T0310041

HANDBOOK OF CONCENTRATOR PHOTOVOLTAIC TECHNOLOGY

HANDBOOK OF CONCENTRATOR PHOTOVOLTAIC TECHNOLOGY

Edited by

Carlos Algora
Universidad Politécnica de Madrid, Spain

Ignacio Rey-Stolle
Universidad Politécnica de Madrid, Spain

WILEY

This edition first published 2016
© 2016 John Wiley & Sons, Ltd

Registered office
John Wiley & Sons Ltd, The Atrium, Southern Gate, Chichester, West Sussex, PO19 8SQ, United Kingdom

For details of our global editorial offices, for customer services and for information about how to apply for permission to reuse the copyright material in this book please see our website at www.wiley.com.

Library of Congress Cataloging-in-Publication Data

Names: Algora, Carlos, editor. | Rey-Stolle, Ignacio, editor.
Title: Handbook of concentrator photovoltaic technology / [edited by] Carlos
    Algora, Ignacio Rey-Stolle.
Description: Hoboken : John Wiley & Sons Inc., 2016. | Includes index.
Identifiers: LCCN 2015039642 (print) | LCCN 2015041278 (ebook) | ISBN
    9781118472965 (cloth) | ISBN 9781118755631 (Adobe PDF) | ISBN
    9781118755648 (ePub)
Subjects: LCSH: Photovoltaic power systems–Handbooks, manuals, etc. | Solar
    concentrators–Handbooks, manuals, etc.
Classification: LCC TK1087 .H345 2016 (print) | LCC TK1087 (ebook) | DDC
    621.31/244–dc23
LC record available at http://lccn.loc.gov/2015039642

A catalogue record for this book is available from the British Library.

Set in 10/12 pt TimesLTStd-Roman by Thomson Digital, Noida, India
Printed and bound in Malaysia by Vivar Printing Sdn Bhd

1   2016

To Lucía, Jara, Violeta, Merche and Carmen

Contents

List of Contributors

Norman Abela
Soitec, Germany

Justo Albarrán
Abengoa Research, Spain

Carlos Algora
Instituto de Energía Solar, Universidad Politécnica de Madrid, Spain

Florencia Almonacid
Centro de Estudios Avanzados en Energía y Medio Ambiente, Universidad de Jaén, Spain

Ignacio Antón
Instituto de Energía Solar, Universidad Politécnica de Madrid, Spain

Kenji Araki
Daido Steel Co., Ltd, Japan
Present address: Toyota Technological Institute, Japan

Stephen Askins
Instituto de Energía Solar, Universidad Politécnica de Madrid, Spain

Shelley Bambrook
Soitec, Germany

Nick Bosco
National Renewable Energy Laboratory, United States

Sebastián Caparrós
Abengoa Research, Spain

Antonio de Dios
Abengoa Research, Spain

Óscar de la Rubia
Instituto de Sistemas Fotovoltaicos de Concentración (ISFOC), Spain

César Domínguez
Instituto de Energía Solar, Universidad Politécnica de Madrid, Spain

Pilar Espinet-Gonzalez
Instituto de Energía Solar, Universidad Politécnica de Madrid, Spain
Present address: California Institute of Technology, United States

Eduardo F. Fernández
Centro de Estudios Avanzados en Energía y Medio Ambiente, Universidad de Jaén, Spain

James Foresi
Suncore Photovoltaics, Inc., Albuquerque, United States

Vasilis Fthenakis
Center for Life Cycle Analysis, Columbia University, United States; and
Photovoltaics Environmental Research Center, Brookhaven National Lab, United States

Iván García
Instituto de Energía Solar, Universidad Politécnica de Madrid, Spain

Tobias Gerstmaier
Soitec, Germany

Andreas Gombert
Soitec, Germany

Maikel Hernández
LPI-Europe, S.L., Spain

Rebeca Herrero
Instituto de Energía Solar, Universidad Politécnica de Madrid, Spain

Sarah Kurtz
National Renewable Energy Laboratory, United States

Ralf Leutz
Leutz Optics and Illumination UG, Germany

Antonio Luque
Instituto de Energía Solar, Universidad Politécnica de Madrid, Spain

Ignacio Luque-Heredia
BSQ Solar, Spain

Pedro Magalhães
Versol Solar, United States

María Martínez
Instituto de Sistemas Fotovoltaicos de Concentración (ISFOC), Spain

David Miller
National Renewable Energy Laboratory, United States

Robert McConnell
Amonix, United States. Present address, CPVSTAR, United States

Rubén Mohedano
LPI-Europe, S.L., Spain

Matthew Muller
National Renewable Energy Lab, United States

Daryl R. Myers
National Renewable Energy Lab., United States

Gustavo Nofuentes
Grupo de Investigación y Desarrollo de Energía Solar,
Universidad de Jaén, Spain

Jerry M. Olson
Consultant, United States

Carl R. Osterwald
National Renewable Energy Laboratory, United States

José A. Pérez
Abengoa Research, Spain

Ignacio Rey-Stolle
Instituto de Energía Solar, Universidad Politécnica de Madrid, Spain

Francisca Rubio
Soitec, Germany

Daniel Sánchez
Instituto de Sistemas Fotovoltaicos de Concentración (ISFOC), Spain

Gabriel Sala Pano
Instituto de Energía Solar, Universidad Politécnica de Madrid, Spain

Gerald Siefer
Fraunhofer-Institut für Solare Energiesysteme ISE, Germany

Sven T. Wanka
Soitec, Germany

Diego L. Talavera
Grupo de Investigación y Desarrollo de Energía Solar,
Universidad de Jaén, Spain

Ignacio Tobías
Instituto de Energía Solar, Universidad Politécnica de Madrid, Spain

Manuel Vázquez
Instituto de Energía Solar, Universidad Politécnica de Madrid, Spain

Marta Victoria
Instituto de Energía Solar, Universidad Politécnica de Madrid, Spain

Tobias Zech
Soitec, Germany

Preface

This volume is the first handbook fully focused on Concentrator Photovoltaic Technology. Essentially, this handbook gathers, in one place, a comprehensive review of all scientific background around Concentrator Photovoltaics (CPV) as well as detailed descriptions of the technology and engineering developed to design, build and manufacture CPV systems and plants. In particular, this book essentially focuses on the current workhorse of the CPV industry: point focus designs based on refractive optics and III-V multijunction solar cells working at concentration levels from some hundreds to over a thousand suns.

In this Preface, we discuss why we believe this handbook is a timely and pertinent endeavor by reviewing the general situation of PV, the key advantages that CPV offers and the history of CPV to conclude with its present status and future prospects.

A Vision of Photovoltaics within the World's Energy Perspective

The Earth receives annually around $1.5 \cdot 10^9$ TWh of solar energy. This overwhelming figure constitutes by far the most abundant energy resource available for mankind heretofore. If adequately harnessed, only a miniscule fraction of this energy would suffice to supply the world's total primary energy demand, which in 2013 was about $1.6 \cdot 10^5$ TWh (i.e. the solar resource on earth is about 10 000 times the energy needs of mankind). The primary energy is processed by the energetic system into different types of readily usable energy forms, among which electricity is considered the key technology for the next decades. Accordingly, the direct generation of electricity from solar radiation (i.e., the production of the preferred consumable form of energy from the richest resource) is a topic of the highest relevance and is the essence of photovoltaics (PV). From the discovery of the photovoltaic effect in 1839 – by French physicist Alexandre-Edmond Becquerel – to the first successful application of photovoltaic panels to power the *Vanguard I* satellite launched in 1958, more than a century went by. Since those pioneering works, many steps forward have been made and the PV industry has evolved from the modest watt-ranged applications of the early days to the giant GW-ranged systems planned today. As a matter of fact, the evolution of photovoltaics over the first decades of the 21st century has been remarkable among all energy technologies. As Figure 1 shows, PV installations have been growing tremendously and by the end of 2015, it is expected to have left behind the non-negligible mark of $200 \, GW_p$ global cumulative installed capacity.

Another sign of ripeness of PV is the size and globalization of the market. As indicated by Figure 2, to reach the cumulated capacities described Figure 1 the photovoltaic industry has

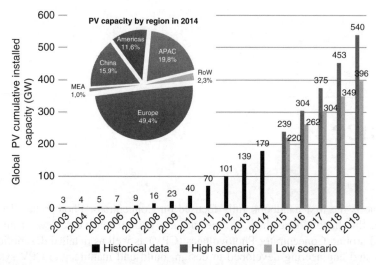

Figure 1 Global PV cumulative installed capacity forecast until 2019.[1] The forecast considers an optimistic (high growth) and a pessimistic (low growth) scenario. The pie chart shows how this capacity is distributed by region as of 2014 (The legend for the pie chart is as follows RoW: Rest of the World; MEA: Middle East and Africa; APAC: Asia Pacific)

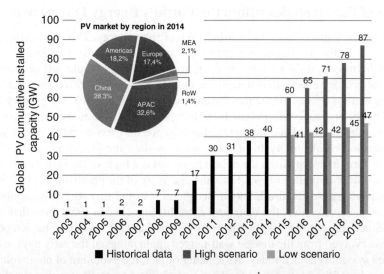

Figure 2 Evolution of yearly global PV installations until 2019.[1] The forecast considers an optimistic (high growth) and a pessimistic (low growth) scenario. The pie chart shows where these installations were made in 2014. (The legend for the pie chart is as follows RoW: Rest of the World; MEA: Middle East and Africa; APAC: Asia Pacific)

---

[1] *Global Market Outlook for Solar Power 2015–2019*, Solar Power Europe, formerly known as EPIA (2015).

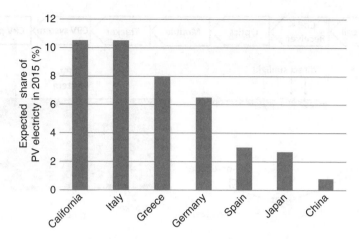

Figure 3 Expected percentage of the electricity demand in 2015 to be produced with photovoltaic power plants in different regions of the world.[2]

maintained almost unprecedented growth rates (annual growth rate of ~44% in installed power from 2003 to 2013). In addition, such growth is no longer concentrated in Europe (as shown in the pie chart included as an inset in Figure 2), but the PV market has become a truly global reality in recent years.

A side effect of the global dimension of the market is the significant penetration that PV is gaining in electricity markets in different regions of the world. Figure 3 shows the expected percentage of the electricity demand in 2015 that will be produced with photovoltaic power plants in different countries or regions. In brief, this figure demonstrates that the expertise to integrate significant fractions of PV electricity in the distribution networks is flourishing in parallel with installations.

In essence, photovoltaics today is a consolidated industry, growing fast worldwide, and gaining relevance in significant electricity markets. All these facts make clear that photovoltaic technology has demonstrated the maturity to become a major source of power for the world. That robust and continuous growth is expected to continue in the decades ahead in order to turn photovoltaics into one of the key players in the pool of technologies involved in generating electricity for the 21st century. The big question for photovoltaic solar energy as of today is not if it will expand, but by how much.

What is CPV? The Role and Advantages of CPV

CPV is one of the PV technologies. Therefore, CPV converts light directly into electricity in the same way that PV does. The difference of CPV regarding PV stands in the addition of an optical system that focuses direct sunlight collected on a large optics area onto a small solar cell. The optics area to the solar cell area ratio is called geometrical concentration or simply concentration level whose dimensionless units are typically referred to as 'suns or ×'.

---

[2] *Snapshot of Global PV 1992–2014*, International Energy Agency - Photovoltaic Power Systems Programme (2015).

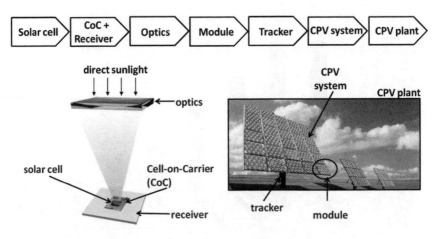

Figure 4   Components and systems of a CPV plant

The CPV approach allows the use of the most efficient cells (although expensive) since the small size of the cell consumes much less semiconductor material. Therefore, CPV replaces costly semiconductor solar cells with cheaper optics. For example, the production of 1 watt of electricity by means of a 40% multijunction solar cell operating at 1000 suns requires 2666 less semiconductor area than if a 15% silicon solar cell without concentration was used for the same purpose. Using much smaller cells promises for lower costs, but CPV systems are more complex than conventional PV systems. The key is if the overcost derived from the complexity of CPV is low enough to be counterbalanced by the savings in semiconductor cell area and the increase in efficiency. In that case, CPV would compete with conventional PV.

Figure 4 shows the different components of a CPV plant. Solar cells used in high concentration CPV systems are typically multijunction solar cells made up of III-V semiconductors. Cells have to be mounted on a carrier (Cell-on-Carrier: CoC) that usually includes a bypass diode. In many designs, CoC is mounted onto a heatsink in order to properly dissipate and remove heat. The optics consists typically of a primary optical element that collects direct sunlight, and may have a secondary element that receives the light from the primary. The assembly of a heatsink, a CoC and a secondary optics (if any) is typically referred to as a *receiver*. By means of the integration of several receivers and primary optics, CPV modules are built. Modules are placed on a sun-tracker structure which allows modules to be pointed towards the sun at all times. The tracker structure together with the modules constitutes the CPV system. Finally, several CPV systems together with inverters, transformers, wiring, etc. form a CPV plant which is able to inject AC electricity to the grid.

Differences in the architectures of PV and CPV plants result in different pros and cons for each technology. Nowadays, silicon-based PV dominates the solar market. When at the end of the 1990s, CPV promised system costs about €3/W_p, PV cost laid at about €6/W_p. Clearly, CPV has accomplished its cost forecast but PV has experienced an unexpected huge price drop. Therefore, if CPV wants to challenge PVs hegemony, it needs to beat PV cost. Nowadays, CPV starts to be cheaper (in terms of LCOE, i.e, Levelized Cost of Energy) than PV in some very hot locations with high direct normal irradiation. However this low cost of CPV electricity has to be widespread by using its advantages summarized in Table 1.

Table 1 Main differences between CPV and PV

Property	CPV	PV
Used sunlight	Direct	Global (direct and diffuse)
Suitable locations	High direct normal irradiation (usually no coastal regions located within 15°–45° of both N and S latitude)	Almost every location within latitude range from 60°N to 60°S
System efficiency	Current values of 30% with much more room for gains. More than two times energy yield per equivalent installation area	Limited to about 15%
Operation temperature	Good performance till 70–90°C	Good performance till 30–50°C
System cost (as of 2014)	~€2/W$_p$	~€1/W$_p$
LCOE (as of 2014)	€c8–20/kWh with room for €c2–3/kWh	€c10–20/kWh with room for €c5–10/kWh
Modularity and scalability	High	Very high
Reliability	Very high (based on a experience of about six years of operation in the field together with accelerated ageing tests suggesting more than 30 years of operation)	Very high (based on a history of about 30 years operation in the field)

Finally, it has to be highlighted than CPV is completely different than CSP (Concentrated Solar Power) which uses heat from the system to generate electricity in a traditional steam engine power plant environment.

History of CPV. Lessons from the Past, Present Status and Expected Future

The use of optical elements to concentrate sunlight, and thus reach higher energy densities, has been known and used by mankind since ancient times. Lighting fires, optical communications or signaling or even setting fire to enemy warships – a legendary feat attributed to Archimedes of Syracuse in the 3rd century BC – are some examples of ancient uses of concentrated sunlight recorded in history books. However, it is not our goal in this preface to present a detailed historical background of such uses of concentrated sunlight; not even of CPV. There are excellent reviews on this topic[3] so here we will focus on some key milestones that –in our opinion– have shaped our short life as a modern electric power industry.

It was the oil crisis in 1973 which spurred the interest on renewable energies in oil-addicted western countries. Solar electricity in general and photovoltaics in particular was a key part of this new wave of interest. Accordingly, it was in the middle 1970s when ambitious development programs were put into practice to develop terrestrial uses of photovoltaics (PV had been

---

[3] See for example, Chapter 1 of *Concentrator Photovoltaics* (eds A. Luque and V. Andreev), Springer Series in Optical Sciences (2007).

used in space to power artificial satellites since 1958). In this context, research on CPV begun as concentration was seen as a natural way to increase the modest efficiency solar cells in those early days.

The first notable effort in the history of CPV technology was the research conducted at Sandia National Laboratories, Albuquerque, New Mexico during the late 1970s. The team at Sandia designed a CPV system operating at 32–40×, based on acrylic Fresnel lenses and silicon solar cells with passive cooling and two-axis tracking. The third generation of this technology (namely SANDIA III) was pilot-produced by Martin Marietta Co. and a power plant of 350 kW$_p$ was installed in the desert of Saudi Arabia by the end of 1981. This plant, namely, the SOLERAS project, operated for more than 18 years in the harsh conditions of the Arabian Desert and was for several years the largest PV installation in the world.

From this seminal milestone to the multi-megawatt CPV plants being deployed today many technological and scientific achievements have occurred over the last 35 years covering the whole value chain of CPV technology.

For example, in the field of solar cells, the early silicon based designs were soon refined into more advanced cell architectures (point contact solar cells) in the mid-1980s. A great leap forward was provided by the move to multijunction solar cells using III-V semiconductors. In 1995, a two terminal monolithic dual junction GaInP/GaAs solar cell was the first solar cell that surpassed the 30% efficiency barrier. By the end of the decade, the addition of a third Ge junction raised the efficiency to over 32%; this design was further optimized and by 2006 the 40% barrier was broken. As of today, we have four-junction solar cells fabricated using wafer bonding techniques with efficiencies in excess of 46%, and with several other architectures (inverted metamorphic, upright metamorphic, dilute nitride lattice-matched, etc.) laying siege to the milestone of 50%.

In the field of optics, a superficial look might give the impression that no such impressive advances have been made since the acrylic Fresnel lenses used in the SOLERAS project are still present in some CPV products today. Moreover, the silicone on glass primary lenses used by some manufacturers also date back to the early 1980s. However, this 35 years have represented also a tremendous advancement in optical technology. The field of nonimaging optics has been intensively explored to optimize the performance of Fresnel lenses and design secondary optics that constitute optical trains with high transmission (>90%); good spatial uniformity (peak to average ratios below 2.5), little chromatic aberration and acceptance angles in excess of ±0.7° for concentrations as high as 1000 suns. And, what is even more important, these accomplishments have been reached while improving the manufacturability, the reliability (UV and weathering resistance) and bringing down the costs for efficient mass production.

Obviously, the field of sun trackers and CPV balance-of-system components has benefited from the tremendous impulse and reduction of costs that microelectronics has experienced over the last three and a half decades.

After two decades in research mode, in the first years of the 21st century, all the progress attained in the different steps of the CPV value chain together with the shortage in silicon supply that was affecting the growth of flat plate PV, brought about a renaissance of CPV technology. Amonix (now Arzon Solar), the CPV industry pioneer founded in 1989, leaped from kW-sized demonstration projects to megawatt ranged power plants. A number of companies fully focused on the CPV business were founded, such as Semprius and Solfocus in the US, Concentrix (later Soitec) in Germany, MagPower in Portugal, Renovalia in Spain, Morgan Solar in Canada, Suncore in China and many others; and companies operating in

Table 2   Ten largest CPV plants as of 2014[4]

Location	Company	Year	Capacity (MW_p)
Goldmud 2 (Quinghai, China)	Suncore	2013	79.8
Goldmud 1 (Quinghai, China)	Suncore	2012	58.0
Touwsrivier (South Africa)	Soitec	2014	44.2
Alamosa (CO, USA)	Amonix	2012	35.3
Borrego Springs (CA, USA)	Soitec	2014	8.9
Villafranca (Spain)	Amonix/Guascor Fotón	2008	7.8
Hatch (NM, USA)	Amonix	2011	5.0
Estoi (Portugal)	Magpower	2012	3.0
Hami (Xinjiang, China)	Soitec	2013	2.6
ISFOC (Puertollano and Almoguera, Spain)	Isofotón, Soitec, Solfocus, Emcore, Abengoa, Arima, Renovalia, Semprius	2008–2013[5]	2.3
Newberry Springs (CA, USA)	Soitec	2013	1.7

conventional PV or other sectors created CPV units, such as Isofoton and Abengoa in Spain, Daido Steel and Sumitomo in Japan or Arima in Taiwan. Many of the aforementioned companies deployed their systems in multi-kW$_p$ sized power plants during the first decade of the 21st century demonstrating module efficiencies approaching 30%.

For CPV technology, the 2010s started with its lights and shadows. On the side of lights, there were a number of CPV companies with mature and reliable products, with record module efficiencies over 30% and orders for tens (sometimes hundreds) of MW$_p$ in the pipeline. On the other hand, regarding shadows, the economic crisis and the astounding price reductions achieved by conventional flat plate PV exerted important pressure on this emerging industry. In this harsh environment a number of the aforementioned companies failed to meet their targets and went out of business. However, we have also witnessed in the last years the commissioning of multi-megawatt CPV power plants (see Table 2). In March 2012, Amonix (now Arzon Solar) completed the installation of a 35 MW$_p$ power plant in Alamosa, Colorado. Soon after, in November 2012, Suncore Photovoltaics commissioned Golmud 1, a 58 MW$_p$ power plant in the Qinghai province in northwest China. This was followed by Golmud 2 in 2013, which added 80 MW$_p$ more to this site. Finally, in 2014 Soitec completed the acceptance tests of two of its largest projects: a 9 MW$_p$ power plant in Borrego Springs, California and a 44 MW$_p$ plant in Touwsrivier, South Africa.

As of today (September 2015), CPV technology has been able to develop the most efficient converters of solar radiation into electricity – the current CPV module record efficiency is slightly above 38% – to prove their reliability and robustness and to deploy about 150 plants around the world with a cumulative installed power of more than 330 MW$_p$ (see Table 2 for the

---

[4] Prepared with data taken from: (a) www.cpvconsortium.com and (b) Philipps SP, Bett AW, Horowitz K, Kurtz S. Current Status of Concentrator Photovoltaic (CPV) Technology, January 2015; CPV Report. TP-6A20-63916 and (c) Soitec.

[5] Depending on the company, first CPV fields were commissioned in 2008 and the last one in 2013.

ten largest plants). However, this power is well below the expectations that, for example, in 2012 predicted the installation of about 1.2 GW$_p$ by 2016.[6] The current installed CPV power is only approximately 0.2% of the total cumulative flat-plate PV power, represented mainly by crystalline silicon.

The aforementioned history of CPV is marked out by successful scientific and technological milestones. In fact, most technology challenges identified at the beginning of CPV development have been already solved. So, why is today's CPV market so weak? Figure 5 shows the different tendencies of the efficiency increase of both concentrator III-V multijunction solar cells and non-concentrator silicon solar cells. In the case of silicon solar cells, the efficiency increase was noticeable at their early development, but from about 1995 their efficiency is almost stagnant and it is precisely from then when silicon PV has experienced its huge deployment. Therefore, it seems that the stillness of silicon solar cells efficiency has conferred the role of 'mature technology' to silicon PV and thus, silicon PV industry has devoted all its efforts to reduce costs in the fabrication of a very well known product by means of learning (i.e. fall in cost as manufacturing ramps up, use of mass production factories, discounts from suppliers, etc.). On the contrary, concentrator multijunction solar cells are experiencing a continuous efficiency increase (as non-concentrator silicon solar cells did at the beginning) which is very similar to those considered as 'Emerging PV technologies' (dye sensitized, perovskite, organic, etc.) in Figure 5. Does it mean that CPV is an emerging and immature technology? The answer is 'no' for the CPV community who prioritizes positive aspects (such as that CPV installations have shown a high reliability with more than six years in the field, that CPV systems integrate very well all of their high technology components – solar cells, optics, modules, trackers, etc.) and emphasizes the advantages of CPV over PV. However, the answer to the same question is probably 'yes' for those out of the CPV community who prioritize negative aspects (such as there is not an only standard CPV product but several ones and CPV products are still in evolution).

The continuous evolution of CPV in the quest for higher efficiency solar cells and modules is because CPV is unable to reach by now the cost of silicon PV. The most efficient way for decreasing cost in the CPV manufacturing process should be by *learning*, taking advantage of the vertical integration together with tens or, even better, hundred MW$_p$ production capacities. Vertical integration allows a better control of both the CPV system performance and total system cost. However, more vertical integration is associated with a higher business risk. This seems to be the case of big companies who have stopped, reduced or reoriented their CPV activity such as Amonix (now Arzon Solar), Solfocus (bankrupted), Soitec (has exited the CPV business), Suncore (has reoriented towards a mixture of concentrator photovoltaics and thermal), etc. Probably, these companies did huge investments in order to be prepared for an expected rising of the market volume that did not arrive on time.

An alternative approach to vertical integration is to subcontract most of the business and thus lower the company's capital needs and at the same time transfer most of the business risk to subcontractors. This approach is advantageous for small companies with limited sources of financing. Besides, this approach would allow these companies to survive while developing more efficient CPV systems in the quest for a disruptive product.

Now, it can be useful to look at one of the great concepts taught at business schools, namely, the 'product life cycle.' A product life cycle consists of an initial product market introduction,

---

[6] http://www.pv-tech.org/news/ims_research_predicts_continued_growth_for_cpv

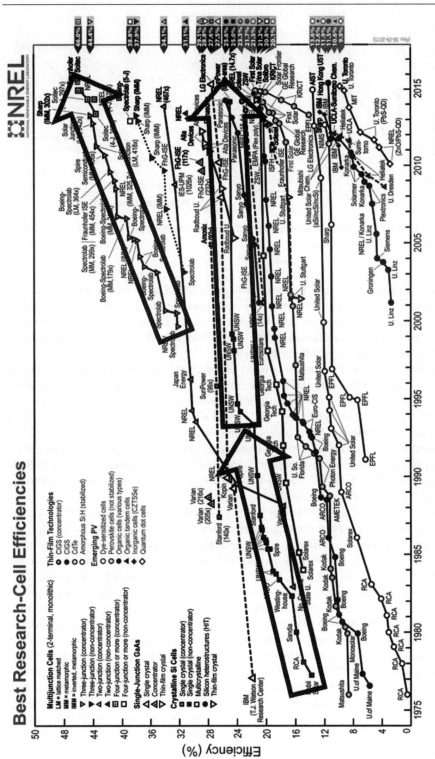

Figure 5 Champion solar cells of different technologies. The top arrow shows the efficiency increase tendency of CPV multijunction solar cells (three and four junctions) while bottom arrows show that of non-concentrator silicon solar cells (left arrow at the early times and right arrow at the maturity stage)

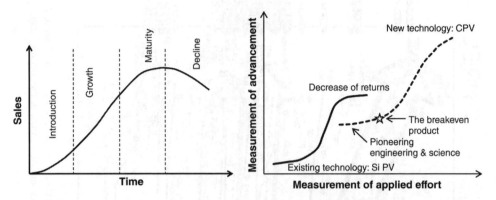

Figure 6 Schematic of a product life cycle (left). 'S-curve' for the technology evolution from Si PV to CPV (right)

then it grows, matures and finally declines (see Figure 6 left). The 'S-curve'[7] allows businesses to predict the rise-fall of new product life cycles within the market-industry. The 'S-curve' means the pattern of revenue growth in which a successful business starts small with a few customers; grows rapidly as demand for the new offering swells, and eventually peaks and levels off as the market matures (Figure 6 right).

High performance companies not only manage to successfully climb the S-curve but before the curve begins to flatten, they quickly shift to the start of the next curve. A paradigmatic example is Osram who being one of the leading makers of light bulbs gave the shift to the incipient (at that time) LED technology and now is the world leader in illumination LEDs.

As we have stated before, Si PV is a very mature technology and because of this, the shift to next S-curve should start. Things often look rosiest just before a company heads into decline. One example: when electronic cash registers went from 10% of the market in 1972 to 90% just four years later, NCR, long the leading maker of cash registers, was caught unprepared, resulting in big losses and mass layoffs.

The main drawback of Si PV is its low and limited system efficiency (about 15%) which is one of the lowest in the present energetic scene. Therefore, technologies with current higher efficiency and with room for further efficiency increase, such as CPV, have already started its own 'S-curve' (see Figure 6 right). Key question is when the shift from Si PV S-curve to a more efficient PV technology, such as CPV, S-curve will happen? The answer is: when a breakeven CPV product will be developed (the star symbol in Figure 6 right). A breakeven CPV product will be that able to produce electricity at a lower cost than Si PV so it will allow CPV companies to achieve a real profit. Such breakeven product can be reached by the current improvement tendency (higher reliability, higher bankability, higher lifetime, cost reduction by learning in the manufacturing, etc.) or by the achievement of a disruptive product (ultra-high efficiency, huge cost reduction, etc.). The time until the appearance of such breakeven CPV product can be a period of upheaval. In fact, CPV has already experienced the creation of its 'bubble' at about the 2000s (with the entering of massive venture capital funds in startup companies, with the inflated forecasts of some reputed consultants, with the massive attendance to CPV

---

[7] Richard N. Foster *Innovation: the attacker's advantage* (1986).

conferences, etc.) and the subsequent pricking the bubble (the fall of many CPV companies, including the death of some big CPV companies that were considered as pillars of the CPV industry development, exit of many venture capitals, etc.). Therefore, CPV is now experiencing (as of mid-2015) a disillusionment-like period as it is usual after a bubble prick. In this catharsis period, CPV has to optimize and reconsider many aspects in order to be strengthened. If proper adjustments are made, CPV will achieve the required breakeven shown in Figure 6 (right) and will outperform Si PV in suitable locations. This Handbook hopes to help the CPV community to reach the required breakeven and to contribute to starting a real golden age for CPV technology.

Structure of the Handbook

As stated above, this handbook gives a comprehensive overview of CPV theory, technology and development status. In order to achieve this, the book is divided into four different building blocks:

- Basic theory (Chapters 1 to 4)
- System design (Chapters 5 to 9)
- System characterization (Chapters 10 to 12)
- Life cycle, costs and market (Chapters 13 and 14).

In the first block about basic theory, the foundations of CPV are covered, namely, direct solar radiation in Chapter 1; multijunction solar cells in Chapter 2 and CPV optics in Chapter 4. For the advanced reader, Chapter 3 reviews emerging solar cell concepts for future CPV systems. In this part of the book a special effort has been put into highlighting the fundamental physics, trying to avoid technology-specific discussions and thus producing a text that will not become obsolete as technology evolves, but, on the contrary, will continue to be of value in the future.

In the second block all topics associated to CPV system design are covered, namely, the impact of temperature in Chapter 5; solar trackers and tracking algorithms in Chapter 6; the integral design and manufacturing of a CPV module in Chapter 7 (including specific case studies of modules and systems from Abengoa, Daido Steel, Soitec and Suncore); the design analysis and operation of a CPV power plant is covered in Chapter 8 (including specific case studies from ISFOC and SOITEC plants); to end with reliability issues (covering the complete value chain from solar cells, optics and modules to CPV power plants) in Chapter 9.

In the third block the different strategies to measure CPV components and systems are covered, namely, the characterization of CPV solar cells is described in Chapter 10; the characterization of CPV optics is described in Chapter 11; and the measurement indoors and outdoors of CPV modules and systems is presented in Chapter 12.

In the final fourth block commercial aspects of CPV technology are covered, namely, a comprehensive life cycle assessment of an exemplary CPV system is carried out in Chapter 13; whilst cost models and analyses for CPV together with market forecast towards grid parity are discussed in Chapter 14.

As a concluding remark, it should be highlighted that this book has been conceived to be used in a versatile way. It can either be read from beginning to end, to acquire a well-structured and comprehensive vision of CPV technology (i.e., as a textbook), or it can be used as reference text to search for data, gain insight into a certain topic or find detailed definitions of key concepts. In this sense, special emphasis has been put in the structuring of the sections and sub-sections to ease the access to specific information from the Table of Contents as well as including sufficient cross-referencing between chapters.

Readership

The prerequisite knowledge of the reader would be a basic background in physics, mechanics and electronics at undergraduate level. The target audience of this book includes CPV specialists (in a wide sense) and thus, this being a specific field, the book will necessarily be a specialist book. However the text is totally understandable and accessible to any reader with a basic background in general physics at an undergraduate level. Therefore, the audience for the book are engineers new to the field and wanting to enter CPV opportunities; specialists in CPV companies and research laboratories; faculty involved in CPV research or teaching; postgraduate students in PV and power engineering; undergraduate students studying PV and CPV; PV specialists wanting to learn more about CPV; commercial institutes and companies engaged in the research, development of elements of CPV systems; electric utilities, etc.

Contributors

The Handbook is authored by almost 50 contributors around the world from academia, industry, research centers, etc. with extensive expertise in each particular field of CPV technology. Several Cherry Award and Becquerel Award recipients have contributed to the Handbook.

Carlos Algora and Ignacio Rey-Stolle
Madrid, September 2015

1

Direct Normal Radiation

Daryl R. Myers
National Renewable Energy Lab., United States

1.1 Concepts and Definitions

The harvesting and conversion of solar radiation by concentrating photovoltaic (CPV) technologies depends explicitly on the quality and quantity of the solar resource that is available, as well as the optical and electrical properties of the photovoltaic technology. This chapter will address the quantitative and qualitative aspects of the solar resource, the direct solar radiation, and briefly, more qualitative discussions of the interaction of the resource with the photovoltaic technologies and system design issues. More quantitative discussion of the latter will be addressed in detail in subsequent chapters.

1.1.1 Orbital and Geometrical Considerations

The Earth orbits a typical star, the sun, which provides energy in the form of optical and thermal radiation that enables and supports life on our planet. A reference for most of the numerical data presented in this section is Allen's Astrophysical Quantities [1].

The sun has a diameter (d_s) of 1 390 000 km (840 000 miles). At the surface of the sun (at radius $R_s = 695\,000$ km from the center) the power flux density emitted is about 6.33×10^7 Wm^{-2}. The Earth's orbit about the sun is an ellipse with an eccentricity of 0.0167. Closest approach of the Earth to the sun (perihelion) occurs on about January 2 or 3, and the furthest distance (aphelion) occurs on about July 4 or 5. The Earth's perihelion, R_p, and aphelion, R_a, distances are about 147.5 million km and 152.6 million km, respectively. That is, the Earth-Sun distance varies from -1.4% to $+2.0\%$ of the average Earth-Sun distance, or a range of 3.4% during the year. The average distance (R_o) between the sun and Earth is 1 Astronomical Unit (AU) of 149 597 870.7 km (92 955 807.273 miles).

Using simple geometry, the apparent angular diameter of the solar disk in degrees at 1 AU is arctangent (d_s/R_o) = arctangent (1.390/149.59787) = 0.532° or 9.28 mrad. The apparent diameter of the solar disk changes by 3.4% as the sun moves from aphelion (arctan (d_s/R_a) = 0.521 =

Handbook of Concentrator Photovoltaic Technology, First Edition. Edited by Carlos Algora and Ignacio Rey-Stolle.
© 2016 John Wiley & Sons, Ltd. Published 2016 by John Wiley & Sons, Ltd.

0.91 mrad) to perihelion (arctan $(d_s/R_p) = 0.539° = 0.94$ mrad). In the absence of an atmosphere, because the solar disk subtends a solid angle of about 0.5°, an observer on the Earth's surface will observe that the rays of sunlight falling on a plane surface with the surface normal (perpendicular) pointed at the center of the solar disk fill a solid angle of the same dimensions. The solar radiation filling the 0.5° cone of rays falling on a surface which is normal (i.e., perpendicular) to the axis of the cone constitute the direct normal radiation, or direct beam irradiance, also called direct normal irradiance, or *DNI*. Note than in the presence a clear, cloudless atmosphere, the actual solid angle of the *DNI* over short periods of time will vary slightly, both in time and physical extent. These tiny variations are due to the effects of turbulence and variations in density of the atmosphere as the direct beam radiation propagates through the atmosphere. The magnitude of these effects is demonstrated by the 'twinkling' of starlight from much more distant and more truly point-source-like stars.

As the sun moves in elevation from the horizon at sunrise, to higher in the sky at noon, to the horizon at sunset, the elevation angle, e, of the solar disk, or angle from the horizon to the center of the disk, is constantly changing. Thus the path length through the atmosphere for the photons (defined as the air mass, m) also changes from long to shorter to longer as the sun moves from sunrise to noon to sunset. The geometrical air mass, m, is defined as approximately $m = 1/\sin(e)$. The complement of the solar elevation angle is the solar zenith angle, z, the angle between the local vertical and the center of the solar disk, thus m is also defined approximately as $m = 1/\cos(z)$.

For a surface or collector to capture the *DNI*, the normal or perpendicular to the surface must point to the center of the solar disk throughout the day. This will keep the incidence angle (the angle between the *DNI* beam and the surface normal, θ) of the *DNI* beam near zero, and requires a mechanism to track the elevation and azimuth of the sun throughout the day. The accuracy of the mechanical system in performing the tracking function is an important aspect of the design of systems for intercepting and concentrating, or focusing the direct beam radiation.

For a stationary horizontal surface the incident angle of the direct beam will vary from 90° at sunrise to the (less than 90°, depending on the latitude of the site) solar elevation angle at noon to 90° at sunset. Because the projected area of the direct beam radiation will vary as the cosine of the incidence angle (known as Lambert's law), the flux density per unit area (I) on an arbitrary surface will decrease at high incidence angles (near sunrise and sunset) and be a maximum at solar noon. That is:

$$I = DNI \cos(\theta), \tag{1.1}$$

where θ is the incidence angle of the *DNI* beam to the surface (in other texts *DNI* is denoted as B). For a horizontal surface, the normal to the surface points to the zenith, or elevation angle of 90°. For this surface, the incidence angle for a *DNI* beam, (I_n), from the disk at solar elevation e is the zenith angle, z, as defined above (i.e. 90°−e, or the complement of the elevation). The *DNI* beam flux on a horizontal surface (I_{bh}) is then:

$$I_{bh} = I_n \cos(z), \tag{1.2}$$

1.1.2 The Solar Constant

The term 'solar constant' was coined when it was assumed that the solar output and thus the intensity of solar extraterrestrial radiation (ETR) at the top of the atmosphere (denoted by I_o)

was indeed constant over time. In the middle of the 19th century, irregular, periodic variations in the appearance and density of sunspots, with a period on the order of 11 years were discovered. This is the so-called 11 year 'sunspot cycle' or 'solar activity cycle'. It has since been determined that irradiance variations on the order of peak-to-peak magnitude of ±0.1% in the ETR are associated with the solar activity cycle [2,3]. Here, the term solar constant continues to be used to denote the average ETR irradiance over several solar cycles, and is denoted by I_o.

Despite the fact that the sun subtends a rather large angle of 0.5° in the sky, it is often treated as a point source of radiation, subject to the inverse square law. The inverse square law states that the flux density of radiation decreases (increases) by the factor $1/r^2$ as the distance r between the source and a detector increases (decreases). If we assume that solar radiation originates at a 'point source' at the center of the Sun, when the optical flux emitted at the surface of the sun quoted above reaches the Earth as the ETR direct beam radiation at 1 AU (I_o), it has been attenuated by a factor of:

$$(R_s/R_o)^2 = (695 \times 10^3/149.5 \; 10^6)^2 = 2.16 \times 10^{-5}, \tag{1.3}$$

where R_s and R_o were defined above in section 1.1.1. The resulting ETR power flux density at the top of the Earth's atmosphere (the solar constant) at the average Earth-Sun distance of Ro is then approximately:

$$I_o = (2.16 \times 10^{-5}) \cdot (6.33 \times 10^7) \, \text{Wm}^{-2} = 1,368 \, \text{Wm}^{-2} \text{ at 1 AU} \tag{1.4}$$

1.1.3 Temporal Variations in Extraterrestrial Radiation (ETR)

Because the Earth-Sun distance varies as described in section 1.1.1, the $1/r^2$ variation in the ETR becomes ±3.3%, theoretically ranging from 1320 Wm^{-2} to 1415 Wm^{-2}. Since 1978, multiple Earth orbiting satellite based broadband radiometers (absolute cavity radiometers) have measured the total solar irradiance with an accuracy of about ±0.5%, or ±7 Wm^{-2}, when corrected to a distance of 1 AU. The actual variations of ±3% in the ETR magnitude to the variations in orbital distance, as well as the approximately 11 year sunspot cycle related variations of ±0.1%, along with very short term solar activity (flares, solar storms, influence of bright faculae, etc.) have been detected by these orbiting sensors. The presently accepted ETR solar constant value based on the 37 year period of record from 1978–2015 is $I_o = 1,366.1$ Wm^{-2} ± 0.6 Wm^{-2} (or ±0.04%) [4]. All uncertainty values quoted for measured data represent one standard deviation about the mean value, unless otherwise noted.

Different satellite sensors have exhibited differences or offsets (biases) between the set of measured data [2,3,5–11]. These differences apparently are dependent upon differing instrument designs. These differences have been analyzed and corrected using various schemes to arrive at a 'composite' standard solar constant value of $I_o = 1366.1$ Wm^{-2} ± 0.04% [4].

As of 2013, the most recent total (solar) irradiance monitor data is that from the US National Aeronautics and Space Administration (NASA) Earth Observing System (EOS) Solar Radiation and Climate Experiment (SORCE) satellite [9–11]. Analysis by Kopp and Lean [12] claim an ETR value at solar minimum of 1,360.8 ± 0.5 Wm^{-2}, 0.4% lower than the solar minimum of 1,365.4 ± 1.5 Wm^{-2} derived from the earlier 1978–2010 observations. The difference, and better accuracy (smaller uncertainty), of the SORCE measurement is attributed to an instrument design that reduces stray light reflected from view limiting apertures and baffles, on-orbit calibrations, and detailed laboratory characterization of the SORCE radiometer. If this lower solar minimum SORCE value is used to correct the accepted 1 AU value of average ETR, the result would be $I_o = 1,361.5 \pm 0.5 \, \text{Wm}^{-2}$. That is, a 1 AU ETR that is 0.34% (4.5 Wm^{-2}) lower than the 1978 to 2010 published values. The investigation of the accuracy of these new measurements are under way as of this writing (2015). The accuracy of the theoretical estimate of I_o from Eq. (1.4) is dependent upon the accuracy of the estimated (theoretically calculated) flux density at the surface of the sun.

Spencer [13] developed a Fourier series expansion for the Earth-Sun distance correction factor R_c as a function of the day of the year, d_n (for Jan 1, $d_n = 1$), using R_o is the mean distance, R is the actual distance, and $d =$ 'day angle' computed from:

$$D = 2\pi(d_n - 1)/365, \tag{1.5}$$

$$\begin{aligned} R_c = (R_o/R)^2 &= 1.000110 + 0.034221 \cos(d) + 0.001280 \sin(d) \\ &+ 0.000719 \cos(2d) + 0.000077 \sin(2d), \end{aligned} \tag{1.6}$$

Multiplying the solar constant I_o by R_c produces the solar extraterrestrial irradiance at the top of the Earth's atmosphere for the day of the year d_n.

Figure 1.1 is a composite time series of corrected and adjusted broadband ETR intensity measurements from space for the period 1975–2008. Descriptions of the history and issues associated with ETR data collection and analysis are provided in references [14–23].

1.1.4 Extraterrestrial Radiation Spectral Power Distribution

Above we discussed the total, or integrated broadband direct beam ETR. This total integrated irradiance is comprised of photons of electromagnetic radiation (as well as energetic atomic particles, such as electrons, protons, neutrinos, etc.). The photons and elementary particles generated by the nuclear reactions deep within the sun eventually propagate outward and escape from the solar surface. The photons (or 'optical radiation') generated range from extremely energetic gamma rays, to X-rays, ultraviolet, visible, infrared, radio and microwave radiation. The distribution of power in the solar emissions with respect to wavelength (or frequency) of the radiation is the solar spectral power distribution. Spectral power distributions are important, in that various photovoltaic technologies respond to or utilize different portions of the solar spectrum to greater or lesser degrees, as will be extensively discussed in Chapter 2 for CPV solar cells.

The presently accepted international standard for a composite solar spectral power distribution, or standard reference solar spectrum, as well as value of I_o, is published by the American Society for Testing and Materials (ASTM) International as ASTM E490-10 [4]. This standard is important to the aerospace community in evaluating the performance of spectrally sensitive components such as materials degradation, detectors and solar cells for satellite remote sensing and power generation applications. Figure 1.2 shows a plot of the 200 nm to 2000 nm spectral region of the extraterrestrial solar spectrum tabulated in ASTM E490. See Chapter 2 for detailed discussion on the influence of terrestrial SPD with respect to CPV solar cell technologies.

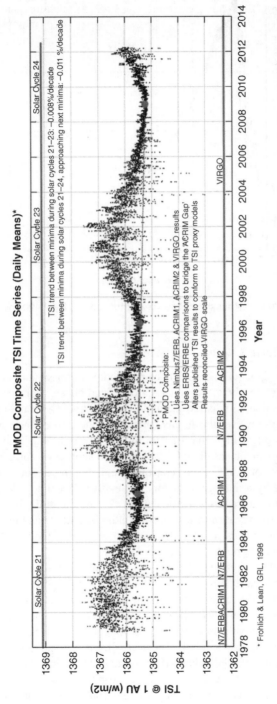

Figure 1.1 Solar constant temporal variations. Source: http://www.acrim.com/RESULTS/Earth%20Observatory/earth_obs_ACRIM_Composite.pdf. Reproduced with permission of Dr. Richard Willson

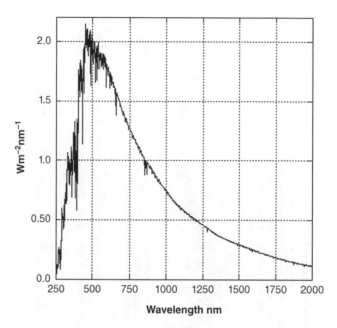

Figure 1.2 Extraterrestrial spectral power distribution. ASTM E490 spectral data

1.1.5 The Atmospheric Filter

So far, we have discussed the ETR, or direct beam radiation, or direct normal irradiance (*DNI*) at the top of the Earth atmosphere. Section 1.1.1 described general considerations regarding solar geometry. This section will discuss the impact of the atmosphere and its effect upon the *DNI* beam radiation as it traverses this highly variable medium. In essence, we could think of the atmosphere as an optical filter that transforms the spectrum and intensity of the ETR. Thereby, the atmosphere attenuates the solar radiation and suppresses some bands as result of the selective absorption of some of its constituents. Even more, such transformation is dynamically influenced by a number of factors that evolve over time – length traversed by the light beam, amount of trace gases, particulate, etc. – and therefore the atmospheric filter changes along the day, from season to season, with altitude, latitude and location. In the next paragraphs, we summarize most of the physical processes that modify the direct normal irradiance as it propagates through the atmosphere.

The gases and particulates present in the atmosphere (or any medium) traversed by the direct beam reflect, absorb, and scatter differing spectral regions and proportions of the direct beam, and thus act as a continuously variable filter. As the narrow cone of *DNI* beam encounters the atmosphere, some photons are reflected by the atmosphere back into space. Some of the remaining photons are selectively absorbed by atmospheric gas molecules, liquid droplets, or particles suspended in the atmosphere [24]. The energy from these photons is either converted to heat (longer wavelength infrared radiation) or re-radiated and 'lost' back to space. Photons with wavelengths approximating the dimensions of atmospheric gas molecules, liquid droplets, or suspended solid particles are preferentially scattered out of the beam and into a broader random radiation field, the diffuse sky radiation. Photons that are not scattered out of the *DNI* beam radiation propagate parallel to the direction of the beam, and are responsible, for instance, for the casting of shadows.

Two different scattering processes based on elastic collision and electromagnetic interactions between photons and atmospheric constituents affect the *DNI*. Rayleigh scattering (named for Lord Rayleigh, who first mathematically described the process) is caused by atmospheric gas molecules or particles smaller than a photon wavelength. Lord Rayleigh derived the strong wavelength dependence of the scattering process named after him, namely that the intensity of the scattered radiation is proportional to the reciprocal of the fourth power of the wavelength (λ^{-4}) and the sixth power of the scattering particle diameter (d^6). Since physical dimensions of atmospheric gas molecules are generally much less (5 Angstroms or 5×10^{-10} m, or the order of hundreds of picometers (pm), or 10^{-12} m), only shorter wavelengths on the order of 300 nm (or 10^{-9} m) to 400 nm or 'blue' photons are scattered the most efficiently and rather uniformly in random directions by the Rayleigh process, hence the apparent blue color of the clear sky.

Mie scattering was described by Danish physicist Gustav Mie by explicitly solving James Clerk Maxwell's equations for the interaction of electromagnetic waves (photons) and particles or molecules larger than the photon wavelengths. Mie scattering is only very weakly dependent on wavelength, but preferentially redirects photons forward, in the direction of the initial path of propagation. Thus, when particles larger than about 500 nm (0.5 μm), are suspended in the atmosphere, a white haze, more extensive and intense as the size and number density of the particles increases, is seen around the solar disk. This Mie scattered radiation is the circumsolar radiation (CSR) or solar aureole. Note that while the CSR is scattered *generally* along the direction of propagation, the photons are scattered into a much larger solid angle, up to several tens of degrees, or a much larger solid angle than the 0.5° of the *DNI* beam. See van de Hulst [24] for detailed technical information on scattering in the atmosphere. CSR will be discussed in more detail in section 1.6.2.

At the Earth surface, where a photovoltaic collector may be located, the combined direct beam radiation (*direct normal irradiance*, DNI) and diffuse sky radiation (*diffuse horizontal irradiance*, DHI) represent the total hemispherical solar radiation on a horizontal surface, usually referred to as the *global horizontal irradiance, GHI*. Thus the relationship of DNI, DHI, and GHI is:

$$GHI = DNI \cdot \cos(z) + DHI, \tag{1.7}$$

It is important to note that diffuse sky radiation photons generally propagate in random, uncorrelated directions because of multiple scattering events and interactions in the atmosphere.

In addition, to the radiation components of Eq. (1.7) there is a fourth one, namely, the *albedo* or reflected component, which accounts for radiation that reaches the target surface after reflection from the ground, buildings, snowy hills or any other reflecting surface.

A general procedure to model *DNI* will be given in section 1.3.1. However, it is also valuable to introduce at this point a few concepts and parameters that are useful in modeling solar radiation and its components. These include the following:

- As has been mentioned in section 1.1.1, the *air mass* (AM or *m*) measures path length through the atmosphere for the photons in the solar radiation. More formally, AM is defined as the ratio of the length of the beam irradiance path through the atmosphere to the vertical length of the atmosphere. Accordingly, in PV terminology AM0 refers to the extraterrestrial irradiance; AM1.5 to an irradiance traversing an atmosphere length 1.5 times its vertical length.

- The key functions to characterize the atmospheric filter are the *atmospheric transmittance T* (*x*) as function of constituent *x*. In essence, *T*(*x*) is the ratio of transmitted direct normal irradiance to that at the top of the atmosphere, as a function of the concentration or intensity of the constituent *x*; with *x* being aerosols, uniformly mixed gases, ozone, water vapor, or the effects of Rayleigh scattering.
- A parameter frequently used to quantify how hazy an atmosphere is the so called *Linke Turbidity*, T_L, which represents the number of clean (aerosol free), dry (free of water vapor) atmospheres necessary to produce an observed attenuation of the extraterrestrial direct normal insolation. Accordingly, T_L is always greater than 1; has a value around 2 for very clean and cold air; and may reach 6 or more for polluted areas.
- Aerosols (small particles of dust, smoke and other materials suspended in the atmosphere) have a deep impact on atmospheric transmissivity. This impact is quantified by means of the *aerosol optical depth*, AOD, usually also denoted as τ or $\tau(\lambda)$, which is calculated as the dimensionless attenuation of incident radiation, I_o, as a function of unit path length through a distance *d* (usually air mass, *m*) of absorbing aerosol. That is, AOD $= \tau = -\ln(I/I_o)$ per unit *m*. In fact, *AOD* varies by wavelength, but is often represented by the AOD at a specific wavelength of 1.0 μm (1,000 nm) or 500 nm. Angstrom's equation for AOD$(\lambda) = \tau(\lambda) = \beta$ $\lambda^{-\alpha}$, where α is the Angstrom exponent, depending on the size distribution of particles, and β is the AOD at the reference wavelength of 1 μm (1000 nm). See the section 1.3.1 below on modeling DNI.
- Finally, in the *clearness index* includes a set of parameters, frequently used to correlate the different components of irradiance. For instance, the *direct normal clearness index*, $K_n = DNI/(R_c\ I_o)$, is the ratio of the terrestrial DNI to the ETR DNI (I_o, corrected for the Earth-Sun distance for the day of the year). Similarly, the *total hemispherical clearness index*, $K_t = GHI/(R_c\ I_o \cos(z))$ is the analogous ratio for global horizontal irradiance.

  All these concepts and parameters will be presented in more detail in section 1.3.1 for the modeling of DNI.

1.2 Measuring Broadband Direct Solar Radiation

Historically, a device to measure solar radiation in general has been called a pyrheliometer, from *pyre* (heat, fire), *helios* (the sun) and meter, or measurement. However, as the total hemispherical solar radiation, diffuse hemispherical radiation, and direct normal (beam) incident radiation components came to be identified and separated, the term pyrheliometer became associated with the DNI beam radiation measurement, as the intensity of the beam radiation from the solar disk was the target measurement. Instruments for the measurement of the hemispherical diffuse and total radiation are now termed pyranometer, or simply 'heat meters' for measuring the combined sky and DNI beam radiation, or total hemispherical sky radiation. See Iqbal [25], Coulson [26], Vignola *et al.* [27] or Stoffel *et al.* [28] for details of solar radiation instrumentation and measurements.

1.2.1 Pyrheliometers

In section 1.1, we described the DNI beam radiation as radiation incident perpendicular to a plane that tracks the position of the solar disk as it moves in azimuth and elevation throughout the day. To obtain only the DNI beam radiation component, one must exclude the sky diffuse hemispherical radiation from intercepting the tracking plane. This is accomplished by aligning

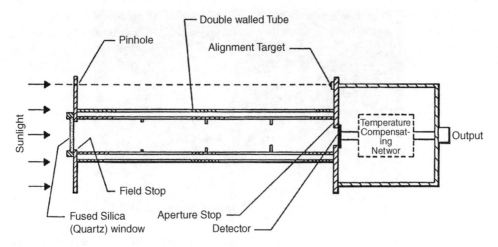

Figure 1.3 Typical pyrheliometer construction. Source: Bahm & Nakos, 1977. Reproduced with permission of the US Department of Energy

a cylindrical tube parallel to the perpendicular to the plane, or along the normal to the plane. Since the normal to the plane is tracking the sun, the tube will be pointed at the sun as well. By situating an opening aperture, a field of view limiting aperture, and light baffles within the tube, a well-defined field of view solid angle can be established (see Figure 1.3). The Russian scientist Alexander Gershun [29] describes such a device in his study of the spatial distribution of sunlight, thusly these view limiting devices are called Gershun tubes.

The construction of a pyrheliometer consists of such a view limiting tube, with a black coating on the internal components, sealed for all weather operation with a detector at the bottom of the tube, and a high transmittance protective window at the end of the tube pointing at the sun. Baffles within the tube limit internal reflections that would generate erroneous (higher) input at the detector. The physical dimensions of the detector, tube, and apertures determine the solid angle field of view at the detector. Figure 1.3 is a sketch of construction of such a device. Figure 1.4 is a photograph of several models of pyrheliometer mounted on a device for pointing the pyrheliometers at the sun, a solar tracker.

The detector is generally a thermopile, comprised of a collection of thermocouples. Thermocouples consist of dissimilar conducting metals in contact with each other at a thermojunction. This junction generates an electrical potential (voltage) dependent on temperature differences between the junction and a similar junction located to sense a reference temperature. Individual thermocouples generate low levels of voltage (microvolts per degree C). Assembling many in series into a 'thermopile', the electrical signal is increased as the individual signals add together. In solar radiometer design, the most common and accurate detectors used are thermopiles in thermal contact with substrate materials coated with black absorbing paints or layers that absorb the photons across the solar spectrum and heat up, transferring thermal energy to the thermopiles. The radiometer output signal is proportional to the difference in temperature between the thermojunctions under the illuminated black absorber and the non-illuminated reference thermojunctions.

A common thermocouple material pair is the type-T thermocouple made from copper bonded to constantan (a nickel–copper alloy). A single type-T thermocouple generates 40 microvolts per degree Celsius (μV/°C) for temperatures between −50°C and +150°C. Thus if

Figure 1.4   Various commercial pyrheliometers on solar tracker. Source: NREL Image Gallery, image 06233, photograph by Thomas Stoffel. Reproduced with permission of the US government

50 thermojunctions are assembled into a thermopile and put in contact with a black absorber that heats up 5°C, and a similar 50 junction thermopile is at ambient temperature, a signal of $40\,\mu V \times 50 \times 5 = 10\,000\,\mu V$, or 10.0 mV is generated. This is a reasonable signal for modern voltage measurement equipment.

Construction of the Gershun tube for early pyrheliometers was based on a simple ratio of the diameter of the front view limiting aperture and the distance between the limiting aperture and the detector of one to ten (1/10). The opening angle, seen from the center of the detector to the edge of the limiting aperture is then the radius of the limiting aperture to the distance from detector to aperture, or 1/20. This simplified manufacturing dimensions, and provided for a relatively large field of view (half opening angle) of arctan $(0.1) = 2.86°$ and a total field of view of 5.7°. Since the pyrheliometer must track the sun throughout the day, and early solar tracking mechanisms were electro-mechanical or clock driven, and not very accurate, the large opening angle permitted some tracking error on the order of $\pm 2°$ or so, without loss of detector illumination.

Since 1978, the World Meteorological Organization (WMO) Commission on Instrumentation, Measurements and Observations (CIMO) has recommended a half-opening angle of 2.5° for primary reference absolute cavity pyrheliometers (see next section) and newer pyrheliometer designs [30]. The resulting total field of view of such a pyrheliometer is then 5.0°, or about 10 times the apparent diameter of the solar disk seen from Earth.

There are two other view angles described in the WMO/CIMO chapter 7 [30] on radiation measurement. These are the slope angle, defined as arctan $(R - r)/L$ where R is the radius of the open aperture at distance L from the detector of radius r, and the limit angle equal to arctan $(R + r)/L$. The former defines the maximum angle at which the edge of the solar disk (accounting for the radius of the solar disk of 0.5°) is just tangent to the field of view opening angle cone. The radiation from the entire solar beam then reaches the detector. The slope angle is considered the useful field of view of the pyrheliometer. The slope angle is generally smaller than the opening angle or nominal field of view. The slope angle for several commercial and primary reference pyrheliometer designs ranges from 20% to 65% of the opening angle. If the center of the solar disk is at an angle greater than the slope angle, a portion of the solar disk is cut off from the point of view of the detector. Thus it is the slope angle of the pyrheliometer that determines the needed tracking, or pointing accuracy of the solar tracking mechanism (see Figure 1.4) required to keep the solar disk (and all direct beam radiation) within the field of view of the pyrheliometer.

All commercial pyrheliometers are provided with an alignment target to indicate the accuracy of the solar alignment. Generally, these consist of a pinhole aperture that produces an image of the solar disk on a target. Target rings are designed to show the approximate angular alignment or misalignment of the pyrheliometer axis with the center of the solar disk. When choosing solar trackers, the tracker accuracy, in terms of angular deviation from perfect alignment with the sun, should be examined to make sure the tracking accuracy is at least as good (comparable to or less than) the slope angle of the pyrheliometer.

The limit angle is the angle beyond which no radiation at all from the solar disk reaches the edge of the detector. There are no internationally accepted standards for the definition of pyrheliometer design other than these WMO CIMO recommendations.

1.2.2 Rotating Shadow Band Radiometers

In section 1.1 we mentioned that the total hemispherical radiation on a horizontal surface or global horizontal irradiance (GHI) results from a combination of the projected DNI beam on the surface, plus the total hemispherical diffuse sky radiation on a horizontal surface (DHI). For a given solar zenith angle (z), DNI can be calculated using Eq. (1.7):

$$DNI = (GHI - DHI)/\cos(z), \tag{1.8}$$

One may use this relationship to derive DNI measurement from the combination of a GHI measurement and a DHI measurement. These measurements are accomplished using a pyranometer, a radiometer with a 180° field of view and a horizontal detector. DHI is measured by blocking the solar disk with an opaque band, ball or disk on a tracking mechanism to track the sun and block the solar disk. These shading devices are designed to subtend the same solid angle (as seen from the pyranometer detector) as the opening angle prescribed for a pyrheliometer.

This method requires two (GHI and DHI) hemispherical irradiance measurements, and a calculated value of the zenith angle. Generally, pyranometer measurements have greater uncertainty than pyrheliometer measurements, and the calculation of z is dependent on accurate time keeping and site longitude and latitude coordinates, as well as an accurate solar position calculation algorithm. Uncertainty in estimating DNI using two independent pyranometers for the GHI and DHI measurements requires combining the uncertainties in the calibration and field performance of each radiometer, increasing the uncertainty in the resulting DNI [28].

It is possible to use only a single pyranometer, alternately unshaded for the *GHI* measurement, then shaded with an appropriate diameter blocking disk or ball at an appropriate distance from the detector for the DHI measurement. The uncertainty in the computed DNI is now dependent on the uncertainty in the one pyranometer, but that uncertainty may be different in each circumstance.

A popular pyranometer design implementing this shade/unshade strategy uses a silicon photodiode detector, which has a spectral response over the limited spectral range of 300 nm to 1100 nm. These radiometers are used in rotating shadow band radiometers (RSR) because their

Figure 1.5 Typical RSR deployed in the field. Source: © David Myers

response time is very short, and the shade/unshade cycle can automated with a rotating band to shade the radiometer during the DHI measurement [31]. Figure 1.5 is a photograph of a typical RSR in the field.

The short wavelength (less than 400 nm) response of these devices is very low, 10% or less of the response in middle of the visible spectrum (550 nm). When shaded, such a pyranometer is receiving radiation from only the shortwave (blue) sky. The detector response of a silicon detector pyranometer is much lower than a pyranometer with a constant responsivity over all wavelengths (such as thermopile in contact with a black absorber). Using a single calibration factor for a silicon photodiode detector, based on the radiation over the entire available spectral response region, to measure both the diffuse sky and total hemispherical (global) radiation will lead to considerable (approaching 50%) errors in the measurement of diffuse radiation. These errors are then carried over into the computation of the DNI. Because the spectral power distribution of the total hemispherical will change throughout the day, the error in computing total hemispherical radiation using a single calibration factor will also vary throughout the day.

There have been efforts to develop correction algorithms to account for spectral, temperature, and cosine response variations for silicon radiometers [32–35], but in general these empirical correlations themselves contain considerable (up to 0.5% or more) scatter. This means that as each correction is applied; an additional ±0.5% random uncertainty is to be combined with the calibration and measurement sources of uncertainty. If we combine three corrections with 0.5% random scatter in each, using the typical root sum of squares approach, an additional 0.87% uncertainty must be combined with calibration and field measurement data uncertainty for the derived DNI [35]. As we will see below, the additional uncertainty due to such corrections approaches the uncertainty in a well calibrated pyrheliometer measurement alone.

In short, as will be shown in the next section, the most accurate measurements of DNI beam (±2%) are accomplished with pyrheliometers using thermopile detectors for direct measurement of the solar DNI. Lower accuracy (±5%) DNI data is generally collected by the RSR radiometers or DNI computed from total hemispherical and diffuse sky radiation measurements. The trade-off between accuracy and cost in terms of equipment and maintenance resources should be evaluated against the impact on the end use or goal of the measurements.

1.2.3 Reference Standards, the World Radiometric Reference (WRR)

Since the design and performance of commercially available pyrheliometers may vary considerably, there is the need for an internationally accepted reference for the calibration of pyrheliometers for measuring DNI beam irradiance. The responsibility for the international reference for the calibration of radiometric instruments rests with the World Meteorological Organization Physical Meteorological Observatory World Radiation Centre (WRC) at Davos, Switzerland [30]. This laboratory maintains a group of highly characterized Absolute Cavity Pyrheliometers (ACP) called the World Standard Group (WSG) that use electrical substitution to calibrate their response. Figure 1.6 is a photograph of the group of reference ACPs in operation at the WRC. These primary reference standard pyrheliometers are absolute in that their response to solar energy is directly traceable to absolute physical and electrical quantities [36–39]. The directly measured dimensions of aperture area, and physical and electrical quantities used to calibrate the response of ACPs, are applied based on the following principles.

Figure 1.6 Working Standard Group (WSG) of reference absolute cavity pyrheliometers at the World Radiation Center in Davos, Switzerland. Source: NREL Image Gallery, image 15505, photograph by Thomas Stoffel. Reproduced with permission of the US government

The ACP consists of a photon trap (cavity) detector, covered with a black absorbing paint or coating, at the bottom of a Gershun tube with no window, with total opening angles of 5.0°, and typical (total) slope angles of 1.5° to 1.75°. Cavity shapes such as cylinders with opening apertures beveled or slanted inward, and pyramidal or inverted cone shapes at the bottom of the cavity are commonly used to enhance the trapping of photons. This trap detector is in thermal contact with thermocouples to measure the temperature rise of the detector when exposed to DNI sunlight.

When the DNI beam is blocked, an electrical current can be supplied to heat the cavity to produce the same temperature rise as that caused by the sunlight. The electrical power required to heat the detector is derived by accurately measuring the electrical current, voltage, and electrical resistance in the circuit that produces the equivalent heating. This electrical power is almost (but not identically) equivalent to the solar DNI beam power that produced the equivalent temperature rise. Slight differences (non-equivalence) in the heat transfer due to the radiative (solar) heating and electrical heating (generally supplied through a different physical pathway and subject to some tiny losses) must be characterized for each individual primary reference ACP design [36–41].

Careful, accurate measurements of the limiting aperture area, efficiency of the trap detector design (absorptivity and reflectivity of coatings, thermal conductivity of the cavity material and coating, etc.), precise properties of the thermocouple junction performance, ohmic losses in conductors, radiative losses during the electrical calibration phase, the above mentioned non-equivalence of electrical and solar heat transfer, and other sources of error in the measurement of electrical current and voltages are required to obtain the highest accuracy ACP radiometric measurements. The process of obtaining the detailed information for the reference ACPs is called characterization, and results in individual correction factors for each individual reference ACP to obtain DNI beam irradiance. Characterization and correction magnitudes generally range from 1 to several hundred part per million (1 ppm to 200 ppm), or 0.001% to 0.020%, depending on the parameter. Table 1.1 shows elements of characterization and the magnitudes involved, from [37]. Figure 1.7 shows sources of uncertainty accounted for in some of the characterization processes.

The WSG instruments are used to establish the World Radiometric Reference (WRR) scale [30]. It is important to note that all the members of the World Standard Group have

Table 1.1 Elements of cavity radiometer comparison

Parameter	Correction factor	3σ uncertainty
Cavity absorptance	1.00115	0.00150
Electrical measurements	1.00000	0.00150
Aperture areas	0.99853	0.00040
Radiative loses	1.00029	0.00300
Coating thermal resistance	1.00007	1.00015
Electrical vs. radiation equivalence	1.00000	1.00015
Conduction from heated aperture	0.99990	0.00015
Internal reflected radiation	0.99991	0.00009
Nonequivalent heat path radiation vs. electrical heat	0.99996	0.00009
Overall correction factor	0.99981	0.00220

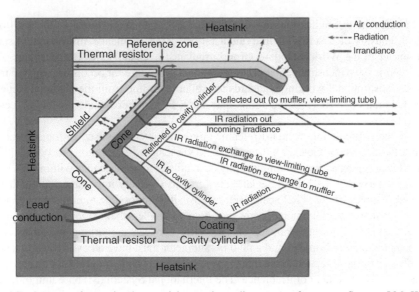

Figure 1.7 Sources of error in characterizing cavity radiometer performance. Source: J.M. Kendall, Personal Communication (1978)

undergone very detailed, individual characterization as described above. They each have individually characterized total field of view opening angles of $5.0° \pm 0.2°$, and slope angles from $1.5°$ to $1.75°$ ($\pm 0.2°$) as recommended by the World Meteorological Organization Commission on Instrumentation Measurements and Observations. The World Radiation Center at Davos, Switzerland is recognized as a National Metrology Institution, and the World Radiation Reference (WRR) scale is recognized by the International Bureau of Weights and Measures (French acronym BIPM), as the international reference standard for solar radiation measurements [41].

Historically, several radiation reference radiometers or scales have been used in meteorology, namely the Angstrom scale of 1905, the Smithsonian scale of 1913, and the international pyrheliometric scale of 1956 (IPS 1956) [30]. Since 1970, the developments in absolute radiometry have very much reduced the uncertainty of radiation measurements. With results of many comparisons of 15 individual absolute pyrheliometers of 10 different types, the WRR was defined in 1975. The old scales can be transferred into the WRR using the following factors:

- WRR = 1.026 Ångström scale 1905, from 1905 to 1913;
- WRR = 0.977 Smithsonian scale 1913, from 1913 to 1956;
- WRR = 1.026 International Pyrheliometric Scale (IPS) 1956, from 1956 to 1975.

During international pyrheliometer comparisons (IPC), organized every five years since 1975, the measuring standards (usually commercially procured ACP or conventional pyrheliometers) of national meteorological centers and laboratories are compared with the World Standard Group, outdoors, over several days. The participating standard pyrheliometer's calibration factors (provided by ACP manufacturers) are used to derive a WRR correction factor to make the test radiometer agree with the WRR. The WRR factor is derived from the statistical analysis of a series of individual ratios of the irradiance from the test standard pyrheliometer to the World Standard Group irradiance over the series of clear days. A very detailed step-by step measurement and computational procedure for accomplishing the transfer of WRR to institutional working standards is available on-line from the National Renewable Energy Laboratory (NREL) publications data base. For further details on this process, see Reda [42].

In turn, the participating standard reference pyrheliometers are used to transfer the WRR scale to network pyrheliometers for use in the field via standard calibration procedures [43–46]. The WRR is accepted as representing the physical units of total irradiance within 0.3% (99.5% confidence interval, or 3 standard deviation uncertainty). A very important point is that since the quoted accuracy of the WRR scale itself is 0.3%, it is impossible to transfer the WRR to a reference standard pyrheliometer with uncertainty smaller than 0.3%. A pyrheliometer calibrated by any additional transfer of the WRR between a working reference ACP or working standard pyrheliometer and a field pyranometer must have an even larger uncertainty as described in the next section. Note that the WRR is based on measurement of electrical units, and not direct measurement of irradiance. The International System (SI) base unit of luminous intensity is the candela.

Recent comparisons in the laboratory between a World Standard Group radiometer and extremely high accuracy (and complex) cryogenic absolute cavity radiometers, used to define the SI laboratory irradiance scale (based upon the candela base unit) show agreement to better than 0.1% between the SI scale and WRR scale at a great many amplitude stabilized laser wavelengths [47]. The World Radiation Center is working to develop a cryogenic version of the Absolute Cavity Pyrheliometer suitable for solar DNI beam measurements. The goal is for uncertainty in the WRR to be lowered to the order of 0.05% and later on to 0.01% if possible; however this is still a work in progress.

1.2.4 Calibration of Pyrheliometers

Periodic comparisons between the World Radiation Center maintained World Standard Group established WRR and national and institutional standard reference ACP and pyrheliometers are performed to transfer the WRR scale to working standards. Similar procedures are used to transfer the WRR from working standard ACP or standard pyrheliometers to field instruments. There are several international consensus standards describing calibrations of field or test pyrheliometer based upon a standard pyrheliometer [43–46]. This transfer process implements the concept of calibration which is defined in the International Vocabulary of Metrology (or VIM for short) [48] as a process that ' . . . establishes a relation between the quantity values with measurement uncertainties provided by measurement standards and corresponding indications with associated measurement uncertainties . . . and uses this information to establish a relation for obtaining a measurement result from an indication.'

The concept of traceability to the WRR scale and reference standard refers to a chain of measurements or comparisons from the standard World Standard Group reference and WRR to the final calibration value. The VIM [27] referred to above states that traceability (pg. 29, definition 2.41) is the 'property of a measurement result whereby the result can be related to a reference through a documented unbroken chain of calibrations, each contributing to the measurement uncertainty.' Note that traceability requires a knowledge of the measurement uncertainty (discussed below) and known sequence of relationships between reference standards and the instruments being calibrated.

The ISO and ASTM consensus standard on calibration procedures [43–46] require sequences of simultaneous measurements of a reference direct beam irradiance, I_{ref}, and signals from the test instruments, usually in voltages, V_{test}, over a period of at least one, and sometimes several days. Each individual ratio of V_{test}/I_{ref} results in a responsivity, Rs, (usually $\mu V/Wm^{-2}$). The collection of all Rs is examined using statistics and graphs of Rs with respect to time and zenith (or elevation) angles to characterize the variability and possible dependence of Rs on time of day or zenith angle, and estimate the uncertainty in the results.

An important note is that all commercially available ACP radiometers used as standard reference radiometers with WRR traceability are designed with the World Meteorological Organization recommended 5.0° full field of view. Commercially available pyrheliometers have different fields of view, ranging from the historical 5.7° full field of view to a 5.0° field of view nominally matching the ACP field of view. Also, the World Standard Group ACP reference pyrheliometers are always operated without a window in place. Reference ACPs compared to the WRR in international and regional comparison experiments generally do not use a window either. It is possible to compare an ACP with a window (say, for all weather operation) with an unwindowed (open aperture) reference ACP to obtain a WRR calibration factor for the windowed unit. However, it has been shown that the addition of the window produces increased random variation in the WRR correction factor due to spectral transmittance and cleanliness issues with the windows.

The final point is that whether fitted with a window or not, and whether the field of view of the test radiometer matches the ACP field of view or not, the correction factors derived for the test pyrheliometer are all relative to the un-windowed, 5° field of view reference ACP radiometer. That is, *the field of view differences and window effects are all relative to the ACP standard configuration and are included in the derived WRR calibration factor.*

Since the un-windowed ACP generally collects, and responds to, a spectral range beyond several tens of micrometers, a pyrheliometer with a window with limited spectral transmittance range is effectively 'corrected' for the presence of the window, and for the difference in field of view, by comparison with the ACP. The derived WRR calibration factor represents the performance of the pyrheliometer and reference ACP under the prevailing conditions at the time of the calibration. Under different conditions than those prevailing at the time the calibrations were conducted, there are additional sources of uncertainty contributing to the uncertainty of pyrheliometers deployed in the field and operating.

1.2.5 Accuracy and Uncertainty

Accuracy is generally defined as the difference between a measurement (which is never perfect), and the value of a 'perfect, zero error' measurement. A series of repeated measurements will always show some natural variation in the measured values. This variation is described as precision, or the distribution of measured values. Uncertainty in a measurement is conceptually a combination of the accuracy (difference from zero error measurement) and precision (scatter in the measurements). Up to this point we have discussed or mentioned several concepts and values related to the accuracy and uncertainty issues applicable to the measurement of DNI beam radiation. Examples are the total uncertainty in the WRR ($\pm0.3\%$), variation in angular field of view ($\pm0.2°$), variation in the diameter of the solar disk ($\pm3.0\%$)

and statistical variation represented by 1-sigma, (σ) and 2σ standard deviations of data distributions. Over the long period of technology development, a wide variety of methods for assessing and (sometimes) reporting accuracy, precision, and uncertainty in measured data have been developed and used. The lack of a standardized methodology often created confusion and even led to engineering errors in many technical fields. Recognizing these problems, during the early 1990s an international community effort produced an international, standardized consensus methodology for expressing uncertainty in measurements.

Since 1995 there has been a formal and explicit recommended standard method for quantifying the uncertainty of measurements in general. This is the *Evaluation of Measurement Data-Guide to the Expression of Uncertainty in Measurements* (JCGM 100:2008), referred to as the GUM [49]. The GUM was developed and published by the Joint Committee for Guides in Metrology (JCGM) of the International Bureau of Weights and Measurements (BIPM). The BIPM, with 54 member countries, headquartered in Sèvres, near Paris, France, was created in 1875 by the International Treaty on the Convention of the Meter, and is the internationally recognized authority for world metrology (the science of measurement). The GUM is also a joint BIPM International Standards Organization (ISO) and International Electrotechnical Commission (IEC) consensus standard, ISO/IEC Guide 98-2008 *Guide to the expression of uncertainty in measurement* (GUM) [50]. A basic summary of the GUM methodology and application to the measurement of direct beam radiation is presented in the following section.

1.2.6 Summary of Guide to Uncertainty in Measurement (GUM) Approach

Basic concepts of the GUM approach are a) the measurement process can be expressed mathematically in a measurement equation dependent on variables leading to the measurement result, an b) there are two fundamental types of sources of uncertainty; Type A, based on statistical analysis of data, and Type B arising from results that are not amenable to statistical analysis, but 'other means'. The latter are represented by things such as a single quoted uncertainty in a standard value or reference, offsets quoted in specifications of measurement equipment (such as $\pm X\,\mu V + 2.5\,\mu V$), or differences between reported values from different sources, etc.

Because statistical processes, such as identifying the standard deviation in a series of measurements have been in use for so long, the term 'standard uncertainty' is used to denote a value that is considered to represent the standard deviation associated with the presumed distribution of errors, or differences between the actual value measured without error and the inevitable spread or distribution of value resulting from many repeated measurements. The steps to be followed for evaluating and expressing the uncertainty of the result of a measurement as presented in the GUM may be summarized as follows:

1. Write down the measurement equation, the mathematical relationship between the measurand Y and the input quantities X_i on which Y depends: $Y = f(X_1, X_2, \ldots, X_N)$. The function f should contain every quantity, including all corrections and correction factors that can contribute a significant component of uncertainty to the result of the measurement.
2. Determine x_i, the estimated value of input quantity X_i, either on the basis of the statistical analysis of series of observations (Type A) or by other means (Type B).
3. Evaluate the standard uncertainty $u(x_i)$ of each input estimate x_i. For an input estimate obtained from the statistical analysis of a series of observations, use the standard deviation

(Type A evaluation of standard uncertainty). For an input estimate obtained by other means, (Type B evaluation of standard uncertainty) $u(x_i)$ is stated, estimated, or evaluated (with an explanation of the derivation).

4. Evaluate the covariances associated with any input estimates that are correlated. This entails computing the sensitivity coefficients for each variable in the function f, defined as the partial derivative of the function with respect to each variable $\partial f/\partial X_i$. The product of the square of the sensitivity coefficient and the square of the estimated standard uncertainty for variable x_i is computed and combined using root sum of squares for all of the products. Variables X_i which are correlated inflate the uncertainty estimates proportional to the cross correlation coefficients

5. Calculate the result of the measurement, that is, the estimate y of the measurand Y, from the functional relationship f using for the input quantities X_i the estimates x_i obtained in step 2.

6. Compute the combined standard uncertainty $u_c(y)$ of the measurement result y from the standard uncertainties (step 3) and covariances (step 4) associated with the input estimates.

7. Compute an expanded uncertainty U, whose purpose is to provide an interval $y - U$ to $y + U$ that may be expected to encompass a large fraction of the distribution of values that could reasonably be attributed to the measurand Y. Multiply the combined standard uncertainty $u_c(y)$ (step 6) by a coverage factor k, typically in the range 2 to 3, to obtain $U = k \cdot u_c(y)$.

8. Select k on the basis of the level of confidence required of the interval.

9. Report the result of the measurement y together with its combined standard uncertainty $u_c(y)$ or expanded uncertainty U. Describe how y and $u_c(y)$ or U were obtained.

The implementation of the GUM approach to estimating the uncertainty in a pyrheliometer measurement in the field will be the result of several successive stages of analysis [51,52]. These stages are summarized in Table 1.2 (with some typical magnitudes indicated):

Finally, the uncertainty estimates for the field data derived from the uncertainties associated with the data collection system component specifications or performance, influence of environmental conditions, specifications and uncertainty in the derived calibration factor for the field pyrheliometer (equipment dependent; ranging from ±1.0% to ±2.0%). See for example the Concentrating Solar Power Best Practices Handbook for the Collection and Use of Solar Resource Data [28].

If summed, assuming a well maintained, regularly cleaned, near perfect tracking mechanism, and no degradation over time, the total accumulated uncertainty in a field measurement of DNI beam radiation is 2.2%. Root sum squaring the uncertainties, we obtain 1.8% for typical

Table 1.2 Stages in the estimation of the uncertainty of a pyrheliometer and typical magnitudes

Source of uncertainty	Typical value
Uncertainty in the characterization of the WSG reference pyrheliometers	±0.1%
Uncertainty in the definition of the WRR derived from the WSG reference pyrheliometers	±0.3%
Uncertainty in the determination of the WRR factor for a standard ACP pyrheliometer during an international or regional comparison	±0.15%
Uncertainty in the derivation of the calibration factor from a standard reference ACP (or secondary reference pyrheliometer) to the field instruments	±0.75%

uncertainties. A year-long comparison of several models and designs of pyrheliometers to evaluate performance in the field reported uncertainties on this order of magnitude [53].

1.2.7 Measurement Data Quality

The quality of measured DNI data is established at the time the data is recorded. Tracking accuracy, window cleanliness, departure of the measurement environment from the environment of the calibration transfer, data recording equipment performance and stability of the radiometer and its components all affect the quality of DNI data. Data quality assessment depends on the amount of effort and resources devoted to examining and understanding the data. Resources can be expended on redundant measurements, such as duplicate instruments, on independent trackers, daily or event driven site visits for cleaning, tracker alignment, visual inspections, and so on [54]. Ancillary data, such as attendant GHI and DHI measurements (so a calculated DNI can be compared with the measured DNI) or environmental data (temperatures below freezing and precipitation leading to icing, for instance) can be leveraged to decide if problems arise.

Given that the basic uncertainty expected in pyranometer measurements is on the order of ±3% to ±5%, the combined uncertainty of GHI and DHI measurements used in the computation of DNI results in an uncertainty of ±3% to ±5%, or two to five times the uncertainty in the DNI data from a pyrheliometer alone. However, one may detect changes greater than the combined uncertainty of 3% to 5% in computed DNI values with respect to measured DNI data. Differences greater than expected measurement uncertainty should be cause for investigating the condition of the pyrheliometer and the tracker used (or the GHI and DHI instruments!)

Other methods of detecting possible problems with the DNI pyrheliometer data is to compare clear sky data with historical clear sky data for similar atmospheric conditions, such as relative humidity, temperature, dew points, etc.; or comparison with mathematically modeled clear sky data. Modeling of clear sky DNI radiation is discussed in the next section. Historical DNI data for periods of clear sky from nearly identical time periods within the year can be compared. Historical, archived clear sky data can be normalized to a selected reference clear sky day, or even the ETR DNI. Comparing similarly normalized measured data with normalized historical data may help identify serious problems with current measurements.

Normalizing measured DNI to the ETR DNI (I_o, corrected for the Earth-Sun distance for the day of the year) is called the *clearness index* (K_n) for the atmosphere [55]:

$$K_n = DNI/(R_c\,I_o), \tag{1.9}$$

Depending on the natural variation in the atmospheric conditions, weather, or climate, the performance of a pyrheliometer may be compared with these reference conditions, and significant deviations (greater than 3%) can alert the operator to investigate the cause of the deviation (poor tracking, contamination, degradation of calibration, etc.).

When DNI measurements are being conducted along with completed and installed concentrating PV (CPV) systems, constant comparison of the DNI data with the system performance data may be used to identify problems with either one or both of the systems [56–58]. The ratio of electrical power out to incident DNI power in (a sort of efficiency only considering DNI) is often used as a performance index for the CPV system [56]. Changes in this ratio greater than the combined uncertainty in the ratio (the measurements of CPV system electrical parameters have their own uncertainty) may be used to alert operators of problems needing attention with either or both systems.

1.3 Modeling Broadband Direct Solar Radiation

Knowledge of the available broadband DNI beam radiation resource data is essential in designing a CPV system. Spectral variations in the DNI beam radiation may affect the performance of a CPV system depending on the solar cell technology used. Because of the expense of measuring equipment, neither of these types of data, and especially spectral data, are generally available. The designer or planner must then resort to surrogate data produced by mathematical models.

There are a relatively few models that have been developed since the 1970s that are in common use at this time (2013) [59–63]. Most are developed from and validated by comparison with measured data, and rely on empirical correlations [64,65]. Typical figures of merit for model performance are *mean bias error* (MBE, the average of differences between measured, preferably from an independent data set, and modeled results), and root *mean square error* (RMSE, the square root of the average of the sum of squares of differences computed in the MBE). That is:

$$MBE = 1/n \sum (y_i - Y_i), \tag{1.10}$$

$$RMSE = \left(\left(\sum (y_i - Y_i)^2 \right)/n \right)^{1/2}, \tag{1.11}$$

where Y_i are estimated from the model, and y_i are measured data, and n is the number of data points compared.

Theoretical physics is sometimes combined with empirical data correlations to produce semi-empirical models. The user of these models must be aware that often the data used to develop the model is used to verify or report the model performance. Verification of the model against independent data sets not used in the model development provides a more accurate estimate of model performance.

Users should keep in mind that it is impossible for empirical models to be more accurate than the measured data use to produce the models. This generally means that claimed model accuracy of better than 5% or so is probably unrealistic [64,65]. Also, measured data inevitably have scatter – represented by the Root Mean Square Error of Eq. (1.11) – about the correlation equations for the model data, which must be combined with the measurement uncertainties to assess the model performance.

1.3.1 Models for Direct Beam Irradiance

Empirical models are based on correlations, functions derived through linear or multi-linear regression analysis (curve fitting). It is assumed that measured solar radiation data can be described as a function of some other independently measured or available variables or parameters. These independent variables can range from simple, single variables such as temperature, to more complex combinations of temperature, relative humidity, day length, cloud cover, sunshine duration in hours, and so on [59–64].

A simple approach is to lump all of the effects of atmospheric transmittance into the one parameter of air mass (m), or path length through the atmosphere, and a 'bulk' attenuation factor [66]. Such a model for computing the direct beam irradiance (DNI) at the surface on a clear day was presented by Meinel and Meinel in 1976 [67]. That model requires only the calculation of the extraterrestrial direct beam radiation, 1366.1 Wm^{-2}, corrected for the Earth-Sun distance; a factor for average clear sky transmittance at air mass $m = 1.0$, equal to 0.7; and an average (exponential) extinction coefficient (0.678) per unit air mass m (at sea level):

$$DNI = 0.7 \, R_c \, I_o^{(0.678 \, m)}, \tag{1.12}$$

If one begins to consider site specific issues, or atmospheric conditions, additional functional terms are needed. In 1970, Laue derived a model involving the altitude, h, in kilometers [68]:

$$DNI = R_c \, I_o[0.7(1.0 - 0.14 \, h)R_c \, I_o^{(0.678 \, m)} + 0.14 \, h], \tag{1.13}$$

These extremely simple models provide only a back-of-the-envelope estimate of clear sky DNI throughout the day, as the air mass decreases and increases.

1.3.2 Atmospheric Component Transmittance

1.3.2.1 Linke's Model

A popular method of estimating the bulk transmittance of the atmosphere is the Linke Turbidity Factor (T_L) proposed by Linke in 1922 [66]. Linke's proposal is based on air mass m, the optical thickness, τ_D, of clean (aerosol free), dry (no water vapor) atmosphere. The transmitted DNI irradiance is expressed as:

$$DNI = R \, I_o \exp(-\tau_D T_L m), \tag{1.14}$$

or, solving for T_L:

$$T_L = \ln(DNI/(R \, I_o))/(-\tau_D \, m), \tag{1.15}$$

with the clean dry atmosphere optical thickness being calculated as:

$$\tau_D = 0.128 - 0.054 \log(m), \tag{1.16}$$

Thus T_L represents the number of clean dry atmospheres necessary to produce the observed attenuation. Databases of *Linke turbidity* are available on line through the European Solar Energy for Professionals website [69].

A 2012 technical report by Reno, Hansen, and Stein [70] reports the accuracy of these simple clear sky models, when compared with measured DNI at several different sites to be on the order of 10%, at best.

1.3.2.2 Comprehensive Transmittance Models

Somewhat better results can be achieved by considering the atmosphere in somewhat more detail. The individual constituent mixed gases (nitrogen, oxygen, carbon dioxide, etc.) water vapor, stratospheric ozone, and aerosols, or small scattering centers suspended in the atmosphere can be considered to have a transmittance (ratio of what impinges at the top of the atmosphere to what remains at the ground level). Each transmittance can be parameterized in terms of the air mass and concentration or amount of a constituent present in the atmosphere. This is the basis of models developed by Bird and Hulstrom [61,62], and earlier work by Watt *et al.* [59], and Atwater and Ball [60]. More recently Gueymard [63] and many others have developed models using a similar approach. The total transmittance of the atmosphere, T, is then calculated as the product of the terms:

- T_r: transmittance due to Rayleigh scattering;
- T_a: transmittance due to aerosol properties;
- T_g: transmittance due to optical properties of gases;
- T_o: transmittance due to ozone (in the stratosphere);
- T_w: transmittance of water vapor.

Therefore, the DNI can be written as:

$$DNI = I_o R_c T_r T_a T_g T_o T_w, \tag{1.17}$$

where I_o is the extraterrestrial direct beam irradiance calculated using Eq. (1.4) and R_c is the Earth-Sun distance correction factor calculated using Eq. (1.6).

The equations defining the above transmittance parameters for the direct beam model of Bird and Hulstrom [61,62] are discussed in the next paragraphs. Note that many of these expressions depend on the air mas (m) which essentially equals $1/\cos(z)$, z being the solar zenith or altitude angle (z). The computation of solar zenith including effects of atmospheric refraction is based on site location (latitude, longitude) and accurate knowledge of local time. Example algorithms are reported in Iqbal [25], Michalsky [71] and Reda [72].

The expression for the Rayleigh transmittance (T_r) is:

$$T_r = \exp(1.0 + M_p - M_p^{1.01})(-0.0903 M_p^{0.84}), \tag{1.18}$$

where M_p is the air mass corrected for station pressure or altitude: $M_p = m$ (station pressure)/ (sea level pressure).

The expression for the mixed gas transmittance (T_g) can be also calculated as a function of M_p:

$$T_g = \exp(-0.0127 M_p^{0.26}), \tag{1.19}$$

The total column ozone amount O_z in atmospheric-cm is the total equivalent depth in cm of all ozone in a vertical column of air, condensed from the atmosphere: $O_m = m \, O_z$. From this magnitude the ozone transmittance (T_o) can be calculated as:

$$T_o = 1 - 0.1611 O_m(1.0 + 139.48 O_m)^{(-0.3035)} \\ -(0.002715 O_m)/(1 + 0.044 O_m + 0.0003 O_m^2), \tag{1.20}$$

Historical total ozone estimates are available from the NASA Ozone and Air Quality website [73].

There is a frequently cited model due to Heuklon [74] to estimate ozone. This model estimates total column ozone based on the day of the year and the latitude and longitude of the site. Since the ozone absorbs strongly in the ultraviolet, and only weakly in the visible part of the spectrum, the impact of T_o on the transmittance of total DNI through the atmosphere is nearly negligible. The difference in total DNI beam irradiance resulting from a low ozone amount of 0.25 atm-cm and a higher amount of 0.35 atm-cm for a mid-latitude continental site, averaged over a year of 4400 daylight hours, is $5 \pm 1.5 \, \mathrm{Wm}^{-2}$ out of an average DNI of $760 \, \mathrm{Wm}^{-2}$, or less than 0.75%.

Water vapor transmittance (T_w) for precipitable water vapor amount PW atm-cm, using: $W = m \, PW$:

$$T_w = 1 - 2.4959 \, W / [(1 + 79.034 \, W)^{0.6828} + 6.385 \, W], \tag{1.21}$$

The aerosol transmittance (T_a) is:

$$T_a = \exp\{(-T_{au}^{0.873})[1.0 + T_{au} - (T_{au}^{0.7088})]m^{0.9108}\}, \tag{1.22}$$

where

$$T_{au} = 0.2758 \, T_{a3} + 0.35 \, T_{a5}, \tag{1.23}$$

and T_{a3} is the aerosol optical depth at 380 nm, and T_{a5} is the aerosol optical depth at 500 nm. Here, optical depth (OD) is the dimensionless attenuation of incident radiation, I, as a function of unit path length through a distance d of absorbing material: $I_{out} = I \exp(-OD \, d)$. That is, $OD = -\ln(I_{out}/I)$ per unit d. Sometimes OD is expressed without units, or as attenuation per unit of measure, as in cm^{-1}, m^{-1}, km^{-1}, etc. In our case, the unit of measure for path length is the dimensionless air mass, m.

Here, T_{au} was based on a rural aerosol distribution of Shettle and Fenn [75], which could be approximated from routine measurements from the National Oceanic and Atmospheric Administration (NOAA) at the time of the model development (1982). Measured aerosol optical depth data is available from the National Aeronautics and Space Administration (NASA) AERONET network described below.

So called *broadband aerosol optical depth* or BAOD may be derived from the attenuation of the *DNI* beam, $I/(R_c I_o) = \exp(-BAOD \, m)$ once the attenuation due to ozone and water vapor is accounted for. Molineaux and Ineichen [76] have shown that BAOD is equivalent to the spectral aerosol optical depth at a specific wavelength, namely 700 nm. Because optical depth is a decreasing function of increasing wavelength, multiplying a BAOD by ~1.5 would be equivalent to spectral optical depth at 500 nm, and $1.8 \cdot BAOD$ is approximately the spectral AOD at 380 nm.

Total precipitable water (the equivalent depth of water in centimeters if condensed out of the entire atmosphere above a location) may be estimated from relative humidity as described by Garrison and Adler [77] using the following sequence:

1. The saturated vapor pressure (E_s) of water vapor, in mbar, can be calculated from temperature, T, in K, using:

$$\log_{10}(E_s) = -8.430 - 1827.178/T - 71208.271/T^2, \tag{1.24}$$

2. The pressure corrected water vapor pressure (E), in mbar, can be calculated from relative humidity (RH) and station pressure (P) with respect to sea level atmospheric pressure of 1013.25 mBar:

$$E = RH\, E_s(P/1013.25), \tag{1.25}$$

3. Finally, the estimated water vapor amount, W, in atmosphere-millimeters (atm-mm) is:

$$W = 1.45\, E + 1.5, \tag{1.26}$$

Aerosol and water vapor data needed for these equations is available from the National Aeronautics and Space Administration AERONET network [78].

1.3.3 Estimating Direct Beam Radiation from Hemispherical Data

Total hemispherical irradiance, GHI, is the most common solar radiation measurement available, because it uses the most simple measurement equipment, a pyranometer on a horizontal surface. Measurements of DNI and GHI have been used to develop correlations between the two measured quantities by many authors. Below we discuss an extremely simple model and a relatively complex model. Both are based on the idea of clearness index, or bulk transmittance of the atmosphere for one or the other or both of the components.

1.3.3.1 Boes Simple Correlation

Boes *et al.* [79] used one year of data from three United States stations with measured *DNI* and *GHI* data. Their correlation produces DNI (in Wm^{-2}) as a function of global horizontal clearness index $K_t = (GHI)/(I_o \cos(z))$ and solar zenith angle, z:

$$DNI = 400\ Wm^{-2}\ \text{for}\ z = 80°\ \text{and}\ K_t > 0.5 \tag{1.27.a}$$

$$= -520 + 1.8 K_t\ Wm^{-2}\quad \text{for}\ 0.3 \le K_t \le 0.85, \tag{1.27.b}$$

$$= 1000\ Wm^{-2}\ \text{for}\ K_t > 0.85 \tag{1.27.c}$$

The original work only utilized K_t values between 0.3 and 0.85; and reported monthly values of clear sky DNI $(K_t > 0.85)$ ranging from $950\ Wm^{-2}$ to $1{,}050\ Wm^{-2}$. They used the value of $1000\ Wm^{-2}$ (a value that certainly can occasionally be exceeded, as a reasonable maximum to simplify the model. Reno, Hansen, and Stein [70] report in a 2012 publication an accuracy of only about $\pm20\%$ for this model.

1.3.3.2 The Maxwell Direct Insolation Simulation Code (DISC) Model

Maxwell [80] at NREL developed his *direct insolation simulation code* (DISC) model using K_n, the direct clearness index $(DNI/(R_c\, I_o))$, rather than the DNI/GHI ratio for the dependent variable. He developed the model using one year of measured data for Atlanta, Georgia, USA. Three United States sites with measured DNI and GHI were used to validate the model. The sites have diverse climates: Brownsville, Texas; Albuquerque, New Mexico; and Bismarck, North Dakota.

Maxwell used parameterizations of K_n with respect to K_t and K_n bins of width 0.05 and air mass, m. He derives a general formulation of K_n versus K_t (from measured GHI data). The model equation produces the change or deviation ΔK_n from clear sky transmittance, denoted by K_{nc}. Modeled DNI is computed as $DNI = I_o\, K_n$, where:

$$K_n = K_{nc} - \Delta K_n, \tag{1.28.a}$$

$$\Delta K_n = a + b \exp(c \cdot m), \tag{1.28.b}$$

The clear sky limit K_{nc} is computed from a polynomial in air mass, m:

$$K_{nc} = 0.866 - 0.122\, m + 0.0121\, m^2 - 0.000653\, m^3 + 0.000014\, m^4, \tag{1.29}$$

To determine the coefficients a, b, and c, the K_t space was partitioned into two parts, $K_t \leq 0.6$ and $K_t > 0.6$.
For $K_t \leq 0.60$:

$$a = 0.512 - 1.56\, K_t + 2.286\, K_t^2 - 2.222\, K_t^3, \tag{1.30.a}$$

$$b = 0.370 + 0.962\, K_t^3, \tag{1.30.b}$$

$$c = -0.280 + 0.932\, K_t - 2.048\, K_t^2, \tag{1.30.c}$$

For $K_t > 0.60$:

$$a = -5.743 + 21.77\, K_t - 27.49\, K_t^2 + 11.56\, K_t^3, \tag{1.31.a}$$

$$b = 41.4 - 118.5\, K_t + 66.05\, K_t^2 + 31.90\, K_t^3, \tag{1.31.b}$$

$$c = -47.01 + 184.2\, K_t - 222.0\, K_t^2 + 73.81\, K_t^3 \tag{1.31.c}$$

The DISC model has bias (mean of differences between measured and model data) errors of about $\pm 50\,\mathrm{Wm^{-2}}$ and random (RMSE) errors of about $\pm 150\,\mathrm{Wm^{-2}}$ depending on the site. There are caveats about the DISC code. One is that the range of zenith angles and air masses for the model development and validation are only for United States continental sites with latitude ranging from 28°N to 45°N.

Perez et al. [81] attempted to improve the DISC model by application of corrections based on water vapor and aerosol estimates, however the improvements were so slight that measured data uncertainty are larger than the reported magnitude of the updated model. Secondly, the model was derived based on hourly average GHI and DNI data. There is no guarantee that the model equations apply equally to data measured at different (higher) time resolution. There is still a need for a more accurate, universal model for converting more widely available GHI data to DNI data for CPV applications [82,83].

1.4 Modeling Spectral Distributions

In sections 1.1.4 and 1.1.5 there was cursory mention of the extraterrestrial solar spectral distribution, the effect of the atmospheric filter, and the issues with spectral sensitivities of PV and optical component materials in CPV applications. Many spectrally dependent parameters for materials can be measured in the laboratory. These laboratory measurements must be combined with solar spectral distributions to compute either detailed or overall CPV system performance. Terrestrial DNI solar spectral power distribution data are very rare, so one must resort to models for generating realistic estimates of CPV performance. As with broadband DNI models, the properties of atmospheric constituents, this time with spectral dependence included, are used to modify an ETR spectral power distribution[1]. We briefly describe here two popular and easy to use spectral models; the simple Spectral 2 (SPCTRL2) model of Bird [84,85], and the moderately complex *simple model of the atmospheric radiative transfer of sunshine* (SMARTS) of Gueymard [86]. The following chapters will discuss the impact of spectral irradiance distribution on concentrating system performance.

1.4.1 Bird Simple Spectral Model (SPCTRL2)

The Bird Simple Spectral Model, SPCTRL2, developed by Bird and Riordan [84,85], at NREL, computes clear sky spectral direct beam, hemispherical diffuse, and hemispherical total irradiances on a prescribed receiver plane – tilted or horizontal – at a single point in time. This model was used as the basis for early (1995 and earlier) and now inactive versions of reference solar spectra standards for PV applications such as ASTM E891/E892 and G159 (see the section on Consensus Standards).

For tilted planes, the user specifies the incidence angle of the direct beam (FORTRAN version) or the tilt and azimuth of the plane (Excel$^@$ and C versions). The wavelength spacing is irregular, covering 122 wavelengths from 305 nm to 4000 nm. Aerosol optical depth, total precipitable water vapor (cm), and equivalent ozone depth (cm) must be specified by the user. No variations in atmospheric constituents or structure are available. There is no separate computation of circumsolar radiation. The direct beam spectral irradiance is assumed to contain the circumsolar radiation within a 5° solid angle.

The equations for the broadband Bird clear sky *DNI* model are re-written in terms of functions of wavelength [84], to generate the spectral *DNI*, $I_b(\lambda)$:

$$I_b(\lambda) = R_c\, I_o(\lambda) T_r(\lambda) T_a(\lambda) T_g(\lambda) T_o(\lambda) T_w(\lambda), \tag{1.32}$$

where $I_o(\lambda)$ is the air mass zero (AM0) or extraterrestrial spectrum; T_r, T_a, T_g, T_o, and T_w are wavelength dependent transmittance functions for Rayleigh scattering, aerosols, uniformly mixed atmospheric gases, ozone, and water vapor, respectively; R_c is the Earth-Sun radius vector correction. Bird used the 1985 Wehrli AM0 spectrum [87] as the starting point for his model. Only 122 irregularly spaced wavelengths at approximately 10 nm intervals from 305 nm to 4000 nm were selected to simplify the spectral calculations. Many absorption features are captured using only three or four wavelengths to identify the shoulders and greatest absorption of the features. The model is easily implemented in about 50 lines of computer programming code and is estimated to be accurate to about 5% to 10% when compared with measured spectra. Source code in FORTRAN, C, and Microsoft Excel$^©$ versions and the ancillary data files (absorption coefficients and ETR spectrum) are available from the NREL Renewable Resource Data Center [88].

---

[1] An extensive catalog of many atmospheric radiative transfer model codes, some very complex and requiring highly esoteric and difficult to locate input parameters, is available at http://en.wikipedia.org/wiki/Atmospheric_radiative_transfer_codes

1.4.2 Simple Model for Atmospheric Transmission of Sunshine (SMARTS)

Gueymard [86] developed a simple model for atmospheric transmission of sunshine, SMARTS, to improve upon the performance of SPCTRL2, improve the existing PV reference standard spectral distribution, and provide more flexible options for solar engineering applications involving solar spectral power distributions. Version 2.9.2 of SMARTS is the basis for the present ASTM and IEC reference spectra ASTM G-173 and ASTM G-177, IEC 60904 [89–91] used for photovoltaic performance testing and materials degradation studies. The model source code is comprised of about 5000 lines of FORTRAN code and is supplied with about 50 ancillary data files for various atmospheric profiles, spectral ground reflectance (albedo), and atmospheric gas absorption. SMARTS computes clear sky spectral irradiances (direct beam, circumsolar, hemispherical diffuse, and total on a tilted or horizontal receiver plane) for specified atmospheric conditions. Users choose one of 10 standard atmospheres or can build their own atmosphere profile. Output for one or many points in time or solar geometries (tilted surfaces, etc.) may be selected.

The algorithms used by SMARTS were developed to match within $\pm 2\%$ the output from the very general MODTRAN complex band models developed by the US Air Force Geophysical Laboratory [92,93]. The algorithms are implemented in compiled FORTRAN code for Macintosh and PC platforms. Source code is included within the model package. The algorithms are used in conjunction with files for atmospheric absorption of atmospheric components and spectral albedo functions. The spectral resolution is 0.5 nm for 280–400 nm, 1 nm for 400–1750 nm, and 10 nm for 1750–4000 nm.

SMARTS users construct text files of 20–30 lines of simple text and numbers to specify input conditions and up to 28 spectral output parameters. Users can specify field-of-view angles for direct-beam computations and a separate computation for the circumsolar component. Gaussian or triangular smoothing functions with user-defined bandwidth can also be specified to compare model results with measurements made with the specified passband. Users can specify only ultraviolet (280–400 nm) computations for erythemal dose, UV index, and similar measurements. Photometric (luminous flux) computations, weighted by a selected photonic response curve, can also be specified. Model output consists of spreadsheet-compatible text files with header information and a record of the prescribed conditions.

Both SPCTRL2 and SMARTS have been validated in many comparisons with measured spectral data [94–100] and generally shown to be within the spectral measurement uncertainty limits, which range from 10% in the ultraviolet to 2%-3% in the visible, and 5% in the infrared, greater than 1200 nm. While more complex, SMARTS is much more flexible than SPCTRL2 and can provide much more detailed information. Given the same atmospheric conditions, however, the two models produce about the same amplitudes of modeled data.

1.4.3 Spectral Distributions from Broadband Data

The SPCTRL2 and SMARTS models both compute clear sky solar spectral power distributions. It is natural to deploy CPV systems in places where the skies are most likely to be clear, to optimize the conversion of DNI beam radiation into energy. At times, clouds of various types and density may intervene and affect both the intensity and spectral content of the direct beam. Should the designer wish to consider these effects upon the CPV system design there very few examples of measured spectral data publically available. One such set is the NREL Spectral Solar Radiation Database [101–103]. The database of over 3000 measured spectra under various conditions at three sites is available on line [104]. The database documentation, including uncertainties associated with the data, is available in Riordan *et al.* [101,102], and Myers [103].

In the absence of measured data, there are published spectra and spectral models for estimating spectral distributions affected by clouds. In 1990, the Commission Internationale de Éclairage (CIE), or International Commission on Illumination published a collection of solar spectra [105] based on the SPCTRL2 model and a model by Justus [106] to account for cloud effects. Since then two models based on the SPCTRL2 model and addressing cloud effects have been published. These are the SEDES (Solar Energy Data Acquisition System, *Erwerb* in German) of the Center for Solar and Hydrogen Science Center (ZSW) in Stuttgart, Germany [107]; and the TMYSPEC model modified version of SEDES developed at NREL [108]. The original SEDES model converts hourly measured broadband *GHI* data to spectral data on a prescribed tilted surface. The TMYSPEC model converts *typical meteorological year* (TMY) or measured GHI data to both GHI and DNI spectral data on a surface of prescribed tilt. The concept of TMY is discussed in more detail in section 1.5.2.

Both SEDES and TMYSPEC models modify the computed clear sky SPCTRL2 model with the ratio of measured broadband data to theoretical clear sky broadband data and empirically derived cloud cover modifiers which are spectrally dependent. The theoretical clear sky broadband data is computed by integrating the modeled clear sky spectral data. For SEDES the resulting modeled spectral power distributions have a spectral range of 300 nm to 1400 nm, and for TMYSPEC the spectral range is 300 nm to 1800 nm. The SPCTRL2 computed spectra are interpolated to 10 nm wavelength intervals. TMYSPEC also generates DNI beam spectra, while the SEDES model only produces the total hemispherical spectra on a prescribed tilted surface. The modeled spectra compare with measured spectra to an uncertainty of about $\pm15\%$ for the total hemispherical spectra, and $\pm20\%$ for the direct beam spectra.

1.5 Resources for Broadband Estimates of CPV Performance

The first choice and best quality data for estimating the performance of a CPV system at a certain location is that measured in place for a long period of record at relatively high temporal resolution, such as hourly average data. For environments with high variability in solar radiation and weather data, higher time resolution data, such as data sampled at 1 minute, 5 minute, or 10 minute intervals is desirable. This permits the study of the impact of input transients on the CPV system. The ASTM 2527 consensus standard [109] was mentioned above describing a methodology for estimating the performance of CPV system after it has been deployed in the field. However, these types of measurements can only be used to validate or evaluate the accuracy of the original design of the system. System design usually includes some modeling of system performance based on broadband DNI beam data, sources of which are described in the next section.

1.5.1 Broadband Direct Beam Radiation Data Resources

The only archive of worldwide measured solar radiation data is maintained at the WMO World Radiation Center (WRC) at St. Petersburg, Russia [110]. Member nations contribute measured data in various formats which are quality assessed and archived at the center. DNI beam radiation is sometimes included from some sites, however only rarely. The units of the archived data are in energy, Joules (or megaJoules, MJ) rather than power. The data is available on line [111]. Data from a station may consist of only a few months of data to many years, depending on the support and operation of the stations by the member nation.

Unfortunately, long periods of recorded solar radiation data, and especially DNI beam radiation, are generally very rare. This means that broadband data for most locations must be modeled. Several sources of data have been assembled slowly over the past 40 years at various national and international levels. In the United States, the National Climatic Data Center (NCDC) in conjunction with Sandia National Laboratory first developed a Solar and Meteorological (SOLMET) database of hourly average measured data for 23 National Weather Service (NWS) solar and meteorological measurement stations [112]. Empirical models were then developed (using the measured data) to produce SOLMET type data for an additional 200 NWS sites where only meteorological data were available. The modeled data were called SOLMET ERSATZ (German for 'substitute') sites [113]. SOLMET/ERSATZ data covered the period from 1952 to 1970. Various original NWS monitoring sites were closed down and either eliminated or re-instrumented between 1975 and 1985. In 1995 the United States solar monitoring network was shut down, and replaced with a 'research' network of only seven stations. See Renné *et al.* [114] for the status of United States measurement networks as of 1999.

In 1990, NREL began developing the first version of the National Solar Radiation Data Base (NSRDB) [115,116]. The NSRDB consists of (95% modeled) solar hourly data for DNI, GHI, DHI, total precipitable water, and broadband aerosol optical depths, along with NCDC reported meteorological data and conditions. NREL developed the Meteorological and Statistical (METSTAT) model to produce the estimated parameters, and algorithms to construct serially complete hourly data files for 239 stations for the period from 1960 to 1990 [117]. The METSTAT model was also modified to produce the Climatological Solar Radiation (CSR) model [118], to produce monthly mean solar radiation data for various renewable energy projects under United Nations Solar and Wind Energy Resource Assessment (SWERA) program[2] and for the King Abdul Aziz City for Science and Technology (KACST) to produce a solar radiation atlas for Saudi Arabia [119].

In 2001, and again in 2012, the NSRDB was updated and expanded [120]. The updates included extending the period of record first from 1990 to 2000, and again from 2000 to 2010. The expansion included the estimation of solar radiation components on a 10 kilometer (approximately) spatial resolution, using Earth orbiting meteorological satellite data (From US GOES, or European METEOSAT platforms) in conjunction with solar radiation estimation models [121]. The quality of meteorological satellite data needed becomes widely available in 1995. The updated NSRDB for 1995 to 2010 now contains approximately 100 000 grid cells, 10 km on a side, with DNI, GHI, and DHI solar radiation estimates representing 1 hour average values. The United States and other nations' satellite conversion models are now used to estimate broadband solar radiation in many countries.

Maps of monthly and annual average solar radiation values for many countries have been developed using these techniques. A few examples are shown in section 1.7 on Direct Solar Radiation Climates. The underlying hourly data for a specific cell may be available from several sources. For the US, NREL provides access through the *Solar Prospector* website [122], and *PVWatts* PV system simulator (for flat plate collectors, but includes solar 2-axis and 1-axis tracking flat plates) [123]. Commercial firms are now providing near real-time (up to

---

[2] (see http://en.openei.org/apps/SWERA/)

'yesterday') estimates of solar radiation resources for specific locations for a fee. These are based on publicly available weather satellite data and their own proprietary satellite conversion algorithms [124,125].

NASA has also prepared long term (22-year) yearly and average satellite based estimates of solar radiation and meteorological data and made it available through the Surface Meteorology and Solar Energy (SSE) website [126][3]. Meteorology and solar radiation for SSE Release 6.0 were obtained from the NASA Science Mission Directorate's satellite and re-analysis research programs [127].

Parameters based upon the solar and/or meteorology data were derived and validated based on recommendations from partners in the energy industry. The 2013 release 6.0 extends the temporal coverage of the solar and meteorological data from July 1983 through June 2010. Most of the data is available as monthly averaged data, but individual daily total radiation are available for each year. The underlying (hourly or sub-daily total) data are not available. The SSE data is based on one degree by one degree cells and is available for single cell, or regions bounded by coordinates provided by the user.

The European Union website PVGIS [128], provides a catalog of, and links to available solar radiation datasets. There is also an online publication containing an inventory of solar databases. Examples of several solar radiation data bases are the *Solar Energy Services for Professionals* (SODA) website where various forms of modeled and measured data are available online [129]. In addition, there is also a European Solar Radiation Atlas produced by the *École des Mines de Paris* [130,131].

A comprehensive world-wide set of combined measured and modeled solar radiation data, including the ability to 'downscale' to higher time resolution (1 minute) from hourly average data is the Swiss METEONORM data set, available from MeteoTest (where some on-line computations are available) [132].

It is important to note that the uncertainty of DNI beam data in any of the models of this type has been shown to be on the order of ±20% for most continental, seasonally variable climate sites, such as Europe, or mid-continent United States. The uncertainty may be expected to be lower for sites suitable for CPV deployment, where consistent clear sky conditions are preferred. The only way to verify this hypothesis is to conduct at least some measurements at a proposed site and evaluate the correlation of the measured and satellite (or other model) derived data sets. This approach is currently described as a 'measure, correlate, predict' (MCP) or 'measure, correlate, evaluate' (MCE) program. See for example Thuman, Schnitzer, and Johnson [133].

1.5.2 Typical Meteorological Year Data for CPV Performance Estimates

The concept of Typical Meteorological Year or TMY was developed by the North Atlantic Treaty Organization (NATO) for heating and cooling applications in 1977 [106]. Data sets for the United States and many international sites have been developed over the years [107–109].

These data are used extensively in the modeling of building heating and cooling loads and building energy performance models. TMY broadband solar and meteorological data have also become popular as input to solar energy conversion system modeling software because of the wide availability of the data.

---

[3] http://eosweb.larc.nasa.gov/sse/

TMY data are serially complete, 8760 hourly records of solar GHI, DNI, and DHI data in conjunction with meteorological parameters such as wind speed, ambient and dew point temperatures, water vapor and broadband aerosol optical depth. The TMY hourly data are selected from a long time series (many years) of hourly data based using a specific statistical analysis and parameter weighting scheme methodology.

For a given month, the methodology is structured so as to find the closest match to an 'average' cumulative frequency distribution of parameters from the collection of all cumulative frequency distributions for the particular month. The single month of hourly values selected from the collection of available months (filtered to remove extreme events such as volcanic events, etc.) is considered 'typical' for the site. The concatenation of the typical months (probably selected from different years) into a typical year is considered 'typical', and somewhere between extremes, for the site in question.

It is wise to remember that since the T in TMY represents 'typical' and that weather and solar conditions vary considerably from typical most of time, that TMY based energy conversion performance predictions may have a great deal of uncertainty, dependent on the inter-annual (and intra-annual) variability of the weather at particular site.

Nonetheless, TMY data sets are very popular input data for modeling PV conversion technology performance for a 'typical' or 'average' year. Many system modeling software packages include these data sets as a library of sites to select from. Such packages also permit 'TMY format' data from other sources to be used as input. The term 'TMY' has become rather generic, but various versions are available, so the user must be knowledgeable of the time frame and formats used. The original TMY [134] was developed from the 1952–1975 SOLMET/ ERSATZ [112,113] data, and based on true solar time, not local standard time at the sites. The format of the earliest TMY data consisted of concatenated (with no spaces), variable length data fields in American Standard Code for Information Interchange (ASCII) text that required parsing or filtering to select the correct values.

The TMY2 [135] developed from the 1960–1990 US NSRDB used the same concatenated variable field length format, but was constructed with slightly different data fields. Local standard time for the site is used. Weighting factors for the solar radiation data, and units for some variables are different and are described in documentation for the TMY2. For a full description, see [136].

The US TMY developed from 1960 to 2010 NSRDB data are denoted as TMY3 files [137], and are available as comma separated variable (CSV) format suitable for spreadsheet import, and are not necessarily compatible with all PV performance models unless they are reformatted. For a full description, see [138].

Because the initial primary application of 'typical year' weather data was the design of buildings to meet indoor environmental needs with respect to the outdoor environment, many international versions of the TMY data are available. The US Department of Energy maintains a library of these (and similar) data sets ('weather year for energy calculation') [139].

The interpretation of results using typical year input data as opposed to multi-year simulation runs may be misleading, especially in highly variable climates. Extreme weather and man-made events such as volcanic eruptions, biomass burning, el-Niño and la-Niña events, etc. may strongly attenuate *DNI*, and are excluded from typical weather year data. Multi-year simulations of performance may provide a much better picture of year to year performance variations that the system may be sensitive to [127]. See section 1.7.3 for a discussion of intra- and inter-annual variability of resources.

1.5.3 CPV Spectral Performance Issues

Because terrestrial solar spectral DNI data is even more complex and sparse than broadband data, model calculations are the tools of choice in studying the sensitivity of a CPV system performance to spectral variations. A very few studies have been made that demonstrate the relative importance of spectral variation in this regard [133,140,141]. This means that the computation of the product of optical properties, PV material spectral response, and spectra under different conditions are required. The spectral distribution of the DNI beam radiation is a strong function of the air mass (as well as atmospheric condition), so correlations of spectral effects with air mass are often used as surrogate for the above complete spectral computations.

Several authors have shown that this approach can lead to erroneous and misleading results [142]. Best practice for evaluation of CPV sites and system design is to collect and process as much information as possible on the parameters that affect spectrally significant performance parameters. Even if there is no desire to perform spectral modeling, at least an estimate of the variation in performance can be obtained from knowledge of the variability of parameters that affect the DNI spectral power distribution.

The most important examples are: variability of the aerosol optical depth, including that of (possibly polluted) upwind sites for the prevailing wind patterns, sand particle size and transport, terrain type (vegetated or not, hard pan desert, such as the American southwest, or loose, fine grained desert surface, such as the Gobi or sub-Saharan deserts), nearby biomass burning, and frequency and type of clouds. If there is concern that these parameters are indeed highly variable, modeling of effects using spectral models for extreme conditions is recommended. See Gueymard [143].

1.6 Sunshape

Recall that section 1.1.1 discussed the geometrical relationship between the Earth and Sun that result in the 0.52° to 0.54° solid angle (cone) of slightly divergent, quasi-parallel bundle of optical rays in the direct solar beam at the top of the atmosphere. Section 1.1.5 on the atmospheric filter described scattering and absorption processes in the atmosphere that attenuate and redistribute the energy in the DNI beam passing through the atmosphere. These effects result in what the CPV community has come to describe as the 'sunshape' as projected on the sky dome, or the conversion device in the CPV system. The desire to optimize the capture of photons in a CPV system in combination with the optical properties of the system design have lead designers to study the field of view and tracking accuracy of CPV systems that depend on sunshape in detail.

1.6.1 The Solar Disk

The sun is essentially a large sphere of extremely high temperature gases whose size is determined by the outward pressure of radiation and opposing gravitational force of the mass of the solar constituents. As seen from Earth above the atmosphere, the solar disk, or photosphere of the sun, appears to have a somewhat ill-defined edge with diminishing brightness. The disk is surrounded by a faint 'corona' or 'solar atmosphere' of ejected, high energy ionized atoms emitting some visible, X-ray, and radio wave energy over a wide spectral range. The photograph of the transit of Venus on June 6, 2012 (Figure 1.8) illustrates the ill-defined edge seen in optical ('white-light') or other spectral ranges. This decrease of intensity in the

Figure 1.8 Solar Limb Darkening. Source: By Brocken Inaglory (Own work) [CC BY 2.5 (http://creativecommons.org/licenses/by/2.5)], via Wikimedia Commons

image of the sun as one moves from the center of the image to the edge is the so-called limb darkening.

Limb darkening results from the combination of the depth dependence of temperature (through the near surface solar atmosphere, and surface layer of the photosphere itself) and the *decreasing* optical depth of the path length as one measures the intensity from the center of the disk to the solar limb. Eddington derived the relationship between optical depth, source function temperature and integrated intensity [144]. A closed form representation of limb darkening as a function of the angle θ from the center of the sun, $(\theta = 0°)$ to the limb $(\theta = 90°)$, using $\mu = \cos(\theta)$ is:

$$I(\mu) = I(0)(3/5)(\mu + 2/3), \tag{1.33}$$

Table 1.3 shows the diminution of the intensity of the radiation as one moves from the center to the limb as computed with the Eddington approximation – using Eq. (1.33) – and the

Table 1.3 Relative intensity of solar disk from center $(\theta = 0°)$ to limb $(\theta = 90°)$

θ (°)	$\mu = \cos(\theta)$	Eddington	Measured at 500 nm
0	1.0	1.00	1.00
37	0.8	0.88	0.88
53	0.6	0.76	0.74
60	0.5	0.70	0.68
66	0.4	0.64	0.64
72	0.3	0.58	0.52
78	0.2	0.52	0.43
84	0.1	0.46	0.32
87	0.05	0.43	0.20
89	0.02	0.41	0.14

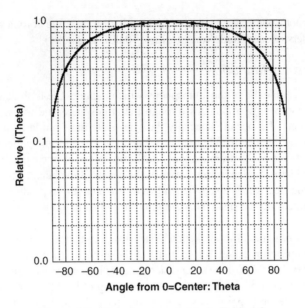

Figure 1.9 Plot of measured data in Table 1.3

measured intensity in the visible, at 500 nm. Figure 1.9 is a plot of a polynomial fit to the measured data in Table 1.3.

Figure 1.10 is a plot of relative intensity as a function of central angle θ for various wavelengths in μm, illustrating that limb darkening is a function of the wavelength of emitted light in the solar spectrum [145]. The shorter the wavelength, the stronger the reduction in intensity as one moves away from the disk center. As soon as the observer line of site moves off

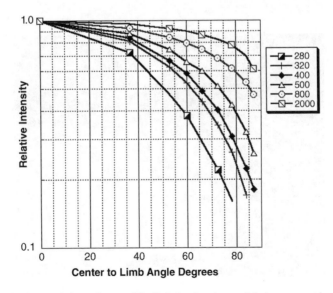

Figure 1.10 Spectral dependence of limb darkening (curve labels are wavelength in nm)

the edge of the photosphere, the emitted optical radiation from the corona falls by more than three or four orders of magnitude (or a factor of 0.001 to 0.0001).

1.6.2 Circumsolar Radiation

As the direct beam at the top of the atmosphere, which includes the limb darkened radiation at the periphery of the beam, propagates through the atmosphere and the scattering processes described in section 1.1.5 occur, especially the Mie scattering from large particles, a circumsolar component of the beam radiation is created. Mie scattering preferentially re-directs photons in the direction of the propagation of the beam, however into a much larger solid angle than the beam and large angles from the axis of the center of the beam. Mie scattering around the solar disk in the clear sky is also independent of wavelength, so produces an appearance of 'white' light surrounding the disk. This circumsolar radiation is also often referred to as the solar aureole.

The radiation from the limb darkened solar disk irradiance is also attenuated by the atmospheric absorption processes. From a site at the surface, if a very narrow field of view instrument (say 0.05°, or 1/10 of the solar disk diameter) is scanned from the center of the solar disk to a few degrees beyond the limit of the solar disk a profile of the intensity of the radiation can be established. Several approaches of experimentally accomplishing this measurement of sunshape under various atmospheric conditions have been developed since the early 1970s.

1.6.2.1 The Lawrence Berkeley Circumsolar Telescope Data

The Lawrence Berkeley Laboratory (LBL) at the University of California in Berkeley, developed a circumsolar telescope with a field of view of 0.025° or 1/20 of the solar diameter [146,147]. A photograph of one of the telescopes rehabilitated by NREL to acquire spectral data is shown in Figure 1.11.

Figure 1.11   LBL Circumsolar Telescope (without data acquisition system). Source: © David Myers

Four copies of the original LBL instrument were deployed at eleven different sites in the United States. The telescopes were configured to automatically perform scans over 6° of arc, with the solar disk at the center, at approximately 10 minute intervals. The scan data was acquired in steps of 1.5 arc seconds (0.0004°). A broadband pyroelectic detector (electrical signal generated proportional to heating by absorption of optical radiation) was exposed to the radiation within the field of view of the telescope. Across the much brighter solar disk, an aperture field of view of 1.5 arc seconds (0.0004°) was used. To account for the large decrease in intensity in the circumsolar region, an aperture with a 4.5 arc second (0.00125°) field of view is used. A reduced version (only a half of the scan data, from 0° to 3.2° away from the disk center, and only about 10% of all the data actually collected) is available online [148].

Table 1.4 is an image of the data format in the reduced data set to illustrate the structure of the data files and assist in their interpretation. The data fields are the site number; telescope number; date; solar time; local time; 2 digit flag status (1 = no errors, 0 = rain flap open); 2 digit line identifier (01 to 07 for broadband data; 21 to 24 for solar disk data; 41 to 47 for circumsolar data). Data lines 01 to 07 contain fields for altitude and azimuth of the sun; Earth-Sun Distance (ESD) in A.U.; flag fields (for instrument problem identification); total hemispherical pyranometer irradiance data for unfiltered and filtered pyranometers on the tracker, normal to the sun, and in horizontal positions; pyrheliometer readings using a filter wheel with clear aperture, filters with wavelength passbands of 0.38–$0.46\,\mu m$, 0.46–$0.54\,\mu m$, 0.54–$0.62\,\mu m$, 0.62–$0.72\,\mu m$, 0.72–$0.85\,\mu m$, 0.85–$1.05\,\mu m$, 1.05–$1.25\,\mu m$, and $>1.25\,\mu m$, and a blank or 'dark' reading; the solar radiation from the disk alone (SolRad); the total radiation within the circumsolar region from $0.30°$ to $3.2°$ from the center (Circum); the circumsolar (C) to total solar plus circumsolar ($C + S$) ratio or $CSR = C/(C + S)$; ACR signal (a Willson type active cavity radiometer [37]); pyrheliometer fractional error; and a correction factor for converting pyroelectric readings for $Wm^{-2}sr^{-1}$. Data fields 21 to 24 contain intensity data in steps of 1.5

Table 1.4 Sample image of data for one NREL LBL circumsolar database scan

```
4 1 77/06/23 12:20 10 01 Time: 13:01 Alt: 78.74 Azi: 204.65 ESD: 1.0163
4 1 77/06/23 12:20 10 02 Flag Field:   00000 00000 11000 00000 00000 0000
4 1 77/06/23 12:20 10 03 Pyranometers. Trk:1068.7 905.5 Hor:1079.1 908.3
4 1 77/06/23 12:20 10 04 Pyrheliometer:  731.1 (clear), filtered next line
4 1 77/06/23 12:20 10 05   45.1  5.0 63.9 42.6 51.4 77.1  6.3 68.5
4 1 77/06/23 12:20 10 06 SolRad: 721.6 Circum: 13.1 C/(C+S): 0.0178815
4 1 77/06/23 12:20 10 07 Misc.  ACR: 0.01293 NIP: 0.01516 CC: 2.811E+07
4 1 77/06/23 12:20 10 21  1.486E+07 1.479E+07 1.462E+07 1.441E+07 1.411E+07
4 1 77/06/23 12:20 10 22  1.356E+07 1.305E+07 1.235E+07 1.089E+07 6.841E+06
4 1 77/06/23 12:20 10 23  1.290E+06 5.535E+04 1.623E+04 1.370E+04 1.236E+04
4 1 77/06/23 12:20 10 24  1.090E+04 9.803E+03 9.182E+03 8.279E+03 7.721E+03
4 1 77/06/23 12:20 10 41  6.697E+03 5.391E+03 4.416E+03 3.689E+03 3.131E+03
4 1 77/06/23 12:20 10 42  2.705E+03 2.371E+03 2.113E+03 1.903E+03 1.730E+03
4 1 77/06/23 12:20 10 43  1.585E+03 1.468E+03 1.367E+03 1.279E+03 1.207E+03
4 1 77/06/23 12:20 10 44  1.143E+03 1.097E+03 1.045E+03 9.999E+02 9.608E+02
4 1 77/06/23 12:20 10 45  9.195E+02 8.804E+02 8.445E+02 8.128E+02 7.893E+02
4 1 77/06/23 12:20 10 46  7.634E+02 7.355E+02 7.175E+02 6.973E+02 6.794E+02
4 1 77/06/23 12:20 10 47  6.641E+02 6.479E+02 6.286E+02 6.071E+02 5.719E+02
4 1 77/06/23 12:20 10 48  5.386E+02 ======= End of Brightness Data ========
4 1 77/06/23 12:20 10 99*********************************************
```

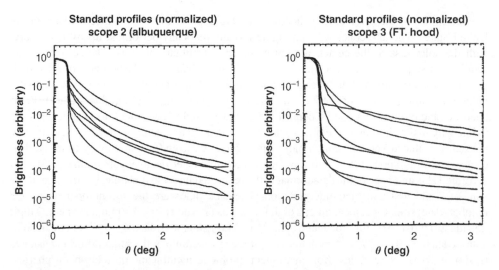

Figure 1.12 Example of sunshape profiles acquired by LBL circumsolar telescope. Source: Hunt et al, 1978 [147]. Reproduced by permission of US Energy Research and Development Administration

arc seconds (reading from left to right and down) from the solar disk center to the limb with a field of view of 1.5 arc seconds. Thus the central intensity is $1.486 \, 10^7$ in relative intensity, and after 10 steps toward the limb (15 arc sec) has fallen to $1.684 \, 10^6$, or 46%. Data lines 41 to 48 contain circumsolar radiation intensity data using a 4.5 arc second field of view. Data line identifier 99 indicates the end of the scan data. Figure 1.12 is a sample of the data from two sites (Albuquerque NM and Fort Hood, TX) with various levels of general atmospheric transparency and atmospheric conditions from reference [147].

The LBL circumsolar data set is the only easily accessible no-cost data available for the investigation of the relationship between circumsolar radiation and broadband radiometric measurements. Figure 1.13 graphically illustrates the relationship between the ratio of DNI beam to GHI and circumsolar radiation. The figure shows the limits of a 5° and 5.7° field of view pyrheliometers, illustrating the difficulty of even measuring quantitatively the impact of circumsolar radiation using broadband pyrheliometers or absolute cavity pyrheliometers.

Recall that pyrheliometer and pyranometer accuracy is at best about 0.5%, or 5 parts out of 1000. Figure 1.13 shows that even if 24% of the disk radiation is scattered into the circumsolar region, (not a likely candidate for site for CPV!). The intensity of the circumsolar radiation at the limits of a 5° field of view cavity pyrheliometer and a 5.7° field of view pyrheliometer is less than a few parts out of 1000, or probably close to the random noise level in such measurements. Under clearer conditions the difference falls to a few parts out of 10 000. This is 100 times the uncertainty in either instrument.

It is also clear that if a CPV system has an acceptance angle or field of view smaller than, or less than the field of view of a pyrheliometer, the pyrheliometer will provide a relative *overestimate* of the radiation available to the CPV system, represented at most (but less than, *as only a small fraction of the circumsolar aureole photons are scattered parallel to the direct beam*) by the integrated circumsolar radiation between the limit of the CPV system acceptance and the limit of the pyrheliometer field of view.

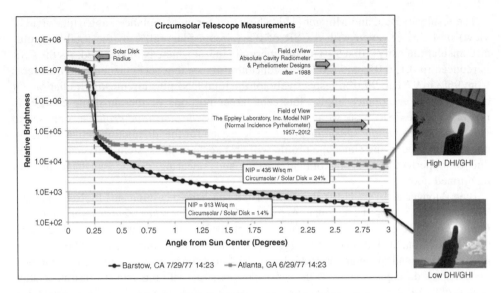

Figure 1.13 LBL circumsolar scan data in relation to instrument fields of view and solar aureole. Note the logarithmic scale of the vertical axis. Source: Thomas Stoffel, US Department of Energy NREL

The spectrally filtered LBL pyrheliometer data itself contains valuable information about how atmospheric conditions affect the spectral power distribution of *DNI* beam radiation. Some of this data is discussed in Buie, Monger and Dey [149] mentioned in the next section. Despite the coarse spectral resolution of this data, it should be possible to correlate the LBL spectral DNI data with clear sky modeled spectra to produce scaling factors to match higher resolution spectral model results with data from spectral regions in the filtered LBL pyrheliometer data set.

1.6.3 Recent Circumsolar Radiation Research

Recent publications on newer instrumentation and results investigating circumsolar radiation have occurred because of the recent current revival of interest in concentrating solar technology and CPV in particular. Using the LBL circumsolar data base, Rabl and Bendl [150] developed a modeled 'standard solar sunshape profile' (figure 7 of [150]) for use in conjunction with optical performance models for concentrating systems. They mention, but do not discuss in detail, the effects of imperfect optical components, scattering of photons off component surfaces, etc., which 'smear out' the image of the sunshape on the conversion device. The profile they selected resembles the 4th curve from the bottom of the Ft. Hood example data plot in Figure 1.12; dropping three orders of magnitude from the central disk intensity at the solar limb, and dropping to 0.0001 of that intensity at 50 mrad from the disk center.

Buie, Monger and Dey [149] and Buie and Monger [151] summarized the LBL circumsolar data set (including some analysis of spectral scan data) and the Rabl and Bendl work mentioned above. They developed their own simulations of the convolution of LBL sunshapes and various concentrating system acceptance angles. They show that as the circumsolar ratio, CSR, [ratio of solar disk/(disk + circumsolar) intensity] decreases or increases that smaller or larger acceptance angles, respectively, are adequate to intercept more beam plus circumsolar energy.

The results of Buie and Monger [151] show that for an acceptance angle (total field of view) of 1 degree, even with a CSR of 0.8 (circumsolar = 80% of the total disk plus circumsolar intensity!) that 94% of the energy in the sunshape is intercepted. As the CSR is reduced to 0.1, 0.05, and 0.02, a 1° concentrating system acceptance angle intercepts from 96% to 97% to 99.0%, respectively, of the total DNI plus circumsolar radiation. Of course, larger acceptance angles intercept larger portions of the sunshape energy no matter what the CSR. Note we use the term *intercept the sunshape profile* as opposed to *collect the energy in the sunshape profile*. The reason for this distinction is that not every photon observed in the sunshape profile will propagate to the system conversion device, due to imperfections in the optical system and smearing of the sunshape profiles mentioned above.

Buie, Monger and Dey [149] also summarized sunshape measurements by Neumann and Witzker [152,153] of the German Aerospace Center (*Deutsches Zentrum fur Luft- und Raumfahrt*, DLR) based on modern charge couple device (CCD) camera system and 12-bit digital resolution. Such devices have a (grayscale) intensity resolution of 0.02% (1 out of 4096 grayscale bins) and dynamic range of about 30 000 to 1; or just over 4 orders of magnitude. They discuss differences between the sunshapes observed with this modern instrumentation and the older LBL circumsolar data sets. They also present a new model, similar to Eq. (1.33), for computing sunshape relative intensity ϕ as a function of θ, in mrad, from the center of the solar disk:

$$\phi(\theta) = \cos(0.326\,\theta)/\cos(0.308\,\theta) \quad \text{for } \theta \leq 4.65 \text{ mrad}(0.265°), \tag{1.34.a}$$

$$\phi(\theta) = \exp(k\,\theta^y) \quad \text{for } \theta > 4.65 \text{ mrad}, \tag{1.34.b}$$

where k and y are derived from the circumsolar ratio (CSR), designated by them as χ:

$$k = 0.9 \ln(13.5\,\chi)\chi^{-0.3}, \tag{1.35.a}$$

$$y = 2.2 \ln(0.52\,\chi)\chi^{0.43} - 1, \tag{1.35.b}$$

The concepts regarding sunshapes described in this section are not definitive, and research in this area continues. However these issues must be taken into consideration in conjunction with a detailed knowledge of limitations or uncertainties associated with the mechanical and optical design of any solar concentrating system.

1.7   Direct Solar Radiation Climates

The discussion of atmospheric parameters affecting DNI beam resources leads one to consider if these parameters produce what might be characterized as direct solar radiation climates. That is, are there large regions where DNI beam resources may be deemed as exceptional, above average, average, below average, or poor for CPV applications? Qualitative adjectives such as 'above average' are only moderately informative. Quantitative values or ranges of values for DNI beam resources provide more information. However, consideration of quantity or magnitude of resources alone may not be sufficient. The quality of the resources, based on parameters such as clearness index for DNI beam (K_n), seasonal patterns for clouds or storms, or ratios such as DNI/GHI or DHI/GHI (if DNI data is not available) provide even more information and should be considered by CPV system designers.

1.7.1 Measurement Networks and Data

Measurement networks and model and measured DNI resources were covered in section 1.5.1 above. We mentioned that the WMO World Radiation Center (WRC) at St. Petersburg, Russia [110] contains measured DNI data when available, however this parameter is rare in that data set. The WMO also archives research data from a volunteer Baseline Surface Radiation Network (BSRN) [154,155]. All sites were last accessed in April, 2015.

The paper by Renné *et al.* [114] describes the sparse and intermittent sets of US solar measurement *n*, which consists mostly of sites oriented toward scientific research and not routine monitoring. A summary list of these networks is shown in Table 1.5. The University of Oregon has operated various stations in the northwest United States since 1978 [156].

The most popular way of presenting solar radiation and DNI beam data is in the graphical form of maps. Figure 1.14 to Figure 1.18 represent example maps of DNI resources derived from either meteorological satellite conversion models such as [121,124], or climatological solar radiation models based on cloud cover and atmospheric parameters such as described in [117,118].

1.7.2 Concentrating Solar Power Site Selection

Selection of a site to optimize the harvesting of direct beam radiation requires knowledge of the solar DNI resources. Many other factors regarding concentrating solar power system design come into play, such as system component design, land availability, terrain types and accessibility, financial resources, etc.; but solar resource information is important for sizing and initial system configurations. Many of these other considerations are addressed in greater detail in the following chapters. Here, we address some approaches and caveats regarding resource evaluation or assessment.

Measured data DNI is a rare, premium product, as previously described. Electronic data files of modeled data available through databases such as the US NSRDB, or the European SODA

Table 1.5   Solar radiation monitoring networks

Network (stations)	Operator	Website
WMO WRC	National Weather Services	http://wrdc-mgo.nrel.gov/
WMO BSRN	Various Research Org.	http://www.bsrn.awi.de/
Australian (16)	Bureau Meteorology Queensland Government	http://www.ga.gov.au/scientific-topics/energy/ resources/other-renewable-energy-resources/ solar-energy/active-solar-ground-stations-across-australia
NOAA ISIS	NOAA CMDL	http://www.esrl.noaa.gov/gmd/grad/isis/
NOAA SURFRAD (8)	NOAA CMDL	http://www.esrl.noaa.gov/gmd/grad/surfrad/ index.html
ARM SGP (30)	US DOE ARM Program	http://www.arm.gov/sites/sgp
U of O Northwest (34)	Univ. Oregon	http://solardat.uoregon.edu/
NREL/MIDC (33)	Various	http://www.nrel.gov/midc/
Texas Solar Radiation DB	Univ. Texas Austin	http://www.me.utexas.edu/~solarlab/tsrdb/ tsrdb.html
Solar Energy Center India (51)	Center for Wind Energy Technology (CWET)	http://natgrp.org/2012/09/10/network-of-solar-radiation-monitoring-stations-in-india/

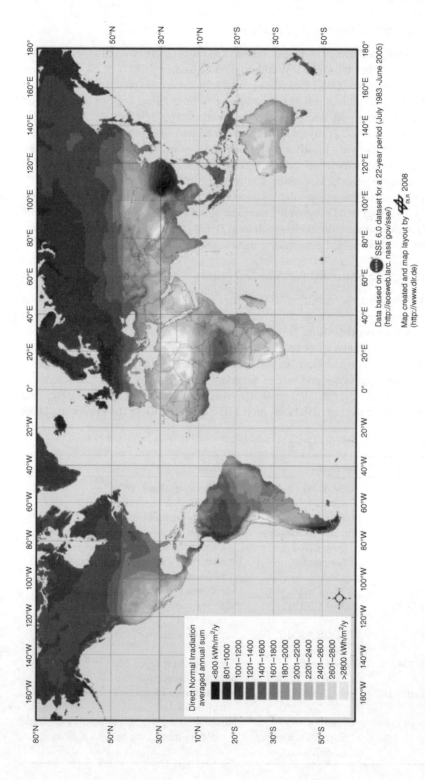

Figure 1.14 Map of worldwide solar radiation. Direct normal irradiance annual average sum, kWhm^{-2} based on NASA SSE model. Source: http://www.dlr.de/tt/Portaldata/41/Resources/dokumente/institut/system/projects/reaccess/ssedni60.jpg. Reproduced with permission of NASA

or European Solar Atlas, worldwide NASA Surface Solar Radiation and WRDC world radiation databases, their availability and limitations, have been described above. Another data source is data manuals summarizing the resources available to various solar collectors, including concentrating PV collectors and solar tracking flat plate PV collectors. One example is the NREL Data Manual for Flat Plate and Concentrating Collectors [157]. This manual presents monthly average radiation resources for direct beam and hemispherical radiation computed from the US NSRDB for various collector configurations using the Perez anisotropic diffuse model [158]. However, the most popular starting point for site selection is an examination of regional, national, continental, or worldwide maps based on combinations of modeled and measured data. Examples, with some caveats, are described in the next section.

1.7.3 Concentrating Solar Power Resource Map Examples

Figure 1.14 presents one version of worldwide CPV resources. The map shows relative yearly average sum, or annual total DNI radiation in $kWhm^{-2}$. The data are based on NASA SSE 22 year model calculations [127]. CPV 'hot spots' are represented by areas exceeding $2400 \, kWh/m^2$ per year total DNI, in lightest tones. This represents about $200 \, kWh/m^2$ total DNI per month per year in these areas. A month of 30 days therefore averages about $6.7 \, kWh/m^2/day$ for total DNI. For an average 12 hour day this means that the average hourly DNI value exceeds $555 \, W/m^2$. These numbers provide a feel for what the lower end of the best (highest) levels of resources are on hourly, daily, monthly and annual DNI time scales.

Often only maps of total horizontal irradiance or hemispherical irradiance on tilted surfaces such as PV panels are available. A rule of thumb for converting hemispherical (for PV panels) to DNI resources can be derived for the sites with the highest resources; namely DNI CPV resources are approximately 1.25 to 1.4 times the hemispherical solar radiation resources for flat plate PV. Similar results can obtained from some of the tabular data base summaries, such as found in NASA SSE monthly result files or the NREL flat plate and concentrating collector data manual [157].

Moving to national and regional scales, Figure 1.15 is a grayscale version of the high resolution color map of annual average DNI resources per day produced by NREL for the Indian Solar Energy Center. The modeled data is produced from seven years of hourly satellite imagery and meteorological input data (estimated aerosol optical depth, water vapor, and ozone) using the satellite base model described of Perez [121] and averaged over the period. The frequency of the DNI resource above $5000 \, Wh/m^2/day$ or $5 \, kWh/m^2/day$ or $1800 \, kWh/$ year total indicates that significant DNI resources for CPV systems are widely available.

These annual summary maps give an indication of the relative magnitude of average resources on an annual basis, but do not help when it comes to evaluating the variability of the solar resource. Variation from month to month and year to year is dependent on regional and national weather and climate patterns can be portrayed in maps covering monthly or seasonal time frames.

An analysis of solar resource variability for both hemispherical and direct normal resources for the United States has been performed by Wilcox and Gueymard [159] and is available online on the NREL Renewable Resource Data Center [160]. These maps show that long term interannual variability in United States DNI resources, computed from the coefficient of variation, or standard deviation divided by the mean (over many years of data) is generally less than 10% for the continental United States, and can be less than 2% for desert regions.

However, even as for inter-annual variability, there is variability from year to year for each month. Figure 1.16, from [159], shows the coefficient of variation of DNI resources for the

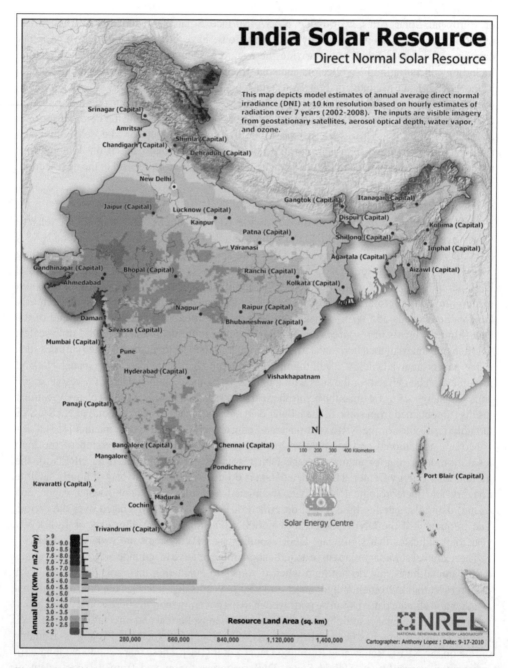

Figure 1.15 Annual average direct normal irradiance derived from meteorological satellite data conversion to solar irradiance. This is a grayscale version of the color original map, available at the NREL database. This figure is included as an example and not for actual use since some artifacts have appeared as a result of the grayscale conversion. Source: http://en.openei.org/w/index.php?title=File: NREL-DNI-Annual.jpg. Reproduced with permission of the US Department of Energy

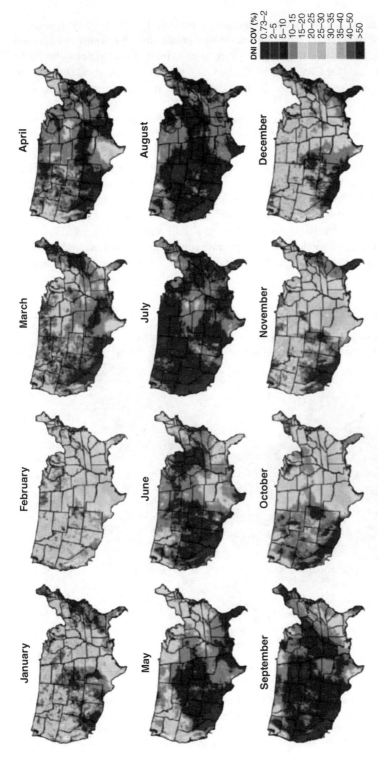

Figure 1.16 Monthly coefficient of variation as percent in DNI resources for the United States from 1998 to 2005. This is a grayscale version of the color original figure from reference [159]. This figure is included as an example only and not for actual use since some artifacts have appeared as a result of the grayscale conversion. Source: Wilcox and Gueymard, 2010 [159]. Reproduced with permission of the US Department of Energy

United States for each of the twelve months of the year, based on eight years of 1998 to 2005. Note how the intra-annual monthly variability decreases greatly in the summer months and increases in the winter months for the mid-continent sites. These considerations should be addressed in the design goals for CPV systems, such as targeting specific seasonal (heating, air conditioning) or long term (refrigeration, lighting) loads and the consistency of the solar resource.

Figure 1.17 illustrates the challenge of spatial resolution in developing national resource maps for very large areas or countries, such as China. This is a map of the DNI annual average resources for China. The western segment of the map was generated using the Perez [121] meteorological satellite conversion model at 10 km resolution. The eastern segment is derived from the NREL 40 km gridded CSR [118] modeled data. Insufficient satellite data or ground based meteorological data may be the cause of the discontinuity in coverage. The discontinuity in magnitudes at about 95° east longitude is probably due to variations in model input data and the factor of 16 in the size of the spatial averages (10 km by 10 m versus 40 km by 40 km) represented.

Whatever the performance of models used to generate such maps, consistent application of the model techniques can be used to establish some indication of the resources for large regions *relative to each other*.

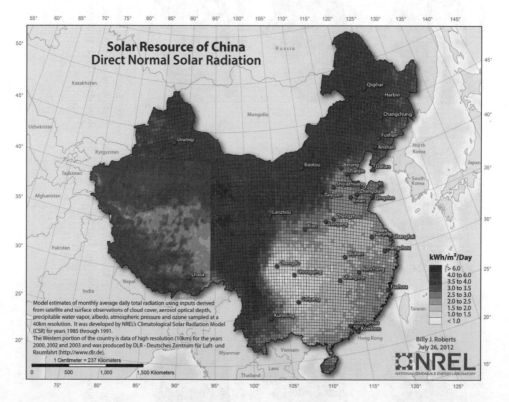

Figure 1.17 Annual average DNI beam for China from meteorological satellite conversion model (Western segment) and 40 km gridded CSR modeled data (eastern segment). This is a grayscale version of the color original map, available at the NREL database. Source: http://en.openei.org/w/index.php? title=File:NREL-China-Solar-CSP-01.jpg. Reproduced with permission of the US Department of Energy

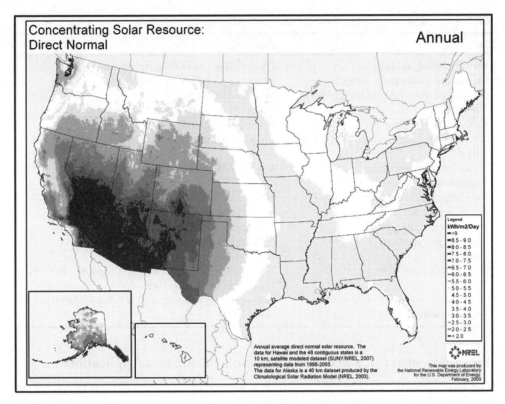

Figure 1.18 United States DNI resource from satellite conversion model. This is a grayscale version of the color original map, available at the NREL database. Source: http://www.nrel.gov/gis/solar.html. Reproduced with permission of the US Department of Energy

This assumption depends on the premise that the input data for the model algorithms is of uniform quality as well. This assumption is usually a weak one, as it is very difficult to obtain some of the required meteorological parameters (especially aerosol data) on a sufficient scale or consistent spatial resolution. Medium, or mesoscale (approximately 100 km to 1000 km per side grid square) and microscale (less than 100 km per side grid square) meteorological parameters are almost impossible to obtain from ground measurements. Satellite based estimates of such input data are dependent on retrieval algorithms, satellite sensor calibration, resolution, quality, and satellite navigational and positional drift issues. Figure 1.18 for the United States are based on satellite imagery and reanalysis meteorological data over 8 years, and 10 km spatial resolution modeling using the model of Perez [121].

1.7.4 Solar Resource Maps and Data Internet Resources

Searching the internet for maps, data, and information pertaining to any subject can be daunting, even with tools such as Google, Ask, or Bing. Even navigating a single website dedicated to renewable, solar, concentrating and photovoltaic power, such as that of the National Renewable Energy Laboratory, can be somewhat frustrating. Listed in Table 1.6 are a few website URLs that have great potential for addressing typical user needs. All sites were last accessed in April, 2015.

Table 1.6   Useful websites to retrieve radiation data and maps

Site	Comment
www.nrel.gov/gis/solar.html	US annual and monthly mean maps
www.nrel.gov/csp/data_resources.html	Listing of general resources for CSP
rredc.nrel.gov/solar/old_data/nsrdb/1961-1990/redbook/atlas/	Low resolution US
www.nrel.gov/csp/maps.html	US State/Region maps, other links
www.nrel.gov/international/global_energy.html	Links to international project map
www.nrel.gov/international/geospatial_toolkits.html	Links to GIS solar toolkits
www.nrel.gov/gis/mapsearch.html	Map search tool for renewables
www.bom.gov.au/	Australian Bureau Meteorology links
www.bom.gov.au/jsp/ncc/climate_averages/solar-exposure/index.jsp	Real time solar radiation exposure maps from the Australian Bureau of Meteorology
www.soda-is.com/eng/services/	A central starting point for general European and African international solar radiation data
http://photovoltaic-software.com/solar-radiation-database.php	A comprehensive listing of solar radiation data bases and services

1.8   Consensus Standards for Direct Solar Radiation Applications

This final section of the chapter provides references to existing (as of March, 2013) national and international consensus standards applicable to CPV applications. As a former participant in consensus standards activities, the author highly recommends the active participation, whenever possible, of industry, academic, and government parties in standards development to help foster the deployment of safe, reliable, efficient, and durable solar energy conversion systems. Note that numerous other consensus standards related to PV performance and characterization in general are in force, but not listed here. Those standards relate to PV reference cell construction, calibration, etc., as well as PV cell and module electrical performance and characterization (spectral response, spectral mismatch calculations, resistance to hail impact, qualification testing, mechanical load testing, etc.). It is highly recommended that engineers and designers become familiar with the widest range of applicable standards relating to general solar radiation applications.

1.8.1   World Radiometric Reference

The internationally accepted WRR reference is defined in the WMO CIMO Guide, Publication No. 8, and recognized by the International Bureau of Weights and Measures as described in section 1.2.3 above; see references [30,41,42]. Some of these standards are cited in the references, however this is a complete compilation as of this writing (2013).

1.8.2   Solar Radiometric Instrumentation Calibration

In the following lists, ISO is the International Organization for Standardization, IEC is the International Electro-Technical Commission, IEEE is the Institute of Electrical and Electronic

Engineers, and ASTM is the American Society for Testing and Materials, now called ASTM International:

- ISO 9846 Solar energy – Calibration of a pyranometer using a pyrheliometer.
- ISO 9847 Solar energy – Calibration of field pyranometers by comparison to a reference pyranometer.
- ISO 9059 Solar energy – Calibration of field pyrheliometers by comparison to a reference pyrheliometer.
- ISO 9060 Solar energy – Specification and classification of instruments for measuring hemispherical solar and direct solar radiation.
- ISO/TR 9901 Solar energy – Field pyranometers – Recommended practice for use.
- ASTM G130 Standard Test Method for Calibration of Narrow- and Broad-Band Ultraviolet Radiometers Using a Spectroradiometer.
- ASTM G138 Test Method for Calibration of a Spectroradiometer Using a Standard Source of Irradiance.
- ASTM G167 Standard Test Method for Calibration of a Pyranometer Using a Pyrheliometer.
- ASTM E816 Test Method for Calibration of Pyrheliometers by Comparison to Reference Pyrheliometers.
- ASTM E824 Standard Test Method for Transfer of Calibration From Reference to Field Radiometers.
- ASTM E927 Standard Specification for Solar Simulation for Photovoltaic Testing.

1.8.3 Spectral Calibration Standards

These standards are used to calibrate instrumentation for measuring spectral distributions or the spectral regions such as the ultraviolet, and optical properties of materials used in solar conversion systems:

- ASTM G130 Standard Test Method for Calibration of Narrow- and Broad-Band Ultraviolet Radiometers Using a Spectroradiometer.
- ASTM G138 Test Method for Calibration of a Spectroradiometer Using a Standard Source of Irradiance.
- ASTM E903 Standard Test Method for Solar Absorptance, Reflectance, and Transmittance of Materials Using Integrating Spheres.
- ASTM E1175 Standard Test Method for Determining Solar or Photopic Reflectance, Transmittance, and Absorptance of Materials Using a Large Diameter Integrating Sphere.

1.8.4 Standard and Reference Spectral Distributions

These standards specify reference solar spectral distributions for comparing the performance of PV materials in terrestrial and extraterrestrial standard reporting conditions:

- ASTM E490 Standard Solar Constant and Zero Air Mass Solar Spectral Irradiance Tables.
- ASTM G173 Standard Tables for Reference Solar Spectral Irradiances: Direct Normal and Hemispherical on 37° Tilted Surface.
- ASTM G177 Standard Tables for Reference Solar Ultraviolet Spectral Distributions: Hemispherical on 37° Tilted Surface.

- ISO 9845-1 Solar energy – Reference solar spectral irradiance at the ground at different receiving conditions, Part 1: Direct normal and hemispherical solar irradiance for air mass 1.5.
- IEC 60904-03 Measurement principles for terrestrial photovoltaic (PV) solar devices with reference spectral irradiance data.

The author hopes these resources provide a foundation for the chapters that follow, which delve into greater detail on many subjects simply mentioned in passing here. The CPV and solar energy conversion industry in general are at a frontier similar to that of the carbon based fossil fuel and electrical generation industry 100 years ago. We hope progress toward a more sustainable and less stressful energy infrastructure will accelerate due to the work of engineers and scientists in pushing this new energy frontier forward.

Glossary

List of Acronyms

Acronym	Description
ACP	Absolute cavity pyrheliometers
AM	Air mass
AM0	Air mass zero
AOD	Aerosol optical depth
ASTM	American Society for Testing and Materials
AU	Astronomical Unit
BAOD	Broadband aerosol optical depth
BIPM	(French acronym) International Bureau of Weights and Measures
BSRN	Baseline Surface Radiation Network
CCD	Charge couple device
CFD	Cumulative frequency distribution
CIE	Commission Internationale de Éclairage
CIMO	Commission on Instrumentation, Measurements and Observations
CSR	Circumsolar radiation, circumsolar ratio, or climatological solar radiation (model)
DHI	Diffuse horizontal irradiance
DISC	Direct insolation simulation code
DLR	German Aerospace Center (Deutches Zentrum fur Luft- und Raumfahrt)
DNI	Direct normal irradiance
DOE	Department of Energy
EOS	Earth observing system
ESD	Earth-Sun distance
ETR	Extraterrestrial radiation
GIS	Geographical information systems
GHI	Global horizontal irradiance
GOES	Geostationary operational environmental satellites
GUM	Guide to the expression of uncertainty in measurement
IEC	International Electrotechnical Commission
IEEE	Institute of Electrical and Electronic Engineers
IPC	International pyrheliometer comparisons
ISO	International Standards Organization

Acronym	Description
JCGM	Joint Committee for Guides in Metrology
KACST	King Abdul Aziz City for Science and Technology
LBL	Lawrence Berkley Laboratory
MBE	Mean bias error
RMSE	Root mean square error
MCE	Measure, correlate, evaluate
METSTAT	Meteorological and statistical
NASA	National Aeronautics and Space Administration
NATO	North Atlantic Treaty Organization
NCDC	National Climatic Data Center
NMI	National Metrology Institution
NOAA	National Oceanic and Atmospheric Administration
NREL	National Renewable Energy Laboratory
NSRDB	National Solar Radiation Data Base
NWS	National Weather Service
PMOD	Physical Meteorological Observatory, Davos
PVUSA	Photovoltaics for utility scale applications
PW	Precipitable water
RMSE	root mean square error
RredC	Renewable Resource Data Center
RSR	Rotating shadow band radiometers
SODA	Solar Energy Services for Professionals
SOLMET	(ERSATZ, German for substitute) solar and meteorological
SORCE	Solar Radiation and Climate Experiment
SPD	(Solar) spectral power distribution
SSE	Surface Meteorology and Solar Energy (NASA data website)
SWERA	Solar and wind energy resource assessment
TMY	Typical meteorological year
VIM	International Vocabulary of Metrology
WMO	World Meteorological Organization
WRC	World Radiation Centre
WRR	World Radiometric Reference
WSG	World Standard Group

List of Symbols

Typical units given in square brackets. If no units are given, variable is dimensionless.

Symbol	Description [Units]
d	Day angle [rad]
DHI	Diffuse horizontal irradiance [Wm^{-2}]
d_n	Day of the year (for Jan 1, $d_n = 1$)
DNI	Direct Normal Irradiance [Wm^{-2}]
d_s	Sun diameter (1 390 000 km) [km]
e	Elevation angle of the solar disk or solar elevation angle [rad]
GHI	Global horizontal irradiance [Wm^{-2}]

(*continued*)

Symbol	Description [Units]
h	Altitude [km]
$I(\theta)$	Direct irradiance component on an arbitrary surface oriented an angle θ to the DNI beam; (sometimes denoted as B) [Wm^{-2}]
I_{bh}	DNI beam flux on a horizontal surface [Wm^{-2}]
I_n	DNI or $I(\theta = 0)$ [Wm^{-2}]
I_o	Solar constant or intensity of solar extraterrestrial radiation (ETR) at the top of the atmosphere ($I_o = 1366.1$ W/m^2 at 1 AU) [Wm^{-2}]
k	Coverage factor for the calculation of the expanded uncertainty
K_d	Diffuse hemispherical clearness index
K_n	Direct normal clearness index
K_t	Total hemispherical clearness index
m	Geometrical air mass or path length through the atmosphere for photons ($m = 1/\sin(e) = 1/\cos(z)$)
MBE	Mean bias error
M_p	Pressure corrected air mass
M_R	Refraction corrected air mass
O_z	Total column ozone amount [atmospheric-cm]
P_o	Sea level pressure (1013.25 mbar) [mbar]
P_s	Site pressure [mbar]
PW	Precipitable water [atmospheric-cm]
R_a	Earth's aphelion distance (152.6×10^6 km) [km]
R_c	Earth-Sun distance correction factor
$RMSE$	Root mean square error
R_o	Average distance between the Sun and Earth (1 AU or 149 597 870.7 km) [km]
R_p	Earth's perihelion distance (147.5×10^6 km) [km]
R_s	Sun radius (695 000 km) [km]
$T(x)$	Atmospheric transmittance
T_{a3}	Aerosol optical depth at 380 nm
T_{a5}	Aerosol optical depth at 500 nm.
T_a	Transmittance due to aerosol properties
T_g	Transmittance due to optical properties of gases
T_L	Linke turbidity
T_o	Transmittance due to ozone (in the stratosphere)
T_r	Transmittance due to Rayleigh scattering
T_w	Transmittance of water vapor
U	Expanded uncertainty
$u(x_i)$	Standard uncertainty of each input estimate x_i
$u_c(y)$	Combined standard uncertainty of the measurement result y
z	Angle between the local vertical and the center of the solar disk; complement of the solar elevation angle ($z = 90 - e$) [rad]
χ	Circumsolar ratio or CSR
$\phi(\theta)$	Sunshape relative intensity ϕ as a function of θ, angle from the center of the sun, ($\theta = 0°$) to the limb ($\theta = 90°$)
θ	Angle between the DNI beam and the surface normal [rad]
$\tau(\lambda)$	Aerosol optical depth or AOD
τ_D	Optical thickness of clean (aerosol free) and dry (no water vapor) atmosphere

References

1. Cox, A.N.E. (2002) *Allen's Astrophysical Quantities*, 4th edn, Springer-Verlag, Berlin.
2. Fröhlich C. and Lean, J. (1997) Total Solar irradiance variations: the construction of a composite and its comparison with models, presented at the 31st ESALB Symposium on Correlated Phenomena at the Sun, in the Heliosphere and in Geospace, ESTEC, Noordwijk, The Netherlands.
3. Fröhlich C. and Lean, J. (2004) Solar radiative output and its variability: evidence and mechanisms, *The Astronomy and Astrophysics Review*, **12**, 273–320.
4. American Society for Testing and Materials, ASTM E490-00a - Standard Solar Constant and Zero Air Mass Solar Spectral Irradiance Tables, ed. West Conshohocken, PA., 2014.
5. Hickey, J.R., Stowe, L.L., Jacobowitz, H. *et al.*, (1980) Initial solar irradiance determinations from nimbus 7 cavity radiometer measurements, Science, **208**, 281–283.
6. J.R. Hickey, B.M. Alton, H.L. Kyle, and D. Hoyt, Total solar irradiance measurements by ERB/Nimbus-7. A review of nine years, *Space Science Reviews*, vol. **48**, pp. 321–334, 1988.
7. Sklyarov, Y.A., Brichkov, Y., Vorobyov, V.A. and Bryantsev, (1991) I.I. Development of a Solar Constant Measurement Programme, *Metrologia*, **28**, 275.
8. Hoyt, D.V., Kyle, H.L., Hickey, J.R. and Maschhoff, R.H. (1992) The Nimbus 7 solar total irradiance: A new algorithm for its derivation, *Journal of Geophysical Research: Space Physics*, **97**, 51–63.
9. Kopp G. and Lawrence, G. (2005) The total irradiance monitor (TIM): instrument design, in *The Solar Radiation and Climate Experiment (SORCE)*, (eds G. Rottman, T. Woods, and V. George), Springer, New York, pp. 91–109.
10. Kopp, G., Heuerman, K. and Lawrence, G. (2005) The total irradiance monitor (TIM): instrument calibration, in *The Solar Radiation and Climate Experiment (SORCE)*, (eds G. Rottman, T. Woods, and V. George), Springer, New York, pp. 111–127.
11. Kopp, G., Lawrence, G. and Rottman, G. (2005) The total irradiance monitor (TIM): science results, in *The Solar Radiation and Climate Experiment (SORCE)*, (eds G. Rottman, T. Woods, and V. George), Springer, New York, pp. 129–139.
12. Kopp G. and Lean, J.L. (2011) A new, lower value of total solar irradiance: Evidence and climate significance. *Geophysical Research Letters*, **38**, L01706.
13. Spencer, J.W. (1971) Fourier series representation of the position of the sun. *Search*, **2**, 172.
14. Hickey, J.R. (1978) A review of solar constant measurements, presented at the Sun: Mankind's future source of energy. Proceedings of the International Solar Energy Congress New Delhi, India.
15. Sofia, S. (1981) Variations of the solar constant, presented at the Proceedings of a Workshop held at Goddard Spaceflight Center, Greenbelt Maryland.
16. Pap, J. (1986) Variation of the solar constant during the solar cycle, *Astrophysics and Space Science*, vol. **127**, 5–71.
17. Crommelynck, D.A., Brusa, R.W. and Domingo, V. (1986) Results of the solar constant experiment onboard Spacelab 1. *Solar Physics*, **107**, 1–9.
18. Fröhlich, C. (1987) Variability of the solar 'constant' on time scales of minutes to years, *Journal of Geophysical Research: Atmospheres*, **92**, 796–800.
19. Lee, R.B., Barkstrom, B.R., Harrison, E.F. *et al.* (1988) Earth radiation budget satellite extraterrestrial solar constant measurements: 1986–1987 increasing trend. *Advances in Space Research*, **8**, 11–13.
20. Lee, R.B., Gibson, M.A., Shivakumar, N. *et al.* (1991) Solar irradiance measurements: Minimum through maximum solar activity. *Metrologia*, **28**, 265.
21. Nikolsky, G.A. (1991) Solar irradiance variability. *Metrologia*, **28**, 281.
22. Gueymard, C.A. (2004) The sun's total and spectral irradiance for solar energy applications and solar radiation models. *Solar Energy*, **76**, 423–453.
23. Gueymard, C.A. (2006) Reference solar spectra: Their evolution, standardization issues, and comparison to recent measurements. *Advances in Space Research*, **37**, 323–340.
24. Hulst, H.C. (1981) *Light Scattering by Small Particles*. Dover Publications, Mineola, NY.
25. Iqbal, M. (1983) *An Introduction to Solar Radiation*. Academic Press, Toronto, ON.
26. Coulson, K. (1975) *Solar and Terrestrial Radiation*. Academic Press, New York.
27. Vignola, F., Michalsky, J. and Stoffel, T. (2012) *Solar and Infrared Radiation Measurements*. CRC Press, Boca Raton, FL.
28. Stoffel, T., Renné, D., Myers, D. *et al.* (2010) *Concentrating Solar Power: Best Practices Handbook for the Collection and Use of Solar Resource Data*. National Renewable Energy Lab, NREL Technical Report

NREL/TP-550-47465 National Renewable Energy Lab, Golden, CO (available at www.nrel.gov/docs/fy10osti/47465.pdf).

29. Gershun, A.A. (1958) Izbrannyye trudy po fotometrii i svetotekhnike (Selected Works on Photometry and Light Engineering). Fizmatgiz, Moscow.
30. World Meteorological Organization (2008) WMO Guide to Meteorological Instruments and Methods of Observation, WMO, Geneva, Switzerland.
31. Michalsky, J.J., Berndt, J.L. and Schuster, G.J. (1986) A microprocessor-based rotating shadowband radiometer, *Solar Energy*, **36**, 465–470.
32. King D.L. and Myers, D.R. (1997) Silicon-photodiode pyranometers: operational characteristics, historical experiences, and new calibration procedures. Conference Record of the 26th IEEE Photovoltaic Specialists Conference, pp. 1285–1288.
33. Michalsky, J.J., Harrison, L. and LeBaron, B.A. (1987) Empirical radiometric correction of a silicon photodiode rotating shadowband pyranometer. *Solar Energy*, **39**, 87–96.
34. Michalsky, J.J., Perez, R., Harrison, L. and LeBaron, B.A. (1991) Spectral and temperature correction of silicon photovoltaic solar radiation detectors, *Solar Energy*, **47**, 299–305.
35. Myers, D. (2011) Quantitative analysis of spectral impacts on silicon photodiode radiometers, presented at the 40th ASES National Solar Conference 2011 (SOLAR 2011), Raleigh, North Carolina.
36. Kendall J.M. and Berdahl, C.M. (1970) Two blackbody radiometers of high accuracy. *Applied Optics*, **9**, 1082–1091.
37. Willson, R.C. (1973) Active cavity radiometer. *Applied Optics*, **12**, 810–817.
38. Crommelynck, D. (1977) Calibration of radiation instruments for the measurement of the radiant flux of an arbitrary source. *Applied Optics*, **16**, 302–305.
39. Brusa R.W. and Fröhlich, C. (1986) Absolute radiometers (PMO6) and their experimental characterization. *Applied Optics*, **25**, 4173–4180.
40. Hickey J.R. and Karoli, A.R. (1974) Radiometric calibrations for the Earth radiation budget experiment. *Applied Optics*, **13**, 523–533.
41. Rüedi I. and Finsterle, W. (2005) The World Radiometric Reference and its quality system, presented at the Technical Conference on Meteorological and Environmental Instruments and Methods of Observation - TECO, Bucharest (Romania).
42. Reda, I. (1996) Calibration of a solar absolute cavity radiometer with traceability to the world radiometric reference. National Renewable Energy Laboratory, Golden CO.
43. American Society for Testing and Materials (2010) ASTM E816 - *Standard test method for calibration of pyrheliometers by comparison to reference pyrheliometers*, ASTM, West Conshohocken, PA.
44. International Standards Organization (1993) ISO 9846:1993 *Solar energy - Calibration of a pyranometer using a pyrheliometer*, ISO, Geneva, Switzerland.
45. International Standards Organization (1990) ISO 9059:1990 *Solar energy - Calibration of field pyrheliometers by comparison to a reference pyrheliometer*, ISO, Geneva, Switzerland.
46. American Society for Testing and Materials (2010) ASTM E824 - *Standard test method for transfer of calibration from reference to field radiometers*, ASTM, West Conshohocken, PA.
47. Romero, J., Fox, N.P. and Fröhlich, C. (1995) Improved comparison of the World Radiometric Reference and the SI radiometric scale, *Metrologia*, **32**, 523.
48. *International Vocabulary of Metrology – Basic and general concepts and associated terms* (VIM 3rd edn) JCGM 200:2012, BIPM, 2012.
49. Evaluation of measurement data — Guide to the expression of uncertainty in measurement JCGM 100. GUM 1995 with minor corrections, Working Group 1 of the Joint Committee for Guides in Metrology (JCGM/WG 1), 2008.
50. International Standards Organization/International Electrotechnical Commission (2008) ISO/IEC Guide 98-3:2008 Uncertainty of measurement - Part 3: Guide to the expression of uncertainty in measurement (GUM:1995).
51. Reda, I., Myers, D. and Stoffel, T. (2008) Uncertainty Estimate for the outdoor calibration of solar pyranometers: a metrologist perspective. *Measure* (NCSLI Journal of Measurement Science) **3**, 58–66.
52. Reda, I. (2011) Method to calculate uncertainties in measuring shortwave solar irradiance using thermopile and semiconductor solar radiometers. National Renewable Energy Lab.
53. Michalsky, J., Dutton, E.G., Nelson, D. *et al.* (2011) An extensive comparison of commercial pyrheliometers under a wide range of routine observing conditions. *Journal of Atmospheric and Oceanic Technology*, **28**, 752–766.

54. Wilcox S.M. and McCormack, P. (2011) Implementing best practices for data quality assessment of the national renewable energy laboratory's solar resource and meteorological assessment project, presented at the 40th ASES Annual Conference (SOLAR 2011), Raleigh, North Carolina.
55. Liu B.Y. and Jordan, R.C. (1960) The interrelationship and characteristic distribution of direct, diffuse and total solar radiation. *Solar Energy*, **4**, 1–19.
56. International Electrotechnical Commission (1998) IEC 61724 - Photovoltaic System Performance Monitoring - Guidelines for Measurement, Data Exchange and Analysis, Geneva.
57. Townsend, T., Whitaker, C., Farmer, B. and Wenger, H. (1994) A new performance index for PV system analysis, in 24th IEEE Photovoltaic Specialists Conference/First World Conference on Photovoltaic Energy Conversion, vol. **1**, pp. 1036–1039.
58. Whitaker, C.M., Townsend, T.U., Newmiller, J.D. *et al.* (1997) Application and validation of a new PV performance characterization method, in *26th IEEE Photovoltaic Specialists Conference*, pp. 1253–1256.
59. Watt, D. (1978) On the nature and distribution of solar radiation. US Department of Energy, Washington, DC, Report HCP/T2552-01.
60. Atwater M.A. and Ball, J.T. (1978) A numerical solar radiation model based on standard meteorological observations. *Solar Energy*, **21**, 163–170.
61. Bird R.E. and Hulstrom, R.L. (1980) Direct insolation models, NREL, Golden CO Solar Energy Research Institute (now National Renewable Energy Laboratory).
62. Bird R.E. and Hulstrom, R.L. Review, (1981) Evaluation, and improvement of direct irradiance models. *Journal of Solar Energy Engineering*, **103**, 182–192.
63. Gueymard, C.A. (2008) REST2: High-performance solar radiation model for cloudless-sky irradiance, illuminance, and photosynthetically active radiation – Validation with a benchmark dataset. *Solar Energy*, **82**, 272–285.
64. Myers, D.R. (2005) Solar radiation modeling and measurements for renewable energy applications: data and model quality. *Energy*, **30**, 1517–1531.
65. Gueymard C.A. and Myers, D.R. (2009) Evaluation of conventional and high-performance routine solar radiation measurements for improved solar resource, climatological trends, and radiative modeling. *Solar Energy*, **83**, 171–185.
66. Linke, F. (1922) Transmissions-Koeffizient und Trubungsfaktor. *Beitraege zur Physik der Atmosphaere*, **10**, 91–103.
67. Meinel A.B. and Meinel, M.P. (1976) *Applied Solar Energy: An Introduction*, Addison Wesley Publishing.
68. Laue, E.G. (1970) The measurement of solar spectral irradiance at different terrestrial elevations. *Solar Energy*, **13**, 43–57.
69. http://www.soda-is.com/eng/services/linke_turbidity_info.html (accessed: April, 2015).
70. Reno, M.J., Hansen, C.W. and Stein, J.S. (2012) Global horizontal irradiance clear sky models: implementation and analysis, Sandia National Labs, Albuquerque, New Mexico.
71. Michalsky, J.J. (1988) The Astronomical Almanac's algorithm for approximate solar position (1950–2050). *Solar Energy*, **40**, 227–235.
72. Reda I. and Andreas, A. (2004) Solar position algorithm for solar radiation applications. *Solar Energy*, **76**, 577–589.
73. http://ozoneaq.gsfc.nasa.gov/ (accessed: April, 2015).
74. Van Heuklon, T.K. (1979) Estimating atmospheric ozone for solar radiation models. *Solar Energy*, **22**, 63–68.
75. Shettle E.P. and Fenn, R.W. (1979) Models for the aerosols of the lower atmosphere and the effects of humidity variations on their optical properties, Environmental Research Paper Air Force Geophysics Lab., Hanscom AFB, MA. Optical Physics Div.
76. Molineaux, B., Ineichen, P. and O'Neill, N. (1998) Equivalence of pyrheliometric and monochromatic aerosol optical depths at a single key wavelength. *Applied Optics*, **37**, 7008–7018.
77. Garrison J.D. and Adler, G.P. (1990) Estimation of precipitable water over the United States for application to the division of solar radiation into its direct and diffuse components. *Solar Energy*, **44**, 225–241.
78. Holben, B.N., Tanré, D., Smirnov, A. *et al.* (2001) An emerging ground-based aerosol climatology: Aerosol optical depth from AERONET. *Journal of Geophysical Research: Atmospheres*, **106**, 12067–12097.
79. Boes, E.C., Anderson, H.E., Hall, I.J., Prairie, R.R. and Stromberg, R.T. (1977) Availability of direct, total and diffuse solar radiation to fixed and tracking collectors in the USA. Sandia Report SAND77-0885, Sandia National Labs.
80. Maxwell, E.L. (1987) A quasi-physical model for converting global horizontal to direct normal insolation. Solar Energy Research Institute (now National Renewable Energy Lab), Golden, CO.

81. Perez, R., Ineichen, P., Maxwell, E., Seals, R. and Zelenka, A. (1992) Dynamic global-to-direct irradiance conversion models. *ASHRAE Transactions Research*, 354–369.
82. Ineichen, P. (2008) Comparison and validation of three global-to-beam irradiance models against ground measurements. *Solar Energy*, **82**, 501–512.
83. Perez, R., Ineichen, P., Seals, R. and Zelenka, A. (1990) Making full use of the clearness index for parameterizing hourly insolation conditions. *Solar Energy*, **45**, 111–114.
84. Bird, R.E., Hulstrom, R.L. and Lewis, L.J. (1983) Terrestrial solar spectral data sets. *Solar Energy*, **30**, 563–573.
85. Bird R.E. and Riordan, C. (1986) Simple solar spectral model for direct and diffuse irradiance on horizontal and tilted planes at the Earth's surface for cloudless atmospheres. *Journal of Climate and Applied Meteorology*, **25**, 87–97.
86. Gueymard, C.A. (2001) Parameterized transmittance model for direct beam and circumsolar spectral irradiance. *Solar Energy*, **71**, 325–346.
87. Wehrli, C. (1985) *Extraterrestrial Solar Spectrum*, Physikalisch-Meteorologisches Observatorium, World Radiation Center (PMO/WRC), Davos Dorf, Switzerland.
88. http://rredc.nrel.gov/solar/models/spectral/ (accessed: April, 2015).
89. American Society for Testing and Materials (2010) ASTM G173 - Standard Tables for Reference Solar Spectral Irradiances: Direct Normal and Hemispherical on 37° Tilted Surface, ASTM International, Conshohocken, PA.
90. International Electrotechnical Commission (2008) IEC 60904-3 Ed.2: Photovoltaic devices - Part 3: Measurement principles for terrestrial photovoltaic (PV) solar devices with reference spectral irradiance data, IEC, Geneva.
91. American Society for Testing and Materials (2010) ASTM G177 - Standard tables for reference solar ultraviolet spectral distributions, ASTM International, West Conshohocken, PA.
92. Anderson, G.P., Wang, J., Hoke, M.L. *et al.* (1994) History of one family of atmospheric radiative transfer codes, pp. 170–183.
93. Anderson, G.P., Kneizys, F.X., Chetwynd, J.J. *et al.*, (1996) Reviewing atmospheric radiative transfer modeling: new developments in high- and moderate-resolution FASCODE/FASE and MODTRAN, pp. 82–93.
94. Jacovides, C.P., Kaskaoutis, D.G., Tymvios, F.S. and Asimakopoulos, D.N. (2004) Application of SPCTRAL2 parametric model in estimating spectral solar irradiances over polluted Athens atmosphere. *Renewable Energy*, **29**, 1109–1119.
95. Gueymard, C.A. (2008) Prediction and validation of cloudless shortwave solar spectra incident on horizontal, tilted, or tracking surfaces. *Solar Energy*, **82**, 260–271.
96. Kaskaoutis D.G. and Kambezidis, H.D. (2008) The role of aerosol models of the SMARTS code in predicting the spectral direct-beam irradiance in an urban area. *Renewable Energy*, **33**, 1532–1543.
97. Michalsky, J.J., Anderson, G.P., Barnard, J. *et al.* (2006) Shortwave radiative closure studies for clear skies during the Atmospheric Radiation Measurement 2003 Aerosol Intensive Observation Period. *Journal of Geophysical Research: Atmospheres*, **111**, D14S90.
98. Gueymard, C.A. (2005) Interdisciplinary applications of a versatile spectral solar irradiance model: A review. *Energy*, **30**, 1551–1576.
99. Myers, D.R., Emery, K. and Gueymard, C. (2004) Revising and validating spectral irradiance reference standards for photovoltaic performance evaluation. *Journal of Solar Energy Engineering*, **126**, 567–574.
100. Utrillas, M.P., Boscá, J.V., Martínez-Lozano, J.A. (1998) A comparative study of SPCTRAL2 and SMARTS2 parameterised models based on spectral irradiance measurements at Valencia, *Spain. Solar Energy*, **63**, 161–171.
101. Riordan, C., Myers, D., Rymes, M. *et al.* (1989) Spectral solar radiation data base at SERI. *Solar Energy*, **42**, 67–79.
102. Riordan, C.J., Myers, D.R. and Hulstrom, R.L. (1990) Spectral solar radiation data base documentation. Solar Energy Research Institute (now National Renewable Energy Lab), Golden, CO.
103. Myers, D.R. (1989) Estimates of uncertainty for measured spectra in the SERI spectral solar radiation data base. *Solar Energy*, **43**, 347–353.
104. http://rredc.nrel.gov/solar/old_data/spectral/ (accessed: April, 2015).
105. International Commission on Illumination (1990) *Solar Spectral Radiation* CIE Technical Report No. 85, 1st edn. CIE Central Bureau, Vienna.
106. Paris M.V. and Justus, C.G. (1988) A cloudy-sky radiative transfer model suitable for calibration of satellite sensors. *Remote Sensing of Environment*, **24**, 269–285.
107. Nann S. and Riordan, C. (1991) Solar spectral irradiance under clear and cloudy skies: measurements and a semiempirical model. *Journal of Applied Meteorology*, **30**, 447–462.
108. Myers, D.R. (2009) Terrestrial solar spectral distributions derived from broadband hourly solar radiation data, presented at SPIE 7410, Optical Modeling and Measurements for Solar Energy Systems III.

109. American Society for Testing and Materials, (2009) ASTM E2527 - Standard test method for electrical performance of concentrator terrestrial photovoltaic modules and systems under natural sunlight. ASTM, West Conshohoken, PA.

110. Tsvetkov, A., Wilcox, S., Renné, D. and Pulscak, M.M. (1995) International solar resource data at the World Radiation Data Center, presented at the ASES Annual Conference (SOLAR 1995), Minneapolis MN.

111. http://wrdc.mgo.rssi.ru/ (accessed: April, 2015).

112. SOLMET Vol 1. Users Manual TD-9724 Hourly Solar Radiation – Surface Meteorological Observations National Climatic Data Center, Ashville, NC.1978.

113. SOLMET Vol 2. Final Report TD-9724 Hourly Solar Radiation - Surface Meteorological Observations, National Climatic Data Center, Ashville, NC.1979.

114. Renné, D., Stoffel, T., Anderberg, M., Gray-Hann, P. and Augustyn, J. (2000) Current Status of Solar Measurement Programs in the US, presented at the ASES Annual Conference (SOLAR 2000), Madison, Wisconsin.

115. NSRDB-Vol. 1 (1992) Users Manual - National Solar Radiation Data Base (1961–1990). Version 1.0., National Renewable Energy Laboratory and National Climatic Data Center, Golden, CO/Asheville, NC.

116. NSRDB-Vol. 2 (1995) Final Technical Report: National Solar Radiation Data Base (1961–1990), National Renewable Energy Laboratory, Golden, CO.

117. Maxwell, E.L. (1998) METSTAT – The solar radiation model used in the production of the National Solar Radiation Data Base (NSRDB). *Solar Energy*, **62**, 263–279.

118. Maxwell, E.L., George, R.L. and Wilcox, S.M. (1998) Climatological solar radiation model, presented at the 1998 American Solar Energy Society Annual Conference, Albuquerque, New Mexico.

119. Energy Research Institute (1998) *Kingdom of Saudi Arabia Solar Radiation Atlas*, 2nd ed. Riyadh, KSA: Energy Research Institute King Abdul–Aziz City for Science and Technology.

120. Wilcox, S. (2010) *National Solar Radiation Database 1991–2010 Update: User's Manual*. National Renewable Energy Laboratory, Golden CO.

121. Perez, R., Kmiecik, M., Moore, K. *et al.* (2004) Status of high resolution solar irradiance mapping from satellite data, presented at the 33rd ASES Annual Conference (SOLAR 2004), Portland, OR.

122. http://maps.nrel.gov/prospector (accessed: April, 2015).

123. http://pvwatts.nrel.gov/ (accessed: April, 2015).

124. Kaku K. and Potter, C.W. (2009) Creating high-resolution solar information from satellite imagery and numerical weather, presented at the Solar09, the 47th ANZSES Annual Conference Townsville, Queensland, Australia.

125. Stein, J.S., Perez, R. and Parkins, A. (2009) Validation of PV performance models using satellite irradiance measurements, a case study, presented at the 38th ASES National Solar Conference (SOLAR 2009), Buffalo, NY.

126. http://eosweb.larc.nasa.gov/sse/ (accessed: April, 2015).

127. Stackhouse, J.P., Whitlock, C.H., Chandler, W.S., Hoell, J.M. and Zhang, T. (2004) Solar renewable energy data sets from NASA satellites and research, presented at the 33rd ASES Annual Conference (SOLAR 2004), Portland, OR.

128. http://re.jrc.ec.europa.eu/pvgis/solrad/index.htm (accessed: April, 2015).

129. http://www.soda-is.com/eng/index.html (accessed: April, 2015).

130. http://catalog.mines-paristech.fr/Files/ESRA11res.pdf (accessed: April, 2015).

131. http://www.pressesdesmines.com/the-european-solar-radiation-atlas-vol-2.html (accessed: April, 2015).

132. http://www.meteotest.ch/en/footernavi/solar_energy/meteonorm (accessed: April, 2015).

133. Thuman, C., Schnitzer, M. and Johnson, P. (2012) Quantifying the accuracy of the use of measure, correlate-predict methodology for long term solar resource estimates, presented at the 2012 World Renewable Energy Forum, Denver, CO.

134. National Climatic Data Center (1981) *Typical Meteorological Year User's Manual TD-9754 Hourly Solar Radiation – Surface Meteorological Observations*. NCDC, Asheville, NC.

135. Marion W. and Urban, K. (1995) *User's Manual for TMY2's Typical Meteorological Years Derived from the 1961–1990 National Solar Radiation Data Base*, National Renewable Energy Laboratory, Golden CO.

136. http://rredc.nrel.gov/solar/old_data/nsrdb/1961–1990/tmy2/ (accessed: April, 2015).

137. Wilcox S. and Marion, W. (2008) *Users Manual for TMY3 Data Sets NREL*, National Renewable Energy Laboratory, Golden CO.

138. http://rredc.nrel.gov/solar/old_data/nsrdb/1991-2005/tmy3/ (accessed: April, 2015).

139. http://apps1.eere.energy.gov/buildings/energyplus/weatherdata_about.cfm (accessed: April, 2015).

140. Muller, M., Marion, B., Kurtz, S. and Rodriguez, J. (2010) An investigation into spectral parameters as they impact cpv module performance, presented at the 6th International Conference on Concentrating Photovoltaic Systems: CPV-6, Freiburg, Germany.

141. Labed S. and Lorenzo, E. (2004) The impact of solar radiation variability and data discrepancies on the design of PV systems. *Renewable Energy*, **29**, 1007–1022.

142. Emery, K., Delcueto, J. and Zaaiman, W. (2002) Spectral corrections based on optical air mass, presented at the Photovoltaic Specialists Conference, 2002, Conference Record of the Twenty-Ninth IEEE, pp. 1725–1728.

143. Gueymard, C.A. (2011) Irradiance variability and its dependence on aerosols, presented at the SolarPACES 2011 - Concentrating Solar Power and Chemical Energy Systems, Granada.

144. Zirin, H. (1988) *Astrophysics of the Sun*, Cambridge University Press, Cambridge UK.

145. Neckel, H.L. (1994) Solar limb darkening 1986–1990 (lambda 303 to 1099 nm). *Solar Physics*, **153**, 91–114.

146. Grether, D., Nelson, J. and Wahlig, M. (1976) Measurement of circumsolar radiation, presented at the Proc. SPIE 0068, Optics in Solar Energy Utilization I.

147. Hunt, A., Grether, D. and Wahlig, M. (1978) Circumsolar radiation data for central receiver simulation presented at the ERDA Solar Workshop on Methods for Optical Analysis of Receiver Systems, Houston, TX.

148. http://rredc.nrel.gov/solar/old_data/circumsolar/ (accessed: April, 2015).

149. Buie, D., Monger, A.G. and Dey, C.J. (2003) Sunshape distributions for terrestrial solar simulations, *Solar Energy*, **74**, 113–122.

150. Rabl A. and Bendt, P. (1982) Effect of circumsolar radiation on performance of focusing collectors. *Journal of Solar Energy Engineering*, **104**, 237–250.

151. Buie D. and Monger, A.G. (2004) The effect of circumsolar radiation on a solar concentrating system. *Solar Energy*, **76**, 181–185.

152. Neumann A. and Witzke, A. (1999) The influence of sunshape on the DLR Solar Furnace beam, *Solar Energy*, **66**, 447–457.

153. Neumann, A., Witzke, A., Jones, S.A. and Schmitt, G. (2002) Representative terrestrial solar brightness profiles. *Journal of Solar Energy Engineering*, **124**, 198–204.

154. Ohmura, A., Gilgen, H., Hegner, H. *et al.* (1998) Baseline Surface Radiation Network (BSRN/WCRP): New precision radiometry for climate research. *Bulletin of the American Meteorological Society*, **79**, 2115–2136.

155. http://www.bsrn.awi.de/ (accessed: April, 2015).

156. University of Oregon (1999) Pacific Northwest Solar Radiation Data, UO SOLAR MONITORING LAB, Physics Department, University of Oregon, Eugene, Oregon.

157. Marion W. and Wilcox, S. (1994) *Solar Radiation Data Manual for Flat-Plate and Concentrating Collectors*. National Renewable Energy Laboratory, Golden, CO.

158. Perez, R., Seals, R., Ineichen, P., Stewart, R. and Menicucci, D. (1987) A new simplified version of the perez diffuse irradiance model for tilted surfaces. *Solar Energy*, **39**, 221–231.

159. Wilcox S. and Gueymard, C. (2010) Spatial and temporal variability of the solar resource in the United States, presented at the 39th ASES Annual Conference (SOLAR2010), Phoenix, AZ.

160. http://rredc.nrel.gov/solar/new_data/variability/ (accessed: April, 2015).

2

Concentrator Multijunction Solar Cells

Ignacio Rey-Stolle,[1] Jerry M. Olson,[2] and Carlos Algora[1]

[1]*Instituto de Energía Solar, Universidad Politécnica de Madrid, Spain*
[2]*Consultant, United States*

2.1 Introduction

Among all energy technologies, the evolution of photovoltaics over the first decades of the 21st century has been remarkable. Photovoltaic (PV) installations have been growing exponentially and by the end of 2016 are expected to have surpassed the non-negligible mark of 200 GW_p cumulative installed capacity in the world. Looking back at the evolution of photovoltaics over the past decades, a range of factors share the responsibility for this success. However, it seems clear that a key milestone in the deployment of PV technology should be attributed to the development of an efficient, robust and reliable device – namely, the crystalline silicon solar cell –, as the cornerstone of the technology.

Obviously, increasing PV demand and solar cell improvement are not independent variables but, on the contrary, reinforce each other through a virtuous cycle: increasing demand allowed for larger research investments, which in turn produced efficiency and reliability improvements in the solar cells, which in the end yielded better products with a positive effect on the demand.

As flat-plate PV is linked to crystalline silicon solar cells, modern CPV systems are undoubtedly tied in with multijunction solar cells. However, in this case the situation is somewhat different since high efficiency multijunction solar cells were originally commercialized for space applications. The unparalleled efficiency, high reliability and good performance at high temperatures of multijunction solar cells riveted the attention of CPV specialist paving the way for the modern renaissance of CPV.

Accordingly, being the cornerstone of CPV technology, this chapter provides a comprehensive view of multijunction solar cells. We will first briefly review, in section 2.2, the

Handbook of Concentrator Photovoltaic Technology, First Edition. Edited by Carlos Algora and Ignacio Rey-Stolle.
© 2016 John Wiley & Sons, Ltd. Published 2016 by John Wiley & Sons, Ltd.

fundamentals of solar cells and how the concept of multijunction solar cell emerges as a response to key loss mechanisms of single junction devices. Then the different ways to implement multijunction solar cells are revised in section 2.3; subsequently, the mathematical models for these devices are presented in section 2.4. The main challenges, particularities and requirements associated to concentrator operation are discussed in section 2.5. Finally, the different multijunction solar cell approaches in the market or under development in research labs are presented in section 2.6.

2.2 Fundamentals

2.2.1 Fundamentals of Photovoltaic Cells

2.2.1.1 A Solar Cell Primer

A solar cell or photovoltaic cell is a device that directly transforms electromagnetic radiation into electrical energy without any thermal cycles, mechanical cycles or chemical reactions [1–3]. Solar cells perform this feat by means of a very specific phenomenon, the so-called *photovoltaic effect*, which was discovered in 1839 by French physicist Alexandre-Edmond Becquerel. The whole process, from incoming solar radiation to electric power delivered to a load, essentially involves the following steps:

1. Electromagnetic radiation needs to be absorbed.
2. Such absorption must excite a pair of charge carriers, positive and negative, from their ground state (i.e. bound state) to a high energy state where they possess some of the energy released by the absorbed photon.
3. While still excited, a separation mechanism must exist to drive apart positive and negative charge carriers, impeding their relaxation to the ground state. Once in the excited state (which must be stable for some time), the carriers store some potential energy that can be delivered to a load.
4. Negative charges (electrons) are directed to a negative contact whilst positive charges move towards the positive contact.
5. Electrons flow out of the negative contact and circulate through the outer circuit delivering power to the attached load.
6. Electrons reach the positive contact where they meet the positive charges returning to the original ground (i.e. bound) state.

The fabrication of devices capable of implementing this process, namely solar cells, was not straightforward. In fact, between the discovery of the photovoltaic effect and the first reasonably successful demonstration of a solar cell – by Daryl Chapin, Calvin Fuller and Gerald Pearson at Bell Labs in 1954 – more than a century went by. In essence, it was only after the advent of modern-day semiconductors that solar cells became a reality. The fact of the matter is that p-n semiconductor junctions make for excellent solar cells (by far the most successful yet). When light is absorbed near a p-n semiconductor junction, photons *with energy greater than the bandgap* induce the photogeneration of charge carriers (electrons and holes) that are efficiently separated by the built-in field of the junction and flow to the outer circuit or load providing electrical power.

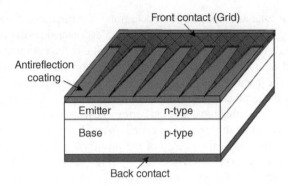

Figure 2.1 Basic structure of a solar cell based on a semiconductor p-n junction

With this background, the basic structure of a solar cell based on a p-n semiconductor junction (depicted in Figure 2.1) becomes evident:

1. The core of the device is a p-n junction made of a semiconductor with adequate bandgap (E_g~1.1 eV for the solar spectrum as will be described in section 2.2.1.2), which will provide absorption of light, carrier generation and charge separation.
2. Two contacts are needed (one for each carrier type). The top contact needs to have some sort of grid structure on the front surface (in order to let most of the light enter the semiconductor) while the back contact can cover completely the rear surface.
3. Finally, as it is done in the lenses of cameras or telescopes, the solar cell front surface is covered with an antireflection coating, to minimize reflection losses.

2.2.1.2 Solar Cell Optimum Bandgap

In the description above, it was highlighted that, in order to create electron-hole pairs in a semiconductor material exposed to light, the photons involved need to have an energy greater than the bandgap energy of the semiconductor. In short, the bandgap represents the energy difference between the top of the valence band and the bottom of the conduction band of the semiconductor. In essence, these are the energy levels associated with the ground state and the lowest excited state of step (2) of the description given for the photovoltaic effect in section 2.2.1.1 above. So a photon with just the energy of the bandgap of a given semiconductor has exactly the energy needed to promote an electron from the top of the valence band, i.e. the ground state, where the electron is bound to an atom, to the bottom of the conduction band, i.e. the excited state, where the electron can freely move and contribute to an electric current. A convenient simplification for solar cells is to assume that photons with energies lower than the bandgap will not be absorbed,[1] since they do not have energy enough to promote the jump between the valence and conduction bands and all intermediate energy levels are forbidden. In other words, the bandgap represents the minimum energy that can be absorbed from a photon to produce a free carrier for conduction. Accordingly, the bandgap is the first key

---

[1] In fact, they can be absorbed by other processes, such as free carrier absorption, but will not contribute to the photovoltaic effect.

parameter in terms of solar cell efficiency as it determines the fraction of the solar spectrum that will not be absorbed, which is all photons with energies lower than the bandgap.

In the light of the discussion above, it could be thought that the optimum bandgap of a solar cell should be small in order to maximize photon absorption. This is not the case (as will be seen in Figure 2.3, to be discussed later). The reason behind that discrepancy is that the bandgap not only determines the number of excited charge carriers (by modulating photon absorption) but also the potential energy that those excited carriers have as compared to their ground state. The key phenomenon behind this process is called *thermalization*. In essence, thermalization means that no matter how high the energy of the absorbed photon is, the resulting excited charge carrier will quickly relax to the base energy level at the band edge losing the excess energy in the form of heat. In other words, the excited carriers cool down fast (within femtoseconds) to an energy level in which they can live stably for much longer times; the relevant energy level for electrons is the bottom of the conduction band and for holes the top of the valence band. Therefore, after a few femtoseconds, excited carriers produced by a very high energy photon are indistinguishable (energy-wise) from excited carriers produced by a photon with the energy of the bandgap. They all form a population of excited carriers with an average potential energy very close to the bandgap energy.[2] Accordingly, the bandgap defines a second key consequence in terms of solar cell efficiency as it determines the potential energy of the charge carriers produced by each absorbed photon.

So, in summary, low bandgap materials can absorb most photons in the solar spectrum (providing a large number of excited charge carriers) but the potential energy per charge carrier is low. On the other hand, high bandgap semiconductors will only absorb the high energy photons in the solar spectrum (providing a small number of excited charge carriers); but after thermalizaiton, the potential energy per charge carrier is high. Therefore, for a given spectral distribution, there must be an optimum bandgap energy. On earth this optimum bandgap energy is $E_g \sim 1.3$ eV as will be shown in Figure 2.3, and represents the band gap energy where the product of the number of generated carriers and energy per carrier is maximized.

2.2.1.3 Basic Solar Cell *I-V* Curve

When exposed to sunlight, a solar cell – i.e. an electronic device as the one in Figure 2.1 – will deliver electrical power to a load through the process described in section 2.2.1.1. The voltage and current levels at which this power is delivered depend on the current–voltage characteristic or *I-V* curve of the solar cell and that of the load. Physical models and subsequent equations for the calculation of the *I-V* curve of a solar cell will be presented in section 2.4.2.2. However, in this section we will present a qualitative approach to *I-V* curves in order to establish a link between the bandgap and the main solar cell electrical parameters.

The *I-V* curve of several solar cells under illumination are schematically depicted in Figure 2.2. Each of these curves has three characteristic points which are: 1) the short circuit current (I_{sc}) which is the current produced by the solar cell at zero voltage; 2) the open circuit voltage (V_{oc}) which is the voltage produced by the solar cell at zero current; and 3) the maximum power point or MPP ($P_{mpp} = V_{mpp} \times I_{mpp}$) which is the point at which the power delivered by the solar cell is maximum.

---

[2] In reality, it is a little lower than the energy of the bandgap, as determined by the separation of the quasi-Fermi levels for electrons and holes. For a detailed explanation see [1].

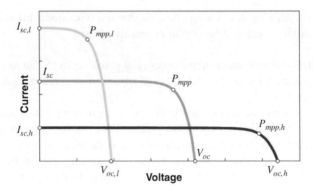

Figure 2.2 Solar cell *I-V* curves for three devices made with semiconductors of different bandgaps, namely, low bandgap (light grey); medium bandgap (dark grey); and high bandgap (black)

Figure 2.2 plots the *I-V* curve of three solar cells made of different semiconductor materials with different bandgaps. Following the discussion in the latter section, it is straightforward to see that 1) the lower the bandgap the larger the short circuit current ($I_{sc,l} > I_{sc} > I_{sc,h}$) since more photons are absorbed in low bandgap materials; and 2) the higher the bandgap the higher the open circuit voltage ($V_{oc,h} > V_{oc} > V_{oc,l}$) since V_{oc} is closely related to the bandgap and thus is a measure of the potential energy of the photogenerated carriers.

2.2.2 Fundamentals of Multijunction Solar Cells

2.2.2.1 Losses in Solar Cells

As will be discussed in Chapter 3, where the thermodynamic efficiency limits for photovoltaic energy conversion will be analyzed in some detail, the maximum efficiency attainable by a solar cell like the one depicted in Figure 2.1 is 30.1% [4]. A slightly higher efficiency of 40.7% (see Table 2.2) can be reached if the device is operated under concentrated light as a result of the premium in voltage provided by concentration, as will be discussed in section 2.5.1.

Analyzing in detail Figure 2.1 and the description of the photovoltaic process given in section 2.2.1.1, it is straightforward to identify the main sources of losses in solar cells:

1. Optical losses. The metallic grid on the front surface will reflect some of the impinging light. In addition, the antireflection coating on the surface will not be perfect for all wavelengths, generating also some undesired reflections in non-metallized areas.
2. Absorption losses. As discussed in section 2.2.1.2, photons with energy below the bandgap will be lost. Moreover, solar cells have a limited thickness and the semiconductor materials they are made of have finite absorption coefficients so even some photons with energies above the bandgap will not be absorbed due to the absence of an infinite absorber.
3. Thermalization losses. As discussed in section 2.2.1.2 as well, the excess energy of photons above the bandgap is lost in the form of heat.
4. Recombination losses. The excited carriers can find ways to release their extra energy and return to the ground state before they reach the solar cell contacts. When this process, called *recombination*, occurs, no power can be delivered to the external load and the extra energy is internally dissipated in the form of light (a photon) or heat.

5. Electrical losses. As a result of current flowing through semiconductor and metal layers of limited conductivity there will be ohmic or resistive losses.

In addition to the list above, more subtle losses exist in solar cells [5] but exert a lower impact on solar cell performance. Losses in solar cells and advanced solar cell designs will be covered in detail in Chapter 3.

All the sources of losses mentioned in the list above are to some extent unavoidable in real solar cells but can be minimized by developing a clever device design and engineering. Table 2.1 summarizes some loss mechanisms in solar cells, their limit value in ideal devices according to Shockley–Queisser's detailed balance calculations [4] (to be extensively discussed in Chapter 3), and some design solutions adopted in real solar cells to approach the ideal limit. This table shows that most losses allow some margin for design *tricks* or material improvements. However, this seems not to be the case for thermalization and below bandgap absorption losses, which can only be tackled in single junction solar cells by the use of materials with optimum bandgap. This fact severely impacts the limit efficiency of single junction solar cells (Figure 2.1) as is clearly shown by Figure 2.3. This figure plots the maximum efficiency as a function of bandgap for a semiconductor solar cell (according to the Shockley–Queisser limit), together with the magnitude of the main losses. For the bandgap energy that optimizes conversion efficiency ($E_g \sim 1.3$ eV), a breakdown of the losses is given.

Table 2.1 Some loss mechanisms in solar cells; their limit value in ideal devices according to Shockley–Queisser's detailed balance calculations [4] and some design solution proposed to approach the Shockley–Queisser limit in real devices

Loss mechanism	Ideal devices	Design solution in real devices
Grid shadowing	Set to zero	Back contacted solar cells (i.e. perform both contacts at the rear surface) Prismatic covers above grid lines to deflect light towards solar cell
Reflection losses	Set to zero	Texture front surface and use double- or triple-layer antireflection coatings
Limited thickness and absorption coefficient	Can be suppressed considering infinite absorbers	Use direct bandgap semiconductor materials with high absorption coefficients Use back-side reflectors to increase effective optical thickness
Limited conductivity in metal and semiconductor	Set to zero	Optimize grid design, contact technology and use thicker metallic grids
Recombination losses	Minimized considering only radiative recombination	Use high quality direct bandgap semiconductor materials limited by radiative recombination
Below bandgap absorption	All photons with energies below the bandgap are lost	Use optimum bandgap
Thermalization	All excess energy above the bandgap is lost	Use optimum bandgap

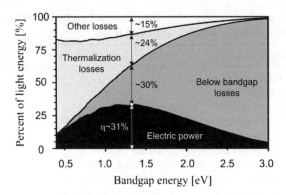

Figure 2.3 Maximum efficiency as a function of bandgap for a semiconductor solar cell plotted together with the magnitude of the main losses. Losses breakdown is given for the optimum bandgap energy ($E_g\sim1.3\,\mathrm{eV}$)

This breakdown demonstrates that thermalization and below bandgap absorption losses take away as much as $\sim54\%$ of the incoming energy, even in the optimum bandgap case. This represents a very large amount of wasted energy and has therefore spurred the quest for novel device architectures, namely, multijunction solar cells, capable of going beyond the Shockley–Queisser limit.

2.2.2.2 Multijunction Solar Cells as a Response to Minimize Losses

The key factor behind the large magnitude of the losses described in Figure 2.3, is that we are pursuing the conversion of the spectrally broad solar resource using a converter which is, in essence, specially suited to deal with monochromatic light: the single junction solar cell. In fact, when exposed to high-intensity well-collimated monochromatic light, solar cells can reach almost full conversion or light into electrical power [6]. In this context, we could turn the tables and assume that what should be adapted to our solar cells is the solar spectrum. Taking this idea to the limit, the solar spectrum could be partitioned into an infinite number of monochromatic components–i.e. infinitesimally narrow energy bands–which could be subsequently trans-formed into electricity by the same number of solar cells each of them with a bandgap energy matched to that of the impinging monochromatic beam. Such an ideal photovoltaic converter would consist of multiple p-n junctions (in fact an infinite number in this ideal case) and therefore is called a *multijunction solar cell*. Correspondingly, each of the junctions in a multijunction solar cell is called *a subcell*. As a result of the multijunction architecture, thermalization and below bandgap absorption losses are minimized and the limiting efficiency for these devices reaches 86.8% for the solar spectrum and an infinite number of junctions [7]. As it will be shown in Table 2.2, multijunction solar cells with much lower number of junctions can reach remarkably high efficiencies too.

In addition, multijunction solar cells present an additional advantage that helps mitigate ohmic losses. For an appropriate combination of bandgaps, the current produced by each subcell in the multijunction device is typically lower than in the case of single junction solar cells. In other words, multijunction solar cells deliver electric power at higher voltage and lower current. Thereby, as ohmic power losses scale with I^2, multijunction solar cells are less affected by electrical resistance losses.

Table 2.2 Optimum bandgap combination for multijunction solar cells with different number of junctions for a blackbody spectrum at 6000 K under concentration [18]

No. of junctions	Type of design	Bandgap combination [eV]	Limiting efficiency [%]
1	Unconstrained	1.11	40.7
	Series-connected	1.11	40.7
2	Unconstrained	0.77/1.70	55.8
	Series-connected	0.77/1.55	55.5
3	Unconstrained	0.62/1.26/2.10	63.8
	Series-connected	0.61/1.15/1.82	63.2
4	Unconstrained	0.52/1.03/1.61/2.41	68.7
	Series-connected	0.51/0.94/1.39/2.02	67.9
5	Unconstrained	0.45/0.88/1.34/1.88/2.66	72.0
	Series-connected	0.44/0.81/1.16/1.58/2.18	71.1
6	Unconstrained	0.40/0.78/1.17/1.60/2.12/2.87	74.4
	Series-connected	0.38/0.71/1.01/1.33/1.72/2.31	73.4

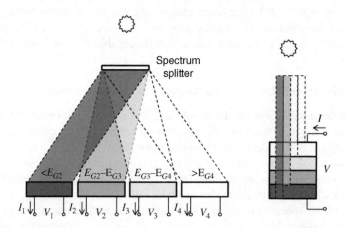

Figure 2.4 Multijunction solar cells, independent subcells with spectrum splitting (left) and monolithic arrangement where spectrum partitioning is achieved by the selective absorption of the upper layers (right)

As will be shown in section 2.3, there are a number of ways to implement multijunction photovoltaic devices. Figure 2.4 shows two basic approximations to achieve the adaptation of the spectrum to subcells with different bandgaps. One option is to split the spectrum by optical means and redirect each band to the appropriate subcell, as in Figure 2.4 (left). Another obvious option, is to exploit the fact that solar cells do not absorb photons with energies below the bandgap. Thereby, subcells in a multijunction cell can be vertically stacked in descending order of bandgaps, as in Figure 2.4 (right).

2.2.2.3 Multijunction Solar Cells and CPV

There are a number of reasons why multijunction solar cells are especially suited for CPV applications and *vice versa*. Among the most prominent are the following:

1. As has been mentioned, the main driver for CPV is the quest for higher PV efficiencies. Thereby, the first obvious advantage of multijunction solar cells is that they can provide much higher efficiencies than single junction solar cells.
2. Multijunction solar cells are structurally more complex and use (as of today) more expensive and scarcer materials. Therefore, the cost per unit area of multijunction solar cells (expressed in €/cm^2 or $/cm^2) is significantly higher than that of single junction devices. However, CPV systems use small area solar cells as semiconductor area needed per cell is roughly reduced by the concentration factor (i.e. 500× to 1300× in current systems in the market). As a result of reduced size, multijunction solar cells for CPV are cost-effective and represent a small fraction of the total CPV system cost (see Chapter 14).
3. As will be explained in sections 2.4 and 2.5, short circuit current scales linearly with concentration. Thus, the impact of ohmic losses experiences a quadratic growth with concentration and hence CPV systems are more demanding in terms of the minimization of ohmic losses. Here, multijunction solar cells have the advantage of delivering its power at a higher voltage and lower current than single junction solar cells.
4. A frequent argument against CPV deals with the impact of high temperatures on solar cell performance. One key advantage of multijunction solar cells stems from their higher efficiency which in turn produces less waste heat to be dissipated. In addition, as will be shown in Chapter 5, the influence of temperature is less significant in multijunction solar cells as compared to single junction devices, exhibiting lower relative voltage degradation as the temperature increases.

On the other hand, there are also some drawbacks for the use of multijunction solar cells in CPV systems associated to the spectral sensitivity of these devices. The fact of the matter is that the solar spectrum does not stay stable but varies during a day, throughout the year, and from location to location. Therefore, in field operation, the partitioning of the spectrum into different bands will be affected accordingly and thus the current produced by each subcell in a multijunction solar cell will suffer fluctuations. This phenomenon, known as *spectral mismatch*, is particularly relevant in series connected multijunction solar cells (the mainstream product in the market today) in which the solar cell current will be limited by the least generating subcell (the topic of current matching will be covered in section 2.3.1.3). On the other hand, different studies (as will be presented in section 2.5.5) and experience in the field (shown in Chapter 8) demonstrate that spectral variations play a minor impact on the energy yield of multijunction solar cells.

In summary, CPV is an enabler for the use of multijunction solar cells for terrestrial applications and at the same time, the high efficiency of multijunction solar cells is a booster for CPV technology.

2.3 Multijunction Solar Cell Structures

In the previous section we have discussed why multijunction solar cell architectures emerge as a natural choice to circumvent the efficiency limits of single junction solar cells and, in addition, are especially suited for concentrator applications. In this section, we go deeper in the discussion by presenting architectures of real devices and design trade-offs for multi-junction solar cells.

2.3.1 Historical Development of Multijunction PV Converters

As a result of its inherent advantages, the quest for a successful implementation of a multijunction solar cell began early in the history of PV research. In this section, we will cover the historical development of multijunction architectures that naturally converge to the monolithic triple-junction two terminal devices that today dominate the market.

2.3.1.1 Spectral Splitting Systems

In section 2.2, we have presented multijunction solar cells as a natural way to adapt our photovoltaic converter to the spectrally broad nature of the solar resource. The idea behind spectral splitting systems is to carry out such adaptation by optical means. A photovoltaic spectral splitting system is a set of optical elements designed to partition the solar spectrum into a set of different bands and redirect each of those bands to a solar cell whose bandgap has been chosen to maximize the photovoltaic conversion of such spectral region [8–10]. Taking this idea to the limit, the perfect spectral splitting system would be an (imaginary) lossless optical system that divides the solar spectrum into an infinite number of monochromatic components (i.e. infinitesimally narrow bands) which are then redirected towards an infinite number of solar cells each of them with a bandgap energy matched to that of the impinging monochromatic beam.

Figure 2.5 shows several possible implementations of spectral splitting systems, consisting of a small, finite number of solar cells. Figure 2.5(a) presents a system based on dichroic filters that reflect a band and let the remaining spectrum through; the reflected beam impinges on a solar cell and the transmitted beam suffers further reflections or reaches the last solar cell. Figure 2.5(b) sketches a system using transmission-diffraction filters in which the main band suffers no change in direction –and the associated solar cell can be placed right below– whilst the secondary band suffers severe diffraction; adjusting the distance between the filters and the solar cells the diffracted beam can be coupled to the appropriate solar cell. Figure 2.5(c) depicts

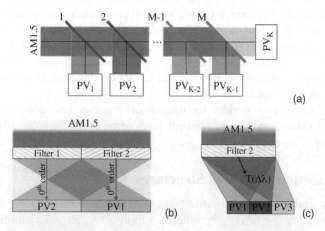

Figure 2.5 Several implementations of spectral splitting systems: (a) cascade configuration using dichroic mirrors; (b) adjacent configuration using transmission-diffraction filters; (c) adjacent configuration based on dispersive filters. Source: Russo et al. 2014 [9]. Reproduced with permission of OSA

a system based on dispersive filters, which act much like prisms spatially spreading the spectrum along a certain distance; placing solar cells along a given length of the dispersion axis a fraction of the spectrum can be collected. All these figures demonstrate a key feature of spectral splitting systems namely the spatial separation of bands which, in turn, allows the separated collection of the light by different (i.e. specially adapted) solar cells.

Cost, losses in optical elements, limitations in the bandgaps available for high quality solar cells and mechanical complexity have posed practical limits to the number of bands (i.e. solar cells) used in spectral splitting systems. The most common combination is two subcells; there are reports of three subcells and it seems impractical to go beyond four bands or subcells. Needless to say that cost; is also a decisive limiting factor in this set-up.

Due to their inherent complexity and elevated costs most of these system have been used for demonstration purposes (i.e. to achieve world records) [11] or for high end applications (i.e. military), and have never achieved commercial application in CPV systems.

2.3.1.2 Mechanical Stack of Solar Cells and Multi-terminal Devices

As shown when introducing the concept of multijunction solar cell in section 2.2, a natural way to implement a multijunction solar cell is simply to stack different solar cells, arranged in descending order of bandgaps, directly exposed to the light beam (see Figure 2.6). This idea is seemingly simple but hides a number of technical challenges. Obviously, it is a must that light not intended to be absorbed in a given subcell can make it through and exit to the next subcell. This means that all subcells (except the bottom subcell) must have a grid both on the front side and the backside to let the unabsorbed light out or at least (in the case of series connected devices) have transparent interconnects. In addition, for some materials, the substrate onto which p-n junctions are fabricated must be etched away to avoid the attenuation of the light by free carrier absorption [12]. Moreover, in order to minimize reflection losses, the back and front grid of adjacent subcells must be aligned, which imposes the need of a certain mechanical precision in the stacking of the subcells. Another challenge deals with the optical coupling of the light exiting one subcell into the next one. To minimize reflection losses antireflection coatings need be deposited on both front a back surfaces and the insertion of an optically dense medium between subcells is highly recommendable. This device configuration is prone to exhibiting two problems under concentrator operation. First, resistive losses may be high as a result of the lateral conduction that takes place in both the front and back contacts. Second, heat dissipation may not be as efficient due to the presence of adhesives or air between the cells (i.e. not all subcells are efficiently connected to the heat sink).

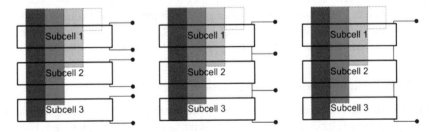

Figure 2.6 Configuration of a mechanically stacked triple-junction solar cell with possible variations for the electrical contacts. (Left) independent contacts; (center) multi-terminal; (right) series connected

However, in addition to the aforementioned difficulties mechanically stacked multijunction solar cells offer a number of advantages. As shown by Figure 2.6, metallic grids are available both at the front and back side of each solar cell in the stack. So, for a series-connected or multi-terminal device (see Figure 2.6), the electrical contact between subcells can be implemented by intermetallic junctions avoiding the need of semiconductor tunnel junctions (which will be introduced in the section 2.3.1.3 about monolithic integration of multijunction solar cells). Having metallic interconnects, it is quite straightforward to give external access to such contacts and turn them into device terminals. Figure 2.6 shows three possible combinations for a triple-junction solar cell to form a two-, four- or six-terminal device. Obviously, having independent electrical access to each subcell in the stack allows for the extraction of the full photocurrent and thus of the full electrical power produced in each subcell, leading to higher efficiencies. In other words, the current matching restriction can be avoided here (it will not be avoided for the case of mechanically stacked series connected devices, as it occurs with monolithically integrated series connected multijunction solar cells; see section 2.3.1.3). Of course, these higher efficiencies at the solar cell level come at the price of complicating the electrical architecture of the CPV module where subcells producing different currents and voltages have to be grouped and connected to form branches with compatible current and voltages. This interconnection has to be engineered carefully to guarantee that energy gains produced at the cell level are not lost in the form of mismatch losses at the module level. It should be noted that these arrangements are not always straightforward since the number of subcells of each type available is fixed, so the possible combinations are limited. Another approach is to have independent connection layouts for the different type of subcells and use separate inverters for each type, which simplifies the connection problem but increases costs.

In brief, mechanically stacked solar cells provide higher efficiencies at the price of 1) having to use a different semiconductor substrate for each subcell; 2) increased subcell processing (front and back grids and ARC needed); 3) having additional issues at high concentration associated to thermal and resistive (lateral conduction) losses.; 4) higher mechanical complexity in solar cell integration (i.e. subcells need to be mounted and aligned in the stack); and 5) sophisticated electrical layouts at the module level. In the 1990s several research groups implemented prototypes that engineered quite elegantly these challenges, producing solar cell efficiencies in the range of 35% [13]. However, these designs never achieved commercial deployment due to their elevated cost and the rapid progress of monolithic two-terminal multijunction solar cells, as will be shown below.

In recent times, a refined version of the concept of mechanically stacked solar cells is gathering momentum, namely, wafer bonded solar cells. In this approach solar cells are bonded at a wafer level (i.e. after epitaxy) and not at a device level. Therefore, once the bond is made, the devices are processed as monolithic two-terminal multijunction solar cells (to be discussed in the next section), and thus current matching applies but are not restricted to lattice matching limitations. A deeper discussion about this approach is held in section 2.6.8.

2.3.1.3 Monolithic Two-terminal Multijunction Solar Cells

In terms of solar cell configuration, monolithic multijunction solar cells are again constituted by stacking different solar cells, arranged in descending order of bandgaps, (see Figure 2.7). The key difference with mechanically stacked solar cells is that the complete solar cell structure – comprising all subcells and interconnecting elements – is grown (i.e. fabricated) in the same

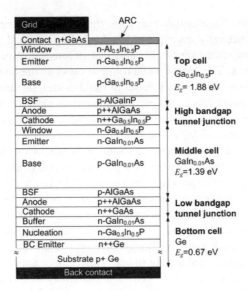

Figure 2.7 Structure (cross-section) of a two-terminal triple-junction GaInP/InGaAs/Ge solar cell

manufacturing process called epitaxy [14]. In this sense, the manufacturing sequence is greatly simplified as compared to mechanically stacked designs. In the monolithic approach there is a single (though longer) epitaxial growth phase where the stacking of the solar cells occurs and subsequent contacts and ARCs are deposited only once.

In Figure 2.7, we show an exemplary design of this solar cell configuration, namely, a monolithic triple-junction GaInP/GaAs/Ge solar cell (see section 2.6.1), which was the first multijunction design massively used in commercial CPV systems (and still is). As can be deduced from Figure 2.7, there are three main challenges to implement such a structure:

1. The semiconductor materials needed for the different subcells have to be structurally compatible so that they can be grown in the same epitaxial process.
2. Electro-optical interconnects between subcells yielding minimum losses have to be implemented also using structurally compatible materials.
3. Since the subcells in the stack are connected in series they should be designed to produce the same current, i.e., they should be *current-matched*.

The requirement of structural compatibility refers to the fact that all semiconductor materials involved in a design as the one sketched in Figure 2.7 need to be grown with high crystallographic perfection, which is a prerequisite to achieve good photovoltaic properties, namely, high minority carrier lifetimes and diffusion lengths. This fact restricts the palette of semiconductors available to *families* which share the same type of crystal lattice and lattice constant. This is the so-called *lattice-matching* requirement for materials in monolithic multijunction solar cells. As will be discussed in section 2.3.2.2 (where materials needs and constraints are assessed in detail) this requirement can be relaxed to some extent allowing some lattice mismatch in the quest for a wider palette of materials (i.e., bandgaps) to implement multijunction solar cells.

In addition to subcells with adequate bandgaps, the monolithic integration of multijunction solar cells requires structurally compatible low-loss electro-optical interconnects between subcells. This again restricts the range of usable materials to lattice-matched (or quasi lattice matched) semiconductors. The most common interconnect is the so-called *tunnel junction*. Tunnel junctions – also known as tunnel diodes or Esaki diodes – are essentially heavily doped p^+-n^+ junctions where quantum tunneling across the junction plays a key role in the conduction mechanism [15]. Such devices were discovered by Nobel-laureate Leo Esaki at the end of the 1950s and were soon proposed for use in multijunction solar cells. The attractive feature of tunnel junctions for monolithic multijunction solar cells is that, as long as quantum tunneling is the dominant conduction mechanism across the p-n junction, they exhibit a linear, resistor-like *I-V* dependence until the so called peak tunneling current (I_p) is reached. This initial ohmic region in the *I-V* curve is ideal to implement low-loss interconnects between the subcells with different energy bandgaps. According to this description, two important requirements can be deduced for tunnel junctions. First, they should provide minimum electrical losses in the interconnection (i.e. low voltage drop) by exhibiting low equivalent resistance and high peak tunneling currents. This requirement is especially demanding in high concentrator applications where photogenerated currents can be on the order of $10 \, A \, cm^{-2}$. Second, they should be transparent (i.e. non-absorbing) to the light travelling to underlying subcells. A detailed discussion of the requirements for the performance of tunnel junctions in CPV applications is presented in section 2.5.3.

The third challenge mentioned in monolithic multijunction solar cell structures is *current matching*. With only two external contacts and several subcells, the current in the multijunction solar cell will be limited by the subcell with the lowest photogeneration, which is referred to as the limiting subcell or junction. Any excess photocurrent produced in non-limiting junctions represents power that cannot be delivered to the external load and thus needs to be internally dissipated in the subcell in question and will only contribute to its heating. Therefore, in order to minimize efficiency losses, in a monolithic multijunction solar cell all subcells are designed to produce the same photocurrent when exposed to the reference spectrum. Of course, in field operation the spectrum changes (daily, seasonally, location-wise, etc.) and therefore a perfect current matching between subcells will seldom occur giving rise to unavoidable losses that monolithic designs should be engineered to minimize (see section 2.5.5).

However, monolithically grown multijunction solar cells do not need to be two terminal devices, as sketched in Figure 2.7. For example, tunnel junctions could be replaced by transparent (i.e., high bandgap) lateral conductive layers onto which contacts could be processed. In this respect, monolithic multi-terminal multijunction solar cells could be manufactured, where the need for tunnel junctions would be avoided and the restriction of current matching would not apply [16]. Despite being a seemingly simple and thus attractive idea, its impact has been quite limited in terms of commercial development for similar reasons as those discussed in multi-terminal mechanically stacked devices, namely:

1. The formation of extra contacts increases the complexity of the manufacturing process and therefore produces increased cost and lower yields.
2. Lateral current extraction to intermediate terminals typically leads to increased resistive losses at high concentration (i.e. lowers efficiency).
3. Multi-terminal devices complicate the electrical interconnection of the solar cells at the module level. In this respect, current mismatch losses are avoided but, to some extent, turned into interconnection losses.

Because of the reasons outlined above, two-terminal devices are preferred as a result of their robustness, easy interconnection and limited impact of mismatch losses for adequately engineered bandgap combinations. For all these reasons the solar cell design that has dominated the CPV industry so far has been the monolithic two-terminal multijunction solar cell or, more specifically, a monolithic two-terminal triple-junction solar cell (Figure 2.7). Alternative structures with more subcells are appearing on the horizon (as will be discussed in section 2.6), but the two-terminal standard seems today unchallenged.

2.3.2 Designing Multijunction Solar Cell Structures

2.3.2.1 Efficiency as a Combination of Bandgaps

Multijunction solar cells (MJSC) split the solar spectrum into different bands that are then converted into electricity by optimized subcells. Such partitioning is shown in Figure 2.8 for three classic solar cell designs used at some point in CPV systems (GaAs single junction; GaInP/GaAs dual-junction; and GaInP/GaAs/Ge triple-junction solar cells). Figure 2.8 illustrates the two aspects in which MJSCs outperform single junction solar cell devices. First, despite the fact that a single junction GaAs solar cell (Figure 2.8 left) and a dual-junction GaInP/GaAs solar cell (Figure 2.8 center) use the same fraction of the AM1.5d solar spectrum, an increase in efficiency is observed in the latter by a gain in V_{oc} attributable to the high bandgap top cell. In other words, the fraction of the spectrum absorbed by the GaInP top cell in the dual-junction design is transformed into electricity with lower thermalization losses compared to the GaAs single junction solar cell. Second, in the transition from a dual GaInP/GaAs (Figure 2.8 center) to a triple GaInP/GaAs/Ge (Figure 2.8 right) solar cell it is the new spectral band accessible by the Ge bottom cell that is responsible for the efficiency increase. The Ge subcell increases the V_{oc} without affecting the current of the multijunction cell. By itself, Ge would have high thermalization losses and a relatively low efficiency. In summary, the use of several bandgaps offers the possibility of reduced thermalization losses and/or access to wider portions of the solar spectrum.

The process for optimizing the efficiency of the multijunction solar cell is a question of determining the number of subcells (i.e. bands) to be used and the bandgap for each subcell. If ideal solar cells are assumed – i.e. total absorption and perfect collection of carriers in their band – the optimum combination of bandgaps and the maximum efficiency of the resulting multijunction solar cell can be calculated as shown in Figure 2.9 (the resulting combination of bandgaps is summarized in Table 2.2) [17–19]. The first obvious observation of Figure 2.9 is that increasing the number of junctions raises the efficiency potential. This idea can be taken to the limit and the absolute limiting efficiency for a multijunction photovoltaic device with an

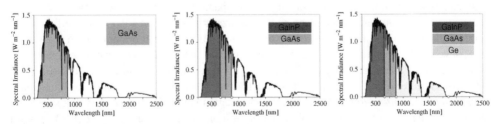

Figure 2.8 Examples of how spectrum is used by different solar cells: (left) Single junction GaAs solar cell; (center) Dual-junction GaInP/GaAs solar cell; (right) Triple-junction GaInP/Ga(In)As/Ge solar cell

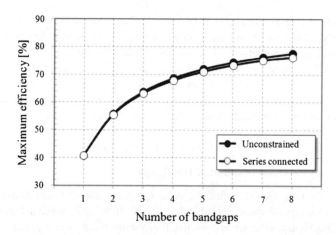

Figure 2.9   Efficiency of an ideal multijunction solar cell as a function of the number of bandgaps under the spectrum AM1.5d. Solid circles correspond to independently connected subcells while empty circles correspond to series connected (i.e. current matched) devices

infinite number of junctions can be calculated to be 86.8% [7]. Figure 2.9 and Table 2.2 also show that the attainable efficiencies for both series connected devices (where current matching applies) and unconstrained multijunctions with independently connected subcells are similar. This implies that the limiting conversion efficiency for multijunction stacks is independent of whether or not it is connected in series, at least for the ideal combination of bandgaps.

As it will be discussed in the next section, the materials available to implement multijunction solar cells are limited and therefore not all bandgap combinations are accessible. Accordingly, a question that naturally arises is 'what is the penalty (in terms of efficiency) if the subcell bandgaps deviate from the optimum values listed in Table 2.2?' Efficiency losses for non-optimum combinations are easily evaluated in a dual-junction solar cell and displayed using contour maps where isoefficiency lines are traced as a function of the bandgaps of top and bottom subcells. An example of an isoefficiency contour map is presented in Figure 2.10(a) [20]. This map is calculated for one-sun AM1.5g spectrum and for not perfectly ideal devices (some losses are allowed for a more realistic result) and thus the maximum efficiency in the map is slightly lower than that of Table 2.2. The analysis of Figure 2.10(a) in the quest for the optimum combination of bandgaps (marked with a black circle) is straightforward. In addition, the penalty paid in efficiency for not using this bandgap combination can be easily ascertained too. Besides, the presence of local maxima (as the one marked with a gray circle) can be revealed offering design alternatives with a minor drop in efficiency as compared to the absolute maximum.

The use of isoefficiency contour maps to optimize triple-junction solar cell is still possible though not so straightforward. As shown by Figure 2.10(b), instead of contours in a plane we now have isoefficiency surfaces in a three-dimensional space since we must now consider as independent variables the bandgaps of the three subcells, namely, top, middle and bottom [21]. The projections of two of these surfaces – for 51% and 52% efficiency – on the Cartesian planes *bottom-middle*; *bottom-top* and *middle-top* define again isoefficiency contours that should be jointly considered to optimize a triple-junction design. For example, Figure 2.10(b) shows a global maximum efficiency for the band gap combination of 1.86, 1.34, and 0.93 eV (marked as a gray ball), and a local maximum at 1.75, 1.18, and 0.70 eV (marked as a gray cone), which are

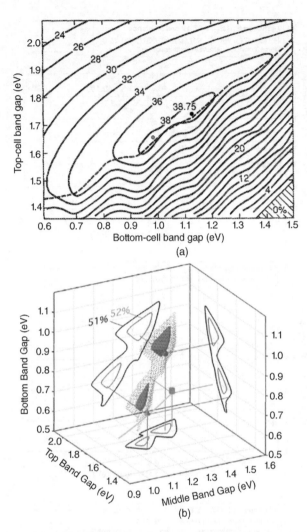

Figure 2.10 (a) Theoretical isoefficiency contour maps for a dual-junction solar cell as function of bottom and top cell bandgap for AM1.5g spectrum, one-sun concentration, and a cell temperature of 300 K. Reproduced with permission from [20]; (b) Theoretical isoefficiency surfaces and their projected contours for series-connected triple-junction solar cells under spectrum AM1.5d at 500 suns concentration and 300 K. Source: Geisz et al. 2008 [21]. Reproduced with permission of AIP

significantly more efficient than the mainstream design of 1.86, 1.39, and 0.67 eV (marked as a gray cylinder), corresponding to the lattice matched GaInP/GaInAs/Ge solar cell to be discussed in section 2.6.1.

The situation for four or more junctions becomes even more complicated, lacking simple graphical representations. Therefore, to study the bandgap sensitivity of designs with 3, 4 or more subcells, it is common to fix all the bandgaps except for two and analyze them in a simple *x-y* isoefficiency contour plot, as shown in Figure 2.11 reproduced from [22]. In Figure 2.11.(a) the top and middle subcell bandgaps in a triple-junction solar cell are studied, with the bottom

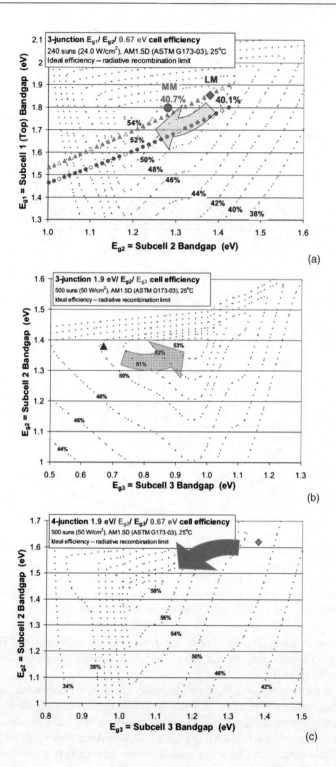

(a)

(b)

(c)

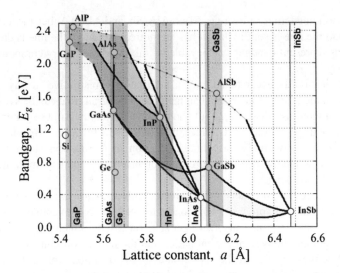

Figure 2.12 Lattice constant vs. bandgap dependence for the most common semiconductor materials used in solar cells. Only materials with the zinc-blende or diamond crystal lattice have been included

cell bandgap fixed to 0.67 eV (i.e. the bandgap of germanium). With this figure, the potential of 3-junction solar cell designs lattice matched to germanium (see section 2.6.1) or upright metamorphic on germanium (see section 2.6.2) can be assessed. In Figure 2.11.(b) the efficiency of a triple-junction solar cell with a 1.9 eV top cell as function of middle cell and bottom cell bandgap is plotted. With this figure, the potential of inverted metamorphic architectures grown on GaAs or Ge (see section 2.6.3) can be assessed. Finally, in Figure 2.11. (c) the efficiency of a quadruple-junction solar cell with a 1.9 eV top cell and a 0.67 eV bottom cell is depicted as function of the two middle cell bandgaps. With this figure, the potential of 4-junction cell designs lattice matched to germanium (see section 2.6.7) or upright metamorphic on germanium can be evaluated.

2.3.2.2 Materials and Constraints

Table 2.2 should establish a clear roadmap to design high efficiency multijunction solar cells of different bandgaps. However, the fact of the matter is that the palette of materials available to implement multijunction solar cells is limited and therefore not all bandgap combinations are accessible. This can be visualized with the aid of Figure 2.12 where the bandgap of the most common semiconductor materials used in PV is plotted against their key structural property: their lattice constant. Circles in this figure represent single or binary compounds. Thick lines connecting two binary compounds that share a constituent atom represent bandgaps achievable

---

Figure 2.11 (a) Theoretical isoefficiency contour maps for (a) a triple-junction solar cell with a 0.67 eV bottom cell as function of middle cell and top cell bandgap; (b) a triple-junction solar cell with a 1.9 eV top cell as function of middle cell and bottom cell bandgap; (c) a quadruple-junction solar cell with a 1.9 eV top cell, a 0.67 eV bottom cell as function of the two middle cell bandgaps. All contours are calculated for AM1.5d spectrum, 500 suns concentration, and a cell temperature of 300 K. Source: King et al, 2009 [22]. Reproduced with permission of WIP

by their corresponding ternary alloy (e.g. AlAs, AlAsSb, and AlSb). When the ternary alloy has a direct bandgap the line is solid; when it has an indirect bandgap, the line is dashed. Finally, shadowed areas represent bandgaps achievable by mixing three binary compounds to yield a quaternary alloy.

In principle, it would appear that Figure 2.12 is not so restrictive. If any circle, line or shadowed area in the figure represents a bandgap (i.e. a material) that can be synthesized, then virtually all combinations in Table 2.2 could be fabricated. Reality is not so generous, at least in order to implement good solar cells, since materials need to be grown not only with the appropriate bandgap but with high structural perfection also. High crystallographic perfection is a prerequisite to achieving good carrier collection properties, namely, high minority carrier lifetimes and diffusion lengths.

Epitaxy is the process most widely used to create semiconductor layers of large area and high structural quality. In this process, semiconductor layers–that eventually will constitute electronic devices–are deposited or *grown* on substrates with which they usually share two key structural features: the type of crystal lattice and the lattice constant, which is in turn the parameter that defines the size of the crystal lattice [14]. In other words, the new layers are deposited on a substrate that is structurally identical in terms of type (crystal lattice) and size (lattice constant). This represents the so-called *lattice-matching* requirement. In this way the substrate acts as a perfect template on which the subsequent growth of the layer can proceed with high *order*. In fact, this is literally what the term epitaxy means, a compound word formed by two ancient Greek roots: *epi* meaning 'on' and *taxy* meaning 'ordering'. Here the word ordering should be understood as a synonym of high crystallographic perfection. Different technologies are used in the semiconductor industry to implement this *ordered* deposit of a material *onto* a substrate. The most common is *metalorganic vapor phase epitaxy* (MOVPE) also known as *metalorganic chemical vapor deposition* (MOCVD). Another technique, *molecular beam epitaxy* (MBE), has become important for multijunction solar cells incorporating diluted nitrides, as will be discussed in section 2.6.

Therefore, if the need for high crystalline quality is only easily attained if there exists a suitable substrate, then an important restriction further limits the palette of semiconductors available for multijunction solar cells. This additional restriction is represented in Figure 2.12 as thin vertical lines that essentially define different *families* of materials that share the same type of crystal lattice and lattice constant (i.e. are lattice-matched). The filled circle in each family represents the substrate material.

The lattice matching requirement can be relaxed to some extent allowing some lattice mismatch in the quest for a wider palette of materials (i.e. bandgaps) to implement solar cells. Unfortunately, this is done at the price of creating some defects in the crystalline structure and therefore losing some photovoltaic quality in the material.

Here emerges the key trade-off for the design of multijunction solar cells: the approach to optimum bandgaps has to be balanced with the attainment of sufficiently high minority carrier parameters (i.e. crystalline perfection) in all subcells forming the multijunction solar cell. This is where the contour maps of Figure 2.10 and Figure 2.11 reveal their usefulness.

In current multijunction solar cell research there are essentially three alternatives to optimize this trade-off:

- *Mechanically stacked multijunction cells.* As already discussed in section 2.3.1.2, here different substrates can be used for the different subcells expanding the choice of bandgaps without sacrificing material quality. Thereby, all materials in the lattice matching lines of Figure 2.12 would be accessible. Advantages and disadvantages have been already discussed in section 2.3.1.2, being cost and complexity the main hurdles for this strategy.

- *Lattice matched solar cells.* Here a single substrate is used to grow all the materials for the different subcells (section 2.3.1.3). In addition, to maximize material quality the choice of bandgaps is limited to lattice matched materials. As a consequence, bandgap combinations are relatively far from the target of optimum values. This is the case of the mainstream triple-junction GaInP/GaInAs/Ge design. Advantages of this design include reduced cost and demonstrated long term reliability resulting from its simplicity and lack of defects.
- *Metamorphic solar cells.* Here a single substrate is also used to grow all the materials for the different subcells (section 2.3.1.3). However, in this case a certain lattice mismatch is allowed in the quest to approach the optimum bandgaps. Advantages of this design include potentially higher efficiencies than lattice matched and moderate costs.

A detailed discussion of various device designs exploring these different routes is given in section 2.6.

2.4 Multijunction Solar Cell Modeling

The computation of the electrical characteristics of solar cells is an essential tool for device design, optimization, analysis and, definitely, also for system integration. For the solar cell designer it is of utmost importance in order to correlate device structural properties with electrical characteristics. For the system integrator it is also important to have a mathematical description of the solar cell to incorporate into higher-level system models. However, an accurate and reliable simulation of high concentrator multijunction solar cells is a challenging task due to their complex configuration, – consisting of stacked p-n-junctions connected by tunnel junctions and involving a large number of semiconductor layers (e.g. see Figure 2.7) – and significant distributed effects. In this section, the different approaches followed to compute electrical characteristics of multijunction solar cells are described, focusing the discussions mostly on the more complex case of series-connected multijunction solar cells (either monolithically integrated or wafer bonded designs). The modeling of multi-terminal devices with independent subcells represents a simpler case where the classic approach for single junction solar cells can be followed.

2.4.1 Numerical Modeling of Multijunction Solar Cell Structures

In the most general approach to the modeling problem multijunction solar cells (as with many other semiconductor devices) can be simulated by solving transport equations for charge carriers along the device structure [23]. The set of equations that govern charge transport in semiconductor devices are the Poisson equation (Eq. (2.1)); the electron and hole continuity equations (Eqs. (2.2) and (2.3)) and the transport equations again for electron and holes. For the particular case of solar cells, carrier transport can be effectively modeled using the drift-diffusion approximation: Eqs. (2.4) and (2.5) [15].

Accordingly, the equations in Table 2.3 provide a general framework for semiconductor device simulation that can be applied to an arbitrary stack of semiconductor layers, and

Table 2.3 Semiconductor equations for numerical modeling of solar cells

Poisson	$\nabla \cdot \varepsilon \nabla V = -q(p - n + N)$	(2.1)
Continuity for electrons	$\nabla \cdot \boldsymbol{J_n} = -q\left(G - R_n - \dfrac{\partial n}{\partial t}\right)$	(2.2)
Continuity for holes	$\nabla \cdot \boldsymbol{J_p} = q\left(G - R_p - \dfrac{\partial p}{\partial t}\right)$	(2.3)
Drift-diffusion for electrons	$\boldsymbol{J_n} = -q\mu_n \nabla(V + V_n) + kT\mu_n \nabla n$	(2.4)
Drift-diffusion for holes	$\boldsymbol{J_p} = -q\mu_p \nabla(V - V_p) - kT\mu_p \nabla p$	(2.5)

therefore to multijunction solar cells, in particular. Within each semiconductor layer the equations in Table 2.3 have to be solved, considering appropriate boundary conditions at each interface. In addition to obvious structural information (doping, thickness, composition, . . .) and operating conditions (temperature, electrical bias, illumination), the final formulation of the equations also needs inputs about material parameters (bandgap, mobilities, optical constants, recombination parameters, etc.) and physical models (generation, recombination, conduction across interfaces, etc.). In the end, this results in a set of non-linear coupled differential equations with respect to time and space derivatives that need to be discretized and numerically solved (i.e. integrated) to obtain the electron density (n), the hole density (p), and the electric potential throughout the structure (V). In general, in solar cells we will be only interested in steady state solutions under an external applied voltage, current and/or illumination so time derivatives disappear slightly simplifying the problem. Despite this simplification, the numerical solution of the equations in Table 2.3 puts stringent conditions on the simulation mesh, which must be adapted to the geometry of the device.

In essence, numerical simulations allow the closest mathematical approximation to rigorous device physics at the price of a higher computational cost. Such fidelity has two key advantages in the field of multijunction solar cells for CPV applications. First, a wealth of internal parameters of the structure is accessible and their influence can be related to the overall performance of the device. This is of key relevance in such complicated solar cell architectures where subtle changes at a point (e.g. passivation at an interface, minority carrier lifetime variation over a layer etc.) can have an impact which would be hard to track in lumped models. Second, numerical simulations allow one to study the effects of parameters with spatial variability, such as a non-uniform, distribution of light, or current flow or temperature, situations commonly associated with CPV applications.

Different software tools have been used over the last decades for the numerical simulation of multijunction solar cells. The most widespread one-dimensional numerical simulators are PC1D [24], developed at the University of New South Wales in Australia, ADEPT [25] developed at Purdue University, AMPS [26] developed at Pennsylvania State University, and AFORS-HET [27] developed at the Helmholtz-Zentrum Berlin für Materialien und Energie, all of them being free software. For two- and three-dimensional calculations commercial semiconductor device simulation software packages include Synopsys SEN-TAURUS [28], Silvaco ATLAS [29] and to a smaller extent Crosslight APSYS [30] and COMSOL Multiphysics [31].

Numerical simulation of MJSCs has only started to receive some attention from the research community over the last decade. This is partly the result of the complexity of the structures involved and the lack of a validated and reliable database of physical models and material parameters. So far remarkable achievements include the numerical simulation of *I-V* and EQE curves of dual- and triple-junction CPV solar cells [32–34]. A particular area where numerical modeling has proven its power to gain insight on device performance is the simulation of tunnel junctions, where the importance of conduction mechanisms such as local and non-local trap assisted tunneling has been demonstrated [35–37]. Finally, 2-D and 3D numerical models are being applied to analyze how MJSC performance changes with illumination intensity and identify the key device layers and interfaces that limit efficiency at high concentration [38,39].

2.4.2   Analytical Modeling of Multijunction Solar Cells

2.4.2.1   Lumped Analytical Models

As in any other electronic device, the goal of an analytical model for a multijunction solar cell (MJSC) is to provide a simple yet precise mathematical description of the device. Ideally, this mathematical model should have the following features:

- Be based on structural parameters of the device and not on phenomenological approaches. In other words, from the structure and configuration of the device we should be able to calculate the model parameters. In this way the model provides a physically meaningful representation of the device and not only a sole mathematical description
- Be easily computable (i.e. analytical)
- Be representative (i.e. provide results close enough to reality).

Obviously, to end up with mathematical models showing the features mentioned above, a number of assumptions and simplifications have to be made. In the following sections we will comment on these assumptions made to deduce simple yet accurate mathematical models to describe single junction solar cells (i.e. subcells) and tunnel junctions. Therefore, we will consider now that individual subcell current density–voltage (*J-V*) curves can be described by an analytical expression in the form $V_i(J)$ for the *i-th* subcell in the MJSC, and that individual tunnel junction current density–voltage curves can be described by an analytical expression in the form $V_{TJi}(J)$ for the *i-th* tunnel junction in the MJSC. Hereinafter, we will discuss how to build up the complete analytical model of a MJSC from those of its constituents.

Figure 2.13 represents a sketch of series-connected MJSC showing internal and external currents and voltages in the device. As this figure anticipates, the analytical model presented in this section will be focused on the case of series connected MJSCs (either monolithically

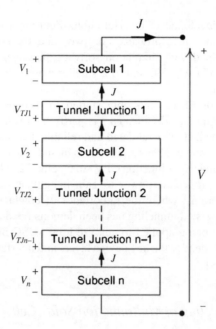

Figure 2.13 Sketch of a series-connected multijunction solar cell showing currents and voltages in the device

integrated or wafer bonded designs). The modeling of multi-terminal devices with independent subcells represents a simpler case where the classic approach for single junction solar cells can be applied for each component cell. The first obvious consequence of Figure 2.13 is that, being a series connection of several devices, the current flowing through the multijunction solar cell will be the same (J), regardless of the fact that photogenerated current might be different in each subcell (i.e. current mismatch). The implications of current mismatch will discussed in section 2.4.2.9. The sign criterion used for the voltages in Figure 2.13 aims to highlight that subcells and tunnel junctions are connected in opposition, so voltage drops have opposite polarity. According to this figure, the voltage of the MJSC will be the sum of all voltages in each subcell minus the voltage drop at each tunnel junction. Hence for an arbitrary MJSC with n subcells and n-1 tunnel junctions, we have:

$$V(J) = \sum_{i=1}^{n} V_i(J) - \sum_{i=1}^{n-1} V_{TJ_i}(J),$$ (2.6)

Equation (2.6) can be further simplified. As will be shown in section 2.4.2.7, a simple approach to model tunnel junctions is to consider them as a resistor (or even a short circuit). In this respect, the role of the tunnel junction can be embedded into the general subcell model as one more component adding to its series resistance. With this simplification we have:

$$V(J) = \sum_{i=1}^{n} V_i(J),$$ (2.7)

2.4.2.2 Modeling Subcells: Single Junction Solar Cells

The simplest circuit model associated with an ideal p-n junction solar cell is the one shown in Figure 2.14.a [1,2]. This circuit consists in the parallel connection of 1) a current source modeling the photogenerated current density (J_L); 2) a diode modeling the dark current density through the p-n junction (J_D). The main assumptions behind this equivalent circuit are that:

- The superposition principle holds for ideal solar cells. In other words, the current flowing in an illuminated ideal solar cell subjected to a forward bias V_j can be calculated by the algebraic sum of the short-circuit photocurrent and the current which would flow at bias V_j in the dark.
- Distributed effects can be neglected and an 'average' operating condition is considered for the whole area of the device. Accordingly, both generation and recombination can be lumped into single circuit elements operating at the same bias V_j.

To account for ohmic effects present in real solar cells two more elements have to be added to the circuit as shown by Figure 2.14.b: 1) a resistor (r_P) in parallel with the elements of the ideal solar cell modeling possible shunts across the junction; and 2) a final series resistor (r_S) modeling ohmic losses in the structure (contacts, semiconductor layers, substrate . . .). Again, the key assumption behind these two new circuit elements is that they can represent in a lumped way all shunts across the p-n junction and all ohmic losses in the device. While for shunts this is a good approximation – in fact in good solar cells the effect of r_P can be neglected ($r_P \to \infty$) – for the series resistance this is seldom the case. Therefore, the assumption of non-distributed ohmic losses will be to a large extent responsible for the lack of accuracy of lumped models for CPV applications.

Consequently, according to the equivalent circuit of Figure 2.14.b, the J-V curve of a real p-n junction solar cell can be calculated using the following expression:

$$J = J_L - J_D(V_j) - J_{rP}(V_j) = J_L - J_D(V + J \cdot r_S) - \frac{V + J \cdot r_S}{r_P}, \qquad (2.8)$$

where J_L is the photogenerated current density in A/cm^2, which to a first approximation is independent on voltage (as will be shown in section 2.4.2.4; J_D is the diode dark current density in A/cm^2 which is a function of the junction voltage ($V_j = V + Jr_s$); r_S is the series resistance in $\Omega \cdot$cm^2; r_P is the parallel or shunt resistance in $\Omega \cdot$cm^2.

In order to derive the expression for the parameters in Eq. (2.8), in the next section we will introduce the layer configuration of a typical subcell (i.e. a single junction solar cell). Once

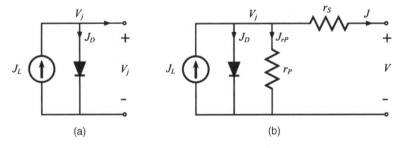

(a)   (b)

Figure 2.14   Equivalent circuit of an (a) ideal solar cell and (b) a real solar cell

known the semiconductor structure over which recombination and generation currents and ohmic losses must be calculated, analytical expressions for J_L, J_D and r_S will be presented in succeeding sections. This derivation plan does not apply to parameter r_p since, being the result of unwanted manufacturing imperfections or poor material quality, it is normally estimated or obtained from a fitting process to experimental *I-V* curves.

2.4.2.3 Layer Structure of III-V Single Junction Solar Cells

To derive the equations for the photocurrent and the recombination currents of a solar cell it is necessary to know its semiconductor layer structure. In this case, we are interested in the typical layer structure of a single junction solar cell made from III-V semiconductors since, as discussed in section 2.3.2.2, III-V compound semiconductors have been the materials of choice to implement high efficiency MJSCs over the last decades.

As can be deduced from the subcells in Figure 2.7 (and has been more clearly depicted in Figure 2.15), the typical layer structure consists of four layers, namely, 1) window; 2) emitter; 3) base; and 4) BSF (Back Surface Field). In essence we have a core p-n junction formed by the emitter and base whose top side is passivated by the window and its rear side by the BSF. From a device modeling standpoint, this set of four layers turns into five when considering the *space charge region* (SCR) formed between the emitter and the base. In Figure 2.15 we have included the whole set of parameters needed for device simulation. An explanation of all the parameters in Figure 2.15 can be found in Table 2.4.

The width of the space charge region (W_{src}) and the resulting effective widths for the emitter (W'_E) and the base (W'_B) can be calculated from basic p-n junction theory [15]:

$$W_{scr} = \sqrt{\frac{2\varepsilon V_{bi}}{q}\left[\frac{1}{N_E} + \frac{1}{N_B}\right]}, \tag{2.9}$$

with V_{bi} being the *built-in* voltage of the p-n junction:

$$V_{bi} = \frac{KT}{q}\ln\frac{N_E N_B}{n_i^2}, \tag{2.10}$$

$$W'_E = W_E - \sqrt{\frac{2\varepsilon V_{bi}}{q} \cdot \frac{N_B}{N_E(N_E + N_B)}}, \tag{2.11}$$

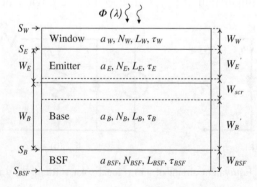

Figure 2.15 Typical layer structure of a III-V single junction solar cell with key parameters

Table 2.4 Set of parameters needed to describe de semiconductor layer structure of a solar cell such as that in Figure 2.15. In all cases, the letter of the subscript indicates the layer, being W = Window, E = Emitter; B = Base; and BSF = Back Surface Field

Structural parameters	
W_W, W_E, W_B, W_{BSF}	Thickness of each layer
W_{scr}	Thickness of the space charge region
W'_E, W'_B	Thickness of the neutral emitter and base (excluding SCR)
N_W, N_E, N_B, N_{BSF}	Doping of the layer
S_W, S_E, S_B, S_{BSF}	Surface recombination velocities at key interfaces (see Figure 2.15)
Minority carrier parameters	
L_W, L_E, L_B, L_{BSF}	Minority carrier diffusion length in each layer
$\tau_W, \tau_E, \tau_B, \tau_{BSF}$	Minority carrier lifetime in each layer
Material parameters	
$\alpha_W, \alpha_{BSF}, \alpha$	Absorption coefficient of window, BSF and of the p-n junction material
ε	Dielectric constant of the p-n junction material
n_i	Intrinsic carrier concentration of the p-n junction material

and analogously:

$$W'_B = W_B - \sqrt{\frac{2\varepsilon V_{bi}}{q} \cdot \frac{N_E}{N_B(N_E + N_B)}}, \tag{2.12}$$

Like most silicon solar cells, III-V solar cells are based on a p-n junction. The most relevant difference, as compared to conventional silicon technology, is how the front and back surface of the p-n junction is passivated (i.e. how surface recombination velocities at those critical interfaces are minimized). The silicon surfaces are generally passivated with a dielectric such as SiO$_2$ or silicon nitride. For III-V PV technology, high-bandgap materials lattice matched to the p-n junction material are used. The front layer is called the window layer and the layer the base layer is called the *back surface field* or BSF layer. The band diagram of the structure in Figure 2.15 is shown in Figure 2.16 and helps clarify the role and requirements for the window and BSF layers:

- Passivation is attained by a favorable band lineup at the critical interfaces. In the minority carrier band a large barrier (i.e. discontinuity) should appear to reflect minority carriers towards the junction. On the other hand, in the majority carrier band no (or a minimum) discontinuity should be present so as not to hinder the transport of majority carriers.
- In order to minimize the presence of crystal defects at those interfaces – and consequently the surface recombination velocity – both the window and BSF should be lattice matched to the p-n junction material (see section 2.3.2.2). In such case, both S_E and S_B can be kept low.

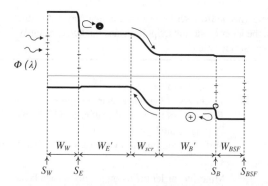

Figure 2.16 Qualitative sketch of the band diagram of the solar cell structure in Figure 2.15 (thick solid lines). Dashed vertical lines delimit the different layers. Short horizontal lines represent trap levels in the bandgap

- In addition, the window layer should be largely transparent to the light (that is why it is called a window) that would be absorbed by the underlying p-n junction. Analogously, the BSF should be transparent for the light that needs to reach the underlying subcells. This highlights again the need for high bandgap materials, preferably indirect, for these layers.

2.4.2.4 Analytical Expressions for the Photocurrent (J_L) of a Single Junction Solar Cell

The classical approach to find the analytical expression for the photocurrent is to solve the minority carrier continuity equations in the structure discussed in the last section (Figure 2.15) when exposed to a given photon flux ($\Phi(\lambda)$, number of photons of wavelength λ entering the solar cell per unit area) that induces a certain generation profile in the layers [3].

For the top cell, the internal photon flux $\Phi_1(\lambda)$ can be calculated from the incident spectra ($E(\lambda)$), the reflectivity of the exposed front surface of the solar cell ($R(\lambda)$) and the *shadowing factor* (*SF*, fraction of the surface area covered by the metal front grid) using the following expression:

$$\Phi_1(\lambda) = E(\lambda) \cdot \frac{\lambda}{hc} \cdot (1 - R(\lambda)) \cdot (1 - SF), \tag{2.13}$$

The reference spectrum $E(\lambda)$ used for CPV applications will be AM1.5d ASTM G303 (see Chapter 1) or, more realistically, $E(\lambda)$ filtered by the primary and secondary optics of the CPV system. For now, we will consider the case of the one-sun reference spectrum and at the end of this section we will extend the results for a solar cell operating in a CPV system. The term λ/hc multiplying the spectrum in Eq. (2.13) is simply the scale factor needed to transform the units of the spectrum (W·cm^{-2} nm^{-1}) to the units of photon flux (number of photons·cm^{-2}s^{-1})

For the *i-th* subcell, its incident photon flux ($\Phi_i(\lambda)$) is calculated from the flux entering the top cell and the transmissivity of the layers (i.e. subcells and tunnel junctions) above ($T_i(\lambda)$):

$$\Phi_i(\lambda) = \Phi_1(\lambda) \cdot T_i(\lambda), \tag{2.14}$$

The simplest way to calculate the attenuation is to use the Lambert–Beer law which states that the internal irradiance or photon flux in the material decays exponentially after passing a distance W. Therefore, assuming that above the i-th subcell there are j uniform layers of thickness W_j each and absorption coefficient α_j each, $T_i(\lambda)$ is given by:

$$T_i(\lambda) = \prod_1^j \exp(-\alpha_j \cdot W_j), \qquad (2.15)$$

Once the internal photon fluxes are calculated, the complete derivation of the photogenerated current can proceed. Full details of this derivation can be found in [3]. In addition to the incident photon flux, J_L depends on the internal quantum efficiency ($IQE(\lambda)$) of the solar cell:

$$J_L = q \int_0^\infty \Phi(\lambda) \cdot IQE(\lambda) \mathrm{d}\lambda, \qquad (2.16)$$

The internal quantum efficiency is the ratio of the number of carriers collected by the solar cell to the number of photons of a given energy that entered the solar cell. Accordingly, the integral (2.16) simply represents the sum of all the carriers collected. The overall $IQE(\lambda)$ of the solar cell is calculated by adding the contributions of all its layers:

$$J_L = q \int_0^\infty \Phi(\lambda) \cdot (IQE_W(\lambda) + IQE_E(\lambda) + IQE_{scr}(\lambda) + IQE_B(\lambda) + IQE_{BSF}(\lambda)) \mathrm{d}\lambda, \qquad (2.17)$$

Expressions for the different components of the IQE is given in Table 2.5 [3]. An even simpler model, can be obtained neglecting the small contributions to the IQE of the window and BSF layers [40], considering them as mere optical filters with no contribution to photocurrent (this approximation is particularly accurate for the BSF).

Some important considerations about the equations in Table 2.5 are:

- In all cases, their dependence with photon wavelength enters through the absorption coefficient (i.e. $\alpha = \alpha(\lambda)$, $\alpha_W = \alpha_W(\lambda)$, $\alpha_{BSF} = \alpha_{BSF}(\lambda)$).
- Except for the window layer, which directly receives the incoming photon flux, the first exponential in all equations represents the attenuation produced by the layers above following a Lambert–Beer dependence, as in Eq. (2.15).
- All carriers absorbed in the space charge region are collected, so that IQE_{scr} only depends on the thickness and absorption coefficient.
- The IQE of the emitter and base are complex functions of minority carrier parameters
- The IQE of the window resembles that of the emitter just changing the corresponding minority carrier parameters and adding a term in the denominator that accounts for the losses in collection suffered by window minority carriers while travelling the emitter to reach the space charge region. An analogous discussion applies for the BSF and the base.

Let us go back to Eq. (2.13) and consider the effect of concentration. If our cell is illuminated in an optical system that takes the reference *one-sun* spectrum – $E(\lambda)$ – and concentrates it

Table 2.5 Equations for the different components of the internal quantum efficiency. The symbols used can be found in Table 2.4

$$IQE_W = \frac{\alpha_W L_W}{(\alpha_W^2 L_W^2 - 1)} \times \left[\left(\frac{S_W \tau_W}{L_W} + \alpha_W L_W \right) - \exp(-\alpha_W W_W) \frac{\left(\frac{S_W \tau_W}{L_W} \cosh \frac{W_W}{L_W} + \sinh \frac{W_W}{L_W} \right)}{\frac{S_W \tau_W}{L_W} \sinh \frac{W_W}{L_W} + \cosh \frac{W_W}{L_W}} - \alpha_W L_W \exp(-\alpha_W W_W) \right] \tag{2.20}$$

$$IQE_E = \exp(-\alpha_W W_W) \frac{\alpha L_E}{(\alpha^2 L_E^2 - 1)} \times \left[\left(\frac{S_E \tau_E}{L_E} + \alpha L_E \right) - \exp(-\alpha W_W) \frac{\left(\frac{S_E \tau_E}{L_E} \cosh \frac{W'_E}{L_E} + \sinh \frac{W'_E}{L_E} \right)}{\frac{S_E \tau_E}{L_E} \sinh \frac{W'_E}{L_E} + \cosh \frac{W'_E}{L_E}} - \alpha L_E \exp(-\alpha W'_E) \right] \tag{2.21}$$

$$IQE_{scr} = \exp(-\alpha_W W_W - \alpha W'_E)[1 - \exp(-\alpha W_{scr})] \tag{2.22}$$

$$IQE_B = \exp(-\alpha_W W_W - \alpha(W'_E + W_{scr})) \frac{\alpha L_B}{(\alpha^2 L_B^2 - 1)} \times \left[\alpha L_B - \frac{\frac{S_B \tau_B}{L_B} \left(\cosh \frac{W'_B}{L_B} - \exp(-\alpha L_B) \right) + \sinh \frac{W'_B}{L_B} + \alpha L_B \exp(-\alpha W'_B)}{\frac{S_B \tau_B}{L_B} \sinh \frac{W'_B}{L_B} + \cosh \frac{W'_B}{L_B}} \right] \tag{2.23}$$

$$IQE_{BSF} = \exp(-\alpha_W W_W - \alpha(W_E + W_{scr} + W'_B)) \frac{\alpha_{BSF} L_{BSF}}{(\alpha_{BSF}^2 L_{BSF}^2 - 1)} \times$$
$$\left[\alpha_{BSF} L_{BSF} - \frac{\frac{S_{BSF} \tau_{BSF}}{L_{BSF}} \left(\cosh \frac{W_{BSF}}{L_{BSF}} - \exp(-\alpha_{BSF} L_{BSF}) \right) + \sinh \frac{W_{BSF}}{L_{BSF}} + \alpha_{BSF} L_{BSF} \exp(-\alpha_{BSF} W_{BSF})}{\frac{S_{BSF} \tau_{BSF}}{L_{BSF}} \sinh \frac{W'_B}{L_B} + \cosh \frac{W'_B}{L_B}} \right] \tag{2.24}$$

onto the cell by a constant factor C, then the new incident spectrum – $E_C(\lambda)$ –could be expressed as:

$$E_C(\lambda) = C \cdot E(\lambda), \tag{2.18}$$

From Eq. (2.18) it is obvious that C – termed the concentration level or concentration factor – is a ratio of two homogeneous quantities (two irradiances), and therefore it should be a dimensionless magnitude. However, since the quantity in the denominator represents a *one-sun* reference condition, the tradition in CPV is to use the symbol × or *suns* as the units for this ratio to highlight its physical meaning.

Therefore, under concentrated light the incident photon flux only changes in a constant factor (C), and thereby Eq. (2.16) transforms now to:

$$J_L = qC \int_0^\infty \Phi^*(\lambda) \cdot IQE(\lambda)\mathrm{d}\lambda = CJ_L^*, \tag{2.19}$$

where Φ^* is the photon flux produced by the one-sun reference spectrum in the subcell of interest and, consequently, J_L^* represents the photogenerated current by the one-sun reference spectrum in the subcell of interest.

Eq. (2.19) is of utmost importance since it sustains one of the key principles in CPV: photogenerated current is proportional to concentration.

2.4.2.5 Analytical Expressions for the Recombination Current of a Single Junction Solar Cell

As with the photocurrent, the calculation of recombination currents in the structure of Figure 2.15 can be developed solving the continuity equations for minority carriers, this time in the absence of luminous excitation. A step-by-step description of this derivation can be found in [3].

In many occasions, for simplicity, only one recombination current (J_D) is considered with the typical Shockley dependence:

$$J_D = J_0 \left[\exp\left(\frac{V_j}{m \cdot V_t}\right) - 1 \right], \tag{2.31}$$

where V_j represents the actual junction voltage (as indicated in Figure 2.14); m the diode ideality factor – sometimes also denoted as n –; J_0 the diode reverse saturation current density (A/cm^2); and V_t is the thermal voltage in V given by the well-known expression:

$$V_t = \frac{kT}{q} \quad \Rightarrow \quad V_t \cong 0.026 \text{ V for T} = 300 \text{ K}, \tag{2.32}$$

where k is the Boltzmann constant $(k = 1.3806 \cdot 10^{-23}$ J/K$)$; q is the elementary charge $(q = 1.6022 \cdot 10^{-19}$ C$)$; and T is the absolute temperature in Kelvin.

Table 2.6 Equations for the different dark recombination currents. Most of the symbols used are those in Table 2.4

Dark current resulting from recombination in neutral regions

$$J_{D,nr}(V) = J_{01}\left[\exp\left(\frac{qV}{kT}\right) - 1\right] \tag{2.25}$$

with J_{01} having a component associated with the emitter (J_{01}^E) and one with the base (J_{01}^B):

$$J_{01} = J_{01}^E + J_{01}^B = \begin{aligned} &q\,\frac{L_E\,n_{iE}^2}{\tau_E\,N_E}\left(\frac{\sinh\dfrac{W_E'}{L_E} + \dfrac{S_E\tau_E}{L_E}\cosh\dfrac{W_E'}{L_E}}{\dfrac{S_E\tau_E}{L_E}\sinh\dfrac{W_E'}{L_E} + \cosh\dfrac{W_E'}{L_E}}\right) \\ &+q\,\frac{L_B\,n_{iB}^2}{\tau_B\,N_B}\left(\frac{\sinh\dfrac{W_B'}{L_B} + \dfrac{S_B\tau_B}{L_B}\cosh\dfrac{W_B'}{L_B}}{\dfrac{S_B\tau_B}{L_B}\sinh\dfrac{W_B'}{L_B} + \cosh\dfrac{W_B'}{L_B}}\right)\end{aligned} \tag{2.26}$$

Dark current resulting from recombination at the space charge region

$$J_{D,scr}(V) = J_{02,scr}2\sinh\frac{qV}{2kT} \approx J_{02,scr}\left[\exp\left(\frac{qV}{2kT}\right) - 1\right] \tag{2.27}$$

with $J_{02,scr}$ being:

$$J_{02,scr} = \frac{\pi}{2}qn_iW_{scr}K\frac{kT/q}{V_{bi} - V} \tag{2.28}$$

with K being a constant related to the abundancy and efficiency of the deep levels in the space charge region.

Dark current resulting from recombination at the perimeter

$$J_{D,per}(V) = J_{02,per}\left[\exp\left(\frac{qV}{2kT}\right) - 1\right] \tag{2.29}$$

with $J_{02,per}$ being:

$$J_{02,per} = qn_iS_0L_S\frac{L_{per}}{A} \tag{2.30}$$

where S_0 is the surface recombination velocity at the perimeter; L_S is the diffusion length at the perimeter; L_{per} is the perimeter length and A is the device area.

However, in the most general situation several recombination mechanisms – occurring in different regions of the device – might coexist in the same solar cell. The most frequent recombination currents found in III-V subcells are:

- *Current associated with the recombination of minority carriers in the neutral regions ($J_{D,nr}$).* This represents the classical injection current in any p-n junction biased at a certain voltage, which is associated with recombination of the excess minority carrier population injected over the potential barrier. As detailed in Table 2.6 by Eq. (2.25), this current has an expression analogous to Eq. (2.31) with $m = 1$.

- *Current associated with recombination in the space charge region ($J_{D,scr}$).* This represents the recombination of carriers traversing the space charge region through deep levels in the gap (i.e. Shockley–Read–Hall processes). As detailed in Table 2.6 by Eq. (2.27), the expression for this current may be simplified to yield a formula analogous to Eq. (2.31) with $m = 2$. Such value for m, results from assuming that the main deep level responsible for the recombination in the space charge region is located right at the middle of the bandgap. If this were not the case or there were multiple levels, different values of m could appear.
- *Current associated with recombination at the junction perimeter ($J_{D,per}$).* This accounts for recombination of minority carriers at the interfacial states of the exposed perimeter of a p-n junction. As detailed in Table 2.6 by Eq. (2.29), the expression for this current may be simplified to yield a formula analogous to Eq. (2.31) with $m = 2$.

Expressions to calculate the different recombination currents described are summarized in Table 2.6. Accordingly, in a solar cell where all these three recombination mechanisms were present, J_D in Eq. (2.8) should be calculated as:

$$J_D = J_{D,nr}(V_j) + J_{D,per}(V_j) + J_{D,scr}(V_j), \tag{2.33}$$

This expression can be rearranged grouping the $m = 1$ term and the $m = 2$ terms:

$$J_D = J_{01}\left[\exp\left(\frac{V_j}{V_t}\right) - 1\right] + (J_{02,scr} + J_{02,per})\left[\exp\left(\frac{V_j}{2V_t}\right) - 1\right], \tag{2.34}$$

2.4.2.6 Series Resistance

The series resistance is a key component influencing the concentrator operation of solar cells. For this reason its role and precise calculation will be thoroughly discussed in section 2.5 *Concentrator requirements*, in particular in subsection 2.5.2. As will be discussed in this section, the lumped series resistance of the solar cell is calculated assuming a simplified picture of the internal current flow in the solar cell (longitudinal in the emitter and transverse in all other layers) and accounting for all ohmic power losses in the different parts of the device (grid, metal-semiconductor contacts, emitter, base, etc.).

2.4.2.7 Modeling Tunnel-junctions

As discussed in section 2.3.1.3, tunnel junctions are thin heavily doped p-n junctions (connected in opposition to the subcell p-n junctions) where quantum tunneling is the main conduction mechanism [15]. The J-V curve of a generic tunnel junction is depicted in Figure 2.17. As this figure shows, the key feature of tunnel junctions for MJSC is that they exhibit a linear *J-V* dependence until the peak tunneling current (J_p) is reached. This initial ohmic region in the *J-V* curve is ideal for low-loss interconnects between the subcells. According to this description, a possible model for a tunnel junction is a resistor-like J-V dependence:

$$J = \frac{1}{r_{TJ}} \cdot V, \tag{2.35}$$

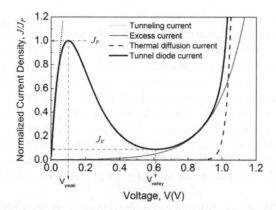

Figure 2.17 Normalized J-V curve of an exemplary tunnel junction. The gray straight dotted line at the left represents the resistor-like model of Eq. (2.35). The characteristic parameters of the J-V curve appear as labels in the figure: peak current density (J_P), peak voltage (V_P), valley current density (J_V) and valley voltage (V_V). Source: Adapted from Espinet-González, 2012 [41]

where r_{TJ} is the tunnel junction equivalent specific resistance in $\Omega \cdot cm^2$. Taking this idea to the limit we could model the tunnel junction (for $J < J_p$) as a short circuit, and therefore no model would be needed.

A more detailed model for tunnel junctions [15], based on semi-empirical considerations, quantifies the J-V curve of the tunnel junction as the sum of three components: a) the tunneling current itself (J_t); b) the excess current (J_x); c) and the classic thermal diffusion current (J_{th}) of any p-n junction (i.e. Shockley's equation):

$$J = J_t + J_x + J_{th},\tag{2.36}$$

$$J_t = J_P(V/V_P)\exp(1 - V/V_P),\tag{2.37}$$

$$J_x = J_V\exp[D(V - V_V)],\tag{2.38}$$

$$J_{th} = J_{0TJ}[\exp(V/kT) - 1],\tag{2.39}$$

The use of Eq. (2.36) to Eq. (2.39) in analytical models will only be needed when the current density in the MJSC exceeds the peak current of the tunnel junction. Therefore, in most cases of practical use for analytical models, the resistor-like model of Eq. (2.35) will suffice.

2.4.2.8 Consolidated Lumped Model for Multijunction Solar Cells

According to what has been described in the preceding subsections, to mathematically describe a multijunction solar cell the schematic representation of Figure 2.13 can be transformed into the equivalent circuit model of Figure 2.18a. In this equivalent circuit:

- The photogenerated current density current for each subcell ($J_{L,i}$) is calculated using Eq. (2.19).
- The two-diode dark current dark for each subcell is calculated using Eq. (2.34).

- The voltage drop in each tunnel junction ($V_{TJ,i}$) is calculated using Eq. (2.36).
- The series resistance of each subcell ($r_{S,i}$) is estimated or calculated following the procedure described in section 2.5.2.
- The parallel resistance of each subcell ($r_{P,i}$) is estimated.
- Finally, the J-V curve of the subcell in question is calculated from this components using Eq. (2.8).

In many solar cell designs the equivalent circuit of Figure 2.18(a) can be further simplified taking into account two facts. First, $r_{P,i}$ is often very large and thus can be neglected.[3] Second, tunnel junctions normally operate in the ohmic region (remember Figure 2.17) and thus their I-V curve can be linearized through their equivalent resistance ($r_{TJ,i}$). Moreover, this resistance can be added as one more component to the series resistance of the subcell right below ($r_{S,i+1} = r_{S,i+1} + r_{TJ,i}$). These two simplifications produce the equivalent circuit of Figure 2.18 (b), where now the series resistance of subcells 2 to n contain the resistive effects of the tunnel junctions. Apart from the change in the calculation of the series resistance, all the other

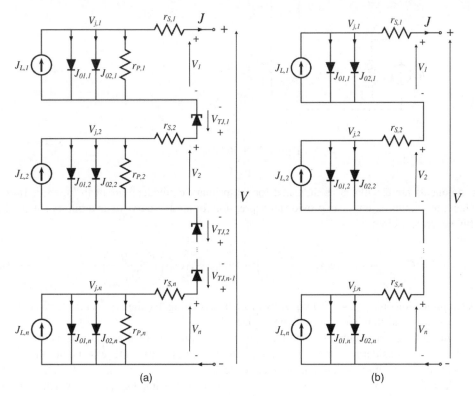

(a)   (b)

Figure 2.18 Equivalent circuit models for multijunction solar cells. (a) Complete two-diode; (b) two-diode without $r_{P,i}$ and tunnel junction diode; Equivalent circuit models for multijunction solar cells. (c) One-diode with lumped r_S; and (d) compact

---

[3] This may not be the case for germanium subcells in triple-junction GaInP/GaInAs/Ge solar cells.

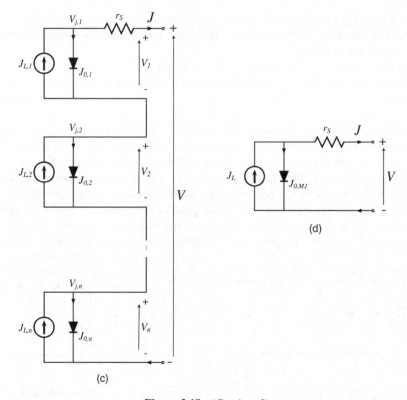

Figure 2.18 (*Continued*)

components are calculated as indicated for the primary equivalent circuit in Figure 2.18(a). Therefore, in this case we can provide an expression for the J-V curve of an arbitrary subcell in this equivalent circuit:

$$J = C \cdot J_{L,i}^* - J_{01,i}\left[\exp\left(\frac{V_i + Jr_{s,i}}{V_t}\right) - 1\right] + J_{02,i}\left[\exp\left(\frac{V_i + Jr_{s,i}}{2V_t}\right) - 1\right], \qquad (2.40)$$

A simplification of Eq. (2.40) is frequently made by assuming that the exponential term is much larger than 1 and thus the latter can be neglected.

Figure 2.18(c) represents an additional step in the simplification of the equivalent circuit that comes from considering again two more circumstances. First, the effects of all series resistances can be lumped into a general series resistance ($r_S = r_{s,1} + r_{s,2} + \cdots + r_{S,n}$). Second, the dark current for each subcell ($J_{D,i}$) can be modeled using Eq. (2.31), which implies considering a single diode with an effective dark reverse saturation current ($J_{0,i}$) and an effective ideality factor (m_i). If, for a given subcell, the $m = 1$ component dominates the dark current then the effective values for $J_{0,i}$ will tend to $J_{01,i}$ and effective ideality factor will tend to 1. Analogously, if, for a given cell, the $m = 2$ component dominates the dark current then the effective values for $J_{0,i}$ will tend to $J_{02,i}$ and effective m_i will tend to 2. On the other hand, if neither component dominates, $J_{0,i}$ will show an intermediate value between $J_{01,i}$ and $J_{02,i}$ and effective ideality will be $1 < m_i < 2$. Therefore, in this case we can provide an even simpler expression for the J-V curve of an arbitrary subcell in this equivalent circuit:

$$J = C \cdot J_{L,i}^* - J_{0,i} \left[\exp \left(\frac{V_i + Jr_{s,i}}{m_i \cdot V_t} \right) - 1 \right], \tag{2.41}$$

It should be noted that for $i \neq 1$, $r_{s,i} = 0$ (see Figure 2.18.c). As occurred in Eq. (2.40), here the -1 after the exponential is frequently neglected also.

So far we have discussed how to calculate the J-V curves for each subcell using equivalent circuits (a) to (c) in Figure 2.18. However, the real goal is to calculate the J-V curve of the whole multijunction solar cell. To get the final voltage on the MJSC at a given current (J), we must add the voltages produced in all the subcells – $V_i(J)$ – at that current, as described in Eq. (2.6) and Eq. (2.7).

Finally, the simplest version of the equivalent circuit for a multijunction solar cell is that of Figure 2.18(d). The fact of the matter is that the circuit Figure 2.18.c is equivalent to that of Figure 2.18(d) if two new circumstances are granted. First, if there is a perfect current matching between subcells (i.e. $J_{L,1} = J_{L,2} = \ldots = J_{L,n} \equiv J_L$) and second, if all recombination currents show the same ideality factor (i.e. $m_1 = m_2 = \ldots = m_n \equiv m$). Under these circumstances, and neglecting the -1 inside the exponential of Eq. (2.41), we have:

$$J = J_L - J_{0,1} \exp \left(\frac{V_1}{m \cdot V_t} \right) = J_L - J_{0,2} \exp \left(\frac{V_2}{m \cdot V_t} \right) = \ldots = J_L - J_{0,n} \exp \left(\frac{V_n}{m \cdot V_t} \right). \tag{2.42}$$

This expression, through simple algebraic manipulations, can be transformed into:

$$J = J_L - (J_{0,1} \cdot J_{0,2} \ldots J_{0,n})^{1/n} \exp \left(\frac{V + Jr_S}{n \cdot m \cdot V_t} \right), \tag{2.43}$$

which is totally analogous to the simplest equation describing a single-junction solar cell (Shockley equation), with the equivalent reverse saturation current density for the multijunction solar cell ($J_{0,MJ}$) being:

$$J_{0,MJ} = (J_{0,1} \cdot J_{0,2} \ldots J_{0,n})^{1/n}, \tag{2.44}$$

and the equivalent ideality factor for the multijunction solar cell (m_{MJ}) being:

$$m_{MJ} = n \cdot m, \tag{2.45}$$

Eq. (2.43) and the equivalent circuit of Figure 2.18(d) look seemingly attractive as a result of their analogy to single junction solar cell equation and equivalent circuit. However, the reader should bear in mind the large collection of assumptions made to reach such simplification. Consequently, their validity – and thus applicability – is very limited in concentrator applications. When it comes to lumped models, the most widely used equivalent circuits are those of Figure 2.18(b) and Figure 2.18(c) –and consequently Eq. (2.40) and Eq. (2.41)–, because of their relative simplicity and reasonable accuracy to represent the cell in a variety of situations. In particular, in the next section, we will use these models to perform a basic analysis of the operation of multijunction solar cells under concentrated light.

2.4.2.9 Multijunction Solar Cell Performance Analysis from Lumped Analytical Models

The preceding sections have developed a framework to obtain an analytical mathematical description of a multijunction solar cell starting from its structural and material parameters. It has been also discussed (and will be further remarked in sections 2.4.3 and 2.5.4) that these models have a limited validity as a result of the distributed effects associated with high irradiances. However, they are certainly useful to gain insight into the basic operation of multijunction solar cells. Accordingly, in this section we will use the models presented –in particular those derived from the equivalent circuit of Figure 2.18.c–to discuss some particularities of multijunction solar cells and to have a first approach to their concentrator performance.

The first topic to be discussed is the J-V curve of a multijunction solar cell. Figure 2.19 represents the case of a lattice matched triple-junction GaInP/GaInAs/Ge solar cell (see section 2.6.1 for more details), including in the same graph the J-V curves of the individual subcells. In this design, as a result of the low bandgap of germanium, the bottom cell produces a lot more photogenerated current as compared to top and middle cells, which are normally fabricated to be current matched (not the case for Figure 2.19 where the top cell is current limiting the device).

With the help of Figure 2.19, it is possible to gain some insight about how the internal biasing of the subcells in the multijunction solar cell operates. Let us consider an arbitrary point in the *flat* part of the *J-V* of the multijunction solar cell (thick black line in Figure 2.19), for example, the point labeled A and marked with a square. The coordinates of the corresponding biasing points of the three subcells are also marked with squares and are $(V_{A,TC}, J_A)$, $(V_{A,MC}, J_A)$ and $(V_{A,BC}, J_A)$ for top, middle and bottom cell, respectively. The fact of the matter is that here (i.e. in the flat region of the *J-V* curve) the overexcited subcells are pined at these biasing points since they are the only operating points at which they can provide the designated current density J_A. If we increase the

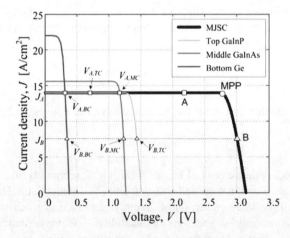

Figure 2.19 J-V curve of a lattice matched triple-junction GaInP/GaInAs/Ge solar cell. The individual *J-V* curves of component subcells are also shown. Calculations made for 1000× AM1.5d. Details about symbols and labels are given in the text

voltage in the MJSC (i.e. move point A towards the maximum power point, MPP) then the voltage of the limiting subcell – the top cell in our example – increases accordingly, whilst the non-limiting subcells stay anchored at $V_{A,MC}$ and $V_{A,BC}$, these voltages being very close to their respective open circuit voltage. This occurs for the *flat* part of the *J-V* roughly until the maximum power point is reached. However, once the voltage is further increased beyond MPP, for example at point B marked with a triangle, then the three cells move their biasing point to meet the required current (J_B in this case) while their voltages are slightly increased towards their V_{oc}. In summary, as long as there exists current mismatch in a multijunction solar cell – and there is always some – it is the limiting subcell that is doing large *voltage* excursions while the non-limiting subcells stay more or less fixed around their V_{oc}.

Another way to look at this is to think that current mismatch curtails the *J-V* curve of the non-limiting subcells making inaccessible the points with current densities above the limiting current. This curtailing, in turn, means that overexcited solar cells contribute to the total multijunction *J-V* curve with artificially high fill factors. This latter fact gives rise to a characteristic evolution of *FF* as a function of current mismatch, which is reproduced in Figure 2.20. In this figure, the effect of the current mismatch between the top and middle cell in a lattice matched triple-junction GaInP/GaInAs/Ge solar cell is analyzed. It should be noted that the spectrum is varied to shift the value of the current matching – defined as the ratio of middle to top cell short circuit current – while keeping the irradiance fixed to 1000× (1 MW/m²). Therefore all points on the curves of Figure 2.20 are receiving the same optical power, with different spectral distributions, though. For instance, when the current matching ratio equals 1, the cell is said to be

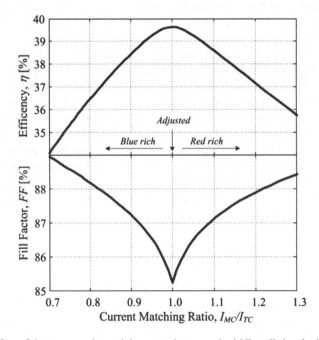

Figure 2.20 Effect of the current mismatch between the top and middle cells in a lattice matched triple-junction GaInP/GaInAs/Ge solar cell. Top graph shows the effect on efficiency; bottom graph shows the effect on fill factor. Calculations made for 1000× (i.e. 10^6 W/m²)

adjusted and (if well designed for reference conditions) this should occur at the AM15.d spectrum. Values of the current matching ratio above 1 indicate a red-rich spectrum which promotes enhanced generation in the middle cell and restricts the number of photons available for the top cell. Exactly the opposite occurs for current matching ratios below 1, indicative of a blue-rich spectrum. Such spectral variations may happen – to a limited extent – in the field but certainly do occur in solar simulators (see Chapter 10), and as indicated in Figure 2.20 cause noticeable effects on *FF* and efficiency. The evolution of *FF* in the lower part of Figure 2.20 can be understood by recalling the fact that current mismatch leads to *J-V* curtailing of non-limiting junctions and thus to enhanced *FF* in the multijunction solar cell. Therefore, as we approach the current matching condition, curtailing of the curves is less and the *FF* approaches that resulting from the addition of the real (i.e. non-curtailed) subcells *J-V* curves. Severe current mismatch can raise *FF* to values close (or above) 90%; so when measuring such high *FF* values – before rushing for champagne – one should seriously consider the possibility of having spectral control issues (i.e. current mismatch).

Obviously, the increase in *FF* is counterbalanced by a severe loss in output current in the multijunction solar cell which is always limited by the subcell with the lowest photogenerated current. This, in turn, means that the maximum efficiency occurs at minimum FF (i.e. under current matching), as shown by the top part of Figure 2.20.

Finally, it should be noted that the equivalent circuit of Figure 2.18(c) and the associated equations can be used to model the complete concentrator response of a multijunction solar cell. The concentrator response is a set of three graphs depicting the evolution of efficiency (η), *FF* and V_{oc} versus concentration and synthesizes very effectively the performance of a CPV solar cell. An example of this is presented in Figure 2.21 for a lattice matched triple-junction GaInP/GaInAs/Ge solar cell including experimental data measured (circles) and the modeling (solid lines). A detailed comment on this type of graphs will be given in section 2.5. Nevertheless, a basic analysis can be made here in terms of the accuracy of the modeling, which seems quite good for $C > 200\times$. This makes sense since the dominant recombination mechanisms at high ($m = 1$ recombination) and low concentrations ($m = 2$ recombination) are different. Therefore, in a single diode per-subcell model only one recombination mechanism can be accurately represented and thus only one region of the curve – either high or low C – can be adjusted. Surely, a better fit could have been obtained with the two-diode per subcell model of Figure 2.18(b). In addition, if the dark *I-V* curve of the multijunction solar cell had to be reproduced also with the same set of parameters that produced the reasonable fit of Figure 2.21 (which in principle should work since structural parameters of solar cell are the same no matter whether there is illumination or not), larger discrepancies would become apparent. These facts highlight the limitations of lumped analytical models and state the need for more elaborate device descriptions as will be described in the next section.

2.4.3 Further Steps: Distributed Circuit-based Modeling

2.4.3.1 Inception of Distributed Circuit-based Modeling

The lumped analytical model is accurate in a restricted range of operating conditions since the assumption of a constant lumped series resistance is only valid when the distributed series resistance effects are small [42]. This is not fulfilled in CPV solar cells due to the high current densities, non-uniform irradiances and chromatic aberrations, which are present under concentrator operation (for detailed explanations on these concepts see section 2.5.4).

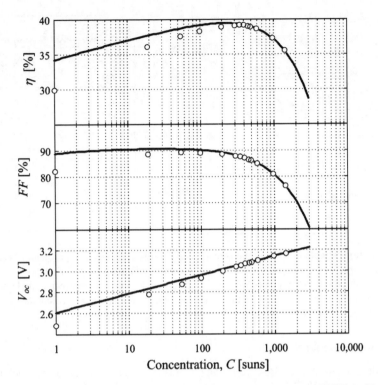

Figure 2.21 Concentrator response of a lattice matched triple-junction GaInP/GaInAs/Ge solar cell. Experimental data is included as circles whilst lines represent the modeling with the equivalent circuit of Figure 2.18(c)

However, this lack of uniformity refers to the whole solar cell area. In fact, if we focus our attention on the micro-scale (e.g. in a region of a few square microns where distributed effects should be insignificant), we may realize that the fundamental assumptions behind the lumped analytical model are essentially still valid. Following this idea the distributed circuit-based simulation of solar cells emerges naturally: the solar cell area is divided into elementary units where the lumped analytical model can be applied and then all such units are interconnected to form a new equivalent circuit for the solar cell. This model should provide an accurate opto-electronic description of the solar cell for any operating condition [41].

The approach of generating a complicated equivalent electronic circuit to model the electrical behavior of the solar cell to account for some distributed effects, was first presented by Nielsen in the early 1980s [43]. Figure 2.22.(a) and (b) show the circuit model used in that work, which was essentially a 2-D representation of current flows in the solar cell which did not consider voltage drops in the grid fingers. Araki and co-workers [44] and Nishioka and co-workers [45] were the first to apply this two-dimensional modeling approach to multi-junction solar cells in 2003 (Figure 2.22.(c)). The extension to 3-D distributed models without any restriction in the current flow – therefore, taking into account any distributed effect that may occur in the solar cell – did not appear until 2005 with the work of Galiana and co-workers [46].

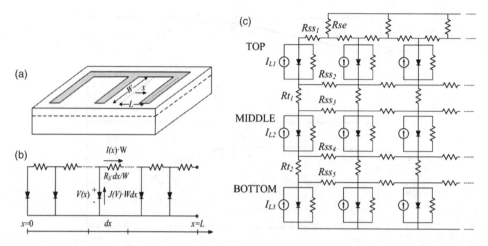

Figure 2.22 (a) Example of a one-dimensional distributed series resistance problem. Sheet current in the top emitter layer of the solar cell is assumed to flow perpendicularly to the grid lines ($W >> L$) in the x-direction. (b) One-dimensional model of the 'elementary unit' (rectangle between dotted lines) in (a) as defined by Nielsen [43]. (c) Modeling of the same 'elementary unit' applied by Nishioka and co-workers to model triple-junction solar cells [45]

2.4.3.2 Approach and Results

The procedure to simulate a MJSC following the distributed circuit-based approach is visually sketched in Figure 2.23. The basic steps to follow are:

1. The solar cell area should be divided into elementary units where the lumped analytical model can be applied. This partitioning will depend on the operating conditions to be simulated and should guarantee that distributed effects are negligible within an elementary unit (i.e. for very high concentrations or highly non uniform irradiances the elementary cells should be smaller than those for moderate concentrations). Figure 2.23 (a) and (b) represent such partitioning for the upper left corner area of a triple junction solar cell with a comb-like grid. To reduce computation overhead, the partitioning could be applied to a so-called geometrically irreducible simulation (GIS) domain, which is the minimum fraction of the area of the device that needs to be simulated according to the symmetry of the grid and the optical excitation. For example, in the case of Figure 2.23 (a) under uniform illumination (or any type of light spot with round symmetry centered in the active area) the GIS domain would be one quarter of the solar cell.

2. The elementary units obtained after the partitioning of the solar cell should be classified into homogeneous types. Since solar cells are highly symmetrical few types of elementary units will be needed to model the whole device. For example, in Figure 2.23 only four different types of elementary cells are found, namely, type A, an elementary unit corresponding to illuminated area; type B, an elementary unit corresponding to metallized area (i.e. under a grid finger or under the bus); type C, an elementary unit corresponding to illuminated area at the perimeter of the device; and type D, an elementary unit corresponding to metallized area at the perimeter of the device. The need to distinguish between bulk and perimeter areas will become apparent in the next point.

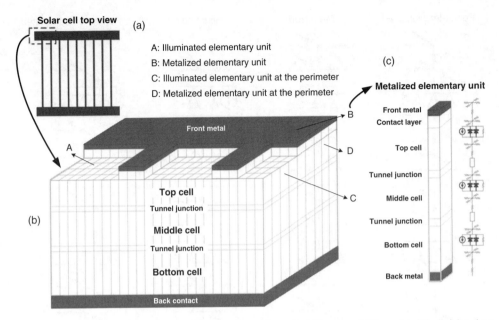

Figure 2.23 (a) Top view of the triple junction solar cell with a comb-like grid; (b) division into elementary units of the area in the dotted rectangle; four different types of elementary units–named A, B, C and D– are needed to describe different areas of the device; (c) Example of the equivalent circuit used to model one of the elementary cells

3. Equivalent circuits should be assigned to each type of elementary unit. For example, in Figure 2.24 the equivalent circuits for the four types of elementary cells obtained in the partitioning in Figure 2.23 are included. In essence, in these equivalent circuits each subcell in the MJSC is described with a shunt resistor and two diodes (one to model recombination in the neutral regions and one to model the recombination in the space charge region); a third diode is added for elementary unit cells at the edge of the device to account for perimeter recombination; illuminated areas have current generators to simulate photo-generation whilst metallized elementary cells lack do not; tunnel junctions are modeled by a specific circuit element whose J-V characteristic can be defined as a mathematical function; finally, a network of horizontal and vertical resistors connect contacts, subcells and tunnel junctions allowing current to flow in any direction (i.e. a 3-D model).

4. The complete equivalent circuit for the GIS domain is obtained by connecting each elementary unit with its neighbors. Values of all the components in the circuit (diodes, current supplies, resistors) need be assigned in relation with the structural or geometrical properties of the solar cell. If a non-uniform irradiance is to be simulated, the values of the currents produced by each current supply will be location dependent. A similar idea would be followed to simulate temperature gradients.

5. The whole circuit is fed into a circuit simulator such as SPICE and the voltage at every node and currents for every branch are obtained for the GIS domain. Following the symmetries of the device, the same information for the complete solar cell is obtained in a straight forward manner.

Perimeter Metalized   Metalized   Illuminated   Perimeter Illuminated

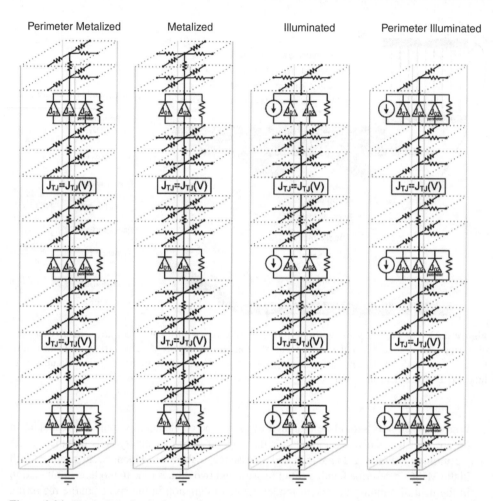

Figure 2.24 Equivalent circuits for the elementary units used in the triple-junction solar cell distributed circuit model of [41]. Source: Adapted from Espinet-González, 2012 [41]

Following the afore-described procedure extremely rich information on the solar cell operation can be extracted. In essence, the exact polarization at any subcell or tunnel junction at any location in the device can be accessed. This fact has spurred the use of this simulation approach to gain insight into the performance of MJSC under real operating conditions in CPV systems. Some notable examples in this direction include the analysis of the impact of non-uniform illumination profiles [41]; the role of tunnel junctions [47,48]; the effect of chromatic aberration [49,50]; the performance under different secondary optical elements [51] or photon coupling [52].

Despite computation times are long (several hours) even for devices in the range of some square millimeters in size – computation times grown exponentially with number of nodes and components – distributed circuit simulation has proven to be accurate, affordable and reliable. In the wait for consistent 3D numerical simulation for device sizes used in MJSCs, distributed circuit-based modeling stands today as the most efficient way to understand the performance of MJSC in CPV systems. A more detailed explanation of how distributed effects affect concentrator solar cell performance will be given in section 2.5.4.

2.5 Concentrator Requirements

2.5.1 High Efficiency

High efficiency solar cells are a must in CPV applications and this has been the basis of the dominance of multijunction solar cells in CPV systems over the last decade. The reasons behind the efficiency improvements associated with multijunction architectures – which apply to both one-sun and concentrated light – have been already discussed in depth in section 2.2.2. Now, we will focus on the benefits of using concentrated light on solar cells – both for single and multijunction – which adds a premium to the unparalleled efficiencies of multijunction devices.

Solar cell efficiency generally increases with concentration as has been incidentally shown in Figure 2.21, over a certain concentration range. As it will be shown in detail, this boost in efficiency is the result of an increase in V_{oc} and to a lesser extent in FF. No efficiency gains are attributable to short-circuit current because in a wide range of concentrations, the photocurrent is simply proportional to the intensity of the impinging radiation. Despite reports of a lack of linearity of photocurrent with the incident light [53,54] – either slightly less than linear (sublinearity), or slightly more than linear (superlinearity) – the linear dependence between irradiance and photocurrent, already described around Eqs. (2.18) and (2.19), is widely accepted and used by the CPV community for current solar cell architectures and normal operating conditions (see Chapter 10 about solar cell characterization).

The increase of V_{oc} (and thus of efficiency) with concentration can be explained in terms of basic semiconductor physics. When a solar is illuminated, the rise of the excited carrier concentrations and their redistribution trigger the splitting of the equilibrium Fermi level into the minority carrier electron quasi-Fermi level and the minority carrier hole quasi-Fermi level. The higher the illumination the larger the quasi Fermi level separation will be. Since the solar cell voltage derives from the quasi Fermi level separation, the increase of light concentration produces an increase of the open circuit voltage [1]. This situation is illustrated in Figure 2.25 for a typical lattice matched GaInP/GaInAs/Ge triple junction solar cell. The one-sun irradiance results in $V_{oc} = 2.70$ V while $V_{oc} = 3.28$ V for 1000×. The V_{oc} increase breakdown by subcell is

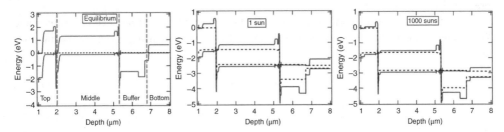

Figure 2.25 Band diagram of a lattice-matched GaInP/GaInAs/Ge triple junction solar cell in equilibrium (left), at 1 sun (center) and at 1000 suns (right). Notice the change of vertical scale in the left figure. Source: Mario Ochoa, Solar Energy Institute, Technical University of Madrid, 2015. Reproduced with permission of Mario Ochoa

13% (from 1.45 to 1.64 V), 20% (from 0.984 V to 1.180 V) and 73% (from 0.26 to 0.45 V) in the GaInP, GaInAs and Ge subcells, respectively.

From a fundamental point of view, the detailed balance theory allows to calculate the variation of the open circuit voltage as a function of concentration [55]:

$$V_{oc} = \frac{E_G}{q} - \frac{kT}{q} \ln\left[\frac{A_o E_G^2 E(W)}{J_L A(W)}\right] + \frac{kT}{q} \ln\left[\frac{C}{r(\theta)}\right] \tag{2.46}$$

where C is the concentration level, $r(\theta)$ is the reflectivity for the cone of light[4] which impinges on the cell from the concentrator as a function of the incidence angle, θ; $A(W)$ and $E(W)$ are the average values for the thickness (W) of absorbance and emittance, respectively, and A_o is a constant. The rest of symbols have their usual meaning. Further details on this Equation and the magnitudes involved can be found in Chapter 3.

An interesting remark is that as V_{oc} depends on the $C/r(\theta)$ ratio, the same effect can be obtained either by increasing C or by decreasing $r(\theta)$. Since θ and C are related by the invariance of radiance:

$$\sin^2 \theta = C \sin^2 \phi_s \tag{2.47}$$

assuming the cell is surrounded by air (of refractive index one) and $\phi_s = 0.267°$ being the angle subtended by the solar disc at the Earth, the maximum V_{oc} occurs for:

$$\frac{C}{\sin^2 \theta} = \frac{1}{\sin^2 \phi_s} \tag{2.48}$$

which is independent on C. This means that the maximum efficiency (i.e, V_{oc}) can be reached at one-sun operation with a highly angularly selective solar cell or with an isotropic cell at the maximum concentration of 46 050 suns [56] or with any other combination in between keeping $C/r(\theta) = 46,050$ [55]. In other words, a V_{oc} related to a given concentration level can be achieved by either increasing photons impinging the cell (i.e. increasing C) or by decreasing the emission of photons from the cell (i.e. increasing photon recycling).

Since FF varies only slightly with V_{oc}, the dependence of efficiency with concentration is almost entirely controlled by V_{oc} in such ideal case [55]. Accordingly, the efficiency of a solar cell system composed with an infinite number of gaps as a function of concentration is presented in Figure 2.26.

Going to a less ideal scenario, if linearity of photocurrent versus irradiance is assumed and the photocurrent at one-sun is denoted by J_L^*, then the photocurrent at a given concentration C (J_L) is:

$$J_L = C J_L^*, \tag{2.49}$$

---

[4] As will commented in Chapter 4, concentrator optics funnel collimated light at the expense of spreading its incidence angle distribution. Accordingly, concentrator solar cells are not illuminated by a collimated low intensity beam but by a cone of high intensity light.

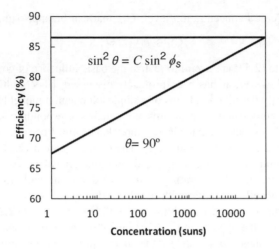

Figure 2.26 Ideal efficiency of a concentrator solar cell with an infinite number of gaps assuming only radiative recombination

which is Eq. (2.19), repeated here for completeness. Assuming the simple one-exponential model of Eq. (2.43) and a negligible (i.e. high enough) shunt resistance, the open circuit voltage is:

$$V_{oc} = m\frac{kT}{q}\ln\left(\frac{J_L}{J_0} + 1\right) \tag{2.50}$$

with m being the diode ideality factor and the rest of the parameters having the common meaning. Therefore, combining (2.49) and (2.50), the open circuit voltage at C suns is:

$$V_{oc} = V_{oc}^* + m\frac{kT}{q}\ln C \tag{2.51}$$

where V_{oc}^* is the open circuit voltage at one sun; Eq. (2.51) assumes also that m and J_0 do not change significantly when concentration level changes and that 1 is negligible as compared to the ratio J_L^*/J_0. Thereby, efficiency as a function of concentration can be approximated by:

$$\eta(C) = \frac{J_{sc}^* \cdot V_{oc}^* \cdot FF(C)}{P_L^*}\left[1 + \frac{mkT}{qV_{oc}^*}\ln C\right] \tag{2.52}$$

where J_{sc}^* and P_L^* are the short circuit current density and illumination power at one sun, respectively. Eq. (2.52) shows that concentration produces a continuous logarithmic increase in efficiency weighted by the FF variation. Again, as we showed before, the increase in efficiency with concentration is limited by the highest concentration achievable on Earth, i.e. 46 050 suns.

In practice, the efficiency increase with concentration is limited by the deleterious effect of the series resistance causing a FF drop together with a V_{oc} increase lower than that predicted by equation (2.51) as we will see below. Accordingly, for a given value of r_S, there is a maximum efficiency achievable at a concentration level, C_M, that can be approximated by [57]:

$$C_M \approx mV_t/r_S J_{sc}^*$$

(2.53)

Equations (2.49) to (2.53) are useful to realize the basic influence of concentration on solar cell performance. However, as has been profusely discussed, III-V multijunction solar cells cannot be conveniently described in terms of a single exponential model because each subcell exhibits several recombination mechanisms with their corresponding ideality factors. An exception is that of the Ge subcell that it is well modeled by $m = 1$ because of the predominance of recombination in the neutral regions [58]. Therefore, equation (2.51) cannot accurately describe the V_{oc} dependence of series-connected multijunction solar cells under concentration. At low concentrations, subcells operating with $m \approx 2$ dominate while from medium to high concentrations the subcells operating with $m \approx 1$ are dominant. At very high concentrations, series resistance is the origin of a V_{oc} sublinearity with ln C (see Figure 2.27). Although in the simplest approximation the open circuit voltage operation involves no current circulation and therefore, no series resistance losses, indeed the photogenerated current is also dissipated in the shaded parts of the solar cell (under the bus bar and grid lines) whereby the series and parallel resistances as well as by recombination diodes [59]).

Finally, as also anticipated by Figure 2.21, the fill factor shows an evolution with concentration somewhat similar to that of efficiency. Initially it increases, as a result of the lower relative impact of recombination losses over concentration. At a given point the effects of the series resistance become apparent and *FF* starts to decline. Such decline anticipates the

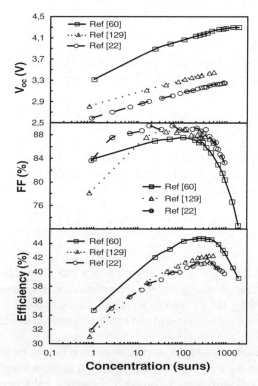

Figure 2.27 Performance of some champion multijunction solar cells as a function of concentration

decay of efficiency which will occur at slightly higher concentration levels, since the increase of V_{oc} can compensate, in a limited concentration range, the loss in *FF*.

In summary, the main limitation of practical solar cells under concentration is the series resistance which is responsible for the efficiency decrease once the increase of open circuit voltage with concentration cannot counterbalance the decrease in fill factor, even in record (as of 2015) solar cells [60]; see Figure 2.27. Therefore, the consideration of series resistance and the ways to minimize its deleterious effect are of key importance and will be analyzed in the next section.

2.5.2 Series Resistance. Grid Designs

Series resistance has been studied by many authors from the theoretical [61–63] and experimental [42,64] points of view. In these works there is no coupling between the semiconductor structure and the ohmic contacts, so that either the semiconductor structure is fixed and the front grid is studied or the ohmic contacts are fixed and the impact of the semiconductor structure is analyzed.

However, the proper analysis of series resistance needs a multidimensional approach with consideration of the whole solar cell. The influence of the different series resistance components, their expressions and their joint optimization for a concentrator GaAs single junction solar cell can be found in [65]. Because of the distributed nature of series resistance a 3D model is compulsory for such study [46]. The extension of this distributed model to a triple junction solar cell including the tunnel junctions is sketched in Figure 2.24 where all the resistive paths can be classified into:

- Horizontal resistances:
  - Metal resistance of the front contact, which depends of the metal sheet resistance, metal thickness and gridline width.
  - Emitter and base resistances, which depend on the corresponding sheet resistances and thicknesses.
  - Anode and cathode resistances of the tunnel junctions, which depend on the corresponding sheet resistances and thicknesses.
- Vertical resistances:
  - Front contact resistance, which depends on the specific metal-semiconductor front contact resistance and on the metal line width.
  - Base and BSF resistances, which depend on the corresponding resistivities and thicknesses.
  - Anode and cathode resistances of the tunnel junctions, which depend on the corresponding resistivities and thicknesses
  - Back contact resistance, which depends on the specific metal-semiconductor back contact resistance.
  - External contact resistance, which depends on the quality of the front wire bonding or tab connection and on the back solder or epoxy joint.
  - Shunt resistances resulting from unwanted shunts created during the solar cell manufacturing process.

In addition to the abovementioned, resistances at the heterojunctions can appear resulting in a deleterious degradation of solar cell performance [66–68].

From the abundant literature in the field, it can be said that the main series resistance components to be taken care of are those of the top cell emitter, front contact and tunnel junctions. Both top cell emitter and front contact resistances are coupled through the front metal grid. In other words, the front metal grid optimization has to take into consideration at least both the practical values of the top cell emitter and the specific front contact resistance [69].

Concentrator solar cells require an ohmic contact or junction between the grid metallization and the underlying semiconductor where the current flows with little parasitic voltage losses and obeys Ohm's law (i.e. the contacts are said to be ohmic as opposed to rectifying or Schottky). Tunneling or field emission is the mechanism responsible for ohmic contacts although in real devices thermionic and thermionic-field can also be present. As the doping level in the semiconductor increases, the width of the potential barrier decreases and tunneling can begin to occur regardless of the barrier height [15]. Therefore, the principal strategy to achieve ohmic contacts is to dope the semiconductor surface sufficiently high to assure that the dominant conduction mechanism is field emission (tunneling). The presence of a highly doped layer (in the range of $10^{19}\,\mathrm{cm}^{-3}$ for n-GaAs) between the metallization and the lower doped semiconductor seems to be a necessary condition for achieving the best ohmic contacts. The influence of metal, deposition technology, contact and alloying fabrication processes on the front contact resistance can be found in text books such as [70].

When optimizing a concentrator solar cell, the main questions are: what series resistance value is sufficient and what technological steps are necessary to achieve that value? Accordingly, some papers deal with the determination of the main factors influencing the series resistance, as well as the establishment of their threshold values for a well-balanced trade-off between cost and performance [65]. Following that philosophy, we present in Figure 2.28 the simulated efficiency at 1000× of a lattice matched GaInP/GaInAs/Ge triple junction solar cell assuming the structure of current commercial devices. Base line values for front contact resistance, metal resistance and top cell emitter resistance are shown in the figure together with the impact of varying the values of one of these resistances while keeping constant the other baseline values. As can be seen, the top cell emitter resistance exhibits the highest impact on efficiency. This result, however, neglects other effects. The top cell emitter resistance is not an independent parameter but depends on the emitter doping and thickness, which, in turn, can influence the internal quantum efficiency in the emitter. Hence the solar cell efficiency cannot be optimized by simply driving the sheet resistance of the emitter to its lowest value.

On the other hand, both the front contact resistance and the metal sheet resistance are independent parameters, whose requested values should be as low as possible. However, there are threshold values from which further improvements do not significantly impact the efficiency. In the case of front contact resistance, once $10^{-5}\,\Omega\cdot\mathrm{cm}^2$ is reached a further improvement (with the subsequent cost) has a negligible effect on efficiency. Regarding the metal resistance, a value of $10^{-3}\,\Omega/\square$ is good enough and further improvements will not affect significantly the efficiency. Again the choice of a number of fingers around the optimum is not a critical point. Of course, these two thresholds are strongly linked to the solar cell size simulated (3×3 mm); larger solar cells would need lower values and smaller devices could withstand higher values without significant loss.

The geometry of the front metal grid is usually circular, square, or rectangular. Circular grid designs are more widespread because of the belief that the concentrated light spot on the cell is circular. However, the shape of the light spot could not be circular depending on the

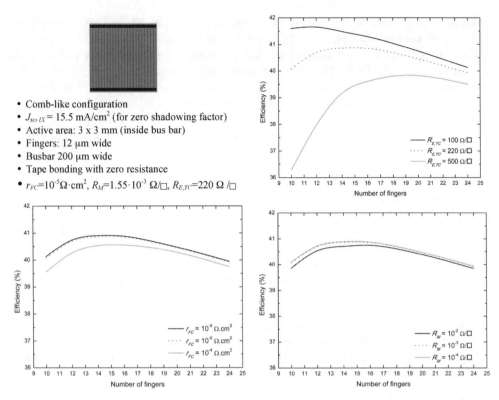

- Comb-like configuration
- $J_{sc,\,1X}$ = 15.5 mA/cm^2 (for zero shadowing factor)
- Active area: 3 x 3 mm (inside bus bar)
- Fingers: 12 µm wide
- Busbar 200 µm wide
- Tape bonding with zero resistance
- r_{FC}=10^{-5}Ω·cm^2, R_M=1.55·10^{-3} Ω/□, $R_{E,TC}$=220 Ω /□

Figure 2.28 Simulation results for a triple junction solar cell structure with the parameters shown in top left corner of the figure and varying some resistance parameters. Simulated efficiency as a function of the number of fingers for several values of emitter resistance of the top cell (top right), front contact resistance (bottom left) and metal resistance (bottom right) for 1000 ×. Source: Vincenzo Orlando, Instituto de Energía Solar, Universidad Politécnica de Madrid, 2015. Reproduced with permission of Vincenzo Orlando

optics used and, in addition, the light spot location onto the solar cell continuously varies depending on the position of the sun allowed by the tracking inside the acceptance angle (see Chapters 4 and 6, devoted to optics and trackers, respectively). Accordingly, any front grid geometry is acceptable when it is chosen considering the optics performance, although, from the series resistance point of view, the inverted square grid pattern performs better assuming the usual finger widths [65].

The size of the solar cells also influences the series resistance. The smaller the size of the solar cell, the lower its series resistance is. However, there are several other factors influencing the optimum solar cell size, as for example, perimeter recombination, heat transfer issues and module manufacturing processes. Each of those will define an optimum size range for the solar cell and a trade-off should be achieved. For instance, when considering the influence of perimeter recombination, an optimum size ranging from 0.1 to 6 mm^2 was determined for GaAs single junction solar cells operating at 1000× showing a sharp decrease in efficiency for sizes larger than 10 mm^2. In order to achieve acceptable temperature increases, sizes no larger than around 10 mm^2 are recommended [59]. Additional cost considerations, derived from module manufacturing processes, suggest an optimum size ranging from 1 to 25 mm^2 [71].

Resistances at the tunnel junctions can have also a big influence. The most important is the one corresponding to the ohmic region of the *I-V* curve (see Figure 2.17). Values between 10^{-4} and 10^{-5} $\Omega \cdot cm^2$, which do not limit the solar cell performance for $C < 1000\times$ or so, have been already achieved [37,72–74]. The other important tunnel resistances are those related to the horizontal components of the anode and cathode. These are not important below the peak current but they can play a decisive role when current is around or higher than the peak current as it is shown in sections 2.5.3 and 2.5.4.

In summary, and as a rule of thumb, for obtaining concentrator multijunction solar cells with a low series resistance the following steps must be followed:

1. Determine the concentration level of operation.
2. Determine the cell size within the range 0.1 to 25 mm^2 (taking into account the whole CPV module: optics, heat extraction, cost, etc).
3. Choose the geometry of the front grid pattern taking into account the optics performance.
4. Optimize the semiconductor structure in order to achieve the highest efficiency taking into consideration the impact of the sheet resistance of the top cell emitter.
5. Verify that the specific resistances of tunnel junctions have a negligible influence on the whole series resistance.
6. Verify that no heterojunction introduces some unexpected series resistance.
7. Measure the metal-semiconductor specific contact resistances and the front metal sheet resistance achieved by the contact technology used.
8. Optimize the number of fingers and the finger pitch of the front metal grid in order to get the highest efficiency at the desired concentration level.

2.5.3 Tunnel Junctions

As it has been shown in section 2.4.2.7, tunnel junctions for multijunction solar cells are characterized by the peak tunneling current density (J_p) and the equivalent resistance (r_{TJ}). For concentration applications, tunnel junctions (TJ) should show sufficiently low r_{TJ} and high J_p so that the performance of the solar cell is not limited by the TJ. The requirements for the equivalent resistance have been described in section 2.5.2. A high peak tunneling current is desirable to: a) be well above the photocurrent generated by the solar cell under concentration; b) withstand peak irradiances on the cell arising from non-uniform illumination profiles created by the optics of the CPV modules (see 2.5.4 and Chapter 4).

The highest peak tunneling currents are in principle obtained with low band gap materials which diminishes the optical transparency of the TJ. In fact, the highest J_p reported to date (10 100 A/cm^2) for a TJ device in the field of III–V MJSCs lattice matched to any of the typical substrates (GaAs or Ge)[5] has a design based on a p^{++}–AlGaAs:C/n^{++}–GaAs:Te heterostructure [74]. This J_p value is much more than enough when considering that the maximum light concentration of about 46 050 suns would produce a short circuit current of about 700 A/cm^2 assuming a short circuit current of 15 mA/cm^2 at 1 sun. Accordingly, a lower J_p could be accepted if high band gap materials were used in the TJ in order to increase its optical transparency.

---

[5] To the best of our knowledge the highest reported J_p is 19,904 A/cm^2 for an MBE-grown p^{++}–GaAsSb:Be/n^{++}–InGaAs:Si tunnel junction that could be suitable for InP/InGaAs tandem solar cells [John C. Zolper, John F. Klem, Thomas A. Plut, and Chris P. Tigges, "GaAsSb-based heterojunction tunnel diodes for tandem solar cell interconnects" Proc. of the 1st World Conference on Photovoltaic Energy Conversion p. 1843-1846 (1994)].

A p^{++}–AlGaAs:C/n^{++}–GaInP:Se TJ with higher optical transparency grown by atomic layer epitaxy was proposed by Jung and co-workers [75]. This device exhibited a J_p of 83 A/cm^2, but this value dropped to 74 and 33 A/cm^2 once annealed for 30 min at 650°C and 750°C, respectively. J–V curves of tunnel diodes for MJSCs grown by MOVPE based on the same heterostructure were later presented with J_p values of 50 A/cm^2 [76], 80 A/cm^2 [37], and 637 A/cm^2 [73]. The effect of the thermal load in the J–V curve was neither shown nor mentioned in any of these works. More recently, a p^{++}–AlGaAs:C/n^{++}–GaInP:Te tunnel junction resulted in a measured peak tunneling current density of 996 A/cm^2 that, after a thermal annealing that emulated the top cell growth, showed a J_p of 235 A/cm^2 (which would correspond to a concentration of about 15 600 suns) [72]. Therefore, the technology of tunnel junctions is well developed at the laboratory level and it does not appear to pose a limitation on solar cell performance neither in optical transparency nor in peak tunneling current.

In addition to the peak tunneling current density and the equivalent resistance, the anode and cathode lateral resistances also play an important role [48] as it is stated below. A widely used method to determine the peak current of the tunnel junction in the multijunction solar cell is to increase the light intensity on the solar cell and monitor the J–V curve until the *dip* in the curve appears [77]. At this point, it can be assumed that the J_{sc} equals J_p. Figure 2.29 (top left) shows the J-V curve of a tunnel junction which is included in a dual junction solar cell which is simulated at different concentration levels in Figure 2.29 (top right): 2900×, which represents a concentration lower than the supposed peak tunneling current of the tunnel junction (3000×); and two concentrations that are slightly higher than the expected peak of the tunnel junction, namely 3090× and 3100×. As expected, at a concentration lower than 3000×, no effects are apparent in the simulated J–V curve; while at higher concentrations (i.e., at 3100×) the expected dip in the J–V curve becomes apparent. However, a surprising result is obtained at concentrations slightly above 3000×: at 3090×, no effect in the J–V is detectable [48]. Therefore, from Figure 2.29 (top right), it can be inferred that the dual-junction solar cell can operate with illumination currents beyond the tunnel diode peak current without exhibiting the dramatic dip-related effects, at least up to approximately 3090×.

The explanation of this result is in Figure 2.29 (bottom and center for 3090× and 3100×) which shows that when the current in the solar cell exceeds the peak tunneling current, the excess current flows laterally until it reaches a dark area (beneath a finger or busbar) where it is able to flow vertically. In other words, dark areas represent an 'unused' tunnel junction area that can sink some photocurrent if lateral current spreads toward those dark areas (when the lateral resistances are not elevated). A direct consequence is that the extra short circuit current that the solar cell can handle before the dip occurs is proportional to the shadowed area accessible to the current spreading, which is typically close to the area covered by the grid [48].

Considering Figure 2.24, if the current density through the vertical resistance in the top cell is larger than the current through the tunnel junction, it means that the excess current from the top cell has flowed away through the anode resistances (r_{anode}). If, on the other hand, at a given position the current density through the tunnel junction is larger than the current through the top cell vertical resistance, then there has been an injection of current from the anode resistances into the tunnel junction. In other words, this area is draining the excess current photogenerated in other regions. Consequently, r_{anode} and $r_{cathode}$ are key parameters on the appearance of the dip in the J–V curve. Higher values of r_{anode} and $r_{cathode}$ would limit current spreading and the dip in the J–V curve would appear at a lower concentration, and *vice versa*.

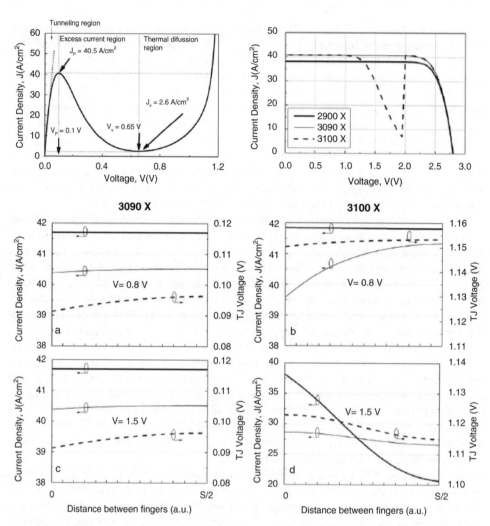

Figure 2.29 (top left) J-V characteristic of the tunnel junction used in the simulations presented in this figure in which the $J_p = 40.5$ A/cm^2 would correspond to 3000x. (top right) J-V curves for a dual-junction solar cell with the tunnel junction shown at the top left under uniform irradiation conditions at different concentration levels. (bottom and center) Longitudinal variation of the current density through the vertical resistance of the top cell, r_{vTC} (thick line, left axis), through the tunnel junction (thin line, left axis) and voltage across the tunnel junction (dashed line, right axis) versus half the distance between two adjacent fingers (where S is the finger pitch) for 3090x ((a) and (c)), for 3100x ((b) and (d)), and for two different dual-junction cell voltages: 0.8 V (before the dip occurs) and 1.5 V (in the dip). Source: Espinet et al, 2011 [48]. Reproduced with permission of Elsevier

Current spreading across the tunnel junction and adjacent layers toward dark regions was also suggested in [78] as a plausible reason for the variations observed in J_p. Also, the same phenomena, a non-negligible current spreading through the layers of the tunnel junction has been observed in simulations of dual-junction solar cells with TCAD models in [38].

2.5.4 Distributed Effects

The operation of solar cells under real conditions with concentration can be far from the standard conditions given by the AM1.5d spectrum with uniform illumination and at a temperature of 25°C [59]. The intense illumination on the cell produced by any optical concentrator generates a 3D temperature gradient. In addition, the most widely used refractive concentrating optics, namely the Fresnel lens, causes flux distributions that are spatially non-uniform, not only in absolute irradiance but also in spectral content (see Chapter 4). Therefore, we describe below the impact on the solar cell performance of gradients in impinging light and temperature.

2.5.4.1 Non Uniformity in Irradiance

Most irradiance patterns produced by CPV optic systems produce an irradiance maximum at the center of the solar cell with decreasing irradiance towards the edges, and this decay is, in most of the conventional CPV imaging optics, well approximated by a Gaussian curve [79]. Moreover, the maximum irradiance can be significantly higher than the nominal (average) concentration level [59]. Therefore, there will be areas of the solar cell where the photocurrent is over the nominal value for which the solar cell is designed. This can affect the solar cell performance causing an efficiency drop.

Some researchers have carried out analysis of the performance of solar cells under non-uniform irradiance using distributed models [44,45,80]. However, in those works the treatment of the distributed effects is restricted to a particular region of the solar cell device and is mostly focused on the series resistance components. The effect of non-uniform light profiles produced by some practical concentrators on the performance of a GaInP top cell was studied in [81]. In that study, the front grid was also optimized for the different light profiles taking into account the asymmetries of the light distribution that can appear because of the variation of the light spot position as a consequence of the sun movement inside the optics acceptance angle.

Regarding dual-junction solar cells, the influence of non-uniform light intensity profiles with a peak-to-average ratio (PAR) as high as 10 on the I_{sc}, V_{oc}, FF and efficiency was studied in [82] by means of the quasi-3D distributed models in which the tunnel junction was modeled as a resistor. A drop below 1% absolute is obtained for a PAR of the Gaussian light profile as high as four and an average concentration of 1000×, due primarily to a reduction in the fill factor and also in the V_{oc}.

A step up in complexity analysed the effect of non-uniform irradiance considering a full description of the tunnel junction (i.e. not only as a mere resistor) [41]. The simulation was carried out with the same Gaussian light profiles in both subcells with three different peak irradiances of: a) 3650×; b) 3700×; and c) 4000×. The average in all cases was 1000×. The three simulated Gaussian light profiles exceed the peak current of the tunnel junction locally as shown in Figure 2.30 (left). I-V curves obtained from the simulations when the dual-junction solar cell is excited by the Gaussian light profiles are depicted in Figure 2.30 (right). Dip-related effects are not observed up to a peak of 3700×, which is a concentration considerably higher than in the case of the uniform light profile (as for 3100× in Figure 2.29 top right). This is because the area which exceeds the peak of the TJ is more localized than in the case of uniform radiation and it is surrounded by areas illuminated with concentrations progressively lower than 3000×. The excess photo-generated current in the TC then spreads laterally through the r_{anode} towards areas of the solar cell with lower concentration in a similar way as that described

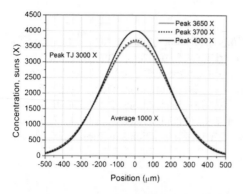 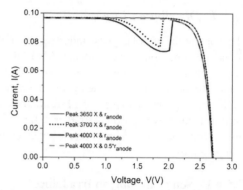

Figure 2.30 Gaussian light profiles used in the simulations with peaks of 3650×, 3700× and 4000× over the average of 1000× in all cases impinging on a 1 mm² (active area) square solar cell (left). *I-V* curves of the dual-junction solar cell under Gaussian light profiles with different PAR. The gray dahsed represents the *I-V* curve of the dual-junction under a Gaussian light profile of PAR 4 but with a low r_{anode} (right). Source: Espinet-González, 2012 [41]. Reproduced with permission of Universidad Politécnica de Madrid

in section 2.5.3. In these areas, the current can flow vertically through the TJ, and then gather back through $r_{cathode}$ towards the areas with higher concentrations in the BC where the current flows through the bottom cell. Therefore, at short circuit current in the case of a Gaussian light profile with peak of 3650× or lower, the tunnel junction works in the low voltage drop region in the whole solar cell for the entire *I-V* curve. Consequently, no-dip related effects appear in the *I-V* curve. However, at higher concentrations (3700× and 4000×) at short circuit conditions the lateral spreading of current density through the TJ resistances is insufficient and the vertical current density through the TJ exceeds 40.5 A/cm² (TJ peak current density) in the center of the solar cell forcing the TJ to work in the high voltage drop region (thermal diffusion regime) and the dip in the *I-V* curve appears. The *I-V* curve of the dual-junction solar cell with low anode resistance ($0.5 \cdot r_{anode}$) under the same Gaussian light profile peaking at 4000× and average 1000× is also depicted in Figure 2.30 (gray dashed line). As can be seen, in the case of low r_{anode} there are still no dip-related effects under these conditions.

This effect, described for dual-junction solar cells, also occurs in triple-junction devices. Again the lateral current spreading through the anode and cathode resistances is the key mechanism for mitigating the loss in efficiency due to non-uniformities [50]. Recently, triple-junction solar cells have been evaluated under real light profiles and their performance has been assessed inside four different optical concentrators with different irradiance profiles, including the effect of misalignments in pointing accuracy of the tracker (i.e. when working on axis and 0.6 degrees off-axis) [51].

2.5.4.2 Chromatic Aberration

Chromatic aberration is a type of distortion in which a lens does not focus all colors to the same convergence point (see more details in Chapter 4). It occurs because lenses have different refractive indexes for different wavelengths of light (the dispersion of the lens). Because single-junction cells are relatively insensitive to spectral variations, the designers of optical concentrators did not worry about chromatic aberration until dual-junction solar cells appeared in the mid-1990s. The introduction of dual-junction solar cells with the need for current matching

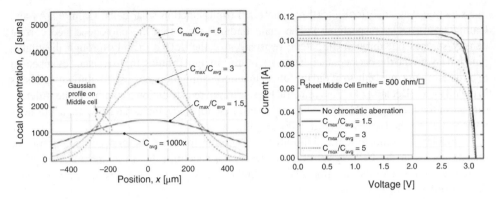

Figure 2.31 Subcell irradiance profiles used (uniform for the top and bottom subcells and Gaussian for the middle cell), which correspond to different chromatic aberration patterns generated by the optics of a hypothetical concentrator assembly (left). Effect of these photocurrent profiles on the triple-junction solar cell I–V curve (right). All profiles with $C_{avg} = 1000$ suns. Source: Iván García, Instituto de Energía Solar Universidad Politécnica de Madrid, 2015. Reproduced with permission of Iván García

promoted the first studies about chromatic aberration [83–85]. For instance, the simulation of the chromatic aberration caused by a Fresnel lens, in combination with a homogenizer, on dual-junction solar cells showed a shunt-resistance-like effect, where the *FF* was affected [86].

Chromatic aberration introduces a spatially non-uniform subcell current mismatch in the multijunction solar cell. Experimental analysis of the effects of chromatic aberration in triple-junction solar cells revealed an efficiency decrease as high as 6.5% (relative) [87]. The simulation of triple-junction GaInP/GaInAs/Ge solar cells by means of a 3D distributed model was carried out in [49] by using C_{max}/C_{avg} ratios from 1 to 5 (see Figure 2.31 left). As the C_{max}/C_{avg} ratio of the middle cell photocurrent distribution increases, the cumulated current mismatch over the solar cell area increases, and thus, a considerable I_{sc} drop and an increased *FF* would be predicted. However, a relatively low I_{sc} drop and a *FF* decrease were observed (see Figure 2.31 right). This unexpected result was caused by the current spreading occurring in the semiconductor structure, which redistributes the excess current generated in some regions and contributes to a lower 'overall' current mismatch in the solar cell. Enabling lateral current flow as much as possible in the solar cell structure can contribute significantly to reducing the effect of chromatic aberration caused by the optics. In other words, the design of the semiconductor structure can help mitigate the effect of chromatic aberration produced by a given optics.

A frequent situation in real CPV systems in the field is the variation of the focal distance of the lens as a consequence of system misalignments and/or temperature impact on the lens (see Chapter 5). This phenomenon is equivalent to moving the cell closer to or away from the lens. On one hand, if the focal distance of the concentrator increases (which would be equivalent to the multijunction solar cell being moved closer to the lens than its design position), then the PAR of the Gaussian light profile of the blue light that is used by the GaInP top cell is higher than the PAR of the flux distribution of the red light used by the Ga(In)As subcell. On the other hand, if the focal distance of the concentrator decreases (which would be equivalent to the multijunction solar cell being moved farther to the lens than its design position), then the spectral content of the flux distribution is the opposite: the Ga(In)As subcell receives a Gaussian light profile with a PAR higher than the PAR of the Gaussian light profile impinging the GaInP top cell [88]. The impact of this situation was analyzed in [50] for dual- and triple-junction solar cells (GaInP/GaAs and GaInP/Ga(In)As/Ge, respectively). The main findings were that the photogenerated current density spreads out through the lateral resistances of the device, mainly through the tunnel junction layers and the back contact. Under non-uniform light profiles, these resistances appeared to be determinant, not only to avoid the tunnel junction's limitation, but also for mitigating losses in the fill factor. Because of the asymmetric lateral current redistribution capability in the structure (higher in the middle cell/bottom cell and lower in the top cell), the deleterious impact of non-uniformities on the top cell is more relevant [50].

As a rule of thumb, in case non-uniformities – due to either the concentrator design or misalignments in the CPV system or changes in the spectrum due to temperature – optical designers should favor the situation where the highest peak of the light profile does not occur on the top cell.

2.5.4.3 Temperature

The illumination of the solar cell produced by any optical concentrator generates a temperature gradient. This gradient is three-dimensional because the nonuniformity of light produces a horizontal temperature distribution and in addition, the different light absorption of the semiconductor layers produces a vertical gradient. Therefore, each point of the multijunction solar cell will be at a different temperature. In a first approach, as the temperature increases, the band gap decreases, so the hottest regions can absorb a larger portion of the incident spectrum.

The net effect of the temperature increase is a reduction in the operational efficiency of the solar cell, and thus a localized current mismatch can occur between various regions of the cell operating at differing temperatures. The thermal energy generation within each subcell in a triple-junction GaInP/GaInAs/Ge solar cell is highest in the Ge bottom cell, and lowest in the GaInAs middle cell [89]. The temperature increase is very dependent on the light uniformity. Simulations of a triple junction cell illuminated at 820× with a Fresnel lens (non-uniform illumination) and a Fresnel lens plus homogenizer (~uniform illumination) suggest that highly illuminated areas may reach temperatures about 40°C higher than the average temperature of the uniform illumination case [90]. The overheating of some parts of the cell can be also due to the appearance of voids in the back solder of the cell which can further promote the so called *thermal runaway* of the cell (see Chapter 9 about reliability).

A 3-D model for the temperature distribution of the multijunction cell requires the accurate illumination profile as the input together with an exhaustive knowledge of the temperature dependence on material parameters such as absorption coefficient, mobility, band gap energy, intrinsic carrier concentration, etc. for all the semiconductors in the multijunction stack. When available, this 3-D modeling will be of great help in determining regions of the solar cell malfunctioning as a consequence of high temperatures.

2.5.5 Atmospheric Spectral Variations and Impact on Energy Yield

Sunlight is absorbed and scattered by the Earth's atmosphere. As explained in Chapter 1, the transmissivity of the atmosphere is governed by a set of components, i.e. *aerosol optical depth* (AOD) at a given wavelength, *precipitable water* (PW), ozone (O_3), etc. and the *air mass* (AM) which is the relative path length through the atmosphere. Therefore, the average efficiencies of solar cells in real operation differ from the efficiencies measured under standard conditions because the spectrum and temperature vary in the field. A pioneering work [91] analyzed the main impact of these variables in single and multijunction cells. The main conclusions were:

- Higher AMs result in a shift of the photon distribution towards longer wavelengths (or lower energies), so high band gap subcells are more strongly affected (see Figure 2.32 left).
- Increased turbidity (AOD) causes more attenuation and scattering of solar irradiance in the UV-visible compared with the NIR region of the spectrum. Therefore, high band gap subcells are more affected than low band gap ones (see Figure 2.32 center).
- The power output of low band gap subcells is more sensitive to water vapor variations than the high band gap ones (see Figure 2.32 right).
- The water vapor effect is less important than the turbidity effect. Turbidity acts more like a broad band filter with a higher attenuation for higher energies. Water vapor affects specific low wavelength bands.
- The AM daily variations produce a current mismatch that is partially compensated for an increase of fill factor.
- Seasonal solar cell performance is a strong function of latitude because the AM distribution functions depend on latitude. The energy produced by solar cells is strongly dependent on latitude for latitudes greater than 30°.

How spectral variations affect solar cells with 3-, 4- and more junctions was analyzed in [92]. The conclusion was that, although variations in atmospheric conditions will affect the generated power in a measurable way, this effect is less than the boost in power obtained by increasing the number of junctions for well-designed cells. Multijunction cells can be optimized to perform best for any set of conditions by changing the thicknesses of the top cells [92], and/or the bandgap combination [93].

A similar conclusion was reached in [95] that modeled the energy production of 4-junction, 5-junction, and 6-junction cells. The effect of changing solar spectrum over the course of the day lowers energy production by only ~1.1% relative in going from 3-junction to 6-junction cells, whereas the efficiency of 6-junction cells under STC is over 23% higher on a relative basis (>23% higher power) than the 3-junction cells [95]. A consideration of the spectral transmissivity of the concentrator optics, which plays a key role in the current matching of the subcells, could change in some extent the above values.

The need to consider the optics for the solar cell optimization [59] must also include variations in the ambient temperature which change the refractive index and even the shape of the optics [96]. For a system based on silicone-on-glass (SoG) Fresnel lenses, if the ambient

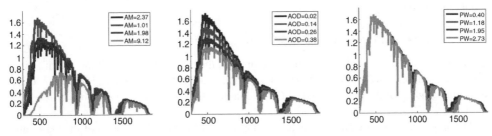

Figure 2.32 SMARTS simulations for Madrid showing the influence of air mass (left), aerosols (center) and precipitable water (right) on the solar spectrum. Source: Núñez at al. 2014 [94]. Reproduced with permission of AIP Publishing

temperature increases the silicone refractive index decreases which makes the lens less converging causing the spectral distribution over the cell to shift to the blue; that is to say, the blue light is more focused and the red light more defocused (equivalent to moving the solar cell closer to the lens, see Chapter 5). On the contrary, if the ambient temperature decreases, the spectral distribution shifts to the red (equivalent to moving the solar cell further away from the lens). In addition, the irradiance is red-shifted in winter and blue-shifted in summer. Therefore, in winter the red light is more focused and simultaneously there is loss of blue light and, conversely, in summer, the blue light is more focused (high temperatures) and simultaneously there is loss of red light. The work of [96] shows that since triple junction solar cells are not symmetric (as design trade-offs result in a top cell with higher lateral resistances than the middle or bottom subcells), higher losses are always caused by higher irradiance peaks in the blue region. The proposed ways to minimize these losses are a) designing the optics to have the highest peak of red light instead of blue light, b) using small solar cells, c) optimizing the semiconductor structure in order to have low lateral series resistances to homogenize or spread the current density.

Jaus and Gueymard compared 37 900 direct spectra from 379 sites worldwide and showed that they are, as an average, more red-rich than the reference spectrum AM1.5d ASTM G173 [97]. So, further optimization of the multijunction cells is possible. This optimization for a given place needs the calculation of an equivalent spectrum for which radiative models such as SMARTS [98] or MODTRAN are used. The generation of the spectrum by these models needs to know the abovementioned atmospheric components (AM, PW, turbidity, etc.) what requires complex and expensive instruments (further details in Chapter 1). Therefore, an alternative model has been proposed [94] to estimate the spectral influence on energy performance of CPV multijunction solar cells.

2.5.6 Temperature Effects

The performance of multijunction solar cells is very influenced by temperature. The corresponding study is presented in Chapter 5.

2.6 Description of Different Cell Approaches

As explained in section 2.3, multijunction solar cells implement high efficiency photovoltaic devices by incorporating materials with different bandgaps trying to get, as close as possible, to the optimum combinations summarized in Table 2.2. In the race to reach a higher number of junctions with the appropriate bandgaps, many materials, device configurations and manufacturing approaches have been explored. This section summarizes the most relevant achievements in this field.

2.6.1 Lattice-matched GaInP/GaAs/Ge

The most widely multijunction cell manufactured on the market today is the monolithic triple junction GaInP/GaAs/Ge device. A general schematic of the cell is shown in Figure 2.7. It is a direct descendent of the original GaInP/GaAs dual junction solar cell [20,99–101] grown on a Ge bottom cell.

The single crystal Ge wafer serves as the starting substrate for the growth of the GaAs and GaInP upper subcells. The most common growth method is metalorganic chemical vapor deposition (MOCVD) [102–106] – and certainly the most relevant for mass production –, but molecular beam epitaxy [107] is also used. The junction in the Ge wafer is usually formed during the initial stages of the deposition process [108–111]. The main advantage of Ge is that it is closely lattice matched to the GaAs and GaInP subcells (see Figure 2.12). This lattice matching helps to ensure the growth of relatively perfect GaAs and GaInP subcells. The band gap of Ge, however, is too low to yield the optimum 3-junction efficiency with GaAs and GaInP. Nevertheless, it does boost the V_{oc} of the device by about 250 mV.

While the lattice matching between Ge and GaAs is close, it is not perfect; the lattice constant of GaAs is slightly smaller than that of Ge. So about 1% indium is usually added to the GaAs so as to more closely lattice match it to the bottom Ge substrate/cell. A comparable amount of In is also added to the GaInP to maintain its lattice match with the GaInAs middle cell. This small amount of In in both cells reduces slightly the V_{oc} but also increases, again slightly, the J_{sc} of the device.

The band gap of the GaInP, at a fixed composition, can be varied from 1.8 eV to 1.88 eV by controlling the degree of Cu-Pt ordering in the material [112–114]. This is typically done with the aid of a surfactant such as Sb [115,116] but can also be accomplished by tuning the growth conditions [117–119], e.g. by changing the growth rate, temperature, PH_3 overpressure, or substrate misorientation.

The optimal top cell band gap energy is around 1.9 eV depending on the spectral content of the incident light [20]. As such, an optically thick GaInP top cell will absorb too much light robbing photocurrent from the GaInAs subcell. One could compensate for this loss of current by lowering the band gap of the middle GaInAs cell by, for example, adding more indium to the middle cell. But now the middle cell is no longer lattice matched to the Ge. One could also increase the top cell band gap by adding Al, but this can introduce other potential problems, associated to the intense affinity between Al and O (i.e. layers with Al content, show high O contents also, which degrade PV properties of the material). A more direct approach is simply to reduce the thickness of the top cell to a point where the top and middle subcells are current matched [20]. For the lowest band gaps (~1.8 eV), the optimum thickness of the top cell is around 500 nm.

The highest efficiency achieved for this device at one-sun is 34.1% [120] and the highest concentrator efficiency is 41.6% at 364× [22]. Although this solar cell has achieved very high efficiencies, its most glaring shortcoming is that there is a large current mismatch between the bottom Ge cell and the upper two cells. In the following we look at several other solar cell structures that address this issue, in an attempt to surpass the performance of this design, which is today the mainstream product of CPV solar cell technology.

2.6.2 Metamorphic GaInP/GaInAs/Ge

As mentioned previously, one way to address the large current mismatch between the Ge bottom cell and the two upper junctions, is to lower the band gap energies of the two top subcells. This can be accomplished by adding indium to both subcells. The addition of indium to both subcells increases their respective lattice constants relative to that of Ge (see Figure 2.12) and the monolithic combination of these lattice-mismatched materials is commonly called a metamorphic structure.

In theory, the efficiency of this device should increase monotonically with In concentration up to the point where the top two subcells are current matched to the Ge bottom cell. Depending on the spectral content of the incident light, this maximum should occur at middle cell In mole fraction of 17% GaInAs [120]. At this point the lattice mismatch between the $Ga_{0.83}In_{0.17}As$ middle cell (and the $Ga_{0.35}In_{0.65}P$ top cell) and the Ge bottom cell is about 1.2%. In practice, however, this has not happened. The lattice mismatch between the Ge subcell and the GaInP and GaInAs subcells causes dislocations to form during the growth process. These dislocations degrade the minority carrier lifetime in the GaInAs and GaInP subcells and reduce the efficiency of the multijunction device [121,122]. For example, the highest efficiency for a $Ga_{0.45}In_{0.55}P/Ga_{0.95}In_{0.05}As/Ge$ triple junction is 41.6% at 484 suns, a number virtually the same as that for the lattice-matched $Ga_{0.5}In_{0.5}P/Ga_{0.99}In_{0.01}As/Ge$ device [22,123].

With a higher In content the highest achieved efficiency for a $Ga_{0.35}In_{0.65}P/Ga_{0.83}In_{0.17}As$ device is 41.1% at 454 suns [120,124]. Considering the large lattice mismatch, this is still a remarkable result and reflects the tremendous progress that has been achieved in this area.

This solar cell growth strategy is termed *upright metamorphic* (UMM or, simply, MM, for short) as opposed to inverted metamorphic designs, which are the subject of the next section.

2.6.3 Inverted Metamorphic GaInP/GaAs/GaInAs

Another option for achieving a current matched triple junction is to increase the band gap energy of the bottom cell. One way to do this is to replace the Ge with a 1-eV $Ga_{0.7}In_{0.3}As$ bottom cell [125]. Adding a 1-eV solar cell can boost the device efficiency but only if the electronic quality of it and the other two junctions remain high. The lattice constant of $Ga_{0.7}In_{0.3}As$ is 0.578 nm or ~2% larger than that of GaAs and is the biggest impediment to achieving high efficiency, particularly if one tries to fabricate the structure in the normal, upright configuration. In this case, threading dislocations generated during the growth of the $Ga_{0.7}In_{0.3}As$ and its buffer layers will degrade the electronic quality of it and any overlying cell structures.

The basic idea of the inverted metamorphic concept (IMM, for short) is to avoid some of the lattice mismatch problems by growing the multijunction solar cell upside down (or inverted). As sketched in Figure 2.33, the growth sequence starts with the growth of the top or high band gap cell followed by the cell with the next lower band gap. By growing the top cell first, followed by the middle cell, the electronic quality of the top and middle subcells, being lattice matched to the substrate, is preserved and relatively unaffected by the subsequent growth of the bottom cell. Usually the growth of the bottom subcell is preceded by the growth of a transparent buffer layer to transition the lattice constant from the GaAs substrate to that of the bottom subcell. With this, lattice defects created in the buffer layer can only degrade the bottom cell. In general, our understanding of the dynamics of dislocation formation and propagation in strained materials has progressed rapidly in the last few decades. In addition, it has been found that GaInAs alloys are less sensitive the presence of dislocations than other III-V alloys. As such, relatively high quality GaInAs alloys, on GaAs-like substrates, can be fabricated. This has led to reports of very high device efficiencies ranging from 42.6% at 327× by the NREL team [126] in 2012 to 44.4% at 302 suns by the team at Sharp in 2014 [127].

The inverted metamorphic approach is not limited to lattice matched GaInP/GaAs top and middle subcells. Geisz and co-workers at NREL fabricated an inverted triple junction device consisting of 1.83 eV $Ga_{0.51}In_{0.49}P$ top junction lattice-matched to the GaAs substrate, a metamorphic 1.34 eV $In_{0.04}Ga_{0.96}As$ middle junction, and a metamorphic 0.89 eV $In_{0.37}Ga_{0.63}As$ bottom junction [21]. An ion beam image and composite 220 dark-field TEM of a cross-section of the device are shown in Figure 2.34. While the efficiency of

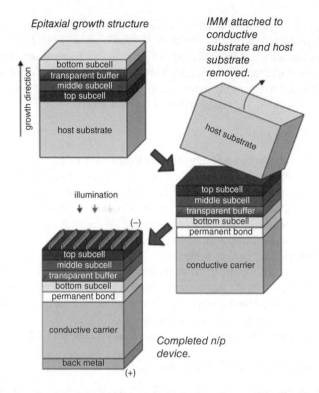

Figure 2.33 Process schematic for concentrator IMM devices. Source: The Boeing Company, Seattle, WA, 2015. Reproduced with permission of Boeing

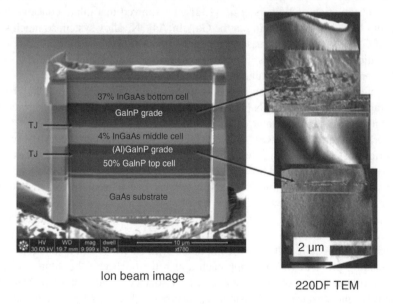

Figure 2.34 Ion beam image and composite 220 dark-field TEM of a FIB cross-section of an unprocessed inverted triple-junction solar cell structure. Source: Geisz et al. 2008 [21]. Reproduced with permission of AIP

this device achieved at that time is only 40.8% at 326 suns, it illustrates the utility of this approach. Other modifications of this device have also been studied [128].

Inverted devices require that the epitaxial device structure be removed from the substrate. The basic process is illustrated in Figure 2.33. This can be accomplished either by destructively removing the substrate or separating the structure from the substrate with the aid of a release layer. The later technique is discussed below in section 2.6.10.

2.6.4 Double Sided Epi

Another way to avoid the problem of threading dislocations from the growth of the 1-eV solar cell propagating into otherwise defect-free junctions is to grow the GaInAs subcell in a separate step on the back side of the GaAs substrate. The GaInP and GaAs subcells are grown on the front side of the same GaAs substrate [129]. This requires at least two MOCVD growth cycles.

The bifacial structure places a number of constraints of the GaAs substrate. First, the GaAs substrate must be optically transparent to wavelengths destined for the bottom cell. The sub-bandgap optical transparency of GaAs substrate – i.e. its free carrier absorption – is only high if it is lightly doped [12]. However, if the doping is too low, electrical resistivity becomes a problem and the efficiency of the device under high concentrations will suffer. Second, both sides of the substrate must be high quality, 'epi-ready' surfaces. Poor quality surfaces generally lead to poor electronic quality of the epitaxial layers.

Despite these problems, this device has achieved an efficiency of 42.3% at 406 suns.

2.6.5 Lattice Matched GaInP/GaAs/GaInNAs

The basic device structure of the GaInP/GaAs/GaInNAs device was conceived by researchers at Sandia National Labs [130] and NREL [131]. The composition of the GaInNAs bottom cell is adjusted to achieve a band gap of 1 eV and a lattice constant equal to GaAs and is based on the original observation of Weyers $et\ al.$ [132] who showed that dilute concentrations of N induce a very large band gap bowing in the $Ga_{1-x}In_xAs_{1-y}N_y$ alloy system so much so that the band gap of GaInAs actually decreases with the addition of N. Since the addition of N and In move the lattice constant of GaAs in opposite directions, adding both in the ratio $y/x = 0.35$ lowers the band gap while maintaining the lattice matching to GaAs.

Unfortunately, it was also soon found that the electronic quality of $Ga_{1-x}In_xAs_{1-y}N_y$ decreases rapidly with increasing N [131,133–137] and was therefore not of sufficient quality for insertion into a 3-junction solar cell.

Later work at Stanford and NREL showed $Ga_{1-x}In_xAs_{1-y}N_y$, prepared by MBE (with the aid of a Sb surfactant), could be suitable for multijunction devices [138]. These ideas have been reduced to practice by the team at Solar Junction, Inc. recently achieving an efficiency of 43.5% at 925 suns [139]. Single junction GaInAsNSb cells have also been grown by MOCVD [140] but appear to be inferior to devices grown by MBE.

The main advantage of this cell is that it is both current- and lattice-matched and can be grown directly on a standard GaAs or Ge substrate. No extended buffer layers nor the removal of the substrate are required. It is also compatible with future 4- and 5-junction devices.

The remaining problem with this approach is that the quality of the dilute nitrides, while sufficient, is still not equal it other conventional 1-eV solar cells and in order to obtain sufficient current density, one must use an intrinsic-layer absorber with low background doping level and wide depletion region. Currently, only MBE can achieve the required depletion layer thickness.

2.6.6 Quantum Dot and Quantum Well Multijunction Solar Cells

Solar cells composed of quantum wells (QW) or quantum dots (QD) have attracted much attention over the last two decades. Much of the attention has been directed towards achieving higher efficiencies for 'single junction' cells such as the 'intermediate band solar cell' [141–143] or the 'hot carrier' solar cell [144]. But another use is to alter or 'engineer' the effective band gap of a solar cell absorber by inserting quantum wells [145–148] or dots [149–154,155]. For example, it is possible to lower the effective band gap or absorption threshold of a silicon solar cell by introducing Ge quantum dots [156] or that of GaAs by introducing InAs dots [155].

This approach has been applied extensively to modify the middle GaAs subcell of a multijunction cell [149,157]. By introducing InAs or GaInAs quantum dots or quantum wells one can lower the effective band edge and increase the middle cell photocurrent.

To compensate for the lattice expansion associated with the In, strain compensating elements such as P can be added to the barrier material (i.e. GaAsP) [158]. By altering the physical width and height of the barrier and confined regions, one can optimize the absorption threshold and transport properties of the middle cell absorber yielding a higher J_{sc} and efficiency for the multijunction device.

When optimized, the results for QW and QD solar cells are in line with what one would expect from a comparable cell with a homogeneous p-i-n structure; the J_{sc} tracks with the absorption threshold, but the V_{oc} of QW and QD solar cells is degraded by higher space charge recombination [159] and, in some cases, higher non-radiative recombination associated with interface recombination and misfit defects.

2.6.7 More Junctions (4, 5, 6)

As shown in previous sections (see Table 2.2), a more or less direct path to higher efficiencies is more junctions. More junctions partition the solar spectrum into smaller chunks where photogenerated carriers suffer reduced thermalization losses. The trick, of course, is to add more junctions without increasing the cost of the device or without introducing other losses. In this regard, the preferred multijunction cell would be a monolithic structure composed of subcells that are of high electronic quality and are lattice- and current-matched. The number of possible structures that fit this bill is clearly limited. It has become clearer, however, that lattice mismatched devices and mechanically stacked structures –in particular wafer-bonded structures –may be able to surmount some of their perceived problems and may be the most cost effective path to higher efficiency and lower cost. In the following we briefly review some the advances in this area.

2.6.7.1 GaInP/GaInAs/GaInNAs/Ge

The nearest approximation to the ideal 4-junction device is the monolithic GaInP/GaInAs/GaInNAs/Ge cell. The device is lattice matched and can be current matched [131,160,161], but the GaInNAs subcell is still less than ideal. Several groups are working on this device, but to date there has been no report of a full 4-junction device with a high efficiency.

2.6.7.2 GaInP/GaAs/GaInAs/GaInAs

The nominal band gap combination for this 4-junction device is 1.8/1.4/1.0/0.7 eV [162,163]. It is a direct extension of the inverted metamorphic 3-junction GaInP/GaAs/GaInAs device described above. The device requires two transparent, graded buffer layers: one to accommodate the strain between the 1.0 eV GaInAs subcell and the GaAs subcell and another to accommodate the additional strain (3.8% relative to GaAs) associated with the 0.7 eV bottom cell. The highest certified efficiency measured for this device is 36.4% at 1-sun, AM1.5g and 43.8% at 327× [162] with a yet to be confirmed report of a 45.7% device at 234×.

2.6.8 Stacked Multijunction Cells

The current world record (Fraunhofer ISE of Freiburg, Germany) for solar cell of any design is the 2-terminal, 4-junction device consisting of a GaInP/GaAs dual junction device 'wafer bonded' to a GaInAsP/GaInAs dual junction grown lattice matched on an InP substrate. It has achieved a certified efficiency of 44.7% at 297× [60] and, at the time of writing (April 2015) a device with a record efficiency of 46% at 508× has been announced by the same lab. The term 'wafer bonded' refers to a process where the two wafers or cells are bonded using only temperature and pressure; the bond is between the constituent atoms at the interface between the two materials and no additional adhesives are used. Wafer bonded cells perform electrically like two-terminal monolithic multijunction cells; cell interconnects are straightforward but the current and efficiency of the device is sensitive to changes in the spectral content of the incident light.

The team at Spectrolab [164] has reported a 'wafer bonded' 2.2/1.7/1.4/1.05/0.73 eV 5-junction cell. The top three junctions are grown on GaAs and the bottom two junctions on InP. The compositions of the various layers were not specified. The NREL certified efficiency for this device was 37.8% at one sun AM1.5g.

There are several other methods for stacking cells. The group at Semprius uses a thin adhesive to bond a 3-junction GaInP/GaAs/GaInNAs top cell to Ge bottom cell. Prior to bonding, the Ge cell is electrically isolated from the top cell, by spin casting a thin layer of As_2Se_3 (~300 nm thick) [165]. Separate contacts (a total of four) are made to the two subcell structures. The As_2Se_3 encapsulant is optically transparent for wavelengths greater than 500nm and has a relatively low thermal resistance and high electrical resistance. The 3-junction cell is detached from the GaAs substrate using a process similar to epitaxial liftoff (to be discussed below). The 3-junction/Ge mechanical stacks are assembled using a technique called transfer printing [166] where an array of detached triple-junction cells are simultaneously transferred to a similar array of Ge bottom cells. Solar cell and module efficiencies (not certified) of 43.9% and 36.5%, respectively, have been reported for this approach.

2.6.9 III-Vs on Silicon

2.6.9.1 Monolithic Epitaxial Structures

Multijunction solar cells based on a silicon bottom junction are very attractive due to the relative low cost of silicon substrates. The ideal multijunction cell would be a monolithic combination of current- and lattice-matched subcells with ideal electronic properties. A dual junction cell with 1.7 eV top junction and silicon bottom cell has a theoretical efficiency of ~38% at 500× [167]. Triple junctions on Si, of course, have the potential for even greater efficiencies. To date, however, there is no Si-based cell that fits this bill. The other approach is to mechanically stack a high quality III-V cell on the Si bottom cell. In the following we briefly discuss some of these structures:

- Monolithic top cell structures using AlGaAs [168,169] and GaAsP [170] have been studied since the 1980s. The problems that limited device efficiency then largely persist today.
- Only GaP and GaPAsN are closely lattice matched to silicon. All other III-V compounds are lattice-mismatched to silicon.
- Independent of lattice mismatch, the heteroepitaxy of a III-V compound with zinc blende structure on a silicon is difficult. The structural differences can lead to the formation of 'antiphase domains' and it is notoriously difficult to prepare a surface of silicon free of oxygen and carbon contaminants, both of which tend to preclude good epitaxy.
- The thermal expansion coefficient of silicon is typically smaller than that of III-V materials. This can lead to cracking of the III-V epilayer when cooling down from growth temperature.

It is difficult to compare the efficiency of devices measured 20 or 30 years ago to devices measured using today's standards, particularly for tandem cells. A better measure of success or progress in this area is the V_{oc} of the top cell relative to its band gap [171]. To first order, the measurement of the V_{oc} of a top cell will only depend on the magnitude of the J_{sc} and not the spectral content of the solar simulator. The V_{oc} s for a number of III-V solar cells grown on Si are tabulated in Table 2.7 below.

The basic trend here is that most top cells grown directly on Si with or without a III-V graded buffer layer, exhibit low a V_{oc} where the $W_{oc} = E_g/q$-V_{oc} is typically 0.6 to 0.8 eV. While for the same top cells grown on GaAs, W_{oc} is 0.5 V to 0.6 V. This reflects some of the added complications of growing III-V solar cells on silicon.

There is one notable exception; the W_{oc} of a GaAs cell grown on a SiGe graded buffer is very close to 0.4 V.

The $GaN_{0.03}P_{0.9}As_{0.07}$/Si tandem solar cell is also unique because it is the only lattice- and (potentially) current-matched Si-based tandem. Similar to the dilute nitride arsenide solar cells described above, if nitrogen and arsenic are added to GaP in the appropriate ratio, the band gap of the alloy can be lowered while maintaining a lattice constant equal to that of Si.

Table 2.7 Summary of some key features of relevant III-V/Si multijunction solar cells reported in the literature

Top cell absorber	Buffer/Substrate	E_g [eV]	V_{oc} [V]	W_{oc} [V]
$GaAs_{0.8}P_{0.2}$ * [170]	GaAsP/GaAs	1.69	1.242	0.45
GaAsP [172]	GaAsP/Si	1.71	0.985	0.72
GaAsP [172]	GaAsP/Si	1.78	1.078	0.70
GaAsP [172]	GaAsP/Si	1.79	1.065	0.72
$GaAs_{0.7}P_{0.3}$ [173]	GaAsP/Si	1.77	0.923	0.79
AlGaAs [174]	Si	1.61	1.04	0.57
GaAs [175]	SiGe/Si	1.40	0.98	0.42
GaPNAs [176]	Si	1.80	1.09	0.71

* This solar cell was grown on GaAs substrate with a lattice mismatch of 0.8%. It is included here for reference, as a benchmark for quality.

Unfortunately, this alloy suffers from the same N-related problems, and the $GaN_{0.03}P_{0.9}As_{0.07}$/ Si is not much better than most other Si-based tandems.

2.6.9.2 Mechanically Stacked Structures

Two-terminal, mechanically stacked structures using Si as the bottom cell have achieved limited success [177,178]. Tanabe [177] and coworkers report an AlGaAs/Si tandem cell where the backside of the AlGaAs top cell, a p+-GaAs layer, is directly bonded to the n+ top surface of the Si bottom cell. Under certain conditions, the p+-GaAs/n+-Si interface forms a tunnel junction. The fill factor and V_{oc} for this device where low.

Most recently, Derendorf and coworkers [178] fabricated a Si-based multijunction cell using a GaInP/GaAs tandem directly bonded to a Si bottom cell with the GaAs/Si forming a conductive interface between the two subcells. Immediately prior to the bonding process the two surfaces are exposed to a 0.4 keV argon beam. This produces amorphous surfaces that facilitate direct bonding without the need for excessive pressures or temperatures [179]. The efficiency of this device was limited to 23.6% at 71 suns by a low J_{sc} in the silicon bottom cell due to a low photocurrent in the bottom cell.

In this space, independently connected, 4-terminal devices appear to have an advantage [180]. Indeed, the highest efficiency silicon-based multijunction solar cells are all mechanically stacked structures. The first solar cell to exceed an efficiency of 30% was a mechanically stacked GaAs/Si device by Gee *et al.* [181]. The top GaAs cell was a p+/n device grown on a lightly doped (and optically transparent to sub-bandgap light) n-type GaAs substrate. Front and back contacts were vertically aligned. Several different bottom cells were studied including a 1.15 eV InGaAs cell fabricated by Varian Research Center and a number of different Si concentrator cells. The top and bottom cells were 'glued' together with an uncured silicone adhesive. The highest measured efficiency was 31% at ~400 suns using an interdigitated back contact Si cell from Stanford University.

2.6.10 Epitaxial Liftoff

High quality, single crystal III-V substrates are expensive, particularly relative to the cost of silicon substrates. As such, most of the mechanically stacked structures and some of the InP-based monolithic structures will only be cost competitive if we can find a way to detach the active device and reuse the substrate in a cost effective way. It has also been shown that the substrate, in most cases, reduces the ultimate efficiency of the solar cell by soaking up the photons from luminescence of radiative recombination events in the device [182].

So it no wonder that epitaxial lift off (ELO) processes are becoming a mainstay of the III-V semiconductor industry. Indeed, many of the devices described above are facilitated by some form of ELO. Small and large area devices are now routinely 'removed' from their original substrate and used in a standalone mode or combined with another to form a mechanical stack.

There are several different versions of ELO. One, called Smart Cut, uses an implanted layer of H^+ ions to create weakened layer below the surface of the active device. A simple cleavage process is then used to separate the device from the substrate.

Perhaps the simplest ELO process is one in which a thin, release layer is introduced between the reusable substrate and the epitaxial layer [183]. For most GaAs-based structures a 10–100 nm thick layer of AlAs is used. Aqueous HF will selectively etch away the AlAs, leaving the rest of the structure untouched. Lateral etch rates as high as 30 mm/hr have been achieved [184,185] and world record GaAs single junction [186] and GaInP/GaAs tandem [187] solar cells have been produced using this process. Release layers suitable for InP-based devices are also available [188].

Acknowledgements

Mr. Mario Ochoa and Mr. Vincenzo Orlando are kindly acknowledged for the preparation of some figures in this chapter.

Glossary

List of Acronyms

Acronym	Description
AM	Air mass
AM1.5d	Reference spectrum for direct radiation
AOD	Aerosol optical depth
BSF	Back surface field
ELO	Epitaxial lift-off
IMM	Inverted metamorphic
MBE	Molecular beam epitaxy
MOCVD	Metalorganic chemical vapor deposition (same as MOVPE)
MOVPE	Metalorganic vapor phase epitaxy (same as MOCVD)
MPP	Maximum power point
MJSC	Multijunction solar cell
NREL	National Renewable Energy Laboratory
PAR	Peak to average ratio
PW	Precipitable water
QD	Quantum dot
QW	Quantum well
SCR	Space charge region
SoG	Silicone on glass
SRV	Surface recombination velocity
TEM	Transmission electron microscopy
TJ	Tunnel junction
UMM	Upright metamorphic

List of Symbols

Typical units given in square brackets. If no units are given, variable is dimensionless.

Symbol	Description [Units]
C	Concentration level or concentration factor [suns or ×]. In other texts also noted as X.
C_M	Concentration level at which maximum efficiency is reached [suns or ×]
$E(\lambda)$	Incident spectra [$\mathrm{Wnm}^{-1}\,\mathrm{cm}^{-2}$]
E_g	Bandgap energy [eV]
FF	Solar cell fill factor [%]

(*continued*)

Symbol	Description [Units]
I	Solar cell current [A]
I_{mpp}	Solar cell current at the maximum power point [A]
IQE_B	Internal quantum efficiency of the base layer [%]
IQE_{BSF}	Internal quantum efficiency of the BSF layer [%]
IQE_E	Internal quantum efficiency of the emitter layer [%]
IQE_{scr}	Internal quantum efficiency of the space charge region [%]
IQE_W	Internal quantum efficiency of the window layer [%]
I_{sc}	Solar cell short circuit current [A]
J	Solar cell current density [Acm^{-2}]
J_0	Generic reverse saturation (dark) current density [Acm^{-2}]
J_{01}	Reverse saturation (dark) current density with $m = 1$ [Acm^{-2}]
J_{02}	Reverse saturation (dark) current density with $m = 2$ [Acm^{-2}]
$J_{0,i}$	Reverse saturation (dark) current density in subcell i [Acm^{-2}]
$J_{0,MJ}$	Equivalent reverse saturation (dark) current density for a multijunction solar cell [Acm^{-2}]
J_D	Solar cell dark current density [Acm^{-2}]
J_L	Photogenerated current density [Acm^{-2}]
J_L^*	Photogenerated current density at one-sun reference conditions [Acm^{-2}]
$J_{L,i}^*$	Photogenerated current density at one-sun reference conditions in the i-th subcell [Acm^{-2}]
J_P	Peak current density in a tunnel junction [Acm^2]
J_{sc}	Solar cell short circuit current density [Acm^{-2}]
J_V	Valley current density in a tunnel junction [Acm^2]
$L_B, L_{BSF},$	Minority carrier diffusion length in each layer of the solar cell structure [cm]
L_E, L_W	(W = window; E = emitter; B = base; BSF = Back surface field)
m	Diode ideality factor; in other texts also noted as n
m_{MJ}	Equivalent diode ideality factor of a multijunction solar cell [Acm^{-2}]
$N_B, N_{BSF},$	Doping level in each layer of the solar cell structure [cm^{-3}]
N_E, N_W	(W = window; E = emitter; B = base; BSF = Back surface field)
n_i	Intrinsic carrier concentration of the p/n junction material [cm^{-3}]
P_{mpp}	Solar cell maximum power or power at the maximum power point [W]
R	Solar cell front surface reflectivity [%]
r_{anode}	Tunnel junction anode horizontal resistance [Ωcm^2]
$r_{cathode}$	Tunnel junction cathode horizontal resistance [Ωcm^2]
r_p	Solar cell parallel or shunt resistance [Ωcm^2]
r_s	Solar cell series resistance [Ωcm^2]
r_{TJ}	Tunnel junction equivalent resistance [Ωcm^2]
$S_B, S_{BSF},$	Surface recombination velocity in each layer of the solar cell structure [cm/s]
S_E, S_W	(W = window; E = emitter; B = base; BSF = Back surface field)
SF	Shadowing factor [%]
T_i	Transmissivity until the i-th subcell [%]
V	Solar cell voltage [V]
V_{bi}	p/n junction built in voltage [V]
V_i, V_j	Voltage drop at the i-th or j-th subcell [V]
V_{mpp}	Solar cell voltage at the maximum power point [V]
V_{oc}	Solar cell open circuit voltage [V]
V_{oc}^*	Solar cell open circuit at one-sun reference conditions [V]
V_P	Peak voltage in a tunnel junction [V]
V_t	Thermal voltage ($k \cdot T/q = 0.0259$ at 300 K) [V]
V_V	Valley voltage in a tunnel junction [V]

Symbol	Description [Units]
W_{oc}	Bandgap to V_{oc} distance ($W_{oc}=E_g/q-V_{oc}$) [V]
W_W	Thickness of the window layer [nm]
W_B, W_E	Geometrical thickness of the base and emitter layers [nm]
W'_B, W'_E	Neutral thickness of the base and emitter layers (excluding SCR) [nm]
W_{scr}	Thickness of the space charge region [nm]
W_{BSF}	Thickness of the BSF layer [nm]
α_{BSF}, α_W	Absorption coefficient of the BSF and window materials [cm^{-1}]
α	Absorption coefficient of the p/n junction material [cm^{-1}]
α_i	Absorption coefficient of the i-th layer in the strcuture [cm^{-1}]
ε	Dielectric constant of the p/n junction material
Φ, Φ_i	Photon flux and photon flux at the i-th subcell [nm^{-1}s^{-1}cm^{-2}]
ϕ_s	Angle subtended by the solar disc at the Earth (0.267°) [°]
η	Solar cell efficiency [%]
θ	Angle of incidence of impinging radiation [°]
λ	Wavelength [nm]
$\tau_B, \tau_{BSF},$	Minority carrier lifetime in each layer [ns]
τ_E, τ_W	(W = window; E = emitter; B = base; BSF = Back surface field)

References

1. Fonash, S.J. (2009) *Solar Cell Device Physics*, 2nd edn, Academic Press, Oxford:.
2. Nelson, J. (2003) *The Physics of Solar Cells*, Imperial College Press, London.
3. Hovel, H.J. 1975 *Semiconductors and Semimetals. Volume 11. Solar Cells*, (eds R. K. Willardson and A. C. Beer), Academic Press, New York.
4. Shockley, W. and Queisser, H.J. (1961) Detailed balance limit of efficiency of p-n junction solar cells. *Journal of Applied Physics*, **32**, 510–519.
5. Hirst, L.C. and Ekins-Daukes, N.J. (2011) Fundamental losses in solar cells. *Progress in Photovoltaics: Research and Applications*, **19**, 286–293.
6. Green, M.A. (2001) Limiting photovoltaic monochromatic light conversion efficiency. *Progress in Photovoltaics: Research and Applications*, **9**, 257–261.
7. Martí, A. and Araújo, G.L. (1996) Limiting efficiencies for photovoltaic energy conversion in multigap systems,. *Solar Energy Materials and Solar Cells*, **43**, 203–222.
8. Imenes, A.G. and Mills, D.R. (2004) Spectral beam splitting technology for increased conversion efficiency in solar concentrating systems: a review. *Solar Energy Materials and Solar Cells*, **84**, 19–69.
9. Russo, J.M., Zhang, D., Gordon, M. *et al.* (2014) Spectrum splitting metrics and effect of filter characteristics on photovoltaic system performance. *Optics Express*, vol. **22**, A528–A541.
10. Mojiri, A., Taylor, R., Thomsen, E. and Rosengarten, G. (2013) Spectral beam splitting for efficient conversion of solar energy – A review,' *Renewable and Sustainable Energy Reviews*, **28**, 654–663.
11. Barnett, A., Kirkpatrick, D., Honsberg, C. *et al.* (2009) Very high efficiency solar cell modules. *Progress in Photovoltaics: Research and Applications*, **17** 75–83.
12. Clugston, D.A. and Basore, P.A. (1997) Modelling free-carrier absorption in solar cells. *Progress in Photovoltaics: Research and Applications*, **5**, 229–236.
13. Fraas, L.M., Avery, J.E., Martín, J. *et al.* (1990) Over 35-percent efficient GaAs/GaSb tandem solar cells. *IEEE Transactions on Electron Devices*, **37**, 443–449.
14. Fornari, R. (2015) 1-Epitaxy for energy materials, in *Handbook of Crystal Growth*, 2nd edn, (ed. T.F. Kuech, North-Holland, Boston, 1–49.
15. Sze S.M. and Lee, M.K. (2012) *Semiconductor Devices: Physics and Technology*, 3rd edn, Wiley, Hoboken, N.J.
16. Flamand G. and Poortmans, (2006) Towards highly efficient 4-terminal mechanical photovoltaic stacks. *III-Vs Review*, **19**, 24–27.

17. Bremner, S.P., Levy, and M.Y. and Honsberg, C.B. (2008) Analysis of tandem solar cell efficiencies under AM1.5G spectrum using a rapid flux calculation method. *Progress in Photovoltaics: Research and Applications*, **16**, 225–233.

18. Brown, A.S. and Green, M.A. (2002) Limiting efficiency for current-constrained two-terminal tandem cell stacks. *Progress in Photovoltaics: Research and Applications*, **10**, 299–307.

19. Tobías I. and Luque, A. (2002) Ideal efficiency of monolithic, series-connected multijunction solar cells. *Progress in Photovoltaics: Research and Applications*, **10**, 323–329.

20. Kurtz, S.R., Faine, P. and Olson, J.M. (1990) Modeling of two-junction, series-connected tandem solar cells using top-cell thickness as an adjustable parameter,' *Journal of Applied Physics*, **68**, 1890.

21. Geisz, J.F., Friedman, D.J., Ward, J.S. *et al.* (2008) 40.8% efficient inverted metamorphic triple-junction solar cell with two independently metamorphic junctions. *Applied Physics Letters*, **93**, 123505.

22. King, R.R., Boca, A., Hong, W. *et al.* (2009) 'Band-gap-engineered architectures for high-efficiency multi-junction concentrator solar cells. 24th European Photovoltaic Solar Energy Conference, Hamburg, Germany, pp. 21–25.

23. Van Roosbroeck, W. (1950) Theory of the flow of electrons and holes in germanium and other semiconductors. *Bell System Technical Journal*, **29**, 560–607.

24. Basore, P.A. and Clugston, D.A. (2003) *PC1D 5.9*. Available at: http://www.pvlighthouse.com.au/resources/PC1D/PC1D.aspx

25. Gray, J.L., Wang, X., Chavali, R.V. *et al.* (2011) *ADEPT 2.1*. Available at: https://nanohub.org/resources/10913

26. Fonash, S., Arch, J., Cuiffi, J. *et al.* (2010) *AMPS-1D 2010*. Available at: http://www.ampsmodeling.org/

27. Stangl, R., Leenderz, C. and Haschke, J. (2010) Numerical simulation of solar cells and solar cell characterization methods: the open-source on demand program AFORS-HET, in *Solar Energy*, (ed. R. D. Rugescu), InTech, Rijeka, Croatia, 319–352.

28. *Sentaurus*. Synopsys, Mountain View, CA, USA. Available: http://www.synopsys.com/

29. *Atlas*. Silvaco, Santa Clara, CA, USA. Available: http://www.silvaco.com/

30. *APSYS*. Crosslight. Vancouver, BC, Canada. Available: http://crosslight.com/

31. *COMSOL Multiphysics*. COMSOL, Inc, Burlington, MA, USA. Available: http://www.comsol.com/

32. Michael S. and Bates, A. (2005) The design and optimization of advanced multijunction solar cells using the Silvaco ATLAS software package, *Solar Energy Materials and Solar Cells*, **87**, 785–794.

33. Baudrit M. and Algora, C. (2008) Modeling of GaInP/GaAs dual-junction solar cells including tunnel junction. 33rd IEEE Photovoltaic Specialists Conference, PVSC '08, San Diego, California, pp. 1–5.

34. Li, Z.Q., Xiao, Y.G. and Li, Z.M. (2006) Modeling of multi-junction solar cells by Crosslight APSYS. Proceedings SPIE 6339, pp. 633909–8.

35. Baudrit M. and Algora, C. (2010) Tunnel diode modeling, including nonlocal trap-assisted tunneling: a focus on III-V multijunction solar cell simulation. *IEEE Transactions on Electron Devices*, **57**, 2564–2571.

36. Hermle, M., Létay, G., Philipps, S.P. and Bett, A.W. (2008) Numerical simulation of tunnel diodes for multi-junction solar cells. *Progress in Photovoltaics: Research and Applications*, **16**, 409–418.

37. Wheeldon, J.F., Valdivia, C.E., Walker, A.W. *et al.* (2011) Performance comparison of AlGaAs, GaAs and InGaP tunnel junctions for concentrated multijunction solar cells. *Progress in Photovoltaics: Research and Applications*, **19**, 442–452.

38. Olson, J.M. (2010) Simulation of non-uniform irradiance in multijunction III-V solar cells. 35th IEEE Photovoltaic Specialists Conference (35-PVSC), Honolulu, Hawai'i, 00201–000204.

39. Kanevce, A., Olson, J.M. and Metzger, W.K. (2010) Numerical simulations of triple-junction GaInP/GaAs/Ge solar cells to provide insight into fill-factor losses at high concentration. 35th IEEE Photovoltaic Specialists Conference (PVSC'10), Honolulu, Hawai'i 002066–002069.

40. Olson, J. M., Friedman, D.J. and Kurtz, S. (2003) High-efficiency III-V multijunction solar cells, in *Handbook of Photovoltaic Science and Engineering* (eds A. Luque and S. Hegedus), Wiley, Hoboken, NJ, Chapter 9.

41. Espinet-González, P. (2012) Advances in the Modeling, Characterization and Reliability of Concentrator Multijunction Solar Cells. PhD Thesis. Instituto de Energía Solar, Universidad Politécnica de Madrid.

42. Araujo, G.L., Cuevas, A. and Ruiz, J.M. (1986) The effect of distributed series resistance on the dark and illuminated Current-Voltage characteristics of solar cells. *IEEE Transactions on Electron Devices*, **33**, 391–401.

43. Nielsen, L.D. (1982) Distributed series resistance effects in solar cells. *IEEE Transactions on Electron Devices*, **29**, 821–827.

44. Araki K. and Yamaguchi, M. (2003) Extended distributed model for analysis of non-ideal concentration operation. *Solar Energy Materials and Solar Cells*, **75**, 467–473.

45. Nishioka, K., Takamoto, T., Nakajima, W. *et al.* (2003) Analysis of triple-junction solar cell under concentration by SPICE. 3rd World Conference on Photovoltaic Energy Conversion, 869–872.
46. Galiana, B., Algora, C., Rey-Stolle, I. and Vara, I.G. (2005) A 3-D model for concentrator solar cells based on distributed circuit units. *IEEE Transactions on Electron Devices*, **52**, 2552–2558.
47. Steiner, M., Guter, W., Peharz, G. *et al.* (2012) A validated SPICE network simulation study on improving tunnel diodes by introducing lateral conduction layers. *Progress in Photovoltaics: Research and Applications*, **20**, 274–283.
48. Espinet, P., García, I., Rey-Stolle, I., Algora, C. and Baudrit, M. (2011) Extended description of tunnel junctions for distributed modeling of concentrator multi-junction solar cells, *Solar Energy Materials and Solar Cells*, **95**, 2693–2697.
49. García, I., Espinet-González, P., Rey-Stolle, I. and Algora, C. (2011) Analysis of chromatic aberration effects in triple-junction solar cells using advanced distributed models. *IEEE Journal of Photovoltaics*, **1**, 219–224.
50. Espinet-González, P., Rey-Stolle, I., Algora, C. and García, I. (2015) Analysis of the behavior of multijunction solar cells under high irradiance Gaussian light profiles showing chromatic aberration with emphasis on tunnel junction performance. *Progress in Photovoltaics: Research and Applications*, **23** (6), 743–753.
51. Espinet-González, P., Mohedano, R., García, I. *et al.* (2012) Triple-junction solar cell performance under Fresnel-based concentrators taking into account chromatic aberration and off-axis operation. *AIP Conference Proceedings*, **1477**, 81–84.
52. Jia, J., Suarez, F., Bilir, T., Sabnis, V. and Harris, J. (2014) 3-D modeling of luminescent coupling effects in multijunction concentrator solar cells. *AIP Conference Proceedings*, **1616**, 3–7.
53. Davis R. and Knight, J.R. (1975) Operation of GaAs solar cells at high solar flux density. *Solar Energy*, **17**, 145.
54. Stryi-Hipp, G., Schoenecker, A., Schitterer, K., Bucher, K. and Heidler, K. (1993) Precision spectral response and I-V characterisation of concentrator cells. 23rd IEEE Photovoltaic Specialists Conference, Louisville KY, USA pp. 303–308.
55. Luque A. and Araújo, G.L. (1990) *Physical Limitations to Photovoltaic Energy Conversion*, Taylor and Francis, Oxford.
56. de Vos, A. (1992) *Endoreversible Thermodynamics of Solar Energy Conversion*. Clarendon Press.
57. Sánchez, E. and Araújo, G.L. (1987) On the analytical determination of solar cell fill factor and efficiency. *Solar Cells*, **20**, 1–11.
58. Espinet-González, P., Rey-Stolle, I., Ochoa, M. *et al.* (2014) Analysis of perimeter recombination in the subcells of GaInP/GaAs/Ge triple-junction solar cells. *Progress in Photovoltaics: Research and Applications*, **23** (7), 874–882.
59. Algora, C. (2007) Very high concentration challenges of III-V multijunction solar cells., in *Concentrator Photovoltaics*, (eds A. Luque and V. Andreev), Springer, pp. 89–111.
60. Dimroth, F., Grave, M., Beutel, P. *et al.* (2014) Wafer bonded four-junction GaInP/GaAs//GaInAsP/GaInAs concentrator solar cells with 44.7% efficiency. *Progress in Photovoltaics: Research and Applications*, **22**, 277–282.
61. Handy, R.J. (1967) Theoretical analysis of the series resistance of a solar cell. *Solid-State Electronics*, **10**, 765–775.
62. de Vos, A. (1984) The distributed series resistance problem in solar cells. *Solar Cells*, **12**, 311–327.
63. Andreev, V.M., Romero, R. and Sulima, O.V. (1984) An efficient circular contact grid for concentrator solar cells. *Solar Cells*, **11**, 197–210.
64. Rumyantsev, V.D. and Rodriguez, J.A. (1990) Method of calculating the distributed and lumped components of the resistance in solar cells. *Solar Cells*, **28**, 241–252.
65. Algora, C. and Díaz, V. (2000) Influence of series resistance on guidelines for manufacture of concentrator p-on-n GaAs solar cells. *Progress in Photovoltaics: Research and Applications*, **8**, 211–225.
66. Galiana, B., Rey-Stolle, I., Baudrit, M., García, I. and Algora, C. (2006) A comparative study of BSF layers for GaAs-based single-junction or multijunction concentrator solar cells. *Semiconductor Science and Technology*, **21**, 1387.
67. Hoheisel, R. and Bett, A.W. (2012) Experimental Analysis of Majority Carrier Transport Processes at Heterointerfaces in Photovoltaic Devices. *IEEE Journal of Photovoltaics*, **2**, 398–402.
68. Olson, J.M., Steiner, M.A. and Kanevce, A. (2011) Using measurements of fill factor at high irradiance to deduce heterobarrier band offsets. 37th IEEE Photovoltaic Specialists Conference (PVSC'11), Seattle WA, USA, pp. 003754–003757.
69. B. Galiana, C. Algora, and I. Rey-Stolle, 'Comparison of 1D and 3D analysis of the front contact influence on GaAs concentrator solar cell performance,' *Solar Energy Materials and Solar Cells*, vol. **90**, pp. 2589–2604, 2006.

70. Williams, R. (1990) *Modern GaAs Processing Methods*. Artech House, Norwood MA, USA.

71. Algora, C., Rey-Stolle, I., Galiana, B. *et al.* (2006) Strategic options for a LED-Like Approach in III-V concentrator photovoltaics. IEEE 4th World Conference on Photovoltaic Energy Conversion, Waikoloa HI, USA, pp. 741–744.

72. Barrigón, E., García, I., Barrutia, L., Rey-Stolle, I. and Algora, C. (2014) Highly conductive p $^{++}$AlGaAs/n $^{++}$GaInP tunnel junctions for ultra-high concentrator solar cells. *Progress in Photovoltaics: Research and Applications*, **22**, 399–404.

73. King, R.R., Fetzer, C.M., Colter, P.C. *et al.* (2002) High-efficiency space and terrestrial multijunction solar cells through bandgap control in cell structures. 29th IEEE Photovoltaic Specialists Conference, New Orleans, USA, pp. 776–781.

74. García, I., Rey-Stolle, I. and Algora, C. (2012) Performance analysis of AlGaAs/GaAs tunnel junctions for ultra-high concentration photovoltaics. *Journal of Physics D: Applied Physics*, **45**, 045101.

75. Jung, D., Parker, C.A., Ramdani, J. and Bedair, S.M. (1993) AlGaAs/GaInP heterojunction tunnel diode for cascade solar cell application. *Journal of Applied Physics*, **74**, 2090–2093.

76. Sasaki, K., Agui, T., Nakaido, K. *et al.* (2013) Development of InGaP/GaAs/InGaAs inverted triple junction concentrator solar cells. *AIP Conference Proceedings*, **1556**, 22–25.

77. Andreev, V.M., Ionova, E.A., Larionov, V.R. *et al.* (2006) Tunnel diode revealing peculiarities at I-V measurements in multijunction III-V solar cells. 4th World Conference on Photovoltaic Energy Conversion, pp. 799–802.

78. Braun, A., Hirsch, B., Katz, E.A., *et al.* (2009) Localized irradiation effects on tunnel diode transitions in multi-junction concentrator solar cells. *Solar Energy Materials and Solar Cells*, **93**, 1692–1695.

79. Herrero, R., Victoria, M., Domínguez, C. *et al.* (2012) Concentration photovoltaic optical system irradiance distribution measurements and its effect on multi-junction solar cells. *Progress in Photovoltaics: Research and Applications*, **20**, 423–430.

80. LaRue, R.A., Borden, P.G. and Gregory, P.E. (1981) A distributed resistance model of an AlGaAs/GaAs concentrator solar cell illuminated with a curved groove Fresnel lens. *IEEE Electron Device Letters*, **2**, 41–43.

81. Garcia, I., Algora, C., Rey-Stolle, I. and Galiana, B. (2008) Study of non-uniform light profiles on high concentration III-V solar cells using quasi-3D distributed models. 33rd IEEE Photovoltaic Specialists Conference (PVSC '08), San Diego CA, USA, pp. 1–6.

82. García, I. (2010) Development of GaInP/GaAs dual-junction solar cells for high light concentrations. PhD Thesis. Instituto de Energía Solar, Universidad Politécnica de Madrid.

83. Kurtz, S.R., Friedman, D.J. and Olson, J.M. (1994) The effect of chromatic aberrations on two-junction, two-terminal, devices on a concentrator system. 1st World Conference on Photovoltaic Energy Conversion – 24th IEEE Photovoltaic Specialists Conference, waikoloa HI, USA, pp. 1791–1794.

84. James, L.W. (1994) Effects of concentrator chromatic aberration on multi-junction cells. 1st World Conference on Photovoltaic Energy Conversion – 24th IEEE Photovoltaic Specialists Conference, pp. 1799–1802.

85. Kurtz S.R. and O'Neill, M.J. (1996) Estimating and controlling chromatic aberration losses for two-junction, two-terminal devices in refractive concentrator systems. 25th IEEE Photovoltaic Specialists Conference, Washington DC, pp. 361–364.

86. Garcia, I., Espinet-González, P., Rey-Stolle, I., Barrigón, E. and Algora, C. (2011) Extended Triple-Junction Solar Cell 3D Distributed Model: Application to Chromatic Aberration-Related Losses. *AIP Conference Proceedings*, **1407**, pp. 13–16.

87. Cotal H. and Sherif, R. (2005) The effects of chromatic aberration on the performance of GaInP/GaAs/Ge concentrator solar cells from Fresnel optics. 31st IEEE Photovoltaic Specialists Conference, Lake Buena Vista FL, USA, pp. 747–750.

88. Victoria, M., Herrero, R., Domínguez, C. *et al.* (2013) Characterization of the spatial distribution of irradiance and spectrum in concentrating photovoltaic systems and their effect on multi-junction solar cells. *Progress in Photovoltaics: Research and Applications*, **21**, 308–318.

89. Sharpe A.M. and Eames, P.C. (2013) Modelling of multijunction cell temperature distributions subject to realistic operating conditions. *AIP Conference Proceedings*, **1556**, 142–146.

90. Ota Y. and Nishioka, K. (2014) Estimation of thermal stress in concentrator cells using structural mechanics simulation. *AIP Conference Proceedings*, **1616**, 25–28.

91. Faine, P., Kurtz, S.R., Riordan, C. and Olson, J.M. (1991) The influence of spectral solar irradiance variations on the performance of selected single-junction and multijunction solar cells. *Solar Cells*, **31**, 259–278.

92. Kurtz, S.R., Myers, D. and Olson, J.M. (1997) Projected performance of three- and four-junction devices using GaAs and GaInP. 26th IEEE Photovoltaic Specialists Conference, Anaheim CA, USA, pp. 875–878.

93. Geisz, J.F., García, I., McMahon, W.E. et al. (2015) Energy yield determination of concentrator solar cells using laboratory measurements. 11th International Concentrating Photovoltaics Conference, Aix-les-Bains, France, pp. 040005-1–040005-8.

94. Núñez, R., Antón, I., Askins, S. and Sala, G. (2014) Atmospheric parameters, spectral indexes and their relation to CPV spectral performance. AIP Conference Proceedings, 1616, 290–293.

95. King, R.R., Bhusari, D., Boca, A. et al. (2011) Band gap-voltage offset and energy production in next-generation multijunction solar cells. Progress in Photovoltaics: Research and Applications, 19, 797–812.

96. Espinet-González, P., Victoria, M., Rey-Stolle, I. et al. (2014) Keys for the joint design of the optics and the solar cell in a CPV system. 40th IEEE Photovoltaic Specialists Conference (40-PVSC), Denver, CO, USA (in press).

97. Jaus J. and Gueymard, C.A. (2012) Generalized spectral performance evaluation of multijunction solar cells using a multicore, parallelized version of SMARTS. AIP Conference Proceedings, 1477, 122–126.

98. Gueymard, C.A. (2001) Parameterized transmittance model for direct beam and circumsolar spectral irradiance. Solar Energy, 71, 325–346.

99. Olson, J.M., Gessert, T. and Al-Jassim, M.M. (1985) GaInP$_2$/GaAs: A current- and lattice-matched tandem cell with a high theoretical efficiency, in Proceedings 18th IEEE Photovoltaic Specialists Conference, Las Vegas 21–25 October, pp. 552–555.

100. Olson, J.M., Kurtz, S.R., Kibbler, A.E. and Faine, P. (1990) A 27.3% efficient $Ga_{0.5}In_{0.5}P$/GaAs tandem solar cell. Applied Physics Letters, 56, 623–625.

101. Bertness, K.A., Kurtz, S.R., Friedman, D.J., et al. (1994) 29.5%-Efficient GaInP/GaAs Tandem Solar Cells, Applied Physics Letters, 65, 989–991.

102. Ikeda, M., Nakano, K., Mori, Y., Kaneko, K. and Watanabe, N. (1986) MOCVD growth of AlGaInP at atmospheric pressure using triethylmetals and phosphine. Journal of Crystal Growth, 77, 380–385.

103. Ohba, Y., Ishikawa, M., Sugawara, H., Yamamoto, M. and Nakanisi, T. (1986) Growth of high-quality InGaAlP epilayers by MOCVD using methyl metalorganics and their application to visible semiconductor lasers. Journal of Crystal Growth, 77, 374–379.

104. Ohba, Y., Nishikawa, Y., Nozaki, C., Sugawara, H. and Nakanisi, T. (1988) A study of p-type doping for AlGaInP grown by low-pressure MOCVD, Journal of Crystal Growth, 93, 613–617.

105. Kibbler, A.E., Kurtz, S.R. and Olson, Carbon J.M. (1991) Doping and etching of MOCVD-grown GaAs, InP, and related ternaries using CCl$_4$. Journal of Crystal Growth, 109, 258.

106. Kurtz, S.R., Olson, J.M., Kibbler, A.E. and Asher, S. (1992) Model for incorporation of zinc in MOCVD growth of $Ga_{0.5}In_{0.5}P$. Proceedings of the 4th International Conference on Indium Phosphide and Related Materials, Newport RI, USA 21–24 April, pp. 109–112.

107. Lammasniemi, J., Kazantsev, A.B., Jaakkola, R. et al. (1998) Characteristics of the first GaInP/GaAs cascade solar cells grown by a production-scale MBE system, in Second World Conference and Exhibition on Photovoltaic Energy Conversion, Vienna, p. 1177.

108. Friedman D.J. and Olson, J.M. (2001) Analysis of Ge junctions for GaInP/GaAs/Ge three-junction solar cells. Progress in Photovoltaics, 9, 179–189.

109. Tobin, S.P., Vernon, S.M., Bajgar, C., Haven, J.V. and Davis, S.E. (1985) MOCVD growth of AlGaAs and GaAs on Ge substrates for high efficiency tandem cell applications, in 18th IEEE Photovoltaic Specialists Conference (PVSC'85), Las Vegas NV, pp. 134–139.

110. Tobin, S.P., Vernon, S.M., Bajgar, C. et al. (1988) High efficiency GaAs/Ge monolithic tandem solar cells, in 20th IEEE Photovoltaic Specialists Conference, Las Vegas 26–30 September, pp. 405–410.

111. Chen, J.C., Ristow, M.L., Cubbage, J.I. and Werthen, J.G. (1991) Effects of Metalorganic Chemical Vapor Deposition Growth Conditions on the GaAs/Ge Solar Cell Properties. Applied Physics Letters, 58, 2282–2284.

112. Gomyo, A., Suzuki, T., Kobayashi, K., et al. (1987) Evidence for the existence of an ordered state in $Ga_{0.5}In_{0.5}P$ grown by metalorganic vapor phase epitaxy and its relation to band-gap energy, Applied Physics Letters, 50, 673.

113. Gomyo, A., Suzuki, T. and Iijima, S. (1988) Observation of Strong Ordering in $Ga_xIn_{1-x}P$ alloy semiconductors, Physics Review Letters, 60, 2645–2648.

114. Kondow, M., Kakibayashi, H. and Minagawa, S. (1988) Ordered structure in OMVPE-grown $Ga_{0.5}In_{0.5}P$. Journal of Crystal Growth, 99, 291.

115. Shurtleff, J.K., Lee, R.T., Fetzer, C.M. and Stringfellow, G.B. (1999) Band-gap control of GaInP using Sb as a surfactant. Physics Review Letters, 75, 1914–1916.

116. Olson, J.M., McMahon, W.E. and Sarah, K. (2006) Effect of Sb on the properties of GaInP top cells, in 4th World Conference on Photovoltaic Energy Conversion, Waikoloa HI, USA, pp. 787–779.

117. Kurtz, S.R., Olson, J.M., Arent, D.J., Bode, M.H. and Bertness, K.A. (1994) Low-Band-Gap $Ga_{0.5}In_{0.5}P$ Grown on (511)B GaAs Substrates. *Journal of Applied Physiccs*, **75**, 5110–5113.

118. Kurtz, S.R., Olson, J.M., Arent, D.J., Kibbler, A.E. and Bertness, K.A. (1993) Competing kinetic and thermodynamic processes in the growth and ordering of $Ga_{0.5}In_{0.5}P$, in *Common Themes and Mechanisms of Epitaxial Growth*, Pittsburgh, PA, pp. 83–88.

119. Kurtz, S.R., Olson, J.M., Friedman, D.J., Kibbler, A.E. and Asher, S. (1994) Ordering and Disordering of Doped $Ga_{0.5}In_{0.5}P$. *Journal of Electronc Matererials*, **23**, 431–435.

120. Bett, A.W., Dimroth, F., Guter, W. *et al.* (2009) Highest Efficiency Multi-Junction Solar Cell for Terrestrial and Space Applications, in 24th European Photvoltaic Solar Energy Conference and Exhibition, Hamburg, Germany, pp. 1–6.

121. Fetzer, C.M., King, R.R., Colter, P.C. *et al.* (2004) High-efficiency metamorphic GaInP/GaInAs/Ge solar cells grown by MOVPE. *Journal of Crystal Growth*, **261**, 341–348.

122. King, R., Fetzer, C., Edmondson, K. *et al.* (2004) Metamorphic III-V materials, sublattice disorder, and multijunction solar cell approaches with over 37% efficiency, in 19th European Photovoltaic Solar Energy Conference and Exhibition, Paris, p. 11.

123. King, R.R., Bhusari, D., Larrabee, D. *et al.* (2012) Solar cell generations over 40% efficiency, *Progress in Photovoltaics: Research and Applications*, **20** (6), 801–815.

124. Guter, W., Schöne, J., Philipps, S.P. *et al.* (2009) Current-matched triple-junction solar cell reaching 41.1% conversion efficiency under concentrated sunlight. *Applied Physics Letters*, **94**, 223–504.

125. Schultz, J.C., Klausmeierbrown, M.E., Ristow, M.L. *et al.* (1993) Development of High-Quantum-Efficiency, Lattice-Mismatched, 1.0-eV GaInAs Solar Cells. *Journal of Electronc Matererials*, **22**, 755–761.

126. Geisz, J.F., Duda, A., France, R.M., *et al.* (2012) Optimization of 3-junction inverted metaphorphic solar cells for high-temperature and high-concentration operation, in 8th International Conference on Concentrating Photovoltaic Systems, Toledo, Spain, *AIP Conference Proceedings*, **1477** (1), 44–48.

127. Green, M.A., Emery, K., Hishikawa, Y., Warta, W. and Dunlop, E.D. (2014) Solar cell efficiency tables (version 43). *Progress in Photovoltaics: Research and Applications*, **22**, 1–9.

128. France, R.M., Geisz, J.F., Steiner, M.A., *et al.* (2012) Pushing inverted metamorphic multijunction solar cells toward higher efficiency at realistic operating conditions, in Photovoltaic Specialists Conference (PVSC), Volume 2, 2012 IEEE 38th, Austin TX, USA 3–8 June, pp. 1–6.

129. Chiu, P., Wojtczuk, S., Zhang, X. *et al.* (2011) 42.3% Efficient InGaP/GaAs/InGaAs concentrators using bifacial epigrowth, in Photovoltaic Specialists Conference (PVSC), 2011 37th IEEE, Seattle WA, USA 19–24 June, pp. 000771–000774.

130. Kurtz, S.R., Allerman, A.A., Jones, E.D. *et al.* (1999) InGaAsN solar cells with 1.0 eV band gap, lattice matched to GaAs. *Applied Physics Letters*, **74**, 729–731.

131. Friedman, D.J., Geisz, J.F., Kurtz, S.R. and Olson, J.M. (1998) 1-eV solar cells with GaInNAs active layer, *Journal of Crystal Growth*, **195**, 409–415.

132. Weyers, M., Sato, M. and Ando, H. (1992) Red Shift of Photoluminescence and Absorption in Dilute GaAsN Alloy Layers. *Japanese Journal of Applied Physics*, **31**, 853–855.

133. Kurtz, S., Geisz, J.F., Friedman, D.J. *et al.*, (2000) Modeling of Electron Diffusion Length in GaInAsN Solar Cells, in 28th IEEE Photovoltaic Specialists Conference, Anchorage, Alaska, pp. 1210–1213.

134. Kurtz, S., Johnston, S. and Branz, H.M. (2005) Capacitance-spectroscopy identification of a key defect in N-degraded GaInNAs solar cells. *Applied Physics Letters*, **86**, 113506.

135. Kurtz, S., Johnston, S.W., Geisz, J.F., Friedman, D.J. and Ptak, A.J. (2005) Effect of nitrogen concentration on the performance of $Ga_xIn_{1-x}N_yAs_{1-y}$ solar cells, in 31st IEEE Photovoltaic Specialists Conference (PVSC'05), Orlando, Florida, pp. 595–598.

136. Kurtz, S.R., Allerman, A.A., Jones, E.D., Klem, J.F. and Seager, C.H. (1999) Minority carrier lifetimes, defects, and solar cells of InGaAsN, lattice-matched to GaAs, in 195th Meeting of the Electrochemical Society, Seattle, WA, pp. 163–169.

137. Kaplar, R.J., Kwon, D., Ringel, S.A. *et al.* (2001) Deep levels in p- and n-type InGaAsN for high-efficiency multijunction III-V solar cells. *Solar Energy Materials Solar Cells*, **69**, 85–91.

138. Jackrel, D.B., Bank, S.R., Yuen, H.B. *et al.* (2007) Dilute nitride GaInNAs and GaInNAsSb solar cells by molecular beam epitaxy. *Journal of Applied Physics*, **101**, 114916-1–114916-8.

139. Derkacs, D., Jones-Albertus, R., Suarez, F. and Fidaner, O. (2012) Lattice-matched multijunction solar cells employing a 1eV GaInNAsSb bottom cell. *Journal of Photonics for Energy*, **2**, 021805-1–8.

140. Kim, T.W., Garrod, T.J., Kim, K. *et al.* (2012) Narrow band gap (1-eV) InGaAsSbN solar cells grown by metalorganic vapor phase epitaxy. *Applied Physics Letters*, **100**, 121120-1–121120-4.

141. Antolin, E., Martí, A., Farmer, C.D. *et al.* (2010) Reducing carrier escape in the InAs/GaAs quantum dot intermediate band solar cell. *Journal of Applied Physics*, **108**, 064513–7.

142. Martí, A., Cuadra, L. and Luque, A. (2001) Partial filling of a quantum dot intermediate band for solar cells. *IEEE Transactions on Electron Devices*, **48**, 2394–2399.

143. Sogabe, T., Shoji, Y., Ohba, M. *et al.* (2014) Intermediate-band dynamics of quantum dots solar cell in concentrator photovoltaic modules. *Scientific Reports*, **4**, 47921-1–47921-7.

144. Nozik, A.J. (2002) Quantum dot solar cells. *Phyisca E*, **14**, 115.

145. Kitatani, T., Yazawa, Y., Watahiki, S. *et al.* (1996) Photocurrent and photoluminescence in InGaAs/GaAs multiple quantum well solar cells. *Japanese Journal of Applied Physics, Pt 1, Japanese Journal of Applied Physics*, **35**, 4371–4372.

146. EkinsDaukes, N.J., Barnes, J.M., Barnham, K.W. *et al.* (2001) Strained and strain-balanced quantum well devices for high-efficiency tandem solar cells. *Solar Energy Materials Solar Cells*, **68**, 71–87.

147. Yang M.J. and Yamaguchi, M. (2000) Properties of GaAs/InGaAs quantum well solar cells under low concentration operation. *Solar Energy Materials Solar Cells*, **60**, 19–26.

148. Browne, B., Lacey, J., Tibbits, T. *et al.* (2013) Triple-junction quantum-well solar cells in commercial production, in 9th International Conference on Concentrator Photovoltaic Systems (CPV-9), Miyazaki Jampan 15–17 April, AIP Conference Proceedings, pp. 3–5.

149. Valdivia, C.E., Chow, S., Fafard, S. *et al.* (2010) Measurement of high efficiency 1 cm^2 AlGaInP/InGaAs/Ge solar cells with embedded InAs quantum dots at up to 1000 suns continuous concentration, in Photovoltaic Specialists Conference (PVSC), 2010 35th IEEE, pp. 001253–001258.

150. Jolley, G., Lu, H.F., Fu, L., Tan, H.H. and Jagadish, C. (2010) Electron-hole recombination properties of In0.5Ga0.5As/GaAs quantum dot solar cells and the influence on the open circuit voltage, *Applied Physics Letters*, **97**, p. 123505.

151. Sablon, K.A., Little, J.W. Olver, K.A. Z. *et al.* (2010) Effects of AlGaAs energy barriers on InAs/GaAs quantum dot solar cells, *Journal of Applied Physics*, **108**, p. 4.

152. Bailey, C.G., Forbes, D.V., Raffaelle, R.P. and Hubbard, S.M. (2011) Near 1 V open circuit voltage InAs/GaAs quantum dot solar cells. *Applied Physics Letters*, **98**, 163105-1–163105-3.

153. Tanabe, K., Guimard, D., Bordel, D. and Arakawa, Y. (2012) High-efficiency InAs/GaAs quantum dot solar cells by metalorganic chemical vapor deposition. *Applied Physics Letters*, **100**, 3905.

154. Polly, S.J., Forbes, D.V., Driscoll, K., Hellstrom, S. and Hubbard, S.M. (2014) Delta-doping effects on quantum-dot solar cells. *IEEE Journal of Photovoltaics*, **4**, 1079–1085.

155. Tomić, S., Sogabe, T. and Okada, Y. (2014) In-plane coupling effect on absorption coefficients of InAs/GaAs quantum dots arrays for intermediate band solar cell, *Progress in Photovoltaics: Research and Applications*, **23** (5), 546–558.

156. Konle, J., Presting, H. and Kibbel, H. (2003) Self-assembled Ge-islands for photovoltaic applications, *Physica E*, **16**, 596–601.

157. Walker, A.W., Thériault, O., Wheeldon, J.F. and Hinzer, K. (2013) The effects of absorption and recombination on quantum dot multijunction solar cell efficiency. *IEEE Journal of Photovoltaics*, **3**, 1118.

158. EkinsDaukes, N.J., Barnham, K.W., Connolly, J.P. *et al.* (1999) Strain-balanced GaAsP/InGaAs quantum well solar cells, *Applied Physics Letters*, **75**, 4195–4197.

159. Lu, H.F., Fu, L., Jolley, G. *et al.* (2011) (2011) Temperature dependence of dark current properties of InGaAs/GaAs quantum dot solar cells. *Applied Physics Letters*, **98**, 3509.

160. Friedman D.J. and Kurtz, S.R. (2002) Breakeven Criteria for the GaInNAs junction in GaInP/GaAs/GaInNAs/Ge four-junction solar cells. *Progress in Photovoltaics*, **10**, 331–344.

161. Kirk, A.P. (2011) High efficacy thinned four-junction solar cell. *Semiconductor Science and Technology*, **26**, 125013.

162. France, R.M., Geisz, J.F., Garcia, I. *et al.* (2014) Quadruple junction inverted metamorphic concentrator devices *IEEE Journal of Photovoltaics*, **5**(1), 432–437, DOI: 10.1109/JPHOTOV.2014.2364132.

163. Miller, N., Patel, P., Struempel, C. *et al.* (2014) Terrestrial concentrator four-junction inverted metamorphic solar cells with efficiency >45%, in 40th IEEE Photovoltaic Specialists Conference (PVSC'14), Denver.

164. Chiu, P.T., Law, D.C. Woo, R.L. *et al.* (2014) Direct Semiconductor Bonded 5J Cell for Space and Terrestrial Applications. *IEEE Journal of Photovoltaics*, **4**, 493.

165. Sheng, X., Bower, C.A., Bonafede, S. *et al.* (2014) Printing-based assembly of quadruple-junction four-terminal microscale solar cells and their use in high-efficiency modules. *Nature Materials*, **13**, 593–598.
166. Carlson, A., Kim-Lee, H.J., Wu, J. *et al.* (2011) Shear-enhanced adhesiveless transfer printing for use in deterministic materials assembly. *Applied Physics Letters*, **98**, 264104-1–264104-3.
167. Wanlass, M.W., Coutts, T.J., Ward, J.S. *et al.* (1991) Advanced high-efficiency concentrator tandem solar cells, in 22nd IEEE Photovoltaic Specialists Conference (PVSC'91), Las Vegas NV, pp. 38–45.
168. Vernon, S.M., Spitzer, M.B., Tobin, S.P. and Wolfson, R.G. (1984) Heteroepitaxial (Al)GaAs structures on Ge and Si for advanced high-efficiency solar cells, in 17th IEEEE Photovoltaics Specialist Conference, Kissimee FL, USA 1–4 May, p. 434.
169. Soga, T., Kato, T., Yang, M. Umeno, M. and Jimbo, T. (1995) High efficiency AlGaAs/Si monolithic tandem solar cell grown by metalorganic chemical vapor deposition. *Journal of Applied Physics*, **78**, 4196–4199.
170. Vernon, S.M., Tobin, S.P., Haven, V.E. *et al.* (1987) Development of high-efficiency GaAsP solar cells on compositionally graded buffer layers, in 19th IEEE *Photovoltaic Specialists Conference* (PVSC'87), pp. 108–112.
171. Yamaguchi M. and Amano, C. (1985) Efficiency calculations of thin-film GaAs solar cells on Si substrates, *Journal of Applied Physics*, **58**, 3601.
172. Geisz, J.F., Olson, J.M., Romero, M.J., Jiang, C.S. and Norman, A.G. (2006) Lattice-mismatched GAASP solar cells grown on silicon by OMVPE, in *4th World Conference on Photovoltaic Energy Conversion*, Hawaii, pp. 172–775.
173. Hayashi, K., Soga, T., Nishikawa, H., Jimbo, T. and Umeno, M. (1994) MOCVD growth of GaAsP on Si for tandem solar cell applications, in 1st World Conf. on Photovoltaic Energy Conversion, Hawaii, pp. 1890–1893.
174. Soga, T., Baskar, K., Kato, T. Jimbo, T. and Umeno, M. (1997) MOCVD growth of high efficiency current-matched AlGaAs/Si tandem solar cell. *Journal of Crystal Growth*, **174**, 579–584.
175. Andre, C.L., Khan, A., Gonzalez, M. *et al.* (2002) Impact of threading dislocations on both n/p and p/n single junction gaas cells grown on Ge/SiGe/Si substrates, in 29th IEEE Photovoltaic Specialists Conference, New Orleans, Louisiana, pp. 1043–1046.
176. Geisz, J.F., Olson, J.M., Friedman, D.J. *et al.* (2005) Lattice-matched GaNPAs-on-silicon tandem solar cells, in *Proceedings of the 31st IEEE Photovoltaic Specialists Conference*, Orlando, Florida, pp. 695–698.
177. Tanabe, K., Watanabe, K. and Arakawa, Y. (2012) III-V/Si hybrid photonic devices by direct fusion bonding. *Scientific Reports*, **2**, 349-1–349-6.
178. Derendorf, K., Essig, S. Oliva. E. *et al.* (2013) Fabrication of GaInP/GaAs//Si solar cells by surface activated direct wafer bonding. *IEEE Journal of Photovoltaics*, **3**, 1423–1428.
179. Essig S. and Dimroth, F. (2013) Fast atom beam activated wafer bonds between n-si and n-gaas with low resistance. *ECS Journal of Solid State Science and Technology*, **2**, Q178–Q181.
180. White, T.P., Lal, N.N. and Catchpole, K.R. (2014) Tandem solar cells based on high-efficiency c-Si bottom cells: top cell requirements for >30% Efficiency. *IEEE Journal of Photovoltaics*, **4**, 208–214.
181. Gee J.M. and Virshup, G.F. (1988) A 31%-efficient GaAs/silicon mechanically stacked, multijunction concentrator solar cell, in Photovoltaic Specialists Conference, 1988., Conference Record of the Twentieth IEEE, vol.1, pp. 754–758.
182. Sheng, X., Yun, M.H., Zhang, C. *et al.* (2015) Device architectures for enhanced photon recycling in thin-film multijunction solar cells. *Advanced Energy Materials*, **5**, 1400919-1–14900919-6.
183. Yablonovitch, E., Gmitter, T., Harbison, J.P. and Bhat, R. (1987) Extreme selectivity in the lift-off of epitaxial GaAs films. *Applied Physics Letters*, **51**, 2222–2224.
184. Schermer, J.J., Bauhuis, G.J., Mulder, P. *et al.* (2000) High rate epitaxial lift-off of InGaP films from GaAs substrates. *Applied Physics Letters*, **76**, 2131–2133.
185. Schermer, J.J. Mulder, P., Bauhuis, G.J. *et al.* (2005) Epitaxial lift-off for large area thin film III/V devices. *Physica Status Solidi (A)*, **202**, 501–508.
186. Kayes, B.M., Nie, H., Twist, R. *et al.* (2011) 27.6% Conversion efficiency, a new record for single-junction solar cells under 1-sun illumination, in Photovoltaics Specialists Conference, Seattle.
187. Kayes, B.M., Ling, Z., Twist, R., Ding, I.K. and Higashi, G.S. (2014) Flexible thin-film tandem solar cells with >30% efficiency. *Journal of Photovoltaics, IEEE*, **4**, 729–733.
188. Aspar, B., Moriceau, H., Jalaguier, E. *et al.* (2001) The generic nature of the Smart-Cut((R)) process for thin film transfer. *Journal of Electronic Materials*, **30**, 834–840.

3

Emerging High Efficiency Concepts for Concentrator Solar Cells

Ignacio Tobías and Antonio Luque

Instituto de Energía Solar, Universidad Politécnica de Madrid, Spain

3.1 Introduction

Solar energy, because of its magnitude, constancy, ubiquity and cleanliness, is certainly the best option for the supply of the ever-growing energetic needs of mankind. A possible drawback is its diluted nature, which would render it impractical at a large scale because of land utilization unless it can be converted with a sufficiently high efficiency. Even at the current efficiency levels of commercial *photovoltaics* (PV) based on silicon flat modules, land occupation is not a fundamental issue [1].

High efficiency, together with low cost, is also important because of economic reasons which used to constitute the main barrier to the penetration of photovoltaics. To overcome this barrier and produce a massive impact of PV on the world energy system, a promising route relies on very high efficiency solar cells in conjunction with very high concentration optics operating in large production plants [2].

This is the realm of ultra-high efficiency concepts, say in the 50% range, using expensive and sophisticated devices. Multijunction solar cells are already a real option, with commercial triple junction solar cells with efficiencies of 39–41%, quadruple-junction devices hitting 45–46% in the laboratory [3], and 50% appearing feasible for 5-junction cells [4]. Beyond adding solar cells to a multijunction device, several concepts have been conceived whose theoretical efficiency is very large even for a single device, though they are still in the early stages of their development and must show that they are able to outperform the single junction cell. However, the high potential of these novel devices has attracted the research effort that may turn them into

Handbook of Concentrator Photovoltaic Technology, First Edition. Edited by Carlos Algora and Ignacio Rey-Stolle.
© 2016 John Wiley & Sons, Ltd. Published 2016 by John Wiley & Sons, Ltd.

a real alternative and that, not less importantly, is producing a wealth of new knowledge in the fields of electronics, energy and nanotechnology.

Nanotechnology indeed offers unique possibilities to control material properties. Technological advances are allowing the fabrication of many new photonic devices based on physical phenomena that until now were not controllable or realizable: optical metamaterials, quantum transport, coherent state manipulation, etc. The impact of nanotechnological developments on photovoltaics will be important too, and, among other things, might bring the ultra-high efficiency PV concepts into practice.

In this chapter, the thermodynamic aspects of PV conversion are briefly recalled first (section 3.2) and the Shockley–Queisser detailed-balance analysis of the single junction cell is presented with more depth (section 3.3). The purpose is to show how to generalize this kind of analysis and to introduce the different approaches able to surpass the single junction limit. The theory explaining the operation of these approaches is then presented, along with an account of their implementation status (section 3.4). Recent developments on other topics important for future high efficiency devices, such as spectrum conversion and light management, are described in section 3.5.

The final part of the chapter (section 3.6) is a review of nanostructures in solar cells, showing how the optoelectronic properties can be tailored, with emphasis on light absorption properties. The objective is to explain how nanostructures can contribute to the implementation of the high efficiency concepts described in the first sections.

3.2 Thermodynamic Efficiency Limits

Enlightening and inspiring studies of the ideal efficiency limits of photovoltaic converters have accompanied the development of PV technology since the beginning. A brief account is given in this and the following section. Table 3.1 summarizes the limiting efficiency values described in this chapter.

It is customary in this kind of ideal analysis to consider the sun as a blackbody at $T_s = 6000$ K and the converter temperature to be $T_c = 300$ K. The effect of more realistic standard spectra can be consulted in [5].

Use will be made of the concept of *étendue* (see Chapter 4 for further details), which is the size of a bundle of rays in phase space which remains invariant as it proceeds along an optical system [6]:

$$\xi = \int d\xi = \int n^2 \cos\theta d\Omega dA, \tag{3.1}$$

where n is the refraction index and the remaining magnitudes are defined in Figure 3.1. The differential power carried by a ray pencil is the radiance times the *étendue* $Rd\xi$.

Figure 3.1 Definition of the differential *étendue* of a pencil of rays. $d\Omega$ is the solid angle and dA area; θ is the angle with the surface normal

Table 3.1 Efficiencies for solar energy conversion. $T_s = 6000$ K, $T_c = 300$ K, $C_{MAX} = 46000$.

	No concentration	Maximum concentration	Comments	Text
Thermodynamic efficiencies				
Carnot efficiency	-	$1-T_c/T_s = 95\%$	Reversible; relative to consumed power.	3.2.2
Solar efficiency (Landsberg&Tonge)	-	$1-4/3(T_c/T_s)+1/3(T_c/T_s)^4 = 93.35\%$	Relative to impinging power.	3.2.2
Solar thermal converter	54.0% (with $E > 0.92$ eV high pass filter and $T_x = 866$ K)	$(1-T_c/T_x)(1-(T_x/T_s)^4 = 85.4\%$ ($T_x = 2544$ K)	Reversible power extraction. Model for thermophotovoltaics. Model for HCSCs in Würfel's form and for $E_G = 0$ MEG solar cells.	3.2.2; 3.4.2.1; 3.4.3.1 Figure 3.2
Photovoltaic (detailed-balance) efficiencies				
System with infinite monochromatic solar cells	68.3%	86.8%	Model for multijunction solar cells; limit for MEG solar cells.	3.3.2; 3.4.3.1 Figure 3.5
Single junction solar cell (Shockley&Queisser)	30.1% ($E_G = 1.31$ eV)	40.7% ($E_G = 1.11$ eV)		3.3.1 Figure 3.3
Hot carrier solar cell	68.1% ($E_G = 0.35$ eV) (Ross&Nozik) 54.0% ($E_G = 0.92$ eV) (Würfel)	86.0% ($E_G = 0.20$ eV) (Ross&Nozik) 85.4% ($E_G = 0$) (Würfel)		3.4.2.1 Figure 3.8
Multi exciton generation solar cell	44.7% ($E_G = 0.77$ eV)	85.4% ($E_G = 0$)	"Maximum" carrier multiplication function.	3.4.3.1 Figure 3.9
Intermediate band solar cell	46.8% ($E_L = 0.92$ eV. $E_H = 1.49$ eV)	63.2% ($E_L = 0.71$ eV, $E_H = 1.24$ eV)		Figure 3.11

Light can go both ways of the ray so that an illumination path is also an emission one. The *étendue* of the sun in the converter ξ_s is then equal or smaller than the total converter emission *étendue* ξ_c. In this case, light from the sky outside the sun disk reaches the converter. In the ideal treatment, diffuse radiation is ignored so that the benefits of concentration will be overestimated.

With reference to the usual plane-parallel geometry of solar cells, the value of ξ_c is πA for a cell of area A radiating isotropically in air, and the sun *étendue* without concentration $\pi A \sin^2 \theta_s$, where θ_s is the semi-angle subtended by the sun. The *étendues* are n^2 times greater if inside a material medium of index n, as with secondaries. The minimum value of $\xi_c/\xi_s = 1$ (if sunlight is not to be lost) is achieved for full concentration obtained when a lossless concentrator casts light on the solar cell with hemispheric isotropy; the maximum value on the Earth surface, considering only the illuminated face, is $(\xi_c/\xi_s)_{MAX} = 1/\sin^2 \theta_s \cong 46\,000 \equiv C_{MAX}$. In a concentrator system, however, the maximum concentration defined as the aperture/cell area ratio or the irradiance ratio is $n^2 C_{MAX}$. If the cell is allowed to emit (as it usually happens) through the non-illuminated (back) face, its *étendue* is to be added to ξ_c.

3.2.1 Disequilibria and Energy Conversion in Solar Cells

Useful work can be extracted from systems in a disequilibrium situation characterized by the difference in an intensive thermodynamic magnitude, be it temperature (thermal disequilibrium), pressure (mechanical) or chemical potential (chemical). This difference drives the work-performing transformation.

The primary source of disequilibrium for solar energy conversion is the temperature difference between the sun and the earth. Since they are in radiation contact, heat is transferred to the converter with which a thermal engine can operate and produce work.

On the other hand, photons do not interact with each other and sun radiation can be seen as a collection of photon systems with different energy and having a different chemical potential and/or temperature [7,8]. For instance, at the converter temperature monochromatic solar radiation can be seen as possessing an energy-dependent chemical potential μ:

$$\frac{E}{k_B T_x} = \frac{E - \mu(E)}{k_B T_c} \Leftrightarrow \mu(E) = E\left(1 - \frac{T_c}{T_x}\right), \tag{3.2}$$

where k_B is Boltzmann's constant and T_x may be an arbitrary temperature, including the sun temperature T_s. This chemical potential drives chemical reactions for transforming and storing solar energy in chemical form. Photosynthesis is the paradigm of chemical solar energy conversion; also photovoltaics can be fruitfully analyzed as a chemical converter.

Solar cells, inspired by thermal or chemical converters, use electrons as the working material and it is interesting to study the different equilibria that electrons inside a semiconductor can establish.

Free electrons can be usually considered equilibrated inside their bands, and thermally between bands, due to carrier-phonon interaction. Their populations can be described by Fermi–Dirac distributions with a common temperature. However, they may be not equilibrated regarding recombination and hence their chemical potentials (quasi-Fermi levels, QFLs) are not equal $\mu_{e1} \neq \mu_{e2}$[1].

---

[1] These are electron quasi-Fermi energies; hole energies are minus the corresponding electron energies.

Electron or hole equilibration (relaxation) inside bands in bulk semiconductors takes place via absorption and emission of phonons. In polar semiconductors, longitudinal optical (LO) phonons are most effective [9]. The relaxation time of band electrons is in the order of several picoseconds.

Assuming the absence of inelastic electron-phonon interaction, an incomplete electron equilibration can also be assumed, so that the electron electrochemical energy depends linearly on the energy itself [8]. The use of Eq. (3.2) applied now to the electron distribution may lead to an equivalent distribution of carriers with a constant electrochemical energy and a temperature above the one of the lattice $T_e > T_c$. In this case we have hot carriers [10]. In fact, it is unthinkable to have the electrons in a solid thermally isolated from the phonons, but the strongest inelastic coupling is produced with the optical photons. Phonons then interact due to the non-harmonicity of the interatomic forces [11] and tend to equilibrate. However, in some materials or nanostructures the energy gap between optical and acoustic phonons is too big and cannot be delivered to a couple of acoustic phonons conserving energy and momentum. That is why it is proposed that the optical phonons become hot [12,13], at the hot electrons temperature.

Moreover, heat is transported through the solid proportionally to the gradient of the phonon dispersion curve, which is very small in optical phonons. Then it is also thought to be possible that heat transmission from the hot electrons to heat sinks situated at the surface of the material is suppressed. The only energy loss mechanism of the hot carriers is to deliver useful work to an external electric circuit in the so called hot carrier solar cell. Indeed, this is an ideal situation, whose approximate accomplishment is under preliminary research.

For complex *densities of states* (DoS) as those found in nanostructured materials, as a rule one should expect thermal and chemical equilibration if there are no gaps, but different QFLs are conceivable if there are energy gaps wider than the LO phonon energy. Different QFLs of conduction electrons and QD bound states have been measured and are the basis of the intermediate band solar cell concept [14].

Electron-electron or electron-hole interaction also tends to equilibrate the distribution function, but the electron system does not lose energy. Scattering with impurities is mainly elastic and does not contribute to energy relaxation, although it does to momentum relaxation.

In a solar cell the interaction of photons with electrons is essential. Absorption and emission of photons are produced with conservation of the energy and the momentum. As compared to the electrons the momentum of the photons is very small so that in practice the electronic momentum is conserved in the transitions. In presence of two gases of electrons of different QFL (the typical situation in the valence and conduction bands) the population of photons is represented by a Bose–Einstein function with a chemical potential that equals the difference of the electronic QFLs [15].

Finally, disequilibrium between mobile electrons in different points in the converter can be found. In fact, the variation of chemical potential is the driving force for the necessary current flow, but if the electron mobilities are high and/or the distance to the contacts is short, this effect (a loss) can be ignored. Contacts are themselves important structures at which external electrodes (metals) are in quasi-equilibrium with the desired electron population. In conventional solar cells, the chemical energy of the electron is transformed into electric energy when traversing the contact [7].

3.2.2 Thermodynamic Efficiencies

The thermodynamics of PV energy conversion and the limits imposed by the second law can be found in a number of works [8,16,17]. Below some important results are collected.

A reversible thermal engine could produce work from the sun's energy at high Carnot efficiency $\eta_1 = 1 - T_c/T_s = 95\%$. This is achievable in a monochromatic cell under maximum concentration (or in any other configuration for which $\xi_c = \xi_s$) and operating very close to open circuit [8]. However the power (work rate) extracted is then negligible. The situation is the same as the one studied in classical Thermodynamics, in which reversible operation requires a very slow transition between the initial and final states, therefore with negligible work production rate.

The definition of efficiency of solar cells is different from the one usually employed in Thermodynamics. In the latter the denominator contains the total power consumed (that is the radiating power input less the power re-radiated) but in solar cells the power re-radiated is not subtracted. With this definition, the highest efficiency was given by Landsberg and Tonge [18] as $\eta_2 = 1 - 4/3(T_c/T_s) + 1/3(T_c/T_s)^4 = 93.35\%$. This requires that no entropy production takes place in the converter and that the emitted radiation is free radiation (with zero chemical potential) at room temperature directed entirely towards the sun. An easy conceptual way of achieving the latter condition is to have the cell under full concentration. The concept is revised and generalized in [19]. Solar cells, even if ideal do not reach this limit and neither do so solar thermal converters even if geared with a Carnot engine, so that some doubt exists whether the Landsberg limit can be accomplished. Green [16], however, based on the work by Ries [20], asserts it can be, at least in principle, by relying on Faraday-effect gyrators that make the directional form of Kirchhoff's law of radiation to fail for the converter.

In an ideal solar cell the QFLs are constant and their difference is the external voltage. Working under monochromatic radiation, Eq. (3.2) can be used to define a cell temperature. In it μ is the voltage, T_c the cell temperature and T_x is the so-called equivalent photon temperature. The radiation emitted by this cell can be either described at the monochromatic energy by a radiation at the cell temperature T_c with the cell voltage as chemical potential, or as a free radiation (zero chemical potential) at T_x. The best monochromatic solar cell has an efficiency:

$\eta_3 = (1 - T_c/T_x)\left(1 - (\xi_c/\xi_s)(T_x/T_s)^4\right)$. This formula is more often applied to solar thermal converters as the one represented in Figure 3.2, because in this case the monochromatic filter is not necessary. The first factor is the reversible, Carnot conversion between a hot reservoir at T_x and room temperature and the second one reflects the loss by radiation back to the sun. In the case of equal *étendues*, the optimum converter temperature is 2544 K giving an efficiency of 85.2%. If $\xi_c > \xi_s$, the achievable efficiency decreases, and in this case it is advantageous to place a high pass optical filter in front in order to reduce radiation losses as suggested in Figure 3.2. The system represented by this formula is the model for thermophotovoltaic converters and hot carrier solar cells.

In the ideal situations described in this section the efficiency is always a decreasing function of the ξ_c/ξ_s ratio, being 1 the minimum (reasonable) value for which efficiency is maximum. In order to approach this situation one can either (i) increase ξ_s with concentration by the concentration factor C or (ii) decrease ξ_c by means of external confinement [21] and they are interchangeable concerning efficiency (see section devoted to concentrator requirements of Chapter 2). In real cases, however, concentration has more effects in PV efficiency.

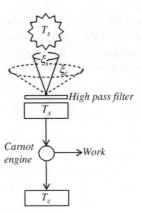

Figure 3.2 A thermal engine with a sun-heated reservoir at T_x. ξ_s, ξ_c are the *étendue* of the sun in the converter and the total converter emission *étendue*, respectively

3.3 Detailed Balance Modeling of Solar Cells

The principle of detailed balance states that in equilibrium every elementary process is balanced by its exact reverse [22], for which initial and final states are swapped. It also applies to energy and matter exchanges between volume elements. Detailed balance is derived from the time reversal invariance of conservative systems, both classical and quantum, known as micro-reversibility. This does not apply in the presence of magnetic fields or macroscopic rotation, and hence detailed balance must be re-examined in those cases.

Detailed balance is related to Onsager's reciprocity relations [23]; these also make use of micro-reversibility for their derivation and ensure positive entropy production for small departures from equilibrium hence guaranteeing the fulfillment of the second law. In the same way, solar cell models complying detailed balance are shown to be consistent with thermodynamic laws [8,19,24]. Detailed balance, along with the constancy of the equilibrium kinetic coefficients under disequilibrium [25], are the roots of a number of interesting reciprocity relationships applicable to solar cells, namely the Van Roosbroeck-Shockley relationship [26] and the quantum efficiency–electroluminescence reciprocity [27], among others.

Kirchhoff's law of radiation states that the emissivity $e(E)$ (emitted power relative to blackbody radiation) and absorptivity $a(E)$ (fraction of incident blackbody radiation absorbed) are equal. Though derived for equilibrium, it is used in non-equilibrium, but then emissivity only refers to spontaneous emission while stimulated emission must be counted as negative absorption [28]. When applied to an electron system with two quasi-Fermi levels separated an amount μ, the important result that is obtained is that, in most cases, the emitted radiation is given by the generalized Planck's law times the absorptivity of the electron system [29]:

$$a(E) \times N(E, T, \mu) = a(E) \times \frac{2}{h^3 c^2} \frac{E^2}{e^{\frac{E-\mu}{k_B T}} - 1}, \tag{3.3}$$

in terms of photons emitted per unit time, energy and *étendue*. Constant c is the speed of light (in vacuum; the effect of the density of the medium would be incorporated in the emission *étendue*). Possible angular and spatial dependences are ignored. This result was arrived at in 3.2.1 by considering equilibrium between photons and electrons.

3.3.1 Shockley–Queisser Model of a Solar Cell

Shockley and Queisser [15] derived a very general model of a solar cell with a single bandgap E_G that represents a true limit for its operation and complies with thermodynamic laws. It has been a most inspiring work from which new devices have been conceived. Detailed balance models have a wide applicability and can be extended to accommodate new developments in nanotechnology and nano-optics.

The initial model has been refined in a number of works [8,16,29]. It is based on the following set of assumptions:

1. One electron-hole pair generated per photon absorbed.
2. Only radiative recombination, one photon per recombination event. Light generation is the necessary process taking place in a solar cell and this assumption is the statement of detailed-balance, since radiative recombination is the reverse of generation. It is the minimum loss mechanism in a solar cell. Not only non-radiative bulk and surface recombination mechanisms must be minimized, but contacts should be perfectly selective too since an electron that exits through the hole contact amounts to a recombination, and contrariwise.
3. Two electron populations with different electrochemical potentials but the same (lattice) temperature due to the thermalizing role of phonons, present in equilibrium concentrations.
4. Constant electrochemical potentials with a separation equal to the external voltage qV. As explained, diffusive quasi-equilibrium allows perfect transformation of the chemical potential into electrostatic potential at the contacts.
5. Perfect absorption by electron-hole generation over the bandgap, and perfect transparency below it.

As it is well known, the steady state current through a two-terminal semiconductor solar cell is the difference between generation and recombination. Assumptions 1 and 2 allow identifying the generation/recombination balance with the photon absorption/emission balance. There is a lot of re-absorption of emission inside the cell (the photon recycling), but the net difference between absorption and emission is equal to the balance of external photons, and this, by assumptions 3, 4 and 5 and the generalized Kirchhoff's law, is written[2] for photons with energy E:

$$\xi_s N(E, T_s, 0) - \xi_c N(E, T_c, qV) = \frac{1}{q}\frac{dI}{dE}, \tag{3.4}$$

---

[2] This formula often includes thermal photons at room temperature arriving through the emission *étendue* not filled with sun rays [16] so that the shortcircuit current would go to zero should the sun be at room temperature.

where I is the cell current and $N(E,T_s,0)$ is defined in (3.3). The power it delivers is maximized with respect to the voltage and divided into the sun power to give the efficiency:

$$\eta = \frac{P}{\xi_s \frac{\sigma}{\pi} T_s^4} = \frac{V \int\limits_{E_G}^{\infty} \frac{dI}{dE} dE}{\xi_s \frac{\sigma}{\pi} T_s^4} \quad \text{with} \quad \frac{\partial P}{\partial V} = 0, \tag{3.5}$$

where σ is the Stefan–Boltzmann constant; it appears divided into π to obtain power per unit *étendue*, and not per unit area. Again, only the ratio of *étendues* enters the efficiency. Full or maximum concentration means $\xi_s = \xi_c$. For the no concentration case, it has been assumed that emission takes place only through the illuminated face ($\xi_c = \xi_s \, C_{MAX}$), or through the back face as well into a substrate with refraction index n ($\xi_c = \xi_s (1 + n^2) C_{MAX}$). The efficiency is plotted in Figure 3.3 as a function of the cell bandgap. The best cell under full concentration is 40.7% efficient with around 1.1 eV bandgap. Without concentration, the optimum gap is shifted to higher values to moderate the emission of low energy photons and the efficiency drops to 31%. Emission into the substrate means another loss greater than 1% absolute.

Figure 3.4 shows the efficiency and the relative amount of different losses as defined in reference [30] for the Shockley–Queisser (SQ) cell as a function of bandgap. The two main losses are due to non-absorption of phonons with energy smaller than the gap and the energy in excess of the gap that is lost through thermalization as heat to the lattice. They have the opposite dependence on the bandgap and have the same magnitude no matter the concentration. The optimum gap is placed where both are similar. The efficiency limit that considers only these two mechanisms is known as Trivich–Finn efficiency [31].

The remaining losses depend on concentration. Emission is very low in both cases so the main loss difference is in the operating voltage (Boltzmann losses in [30]), which varies in an amount approximately $k_B T_c \ln \xi_c / \xi_s$. This is very similar to the familiar open circuit voltage enhancement achieved by concentration $k_B T_c \ln X$.

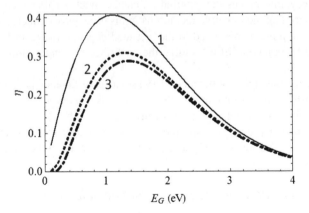

Figure 3.3 Limiting efficiency of the single-gap solar cell as a function of the bandgap energy, E_G, for maximum concentration (1) and no concentration with (3) and without (2) back emission into a substrate $n = 3.6$

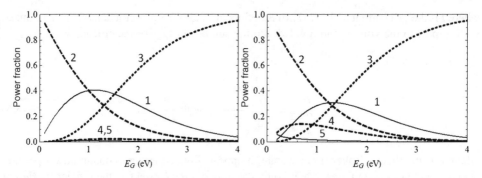

Figure 3.4 Power fraction taking into account single power-loss mechanisms and output of the single gap SQ solar cell as a function of bandgap according to [30]. Left: maximum concentration; right: no concentration. 1: power output. 2: no absorption. 3: thermalization. 4: voltage lower than the bandgap. 5: emission

There is a difference however in the effect of concentration in solar cells limited by radiative recombination or by non radiative mechanisms. In the former case, voltage enhancement can be achieved through concentration and through external confinement, since it is the ratio and not the absolute value of the *étendues* that matters, though the energy density is increased only by concentration. In non-radiative cells, confinement is used to increase absorption, but only marginally voltage for which concentration is needed.

The best single-junction cell at one sun today (as of April 2015) is an impressively 28.8% efficient (with the standard AM1.5 Global) GaAs solar cell (1.42 eV) by Alta Devices [3]; it is near the 90% of the SQ efficiency limit for this spectrum. A high material quality is responsible for suppressed parasitic recombination and high internal emission efficiency, which is the ratio of radiative to total recombination events. The quantity that informs about how near the device is to the SQ limit is the external emission quantum efficiency, i.e., the fraction of the recombination current that is due to emission out of the cell, unity in the SQ model. Photon recycling makes the external efficiency much smaller than the internal one. Then, it is necessary that the radiative recombination is efficiently extracted, as in LEDs [32]. Moreover, emission must be restricted to the intended *étendue*; the record cell is a thin film device fabricated by epitaxial lift-off so that the back face no longer emits into an optically dense substrate.

On the other hand, the highest silicon (1.12 eV) solar cell efficiency is 25.6% [3] in a device with extraordinarily high recombination lifetimes; it makes even a very good light emitting diode for silicon, but this material is intrinsically a poor emitter (and consequently a weak absorber for near bandgap energies) and the unavoidable Auger recombination will limit its performance below the SQ levels. Reference [33] presents a study of the emission efficiency of record solar cells.

3.3.2 The System with Infinite Monochromatic Solar Cells

In order to avoid both non-absorption and thermalization losses, radiation of a given photon energy E should interact with electrons only within a narrow energy band. This can be accomplished in a system featuring two discrete electron energy levels, or in a conventional semiconductor of bandgap E by using a narrowband filter restricting emission and absorption to this energy. The extracted power from such a monochromatic cell [8] is:

$$qV(E)\{\xi_s N(E, T_s, 0) - \xi_c N[E, T_c, qV(E)]\} = \frac{dP}{dE},\tag{3.6}$$

The thermodynamic efficiency would be found by dividing into the net consumed (absorbed minus emitted) power, giving $qV(E)/E$. This efficiency is maximum at open circuit and equal to the Carnot efficiency for the ideal case $\xi_s = \xi_c$. A collection of these cells would realize the perfect chemical converter with Carnot efficiency commented on in section 3.2.

However, solar efficiency calculated with respect to the monochromatic impinging power $q\xi_s N(E, T_s, 0)$ is obtained by optimizing the voltage function $V(E)$. The result can be seen in Figure 3.5 with and without concentration.

If the whole spectrum is thus converted in separate monochromatic cells, the efficiency would be the integral of (3.6) divided into the solar spectrum, amounting to:

$$\eta = \frac{\displaystyle\int_0^\infty qV(E)(\xi_s N(E, T_s, 0) - \xi_c N(E, T_c, qV(E)))dE}{\xi_s \dfrac{\sigma}{\pi} T_s^4},\tag{3.7}$$

$$= \begin{cases} 0.868\ (\xi_c = \xi_s) \\ 0.682\ (\xi_c = X_{MAX}\xi_s) \end{cases}$$

which is the maximum PV efficiency, at least without the use of non-reciprocal optical elements in whose case it could be possible to achieve Landsberg's efficiency.

A summary of the concepts presented in this section is shown in Table 3.1.

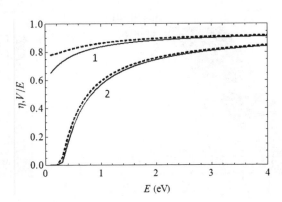

Figure 3.5 Efficiency (solid lines) and optimum voltage into photon energy (dashed) of monochromatic solar cells for maximum concentration (1) and for no concentration (2)

3.4 Solar Cell Concepts Exceeding the Single Junction Shockley–Queisser Limit

3.4.1 Multijunction Solar Cells

The ideal PV converter presented in the last section 3.3.2 can be approached with a series of solar cells with different bandgaps $E_{G1} < E_{G2} < E_{G3}$ that work with the part of the spectrum best adapted to it, i.e., the k-th solar cell should interact with photons of energy E such that $E_{Gk} < E < E_{Gk+1}$. The multijunction solar cell is extensively described in Chapter 2 so here only some results are given for reference and comparison with other high efficiency approaches.

The detailed-balance efficiency of an N_J-junction device, provided that each cell in it can be independently biased, is obtained by optimizing the following expression with respect to the N_J voltages [29]:

$$
\eta = \frac{\sum_{k=1}^{N_J} \int_{E_{Gk}}^{E_{Gk+1}} qV_k[\xi_s N(E, T_s, 0) - \xi_c N(E, T_c, V_k)]dE}{\xi_s \dfrac{\sigma}{\pi} T_s^4}, \tag{3.8}
$$

where $E_{GNJ+1} \to \infty$. If N_J is increased to infinity and the lowest bandgap decreased to zero, the maximum efficiencies stated in equation (3.7) are obtained.

In a direct implementation of the concept, energy selectivity can be accomplished by the use of optical devices splitting the spectrum in the desired ranges, as suggested in Figure 3.6. There have been several practical realizations of this system [34,35] (see Chapter 2 for further details), but by far the most successful realization of the multijunction solar cell, producing the highest practical PV efficiencies, is the series-connected monolithic solar cell shown also in the Figure 3.6. III-V semiconductors allow many bandgap combinations and loss-less tunnel

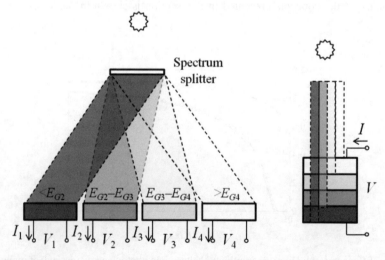

Figure 3.6 Multijunction solar cells, independent cells with spectrum splitting (left) and monolithic arrangement (right)

junctions for series connection, even at high concentration. Energy selectivity is attained because higher gaps on top of the device absorb in turn the higher energy ranges and each photon is automatically absorbed by the adequate solar cell.

The series connection introduces N_J-1 constraints, namely, that all currents be the same $I/q = \int_{E_{Gk}}^{E_{Gk+1}} [\xi_s N(E, T_s, 0) - \xi_c N(E, T_c, V_k)]dE$, and as a result the achievable ideal efficiency is somewhat lower. However, the difference decreases with the number of cells, and for $N_J \to \infty$ the maximum efficiency is also obtained [36,37].

3.4.2 Hot Carrier Solar Cells

As apparent in Figure 3.4, a large portion of the energy lost in the ideal single-junction solar cell is due to thermalization of carriers, i.e., the photon energy in excess of the bandgap is given to the lattice as heat through the emission of phonons. If carriers could be extracted to the external circuit before they had enough time to lose this energy, the efficiency could be much higher. The *hot carrier solar cell* (HCSC) has been proposed to work under this principle.

3.4.2.1 Detailed Balance Model

The first detailed-balance analysis of this hypothetical device is due to Ross and Nozik [38]. High-energy electron hole pairs are created in a semiconductor upon absorption of solar radiation. They are supposed not to exchange energy with the lattice by inelastic phonon processes. Carrier-carrier collisions however equilibrate them to a Fermi–Dirac distribution with temperatures $T_e, T_h > T_c$ and different chemical potentials μ_{HCe}, μ_{HCh}. If electron-hole scattering is also present, carrier temperatures are equal $T_e = T_h \equiv T_{HC}$ although the formalism can be extended to different temperatures in both carrier bands [39]. In these distributions, there are high energy carriers that will be extracted to the external circuit at defined energies E_{xe}, E_{xh} through a special contact system [10] (see Figure 3.7 for a schematic energy diagram of the

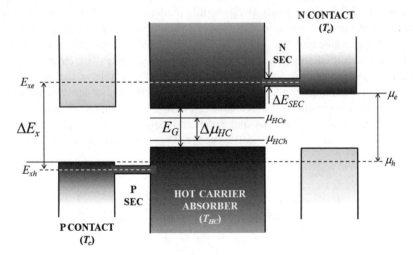

Figure 3.7 Schematic representation of a hot carrier solar cell (after reference [10])

HCSC). Since only radiative recombination is assumed, carrier balance applies as for the single junction SQ cell case, except for the carrier temperature:

$$\frac{I}{q} = \frac{2\xi_s}{h^3 c^2} \int_{E_G}^{\infty} \frac{E^2 dE}{e^{E/k_B T_s} - 1} - \frac{2\xi_c}{h^3 c^2} \int_{E_G}^{\infty} \frac{E^2 dE}{e^{(E-\Delta\mu_{HC})/k_B T_{HC}} - 1}, \quad (3.9)$$

where I is the extracted current and $\Delta\mu_{HC} = \mu_{HCe} - \mu_{HCh}$. Besides, since there is no power loss to the lattice, the energy balance has the same terms:

$$\frac{I\Delta E_x}{q} = \frac{2\xi_s}{h^3 c^2} \int_{E_G}^{\infty} \frac{E^3 dE}{e^{E/k_B T_s} - 1} - \frac{2\xi_c}{h^3 c^2} \int_{E_G}^{\infty} \frac{E^3 dE}{e^{(E-\Delta\mu_{HC})/k_B T_{HC}} - 1}, \quad (3.10)$$

This is not the useful power yet. Carriers must be cooled down since the external circuit is at ambient temperature in the selective energy contact, which is a material region between the hot carrier absorber and a conventional semiconductor at ambient temperature that carriers can only pass through in a very narrow energy range. Transport equilibrium between regions can be established at this precise energy by equating the Fermi–Dirac distribution occupancy:

$$\left. \begin{aligned} \frac{E_{xe} - \mu_{HCe}}{k_B T_{HC}} &= \frac{E_{xe} - \mu_e}{k_B T_c} \\ \frac{E_{xh} - \mu_{HCh}}{k_B T_{HC}} &= \frac{E_{xh} - \mu_h}{k_B T_c} \end{aligned} \right\} \Rightarrow qV = \mu_e - \mu_h = \Delta\mu_{HC}\frac{T_c}{T_{HC}} + \Delta E_x\left(1 - \frac{T_c}{T_{HC}}\right), \quad (3.11)$$

where ΔE_x is fixed. Given the voltage V, this set of three equations allows to calculate I, T_{HC} and $\Delta\mu_{HC}$ and then the output power IV and the efficiency $\eta = IV/\xi_s\sigma T_s^4$. It is given in Figure 3.8 as a function of the absorber bandgap E_G. In this model the quasi-Fermi level separation is negative inside the converter to limit emission and the extraction energy ΔE_x has no effect on efficiency. A large value will lead to low current, high voltage cells and contrariwise.

Würfel [40] modified the preceding analysis by considering that electron-hole Auger recombination and its reverse, impact ionization, are unavoidable in the hot carrier solar cells. Besides, some strange results of the Ross–Nozik model (higher carrier temperature than sun temperature in some cases) are avoided. In the limiting situation, Auger processes will force equilibrium between electrons and holes and $\Delta\mu_{HC} = 0$. This equation replaces the particle balance equation (3.9) in the analysis above. Then, the cell emits as a blackbody at T_{HC} and it is obtained:

$$\eta = \frac{\left(\dfrac{2\xi_s}{h^3 c^2} \displaystyle\int_{E_g}^{\infty} \dfrac{E^3 dE}{e^{E/k_B T_s} - 1} - \dfrac{2\xi_c}{h^3 c^2} \displaystyle\int_{E_g}^{\infty} \dfrac{E^3 dE}{e^{E/k_B T_{HC}} - 1}\right)\left(1 - \dfrac{T_c}{T_{HC}}\right)}{\xi_s \dfrac{\sigma}{\pi} T_s^4}, \quad (3.12)$$

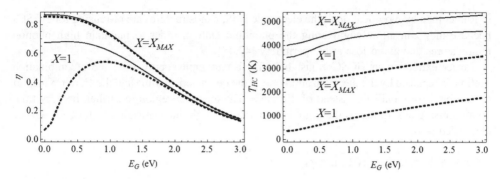

Figure 3.8 Efficiency (left) and carrier temperature (right) as a function of bandgap for extreme values of concentration factor. Solid lines stand for the Nozik–Ross model and dashed ones for the Würfel model

with T_{HC} to be optimized for maximum output. This is equivalent to the thermal converter in Figure 3.2 with E_G playing the role of the high pass filter mentioned there. The efficiency in this model is lower, particularly with no concentration, due to the loss of a degree of freedom.

3.4.2.2 Hot Carrier Absorbers

Free carriers are strongly coupled to the lattice and the loss of energy by phonon emission is unavoidable. Luque and Martí [41] investigated how good HCSCs made on common semiconductors could be. Phonons at the lattice temperature are not in equilibrium with hot carriers and so these lose energy by net phonon emission. Using mobility data, which is controlled by inter-valley scattering phonon emission in indirect gap semiconductors (Si and Ge) and by *longitudinal optical* (LO) phonon emission in polar semiconductors (GaAs, InAs, . . .), the thermalization loss is introduced in Würfel's formulation by adding a phonon term in equation (3.10). Lattice heating severely limits the ideal efficiency of HCSCs, but it can still be higher than the SQ limit. Losses increase very much with carrier concentration, that must be kept as low as possible. A high bandgap would be thus preferable save, of course, for limited absorption. Polar semiconductors (such as GaAs) are worse than non-polar semiconductors (Si, Ge). Thermalization times in the order of ps are deduced, in agreement with measurements in bulk semiconductors.

When considering finite extraction times, i.e., the average time needed for a photogenerated carrier to exit through the contact, much higher thermalization times on the order of 100 ps should be achieved [42,43]. In these references, the effect of Auger recombination is included in the particle balance equation and the authors conclude that Ross and Nozik's model rather than Würfel's is valid, except near open circuit.

A low optical phonon energy would help [44] in delaying carrier cooling by increasing the number of emission events needed and hence decreasing the cooling rate. Another option is to keep the phonons hot, as explained in section 3.2.1. This can be favored if there is a significant energy gap between zone-center optical phonons and high-k acoustic ones. Different materials have been investigated with regard to the potential for maintaining a hot-phonon population, such as GaAs and InP [45], but with InN [46] being one of the most promising.

Rather than by modification of the electronic DoS, nanostructures are promising for HCSCs because they can allow engineering the phononic DoS in order to maintain high phonon temperatures and avoid loss at the periphery [47,48].

Carrier extraction in HCSCs must be very fast to limit energy loss: the recombination lifetime is replaced by the thermalization one as the characteristic time and the carrier transport should be mainly ballistic instead of diffusive. The absorber region must then be very thin, compromising absorption so that light confinement or light intensification techniques will be demanded [49].

3.4.2.3 Selective Energy Contacts

Ultrafast carrier extraction, such as at heterointerfaces, is not sufficient and thus the need for energy selectivity was soon realized. The broader the carrier energy range allowed going through the contact, the faster the extraction, but the greater the loss. These contacts could be realized with quantum dots that select the carriers resonant with their localized energy levels in a quantum transport process. Silicon on silicon nitride and oxide systems [50,51] are being investigated and resonant diode behavior has been observed. An interesting structure based on quantum dot superlattices incorporates energy selectivity to the absorber region [52].

3.4.2.4 Prospects for Hot Carrier Solar Cells

In spite of recent advances in theory and practice, these devices are still very far from practical application. The absorber material is unknown and the contact design is at an early stage. It seems however that nanostructures may hold the key to HSCSs. In spite of the uncertainty regarding the final incarnation of HSCSs, they constitute a high concentration, high efficiency concept. Since the operating conditions of real devices are not known, it cannot be predicted the impact of concentration outside the detailed balance limit, which certainly will not be attained. It depends on how fast thermalization and limited contact conductance losses grow with concentration.

3.4.3 Carrier Multiplication or Multi-Exciton Generation Solar Cells

From the point of energy balance, high energy photons can produce more than one electron hole pair, instead of being lost by thermalization. Carrier multiplication or multi-exciton generation (MEG) solar cells were suggested in [53,54] after observing higher than 100% external quantum efficiency in silicon solar cells in the ultraviolet. In this concept, thermalization loss is again circumvented.

3.4.3.1 Detailed Balance Modeling

In order to build a thermodynamically acceptable model [55,56], the equilibrium between photons and electron-hole carriers must be revised. If a photon with energy E produces $M(E)$ pairs, the relationship between the photon μ_{ph} and the electron hole pair μ_{eh} chemical potentials that guarantees thermodynamic consistency is [8,19]:

$$\mu_{ph}(E) = M(E)\mu_{eh} = M(E)qV, \tag{3.13}$$

Then, the particle balance includes the factor $M(E)$ in the generation and radiative recombination terms, and in the chemical potential of the emitted photons so that the efficiency is written:

$$\eta = \frac{\displaystyle\int_{E_G}^{\infty} qV \times M(E)[\xi_s N(E,T_s,0) - \xi_c N(E,T_c,qV \times M(E))]dE}{\xi_s \dfrac{\sigma}{\pi}T_s^4}, \tag{3.14}$$

The ideal $M(E)$ variation is a staircase function so that as many pairs are generated as is possible with the photon energy E, up to a maximum value M_{max}:

$$M(E) = \min\left(\lfloor E/E_G \rfloor, M_{max}\right), \tag{3.15}$$

where the brackets mean the integer part. In Figure 3.9 this function is presented for $M_{max} = 1$ (normal one-pair-per-photon situation) and $M_{max} \to \infty$. The linear function corresponds to measured multiplication factors and will be commented later. In the same figure, the efficiencies belonging to these functions are given as a function of E_G. The concentration-induced efficiency increase for this concept is especially large.

This formulation contains interesting limiting cases as well:

- In the case $E_G = 0$, $M_{max} \to \infty$, the MEG solar cell transforms into a zero-bandgap hot carrier device, i.e., the thermal converter in Figure 3.2 again.

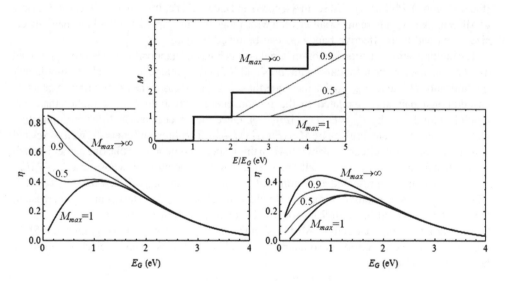

Figure 3.9 Efficiencies with maximum (left) and with no (right) concentration for MEG cells with the multiplication functions shown in the upper inset. Apart from the ideal cases for $M = 1$ and $M \to \infty$ linear multiplication functions $M = (E/E_G - 1)\eta_{EHPM}$ (see section 3.4.3.2) correspond to measured trends [63], which are labeled with the values 0.5 and 0.9 for η_{EHPM}

- Eq. (3.14) is formally identical, for $E_G = 0$, to (3.7) with $V \times M(E) \to V(E)$. If the multiplication factor could be adjusted so that $V \times M(E)$ equals the optimum $V(E)$ function shown in Figure 3.5, the maximum photovoltaic efficiencies of the ideal converter will be obtained. For $E_G \neq 0$, it is similar to a multijunction solar cell. Then, the 'ideal' staircase multiplication does not produce maximum efficiency due to larger luminescent loss.

3.4.3.2 Carrier Multiplication/Multi-Exciton Generation Materials

Bulk semiconductors (Si, Ge and SiGe alloys) were considered first for PV conversion purposes [53,57]. Impact ionization is the commonly accepted mechanism to produce carrier multiplication. The reverse process, Auger recombination, will then be an unavoidable loss [19]. For high multiplication factors to take place, impact ionization should be faster than energy loss to phonons [58]. Moreover, momentum conservation in bulk materials introduces an energy threshold for impact ionization that is much larger than the bandgap. These facts make the multiplication rate in bulk semiconductor materials to be small and to occur at deep UV wavelengths which are not very interesting for photovoltaics [59].

The concept, renamed as *multi exciton generation* (MEG), was revitalized by the developments in semiconductor nanocrystals or quantum dots [60]. It was predicted that in nanocrystals the generation of more than one carrier pair per photon could be more efficient due to the enhanced Coulomb interaction between confined electrons, the lack of momentum conservation selection rules and the possibility of delayed carrier thermalization in discrete DoS. The multiplication mechanism involves the formation of multiexcitonic coherent states [61]. During some time, the MEG effect in nanocrystals was overestimated due to misinterpretation of experimental results. However, it is now well established that quantum yields for carrier generation higher than 100% are produced in colloidal quantum dots of PbSe (the first report), PbS, PbTe, CdSe, InAs, core/shell CdSe/CdTe, InP, etc.(see [62] for a review of MEG materials). These materials cover a wide range of bandgaps. The QDs have nanometer-sized radii and their effective bandgaps can be tuned by them.

The multiplication function in most materials –both bulk and nanoparticles– can be represented as a one-parameter linear function with threshold [63]. The parameter is the electron-hole pair multiplication efficiency η_{EHPM} that measures the energy in excess of the gap required to generate an additional pair and is related to the photon energy threshold E_{th} for higher than one multiplication factor by $E_{th} = E_G (1 + 1/\eta_{EHPM})$. Once the onset is reached, the multiplication factor (or quantum yield) is given by $M = (E/E_G - 1)\eta_{EHPM}$. The threshold energy is well correlated with the asymmetry of effective masses [64]. In this respect, it is beneficial that one carrier is much lighter than the other one and takes all excess energy upon generation, in whose case $E_{th} \to 2E_G$.

This function is represented in Figure 3.9 for $\eta_{EHPM} = 0.5$ ($E_{th} = 3 E_G$) and $\eta_{EHPM} = 0.9$ ($E_{th} = 2.1 E_G$), representative of measured values in colloidal QDs. The efficiency increase in these materials would be significant for low bandgaps, as shown in the same figure.

Another interesting material for MEG PV in the $E_G = 0$ limit seems to be graphene [65]. In this case, the role of electrons in conduction and valence bands is taken by electrons above or below the Fermi energy.

3.4.3.3 Prospects for Carrier Multiplication Solar Cells

Most MEG measurements have been made on nanoparticles in colloidal solution which bear no interest for practical solar cells. To make PV devices, the photogenerated carriers must be extracted from the nanocrystal (an ultrafast process) and transported with sufficient mobility to selective contacts (which do not need to be energetically selective). Three configurations are envisaged [66] and experimentally tested:

- QD films as MEG absorbers. Conduction of photogenerated carriers takes place by hopping between neighboring QDs. Shottky-contact solar cells based on QD films have been fabricated with sizeable efficiencies. Higher than 100% quantum efficiency has been verified in solar cells based on QD films [67], thus confirming the feasibility of the MEG approach.
- QDs replacing the dye as sensitizers in *dye-sensitized solar cells* (DSSCs), with the possibility of efficiency increase by MEG.
- QDs in semiconducting-polymer blends which provide carrier separation and transport.

In spite of the strong influence of concentration shown in Figure 3.9, these solutions fit rather the low cost, low efficiency approach to PV conversion than the high concentration, high efficiency option which is the subject of this chapter, for which other implementation modes should be developed.

3.4.4 Intermediate Band Solar Cells

In order to avoid the loss of non-absorbed photons without the decrease of voltage that would imply using a lower bandgap, the possibility of generating one electron-hole pair using more than one low-energy photon is put up. An intermediate electronic level in the bandgap is needed so that this process has a sizeable probability by taking place in two steps, as schematized in Figure 3.10. The first works on this topic aimed at deep impurity levels [68]; but the conditions under which deep levels would increase instead of degrade the efficiency were not recognized.

The concept was reformulated in the detailed-balance formalism by Luque and Martí [69] contemplating different quasi-Fermi energies for electrons in bands and in the intermediate levels. Instead of localized, the intermediate levels are extended (and hence the name Intermediate Band, IB) to suppress non radiative recombination [70], and the contacts are made only to CB and VB through doped regions without IB (the emitters), the IB being kept open circuited. Figure 3.10 illustrates the band diagram of an *IB solar cell* (IBSC). In it, the IB

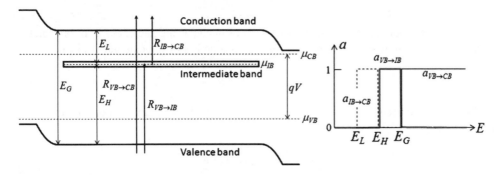

Figure 3.10   Left: idealized IBSC band diagram showing p and n emitters; the IB QFL is shown inside it. Right: ideal absorbances for the three transitions

is placed closer to the CB, but at this ideal stage the device is symmetric with respect to the IB position from midgap. The threshold energies for the three transitions are E_L (IB → CB) < E_H (VB → IB) < E_G (VB → CB), with $E_G = E_L + E_H$. Recent reviews of the IBSC can be found in [71,72].

3.4.4.1 Detailed Balance Modeling of IBSCs

The assumptions for the detailed balance modelling of IBSCs are:

- The existence of three separate, constant quasi-Fermi levels for electrons in the CB, the IB and the VB. CB and VB are in diffusive equilibrium with the emitters so that the external voltage is the difference $qV = \mu_{CB} - \mu_{VB}$. This allows the external voltage to exceed the energies of the intermediate transitions while fulfilling detailed balance.
- The only electronic transitions taking place between bands are radiative. As indicated in Figure 3.10, every photon must be absorbed in the most favorable transition, and absorption is complete.

This and the preceding assumption allow the net transition rates R (s^{-1}) to be written as:

$$R_{VB \to CB} = \int_{E_G}^{\infty} [\xi_s N(E, T_s, 0) - \xi_c N(E, T_c, qV)] dE$$

$$R_{VB \to IB} = \int_{E_H}^{E_G} \left[\xi_s N(E, T_s, 0) - \xi_c N(E, T_c, \mu_{IB} - \mu_{VB}) \right] dE$$

$$R_{IB \to CB} = \int_{E_L}^{E_H} \left[\xi_s N(E, T_s, 0) - \xi_c N(E, T_c, \mu_{CB} - \mu_{IB}) \right] dE =$$

$$= \int_{E_L}^{E_H} \left[\xi_s N(E, T_s, 0) - \xi_c N(E, T_c, qV - (\mu_{IB} - \mu_{VB})) \right] dE$$

(3.16)

Since the IB is not connected, the rate of electrons in equals the rate of electrons out and the balance equation allows calculating $\mu_{IB} - \mu_{VB}$ as a function of the external voltage:

$$R_{IB \to CB} = R_{VB \to IB}, \qquad (3.17)$$

The external current is the sum of net VB to CB and IB to CB transition rate times the electron charge:

$$\eta = \frac{IV}{\xi_s \dfrac{\sigma}{\pi} T_s^4} = \frac{qV(R_{IB \to CB} + R_{VB \to CB})}{\xi_s \dfrac{\sigma}{\pi} T_s^4}, \qquad (3.18)$$

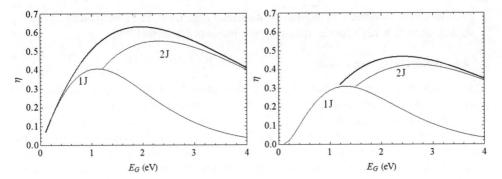

Figure 3.11 Thick lines: IB efficiency as a function of the largest gap, E_G, with optimized IB position for maximum (left) and one sun concentration (right). Label '1J' indicates single junction cells and '2J' a series connected, double junction device for which E_G is the sum of the two cell bandgaps

The efficiency of IB solar cells in the detailed balance limit is given in Figure 3.11 as a function of the largest bandgap E_G, with optimization of the IB position. 63.2% is the maximum value under concentration (very similar to a triple junction device [29]), obtained for $E_G \cong 1.95$ eV, $E_L \cong 0.71$ eV, $E_H \cong 1.24$ eV for the case of maximum concentration.

From the electrical point of view, the IB solar cell contains three cells, two in series and these in parallel with the largest gap, and this makes its efficiency very similar to a triple junction device. Besides, thanks to the parallel connection the IB cell has the advantage of being more tolerant to spectral mismatches. In Figure 3.11, the efficiency of a 1 J cell and a series-connected 2 J device (such that the sum of the bandgaps equals the E_G of the IBSC) are also drawn. It can be seen that the IBSC is as good as the 1 J cell for low E_G, as the 2 J device for high E_G and significantly better than both at medium E_G.

The theory of the IBSC has been developed in a number of works in the frame of detailed balance, circuit theory or numerical device simulation; among the topics addressed:

- The effect of concentration. As shown in Figure 3.11, the additional recombination path can make the IBSC less efficient than the single cell with the same bandgap even in the detailed balance model. However, with increasing concentration this additional recombination loses importance while the IB generation remains, boosting the efficiency.
- The absorption coefficients overlap. The energy selectivity depicted in Figure 3.10 does not seem feasible, but failing to achieve it does not necessarily imply a large loss. Selectivity for higher energy photons is automatically incorporated with the scheme in the figure, since they will be absorbed in the illuminated emitter before reaching the IB region. Suggested remedies to the absorption overlap issue include having very different absorption coefficients with an effective light trapping scheme [73] and varying the position or the electron filling of the IB [74].
- The energy width of the IB and the transport along it. Finite width bands supporting transport are not really needed to realize the potential of the IB concept, though the homogenizing effect can be beneficial [75].
- The filling of the IB. For both transitions to be possible, in the IB there must exist both filled and empty electron states. If a true band, it must present a metallic character. If obtained by

discrete levels associated to defects or nanostuctures, these must be partially filled by doping, though at high concentration doping may not be needed [76].

3.4.4.2 IBSC Characterization

A device operates as an IBSC instead of as a conventional cell if a) external current can be produced by two-step processes with sub-bandgap photons and b) the open circuit voltage can exceed the IB transition threshold, being limited by the host material bandgap [77]. Due to thermal coupling that holds together the quasi-Fermi levels of the IB and the neighboring band, to obtain this behavior it can be necessary to work at a low temperature. To verify current generation through the IB, conventional spectral response measurements are not sufficient and two photon experiments are performed. For the determination of the voltage limit, besides lowering temperature, concentration is used. The reason is that two-step recombination rates increase more slowly with concentration than band-to-band processes, and the voltage at very high concentration becomes limited by the high bandgap (voltage recovery).

3.4.4.3 Quantum Dot Intermediate Band Solar Cells

After the suggestion of using the confined electronic levels of QDs to form the IB [78], the first devices were fabricated with InAs QDs in GaAs, the best known system [79]. Delta-doping with Si in the barrier region with an areal density similar to the QD density is used to partially fill the ground state of the dots.

The corresponding electron energy diagram will be shown in Figure 3.16 for commonly found dot dimensions. As shown by the figure, it is very far from the optimum values revealed by detailed balance analysis. E_G corresponds to the distance between electron and hole confined ground states; hole states are quasi-continuous and hence are in equilibrium with the valence band. Besides, the *wetting layer* (WL) introduces a quasi 2-D electron distribution close to the dots that further reduces the effective bandgap. With the IB constituted by the ground QD state and several levels inside the well but very near the CB, this material features $E_H \cong 0.9$–1.0 eV, $E_L \cong 0.1$–0.2 eV. Many groups have fabricated solar cells based on this QD system [80–83].

These cells show sub-bandgap photocurrent in the spectral response but the contribution to the total current is low. A first step towards enhancing absorption by the IB is to build a large number of QD layers. The accumulated strain inherent to the Stranski–Krastanov method produces dislocations and severely degrades the solar cell but several strain compensation techniques using layers with the opposite stress character have been developed and tens of layers can be grown without introducing defects [84,85].

The subject of absorption of light by QDs has been studied with the simplified model explained in section 3.6 to ascertain how to increase the IB current [86,87]. The conclusion is that it would be required to increase the in-plane QD density, which seems possible by performing growth on high-index crystallographic planes such as (311) [88].

On the other hand, many practical devices feature open circuit voltages substantially lower than the reference cells without QDs. It is the consequence of thermal contact among the IB (the first confined level) and the conduction band due to the small energy separation and the presence of other confined states that facilitate thermal transitions in a ladder fashion [89]. At room temperature, there are a lot of thermal photons in this energy range that might by themselves or with assistance of phonons explain the reduced voltage since they maintain in equilibrium the IB and CB, i.e., no Fermi-level separation is possible and the gap is limited by E_H [90,91]. Decreasing the temperature uncouples the states from the band. In this way, the first measurement of a voltage limited by E_G in a sample showing sub-bandgap light generation was obtained. It was also found that field-assisted escape in the space charge regions shortcircuits IB and CB and cannot be avoided by decreasing the temperature. This was solved by increasing the separation between QD layers [92].

To improve the voltage of QD-based devices it is needed to isolate the ground state from excited states so that quasi-Fermi level separation is permitted. Using smaller dots, with 10 nm size instead of the current 16 nm, could help by removing all excited states into the CB. Better material systems with larger E_G and E_L, closer to the IBSC optimum, are being sought [93] with InGaAs/AlGaAs appearing as a promising alternative.

3.4.4.4 Bulk Intermediate Band Materials and Solar Cells

Many *ab initio* calculations have been performed [94] in the search of a material that can present the IB in the bulk bandgap structure. Several compounds have been identified among which $V_{0.25}In_{1.75}S_3$ [95] has been synthetized and experiments are very promising.

Introducing large amounts of a deep level impurity in a conventional semiconductor can be a method to produce an IB from the extended states formed by the overlapping of the impurity-bound wavefunctions. It is argued that Shockley–Read–Hall recombination will then be suppressed, due to the weaker coupling of the extended wavefunctions with the lattice distortion that launches multi-phonon relaxation [70] (in QDs, the confined wavefunctions are in fact many atomic distances wide and, in this sense, they are also extended, so that multiphonon recombination is not favored and, for this reason, it is not needed that they form a band). Ion-implanted Ti in Si [96] and Ti and Fe in low-temperature MBE GaAs have been experimentally investigated and solar cells fabricated, in some cases showing voltage recovery under concentration [97].

Highly mismatched alloys of III-V (InGaNAs) [98] or II-VI (ZnTeO) [99] obtained by MBE form intermediate bands by the so-called band anticrossing mechanism. In both cases, solar cells have been fabricated that present very promising IB behavior.

3.4.4.5 Prospects for Intermediate Band Solar Cells

IBSCs have taken off towards practical development with several materials, both nano-structures and alloys, and technologies having demonstrated the operating principles. Efficiencies are still low, with 18.3% at one sun being the highest value recorded [81] in a QD device, but below the reference cell in the experiment without quantum dots. The IBSC is a high efficiency concept that will surely be used under concentration because of both material cost considerations and operational benefits. In the QD implementation, they could be incorporated to tandem devices drastically cutting down the number of cells needed for ultra-high efficiency with complete fabrication compatibility [100].

3.5 Other Concepts

Once section 3.4 has described the main concepts being investigated to achieve very high efficiency solar cells, in this section we present developments on other topics that could also contribute to enhancing the efficiency of solar cells.

3.5.1 Light Management for High Efficiency Photovoltaics

Materials with a direct bandgap have a high absorption coefficient and, according to Van Roosbroeck–Shockley reciprocity [26], high radiative emission too. The internal emission efficiency can then be high and these materials approach SQ conditions. Light management is needed to: (1) increase absorption, because after equations (3.3)–(3.5) SQ efficiency is almost proportional to the spectrum-weighted absorptivity; (2) extract the emission out of the sample (increase the external efficiency), because at each photon emission-reabsorption-reemission cycle non-radiative recombination introduces a loss; and (3) steer the emission towards the sun, i.e., match the emission *étendue* to the sun *étendue*, as explained in sections 3.2 and 3.3. Stated in other terms [101], the number of supported light modes (loosely speaking, photon states) in the absorber must be increased and all the modes must be put in contact with the external modes in the sun *étendue*.

Indirect gap materials are poor absorbers and emitters near the bandgap and are farther from their SQ limit. With them the objective is to increase absorption by light trapping, but the same prescription is necessary: to increase the number of internal modes and to put them in contact, i.e., to illuminate them, with the external modes from the sun. Hence, emission extraction and absorption are closely related phenomena by virtue of reciprocity relationships [102,103]. Light trapping is becoming increasingly important due to the interest in very thin solar cells.

Thick solar cells are usually described in the ray optics approximation. The modes considered are propagating plane waves described as rays carrying radiance. In planar samples, the illuminated rays are within the total internal reflection cone. Textures, which are front and/or back layers able to spread the light directions after transmission and reflection, are used to connect these modes with the remaining modes outside the cone. The light trapping limit for very weak absorption, which is the absorption enhancement with respect to the planar cell without mirror, is attained when all modes are filled with the sun radiance, and is the celebrated $4n^2$ in the case of not limiting the emission cone, and $4n^2/\sin(\theta_x)^2$ in the case of matching the sun *étendue*, with θ_x the semiangle of the light cone from the concentrator [21].

Besides chemical or physical etching, metal [104] and dielectric [105] nanoparticle layers have been used as light-spreading layers. Then, back mirrors, external optics (concentrators or external light confining devices [21]) or angularly-selective filters [106] match the emission and the sun *étendues*. Diffraction gratings can serve both purposes of connecting all internal modes between themselves and with the sun *étendue* [107].

In order to increase the density of modes, those not accounted for in the ray optics regime must be considered. In reference [101] a number of options to increase it are presented, based on using nanostructures or other materials in the vicinity of the absorber. Physical optics must be used to study these structures but the very general principles of detailed-balance and the main conclusions of the SQ analysis are still valid. Absorption and emission rate (and thus internal efficiency) increase with the density of states by the Purcell enhancement effect [108]), but are still linked through reciprocity relations.

For instance, evanescent, guided waves can be excited in the absorber surface if sandwiched between higher refractive index materials: these modes can contain electromagnetic energy and are not accounted for in the ray optics description. The evanescent modes associated to metal-dielectric planar interfaces (surface plasmon polaritons, SPPs [109]) can also be used; a diffraction grating or other structure must couple them to illuminated modes.

Metal nanoparticles in a dielectric medium (sustaining resonant localized SPPs) also support evanescent fields. If these modes are 'filled', significant enhancements of the electric field that drives absorption/emission can be obtained [110,111]. Due to the short range, evanescent nature of these modes, the nanoparticles must be immersed in the absorber.

In general, nanostructures with different refraction index excite local near fields. Dielectric particles in the range of the wavelength acting as nanolenses have been proposed too [112]. Note that nanostructures are used for light management in PV in two different regimes: placed on the surfaces, as light scatterers in the far field, and immersed in the absorber to excite near fields and increase the mode density [113].

If the absorber dimensions are in the order of the wavelength, the density of modes is altered with respect to the bulk value and becomes sensitive to its environment, for instance in ultra-thin solar cells. Absorption is greatly enhanced near resonances and small volumes can collect a high quantity of light [114]. In *photonic bandgap* (PBG) structures, the absorber is periodically structured so that light bands are formed. Near the edges the density of states greatly increases, but is reduced at other frequencies.

In the physical optics regime, for a given absorber volume, the ray optics limit can be overcome, at least in a certain wavelength range [115]. A controversial point is whether in thin solar cells the absorption reduction due to decreased volume can be compensated by resorting to guided resonances.

3.5.2 Spectrum Conversion

It has been shown in section 3.3.2 that a solar cell can convert bandgap quasi-monochromatic light with a very high efficiency. The losses quantified by the SQ analysis are due to the broad-band nature of sunlight and so it is natural to consider as a first stage in PV conversion to redistribute the sun energy in a narrower spectral range able to be efficiently converted by even conventional solar cells. This spectrum conversion process is however subject to the same thermodynamic and reciprocity constraints so that it brings energy losses by itself. Several schemes, depicted in Figure 3.12, have been proposed and in some cases experimentally realized:

- *Thermophotovoltaics* (TPV) is based on the converter in Figure 3.2 with the Carnot engine replaced by a solar cell. Concentrated sunlight heats a refractory material receiver that emits (with the assistance of energy-selective emitters or band-pass filters to narrow the spectrum)

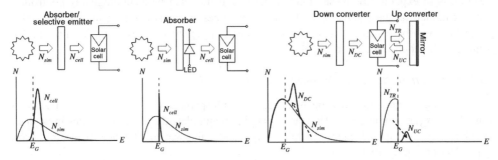

Figure 3.12 Spectrum conversion systems; from left to right: thermophotovoltaics, thermophotonics and down- and up-conversion. Photon spectra are shown to illustrate the processes

towards the solar cells, arranged around it to increase the useful *étendue*. In spite of the high theoretical efficiency limit, experimental efficiencies are modest at the moment [116,117].

• Termophotonics is a development of TPV aimed at decreasing the operating temperature, thus avoiding heat losses and simplifying material issues, etc [118]. Sunlight heats a receiver that has an LED diode attached; the hot LED, with the same bandgap as the converter solar cell to which its luminescence radiation is directed, works under forward bias so that some electrical energy is consumed, but can very efficiently convert heat to luminescent, narrow band radiation. Ideal efficiencies are very high, but practical LEDs with the extremely high external quantum efficiency needed are not available yet.

• Up- and down-conversion is based on quantum luminescent conversion at room temperature to modify parts of the spectrum. Up-converters should produce useful photons, i.e., with energy higher than the bandgap E_G of the solar cell, out of low energy photons that would pass through unabsorbed, and for this reason they can be placed on the back of the cell. Down-converters on the illuminated face of the cell 'cut' photons with energy higher than $2E_G$ into two near-bandgap photons and thus decrease thermalization loss.

Luminescent spectrum converters have been modeled as three-band materials [119,120]. They are very similar in operation and detailed balance efficiency to IB and MEG converters, respectively, the advantage of spectrum conversion lying on its opening the range of useful materials by separating the functions of spectrum modification and PV conversion.

Their practical implementation has resorted to nanocrystals [121] and, specially, to rare-earth elements in oxide powders. There are a large number of possible material combinations with a rich variety of phenomena for up- and down-conversion (see [122] for a recent review). A common disadvantage is that the response tends to be narrow band.

The many experiments performed use spectrum conversion as a way of improving the efficiency of flat-plate modules rather than as a high efficiency, high concentration option, though the efficiency of photon conversion processes usually increases with light intensity.

3.6 Nanostructures in Solar Cells

In the preceding sections it has been shown that scientists have frequently resorted to nanostructures to realize the new PV concepts overcoming the SQ limit. Nanostructures offer the possibility of obtaining materials that show the demanded properties: slowed-down carrier cooling, engineered density of states, enhanced carrier multiplication, etc. The objective of this section is to explain how these effects emerge and how can be technologically controlled. Simple models for the calculation of the electronic states in nanostructures and their interaction with light will be described, as these are the most important features that will allow understanding the application to solar cells.

3.6.1 Electron States in Nanostructures

Semiconductor nanostructures consist of regions with spatial variation of the materials at very small scales (i.e. nanometers). They are created to manipulate the electronic states in devices: their density, their spatial extent and their energy, and in this way to influence the electronic and optoelectronic properties.

To affect the electronic states the dimensions of these regions must be in the nanometer range, in the order of the electron wavelength in the semiconductor. If they were much greater the bulk states would not be affected. On the other hand, since the interatomic distance is around 0.5 nm, nanostructures contain many atoms and electrons inherit many properties of the bulk states. The existence of these two scales inspires the modeling tools used most often that will be briefly described in the following.

Almost all the results presented are based on the one-electron, empty-lattice approximation. Although many-body effects in nanostructures constitute one of the most active research fields, they are perhaps not very important for photovoltaics and, at any rate, are completely out of the reach of this work. Likewise, spin is not changed during light absorption, so it will be generally ignored (except for the density of states) unless otherwise specified.

3.6.1.1 The Single Band, Effective Mass Approximation

A widely used model to calculate energy levels in semiconductor nanostructures is the so called single band, effective mass approximation [123]. In an infinite crystal, the electron or hole wavefunction is given by a Bloch function consisting of a periodic function varying at the atomic scale times a plane wave with wavenumber $\mathbf{k}$:

$$\Xi_{\mathbf{k}\nu}(\mathbf{r}) = e^{i\mathbf{k}\cdot\mathbf{r}} u_{\mathbf{k}\nu}(\mathbf{r})/\sqrt{\Omega}, \tag{3.19}$$

where $\mathbf{r}$ is the position vector, ν denotes the band index, $\mathbf{k}$ is restricted to the first Brillouin zone and Ω is the full volume where the wavefunctions are confined.

In the single band, *effective mass approximation* (EMA) the wavefunction is written as the product of a slowly varying envelope function $\Psi(\mathbf{r})$ and the periodic part of the Bloch function belonging to the perfect crystal solution for a given $\mathbf{k_0}$ value in the band considered, $u_{\mathbf{k}0\nu}(\mathbf{r})$:

$$\Xi(\mathbf{r}) = \Psi(\mathbf{r}) u_{\mathbf{k}0\nu}(\mathbf{r}), \tag{3.20}$$

This is a good approximation to the true wavefunction if: (1) electrons or holes near $\mathbf{k_0}$ are considered, (2) only one band is involved (which discards valence-band wavefunctions in zinc-blende or diamond crystals, and staggered heterojunctions), (3) the Bloch wavefunctions are very similar in the materials involved and (4) the perturbing potential varies slowly in a crystal unit cell (see [123] for a thorough discussion of the validity of the approximation).

The envelope function then fulfills the effective mass equation:

$$E\psi(\mathbf{r}) = -\frac{\hbar^2}{2m^*}\nabla^2\psi(\mathbf{r}) + V(\mathbf{r})\psi(\mathbf{r}), \tag{3.21}$$

where $V(\mathbf{r})$ is the perturbing potential, in the case of a heterojunction the variation of the energy of the $\mathbf{k_0}$ point, and m^* the effective mass that contains the effect of the periodic crystal potential. This equation would also hold for hole wavefunctions near the maximum of a valence band, with the energy increasing away from the maximum. For abrupt material changes, as normally considered in the study of heterostructures [124], the approximation holds inside each region and solutions must be connected across boundaries. The boundary conditions normally used are: (1) the continuity of the envelope function and (2) of its normal derivative divided into the effective mass, which is related to the continuity of the particle flux through the interface. Since the effective mass equation is an approximation, there is still some discussion on this second condition [124,125] and its implications in the approximation.

Notice that $u_{k0_L}(\mathbf{r})$ is a periodic function and that $\Psi(\mathbf{r})$ varies slowly as compared to it. A general property widely used in the mass equation approximations is the following: if $f(\mathbf{r})$ is a periodic function of very short period, repeating itself in each crystal unit cell, and $g(\mathbf{r})$ is a function that varies slowly, then:

$$\iiint f(\mathbf{r})g(\mathbf{r})\mathrm{d}^3r \cong \iiint_{\text{unit cell}} f(\mathbf{r})\mathrm{d}^3r \iiint g(\mathbf{r})\mathrm{d}^3r, \tag{3.22}$$

3.6.1.2 Electron-Photon Interaction Matrix Element

A very important application of the calculation of the quantum states in nanostructured solar cells, and the main reason to bring it here, is the understanding of the absorption and emission of photons. This topic will be considered later in this section but it is opportune to introduce it now since it requires the calculation of the matrix element of this interaction, which is given in quantum mechanics [126] as:

$$\langle \Xi | \boldsymbol{\varepsilon} \cdot \mathbf{r} | \Xi' \rangle = \iiint \Xi^*(\boldsymbol{\varepsilon} \cdot \mathbf{r})\Xi' \mathrm{d}^3r, \tag{3.23}$$

where Ξ and Ξ' are the initial- and final-state wavefunctions respectively and $\boldsymbol{\varepsilon}$ is the light polarization vector. This matrix element is valid when at least one of the wavefunctions is bound. It is not valid for photon interaction with initial and final extended states (see e.g. [127]). By applying the property expressed in equation (3.22) and assuming that $u_{k0_L}(\mathbf{r})$ is normalized in the unit cell and that Ξ and Ξ' belong to the same band:

$$\langle \Xi | \boldsymbol{\varepsilon} \cdot \mathbf{r} | \Xi' \rangle = \iiint \Psi^*(\boldsymbol{\varepsilon} \cdot \mathbf{r})\Psi' \mathrm{d}^3r, \tag{3.24}$$

which depends only on the envelope function.

When the initial and final wavefunctions cannot be approximated in a single band, Eq. (3.24) is not applicable and we shall be forced to use a multiband approach. This will be presented later in this section.

3.6.1.3 The Rectangular Step in One Dimension

A convenient, simple model for approaching the electronic configuration of nanostructures is the problem of a 1-D rectangular potential step. In heterojunctions, the band discontinuities are very abrupt and develop within a few atomic planes if the material compositional change is abrupt. Band line-up is determined by the differences in work function, interface dipoles, etc. In addition, semiconductor nanostructures are often grown under compressive or tensile stress which leads to changes in the band structure with respect to the bulk material [128].

The 1-D rectangular potential can describe the variation of the band-edge energies in quantum wells normal to the interface planes, taken as the z direction. The time independent equation:

$$\frac{d^2\Psi(z)}{dz^2} + \frac{2m_1^*}{\hbar^2} E\Psi(z) = 0 \quad \text{for } |z| < c$$

$$\frac{d^2\Psi(z)}{dz^2} + \frac{2m_2^*}{\hbar^2}(V - E)\Psi(z) = 0 \quad \text{for } |z| > c \tag{3.25}$$

where different effective masses have been assumed in the well (m_1^*) and the barrier (m_2^*) material. Equation (3.25) has to be supplemented with the four boundary conditions demanding continuity of Ψ and $1/m^* d\Psi/dz$ at the interfaces $z = \pm c$.

Since the potential is symmetric, the solutions have a definite parity, i.e., there are odd and even solutions. Solutions to this textbook problem fall into two categories:

Bound states. These are found for $0 < E < V$ and they are sine (for odd solutions) or cosine (for even solutions) functions with wavenumber $k = (2m_1^*E)^{1/2}/\nabla$ inside the well and evanescent, exponentially decaying functions in the barrier material with decay constants $\alpha = (2m_2^*(V-E))^{1/2}/\hbar$.

Some properties of the solutions follow. The higher the effective mass, the closer the energy levels. The state with the lowest energy is even and features no nodes, the second one is odd and has one node, and so on. The number of states is the integer part of $1 + 2c/\pi\hbar \times (2m_1^*V)^{1/2}$. Hence, there is always at least one confined state and their number increases with the well thickness. Figure 3.13(a) shows the wavefunctions corresponding to the well described in the caption, while Figure 3.13(b) gives the variation of the energy levels with well thickness. It is this possibility of controlling the energy spectrum of electrons that is used in photovoltaic devices.

Every state can accommodate two electrons with opposite spin. The derived energies are valid for an empty well. If a second electron were to be admitted in, in the same state or in another one, the electrostatic energy of the one already present should be incorporated to the equation. Again, the effect is small in many cases.

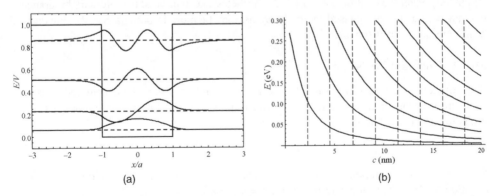

(a)   (b)

Figure 3.13 (a) Bound wavefunctions in the potential well. The wavefunctions are real sine and cosine. The horizontal dashed lines indicate the energy of the state. The absolute amplitude of the functions bears no significance. (b) The energy levels as a function of the well half-thickness. Data for the calculation: $c = 8$ nm; $V = 0.3$ eV; $m^* = 0.06$ m_0, representative of the conduction band in InAs/GaAs quantum dots

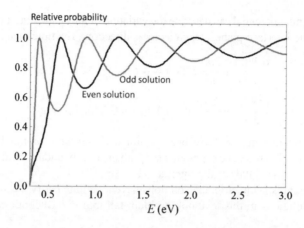

Figure 3.14 Probability of finding the electron inside the well (relative to a plane wave) for extended states as a function of energy

Extended states. For $E > V$, inside the well, as before, there are odd sine solutions and even cosine ones with wavenumber $k = (2m_1^* E)^{1/2}/\hbar$. In the barrier, the wavefunction is also of oscillatory character with a lower wavenumber $k = [2m_2^*(E - V)]^{1/2}/\hbar$. All energies are allowed and, in this representation, the k values are twice degenerate (for odd and even solutions).

The effect of the well on extended waves can be understood as the multiple interference between reflected waves at the well edges. When definite phase relationships are fulfilled, as in an interference optical filter, the wavefunction is more confined in the well. In Figure 3.14 the probability of finding the carrier in the well as a function of energy is presented, relative to the plane wave case encountered for very high energies. States with accumulated probability are important for light absorption from confined levels because of the spatial overlap with true bound states. They are important for carrier transport too; the well is an obstacle for carrier propagation and these effects influence the transmission probability. The resonance is more abrupt the greater the confining energy V. Nevertheless, keep in mind that ordinary hindrance in the transport, as the one caused by the phonons, may be more important than the one derived from pure quantum mechanics.

3.6.1.4 Quantum Wells

Real structures are three-dimensional, so that coordinates x, y must be considered. Quantum wells are invariant under continuous transverse translation so that the eigenfunctions of the problem can be written as:

$$\Psi(x, y, z) = \zeta(z)e^{i\mathbf{k_t}\cdot\mathbf{r}}, \tag{3.26}$$

Where $\mathbf{k_t}$ has no z component. The functions are extended in x,y. The envelope Schrödinger equation lends itself to the 1-D form; if $m^* = m_1^* = m_2^*$:

$$-\frac{\hbar^2}{2m^*}\frac{d^2\zeta(z)}{dz^2} + V(z)\zeta(z) = \left(E - \frac{\hbar^2 k_t^2}{2m^*}\right)\zeta(z), \tag{3.27}$$

where $\zeta(z)$ is a function of z only. The rectangular step equation is obtained, except for the transverse kinetic energy. If the effective mass is not the same in both materials an average value is to be used [124], but this is usually a small effect. The energy of eigenfunctions (see equation (3.27)) is the energy of the 1-D problem in z plus the transverse kinetic part:

$$E_{n\mathbf{k}_t} = E_n + \frac{\hbar^2 k_t^2}{2m^*},$$
(3.28)

The density of states (DoS) for a quantum well structure is a step function for each bound level of the 1-D problem:

$$g_{2D}(E) = \frac{2}{L_z} \times \frac{m^*}{2\pi\hbar^2} \sum_{n=1}^{N} U(E - E_n),$$
(3.29)

where

$$U(t) = \begin{cases} 0 \text{ for } t < 0 \\ 1 \text{ for } t \geq 0 \end{cases},$$
(3.30)

$U(t)$ is the Heaviside unit step function, N the number of bound solutions and a factor 2 is put at the front to account for spin degeneracy. L_z is the length of the well structure in the z direction and has no physical effect other than transforming the 2-D DoS of the quantum well into a density per unit volume. If we have a *multi quantum well* (MQW) structure, L_z is chosen as the distance between consecutive wells. The DoS is continuous for $E > E_1$.

For $E > V$ the continuum of propagating states in the three dimensions is found; its density is given by the well-known expression:

$$g_{3D}(E) = 2 \times \frac{\sqrt{2}(m^*)^{\frac{3}{2}}}{2\pi^2\hbar^3} \sqrt{E - V}\, U(E - V),$$
(3.31)

Here the effective mass should be that of the barrier. Though the density of 3-D states is the same as for the homogeneous material, the wavefunctions are distorted by the presence of the well [129], as discussed before. On the other hand, there are 2-D states in (3.29) with total energy greater than the barrier that become immersed in the continuum. These states are evanescent modes in the z direction and cannot be excited in a perfect infinite crystal: they appear due the presence of the well. Both effects, the modification of the continuum wavefunctions and the presence of the virtually bound ones, are important for light absorption.

3.6.1.5 Quantum Dots

Self-assembled quantum dots can be modeled as parallelepipeds of dimensions a, b, c. The solution can be readily got from the analysis of the 1-D potential step [130] if the true three dimensional well is approximated by the sum of three one-dimensional ones as illustrated in Figure 3.15. In the corner and edge regions the assumed potential is larger than the intended one, but it can be shown using perturbation theory that the effect is small [125,131], at least for the

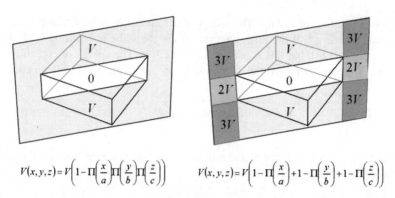

$$V(x,y,z) = V\left(1 - \Pi\left(\frac{x}{a}\right)\Pi\left(\frac{y}{b}\right)\Pi\left(\frac{z}{c}\right)\right) \qquad V(x,y,z) = V\left(1 - \Pi\left(\frac{x}{a}\right) + 1 - \Pi\left(\frac{y}{b}\right) + 1 - \Pi\left(\frac{z}{c}\right)\right)$$

Figure 3.15 Three dimensional potential well and approximation for separation of variables solution. $\Pi(t) = \begin{cases} 1 \text{ for } |t| \leq 1 \\ 0 \text{ for } |t| > 1 \end{cases}$. The graphs show the potential in a meridian, diagonal plane

low energy states. Separation of variables for this approximate potential allows expressing the solution as the product of three functions of the one-dimensional step, bound or extended.

For solutions confined along the three directions, the energy, labelled by the order number of the three solutions $\{n_x, n_y, n_z\}$, is:

$$E_{n_x n_y n_z} = E_{n_x} + E_{n_y} + E_{n_z}, \tag{3.32}$$

The DoS of confined states is a series of $N_x N_y N_z$ deltas:

$$g_{0D}(E) = \frac{2}{L_x L_y L_z} \sum_{n_x=1}^{N_x} \sum_{n_y=1}^{N_y} \sum_{n_z=1}^{N_z} \delta\left(E_{n_x} + E_{n_y} + E_{n_z}\right), \tag{3.33}$$

Some of these states (called virtual bond states) can have an energy greater than the barrier and thus are merged into it, but, representing regions of accumulation of the probability density, they have a role in absorption of light, as already mentioned.

Taking one or two of the one-dimensional functions to be extended, one finds wire- and well-like states (i.e. one-dimensional and two-dimensional extended states, respectively) with energies greater than V and evanescent in one or two dimensions. The importance of these states for absorption has been addressed [131]. Besides, there is the 3-D continuum of the barrier material with density given in equation (3.31).

3.6.1.6 Spherical Quantum Dots

While box shaped quantum dots are a reasonable approximation for epitaxially grown structures, colloidal quantum dots –also known as *nanoparticles* (NP) – are better simulated by a sphere. The nanoparticle surface is usually covered with a polymer or with a wider bandgap semiconductor, and they are immersed in the host material. The potential is then constant inside the dot and a barrier appears at the surface. This barrier can be very large if no charge transfer is intended between the dots and the surrounding medium, as is the case in photoluminescence applications.

Applying separation of variables to the envelope function, spherical harmonics are obtained for the angular part while the radial part fulfills the spherical Bessel equation [132]. There are both a discrete spectrum (for which an eigenvalue problem must be solved) and a continuous spectrum. The energy of the confined states is largely determined by the dot radius, so that the optoelectronic properties can be easily tuned.

There seem to be important differences though with the parallelepiped dots described above since there are no bound states with energies larger than the barrier and no analogues to the 1D and 2D states that appear surrounding the parallelepipedic dots [131], perhaps implying different light absorption properties related to the shape of the nanostructures. Resonant states within the continuum are also found [132].

3.6.1.7 Superlattices

The penetration of the confined solution into the barrier region is given by $1/\alpha = \hbar/\sqrt{2m^*(V-E)}$, in the nanometer range. If the structures are closer than that, the overlap of the wavefunctions makes the energies to separate. In periodic arrangements called super-lattices a miniband is formed from the interacting, bound states. The envelope function equation mimics at a greater scale the electron Schrödinger equation for a crystal; in case of the abrupt potential steps assumed here, the Kronig–Penney [133] model is applicable.

Only in the case of quantum dot superlattices an isolated energy band forms. With wells and wires, the DoS of the isolated structure is already continuum so that no gaps are found in the superlattices either.

3.6.1.8 Multiband k·p Models

The most important solar cell mechanism, which is the pumping of electrons from the valence band to the conduction band cannot be treated with the formalism so far presented because only photonic transitions within one band have been considered. Letting aside *ab initio* calculations, almost impossible with nanostructured semiconductors because they require to deal with over 10 000 atoms, in nanostructured cells more suitable tools are the so called k·p models. This topic is also very vast and only the method based in the so called *empirical k·p Hamiltonian* (EKPH) [134,135] will be considered here. It uses four bands which are the conduction band (*cb*), and the three valence bands in zinc-blende crystals: heavy holes (*hh*), light holes (*lh*) and split off (*so*) bands. In this model the bands are considered spin-degenerated but in fact there are separated spin up and spin down bands for each one of them. In this chapter the model will not be justified (further reading is available for example in [134,135]), and only its use will be explained.

The periodic functions $u_{k0\nu}(\mathbf{r})$ corresponding to the EKPH model are denoted $|X\rangle, |Y\rangle, |Z\rangle, |S\rangle$, the three first corresponding to the $k=0$ point of the valence band and the fourth corresponding to the $k=0$ point of the conduction band. A general wavefunction, represented in the one band model by equation (3.20) is represented in the EKPH model by:

$$\Xi(r) = \Psi_X(\mathbf{r})|X\rangle + \Psi_Y(\mathbf{r})|Y\rangle + \Psi_Z(\mathbf{r})|Z\rangle + \Psi_S(\mathbf{r})|S\rangle, \tag{3.34}$$

where $\Psi_X(\mathbf{r}), \Psi_Y(\mathbf{r}), \Psi_Z(\mathbf{r})$ and $\Psi_S(\mathbf{r})$ are the envelopes of $|X\rangle, |Y\rangle|Z\rangle$ and $|S\rangle$, respectively. They represent the probability of the electron being projected on one of the Bloch periodic function; thus $\iiint \Psi_X^* \Psi_X \mathrm{d}^3 r$ is the probability of being projected on $|X\rangle$. Within this framework, assuming that $\Xi(\mathbf{r})$ is normalized to one, the one-band approximation is valid for as far as $\iiint \Psi_S^* \Psi_S d^3 r$ is close to unity, leaving small values for the norm of the other envelope functions. To calculate the envelope functions the following four steps are to be followed:

a. calculate the diagonalized EKPH eigenfunctions;
b. perform their Fourier transform;
c. multiply by the pertinent matrix element;
d. perform the inverse Fourier transform.

a. *Calculating the diagonalized EKPH eigenfunctions*
   The diagonalized Hamiltonian for a given band is the same used for the single band effective mass approximation, given in equation (3.21). The effective mass is the one corresponding to the band in consideration: *cb, hh, lh,* or *so*. For bound functions, it seems more accurate to use the value corresponding to the nanostructure. The use of averaged values of effective masses or modified Hamiltonians has to be considered carefully. For the valence bands (*hh, lh* and *so*) the effective masses are negative. The potential is the corresponding band offset. For valence bands, Eq. (3.21) may be changed of sign so that pedestals in the band offsets behave like wells and bind the wavefunctions, but the eigenenergies obtained are changed of sign. Putting together the full energy spectrum, both in the valence and the conduction band, requires the proper consideration of the potential origin of each single band Schrödinger equation and which energies go upwards from the potential origin and which downwards. An example, corresponding to the InAs/GaAs QDs labeled SB in [92], is given in Figure 3.16. The results from the separation of variables method explained above constitute the zero order approximation. The first order approximation [131] is not considered in this work, but is very close to the exact values and also drawn for comparison purposes. Note that the differences are very important for the high energy bound levels within the conduction band. However the zero order approximation is very accurate for the levels located inside the bandgap.
   The obtained eigenfunctions are $\Phi_{cb}^{(n_x,n_y,n_z)}$, $\Phi_{lh}^{(n_x,n_y,n_z)}$, $\Phi_{hh}^{(n_x,n_y,n_z)}$ and $\Phi_{so}^{(n_x,n_y,n_z)}$. We present in Figure 3.17, as an example, diagonalized (1,2,1) eigenfunctions for the *cb*, the *hh* band and the *lh* band. Note how due to the higher effective mass the heavy hole state is more confined into the dot.

b. *Performing their Fourier transform*
   The wavefunctions obtained above must now be developed in plane waves. Mathematically speaking, this means obtaining the 3D Fourier transform of a function in the three space variables (x, y, z) to the variables (k_x, k_y, k_z). Depending on the case, this transform may be a Fourier integral or a discrete Fourier transform. In few cases the Fourier integral may be obtained analytically, but usually the required numeric integration is rendered identical to the discrete Fourier transform with simple algorithms. These transforms may accept multiple definitions that differ in a constant of proportionality, in the variables $\mathbf{k}$, which may also differ in a constant of proportionality, and in the field of variation of $\mathbf{k}$ that can extend, or not, symmetrically, around $k = 0$. The symmetric case is used in this text and this might require performing permutation of terms when calculating it with standard Fourier transform subroutines. The exact definitions of the plane wave developments and their exact relationship with the Fourier transforms (integral and discrete) can be found in [127].

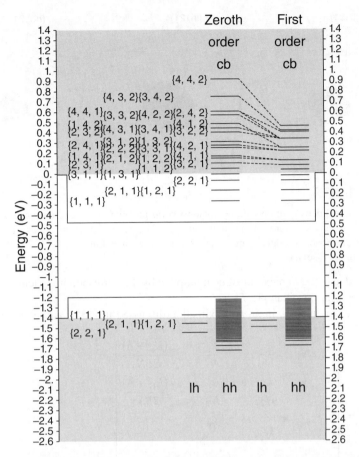

Figure 3.16 Energy spectrum calculated for the *cb*, *hh* and *lh* bands (for *so* band the energies are not calculated because they are too low and do not play a significant role in the subbandgap photon absorption). Separation of variables calculation: the zeroth and first order calculations are shown in parallel

c. *Multiplying by the pertinent matrix element*

The diagonalized EKPH eigenfunctions must be converted into the envelopes through a transformation matrix:

$$[T] = \begin{bmatrix} T_{cb}^X & T_{cb}^Y & T_{cb}^Z & T_{cb}^S \\ T_{lh}^X & T_{lh}^Y & T_{lh}^Z & T_{lh}^S \\ T_{hh}^X & T_{hh}^Y & T_{hh}^Z & T_{hh}^S \\ T_{so}^X & T_{so}^Y & T_{so}^Z & T_{so}^S \end{bmatrix}, \tag{3.35}$$

To obtain one of the envelopes, e.g. the *Y* envelope function, the Fourier transform of the diagonalized EKPH eigenfunctions obtained with the effective mass and the offset of a

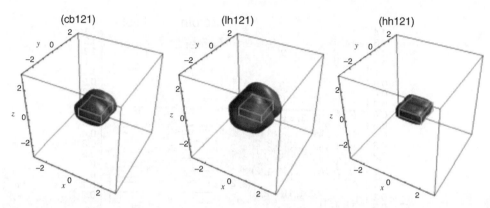

Figure 3.17 Equal density of probability contours corresponding to 0.1% of their maximum value for diagonal EKPH eigenfunctions of quantum numbers (1,2,1) for different bands. The central box represents the QD. Space coordinates are in unities of 10 nm. Source: Luque et al, 2012 [135]. Reproduced with permission of Elsevier

given band, e.g. the *lh* band, must be multiplied by the corresponding element of matrix in the equation, in our example T^Y_{lh}.

Values of the matrix, in an approximation discussed in [135], are given now:

$$[T] = \begin{bmatrix} k_x/N_1 & \dfrac{k_y}{N_1} & \dfrac{k_z}{N_1} & \dfrac{(E_G + A)}{2BN_1} \\[2ex] k_x/N_2 & \dfrac{k_y}{N_2} & \dfrac{k_z}{N_2} & \dfrac{(E_G - A)}{2BN_2} \\[2ex] -k_y/N_3 & \dfrac{k_x}{N_3} & 0 & 0 \\[2ex] k_x k_z/N_4 & \dfrac{k_y k_z}{N_4} & -\dfrac{\left(k_x^2 + k_y^2\right)}{N_4} & 0 \end{bmatrix}$$

$$A = \sqrt{E_G^2 + 4B^2\left(k_x^2 + k_y^2 + k_z^2\right)} \qquad , \qquad (3.36)$$

$$N_1 = \sqrt{8B\left(k_x^2 + k_y^2 + k_z^2\right) + 2E_G\left(E_G + \sqrt{E_G^2 + 4B^2\left(k_x^2 + k_y^2 + k_z^2\right)}\right)}\Big/2B$$

$$N_2 = \sqrt{8B\left(k_x^2 + k_y^2 + k_z^2\right) + 2E_G\left(E_G - \sqrt{E_G^2 + 4B^2\left(k_x^2 + k_y^2 + k_z^2\right)}\right)}\Big/2B$$

$$N_3 = \sqrt{\left(k_x^2 + k_y^2\right)}; \quad N_4 = \sqrt{\left(k_x^2 + k_y^2\right)\left(k_x^2 + k_y^2 + k_z^2\right)};$$

$$B = \sqrt{E_G(m/m^*_{cb} - 1)\hbar^2/2m}$$

Since the *k*-values used above correspond to the plane-wave development often used in quantum mechanics (they are the coordinates of the reciprocal space in which the energy bands are often represented), the Fourier transforms for the diagonalized EKPH eigen-functions must be compatible with this development.

d. *Obtaining the envelope functions*

The envelope functions are now obtained by calculating the inverse Fourier transforms of the (k_x, k_y, k_z)-functions obtained above. In particular, for the envelope functions corresponding to a given energy level pertaining to a given band the procedure above must be repeated with the product of the elements of the band line of the [T] matrix and the Fourier transform of the diagonalized EKPH eigenfunction for this band. Several examples are given in Figure 3.18. They correspond to envelopes derived from the EKPH eigenfunctions drawn in Figure 3.17.

3.6.1.9 Excitons

Excitons are two-particle solutions to the Schrödinger equation representing an electron-hole pair bound by their coulombic attraction, in a similar fashion to electron states introduced by shallow dopants. Bulk excitons have in general low binding energy (less than 10 meV). In quantum wells the energy is higher because carrier confinement brings the carriers closer to one another, but still low enough for them to be dissociated at room temperatures [125].

Excitons are visible as absorption peaks in quantum-well solar cells [136]. In quantum dots they manifest themselves by a lowering of the energy of electron-hole transitions and by different recombination kinetics [137].

3.6.2 Light Absorption by Nanostructures

According to [134], the absorption coefficient (dimension: the inverse of a length) for a beam of photons incident on a material with QDs pumping electrons form a lower level to an upper level separated by an energy E_{line}, is:

$$\alpha^{max} = 2 \times \frac{2\pi^2 e^2 E}{n_{ref} ch\varepsilon_0} \frac{|\langle \Xi | \mathbf{r} \cdot \mathbf{\varepsilon} | \Xi' \rangle|^2}{4ab} FN_l \delta(E - E_{line}), \tag{3.37}$$

where ab is the area of the QD base, n_{ref} is the index of refraction of the material, F is the fraction of semiconductor area occupied by QDs and N_l is the number of QD layers per unit length. Universal constants also appear in the formula with their usual symbols. This type of transitions between bound states, which are the most important, have the shape of a Dirac delta. Due to non-homogeneities, the Dirac delta can be replaced by some of its approximations, such as a Gaussian.

In the case of a single band approximation the matrix element appearing in the absorption can be applied straightforwardly, by using Eq. (3.24). Usually the incoming light impinges as electromagnetic waves along the z-direction and therefore, the polarization vector can only be in the x- or y-direction. Therefore the matrix elements are $\iiint \Psi^* x \Psi' d^3 r$ or $\iiint \Psi^* y \Psi' d^3 r$. For non-polarized radiation the factor to use is $\left(|\iiint \Psi^* x \Psi' d^3 r|^2 + |\iiint \Psi^* y \Psi' d^3 r|^2 \right)/2$. The 2 before the multiplication sign of Equation (3.37) refers to the two spins involved; transitions conserve the spin. Finally the super-index '*max*' refers to the case where the initial state is full of electrons and the final state is empty of them. In the general case, the absorption coefficient includes the factor $f_i(1-f_f)$ where f_i and f_f are the electron filling factor of the initial and final

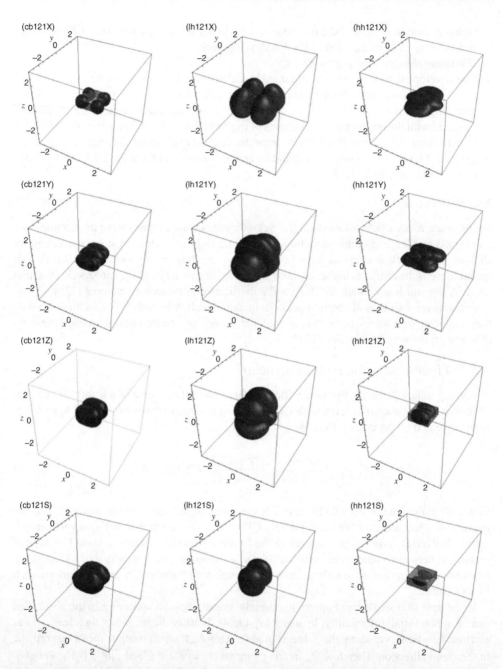

Figure 3.18 Equal density of probability contours corresponding the same values as in Figure 3.17 for the states (1, 2, 1). The band corresponding to each picture (*cb*, *lh* and *hh*) is indicated in the labels, together with the (*X*-, *Y*-, *Z*- and *S*) envelope it corresponds to. Space coordinates are in unities of 10 nm. Source: Luque et al, 2012 [135]. Reproduced with permission of Elsevier

states, respectively. In equilibrium these are Fermi factors but often they have different quasi Fermi level.

This procedure can be used to obtain the absorption coefficients from a given cb state within the bandgap, e.g. cb(1,1,1) to other states also within the bandgap or within the conduction band (virtual bound states). For instance, we may be interested in the transition cb(1,1,1)→ cb(1,2,1) between two states within the bandgap or we may be interested in the transition cb(2,2,1)→cb(3,2,1), the first being within the bandgap and the second being a virtual bound state. The virtual bound states constitute the best path for transitions from bound states within the bandgap to the conduction band. By summing up the absorption coefficients of all the transitions to virtual bound states, the absorption from a bound state within the bandgap to the whole conduction band is obtained. In box shaped QDs, only transitions between states differing in a single quantum number, both of different parity, are permitted. Only transitions relating adjacent quantum numbers are strong.

The transitions we have been referring to are called intraband transitions because they are all are related to the $|S\rangle$ periodic part (in the conduction band) of the Bloch functions.

In the case of the EKPH approximation, the expression is the same as in Eq. (3.37) but:

$$\langle \Xi | \mathbf{r} \cdot \mathbf{\epsilon} | \Xi' \rangle = \iiint \Psi_X^* \mathbf{r} \cdot \mathbf{\epsilon} \Psi'_X d^3 r + \iiint \Psi_Y^* \mathbf{r} \cdot \mathbf{\epsilon} \Psi'_Y d^3 r + \\ + \iiint \Psi_Z^* \mathbf{r} \cdot \mathbf{\epsilon} \Psi'_Z d^3 r + \iiint \Psi_S^* \mathbf{r} \cdot \mathbf{\epsilon} \Psi'_S d^3 r \tag{3.38}$$

Being the latter expression a consequence of the rule in Eq. (3.22) and of the ortho-normal property of the $|X\rangle$, $|Y\rangle$ $|Z\rangle$ and $|S\rangle$ periodic part of the Bloch functions. To calculate the absorption coefficient for a vertical beam of non-polarized photons, the expression in equation (3.38) has to be calculated for $\mathbf{r} \cdot \mathbf{\epsilon} = x$, and then obtain the squared of its absolute value. The same has to be done for $\mathbf{r} \cdot \mathbf{\epsilon} = y$ to finally obtain the average value of both terms.

Limiting ourselves to transitions between bound states, which as it has been said are the most important, the absorption to a given state, e.g. cb(1,1,1), is obtained by summing the absorption coefficients of the transitions of all the lh, hh and so states to this specific state. These transitions are called interband transitions.

The absorption coefficients for a vertical beam of non-polarized photons are represented in Figure 3.19.

According to the legend, in the upper left panel the transitions from all the valence band states into the three bound states within the bandgap (IB states) existing in this specific quantum dot are represented. In the lower left panel, the transitions from the bandgap bound states into the conduction band states are shown. In the upper right panel, transitions between bandgap bound states are presented. Finally in the lower right panel the sub-bandgap transitions between bound states near the valence band and bound states within the conduction band are represented. It is observed that the interband absorptions are rather weak. The initial states of these transitions are states with important X- Y- or Z-envelopes but weak S-envelopes; they are essentially valence band functions. On the contrary, the final function, of the cb, has an important S-envelope (whose norm is almost one) but they have small X- Y- or Z-envelopes. In consequence, all the four terms in Eq. (3.38) are small leading to a small absorption coefficient. The lines corresponding to IB-CB transitions, then, are revealing stronger absorption because initial and final states have a strong S-envelope.

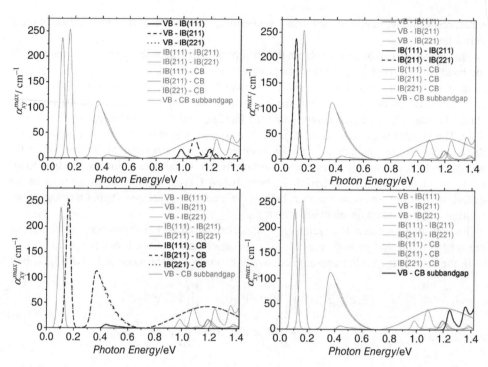

Figure 3.19 Absorption coefficients for non-polarized photons travelling in the z-direction, for the different transitions in the SB prototype of [92] before modification by the Fermi filling factors. Each curve represents a different electronic transition between all the bands and levels shown in Figure 3.16. Source: Mellor et al, 2014 [144]. Reproduced with permission of Wiley

3.6.3 Relaxation, Capture and Recombination in Nanostructures

Relaxation, or thermalization, in a system is the process of recovering equilibrium after some event has disturbed it. High energy light generation creates electrons and holes in states that are barely occupied in equilibrium, so that they lose energy by interaction with the environment. Electron relaxation is faster when the density of states is continuous in energy than when a significant energy gap is to be bridged, as between conduction and valence bands, in whose case the term is recombination. Their mechanisms and rates are very different. Relaxation from a continuum band to the localized states of nanostructures is often referred to as capture, as with defect levels.

Relaxation –or capture and its counterpart, escape– from the conduction band to nano-structure states is affected by carrier confinement effects [60]. The phonon structure is not very much altered in common III-V material systems [138,139] (though its modification by nanostructuring is being explored as commented in section 3.4.2 when dealing with hot carrier solar cells). With quantum wells there is a continuum of states and the situation is similar to the bulk's, but with quantum dots, even if superlattices are formed, there are open gaps between the continuum bands and the bound states, so that the picture can be different. In this respect, hole relaxation is much easier because of the closer level spacing related to the higher effective mass.

If the gap energy is greater than the longitudinal optical (LO) phonons, it is predicted that relaxation will be much slower because less probable, multiphonon events must be invoked: this effect is known as phonon bottleneck [140], and capture times in the order of 1 ns could in principle be expected. In practice, however, and though there are varied experimental results [141,142], times are shorter. There are several reasons, such as polaron formation [143] and carrier-carrier interaction, more important at high carrier densities. Interaction with thermal photons, using the excited bound states of the quantum dot within the bandgap is also able to produce a fast relaxation, at least at room temperature [91,92]. This mechanism has explained well the temperature decrease of the photocurrent in IB solar cells described in 3.4.4 [144], since in these cells capture is seen as a recombination event.

An external electric field, besides altering the energies of confined states [124], makes it possible for carriers to pass through energy barriers by tunneling and tunneling-assisted thermionic effect and increases capture/escape rates. High electric fields are found in the space charge region of solar cells.

Modeling of capture and escape processes is frequently done with a rate equation formalism, in which the rate of a given process is proportional to the concentration of electrons in the initial states, the availability of empty states in the final set and the inverse characteristic time [145,146].

Relaxation to the ground state of nanostructures is considered as part of a recombination event. Recombination in both bulk materials and nanostructures proceeds through defect states via the Shockley–Read–Hall mechanisms, through light emission or through Auger recombination.

3.6.4 Nanostructures for Multijunction Solar Cells

Monolithic multijunction solar cells are the most efficient devices today and are the choice for space and concentrator applications. In series connected stacks, it is very important that the cell bandgaps are well adapted to the working solar spectrum because the series connection severely penalizes current mismatches. Lattice matched material systems offer good photovoltaic quality and high lifetimes, but not all bandgap combinations are available. Metamorphic solar cells allow some lattice mismatch for achieving a better spectral adaptation [4,147,148], but this entails a reduction of the crystal quality. It is not clear then which approach is better with experimental current efficiencies very similar (see Chapter 2 for further details about record efficiencies as of April 2015).

The use of low dimensional nanostructures in the absorber region of a solar cell could retain the best of both approaches and thus push the efficiency forward, especially for multijunction solar cells [149]. Multi-quantum wells can be grown with little lattice distortion and there is an ample freedom for adjusting the energy levels by tuning the well width, as explained in the preceding paragraphs. The quantum well states operate as an extension of the bands, with which they are in equilibrium, so that the effective bandgap is controllable by the well width. If absorption is not complete, the short-circuit current depends on the number of wells. Also, defects are possibly introduced that affect mainly to the open circuit voltage, but also to the short circuit current. In this respect, MQW cells are conventional single junction devices with the usual compromise between short-circuit current and open circuit voltage [150]. GaAsP/InGaAs MQW solar cells have attained 28.3% efficiency under concentrated light [151]. Quantum dots can also been used with this purpose and double and triple junction solar cells have been fabricated [152].

Silicon nanostructures are being investigated [153] in order to develop a silicon-based technology in which high efficiency concepts such as multi-junction, monolithic solar cells are possible.

3.6.5 Fabrication Techniques

Quantum wells are formed by *molecular beam epitaxy* (MBE) or *metal organic chemical vapor deposition* (MOCVD). Both methods are employed for the fabrication of III-V devices with extremely tight control of the composition and the thickness of the deposited layers, requisites for feasible quantum well formation.

Quantum dots of III-V compounds are obtained by MBE and MOCVD too, mainly in the Stransky-Kastranov growth mode [154]. The dot material must present a larger lattice constant than the barrier so that, when a few atomic layers are deposited, three-dimensional islands form to minimize the internal energy associated to strain. These islands can have different shapes, lentil-like or truncated square-based pyramids. Typical InAs/GaAs quantum dots obtained by this technique are shown in Figure 3.20. They are in the 10 nanometer range and shorter in the growth direction than in the normal ones. Before islands form, a monolayer of dot material grows: this is called the Wetting Layer (WL) and in fact constitutes a quantum well lodging 2D electron states that alter the band diagram of the QDs [92]. In-plane densities for growth on (100) planes are in the mid 10^{10} cm^{-2} range. Consecutive QD layers can be stacked by growing barrier layers and then repeating the QD formation cycle. If the barrier is not too thick, the strain propagated from the underlying dots favors nucleation just above them, producing the columnar structures shown in the figure.

Quantum dots can also be formed by chemical reaction in solution (*colloidal quantum dots*, CQD) [60]. These techniques present clear advantages regarding throughput, cost and applicability of the QDs with respect to epitaxial growth. Different methods exist that produce nm-diameter particles with controllable size, very good homogeneity and crystal perfection. II-VI (CdS, CdTe, CdSe) and IV-VI (PbTe, PbS, PbSe) are well-developed QD materials with

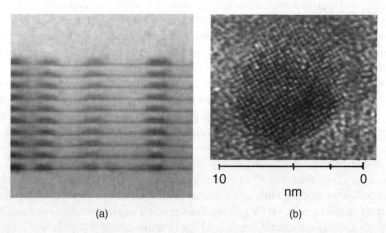

(a)   (b)

Figure 3.20 Transmission electron microscopy images of (a) several layers of InAs/GaAs QDs grown by MBE. Source: Martí et al, 2007 [155]. Reproduced with permission of AIP. (b) a colloidal PbS QD. Source: Bakkers at al, 2000 [156]. Reproduced with permission of Elsevier

a growing spectrum of applications. The large ratio of surface to bulk atoms in these dots makes it very important the role of surface states that can dominate electron recombination and relaxation. Surface passivation is possible with core-shell structures, in which an outer layer with higher gap is grown, and organic molecules are used for terminating surface bonds. Another issue with CQDs is the transfer of electrons to and from the surrounding medium, something necessary to make photovoltaic devices.

Glossary

List of Acronyms

Acronym	Description
CB	Conduction band
CQD	Colloidal quantum dot
DoS	Density of states
EKPH	Empirical k.p Hamiltonian
EMA	Effective mass approximation
HCSC	Hot carrier solar cell
IB	Intermediate band
IBSC	Intermediate band solar cell
LO	Longitudinal optical
MBE	Molecular beam epitaxy
MEG	Multi exciton generation
MOCVD	Metal organic chemical vapor deposition
MQW	Multi quantum well
PBG	Photonic bandgap
PV	Photovoltaic
QD	Quantum dot
QFL	Quasi-Fermi level
SPP	Surface plasmon polariton
SQ	Shockely-Queisser
VB	Valence band
WL	Wetting layer

References

1. Luque, A. and Hegedus S. (eds.), (2011) *Handbook of Solar Energy Engineering*, 2nd edn, Wiley, New York, Chapter 1.
2. Luque, A. (2011) Will we exceed 50% efficiency in photovoltaics? *Journal of Applied Physics* **110**(3), 031301 1–19.
3. Green, M.A., Emery, K., Hishikawa, Y., Warta W. and Dunlop, E.D. (2015) Solar cell efficiency tables (Version 45). *Progress in Photovoltaics: Research and Applications* **23**, pp. 1–9.
4. King, R.R., Boca, A., Hong, W., Liu X.Q. *et al.* (2009) Band-gap-engineered architectures for high-efficiency multijunction concentrator solar cells. Proceedings of the 24th European Photovoltaic Solar Energy Conference, WIP, Munich, 21–25.
5. Green, M.A. (2012) Limiting photovoltaic efficiency under new ASTM International G173-based reference spectra. *Progress in Photovoltaics*, **20**, 954–959.
6. Welford, W. and Winston, R. (1978) *The Optics of Nonimaging Concentrators*, Academic Press, New York, Chapter 2.

7. Würfel, P. (2005) *Physics of Solar Cells*, Wiley VCH–Verlag GmbH andKGaA, Weinhem, Germany, Chapter 4.

8. Luque A. and Martí, A. (2011) in *Handbook of Solar Energy Engineering*, 2nd edn (eds A. Luque and S. Hegedus), Wiley, New York, Chapter 4.

9. Yu P.Y. and Cardona, M. (1999) *Fundamentals of Semiconductors*, 2nd edn, Springer, Berlin.

10. P.T. Würfel, (1997) Solar energy conversion with hot electrons from impact ionisation. *Solar Energy Materials and Solar Cells*, **46**, 43–52.

11. P.G. Klemens, (1966) Anharmonic decay of optical phonons. *Physical Review*, **148**, 45–848.

12. Pötz W. and Kocevar, P. (1983) Electronic power transfer in pulsed laser excitation of polar semiconductors. *Physical Review B*, **28**, 7040–7047.

13. Goodnick, S.M. and Honsberg, C. (2011) Ultrafast carrier relaxation and nonequilibrium phonons in hot carrier solar cells. *Photovoltaic Specialists Conference (PVSC), 2011 37th IEEE* 002066–002070.

14. Luque, A., Martí, A., López, N., Antolín E. *et al.* (2005) Experimental analysis of the quasi-Fermi level split in quantum dot intermediate-band solar cells. *Applied Physics Letters*, **87**, 083505 1–3.

15. Shockley W. and Queisser, H.J. (1961) Detailed balance limit of efficiency of p-n junction solar cells. *Journal of Applied Physics*, **32**, 510–519.

16. Green, M.A. (2006) *Third Generation Photovoltaics*. Springer-Verlag, Berlin, Chapter 3.

17. Markvart, T. (2007) Thermodynamics of losses in photovoltaic conversion. *Applied Physics Letters* **91** 064102 11–3.

18. Landsberg P.T. and Tonge, G. (1980) Thermodynamic energy conversion efficiencies. *Journal of Applied Physics*, **51**, R1–R20.

19. Luque A. and Martí, A. (1997) Entropy production in photovoltaic conversion. *Physical Review B*, **55**, 6994–6999.

20. Ries, H. (1983) Complete and reversible absorption of radiation. *Applied Physics B*, **32**, 153–156.

21. Miñano J.C. (1990) "Optical confinement in Photovoltaics", in *Physical Limitations to Solar Energy Conversion*, (eds A. Luque, G. L. Araújo), Adam Hilger, Bristol, 50–83.

22. Kondepudi D. and Prigogine, I. (1998) *Modern Thermodynamics*, John Wiley and Sons, Chichester (UK).

23. Onsager, L. (1931) Empirical relations in irreversible processes. *Physical Review*, **37**, 405–426.

24. Luque, A., Martí A. and Cuadra, L. (2002) Thermodynamics of solar energy conversion in novel structures. *Physica E*, **14**, 107–114.

25. Rau U. and Brendel, R. (1998) The detailed balance principle and the reciprocity theorem between photocarrier collection and dark carrier distribution in solar cells. *Journal of Applied Physics*, **84**, 6412–6418.

26. van Roosbroeck, W. and Shockley, W. (1964) Photon-radiative recombination of electrons and holes in germanium. *Physical Review*, **94**, 1558–1560.

27. Kirchartz, T., Helbig, A., Reetz, W. *et al.* (2009) Reciprocity between electroluminescence and quantum efficiency used for the characterization of silicon solar cells. *Progress in Photovoltaics*, **17**, 394–402.

28. H.P. Baltes in E. Wolf, Ed., *Progress in Optics* **13**, Elsevier, 1976. pp. 1–25.

29. Araújo G.L. and Martí, A. (1994) Absolute limiting efficiencies for photovoltaic energy conversion. *Solar Energy Materials and Solar Cells*, **33**(2), 213–240.

30. Hirst L.C. and Ekins-Daukes, N.J. (2011) Fundamental losses in solar cells. *Progress in Photovoltaics*, **19**, 286–293.

31. Green, M.A. (2012) Analytical treatment of Trivich–Finn and Shockley–Queisser photovoltaic efficiency limits using polylogarithms. *Progress in Photovoltaics*, **20**, 127–134.

32. Miller, O.D., Yablonovitch E. and Kurtz, S.R. (2012) Strong internal and external luminescence as solar cells approach the Shockley–Queisser limit. *IEEE Journal of Photovoltaics*, **2**(3), 303–311.

33. Green, M.A. (2012) Radiative efficiency of state-of-the-art photovoltaic cells. *Progress in Photovoltaics*, **20**(4), 472–476.

34. Barnett, A., Kirkpatrick, D., Honsberg, *et al.* (2009) Very high efficiency solar cell modules. *Progress in Photovoltaics*, **17**(1), 75–83.

35. Martí A., Davies, P.A., Oliván, J. *et al.* (1993) High efficiency photovoltaic conversion with spectrum splitting on GaAs and Si cells located in light confining cavities. *Photovoltaic Specialists Conference, 1993, Conference Record of the Twenty Third IEEE*, Louisville, Kentucky, USA, 768–773.

36. Brown A.S. and Green, M.A. (2002) Limiting efficiency for current-constrained two-terminal tandem cell stacks. *Progress in Photovoltaics*, **10**(5), 299–307.

37. Tobías I. and Luque, A. (2002) Ideal efficiency of monolithic, series-connected multijunction solar cells. *Progress in Photovoltaics*, **10**(5), 323–329.

38. Ross R.T. and Nozik, A.J. (1982) Efficiency of hot-carrier solar energy converters. *Journal of Applied Physics*, **53**, 3813–3818.
39. Würfel, P. (1982) The chemical potential of radiation. *Journal of Physics C: Solid State Physics*, **15**(18), 3967.
40. Würfel, P., Brown, A.S, Humphrey, T.E. and Green, M.A., Particle article conservation in the hot-carrier solar cell. *Progress in Photovoltaics* **13**, pp. 277–285 (2005).
41. Luque A. and Martí, A. (2010) Electron-phonon energy transfer in hot-carrier solar cells. *Solar Energy Materials and Solar Cells*, **94**(2), 287–296.
42. Takeda, Y., Ito, T., Motohiro, T. *et al.*, (2009) Hot carrier solar cells operating under practical conditions. *Journal of Applied Physics*, **105**(7), 07490 5–10.
43. Takeda, Y., Ito, T., Suzuki R., et al., (2009) Impact ionization and Auger recombination at high carrier temperature. *Solar Energy Materials and Solar Cells*, **93**(6–7), 797–802.
44. König, D., Casalenuovo, K., Takeda, Y. *et al.*, (2010) Hot carrier solar cells: Principles, materials and design, *Physica E: Low-dimensional Systems and Nanostructures*, **42**(10), 2862–2866.
45. Clady, R., Tayebjee, M.J.Y., Aliberti, P. *et al.*, (2012) Interplay between the hot phonon effect and intervalley scattering on the cooling rate of hot carriers in GaAs and InP. *Progress in Photovoltaics*, **20**(1), 82–92.
46. Aliberti P. *et al.*, (2010) Investigation of theoretical efficiency limit of hot carriers solar cells with a bulk indium nitride absorber. *Journal of Applied Physics*, **108**, 094507 1–10.
47. Conibeer, G.J., König, D., Green, M.A. and Guillemoles, J.F. (2008) Slowing of carrier cooling in hot carrier solar cells. *Thin Solid Films*, **516**, 6948–6953.
48. Conibeer, G.J., Ekins-Daukes, N., Guillemoles, J.F. *et al.* (2009) Progress on hot carrier solar cells. *Solar Energy Materials and Solar Cells*, **93**, 713–719.
49. Le Bris A., Rodière J., Colin C. *et al.*, (2012) Hot carrier solar cells: Controlling thermalization in ultrathin devices. *IEEE Journal of Photovoltaics*, **2**(4), 506–511.
50. Conibeer, G.J., Jiang, C.W., König, D. *et al.* (2002) Selective energy contacts for hot carrier solar cells. *Thin Solid Films*, **516**, 6968–6973.
51. Veettil, B.P., Patterson, R., König, D. *et al.* (2011) Optimized resonant tunneling structures with high conductivity and selectivity. *EPL (Europhysics Letters)*, **96**, 57006 1–94.
52. König, D., Takeda, Y. and Veettil, B.P. (2012) Lattice-matched hot carrier solar cell with energy selectivity integrated into the hot carrier absorber. *Japanese Journal of Applied Physics*, **51**, 10ND02 1–4.
53. Kolodinski, S., Werner, J.H., Wittchen, T. and Queisser, H.J. *et al.* (1993) Quantum efficiencies exceeding unity due to impact ionization in silicon solar cells. *Applied Physics Letters*, **63**(17), 2405–2407.
54. Werner, J.H., Kolodinski, S. and Queisser, H.J. (1994) Novel optimization principles and efficiency limits for semiconductor solar cells. *Physical Review Letters*, **72**(24), 3851–3854.
55. Spirkl, W. and Ries, H. (1995) Luminescence and efficiency of an ideal PV cell with charge multiplication. *Physical Review B*, **52**, 11319–11325.
56. De Vos A. and Desoete, B. (1998) On the ideal performance of solar cells with larger-than-unity quantum efficiency. *Solar Energy Materials and Solar Cells*, **51**, 413–424.
57. Wolf, M., Brendel, H., Werner, R.J. and Queisser, H.J. (1998) Solar cell efficiency and carrier multiplication in $Si_{1-x}Ge_x$ alloys. *Journal of Applied Physics*, **83**, 4213.
58. Green, M.A. (2006) *Third Generation Photovoltaics*, Springer-Verlag, Berlin, Chapter 7.
59. Schaller R.D. and Klimov, V.I. (2004) High efficiency carrier multiplication in PbSe nanocrystals: implications for solar energy conversion. *Physical Review Letters*, **92**, pp. 166601 1–4.
60. Nozik, A.J. (2001) Spectroscopy and hot electron relaxation dynamics in semiconductor quantum wells and quantum dots. *Annual Review of Physical Chemistry*, **52**(1), 93–231.
61. Nozik, A.J. (2008) Multiple exciton generation in semiconductor quantum dots. *Chemical Physics Letters*, **457**, 3–11.
62. Nozik, A.J., Beard, M.C., Luther, J.M. *et al.* (2010) Semiconductor quantum dots and quantum dot arrays and applications of multiple exciton generation to third-generation photovoltaic solar cells. *Chemical Reviews*, **110**, 6873–6890.
63. Beard, M.C. (2010) Multiple exciton generation in semiconductor quantum dots. *The Journal of Physical Chemistry Letters*, **2**, 1282–1288.
64. Klimov, V.I. (2006) Detailed-balance power conversion limits of nanocrystal-quantum dot solar cells in the presence of carrier multiplication. *Applied Physics Letters*, **89**, 123118 1–3.
65. McClain J. and Schrier, J. (2010) Multiple exciton generation in graphene nanostructures. *Journal of Physical Chemistry C*, **114**(34), 14332–14338.

66. Nozik, A.J. (2002) Quantum dot solar cells. *Physica E* **14**(1–2), 115–120.

67. Beard, M.C., Luther, J.M., Semonin, O.E. and Nozik, A.J. (2013) Third generation photovoltaics based on multiple exciton generation in quantum confined semiconductors. *Accounts of Chemical Research*, **46**(6), 1252–1260.

68. Wolf, M. (1960) Limitations and possibilities for improvement of photovoltaic energy converters. Part I: considerations for earth's surface operation. *Proceedings IRE*, **48**, 1246–1263.

69. Luque A. and Martí, A. (1997) Increasing the efficiency of ideal solar cells by photon induced transitions at intermediate levels. *Physical Review Letters*, **78**(26), 5014–5017.

70. Luque, A., Martí, A., Antolín E. and Tablero, (2006) Intermediate bands versus levels in non-radiative recombination. *Physica B*, **382**(1–2), 320–327.

71. Martí, A., Antolín, E., García-Linares, P. *et al.* (2013) Six not so easy pieces in intermediate band solar cell research. *Proceedings SPIE* 8620, Physics, Simulation, and Photonic Engineering of Photovoltaic Devices II, San Francisco, California, USA, p.86200J.

72. Luque, A., Martí A. and Stanley, C. (2012) Understanding intermediate-band solar cells. *Nature Photonics* **6**, 146–152.

73. Cuadra, L., Martí A. and Luque, A. (2004) Influence of the overlap between the absorption coefficients on the efficiency of the intermediate band solar cell. *IEEE Transactions on Electron Devices* **51**(6), 1002–1007.

74. Lin A. and Phillips, J. (2012) Decoupling spectral overlap intermediate band solar cells using low-high state filling. *Proceedinghs of the 38th IEEE Photovoltaic Specialists Conference (PVSC)*, Austin, Texas, USA, 73–77.

75. Martí, A., Cuadra L. and Luque, A. (2002) Quasi drift-diffusion model for the quantum dot intermediate band solar cell. *IEEE Transaction on Electron Devices*, **49**(9), 1632–1639.

76. Strandberg R. and Reenaas, T.W. (2011) Drift-diffusion model for intermediate band solar cells including photofilling effects. *Progress in Photovoltaics*, **19**(1), 21–32.

77. Martí, A., Antolín, E., Linares P.G. and Luque, A. (2012) Understanding experimental characterization of intermediate band solar cells. *Journal of Materials Chemistry*, **22**, 22832–22839.

78. Martí, A., Cuadra L. and Luque, A. (2000) Quantum dot intermediate band solar cell. *Proceedings of the 28th IEEE Photovoltaics Specialists Conference*, Anchorage, Alaska, USA, 904–943.

79. Luque, A., Martí, A., Stanley, C. *et al.* (2004) General equivalent circuit for intermediate band devices: Potentials, currents and electroluminescence. *Journal of Applied Physics*, **96**, 903–909.

80. Zhou, D., Sharma, G., Thomassesn, S.F. *et al.* (2010) Optimization towards high density quantum dots for intermediate band solar cells grown by molecular beam epitaxy. *Applied Physics Letters*, **96**, 061913 1–3.

81. Blokhin, S.A., Sakharov, A.V., Nadtochy, A.M. *et al.* (2009) AlGaAs/GaAs photovoltaic cells with an array of InGaAs QDs. *Semiconductors*, **43**, 514–518.

82. Bailey, C.G., Forbes, D.V., Raffaelle R.P. and Hubbard, S.M. (2011) Near 1 V open circuit voltage InAs/GaAs quantum dot solar cells. *Applied Physics Letters*, **98** 163105.

83. Guimard, D., Morihara, R., Bordel, D. *et al.* (2010) Fabrication of InAs/GaAs quantum dot solar cells with enhanced photocurrent and without degradation of open circuit voltage. *Applied Physics Letters*, **96**, 203705 1–3.

84. Hubbard, S.M., Cress, C.D., Bailey, C.G. *et al.* (2008) Effect of strain compensation on quantum dot enhanced GaAs solar cells. *Applied Physics Letters*, **92**, 123512.

85. Oshima, R., Takata A. and Okada, Y. (2008) Strain-compensated InAs/GaNAs quantum dots for use in high efficiency solar cells. *Applied Physics Letters*, **93**, 083111.

86. Luque, A., Martí, A., Mellor, A. *et al.* (2013) Absorption coefficient for the intraband transitions in quantum dot materials. *Progress in Photovoltaics*, **21**, 658–667.

87. Luque, A., Mellor, A., Ramiro, I. *et al.* (2013) Interband absorption of photons by extended states in intermediate band solar cells. *Solar Energy Materials and Solar Cells*, **115**, 138–144.

88. Akahane, K., Kawamura, T., Okino, K. *et al.* (1998) Highly packed InGaAs quantum dots on GaAs(311)B. *Applied Physics Letters*, **73**, 3411–3413.

89. Luque, A., Linares, P.G., Antolín, E. *et al.* (2010) Multiple levels in intermediate band solar cells. *Applied Physics Letters*, **96**(1), 013501 1–3.

90. Linares, P.G., Martí, A., Antolín, E. *et al.* (2012) Voltage recovery in intermediate band solar cells. *Solar Energy Materials and Solar Cellsi*, **98**, 240–244.

91. Luque, A., Martí, A., Antolín, E. *et al.* (2011) Radiative thermal escape in intermediate band solar cells. *AIP Advances* **1**, 022125 1–6.

92. Antolín, E., Martí, A., Farmer, C.D. *et al.*, (2010) Reducing carrier escape in the InAs/GaAs quantum dot intermediate band solar cell. *Journal of Applied Physics*, **108**, 06453 1–7.

93. Linares, P.G., Martí, A., Antolín, E. and Luque, A. (2011) III-V compound semiconductor screening for implementing quantum dot intermediate band solar cells. *Journal of Applied Physics*, **109**(1), 014313 1–8.

94. Wahnón P. and Tablero, C. (2002) Ab initio electronic structure calculations for metallic intermediate band formation in photovoltaic materials. *Physical Review B*, **65**, 165115 1–10.

95. Palacios, P., Aguilera, I., Sanchez, K. *et al.* (2008) Transition metal-substituted indium thiospinels as novel intermediate-band materials: Prediction and understanding of their electronic properties. *Physical Review Letters*, **101**, 046403.

96. Antolín, E., Martí, A., Olea, J. *et al.* (2009) Lifetime recovery in ultrahighly titanium-doped silicon for the implementation of an intermediate band material. *Applied Physics Letters*, **94**, 042115 1–3.

97. Linares, P.G., Martí, A., Antolín, E. *et al.* (2013) Extreme voltage recovery in GaAs:Ti intermediate band solar cells. *Solar Energy Materials and Solar Cells*, **108**, 175–179.

98. López, N., Reichertz, L.A., Yu, K.M., Campman, K. and Walukiewic, W. (2011) Engineering the electronic band structure for multiband solar cells. *Physical Review Letters*, **106**, 028701 1–4.

99. Wang, W., Lin A. and Phillips, J.D. (2009) Intermediate-band photovoltaic solar cell based on ZnTe:O. *Applied. Physics Letters*, **95**, 011103 1–3.

100. Antolín, E., Martí, A., Linares, P.G. *et al.* (2010) Raising the efficiency limit of the GaAS-based intermediate band solar cell through the implementation of a monolithic tandem with an AlGaAs top cell. *Proceedings of the 21st European Photovoltaic Energy Conference*, 12–415.

101. Callahan, D.M., Munday J.N. and Atwater, H.A. (2012) Solar cell light trapping beyond the ray optic limit. *Nano Letters*, **12** 214–218.

102. Martí, A., Balenzategui, J.L. and Reyna, R.F. (1997) Photon recycling and Shockley's diode equation. *Journal of Applied Physics*, **82**, 4067–4075.

103. Kosten, E.D., Atwater, J.H., Parsons, J., Polman A. and Atwater, H.A. (2013) Highly efficient GaAs solar cells by limiting light emission angle. *Light: Science and Applications*, **2**(e45), 1–6.

104. Nakayama, K., Tanabe K. and Atwater, H.A. (2008) Plasmonic nanoparticle enhanced light absorption in GaAs solar cells. *Applied Physics Letters*, **93**, 121904 1–3.

105. Chang, T.H., Wu, P.H., Chen, S.H. *et al.* (2009) Efficiency enhancement in GaAs solar cells using self-assembled microspheres. *Optics Express*, **17**, 6519–6524.

106. Peters, M., Goldschmidt, J.C., Kirchartz T. and Bläsi, B. (2009) The photonic light trap – Improved light trapping in solar cells by angularly selective filters. *Solar Energy Materials and Solar Cells*, **93**(10), 1721–1727.

107. Mellor, A., Tobías, I., Martí, A., Mendes M.J. and Luque, A. (2011) Upper limits to absorption enhancement in thick solar cells using diffraction gratings. *Progress in Photovoltaics*, **19**(6), 676–687.

108. Purcell, E.M. (1946) Spontaneous emission probabilities at radio frequencies. *Physical Review*, **69**, 681.

109. Maier, S.A., (2007) *Plasmonics: Fundamentals and Applications*, Springer, New York.

110. Pillai S. and Green, (2010) Plasmonics for photovoltaic applications. *Solar Energy Materials and Solar Cells*, **94**(9), 1481–1486.

111. Mendes, M.J., Luque, A., Tobías, I. and Martí, A. (2009) Plasmonic light enhancement in the near-field of metallic nanospheroids for application in intermediate band solar cells. *Applied Physics Letters*, **95**, 071105.

112. Mendes, M.J., Tobías, I., Martí A. and Luque, A. (2011) Light concentration in the near-field of dielectric spheroidal particles with mesoscopic sizes. *Optics Express*, **19**(17), 16207–16222.

113. Atwater H.A. and Polman, A. (2010) Plasmonics for improved photovoltaic devices. *Nature Materials*, **9**, 205.

114. Wallentin, J., Anttu, N., Asoli, D. *et al.* (2013) InP Nanowire Array Solar Cells Achieving 13.8% Efficiency by Exceeding the Ray Optics Limit. *Science*, **339**, 1057–1060.

115. Yu, Z., Raman A. and Fan, S. (2010) Fundamental limit of nanophotonic light trapping in solar cells. *Proceedings of the National Academy of Sciences USA* **107**(41), 17491–17496.

116. Datas A. and Algora, C. (2012) Global optimization of solar thermophotovoltaic systems. *Progress in Photovoltaics*, **21**, 1040–1055.

117. Datas A. and Algora, C. (2012) Development and experimental evaluation of a complete solar thermophotovoltaic system. *Progress in Photovoltaics*, **21**, 1025–1039.

118. Harder, N.P. and Green, M.A. (2003) Thermophotonics. *Semiconductor Science and Technology*, **18**(5), S270.

119. Trupke, T., Green M.A. and Würfel, P. (2002) Improving solar cell efficiencies by down-conversion of high-energy photons. *Journal of Applied Physics*, **92**(3), 1668–1674.

120. Trupke, T., Green M.A. and Würfel, P. (2002) Improving solar cell efficiencies by up-conversion of sub-band-gap light. *Journal of Applied Physics*, **92**(7), 4117–4122.

121. van Sark, W.G., Meijerink, A. and Schropp, R.E. (2102) Solar spectrum conversion for photovoltaics using nanoparticles in *Third Generation Photovoltaics*, (ed. V. Fthenakis) ISInTech, DOI: 10.5772/39213. Available from: http://www.intechopen.com/books/third-generation-photovoltaics/solar-spectrum-conversion-for-photovoltaics-using-nanoparticles

122. Huang, X., Han, S., Huang W. and Liu, X. (2013) Enhancing solar cell efficiency: the search for luminescent materials as spectral converters. *Chemical Society Reviews*, **42**, 173–201.

123. Datta, S. 1989 *Quantum Phenomena*, Addison-Wesley, Reading, MA, USA.

124. Bastard, G. 1988 *Wave Mechanics Applied to Semiconductor Heterostructures*, Les Éditions de Physique, Paris.

125. Harrison, P. 2005 *Quantum Wells, Wires and Dots*, 2nd ed., John Wiley and Sons, New York.

126. Messiah, A. (1960) *Mécanique Quantique*, Dunod, Paris.

127. Luque, A., Antolín, E., Linares, P.G. *et al.* (2013) Interband optical absorption in quantum well solar cells. *Solar Energy Materials and Solar Cells*, **112**, 20–26.

128. Adachi, S. (1992) *Physical Properties of III-V Compounds*, John Wiley and Sons, New York.

129. Davies, J.H. (1998) *The Physics of Low Dimensional Semiconductors*, Cambridge University Press, Cambridge, UK.

130. Luque, A., Martí, A., Antolín, E. and García-Linares, P. (2010) Intraband absorption for normal illumination in quantum dot intermediate band solar cells. *Solar Energy Materials and Solar Cells*, **94**, 2032–2035.

131. Luque, A., Mellor, A., Tobías, I. *et al.* (2013) Virtual-bound, filamentary and layered states in a box-shaped quantum dot of square potential form the exact numerical solution of the effective mass Schrödinger equation. *Physica B: Condensed Matter*, **413**, 73–81.

132. Buczko, R. and Bassani, F. (1996) Bound and resonant electron states in quantum dots: The optical spectrum. *Physical Review B*, **54**, 2667–2674.

133. Kittel, C. (2005) *Introduction to Solid State Physics*, 8th edn, John Wiley and Sons, NewYork.

134. Luque, A., Martí, A., Antolín, E. *et al.* (2011) New Hamiltonian for a better understanding of the quantum dot intermediate band solar cells. *Solar Energy Materials and Solar Cells*, **95**, 2095–2101.

135. Luque, A., Mellor, A., Antolín, E. *et al.* (2012) Symmetry considerations in the empirical k.p Hamiltonian for the study of intermediate band solar cells. *Solar Energy Materials and Solar Cells*, **103**, 171–183.

136. Kailuweit, P., Kellenbenz, R., Philipps, S.P. *et al.* (2010) Numerical simulation and modeling of GaAs quantum-well solar cells. *Journal of Applied Physics*, **107**, 064317 1–6.

137. Dawson, P., Rubel, O., Baranovskii, S.D. *et al.* (2005) Temperature-dependent optical properties of InAs/GaAs quantum dots: Independent carrier versus exciton relaxation. *Physical Review B*, **72**, 235301 1–10.

138. Tsai, C.Y., Eastman, L.F., Lo Y.H. and Tsai, C.Y. (1994) Breakdown of thermionic emission theory for quantum wells *Applied Physics Letters*, **65**, 469–471.

139. Register, L.F. (1992) Microscopic basis for a sum rule for polar-optical-phonon scattering of carriers in heterostructures. *Physical Review B*, **45**, 8756–8759.

140. Benisty, H., Sotomayor-Torrès, C.M. and Weisbuch, C.H. (1991) Intrinsic mechanism for the poor luminescence properties of quantum-box systems. *Physical Review B*, **44**, 10945–10948.

141. Sosnowski, T.S., Norris, T.B., Jiang, H. *et al.* (1998) Rapid carrier relaxation in In0.4Ga0.6As/GaAs quantum dots characterized by differential transmission spectroscopy. *Physical Review B*, **57**, R9423–R9426.

142. Zibik E. *el al.* (2007) Fast intraband capture and relaxation in InAs/GaAs self-assembled quantum dots, in *Quantum Electronics and Laser Science Conference, 2007. QELS '07*, Optical Society of America, 6–11.

143. Zibik, E.A., Wilson, L.R., Green, R.P. *et al.* (2004) Intraband relaxation via polaron decay in InAs self-assembled quantum dots' *Physical Review B*, **70**, 161305 R1–R4.

144. Mellor, A., Luque, A., Tobías I. and Martí, A. (2014) Realistic detailed balance study of the quantum efficiency of quantum dot solar cells. *Advanced Functional Materials*, **24**, 339–345.

145. Tessler N. and Eisenstein, G. (1993) On carrier injection and gain dynamics in quantum-well lasers. *IEEE Journal of Quantum Electronics*, **29**(6), 1586–1595.

146. Luque, A., Martí, A., López, N. *et al.* (2006) Operation of the intermediate band solar cell under nonideal space charge region conditions and half filling of the intermediate band. *Journal of Applied Physics*, **99**, 094503 1–9.

147. Guter, W., Schone, J., Philipps, S.P. *et al.* (2009) Current-matched triple-junction solar cell reaching 41.1% conversion efficiency under concentrated sunlight. *Applied Physics Letters*, **94**, 223504 1–3.

148. Lumb, M.P., González, M., Vurgaftman, I. *et al.* (2012) Simulation of novel InAlAsSb solar cells. *Proceedings SPIE 8256, Physics, Simulation, and Photonic Engineering of Photovoltaic Devices*, 82560 S1–13.

149. Bushnell, D.B., Tibbits, T.N., Barnham, K.W. *et al.* (2005) Effect of well number on the performance of quantum-well solar cells. *Journal of Applied Physics*, **97**, 124908 1–4.

150. Tobías, I., Luque, A., Antolín, E. *et al.* (2012) Realistic performance prediction in nanostructured solar cells as a function of nanostructure dimensionality and density. *Journal of Applied Physics*, **112**, 124518–124524.
151. Adams, J.G., Browne, B.C., Ballard, I.M. *et al.* (2011) Recent results for single-junction and tandem quantum well solar cells. *Progress in Photovoltaics*, **19**(7), 865–877.
152. Kerestes, C., Polly, S., Forbes, D. *et al.* (2014) Fabrication and analysis of multijunction solar cells with a quantum dot (In)GaAs junction. *Progress in Photovoltaics*, **22**(11), 1172–1179.
153. Cho, E.C., Green, M.A., Conibeer, G.J. *et al.* (2007) Silicon quantum dots in a dielectric matrix for all-silicon tandem solar cells. *Advances in OptoElectronics*, **2007**, 69578 1–11.
154. Bimberg, D., Grundmann M., and Ledentsov, N.N. (1999) *Quantum Dot Heterostructures*, John Wiley and Sons, New York.
155. Martí, A., López, N., Antolín, E. *et al.* (2007) Emitter degradation in quantum dot intermediate band solar cells. *Applied Physics Letters*, **90**(23), 233510 1–4.
156. Bakkers, E.P., Kelly J.J. and Vanmaekelbergh, D. (2000) Time resolved photoelectrochemistry with size-quantized PbS adsorbed on gold. *Journal of Electroanalytical Chemistry*, **482**(1), 48–55.

4

CPV Optics

Rubén Mohedano[1] and Ralf Leutz[2]

[1]*LPI-Europe, S.L.; Spain*
[2]*Leutz Optics and Illumination UG, Germany*

4.1 Introduction

A CPV power plant operates at increased profitability firstly when it is able to produce electricity at a high rate, regularly and for a long period of time (i.e. it is efficient, reliable and durable) and secondly when is based on inexpensive modules. An optical design can help meet these two goals by, for instance, using as few elements as possible (making sense for reducing optical losses), or by choosing an architecture that makes performance less sensitive to tracking, assembly and alignment errors. Efficiency also involves optics producing a 'white' spot on the solar cell without chromatic aberration. Optics might be suitable for large-scale and low-cost production if it shows high manufacturing and aiming tolerances, but always maintaining high concentration (>500×), to minimize the usage of high performance and high cost solar cells.

Sunlight passes optical elements on its way into the solar cell. The succession of optical surfaces and materials is termed the optical train. The optical train has to concentrate the incoming sunlight: take a beam of nearly parallel rays falling at its entry aperture and carry as much as possible of it onto a small (compared to the entry aperture area) solar cell, where the beam is necessarily wider angularly, as we demonstrate below. The different steps involved in this task will be revised in the present chapter. Basic optical concepts and their connection to CPV will be covered in section 4.2, to be later explained in physical detail in Section 4.3. We will look at the way optical designs are created in the imaging and non-imaging traditions of optical design in Sections 4.4 and 4.5, respectively. On the one hand, Section 4.4 is devoted to the calculation of the optical parts of a CPV system, including examples of solar optics meeting practical restraints. On the other hand, section 4.5 deepens the analysis of designs using ray tracing, which is actually a key part of the design process, when it goes through an optimization process. Finally, the manufacturing options for CPV optics will be treated in Section 4.6. Some final thoughts on optics for CPV are given in Section 4.7.

Handbook of Concentrator Photovoltaic Technology, First Edition. Edited by Carlos Algora and Ignacio Rey-Stolle.
© 2016 John Wiley & Sons, Ltd. Published 2016 by John Wiley & Sons, Ltd.

For the interested reader, some classical theoretical topics and illustrative design examples are included in the Annexes 4-I to 4-III.

4.2   Light, Optics and Concentration

The optical designer needs to understand the nature of light and optics well, and the technology of the solar cell as well as possible. The reason is that solar cells only convert light into electricity efficiently in a limited spectral range: the optics must be as efficient as possible at least in that range. The question of finding optimal optical systems became trickier with the success of multijunction solar cells: with the junctions acting as series-connected cells, the challenge is keeping high efficiencies constant throughout the wavelengths that affect each junction independently, rather than keeping a high average value. A sudden transmission loss in a quite narrow spectral range can lead to noticeable photocurrent drops in one of the subcells, limiting the intensity and therefore the power extracted from the whole multijunction cell. In the frequent case of triple-junction 3J solar cells, this problem is more often noticed in the top (blue) and middle (green) cells (sensitive to light in the ranges 380–630 nm and 600–920 nm, respectively), since these subcells are designed with the bottom (red) cell (sensitive to light within 900–1800 nm) producing an excess of photocurrent (compared to that of the other two subcells) for typical sun spectrums. For such cells, the optical train could afford to give up some transmittance in the near infrared region.

An indication of the sensitiveness of the performance of the solar concentrator on solar direct spectrum G 173, the external quantum efficiency (EQE) of the device (here a triple-junction device), and the spectral transmission of the concentrator primary optics' material (here PMMA) is given in Figure 4.1. Note the windows created by absorption of various materials, which will have an influence on the design of subcells.

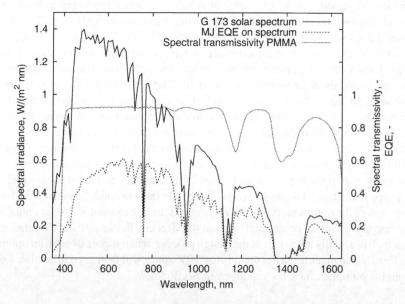

Figure 4.1   Direct solar spectrum G 173, spectral response of a triple-junction device weighted by its external quantum efficiency (EQE); and spectral transmission of PMMA, often the primary optics (Fresnel lens) material

Another noteworthy effect is the diminishing EQE in the ultraviolet (UV) region of the spectrum. Using high UV-transmission materials will increase the performance of the multijunction solar cell only very moderately, while there is a risk of introducing reliability issues, such as solarization of some glass secondary optics [1,2], or deterioration of the adhesive used between cell and secondary.

As multijunction cells with more than three subcells (and therefore more efficient) become available, spectral issues would become a major limiting performance factor, and the concentrating optics must be designed with a close view on the effects that concentrated light may have on the performance of the multijunction solar cell device.

4.2.1 Light and Optics

Think of light as an incompressible medium, as being bundles of rays of energy. Picturing light as rays lets us simplify optics into geometrical optics. Light rays become straight lines. We need neither be concerned with physical optics, nor with the nature of light, its occurrence as wave and particle and its non-linear interference with matter.

Light is made up of particles of various wavelengths, or energies. This distribution is called the spectrum. Thermal sources of light emit spectra that follow a Planck distribution specific to the temperature of the emitter.

Optical elements handle light. In the understanding of geometrical optics, light can be reflected or refracted when impinging at the interface between two optical materials. Light can be absorbed or dispersed when traveling within an optical material. An optical material can most easily be described by its refractive index, which is the coefficient of the speed of light in vacuum over the speed of light in that material, and by its transmittance to light. Refractive index and transmission are wavelength-dependent.

Optical elements include mirrors and lenses, based on the working principles of reflection and refraction, respectively. Reflection in even the best mirrors is not complete, but some light is being transmitted or absorbed. This reflectivity varies greatly over wavelength. Likewise, lenses are not only refracting transmitted light, but there is some degree of reflection at every surface. These surface reflections are termed Fresnel reflections.

Regarding geometrical optics, Augustin Fresnel (1788–1827) had in 1822 designed a lens made of prism slabs to be used in collimating the beams of lighthouses. After him, this type of step lens composed of prisms is called a Fresnel lens. Fresnel understood that for proper refraction of light only the slope of the surface of the lens needs to be considered, not the bulk of the material of the lens. Local slope and global shape of the lens were decoupled, offering an additional degree of freedom for the designer. The degree of freedom was used to create a globally thin and inherently aspheric lens.

Light paths are reversible. A collimating lens becomes a concentrating lens when source and target are exchanged. The collimator creates parallel light from a small source, the concentrator focuses parallel light onto a small target.

Lavoisier [3] had used in 1772 a conventional bulky large glass lens to burn a diamond at 3500°C forming carbon, using the power of the sun. Solar power became *en vogue*. In 1866 Mouchot [4] presented a paraboloidal mirror concentrating sunlight onto a linear absorber aligned with the optical axis of the mirror where steam was generated, in turn driving a printing press. This was done at the world exhibition in Paris.

Paraboloidal mirror concentrators had been evaluated much earlier. The Greek ignited the fires at their oracles by sunlight. It had been found that paraboloids provided a much better focus than their spherical relatives. Spherical aberrations were known and were calculated more than 2000 years ago by Diocles [5].

Conic sections (those shapes than can be created by cutting a cone using a plane) became the elements for imaging optical design. In imaging optics, the goal is to create an image of the source on the target. The efforts of the designer go into eliminating aberrations. Spherical lens surfaces are still used widely, partly due to ease of their manufacturing. Statistical polishing strategies will result in a very accurate spherical surface. Astronomical telescopes explored combinations of conic shapes, such as Cassegrain optics, where light reflected up by paraboloidal primary mirror is reflected down by a hyperboloidal secondary mirror through a hole in the primary onto the target. There are telescopes with spherical primary mirrors and secondaries correcting the aberrations introduced by the primary.

Imaging optics deals with small (ideally: point) sources and parallel, paraxial light. Non-imaging optics, on the other hand, offers prescriptions on designing optics for extended sources and spreading beams of light. The sun is a good example for a borderline case: the sun appears small at 4.5 mrad or 0.265° half-angle. The sun qualifies as a point source, and the light can be regarded as parallel in some design cases. However, practical solar optics for CPV needs to consider tolerances as optics and sun may not be perfectly aligned along the optical axis.

Designers of non-imaging optics are not concerned with the quality of an image, but rather with the efficient guidance of a quantity of light through the optical system. Non-imaging optics had been invented in 1966 for collecting Cherenkhov radiation [6–8]. The device, a mirror called *compound parabolic concentrator* (CPC) funneled incoming radiation onto a target. In its linear (2D) version, it is an ideal concentrator, all light incident within a so-called acceptance half-angle gets directed to the target. The rotationally symmetric (3D) version is not ideal, and suffers from skew rays which may be reflected out of the concentrator, even though they were incident within the acceptance half-angle. Edge-rays mark the acceptance half-angle. The designer of solar optics may use them, among other concepts to be explained.

Theorems in optics hold equally true for imaging and non-imaging optics: one is the brightness theorem that states that no image can be brighter than its source. In another interpretation of the brightness theorem, multiplication of the cross-sectional area of the light beam and the solid angle of that same light beam yields the *étendue* (throughput) of light. This property is conserved along the passage of the beam through an optical system. Which brings us back to understanding light as incompressible medium.

4.2.2 Optics for Concentration Photovoltaics

The initial idea behind CPV was to replace costly silicone cell area by less costly optics area. The first CPV systems of scale were built in the early 1980s in Saudi-Arabia, the USA and Japan, with remains still existing.

Indeed, the first CPV systems were using silicon single-junction cells with efficiencies just over 20%. The lenses were made of PMMA, though at the time the first silicone-on-glass lens had been developed [9].

When multi-junction cells with efficiencies above 35% became available as space cells were modified for concentration terrestrial use, the lower usage of solar cell area became actually a second order argument when compared with the peak power per unit area, clearly surpassing that of flat plate technologies, and literally dozens of CPV systems were developed in the 2000s. Few survived the industry consolidation after 2010. IEC standards (IEC 62989 will be a

Figure 4.2 Typical CPV-module with Fresnel lens parquet (5 × 4 lenses), twenty matching secondary optical elements, and cells at focal distance, side walls not shown. Explosion drawing

technical specification on primary optics in CPV, see Chapter 9) were developed and a typical CPV module emerged: multiple Fresnel lenses made as parquet, and placed at a focal distance with or without secondary optical elements (SOE) coupled to the cell. All in a box, as shown in Figure 4.2 (with SOE).

Non-imaging optics suits CPV optics, as its key challenges are energy efficiency and specific distributions of light (including concentration ratio, acceptance angle and spectral and local uniformity), rather than the reproduction of the source (the sun). Optics for CPV may be characterized by material properties, design solutions and manufacturing processes.

Optical-grade materials must be highly transmissive (lens efficiency approaching 90% including absorption and Fresnel surface reflections) or highly reflecting (95% for front surface mirrors). Surfaces should, if manufactured properly, exhibit a roughness measured in one-digit nanometers, or just above.

Optical designs for CPV are at the core of CPV as a whole. Modules can only be constructed meaningfully because the design of the optics connects the mechanical size and structure of the module with the cell requirements in local, specular and spectral irradiance uniformity for minimized current mismatch over subcells. Typically, sunlight is focused by the primary optical element (POE) onto a secondary optical element (SOE); both elements are responsible for the concentration ratio, acceptance angle, irradiance uniformity and ultimately the efficiency of the module. The geometrical sizes of the optical elements scale with cell size and concentration ratio. Module geometry should depend on optical decisions; the module is a mechanical container for optics. Semiconductor physics of the cell and dissipation of heat are subsets of optics, too, but the focus of this Chapter is on geometrical optics (see Chapters 5 and 7 for the other commitments).

Lenses and mirrors manufactured for CPV must be reproducible in high quality at minimum cost. Quality refers to the limits of product specifications. The service life of the optics must exceed 20 years outdoors (see Chapter 9) with minimum degradation of efficiency (<10%). Developing manufacturing processes is crucial in achieving lowest costs for optics the size of square meters, which earlier had been available only for astronomy, at high prices.

4.3   Optical Background

A system with a surface where the *irradiance* (the power of electromagnetic radiation per unit area incident on a surface) has increased with regards to that at its entry aperture is called a *concentrator*. Basically, the optical train of a CPV system has to transfer as much as possible of the radiation power falling at its *entry aperture* to the solar cell in the best possible conditions. This implies, as explained above, that radiation power is transferred with minimum losses and in such a way that the long-term production of energy is assured (i.e. the system has certain aiming tolerances and the power hits the solar cell in good conditions –concentrated and uniformly, both spectrally and irradiance-wise). Some key-concepts in optics of particular interest in CPV are shown in the next sections. Along with *Fermat's Principle* and the definition of the *Étendue Invariant*, the *Edge Ray Theorem* that is the basis for advanced *Nonimaging* designs requires special attention.

4.3.1   Basic Concepts in Geometrical Optics

Geometrical optics is the branch of optics that is characterized by a disregard for the wavelength, since this approaches zero in the electromagnetic field of high frequency associated with light in a wide range (including visible and infrared light). In this approach, the laws may be formulated in the language of geometry (hence the name), and we can consider the energy to be transported along certain curves, called *light rays* [10].

4.3.1.1   Fermat's Principle. Rays and Wavefronts. Refraction, Reflection and Total Internal Reflection

Fermat's principle states that the trajectory of a ray travelling from P to Q, two points located at two different media with refractive indexes n_i and n_t, respectively, is such that minimizes the optical path length between P and Q. The optical path length is the sum of the distances covered by the ray when travelling from P to Q multiplied by the correspondent refractive index at each media they cross (see Figure 4.3). In a more formal way, this principle states that a ray is the extremal curve [11] that minimizes the following curvilinear integral:

$$L_{opt} = \int_P^Q n(x, y, z)ds, \qquad (4.1)$$

where $n(x, y, z)$ is the *refractive index* of the current medium at point (x, y, z) and ds stands for the differential length along the integration path between points P and Q. The value L_{opt} is calculated along the extremal curve (which is the trajectory of the ray, as mentioned above) between points P and Q and is called the *optical path length* between these points. In practice, this principle states that light will follow the shortest time-trajectory between points P and Q, considering the variation of light speed with $n(x, y, z)$.

A wave front (ζ in Figure 4.3, right) is the surface that is normal to all the rays of a bundle at different points through the propagation, and can be determined by the optical path length travelled by all the rays of the wavefront through the optical system, which is identical for all of them. The constancy of optical path length for all rays of the bundles at the entry surface till they reach the solar cell is a very common design tool to calculate the surfaces of an optical system.

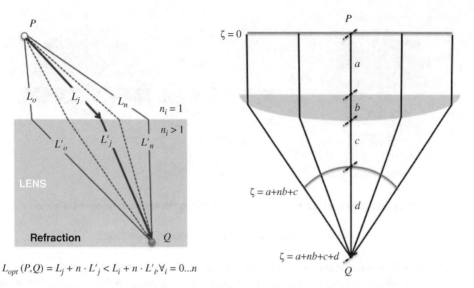

$$L_{opt}(P,Q) = L_j + n \cdot L'_j < L_i + n \cdot L'_i, \forall_i = 0...n$$

Figure 4.3 (Left) One-time refraction example: Fermat's principle states that the trajectory of a ray travelling from P to Q, is such that minimizes the optical path length between P and Q. In this example, segments L_j and L'_j form the correct ray trajectory. (Right) Two-times refraction example: a wave front ζ is the surface that is normal to all the rays of a bundle at different points through the propagation, and can be determined by the optical path length travelled by all the rays of the wavefront through the optical system, which is identical for all of them

The *refraction* and *reflection* laws, the *reversibility* of the trajectories of rays, and the equality of optical path length for all the rays belonging to a continuous bundle linking two wave fronts can be deduced straightforwardly from this principle. These principles are used one way or another in all optical designs.

The CPV optical train should get the light falling in a large area and take it to a small one: this implies the trajectories of most rays should change (more than once typically) to reach the cell. From the deflection angle standpoint, there are two main ways to change the path of a ray in a CPV system, refraction and reflection (that can be divided into metallic and total internal reflection TIR).

Refraction occurs when a ray crosses the boundary separating two media having different *refractive indexes* n_i and n_t, and consists of a sudden deviation of the ray trajectory that depends on the slope of the boundary and both refractive indexes (Figure 4.4). Optical devices changing the direction of the rays by refraction are usually referred to as '*lenses*'. *Snell's law* states:

$$n_i \cdot \sin \theta_i = n_t \cdot \sin \theta_t, \tag{4.2}$$

where θ_i and θ_t are the angles between rays hitting (*i*) and leaving (*t*) point P of the lens and the lens normal at such point, while n_i and n_t are the refractive indexes of media for incident and transmitted (refracted) rays (see Figure 4.4). Since the index of refraction is a function of the wavelength $n(\lambda)$, the beam deviation also depends on the spectrum.

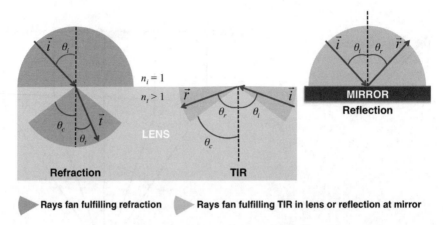

Rays fan fulfilling refraction   ▶ Rays fan fulfilling TIR in lens or reflection at mirror

Figure 4.4   Refraction, TIR and metallic reflection

For normal incidence, $\theta_i = \theta_t = 0$ and there is no deviation in the trajectory of the ray. Notice the change in direction is somewhat subtle (especially when compared with reflection, as we will show later) and rises when the difference between the values of n_i and n_t rises.

In practice, for CPV applications one of the media is air, and therefore either n_i or $n_t = 1$. The optically-dense media has $n > 1$ typically ranging from 1.4 to 1.6. According to Table 4.1, glass lenses have more power to deviate the light thanks to their higher index of refraction (compared to plastics and silicone) in the wavelengths of interest.

When $n_i = 1$, and $n_t > 1$, the maximum deviation of the refracted ray is achieved when $\theta_I = 90°$, and in this case θ_t is:

$$\theta_t = \sin^{-1}\left(\frac{1}{n_t}\right) = \theta_c, \tag{4.3}$$

This angle θ_c is called '*critical angle*' and is the origin of another way to change the direction of a ray, the so-called '*total internal reflection*' or TIR. Indeed, if the ray now comes from the optically dense media and hits the lens surface with an angle with respect to the normal $\theta_i > \theta_c$, it cannot be refracted anymore, and remains trapped within the lens, after being reflected. As in all canonical (metallic) reflections, the angle between the surface normal and the reflected ray r coincides with that of the impinging ray i and surface normal ($\theta_I = \theta_r$, see Figure 4.4).

TIR is a very attractive way to bounce the light, because it has the bending power of a metallic reflection but does not need one additional technological step (mirror deposition),

Table 4.1   Values of the refractive index for a few optical materials utilized in CPV ($\lambda = 590$ nm)

Optical material	n ($\lambda = 590$ nm)
Glass (BK7)	1.52
Silicone rubber	1.41
PMMA	1.49
Sol-gel	1.45

leading to potential cost cuts. Additionally, a TIR is 100% efficient, while a metallic reflection always implies some loss (absorption or transmission) in the metal (see section 4.5.1). The complexity of designing TIR optics it that these should observe two constraints at the same time: directing the light towards the proper direction and making sure all design rays hit the surfaces with angles above the critical angle. The latter lowers with increasing refractive index and therefore materials with high n are preferred for TIR surfaces.

The final option in geometrical optics to change the path of a ray is the metallic or mirror 'reflection'. In this case, the law $\theta_i = \theta_r$ holds independently on θ_i. The maximum deviation of a ray is achieved when the ray hits the surface of the mirror perpendicularly, and both $\theta_I = \theta_r = 0$, meaning that the impinging ray reverses its direction.

4.3.1.2 The Phase Space. *Étendue*

In the *Hamiltonian formulation* of optics [12] the ray that passes through point (x, y, z) with the direction that is specified by the unitary vector v is represented as (x, y, z, p, q, r), where $(p, q, r) = n(x, y, z), \cdot v$ are the *optical direction cosines* of the ray with respect to the X, Y and Z axes, respectively.

If n is known, and one of the direction cosines is always positive, a ray is a five-parameter entity defined by a point and two direction cosines. This five-dimension space is called *extended phase space*. A *ray-bundle* M_{4D} (or ray-manifold) is a four-parameter entity, topologically, a closed set of points in the extended phase space, with each point representing a different ray (i.e., two different points cannot correspond to the same ray at two different instants). Often, a ray manifold M_{4D} is defined at its intersection with a reference surface Σ_R, which must observe the condition of intersecting only once the trajectories of the rays belonging to M_{4D}. This reference surface defines a four-parameter variety called *Phase Space*. For instance, if the reference surface is a plane $z = 0$, the phase space is (x, y, p, q). In 2D geometry, all these concepts can be defined similarly. For instance, the extended phase space is the three-dimensional variety defined by $p^2 + q^2 = n^2(x, y)$ in the four-dimension space of coordinates x-y-p-q. The reference surfaces become curves in the xy plane and a ray-bundle m_{2D} shall be a two-parameter entity.

Each ray crossing the reference surface Σ_R is represented by a single point in the phase space, and the entire bundle M_{4D} occupies certain volume in the same space. The volume occupied by the ray bundle in the phase space increases with its angular span and the size of the spot produced at the evaluation surface, and has a somewhat energy-related meaning, since it accounts for the number of rays included in the bundle. Such volume is called *étendue* [13], and can be calculated as described in 4-I Annex.

Bundles having a wide angular span or crossing large surface areas have more étendue, as Figure 4.5 shows.

The *étendue* value is constant along the propagation of the ray bundle, no matter the surface of evaluation, according to the *Brightness Theorem*. Checking the matching of the *étendue*s of the input and output bundles connected in an optical design is a reliable verification of whether the design was carried out properly. However, the main implication of the *étendue* constancy is the statement of an upper bound, which can also be derived from the *2nd principle of thermodynamics*, for the transfer of light between a source and a target. This upper bound shall be reviewed in Section 4.3.2.3 for the particular case of concentrators design, a case in which a narrow bundle hitting a large area (the concentrator's entry aperture) is transformed into a wide-angle bundle hitting a small area (the solar cell), preserving the *étendue*.

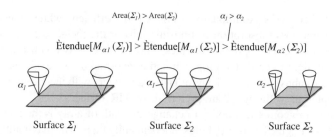

Figure 4.5 Compared decreasing *étendue*s of three different bundles, from left to right: (1) rays hitting surface Σ_1 with angles $\leq\alpha_1$; (2) rays hitting surface Σ_2 (*area Σ_2 < area Σ_1*) with angles $\leq\alpha_1$; (3) rays hitting surface Σ_2 with angles $\leq\alpha_2$ ($\alpha_2 < \alpha_1$)

4.3.2 Basic Concepts in Nonimaging Optics

The research field of *nonimaging optics* was informally founded in the 1960s with the invention of the *compound parabolic concentrator* (CPC). This simple design avoided image formation (a condition that is not needed in principle when we only need an efficient transfer of radiative power from a source to a target) to achieve the thermodynamic upper bound of concentration in 2D. The design-key was the so-called *Edge Ray Theorem*, which supports most nonimaging designs, and is explained in a simplified way in section 4.3.2.2. The devices that came afterwards[1] showed suitability in several fields, especially in the illumination and solar energy contexts.

Although imaging optics, (as opposed to nonimaging optics) does not particularly specialize in problems of power transfer, its devices and design tools can lead to optics doing at least part of the job just as good. Apart from showing low acceptance values, imaging devices in this application produce a concentrated image of the sun onto the solar cell, which is bad for both the cell efficiency and lifetime if the concentration is high, and systems using some imaging stages should provide additional nonimaging stages to prevent high irradiance peaks. The main examples are those based on Köhler-based imaging optics, which do not produce an image of the sun onto the solar cell, but an image of the primary optics instead. Imaging solutions are simple and most effective in paraxial problems, where all the rays that are involved in the problem are close to the optical axis, and hence form a narrow angle with respect to it. This seldom happens in the CPV framework, where the constraint of image formation should be avoided and replaced by other alternative constraints leading to better performance features.

The aim in a nonimaging device is often the efficient transfer of light power between a source and a receiver, disregarding image formation (see Figure 4.6), so that these systems are in essence ray-bundles transformers from the point of view of geometrical optics. However, designing with nonimaging tools does not necessarily imply that image formation cannot be achieved (but rather that it is not forced). In fact, there exist nonimaging devices that show good imaging properties [15].

---

[1] A nice overview on the works developed in this field along 30 years can be seen in the compilation *Selected Papers on Nonimaging Optics* [14]. To review the current status one should take a check the latest contributions to the proceedings of the SPIE conference *Nonimaging Optics: Maximum Efficiency Light Transfer*, Roland Winston, editor.

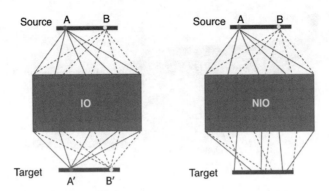

Figure 4.6 Nonimaging Optics (NIO in the drawing) do not necessarily focus points of the source onto points at the target (image) like an imaging optics (IO in the drawing) system does. Other types of contour conditions are fixed in this branch on Geometrical Optics to assure all light coming from the source hits the interior of the target

4.3.2.1 Common Ray-Bundles in CPV Designs

Therefore, in essence, a nonimaging device connects two ray-bundles that we call input (M_i) and output (M_o) bundles. Most nonimaging designs begin with their definition. Let M_i be the ray-bundle emitted by the source (the sun) hitting the entry aperture of the concentrator and let M_o be the output bundle, which in principle is the entry aperture as subtended by the receiver (solar cell). This bundle is a subset of all the rays that may impinge on the receiver, M_R (in a plane receiver with a single active face, M_R consists of all rays coming from the hemisphere faced by that face). The bundle of collected rays M_c is the set formed by the rays common to M_i and M_o. When the bundle of collected rays, M_c, coincides with M_i, the design is *loss-free*. When a design perfectly couples the bundles M_i and M_o (i.e., the $M_i = M_o = M_c$) it is called *ideal*. When $M_c = M_o = M_R$ the device is said to be *maximal*, and the target is illuminated from all the available directions. A device that fulfills both the ideality and maximal conditions is called *optimal* [16]. In practice, nonimaging designs do not necessarily aim to be maximal or ideal. The concepts of *optimal*, *ideal* and *maximal* concentrators apply for 2D devices also, the definitions dealing with bundles m_i, m_o, m_R and m_o. Notice we can refer to 2D bundles using lower-case letters to differentiate them from their 3D counterparts, denoted by upper-case.

There is one type of bundle that shows up very often when designing CPV optics in the nonimaging framework. Often referred to as *infinite-source* type, this input bundle can be defined as formed by the rays hitting a given surface (which could be the entry aperture of a concentrator) with angles below the so-called *acceptance* (or *semi-acceptance*) angle, as shown in Figure 4.7. Hence, the acceptance angle can be defined as the angle subtended by the 'design source' when this is at infinity. Notice this design source should be larger than the sun itself, which has to be inscribed into it: the larger the difference, the more excess of acceptance angle (sometimes referred to as 'acceptance (or tolerance) budget' [17] to counter (or exploit from a cost-perspective) manufacturing and operation tolerances.

It is obvious that the 2D equivalents for these bundles are defined over curves rather than surfaces. They have also been depicted in Figure 4.7, in the diagonal of the square entry aperture.

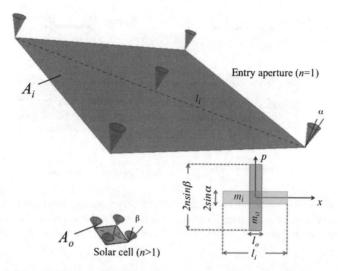

Figure 4.7 Typical ray bundles in nonimaging CPV designs at reference surfaces (the entry aperture and the cell) of a design when the design sources defining the input and output bundles are at infinity, in the 3D and 2D cases. The design sources subtend an angle 2α at the entry aperture in this case, and 2β at the solar cell. The *étendue*s of the input and output bundles are the same if the optical system is ideal. Owing to *étendue* conservation, and owing to concentration, since $A_i > A_o$, results in $\alpha < \beta$. The *étendue* conservation can be depicted in 2D (bottom right), looking at the optical system diagonal: notice the area of the phase space charts of the bundles m_i and m_o are the same

When the source is at infinity, the *étendue* of these infinite-source bundles (very common in the CPV field) can be calculated [18], in the 3D case, as:

$$E_{M_j} = \pi A_j n_j^2 \sin^2(\alpha), \tag{4.4}$$

where j can be either i or o, depending on whether we are calculating the *étendue* at the entry aperture or at the solar cell, respectively. In two dimensions:

$$E_{m_j} = 2l_j n_j \sin(\alpha), \tag{4.5}$$

notice that $E_{m_j} = (4/\pi)\sqrt{E_{M_j}}$ in both cases.

4.3.2.2 The Edge-Ray Principle. Point-Source Approximation Design

Let δM_i and δM_o be the boundaries of the regions defined by M_i and M_o in the phase space introduced above, i.e., the *edge-ray* (or *extreme-ray*) subsets of such bundles. The edge-ray subsets have one dimension fewer than M_i and M_o, i.e., $dim(\delta M_i) = dim(\delta M_o) = 3$. The edge-ray principle states that, in order to couple the bundles M_i and M_o by means of an optical system, it suffices to connect the edge-ray subsets δM_i and δM_o. Conversely, a perfect matching between bundles M_i and M_o implies the coupling between their edge-rays. Obviously, this principle also holds for the two-dimensional bundles m_i and m_o.

The edge-ray principle is the design key in most nonimaging designs, and shows the benefits that arise from the elimination of the imaging requirement (notice that, roughly speaking, the design deals with a considerably smaller number of rays).

Rays either emerging (or impinging) from (or at) an edge of the source (or receiver) or tangent to it are edge-rays. For instance, in 2D, for a source placed at infinity, the edge-rays of the input bundle m_i are:

- those impinging on the straight line of length l_i tilted at an angle $\pm\alpha$ with respect to the vertical;
- all rays within $\pm\alpha$ at the edges of said line.

That is, in Figure 4.7, bottom right, the boundary of the horizontal rectangle ($p = sin\alpha$, $p = -sin\alpha$, $x = -l_o/2$, $x = l_o/2$). In the same figure, the boundary of the vertical rectangle representing the output bundle m_o gathers the edge rays of such bundle. A simple utilization of the *edge-ray principle* can be found in [13], where the design of the CPC is detailed.

One optical surface needs to be designed to connect a pair of input and output bundles. Nonimaging CPV systems are designed with four bundles defining the edges of the solar cell and the angular limits (to the left and to the right about the sun's direction) of the acceptance angle, and therefore need two surfaces designed specifically to connect these bundles.

A simpler approach that does not follow nonimaging design rules consists of the calculation of optical systems connecting only one pair of design bundles, for instance, the plane wavefront defined by sun's central point and the spherical wavefront defined by solar cell central point. Such simplified designs, that can comprise only one designed surface, are actually making a 'point source approximation' of both the light source and target, and perform far from the thermodynamic upper bounds discussed in the next section.

4.3.2.3 Geometrical Concentration and Angular Acceptance. Thermodynamic Trade-Off and Concentration-Acceptance-Product (CAP)

We all know that light can be concentrated. A magnifying glass, held into the sun, can concentrate sunlight into a focus hot enough to light a fire. The glass shadows the area around the focus, which has lost the redirected light. 2D, or linear, concentrators concentrate light in one plane, 3D, concentrators do so in two planes in space. The *geometrical concentration C_g* is the ratio between the area (width) of the entry aperture to the area (width) of the exit aperture, for 3D and 2D concentrators, respectively. A times symbol, $\times$ follows the value of the current geometrical concentration of a device: hence, a concentrator whose entry aperture area is 500 times the area of the target will be referred to as a 500$\times$ concentrator.

Since light is incompressible, and the *Brightness Theorem* introduced above dictates that the image of a source cannot be brighter than the source itself, the area cannot be the only parameter changing during concentration. The *étendue* of concentrated light between entry and exit apertures remains unchanged:

$$n_{entry}^2 A_{entry} \Omega_{p,entry} = n_{exit}^2 A_{exit} \Omega_{p,exit}, \tag{4.6}$$

where the projected solid angle Ω_p measures the beam spread. Therefore, concentrated light at the solar cell has always a wider angular span than the un-concentrated sunlight hitting the

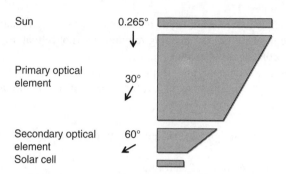

Sun   0.265°

Primary optical   30°
element

Secondary optical   60°
element
Solar cell

Figure 4.8 Geometrical concentration (here schematically shown in 2D) reduces the width of the beam of light between entry aperture (top) and exit aperture (bottom), but conserves *étendue* by increasing the solid angle of beam hitting the solar cell

concentrator's entry aperture. In concentration, as the exit area becomes smaller, the solid angle changes from the one of near-parallel sunlight to one that fills 90° in theory for the maximum of concentration. In practice, the angle of acceptance on the target, e.g. a solar cell is often smaller than 90°. Figure 4.8 shows a schematic of the dependence of area and angle.

The definition of acceptance angle α is linked to that of the *design source* placed at infinity. In that approach, the size of the source must be measured in terms of angle. For instance, the sun is a light source that subtends[2] an angle of $\theta_{sun} = \pm 0.265°$.

Following the nomenclature of Figure 4.8, let M_i be a bundle defined by the acceptance angle α and by an entry aperture whose area is A_i surrounded by a medium of refractive index n_i. Let M_o be the bundle at the target, defined by its angular spread β, the area of the receiver A_o and the refractive index of its surroundings n_o. If the concentrator perfectly couples these two bundles, the equality of *étendue*s is also observed, thus, according to eq. (4.6):

$$E_{M_i} = A_i n_i^2 \pi \sin^2 \alpha = A_o n_o^2 \pi \sin^2 \beta = E_{M_o}, \tag{4.7}$$

The maximum concentration is achieved when the bundle M_o matches M_R (this is the maximal design defined above), i.e., the receiver is illuminated isotropically ($\beta = 90°$). Since, in general, $E_{M_o} \leq E_{M_R}$ we may write:

$$A_i n_i^2 \pi \sin^2 \alpha \leq A_o n_o^2 \pi, \tag{4.8}$$

an expression that, re-written in the correct way, enables the appearance of an upper bound $C_{g,max}^{3D}$ for the geometrical concentration:

$$C_g^{3D} = \frac{A_i}{A_o} \leq \frac{n_o^2}{n_i^2 \sin^2 \alpha} = C_{g,max}^{3D}, \tag{4.9}$$

[2] In this chapter, finite sun refers to an extended source type having such angle, as opposed to a plane wavefront (point source located at infinity) used elsewhere as an approximation to model sunlight.

Let us look more deeply into this expression, focusing on the fairly frequent case in which the entry aperture faces the air $n_i = 1$ and the receiver is embedded in a medium of refractive index n. It is obvious that:

$$n^2 = C_{g,max}^{3D} \sin^2 \alpha, \tag{4.10}$$

and, therefore, we infer that there exists a trade-off between the achievable concentration and the angular spread of the input source. For instance, if we want to increase the irradiance on a receiver above $C_{g,max}^{3D}$ times that existing on the entry aperture, the maximum achievable acceptance angle is $\alpha \le \sin^{-1}\left(\sqrt{n^2/G_{g,max}^{3D}}\right)$.

The *concentration–acceptance* product [19] CAP $= \sin(\alpha) \cdot C_g^{1/2}$ gives an idea on how close a design gets to the thermodynamic limit and therefore, is a number that rates a given optical train. The maximum value is $CAP_{max} = n \approx 1.5$. The CAP depends only on the concentrator architecture. Again, the 2D analysis leads to similar expressions, now dealing with lengths instead of areas:

$$C_g^{2D} = \frac{L_i}{L_o} \le \frac{n_o}{n_i \sin \alpha} = C_{g,max}^{2D}, \tag{4.11}$$

Again, according to *étendue* conservation, concentration is feasible because the beam falling on the concentrator's entry aperture has a much narrower span ($\theta_{sun} = \pm 0.265°$, as mentioned above) than that reaching the solar cell, and this way the entry aperture area can be much larger that the solar cell area.

It is straightforward that for high concentration systems the acceptance is relatively small (see Figure 4.9). The thermodynamic limit tells the maximum concentration achievable for a system only collecting rays within $\pm \theta_{sun}$ is $C_g = 105,181\times$, for $n = 1.5$, in 3D (in 2D -linear systems-, the limit is $324\times$). A solar concentrator with $C_g \ge 100\,000\times$ has been designed and made by Winston [20], which is only interesting from the theoretical standpoint, since a solar

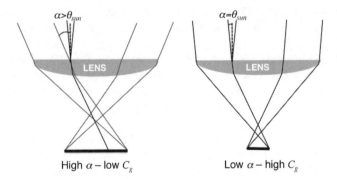

Figure 4.9 The collection of rays when a small cell is placed under certain lens is limited to a narrow fan. This drawing shows the trade-off between acceptance angle and geometrical concentration, directly linked to *étendue* conservation: for a given optical train, if we mean to collect a wider beam (increase acceptance), the receiver must increase its size with respect to the optics, and therefore concentration is reduced

cell cannot manage such concentration levels and moreover the CPV systems cannot be perfectly manufactured and cannot be perfectly aimed at the sun continuously. Therefore most CPV optics are designed to be able to collect rays beyond this minimum θ_{sun}. This is the reason why the rays used in the design of CPV optics are of infinite source type with beam span clearly $>\theta_{sun}$. For ultra high concentration levels ($>1000\times$), two-axis tracking and a secondary optics are compulsory. For instance, for $n = 1.5$, isotropic illumination of solar cell, and $C_g = 1000\times$, the thermodynamic limit for α is 2.7°.

The excess of acceptance angle over θ_{sun} can be understood as an 'acceptance budget' (see Section 4.6.2). The benefits of a wide acceptance budget are:

- More manufacturing tolerances that can help reducing costs:
  - Easier alignment/assembly of parts;
  - Lighter trackers;
  - Optical and mechanical parts can be made less accurately (lighter housing, poorer optical surfaces -roughness and shapes-, panels warp etc.);
- More operation tolerances, that improve the energy production:
  - Lower series-connection mismatches (reduced risk of having one series-connected cell out of focus, compromising performance of the entire array [21]);
  - Insensitivity to slight tracker misalignments;
  - Insensitivity to thermal effects on SoG lenses based-systems;
  - Insensitivity to wind loads;
  - Collection of circumsolar radiation.

4.4 Design of the Optical Train: Calculation of Surfaces

Designing directly in 3D is a difficult task. For sources at infinity, only complex optical systems with non-homogeneous media have been designed [22–24] attaining ideality in 3D. In this chapter, two simpler approaches to 3D designs will be explained in 4.4.2.3.

Designing in 2D with the subsets (associated with an invariant of the symmetry, for instance, p or h) m_i and m_o of the bundles M_i and M_o permits an easier treatment of optical problems (see 4-II Annex). The design of a cross-section of a device that is obtained afterwards by extending that section along or around an axis suits several applications, including CPV.

Although the 2D methods may lead to devices that are ideal in 2D, many 3D rays are not under designers' control, and their behavior has to be analyzed later by means of *ray tracing*. This does not imply that 3D ideality is forbidden in such designs: indeed, there exists a device [25] (the *Flow Line Concentrator*, FLC) that just happens to attain ideality in both 2D and 3D.

The sun subtends a small angle ($\pm0.265°$) and tracking systems achieve accuracies below 0.1° (see Chapter 6). At first sight it might seem needless to go to acceptance angles beyond 0.4° for instance. Therefore, many concentration strategies use simple optical devices such as parabolic mirrors or Fresnel lenses that sometimes work in cooperation with secondary optics to slightly increase the CAP (acceptance or concentration or both). The majority of these devices are far from achieving the maximum acceptances attainable for a given concentration level and their sensitivity to manufacturing, assembly and alignment errors, wind loads, etc., vary.

Along with the two major performance features, optical efficiency and the CAP, the devices shown in this section will also be discussed taking into account a few other aspects that matter as well:

- Compactness (the ratio of their optical depth over the entry aperture diagonal, connected to the so called *f*-number –i.e. focal length over lens diameter– in Fresnel systems), which has an impact on housing and transportation costs.
- Ability to produce irradiance uniformity on the solar cell (and spectral dependence of the latter), which has a beneficial impact on cell efficiency (see Chapter 2).
- Main technological issues (type of optical materials used in the manufacturing of parts, SOE-to-cell bonding), which have an impact on the manufacturing costs and reliability/ durability.

4.4.1 Types of Concentrators as a Function of Concentration Level

At low concentration ratios (<10×), the size of the cell is in the order of that of the optics entry aperture, and the acceptance angle can be relatively large: so large that nearly covers all the sky points occupied by the sun along the year, and therefore the concentrator does not need to track the sun in order to cast its rays. If the concentrator does not track the apparent movement of the sun, the concentrator is called a *static, or stationary concentrator* [26].

Unlike tracked systems, static concentrators share many of the advantages of the flat plate, such as the feasibility of collecting diffuse light (which may reach more than 60% of total irradiance available troughout the year in cloudy areas). Static concentrators are often designed in 2D (in a vertical plane including the north to south axis) to collect a wide angular range (at least 50°, full angle) of sunrays covering all the declinations between the winter and summer solstices. Subsequently, the designed 2D cross-section is extended longitudinally to achieve a linearly symmetric device. With this maneuver, it is assured that the majority of rays impinging on the concentrator's entry aperture throughout the day can strike the solar cell. Static concentrators are often based on conventional silicon cells. It can be shown that a truly static concentrator whose entry aperture is tilted an angle equaling the site latitude has a maximum geometrical concentration ratio of $C_{staticMAX} = n/\sin \delta$, where $\delta = 23.45°$ is the declination of the earth's path around the sun. Therefore, $C_{staticMAX} = 3.77$ (considering the refractive index of the material of the concentrator, $n = 1.5$). One may design concentrators that track seasonally, or are tilted at appropriate times in order to achieve geometrical concentration ratios above the one for the static concentrator, reaching 10×. One trick involves the use of bi-facial cells, which due to their double-sided nature have double the surface in one unit. With some goodwill, we may calculate the geometrical concentration ratio with the area of the cell, not with the combined areas of both surfaces. Then, the concentration ratio appears to have doubled.

Theoretical concentration limits also grow when the static collectors are tilted angles different from the latitude (for instance, if these are integrated in building facades), as [27] showed. Owing to the cosine factor in the sun beam hitting the system's entry aperture, the 'clearest' part of the sky (comprising the points where the sun has been some time along the year) subtends a narrower angle, and concentration can rise, as we showed in Figure 4.9.

There is another way to increase the concentration for stationary concentrators by redirecting light incident in the plane of concentration into the perpendicular plane on the optical axis, using microstructures [28].

Figure 4.10 The EUCLIDES: a medium concentration (~30×) 1-axis tracking (line-focus) CPV System. Photo courtesy of Ignacio Antón (IES-UPM)

Medium concentration systems ($10\times < C_g < 100\times$), conceptually very similar to linear solar collectors widely used in thermal *Concentrated Solar Power* (CSP) [29] are found more easily. Typically, these devices have a trough shape [30–32] (both the cells and optics have linear symmetry) and therefore, only concentrate the light in one transversal plane (see Figure 4.10).

In the longitudinal direction, optics and cells have similar lengths, thus there isn't any concentration. Thanks to this fact, and the properties pointed out in Section 4.4.1, such systems only need to azimuthally track the sun rotating around one single axis, parallel to the symmetry axis, which is typically horizontal (horizontal axis tracking) or tilted an angle equaling the latitude of the installation site (polar axis tracking) [33].

The major part of research, industrial and commercial activities in the CPV world have been devoted to higher concentration systems ($C_g > 100\times$), though. The reason is the rapid evolution of high concentrator solar cells, whose efficiency, reliability and availability have the type of potential to make CPV cost-effective. In fact, as of 2015 the last efficiency records for solar cells have been achieved at high concentration (a few hundred suns typically) by III-V multijunction cells [34] with efficiencies ranging from 40% to 46%. Light collection at high concentration levels is only feasible with the sun aiming perpendicularly at the concentration entry aperture, which implies the need of two-axis tracking.

4.4.2 Design Examples

The design of the optical train of a CPV system can be as simple as finding the shape of the surfaces of a focusing lens. More advanced designs, for instance using nonimaging principles and (or) taking into account all the special features that help attaining better performance features (high CAP values, uniform irradiances no matter the spectrum of light) or practical advantages (smaller secondary optics, easier to glue lenses etc.) eventually lead to more efficient CPV systems. This section discusses the characteristics of a few of the most common high-concentration CPV systems that can be found nowadays, and mentions the design principles utilized in each case.

4.4.2.1 The Fresnel Lens

To perform the concentration function, conventional lenses tend to be bulky (specially for large collection areas) for reasonably short focal lengths. Such lenses compromise system costs and weight. An 1822 invention, the Fresnel lens, solves this problem by dividing the lens collection area into rotationally-symmetric prism sectors, which refract the light as needed. Most Fresnel lenses have flat prism faces, which simplifies manufacturing. A simple focusing Fresnel lens can be used as solar collector [35], although its performance tends to be poor, especially in terms of both acceptance angle (at concentrations above 400× the latter is too close to sun's opening angle) and irradiance uniformity.

For a lens with the grooves facing the solar cell, the focusing Fresnel lens solution is analytical [36], and each prism angle θ can be calculated as follows:

$$\tan \theta = \frac{\rho}{n\sqrt{\rho^2 + f^2} - f},\qquad(4.12)$$

where n is the lens material refractive index, ρ is the radial dimension and f the lens focal length (distance from top flat surface to focal point).

As we will show below, the Fresnel lens is the core of most CPV optical trains found nowadays, which use it as primary optical element and provides the major part of the concentration factor, leaving for the secondary optical element the responsibility of providing a slight CAP increase (typically, both the acceptance and concentration) and irradiance uniformity to the overall system.

4.4.2.2 CPV Systems Designed in 2D

The optical trains of concentrators are designed in two dimensions in most cases: the calculated cross section is either rotated or extended along a preferred axis, depending on the type of concentrator, to get its actual 3D shape.

The simplest possible CPV design utilizes the point-source approximation explained above. It consists of the calculation of one single optical surface (either a lens or a mirror) typically focusing the sun's central point at the cell center. In the case of a mirror, the shape that does that is the parabola, a shape widely used in CSP owing to simplicity and historically proven performance.

Fresnel lens designs involve Snell's law introduced above (see equation (4.2)). Fresnel lenses can be designed flat or shaped, most solutions to the geometrical problems are closed, others require numerical approaches. Simple Fresnel lenses are flat lenses, decoupling global slope (the slope of a plano-convex lens) and local slopes (prisms), as shown in Figure 4.11. A generic imaging Fresnel lens can be designed by calculating prism by prism. Any prism can be designed independently of its neighboring prisms. This degree of freedom allows for the design of nonimaging Fresnel lenses: not all prisms necessarily refract the impinging sunlight into the same focal spot. Distributions of the irradiance on the target are possible within the bounds of dispersion. Fresnel lenses are inherently corrected for spherical aberrations. The reader is referred to Leutz and Suzuki's monograph on nonimaging Fresnel lenses [37].

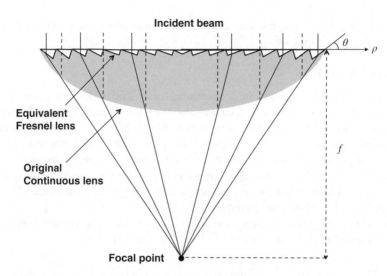

Figure 4.11 Each prism in a Fresnel lens is calculated to refract the light towards a preferred direction. The prism angle θ is a function of its radial location ρ, the refractive index n and lens focal length f

The dispersion (variation of refractive index with wavelength, which will be discussed in 4.5.1) in Fresnel lenses is limiting the CAP. Dispersion can be corrected using diffractive gratings [38] on the prisms. Another option is to satisfy the achromatic condition, which is impossible with plastic materials, but is fulfilled by cementing crown and flint glasses in its application to Fresnel lenses [39]. All these solutions are impractical due to the current status of manufacturing, efficiency issues or fundamental problems with thermal expansion. Note that diffractive structures on large-area Fresnel lenses have been designed, diamond-turned and manufactured in hot embossing.

Dispersive beam spread requires 'slower' lenses with higher aspect ratios (longer focal lengths), unless a secondary is present. On the other hand, when the focal length becomes long, the increased lever of that beam makes the system sensitive to tracking errors (remember the CAP of a concentrator rises with wider beams onto the solar cell). The optical designer needs to be aware of these effects, and design the optical train accordingly. The goal must be achieving high optical efficiency and high geometrical concentration, within the boundaries of the tolerance budget. The higher the latter, the easier is the achievement of high efficiency, especially at array level, where the probability of sunlight spillage in every single cell must be reduced to a minimum (series connection mismatch) without increasing the system complexity.

The optical train of Fresnel lens based CPV systems consists of a Fresnel lens and, more often than not, of a secondary optical element. The secondary may be a solid piece of glass, or can be a reflector. Among those that prefer the second option [40], the one using a truncated inverted pyramid is the most common. In this type of device, which has been named Fresnel XTP elsewhere [41], the Fresnel lens is designed in such a way that, at normal incidence, when the system is perfectly manufactured, the light hits the solar cell without using the secondary mirror. Since such perfection is seldom attainable at reasonable costs, a 'preventive' secondary mirror is installed to be able to collect part of the light that otherwise would miss the cell. Since the primary and secondary optics in this XTP design are not

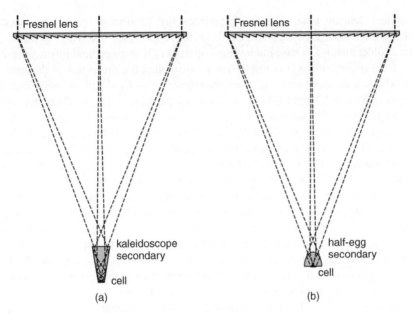

Figure 4.12 Cross-section of the optical trains of Fresnel lenses with refractive secondaries. The optical train (a) has a kaleidoscope-type glass secondary working on the principle of total internal reflection. The optical train (b) has a half-egg glass secondary

designed together, we can consider this an example of a classic point-source approximation design, in which one optical surface is designed at a time observing one design constraint only: for instance, the Fresnel lens could focus the sun central point (m_i is a plane wavefront hitting the Fresnel lens plane surface perpendicularly) onto the cell central point (m_o is a spherical wavefront centered at cell centre). This rather low CAP (0.36) device attains a concentration of about 425× for an acceptance angle of ±1°.

From Eq. 4.10 the refractive index n^2 in the numerator gives a clear advantage of 1.5^2 instead of 1 in either geometrical concentration (or acceptance angle) for glass against air (which is the case of reflectors like those of the XTP). There are several possible types of secondaries, two of the most typical are pictured in Figure 4.12.

The kaleidoscope-type secondary (on the left of Figure 4.12) works on the principle of total internal reflection (TIR) and has a tapered section that ends on top of the solar cell and bonds to it. Notice the prism has two functions: increasing the CAP and providing a more uniform irradiance on top of the solar cell. This device has been referred to as F-RTP elsewhere. Despite having two optically active stages (primary and secondary), their design is mostly unconnected, and does not follow general nonimaging principles (Edge Ray Theorem). For instance, the Fresnel lens can focus the plane input wavefront m_i (sun central point) at the prism entry-surface center, with the only major designer's choice being the focal length. In the design of the prism the length will determine the irradiance uniformity achievable, while the prism angle will fix the CAP, which is also influenced by the beam angles at the entry of the prism (defined by the focal length mentioned before). This device shows CAP values of about 0.45, which implies it can achieve concentrations of about 677× and still keep an acceptance angle of ±1°.

Two optical elements (lenses, or simply surfaces) that form an image of a preferred object (particularly, the other optical element) onto a preferred location yield what is called a Köhler integrator. Köhler integrators have interesting properties [42], such as the ability to transfer some features of the preferred object onto the target, rather than the properties of the light source involved. The Köhler approach brings interesting benefits for CPV optical trains. The half-egg secondary (or SILO, on the right of Figure 4.12) is a simple Köhler-design, where an image of the Fresnel lens (uniformly illuminated by the sun) is projected onto the target (the solar cell) by the secondary stage, and was proposed [43] by Sandia Labs in the late 1980s. In this case, the design of the secondary stage is more directly linked to the primary design, although in the general case it is not yet a nonimaging design. With CAP values ranging 0.3 in the best cases, this architecture attains concentrations of about 248× for an acceptance angle of ±1°.

In the RTP secondary, rays are reflected multiple times in a statistical fashion, whereas the half-egg secondary refracts the incoming light only once in a deterministic way. The kaleidoscope offers a better CAP and a better uniformity of the irradiance on the target if it is long enough, but is difficult to assemble (glue leakage leads to important power losses at the cell-prism interface) and is very sensitive to surface roughness. In contrast, the half-egg Köhler can be easily placed onto the cell, and can be molded in manufacturing with a surface quality that suffices for refraction (but would be poor for TIR).

The acceptance half-angle of the optical train is the parameter to be assessed in the framework of the tolerances set by manufacturing and operation (tracking). An example shows the relation of optical efficiency, optical concentration ratio, aspect ratio, and acceptance half-angle for a lens of a geometrical concentration ratio of 1000×, in Figure 4.13. The optical train has a kaleidoscope-type secondary. Results are found by

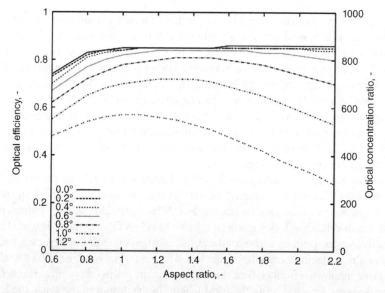

Figure 4.13 Acceptance half-angle, aspect ratio, optical efficiency and optical concentration ratio for an optical train of a concentrator consisting of a flat Fresnel lens and a refractive kaleidoscope-type secondary (F-RTP) with a geometrical concentration ratio of up to 1000×. The system is not optimized, values are result of ray-tracing simulations. Example reading: at an aspect ratio of 1.4, the concentrator has an acceptance half-angle of 0.8°, reaching an optical efficiency of 80%, and an optical concentration ratio of 800×

ray-tracing. The system is not optimized, but the drawing helps understanding the working principles of devices of this kind. It appears to be challenging to reach an efficiency of 80%, and still catching all light within a cone of half-angle of 0.8°. The system reacts quite sensitive to the aspect ratio, which should be around 1.2, even if the additional module height causes additional cost.

4.4.2.3 CPV Systems Designed in 3D

Some high-concentration systems lack of either rotational or linear symmetry and are designed directly in three dimensions. This strategy can help achieving enhanced performance characteristics (CAP and/or irradiance uniformity) and other practical advantages. Two examples of CPV optics designed in three dimensions are shown in this section.

The first one is the so-called Fresnel-Köhler (FK) concentrator [44]. The FK design combines general nonimaging and *Köhler* integration principles introduced above. This two-stage device comprises a Fresnel lens POE divided into a set of off-axis folds (typically square in shape) working together with a SOE (divided into folds as well), where each sector images a POE sector onto the solar cell. In its most common version, the FK comprises four square-folds sectors, like the one shown in Figure 4.14. According to the Köhler principles, each point at the POE is imaged onto a point at the cell (see the trajectories of white rays in the figure), and, at the end, the optical train produces a square beam that is the overlapping of the four square POE folds imaged by the SOE onto the cell. The design recipe, which is based on the *Edge Ray Theorem*, has been described elsewhere [41]. The final result is a POE lens comprising four identical sets of off-axis rotationally symmetric Fresnel lenses, on top of a SOE

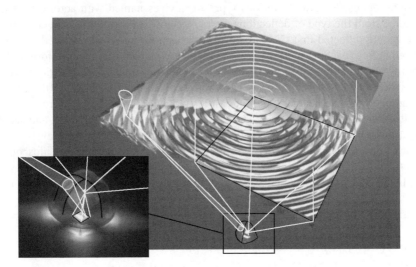

Figure 4.14 The generic FK concentrator is divided into four Köhler channels designed in such a way that a square irradiance is achieved on top of the solar cell, thanks to the imaging properties of the 4-quadrant SOE (notice each point at the POE quadrant is imaged onto a specific point of the cell so the square POE perimeter is imaged onto the cell perimeter). The Fresnel facets, beam angles and perspective are exaggerated in this drawing for the sake of clarity. Courtesy of LPI

Figure 4.15 Close-up rendering of the center of the FK Fresnel lens, showing the four sectors (left). Photograph of a FK SOE made by glass molding with a ring to be used as holder (right). Courtesy of LPI

lens comprising four identical free-form sectors (see Figure 4.15). The design is symmetrical with respects to planes $X = 0$ and $Y = 0$, with Z being the optical axis. The FK device is inspired by the SILO, and can be considered an evolution of it: along with relatively high CAP (up to 0.63, which enables concentrations up to 1057× for an acceptance angle of $\pm 1°$) this device attains excellent irradiance uniformity no matter the wavelength of the light hitting the concentrator [45], avoiding power drops linked to chromatic aberration [46]. This performance holds for a wide range of *f-numbers*, starting at $f\# = 0.85$, so the FK can be also quite compact, compared to other Fresnel-lenses approaches.

The second 3D design example is the so-called XR concentrator [47], a compact device (depth over width < 0.5) based on a primary mirror and a secondary lens (see Figure 4.16). Its name comes from the *Simultaneous Multiple Surface* (SMS) design method [48] of nonimaging optics, that yields various types of devices named with acronyms where the letters come from the type of deflections the rays follow from source to target. In this case, the sunrays are reflected first (X is assigned to reflection) and then refracted (R stands for refraction) before reaching the solar cell. A detailed explanation of the free-form XR design procedures is available in Annex 4-III. Unlike other rotationally symmetric mirror-based devices (such as those based on Cassegrain mirrors [49] or on-axis XR [50]), the receiver (SOE and cell) and the heat sink do not cast a shadow on the primary mirror, maximizing light collection per system unit area. The CAP of the XR is fairly high (≈ 1), and acceptance angles in the range of $\pm 1.81°$ have been reported [51] for concentration levels of 1000×. Irradiance uniformity in this device is achieved by means of a short kaleidoscope prism block at the bottom of the SOE, and on top of the cell, which creates good uniformity despite its length, thanks to the wide angular spread of the beam entering it (as corresponds to a high CAP device working close to the thermodynamic upper bound). A summary of all these concentrators main features can be found on Table 4.4.

4.4.3 Secondary Optical Elements: Design Details

The optical train of the CPV system is often designed with secondary optical elements (SOEs) matching the primary optical elements (such as the Fresnel lens parquet shown in Figure 4.22). The secondary can improve the optical performance of the optical train and, more important, can improve the electrical efficiency of units comprising several solar cells (modules and systems).

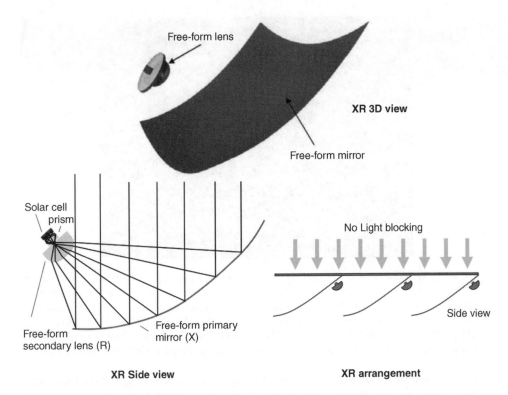

Figure 4.16 Views of the free-form XR concentrator. Showing their 3D shape (top), how rays are collected in a 2D cross section (bottom left) and how different units can be arranged in a panel without the receivers casting any shadow on the primary optics, and therefore maximizing the footprint usage

4.4.3.1 Purposes of Secondary Optical Elements

Secondary optical elements (SOE) are used in the optical train of CPV systems [52,53] with primaries in order to:

- Increase the optical system efficiency by catching refracted/reflected light that otherwise would miss the receiver by using a SOE with entry aperture greater than the exit aperture.
- Provide additional concentration of n (2D) and n^2 (3D) via adopting SOE of refractive indexes greater than that of air, assuming a fixed acceptance angle.
- Widen the acceptance angle, for a fixed concentration ratio. The acceptance angle may also be named tolerance angle, which needs to accommodate light incident off its designed path due to material properties, manufacturing processes, assembly, tracking and operating conditions (e.g. spectral, DNI, temperature, humidity, etc.). Wide acceptance angles can enlarge the tolerances of component manufacture, module assembly, system installation and tracking, in turn lowering the cost of generating electricity.
- Homogenize the irradiance and the spectral flux distributions over the cell by smoothing the irradiance peaks down to a peak-to-valley ratio (p-v value) smaller than 1.1 and redirecting or mixing the concentrated light from the primaries uniformly onto the cell. The irradiance peaks created by the sun image can cause thermal stresses which could damage the solar cell.

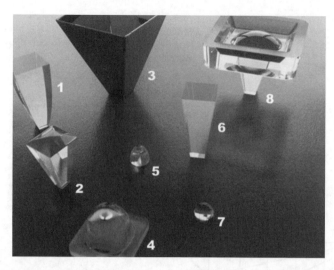

Figure 4.17 Secondary optical elements (SOE): (1) molded kaleidoscope with flat top and compound parabolic concentrator (CPC) shaped walls; (2) molded kaleidoscope with crossed top (Circadian Solar, Concentrator Optics); (3) reflective inverted truncated pyramid or *reflective catcher* (Amonix); (4) molded half-egg with cavity at the bottom (Concentrator Optics, Isuzu glass, Azur Space Solar Power); (5) small half-egg (Moulded Optics); (6) polished kaleidoscope (Moulded Optics); (7) molded hemispherical secondary (Moulded Optics) and (8) molded kaleidoscope with spherical top (Concentrator Optics, Isuzu glass)

Multijunction solar cells may suffer performance losses when, owing to chromatic aberration, tunnel junctions locally operate out of their ohmic region (see Chapter 2).
- Transform the beam cross section into a desired shape (e.g. from circular to square cross-sectional shape to match the cell shape).

For solar concentrators of medium-high concentration ratios (less than 400 suns), SOEs are not always required (although recommended). For higher concentration CPV, considering the above-mentioned advantages, secondaries are almost compulsory in practice. The most common SOEs used in CPV systems can be seen in Figure 4.17.

4.4.3.2 Optical Principles

Commonly used SOEs can be classified into three types (listed in Table 4.2) according to the optical functions of the surfaces used to fulfill the purposes of the SOE. These surfaces interact

Table 4.2 Types of secondaries in CPV systems

Type	Examples
Reflective	Reflective truncated pyramid (or reflective kaleidoscope), reflective truncated cone, reflective compound parabolic concentrator (CPC)
Refractive	Spherical or aspherical dome (half-egg shape), Köhler SOE, ball lens
Dielectric total internal reflection concentrator (DTIRC)	Kaleidoscope, domed kaleidoscope, circular-into-square (CISQ) homogenizer, kaleidoscope with shaped (e.g. freeform) walls

with the impinging light; they can be reflective (either mirrored or TIR) and refractive. The geometries of the different refractive, TIR and reflective types of SOEs are illustrated in Figure 4.21. The geometry of the SOE may be designed in such a way that the optical principles of all surfaces are combined. The design results in secondaries that fulfill the purposes listed in section 4.4.3.1 to a higher or lower degree, in operation.

4.4.3.3 Operating Principles

Notwithstanding the optical performance of the secondary, we distinguish two operating principles basic to its design, namely, deterministic and statistic operation. This refers to the way incident light is treated. For the deterministic case, the light is reflected or refracted any number of times determined in the design of the secondary, e.g. by the number and slope of surfaces the impinging beam interacts with. The loci of interaction are predicted within narrow limits. An example is the half-egg, and other Köhler designs.

For the statistic case, the number of interactions of each ray with the surfaces of the SOE remains undefined. Unless a Köhler approach is used [54], for purposes of mixing chromatic aberrations induced by a primary lens, or for homogenizing the irradiance distribution on the target, rays need to interact with the surfaces of the secondary as often as possible, on loci only roughly predictable. The surfaces need to reflect lossless and should be designed as to break any symmetry that could create unintended flux distributions. The TIR kaleidoscope in the F-RTP approach is the best example for this operating principle.

Note that in the F-XTP approach, the metal mirror secondary is often used as catcher only. During most operating conditions of the CPV module, the reflecting surface does not intercept any light. When light threatens to miss the receiver, the mirror catches and redirects it to the cell.

A Köhler design SOE with a large number of lobes would probably be a good approximation of the statistic case. The discussion should be understood the same way as discussion on the distinctions of imaging versus nonimaging optics, in some limit, the differences become small and the concepts become governed by identical mathematical descriptions.

4.5 Performance Analysis and Optimization of the Optical Train

The design of an optical train needs some feed-back information for its optimization. The process of analyzing stochastically what happens with rays travelling through an optical system is called ray tracing. Although in the simplest approach the ray tracing consists only of a statistical calculation of the number of rays reaching the target with respect to the total traced rays in an ideal loss-free system, the calculation of losses due to different mechanisms, (namely, *absorption*, inefficient reflections, *Fresnel losses*, etc.) is also present in more accurate analyses. There are a few different software tools specialized in this type of analysis, which typically needs either a CAD model of the optical train or a mathematical description of it. The forecast of how a CPV will perform in real life is as accurate as the ray trace model and simulation used in the performance evaluation: this requires not only knowledge of material properties and how losses can occur in the optical train, but certain experience with the mathematics and metrics of a ray trace simulation. This section gives an introduction to both aspects.

The *total transmission, optical efficiency* or *collection efficiency* [55] is the ratio of the power of the collected rays bundle $P(M_c)$ to the power of the input ray bundle $P(M_i)$:

$$\eta_{opt} = \frac{P(M_c)}{P(M_i)}, \tag{4.13}$$

where $P(M_i)$ is defined at the concentrator's entry aperture. Some designs start with a noticeable portion of the entry aperture that is optically inactive, such us the Cassegrain mirrors (see above), the on-axis XR [56] or even the RXI [57]. These devices would prefer a definition of the efficiency based on the optically-active area, for obvious reasons. The fact that part of their entry aperture cannot collect light has an influence on the systems footprint, but this issue affects all designs up to a certain degree. For instance, all modules need an outer frame around the optics to hold it in place on top of the housing, and several manufacturers take that into account for their efficiency/m^2 specifications.

Obviously, $\eta_{opt} \leq 1$ (only a loss-free design achieves $\eta_{opt} = 1$). When all rays convey the same amount of power, the collection efficiency becomes a concept that deals only with collected vs un-collected rays.

As explained above, in most concentrators the light collection is only possible when the optics is aiming at the sun with a certain degree of accuracy. In order to do that efficiently, the concentrator has to accurately track the sun on its apparent path around the earth. When M_i corresponds to a source placed at infinity, we can compute the collection efficiency associated with a given point of the source (i.e. to a given impinging angle θ), and in this case we define the *angular transmission* $T(\theta)$ as a plot containing the fraction of direct irradiance available at the entry aperture reaching the solar cell as a function of the angle θ between the rays coming from the sun central point and the concentrator entry aperture normal:

$$T(\theta) = \frac{dP[M_c(\theta)]}{dP[M_i(\theta)]}, \tag{4.14}$$

where $dP(M_i(\theta))$ stands for the infinitesimal power transported by rays entering the aperture tilted at an angle θ with respect to the vertical and $dP(M_c(\theta))$ is the same value evaluated for the same rays at the target. The *optical efficiency* η_{opt} is the maximum $T(\theta)$ achieved. Often, this occurs at normal incidence, $\theta = 0°$.

A theoretical definition of the acceptance angle has been given above, when introducing the *étendue*. In practice, the *acceptance angle* $\alpha(°)$ is the maximum tilt of rays with regard to the entry-aperture normal, such that at least a 90% of them (compared to normal incidence) will eventually strike the solar cell. In the transmission curve, the acceptance is the angle α such that $T(\alpha) = 0.9T(0°)$.

The product of the collection efficiency and geometrical concentration $C_o = C_g \eta_{opt}$ is called *optical concentration*, and gives an idea of the true irradiance gain (irradiance at the cell over irradiance at the entry aperture) achieved by a concentrator. The optical concentration is measured in *suns*, since it is related to a reference sun's irradiance (typically, 1000 W/m^2).

When the word 'concentration' is mentioned alone in the CPV field, it often means geometrical concentration. The main physical laws that apply to concentration optics deal with C_g, which is a purely geometrical concept. Note that $C_o < C_g$, since $\eta_{opt} < 1$.

4.5.1 Efficiency. Sources of Losses

Only a fraction of the DNI hitting the entry aperture of a concentrator reaches the solar cell, and can potentially be converted into electricity. The ratio of light power at the solar cell over the direct irradiance available at the aperture is called *optical efficiency,* as anticipated above. This fraction depends on many factors, particularly the angle of incidence of the beam at the concentrator entry, but when a manufacturer mentions *optical efficiency* usually refers to normal incidence, probably because in that situation most concentrators achieve a maximum. The *optical efficiency* should be calculated spectrally throughout the entire spectrum of interest (i.e. starting at 300 nm and up to 1900 nm to cover the EQE of current triple-junction solar cells). Some manufacturers refer to the efficiency at one single wavelength or at a limited spectral range, tough. It is the combination of high optical efficiency and high electrical efficiency over the year what makes a CPV system cost-effective.

Multi-junction cells are designed to perform optimally under certain types of solar spectra. The use of improper materials in the optical parts or an incorrect design can reduce the delivered power. The optical designer should take care of maximizing the transmission of light that is converted into electricity in every junction of the cell. For current triple-junction solar cells, this is especially relevant for top and middle, since the bottom cell (germanium) typically produces an excess of photocurrent, and there is some room for being slightly 'inefficient'.

There are different sources of losses that compromise the maximum optical efficiency achievable by a concentrator (see the list included in Table 4.3), and this holds even for an optical train manufactured with the highest accuracy and the best optical materials.

Here we classify the types of losses into two: those that depend on the optical material properties and those that depend on the optical surfaces.

Table 4.3 Summary of mechanisms for losses in CPV optical trains

Type	Sub-type	Comments
Fresnel surface losses		Refraction physics. Increase with refractive index and incidence angle
Absorption	In lenses	Propagation physics (Beer's Law). Grow with ray path length inside solid optical parts
	In mirrors	Material reflectivity. Silver mirrors show typical reflectivities above 95%
Dispersion		Refraction physics. Negligible in high-CAP optics, but limiting concentration ratio in low-CAP optics
Geometrical	Architecture limitations	Secondary stages cast a shadow on the system
	Draft angles and tip and groove radii	Fresnel lenses manufacturing limitations
	Glue excess on kaleidoscope-type (prism) SOEs	Assembly, bonding SOE to cell
	Form factor	Manufacturing limitations. Lower in high-acceptance optics
	Alignment of parts	Assembly. Lower in high-acceptance optics
	Scattering	Refraction/reflection physics. Grow with surface roughness. Lower in high-acceptance optics

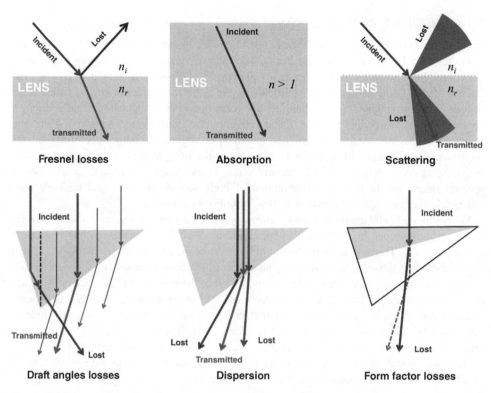

Figure 4.18 Different sources for losses in a concentration system comprising lenses. Absorption, scattering and form factor losses also apply to mirrors

4.5.1.1 Losses Depending on the Properties of the Optical Materials

When it comes to a CPV system, the main sources for losses that depend on material properties are (see Figure 4.18):

Fresnel Surface Losses

In a lens, a certain percentage of light is reflected at each refraction at a surface, with the direction following the laws of reflection. These Fresnel-reflection surface losses grow with the difference between the refractive indexes of the two media involved (and therefore depend on λ) and with the incidence angle, as the equation for the Fresnel reflection coefficient shows:

$$R_f = \left| \frac{n_i \cos(\theta_i) - n_t \cos(\theta_t)}{n_i \cos(\theta_i) + n_t \cos(\theta_t)} \right|^2 , \qquad (4.15)$$

where n_i and n_t are the refractive indexes of the incident and transmitted (refracted) ray, respectively. At normal incidence, when the transition is from air to an optically dense media ($n > 1$), the Fresnel surface losses are about 4–5%, depending on the material. There are anti-reflective (AR) coatings that can reduce the Fresnel losses to about a 1–2%, depending on the wavelength. The good performance of such AR layers depends on the substrates (glass being a typical substrate) and on the incidence angle of light. Therefore, AR coatings, which are rare, are more often found on the flat entry aperture of Fresnel-lenses based systems. Their application on SOE lenses is more complex and need a careful design of the layers, in collaboration with the optical designers in charge of the optics. When thinking on the usage of AR coatings, the CPV system designer should take into account their costs and the actual electrical efficiency increase they produce. There are different types of AR coatings: a) interference layer coatings; b) nano-structured layers on a substrate (e.g. moth-eye structures) and c) wet-etched modifications of the surface of the denser medium.

All AR coatings present the impinging light with a gradually increased refractive index, as the mixture of air and material the incident light 'sees', gradually increases. Thus, the refraction happens gradually, and the associated losses are 'forgotten'. Challenges with all AR coatings include lifetime and applicability to polymers, as well as spectral bandwidth.

Spectral Absorption and Reflectivity

Optical materials that transmit the light are neither perfectly transparent (in the case of lenses) nor perfectly reflective (for mirrors): part of the light is always absorbed inside the material. In the case of lenses and dielectric bodies, the absorption coefficient α_{abs} depends on λ, and is measured in mm^{-1}. Absorption grows with α_{abs} and lens thickness (ray path length L inside optics). A ray having a power P_{in} has a power P_{out}, after having travelled L mm through the dielectric body, given by the Beer's law:

$$P_{out} = P_{in}e^{-\alpha_{abs}L} \ , \tag{4.16}$$

In a mirror, a fraction of the light is absorbed at the metal coating (typically, 10% for aluminum mirrors and 5% for silver mirrors). The mirror reflectivity, often found in the literature, which is the inverse of this absorption, also varies with incidence angle in some materials.

Dispersion

In a dielectric body, light is refracted towards different directions depending on its color, despite striking the same point, owing to the variation of the index of refraction as a function of λ. The *dispersion* is the mechanism responsible for the rainbow, and the color separation when a white beam passes through a solid transparent prism. Dispersion produces deviations for the rays of certain wavelengths that might cause them miss the solar cell if the design is not tolerant enough. Most designs are carried out using one single value for n and therefore, can be considered purely monochromatic, becoming polychromatic later at the ray trace analysis stage.

4.5.1.2 Losses Depending on the Surface Shape and Finish

The quality and shape of an optical surface have a strong influence on the ability of surfaces to properly serve their purposes. The main types of losses linked to optical surface properties are (see Figure 4.18):

Scattering

Rays do not exactly follow the design direction unless the optical surfaces are optically smooth. Light is scattered (forwards and backwards in the case of lenses) around the chief directions (refracted ray and Fresnel reflected ray in lenses, reflected ray in mirrors) when surfaces are rough [58,59]. Losses increase with increasing roughness. Both the fraction of power scattered and the angular span of the scattered light depends on the roughness pattern.

The same roughness pattern provokes more severe ray deviations in mirrors and TIR surfaces than in refractive surfaces, owing to the nature of reflection. Note that a slope deviation of γ degrees provokes a ray deviation of 2γ degrees, while refraction calculates through Snell's Law, causing the deviation to remain roughly constant (see [37]). Scattering in mirrors basically behave like back scattering in lenses angularly.

The actual power loss depends on the roughness pattern and on the ability of the optics to collect some of the scattered light. The maximum roughness permitted in the optics of a CPV optical train is not easy to determine and depends on the architecture of the system. Apart from the differences between reflective and refractive surfaces just mentioned, POEs typically allow for less roughness than SOEs, owing to their smaller acceptance angle. In secondaries made of molded glass, the surface tension of the cooling glass can supply a smooth surface. The rule of thumb is that the roughness of the surface should be one order of magnitude smaller than the minimum wavelength of the light it handles (therefore, around 30–35 nm in the case of CPV). One needs to distinguish local and global surface errors. Surface roughness is a local error while shape distortions or visible waviness are global errors, or form-factor errors, as explained below.

Geometrical Losses

This general term geometrical loss refers to cases in which the cell is not clearly visible[3] from certain points of the optics. Geometrical losses sometimes may deal with the optical architecture approach itself, when some of its parts cast shadows onto the receivers, examples being secondary lenses and mirrors in the Cassegrain designs and on-axis XR approaches. In other cases, geometrical losses deal with manufacturing constraints or errors, and even with design errors/limitations. Some of them can be minimized in systems having wide acceptance.

An example of a manufacturing constraint in a Fresnel lens are losses that come from tips-rounding and draft angles on the (no longer) vertical facets needed to withdraw the lenses from the mold. Other examples of this type of errors are lens warp [60] (in PMMA POEs) and prism deformation (in SOG POEs [61]) linked to manufacturing processes.

A typical manufacturing error is found in F-RTP prisms, where the optical glue leaking at the bottom of the prism might cause some light spillage because rays hit the glue meniscus shapes with angles below the critical angle, forcing them to be refracted, rather than being reflected. Other manufacturing error deals with surfaces having shapes slightly different from those calculated at the design stage ('form factor' errors). Rays may hit these surfaces at points where neither the coordinates nor the surface normals (slopes) are as designed, and therefore may be refracted (or reflected) with a slight deviation. The same slope error produces a larger deviation in the case of reflections, as pointed out in the scattering section, and therefore, mirror/TIR surfaces are more sensitive to surfaces errors and require more accuracy. Errors in the form factor of a lens are less likely to affect the concentrator performance than similar errors in a mirror [62].

Poor alignment between optical parts is another source for geometrical losses in CPV systems. These are typically more tolerant to misalignments along the optical axis than to lateral misalignments. Rotations of the optical parts with regard to each other, or the cell with regard to the SOE may provoke significant efficiency drops as well.

Figure 4.19 shows how losses typically occur in a Fresnel + refractive SOE system. For this example, we have selected a F-RTP optical train and have considered the system to be perfectly

---

[3] This is actually true: when there are geometrical losses of certain size, they can be noticed by moving the eye over the surface of the entry aperture of the primary optics, looking at the receiver: the latter disappears in areas where geometrical losses occur.

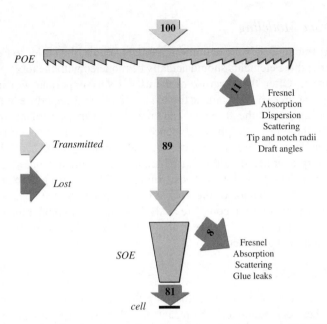

Figure 4.19 Drawing showing how losses (figures are approximate) occur in the different stages of an optical system based on Fresnel lenses and a transparent solid secondary stage (RTP, half-egg, FK). In the example, 81% of the incident power hits the solar cell if the system is perfectly aligned and aimed at the sun

aligned: therefore the losses listed in the figure (quantities may vary) can be considered unavoidable (those at the glue leaks and those dealing with scattering can be minimized, though). Figure 4.20 shows a similar diagram to illustrate the losses in a Cassegrain type concentrator.

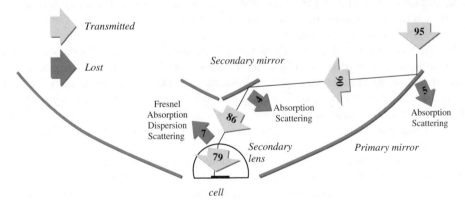

Figure 4.20 Drawing showing how losses (figures are approximate) occur in the different stages of an optical system based on two mirrors and a lens (Cassegrain type). Note that the secondary mirror casts a shadow whose area, projected onto sun-beam direction, covers 5% of the total primary mirror area leading to geometrical losses. In the example, 79% of the incident power hits the solar cell, if the parts are perfectly aligned and the system is perfectly aimed at the sun. If the system has a cover glass the losses are 4–9% larger, depending on whether the glass has an antireflective coating

4.5.2   Ray Trace Modeling

The simplest ray tracing needs the definition of a light source, the optics to be analyzed and the receiver. These three elements should actually be modeled with the tools provided by the ray tracing software. Often the expertise of the user is very important. For instance, in the CPV field the light source should emulate the sun as seen from the aperture of the CPV panel. In a first rough approach, the designer can model the sun as a plane monochromatic wavefront hitting the optics at a given incidence angle. However, this is quite incomplete. The sun, as seen from the earth, subtends a small angle ($\pm 0.265°$) so the rays at the concentrator's entry aperture should hit at all angles within the angular size of the sun, at least. The sun emits sunlight with Planck spectrum at 5777 K, starting in the UV and ending in the far infrared. The behavior of the optical parts depends on the spectrum of light, as explained before, and therefore a complete analysis should try to model the sun throughout the entire spectrum. Similar considerations apply to the optical parts and receivers, and will be discussed in the next sub-sections.

4.5.2.1   Light Source Model

Since rays coming from the sun, when reaching the entry aperture of a concentrator, are almost parallel, these can be modeled by means of a plane surface perpendicular to the rays or by a point source located at infinite (very far away from the optical model, which makes this option somewhat impractical). Some programs have specific tools to model the sun. If not, most ray tracing software allow for the definition of a set of rays stemming from any surface (particularly, a planar one) following certain directions, for instance following the normal to the surface. The point source (spherical wavefront) is another type of source that is always available in ray tracing software. In order to model the small angular size of the sun, as seen from the earth, the easiest way is selecting a limited angle ($\pm 0.265°$) beam stemming from a plane surface. The position (particularly, the rotation) of either the point source of the plane surface with regards to the optics defines the incidence angle. The analysis of a set of different relative positions of these elements permits the calculation of the *angular transmission* and *acceptance angle*.

Apart from the geometrical beam characteristics, the user can define the beam power. Since the analysis of a concentrator often looks for the ratio of light power reaching the cell over the light power available at the entry aperture, what really matters is getting to know the actual power available at the aperture, so all calculations can be referred to it. To avoid any doubts, the user can locate a flux sensor having just the same shape of the concentrator's entry aperture on top of it, making sure it has no optical properties and do not interfere on the natural trajectories of the rays.

Another important user-selected value is the number of rays. Once the beam power and the number of rays are selected, each ray within the analysis ray set will have a starting direction and power (the sum of rays individual power is the power of the beam). On their way through the optical system, each ray can potentially change its direction and have is power reduced. The number of rays is a very important value: considering the ray tracing is a statistical calculation, it has a direct impact on the accuracy of the results obtained at the end. It has also an important influence on the calculation time. A reduced number of rays are traced very rapidly and results are available very soon, but these might be misleading. The user has to determine the number of rays needed for each type of simulation so the errors are minimized. For instance, efficiency calculations need a small amount of rays (ten to one hundred thousand typically) compared to irradiance patterns ray traces (five hundred thousand to one million rays) needed to analyze in detail the maximum and minimum local irradiance peaks at the cell.

In principle, the number of rays n_{rays} required relates to the confidence interval σ, or statistical error, as:

$$\sigma = 1/\sqrt{n_{rays}} , \tag{4.17}$$

With 100 rays, the error is 1/10, or 10%. Except for collected-flux calculations, where a limited number of rays might suffice to get an accurate impression of the concentrator performance, receivers in ray tracing are often segmented into bins, fields of mostly equal area. A square cell may have been segmented into 10×10 bins, in order to detect and visualize the inhomogeneity of the irradiance, for instance. The error for each bin will be 10% for 100 rays. With 100 bins present, 10 000 rays are required. It often makes sense to employ simplifications, such as symmetries to minimize the number of rays. Algorithms may 'learn'; new rays may be traced into areas where the expected marginal information increase [63] is greatest.

If a full spectral analysis of the optical train is needed, the light source should be described as a function of wavelength, and local/angular intensity variations if it is an extended source such as the solar disc.

4.5.2.2 Optics Model

Ray tracing programs [64] accept CAD files with a 3D description of the solids to be traced (the CAD format depend on each case, *sat*, *stp* and *igs* files are compatible with most software tools). Some programs offer design tools oriented to illumination problems in general, which permit the calculation of simple optical trains (one-stage mostly).

The first step in modeling the optical train requires loading the CAD file and aligning it with the correct positions of light source and solar cell. Then the material for each solid needs to be defined. Depending on the software, the properties can be applied to the entire solid, to the surfaces that form the solid or to both. This step is critical, especially when there are different materials together in optical contact. The most typical case is the glue bonding secondary optics and cell. The accurate definition of such interfaces is very important if we want to know the actual performance of the optical train. The possibility of applying special properties to the surfaces of a solid is important, since this enables modeling AR coatings, for instance, which are often applied to only one surface of a lens.

All ray tracing tools permit the definition of basic materials properties such as 'perfect refractive' (only defined by a refractive index), 'reflective' (defined by a reflectivity), perfect absorber, among others. Some tools have a database with common optical materials such as different commercial optical grade PMMA, Polycarbonates, silicones and glasses. Apart from such databases, the programs permit user-defined materials, introducing values such as refractive index (constant or wavelength-dependent), absorption coefficients for the bulk materials, and spectral reflectivity or transmission characteristic for the thin layers. AR coatings can be modeled by means of a file describing their spectral response or 'ideally' by just defining the corresponding optical surface as ideally refractive, without Fresnel losses.

Scattering properties of surfaces are more easily modeled applying scattering properties to selected surfaces, whose overall shape can be kept smooth. Commercial software tools provide different types of models such as Mie [65] and Henyey–Greenstein [66]).

One way to find out the contribution of the different sources of losses to the overall efficiency is defining perfect surfaces and loss-free materials, except for the parts whose contributions need to be analyzed. The process can be repeated with every material and stage, to understand how the optical train can be enhanced. This should be understood as the standard procedure when conducting a sensitivity analysis. Occasionally, modeling of the parts may be difficult, as the ray-tracing package isn't equipped with the necessary functionality. In such cases, experience and laboratory experiments may help to introduce a black box into the optical model that adds the desired modifications. An example is a statistical scattering, which isn't introduced as geometrical model of the surface, but as statistical change of direction of the rays passing through a black box instead.

4.5.2.3   Receiver (Target) Model

The solar cell can be modeled with a simple surface having the shape of the active area, defined as an absorber. Some tools do not even need a specific definition of features for such surface: just its proper position, and immersion into the secondary optics, if that is the case.

When it comes to the solar cell, all attention has to be paid to the correct geometrical modeling and the proper definition of surrounding materials.

If some more specific details about the cell performance need to be analyzed, a more accurate modeling can be tried. This can include the addition of grid lines (with shape, size, reflectivity and scattering properties) and the absorption/reflectivity characteristics of the active areas. The spectral response of the solar cell cannot be added in current commercial ray tracing software tools, but can be modeled separately if the power and spectrum of incident light is known (see Figure 4.1 where the EQE has been calculated).

When the goal of the analysis is getting to know the irradiance on the entry aperture of the solar cell, one has to make sure the receiver surface has a resolution sufficient to resolve sharp irradiance peaks. Most tools let the user define the number of bins within the receiver surface.

4.5.2.4   Analyzing the Results

Once the model is completed, a ray trace analysis can be run. The calculation time depends on the complexity of the model but most of all in the number of rays and type of ray modeling (whether Monte Carlo or ray splitting are selected, whether information of all rays traced is stored or only those going through preferred surfaces).

The set, or better, a subset, for a clear interpretation of results of rays traced through the optical set up can be made visible in most software packages. This allows the designer to take a first look at the performance of the system, and visually confirm the alignment of parts. In a closer look at results, values such as optical efficiency or irradiance available at the solar cell for one particular sun beam incident angle can be determined. Such values are not directly available: the fluxes available at the cell and at the entry aperture are, and the optical efficiency is the ratio of them both. There are means to trace automatically different variations of the same set up if the goal is finding the angular transmission and acceptance curve, for instance. One is preparing a script program that runs several iterations of the same model with a set of different system orientations with regards to the light source, and stores the results obtained at each iteration. The results can be processed later to draw the angular transmission and deduce the acceptance angle. Other important characteristics that can be studied through ray tracing are part of a sensitivity analysis:

- Determining the tolerance of the optical train to some shifts or rotations between the optical parts
- Finding the actual contributions to losses of some particular elements
- Comparing the performance of alternative materials
- Simulating the light in radical off-axis situations, where the beams do not hit the solar cell and can potentially damage concentrator cell assembly (CCA) and housing features.

Visualization is probably the most important sanity check on the optical system. Geometrical changes may be introduced to tweak the optics. The ray-tracing analysis results in a list of error contributions with sensitivities and dependencies for each optical element. The metric to be chosen for easy comparison and evaluation must be a tolerance angle, in order to compare the errors with the acceptance angle of the optical system. Now, the optical analysis begins with an ideal system. Manufacturing of components, their assembly, and operations under specified conditions make any sensitivity analysis a complex and time-consuming task. Most tolerances for manufacturing, to give an example, will only be known once high-volume production has begun. This is even more obvious for high-volume, automated (as opposed to manual) assembly. Errors due to operations, too, will likely be known only once the environmental setting of the CPV power plant is understood for specified conditions, and a number of CPV systems have been installed and monitored. Only then the tolerance budget required can be compared to the tolerance budget provided by the optical design. Discussing the manufacturing of the components of the optical train of the CPV module provides the inroads on understanding tolerances.

4.5.2.5 Design Summary

Figure 4.21 shows the cross sections of the optical trains discussed above. Notice all the devices have the same entry aperture area, and the depth has been drawn to scale, so the drawing gives an idea on how deep these systems must be with respect to each other to perform well. Notice the single Fresnel lens approach is, by far, the deepest device, while the XR is the most compact of all. For a constant 800× concentration, Table 4.4 compares these devices in terms of

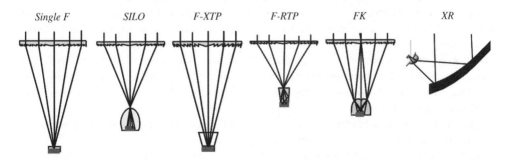

Figure 4.21 Main concentrators explained in Section 4.4. They all share the same entry aperture area and depths are scaled with respect to these. Cell and SOE are oversized, for clarity purposes

Table 4.4 Compared features of the different devices introduced in section 4.4. For the calculation of the acceptance angle, we have assumed $C_g = 800\times$ in all cases. Note that the optical efficiency of a perfectly manufactured concentrator is higher in the case of the single F and F-XTP, owing to the lack of secondary stage. However, for the same reason, and the correspondent reduction in the acceptance angle, these systems tend to attain lower electrical efficiencies in large arrays, where their high sensitivity to various tolerances become evident

	Single F	SILO	F-XTP	F-RTP	FK	XR
C_g	800×	800×	800×	800×	800×	800×
α	±0.36°	±0.6°	±0.73°	±0.91°	±1.28°	±2°
$f\#$	>1.4	>1	>1.4	>0.85	>0.85	??
Irradiance uniformity on cell	Bad	Good	Bad (or medium at the expense of some CAP)	Medium	Excellent	Medium - good
η_{opt}	85–88%	82%	85–88%	82%	82%	82%
Technology	Fresnel	Fresnel + lens SOE	Fresnel + mirror SOE	Fresnel + lens SOE	Fresnel + lens SOE	Mirror + lens SOE

acceptance angle, f-number ($f\#$), irradiance uniformity on the solar cell, optical efficiency and technology used to manufacture them.

4.6 Optics Manufacturing

4.6.1 Optical Materials for CPV

Optical properties, durability, and manufacturability are determining the suitability of optical materials for CPV in the real world. CPV is installed in predominantly arid climate zones where the DNI is high. Optical elements will be exposed to temperatures ranging from below zero degrees centigrade to above 50°C. Humidity can be high, the optics may come in contact with rain, hail and snow from the outside and harbor condensation inside. Wind and thermal cycling can dynamically stress the optics to the point of their failure in the so-called environmental stress cracking (ESC); hail poses the risk of destruction. Sandstorms, soiling and cleaning cause surface abrasion and may set the stage for corrosion processes such as ESC. See Chapter 9 for further details about optics reliability.

In this chapter, the terms durability, reliability and longevity of optics are used synonymously although they strictly express different meanings (see also Chapter 9).

Optical materials are exposed to outdoor levels of ultraviolet (UV) radiation. Irradiance values are higher than average, as CPV plants may experience many sunlight hours; additionally, installations may be at high elevations. UV radiation may break organic bonds endangering the integrity of plastic optics. Yellowing is a first indication of ESC, and, ultimately, failure. Inorganic optics may contain elements reacting to high ultraviolet flux, one example is the solarisation of glass, where manganese oxide or other material traces are oxidized by UV radiation.

The most commonly used optical materials in CPV are listed in Table 4.5. Some of the characterizations are based on experience.

Glass is in many CPV modules used for the front plate. All light to be concentrated passes the front plate, directly affecting module efficiency. Losses are caused by Fresnel surface reflections. Absorption within the glass material is minimized by using thin iron-free glass.

Table 4.5 Common optical materials for CPV optics: Function, optical properties and durability

	Function	Material type	Optical properties	Durability
Glass	Cover glass	Soda-lime, iron-free	Excellent	Excellent
	Mirror glass	Soda-lime, iron-free	Good	Good
	Secondary Optical Element (SOE)	BK7, B270	Good	Good
Sol-gel	SOE	Sol-gel	Good	Good
Silicones	Silicone-on-glass Fresnel lens	Optical silicones, glass	Good spectral transmissivity, large thermal expansion difference between silicone and glass	Good (data available from 2005)
	Glue between SOE and cell	Silicone	Good	Good (shorter data history than for SoG)
PMMA	Primary Optical Element (POE)	PMMA	Good	Good (warranted and well known)
Coatings	Antireflective (AR) coatings such as interference, nanoscale coatings, or etched surfaces	SiO_x for moth-eye type nanoscale coatings	Good	Medium
Aluminum	Mirror coatings	Al	Good	Good, if properly sealed

Only approximately 2% of all glass manufactured worldwide is used for solar applications. There are a very limited number of purely solar glass lines, may it be float glass or the less common rolled glass. Glass manufacturers are running campaigns drawing iron-free glass. The glass sheet should be smooth and flat with its edges beveled. Beveling reduces the risk of injury during handling and also reduces the risk of tempered glass breaking due to the tension release triggered by micro-cracks on the edges.

Modern glass lines automatically test glass ensuring it being free of defects. During the transport of glass, surface corrosion may be introduced when water is present in the space between glass plates.

Glass forms the superstrate of the silicone-on-glass (SOG) Fresnel lens parquet. Ideally, the Si present in both the glass and the silicone forms chemical bonds. Glass sheets must be clean and dry. For the sake of optical efficiency, the local waviness of the glass should be minimized. The difference between highest and lowest points in such waviness (the peak-valley-value, or pv-value) must be smaller than the thickness of the base layer of silicone on the glass. Otherwise, prism valleys would be on the glass, lowering efficiency and potentially damaging the tools used to shape the optical structures of the lenses. Global shape distortions, such as dishing or warping need to be kept small, in order not to compromise the optics or handling suitability of the sheet.

Float glass is tempered in ovens, the more modern of which use controlled air valves to float and selectively cool the glass sheet aiming at a flat shape in the tempering process. There is tempered safety glass and heat-strengthened glass. In most cases, heat-strengthened glass is used in CPV. If the glass breaks, the resulting shards are larger and have less blunt edges in heat-strengthened glass than in tempered safety glass. The thickness of float glass is typically 3.2 mm or 4.0 mm, with 2 mm glass appearing on the market.

PMMA is equally used for the primary optics of CPV modules. PMMA ($80 \cdot 10^{-6}$ m/(m K)) has a lower coefficient of thermal expansion than glass or silicone ($5-10 \cdot 10^{-6}$ m/(m K)), or ($200-500 \cdot 10^{-6}$ m/(m K), respectively). PMMA is one of the few organic materials that are optically clear and can be modified to withstand outdoor UV exposure for more than twenty years, sometimes warranted by suppliers.

4.6.2 Tolerance Budget

The tolerance budget of a solar concentrator is limited by the concentration ratio of the optics. The higher the geometrical concentration ratio C_g, the lower the acceptance half-angle α (as explained above in Eq. (4.10), $C_g = n^2/\sin^2(\alpha)$). In Figure 4.13, an example of an acceptance half-angle of 0.8° is given for a HCPV system. These 0.8° can be considered the limit of the tolerance budget of this particular concentrator, in the sense that for higher incidence angles, the losses increase significantly and lower the optical concentration ratio of the module. Assume that the system should perform according to design. System assemblers must allocate individual tolerances to the system components, their assembly, and operations. The overall metric should be in the unit of the acceptance half-angle, but this can easily be transformed into geometrical lengths. A recipe for the calculation of the overall tolerance budget with examples and details has been given earlier [67,68].

The tolerances of manufacturing Fresnel lens parquets are a simple example of statistics in production. Assume that a monolithic master mold has been made for the production of silicone-on-glass lenses; tolerances under consideration are length variations that occur in the tooling (alignment issues), and the process (laying of the glass, thermal expansion issues). Assume further that the variations within each tolerance data set follow a normal distribution (values are found symmetrically and relatively close to the expected value, μ). The data sets shall be independent of each other. There is an equal probability that statistical errors increase or decrease a value. This statistical compensation of positive and negative errors is the Gaussian Error Propagation yielding the propagated standard deviation s_y of a number of data sets x_i [69]:

$$s_y = \sqrt{\left(\frac{\partial y}{\partial x_1}\right)^2 s_{x1}^2 + \left(\frac{\partial y}{\partial x_2}\right)^2 s_{x2}^2 + \ldots + \left(\frac{\partial y}{\partial x_i}\right)^2 s_{xi}^2}, \qquad (4.18)$$

The partial derivatives are a measure for the slope or sensitivity of the tolerance. All tolerances are treated as variations in length, caused by approximately linear effects such as thermal expansion well away from any phase changes. Therefore,

$$s_y = \sqrt{s_{x1}^2 + s_{x2}^2 + \ldots + s_{xi}^2}, \qquad (4.19)$$

If sufficient volumes have been already produced, the sample standard deviations s_x are known; likely even the propagated standard deviation s_y of the product has been measured and calculated.

Table 4.6 Process Capability Index (C_{pk}) and sigma level for normal distributions

Process Capability Index C_{pk}	Sigma level	Area under probability density function (normal distribution)	Process yield, %
0.33	1	0.6826894921	68.3
0.67	2	0.9544997361	95.4
1.00	3	0.9973002039	99.7
1.33	4	0.9999366575	
1.67	5	0.9999994267	
2.00	6	0.9999999980	

The propagated standard deviation can be estimated in advance, as the sample standard deviations s_x can be predicted based on the so-called *Process Capability Index, C_{pk}* [70].

In normal distributions, the empirical rule states that approximately 68% of all data values are within one standard deviation of the expected value μ, mathematically $\mu \pm \sigma$. 99.73% of all products are within $\mu \pm 3\sigma$, hence the well-known $3 - \sigma$ level in quality assurance. The relation between C_{pk}, sigma level and process yield is given in Table 4.6.

Assuming that any tooling accuracy or process instability can be controlled on the $3 - \sigma$ level:

$$s_x = \frac{T/2}{3\,c_{pk}}, \tag{4.20}$$

where T is the tolerance, here symmetrical on the specified length l_{-T}^{+T}.

The tolerance budget is a critical issue in designing and manufacturing large-area optical elements for CPV. One should be aware of material properties such as thermal expansion values and process limitations.

4.6.3 Manufacturing of Primary Optical Elements

A few decades ago, large area optics meant telescope optics; optics for astronomy at astronomical prices, as has been quipped. Large displays based on backlight screens required the production of single large-area Fresnel lenses, with diameters up to five meters. With diamond-turning allowing for the manufacturing of optical surfaces, molds could be made. Molds could then be replicated. Assemblies of single lens molds could then be replicated and thus monolithic molds for Fresnel lens parquets can be created. It becomes possible to base mass manufacturing of parquets on a single master mold. This splits optical tolerances into systematic ones present in the master mold and statistical ones introduced in the manufacturing process of the product itself. Systematic tolerances are usually known and often CPV modules can be designed in ways to accommodate systematic tolerances. An example is the systematic lens center distance variations present in the master mold; once known, receivers can be placed accordingly on the back plate of the module.

We focus now on the example of parquets, i.e. assemblies of macroscopic single optical elements into large-area primary optics for CPV. Parquets can be characterized by systematic

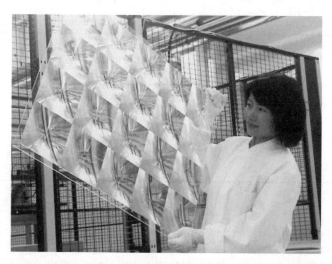

Figure 4.22 Primary optical element (POE): Fresnel lens parquet made by hot embossing

tolerances; they usually combine module cover and primary optics, thus potentially increasing efficiency and ease of module assembly. Though parquets of reflective optics are theoretically possible, Fresnel lens parquets represent the vast majority of CPV primary optics.

Regarding mold manufacturing, diamond-turning reliably reproduces optical structures such as prisms or diffractive elements with a size of 10 nm. Any optical surface should have a local surface roughness at least one (better two) order(s) of magnitude below the wavelength of light. This does represent a challenge for single-point diamond turning where a diamond with a small tip radius (typically 4–8 μm) cuts the surface of the rotating mold. The feed of the machine and the rotation of the mold create adjacent paths cut by the diamond separated by 'wave' crests. Diffraction should be negligible and typically >99% of light is contained in the 0th-order maximum.

Prism heights are often in the range of 0.2–0.7 mm, draft angles 0–2°, tip and groove radii 5–15 μm. The surface roughness may be measured in white light interferometry with good values being around 10 nm.

Diamond-turned molds can be assembled into a parquet, and replicated galvanically into a monolithic master mold. Copies can be drawn and the actual manufacturing process of the Fresnel lens parquet may begin.

Parquets are often 0.2–1.5 m^2 in area, mostly depending on the overall expansion of materials and potential mismatch between front aperture and back plate, as well as on mechanical stability.

Fresnel lens parquets are often made in hot embossing or casting processes. This involves the heating and cooling of a lens parquet material to cure or fix the shapes of the prisms. Fresnel lenses in PMMA (Table 4.7; Figure 4.23 and Figure 4.24) are hot embossed into pre-fabricated sheets or foils. The sheet is heated some 50°C above glass point (110°C), the mold closes embossing the lens features into the PMMA. Subsequently, the mold and product are cooled down under pressure to a temperature below glass point. Finally, mold and product are separated.

Table 4.7 Common optical materials for CPV: processes and manufacturability

	Function	Process	Manufacturability
Glass	Cover glass	Float line	Mature
	Mirror glass	Casting	Good
	Secondary Optical Element (SOE)	Polishing, molding	Mature, good (respectively)
Sol-gel	SOE	Sintering	Good
Silicones	Silicone-on-glass Fresnel lens	Casting	Good
	Glue between SOE and cell	Casting	Good
PMMA	Primary Optical Element (POE)	Hot embossing	Mature
Coatings	Antireflective (AR) coatings such as interference, and nanoscale coatings	Deposition	Medium
Aluminum	Mirror coatings	Deposition	Medium (high aspect ratio of part complicates shaping in two dimensions)

There are variations to the batch process of hot embossing. Manufacturers are embossing optics on PMMA foil passing between two hot calenders, one of which contains the Fresnel lens mold. The process is called roll-to-toll. The foil is then cut to size and laminated onto PMMA sheets, creating the Fresnel lens parquets.

Silicone-on-glass (SOG, Table 4.7, Figure 4.24) Fresnel lens parquets are made in a casting process where silicone is applied to the cleaned glass. The two-part silicone must be evacuated and mixed. Once the mold is embossed into the silicone, the composition is heated, leading to the silicone curing and chemically bonding itself to the glass.

Glass and silicone have very different thermal expansion coefficients, as given above. The bonding of the silicone onto the glass results in the silicone being restricted in its expansion or contraction when the operating temperature of the SOG composite differs from the curing temperature of the silicone. The effect leads to a third-order distortion of the prism faces, in turn compromising the lens' optical efficiency, which will significantly depend on typical operating and manufacturing temperatures [71]. In fact, the distorted prism face creates two focii, both visible in the experiment and in the ray-tracing simulation after a finite element analysis.

The impact of the differential thermal expansion coefficients for SOG can be analyzed by finite element analysis and moderated by including a suitable shrinkage compensation in the mold. Thus, one may cure at a relatively higher temperature, which reduces cycle times and cost. There remains the differential between highest and lowest operating temperatures, with related efficiency reductions. Final system design should include the module and intended location's temperature profiles, as well as the concentration ratio and acceptance half-angle, or tolerance of the optical train. Systems with more acceptance and tolerances will more easily counterbalance this thermal effect.

Products are characterized and cleared by quality inspection. Each parquet should be marked or labelled allowing for tracing of each individual parquet over its warranty period (>20 years).

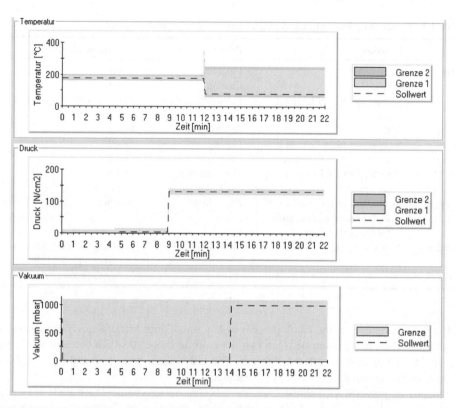

Figure 4.23 Typical process steps in hot embossing of Fresnel lens parquets. From top to bottom, temperature, pressure and vacuum settings over the process time are shown. Dashed lines are intended values, limits ('Grenzen') are shown. Screenshot of the control terminal of a Maschinenfabrik Lauffer GmbH and Co. KG vacuum lamination press

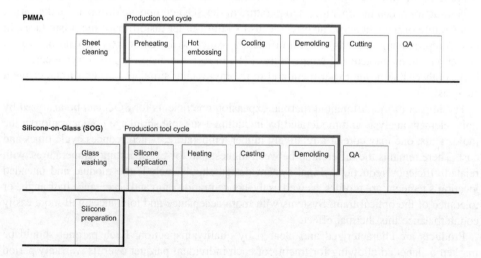

Figure 4.24 Hot embossing of PMMA and casting of silicone-on-glass (SOG). Typical cycle steps

4.6.4 Manufacturing of Secondary Optic Elements

4.6.4.1 Materials and Manufacturing Processes

Materials for secondary optical elements are given in Table 4.8. Metal films are used for reflective (mirror) SOEs. These films need to be highly reflective, which can be a challenge for the applicable solar spectrum and grazing angles of incidence. All materials must withstand high fluxes of several thousand suns locally in the focus of the primary, including UV, elevated temperatures humidity and possibly other accelerators like glues.

The glasses must not show solarization. Stress in molded glass can be an issue. Other materials than those listed in Table 4.8 have been tested with negative results; thermal stability is absolutely essential as dust and the high flux can easily lead to a runaway reaction destroying the optical element. Transmissivity of materials for secondaries can become an issue only if the paths of the light inside the lenses are excessively long (more than tens of millimeters).

Materials define manufacturing processes. Metal foils have a limiting radius for bending. They cannot be bent in more than one dimension and stretching is very limited. Glasses are polished or molded. In polishing, flat or spherical surfaces are standard, aspheres are possible. Molding allows for free-form surfaces, as long as the negative mould can be manufactured. For demolding, a minimum draft angle of $2°$ is typically necessary. Molding is an art and only very recently automated processes have begun to add to the traditional ways. For mass production, molten glass is directly used instead of pre-cast glass gobs or white-hot rods (which are manually molded and thus only feasible at the prototype stages). This is necessary to cope with the sheer number of optical elements required in CPV.

The number of glass lenses per unit-power obviously depends on the size of the solar cell, the concentration level and module efficiency. Assume that a module carries 100 SOEs per square meter (in a $C_g = 1100×$ CPV module comprising $9\,mm^2$ solar cells), or about 370 000 SOEs/MW$_p$, (assuming 30% module efficiency and $900\,W/m^2$ irradiance for concentrator standard *operating* conditions, CSOC, see chapter 12) and therefore, 370 million SOE's glass pieces for one GW$_p$ in CPV.

Surface quality of glass, whether polished or molded is often good. The high surface tension of glass generally prevents surface imperfections, though some visible surface roughness remains, as seen in Figure 4.25. Such roughness is unlikely to affect performance owing to the wide local acceptance angle of secondary elements, which often work with wide beams coming from the POEs.

Table 4.8 Materials for secondary optical elements

Type	Examples
Aluminum for reflective SOE	Alanod MIRO-SUN®
Glass for polishing process	BK7, Schott: $n_d = 1.5168$, $v_d = 64.2$;
	K9, China: $n_d = 1.5163$, $v_d = 64.1$;
	BSC7, Japan: $n_d = 1.5168$, $v_d = 64.2$
Glass for molding process	B270, Schott: $n_d = 1.523$, $v_d = 58.64$;
	H-K51, $n_d = 1.523$, $v_d = 58.64$;
	LIBA2000, $n_d = 1.521$, $v_d = 62.0$;
	Schott BOROFLOAT® 33, $n_d = 1.4714$, $v_d = 66.02$;
	Swarovski C5 HT, $n_d = 1.5608$, $v_d = 61.33$
Sol-gel method	Evonik Savosil®, $n_d = 1.458$, $v_d = 71.0$

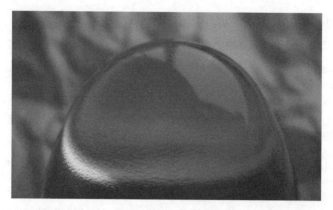

Figure 4.25 Surface roughness of experimental molded secondary optical element (SOE). Approximate diameter 20 mm

4.6.4.2 Mounting Secondary Optical Elements

Secondaries need to be mounted onto the solar cell. Glass and cell can be bonded securely and optically efficient with silicones. Once the secondary becomes heavy and tall, its leverage threatens to break the cell in case of mechanical shock during transport or operation. Just like mirror secondaries, the SOEs made of glass need a support structure, which adds to the complexity and cost of the module. Short secondaries are easier to glue and hold. If the base area of the secondary is larger than the area of the cell, encapsulation can be an additional purpose of the SOE. One major issue in this type of encapsulation is the appearance of air bubbles on top of the cell, which compromise the optical efficiency. Kaleidoscope secondaries based on TIR must use exact amounts of glue, as any excess leads to the spillage of light at its critical base where the flux density is highest.

4.7 Impact of CPV Optics in a Nutshell

The optical train nature and performance has an influence on the CPV system that equals that of the solar cell itself. It has a deep impact on the final system efficiency and moreover on its maintenance throughout the year (energy production) but it also determines part of the capital expenditures (CAPEX), those of the optical parts being obvious, but also those of additional parts whose tolerances and weight (to mention only two features) depend on how wide the acceptance of the optical train is.

Table 4.9 gives some examples of how the optics can influence CPV systems in terms of costs, looking at the different parts of the system, from module to plant.

The importance of the optical parts from the costs standpoint starts with the solar cell: on the one hand the optics determine the concentration achievable and therefore, the usage of solar cell (linked to the system bill of materials), but also the cell efficiency and reliability (through the ability of the optics to produce uniform irradiance free of chromatic effects).

The CPV optics needs to be simple, efficient (a reduced number of optical stages help to meet both goals) and must come with wide acceptance angles and good means to produce uniform irradiance onto the cells. The wide acceptance gives the CPV system a large 'tolerance budget', that should be partly consumed in an optimal way (for instance, loosing manufacturing tolerances or plant installation requirements, or reducing tracker weight) and partly reserved, to

Table 4.9 Optical train influence on the total system costs (capital expenditures, CAPEX) and also the energy production. This table shows how this influence can occur looking at all the parts of a CPV system

	CPV module			CPV system	CPV plant
	Solar cell	Optical train	Housing and module assembly		
Less CAPEX	Reduce cell area through higher concentration	Reduced number of parts Compatible with mass production	Loose manufacturing and assembly tolerances	Lighter supporting structure and simpler tracker	Rapid on-site installation Reduced maintenance
More Generation	Increase conversion efficiency and assure long lasting through uniform irradiance	Maximize transfer of light	Lower mismatch losses between the cells of a module (alignment)	Lower mismatch losses between the modules of a system (aiming)	Reduced wind and soil effects

assure the system will perform efficiently even with tracker misalignments and windy days, or when the modules are a little dusty.

Optics lacking adequate tolerance budget and with tight manufacturing tolerances may generate less electricity than expected at array level, no matter how well a single-cell prototype works. These systems can produce electricity clearly below expectations, compromising the plant profitability. Such type of systems would perform according to expectations by means of a higher quality and higher cost manufacturing or by changing the optical approach to one showing better tolerances.

The last and not the least important feature one should expect from the optics refers to the rest of the parts of the CPV system, and is reliability and long term durability. Considering these systems need a lifetime of over 20 years to be competitive in terms of costs, these goals are compulsory. In other rare approaches [72], there's some room for inexpensive optical materials, provided that the CPV system is designed in such a way that the optical parts can be easily replaced from time to time, without large investments. In such cases the durability of optical materials is not a must.

Acknowledgements

The contribution of Ling Fu on the discussion of secondary design and manufacturing is gratefully acknowledged

Glossary

List of Acronyms

Acronym	Description
2D	Two-dimensional
3D	Three-dimensional
AR	Anti-reflective coating
CAP	Concentration-acceptance product

(continued)

Acronym	Description
CAPEX	Capital expenditures
CCA	Concentrator cell assembly
CPC	Compound parabolic concentrator
CSP	Concentrated solar power
DNI	Direct normal irradiance (W/m^2)
EQE	External quantum efficiency
ESC	Environmental stress cracking
FF	Fill factor
FK	Fresnel Köhler concentrator
FLC	Flow line concentrator
F-RTP	Concentrator consisting on a Fresnel lens primary optic and a solid pyramid solid TIR mirror as SOE
F-XTP	Concentrator consisting on a Fresnel lens primary optic and a hollow pyramid mirror as SOE
IO	Imaging optics
NIO	Nonimaging optics
PC	Polycarbonate
PMMA	Polymethyl- methacrylate
POE	Primary optical element
RXI	SMS-designed concentrator consisting of a combination of refractive, TIR and reflective surfaces
SILO	Concentrator consisting on a Fresnel lens primary optic and a lens SOE designed as a Köhler pair
SMS	Simultaneous multiple surfaces design method
SOE	Secondary optical element
SOG	Silicone-on-glass, material combination for Fresnel lens parquets
TIR	Total internal reflection
UV	Ultraviolet
XR	SMS-designed concentrator consisting of a mirror primary optics and a refractive secondary optics

List of Symbols

Typical units given in square brackets. If no units are given, variable is dimensionless.

Symbol	Description [Units]
λ	Wavelength [nm]
δM_i	Edge rays associated with *3D* bundle M_i. These comprise rays at the boundary of the region describing bundle M_i in the *Phase Space*.
α	Acceptance angle [°]
α_{abs}	Absorption coefficient
η_{opt}	Optical efficiency
θ_c	Critical angle [°]
θ_{sun}	Angle subtended by the sun, as seen from the earth [°]
$\theta_I, \theta_r, \theta_t$	Angles formed by the normal of a surface with respect to incident, reflected and transmitted (refracted) rays [°]

Symbol	Description [Units]
ζ	Wavefront
θ_c	Critical angle [°]
θ:	Prism angles in a Fresnel lens [°]
ρ	The radial dimension in the design of a Fresnel lens [μm]
Ω_p	Projected solid angle [°]
γ	Surface slope deviation
σ	Confidence interval or statistical error in a ray trace
δM	Boundary of ray bundle M in the phase space
A_{entry}	Area of the entry aperture of a concentrator or optical system (mm^2, cm^2, etc.)
C_g	Geometrical concentration
C_o	Optical concentration
C_{pk}	Process capability index
$C_{staticMAX}$	Maximum concentration for static concentrators
E_M	*Ètendue of bundle M*
f	Lens focal length [mm]
$f\#$	*f-number.* In a circularly-trimmed imaging lens, ratio of focal length over lens diameter. In a *CPV* system, optical system depth over *POE* diagonal length, when the latter as a square outer shape
I_{sc}	Short circuit photocurrent [A]
L_{opt}	Optical path length [cm]
M	Ray bundle in 3D
m	Ray bundle in 2D
$n(\lambda)$	Spectral refractive index
n_i	Refrative index of media i
n_{rays}	Number of rays hitting a sensor bin
$P(M)$	Power of ray bundle M
p, q, r	direction cosines of a ray with respect to the X, Y and Z axes in 3D space
R_f	Fresnel reflection coefficient
s_y	Propagated standard deviation
$T(\theta)$	Angular transmission
x, y, z	Coordinates of a point in 3D space

References

1. Long, B.T., Peters, L.J. and Schreiber, H.D. (1998) Solarization of soda-lime-silicate glass containing manganese. *Journal of Non-crystalline Solids*, **239** (1–3), 126–130.
2. Biertümpfel, R., Engel, A. and Reichel, S. (2012) Solarization of optical filter glass. *DGaO-Proceedings* (available at: http://www.dgao-proceedings.de; urn:nbn:de:0287-2012-B010-1).
3. Lavoisier, A.L. (1865) *Oevres de Lavoisier (Tome III)*, Imprimerie Imperial, Paris.
4. Mouchot, A.B. (1869) *La Chaleur Solaire et ses Applications Industrielles*, Gauthier-Villars.
5. Toomer, G.J. (2012) *Diocles, On Burning Mirrors: The Arabic Translation of the Lost Greek Original*, Springer, Berlin.
6. Hinterberger, H. and Winston, R. (1966) Efficient light coupler for threshold Cerenkov counters. *Review of Scientific Instruments*, **37**, 1094.
7. Baranov, V.K. and Melnikov, G.K. (1966) Study of the illumination characteristics of hollow focons. *Soc. J. Opt. Technol.* **33**, 408.
8. Ploke, M. (1967) Lichtführungseinrichtuger mit starker Konzentrationswirkung, *Optik* **25**, 31.

9. Sala G. and Lorenzo, E. (1979) Hybrid silicone-glass Fresnel lens as concentrator for photovoltaic applications, *Proceedings 2nd EUPVSEC*, Berlin.
10. Born, Max and Wolf, Emil (1980) *Principles of Optics*, Pergamon Press, New York, p. 109.
11. Dacorogna, B. (2014) *Introduction to the Calculus of Variations*, 3rd edn, World Scientific Publishing.
12. Luneburg, R.K. (1975) *Mathematical Theory of Optics*, Pergamon Press, Oxford.
13. Winston, R., Miñano, J.C. and Benítez, P. (2005) *Nonimaging Optics*, Academic Elsevier.
14. *Selected Papers on Nonimaging Optics* (1995) SPIE Milestone Series, volume MS106.
15. Benítez, P. and Miñano, J.C. (1997) Ultrahigh-numerical-aperture imaging concentrator. *Journal of the Optical Society of America A*, **14**, 1988–1997.
16. Mills, D.R. (1995) Two-stage collectors approaching maximal concentration. *Solar Energy* **54**, 41–47.
17. Mohedano R., Cvetkovic A., Benítez P. *et al.* (2010) Compared performance of Fresnel-based concentrators at array level. *Proceedings of the 25th EU PVSEC/WCPEC-5*, Valencia, 913–917.
18. Chaves, J. (2008) *Introduction to Nonimaging Optics*, CRC Press.
19. Benítez, P., Miñano, J.C., Zamora, P. *et al.* (2010) High performance Fresnel-based photovoltaic concentrator. *Energy Express, Optical Society of America*, **18** (S1), A25–A40.
20. O'Gallagher, J., Welford, W.T. and Winston, R. (1987) Axially symmetric nonimaging flux concentrators with the maximum theoretical concentration ratio. *Journal of the Optical Society of America, A* **4** (1), 66–68.
21. Vorster, F.J. and van Dyk, E.E. (2005) Current-voltage characteristics of high-concentration, photovoltaic arrays. *Progress in Photovoltaics: Research and Applications*, **13**, 55–66, DOI: 10.1002/pip.563.
22. Miñano, J.C. (1985) Design of three-dimensional nonimaging concentrators with inhomogeneous media. *Journal of the Optical Society of America*, **3** (9), 1345–1353.
23. Miñano, J.C. (1985) Refractive index distribution in two-dimensional geometry for a given one-parameter manifold of rays. *Journal of the Optical Society of America A*, **2**, 1821–1825.
24. Miñano, J.C. (1993) Poisson brackets method of design of nonimaging concentrators: a review in Nonimaging Optics: Maximum-Efficiency Light Transfer II, (eds R. Winston, R.L. Holman), *Proceedings SPIE 2016*, 98–108.
25. Winston, R. and Welford, W.T. (1979) Geometrical vector flux and some new nonimaging concentrators. *Journal of the Optical Society of America*, **69** (4), 532–536.
26. Luque, A. *et al.* (1989) *Solar Cells and Optics for Photovoltaic Concentration*, Adam Hilger, Bristol, pp. 305–352.
27. Mohedano, R., Benítez, P. and Miñano, J.C. (1998) Cost reduction of building integrated PVs via static concentration systems. *Proceedings of the 2nd World Conference and Exhibition on Photovoltaic Solar Energy Conversion*, Vienna, 2241–2244.
28. Leutz, R. and Annen, H.P. (2007) Reverse ray-tracing model for the performance evaluation of stationary solar concentrators. *Solar Energy*, **81**, 761–767.
29. Norton, B. (2013) *Harnessing Solar Heat*, Springer.
30. Sala, G., Arboiro, J.C., Luque, A. *et al.* (1998) 480 kW peak EUCLIDES concentrator power plat using parabolic troughs. *Proceedings of the 2nd World Conference and Exhibition on Photovoltaic Solar Energy Conversion*, Vienna, Austria, 6–10 July, 1963–1968.
31. Luque, A., Sala, G., Arboiro, J.C. *et al.* (1997), Some results of the EUCLIDES photovoltaic concentrator prototype. *Progress in Photovoltaics: Research and Applications*, **5**, 195–212. DOI: 10.1002/(SICI)1099-159X (199705/06)5:3<195::AID-PIP166>3.0.CO;2-J.
32. Benítez, P., Mohedano, R. and Miñano, J.C. (1997) DSMTS: A novel linear PV concentrator. *Proceedings 26th IEEE Photovoltaic Specialists Conference*, Anaheim, California, 1145–1148.
33. Finot, M. and MacDonald, B. (2011) Significant cost reduction through new optical, thermal, and structural design for a medium-CPV system. *Proceedings SPIE 8108, High and Low Concentrator Systems for Solar Electric Applications VI, 81080B*, September 19, DOI: 10.1117/12.894187; http://dx.doi.org/10.1117/12.894187
34. Green, M.A., Emery, K., Hishikawa, Warta Y.W. and Dunlop, E.D. (2015) Solar cell efficiency tables (Version 45). *Progress in Photovoltaics: Research and Applications*, **23**, 1–9.
35. Bett, A.W. and Lerchenmüller, H. (2007) The FLATCON system from Concentrix Solar, in *Concentrator Photovoltaics* (eds Lugue, A.L. and Andreev, V.M.) Springer, Berlin, (Springer series in optical sciences 130), pp. 301–319.
36. Tver'yanovich, E.V. (1984) Profiles of solar-engineering Fresnel lenses. *Geliotekhnika*, **19** (6), 31–34. Translated into English in *Applied Solar Energy* **19** (6), 36–39.
37. Leutz R. and Suzuki, A. (2001) *Nonimaging Fresnel Lenses: Design and Performance of Solar Concentrators*, Springer Verlag.

38. Languy, F., (2012) Achromatization of nonimaging Fresnel lenses for photovoltaic solar concentration using refractive and diffractive patterns. Doctoral thesis. Hololab, Faculty of Sciences, University of Liege, Belgium.
39. Leutz, R., Fu, L. and Ries, H. (2003) Secondary optics for solar concentrators: concentration, beam shaping, and illumination uniformity, in *Proceedings of the 2nd International Solar Concentrator Conference for the Generation of Electricity or Hydrogen*, Alice Springs, Australia, November. U.S. Department of Energy, NREL, CD-520-35349. Invited presentation, slides only.
40. Gordon, R., Slade, A. and Garboushian V. (2007) A 30% efficient (>250 Watt) module using multijunction solar cells and their one-year on-sun field performance. *High and Low Concentration for Solar Electric Applications II* 6649. DOI: 10.1117/12.732700.
41. Benítez, P., Miñano, C., Zamora, P. *et al.* (2010) High performance Fresnel-based photovoltaic concentrator. *Optics Express* **18**, A25–A40.
42. Cassarly, W. (2001) Nonimaging optics: Concentration and illumination, in *The Handbook of Optics*, 2nd ed., McGraw-Hill, New York, pp. 2.23–2.42.
43. James, L.W. (1989) Use of Imaging Refractive Secondaries in Photovoltaic Concentrators, Contractor Report SAND89-7029.
44. US Patent 8 000 018 Fresnel-Köhler (FK), 2011.
45. Espinet-González, P., R. Mohedano, I. García, *et al.*, (2012) Triple-junction solar cell performance under Fresnel-based concentrators taking into account chromatic aberration and off-axis operation. *Proceedings CVP-8*, April 16–18, Toledo, Spain.
46. Kurtz, S., O'Neill, M.J. (1996) Estimating and controlling chromatic aberration losses for two-junction, two-terminal devices in refractive concentrator systems. *Proceedings 25th PVSC*, 361–367.
47. Patent Optical Concentrator, Especially for Solar Photovoltaics, Priority Number US20070894896P 20070314.
48. Three Dimensional Simultaneous Multiple-Surface Method and Free-Form Illumination-Optics Designed Therefrom, US Patent 7,460,985 B2, Dec. 2, 2008.
49. Hernández, M., Benítez, P., Miñano, J.C. *et al.* (2007) The XR nonimaging photovoltaic concentrator. *Proceedings SPIE Nonimaging Optics and Efficient Illumination Systems II*, (eds Winston R. and Koshel, J.), San Diego, USA.
50. Horne, S., Conley, G., Gordon, J. *et al.* (2006) A solid 500 sun compound concentrator PV design. *IEEE 4th World Conference on Photovoltaic Energy Conversion*, **1** and **2**, 694–697.
51. Cvetkovic, A., Hernandez, M., Benitez, P. *et al.* (2008) The free form XR photovoltaic concentrator: a high performance SMS3D design. *Proceedings SPIE* **7043**, High and Low Concentration for Solar Electric Applications III, 70430E, September 09, DOI: 10.1117/12.793714.
52. Fu, L., Leutz, R. and Annen, H.P., (2010) Secondary optics for Fresnel lens solar concentrators. *Proceedings SPIE* **7785**, Nonimaging Optics: Efficient Design for Illumination and Solar Concentration VII, 778509, doi: 10.1117/12.860438; http://dx.doi.org/10.1117/12.860438
53. Victoria, M., Dominguez, C., Anton I. and Sala, G. (2009) Comparative analysis of different secondary optical elements for aspheric primary lenses. *Optics Express* **8**, 6487–6492.
54. Zamora, P., Espinet-González, P., Benitez, P. *et al.* (2013) Experimental confirmation of FK concentrator insensitivity to chromatic aberrations. *AIP Conference Proceedings* **1556**, 88 http://dx.doi.org/10.1063/1.4822206, 15–17 April 2013, Miyazaki, Japan.
55. A. Luque *et al.*, (1989) *Solar Cells and Optics for Photovoltaic Concentration*, Adam Hilger, Bristol.
56. Hernández, M., Benítez, P., Miñano, J.C. *et al.* (2007) The XR nonimaging photovoltaic concentrator. *Proceedings SPIE* **6670**, Nonimaging Optics and Efficient Illumination Systems IV, 667005 (21 September 2007), DOI: 10.1117/12.736897.
57. Miñano J.C. and González, J.C. (1992) New method of design of nonimaging concentrators. *Applied Optics* **31**, 3051–3060, http://www.opticsinfobase.org/abstract.cfm?URI=ao-31-16-3051
58. Kerker, M. (1969) *The Scattering of Light*. Academic, New York.
59. Stover, J.C. (1995) *Optical Scattering: Measurement and Analysis*, DOI: 10.1117/3.203079.
60. Cvetkovic, A., Mohedano, R., Gonzalez, O. *et al.* (2011) Performance modeling of Fresnel-based CPV systems: effects of deformations under real operation conditions. *AIP Conference Proceedings* **1407**, 74, http://dx.doi.org/10.1063/1.3658298
61. Askins, S., Victoria, M., Herrero, R. *et al.* (2011) Effects of temperature on hybrid lens performance. *AIP Conference Proceedings* **1407**, 57, http://dx.doi.org/10.1063/1.3658294
62. Benitez, P., R.M. Arroyo, Minano, J.C. (1997) Manufacturing tolerances for nonimaging concentrators. *Proceedings SPIE* **3139**, Nonimaging Optics: Maximum Efficiency Light Transfer IV, 98 DOI: 10.1117/12.290214.

63. Schmidt, T.C. and Ries, H. (2010) Optimization of secondary concentrators with the continuous information entropy strategy. *AIP Conference Proceedings*, **1277** (1), 101.
64. Fred (http://photonengr.com/software/), TracePro (http://www.lambdares.com/features) LightTools (http://optics.synopsys.com/lighttools/) are some examples
65. A modern formulation of the Mie solution to the scattering problem on a sphere can be found in many books, e.g., in Stratton, J. A. (1941) *Electromagnetic Theory*, McGraw-Hill, New York.
66. Henyey, L.G. and Greenstein, J.L. (1941) Diffuse radiation in the galaxy. *Astrophysical Journal*, **93**, 70–83.
67. Leutz, R., Annen H.P. and Fu L. (2010) Optical design for reliability and efficiency in concentrating photovoltaics. *Proceedings SPIE* **7773**, Reliability of Photovoltaic Cells, Modules, Components, and Systems III, 777304, DOI: 10.1117/12.861156; http://dx.doi.org/10.1117/12.861156
68. Hornunga, T., Bachmaiera, A., Nitza P. and Gombert, A. (2010) Temperature dependent measurement and simulation of Fresnel lenses for concentrating photovoltaics. *AIP Conference Proceedings*, **1277**, 85, http://dx.doi.org/10.1063/1.3509239
69. Ku, H.H. (1966) Notes on the use of propagation of error formulas. *Journal of Research of the National Bureau of Standards* **70C** (4), 262, DOI: 10.6028/jres.070c.025, (accessed 3 October 2012).
70. http://en.wikipedia.org/wiki/Process_capability_index).
71. Annen, H.P., Fu, L., Leutz, R., González L. and Mbakop J. (2011) Direct comparison of polymethylmetacrylate (PMMA) and silicone-on-glass (SOG) for Fresnel lenses in concentrating photovoltaics (CPV). *Proceedings SPIE* **8112**, Reliability of Photovoltaic Cells, Modules, Components, and Systems IV, 811204, DOI: 10.1117/12.893884, http://dx.doi.org/10.1117/12.893884
72. http://www.renewableenergyworld.com/rea/news/article/2008/12/cool-earth-is-scaling-up-solar-energy-generation-54228

4-I Annex

Étendue Calculation

The *étendue* of a bundle M_{4D} can be calculated at surface Σ_R as follows:

$$E(M_{4D}) = \int_{M_{4D}(\Sigma_R)} dxdydpdq + dxdzdpdr + dydzdqdr, \qquad (4\text{-}I.1)$$

Let Σ_i and Σ_o be two plane reference surfaces with $z = constant$. In this case, since the *étendue* of the bundle M_{4D} remains the same after propagation, we may write:

$$\int_{M_{4D}(\Sigma_i)} dxdydpdq = \int_{M_{4D}(\Sigma_o)} dx'dy'dp'dq', \qquad (4\text{-}I.2)$$

where primed symbols have been used to identify the local variables at plane Σ_o. Notice $dz = dz' = 0$. Notice p, q, p' and q' are the optical direction cosines and are multiplied by the current refractive index n at surfaces Σ_i and Σ_o (see Figure 4-I.1).

When the trajectories of the rays are included in a plane, (e.g., $y = constant$), the problem becomes a two-dimensional one and there exists another magnitude that is also called *étendue*, and for a bundle m_{2D} is:

$$E(m_{2D}) = \int_{m_{2D}(\Sigma_R)} dxdp + dzdr + dydq, \qquad (4\text{-}I.3)$$

This invariant equals the *Lagrange invariant* [1]. On this occasion, if we evaluate the bundle in two surfaces $z = constant$, and since the y and y' axes are orthogonal to the plane where the rays are included ($dy = dy' = 0$), the *étendue* conservation can be expressed as:

$$\int_{\Sigma_i} dxdp = \int_{\Sigma_o} dx'dp'. \qquad (4\text{-}I.4)$$

Handbook of Concentrator Photovoltaic Technology, First Edition. Edited by Carlos Algora and Ignacio Rey-Stolle.
© 2016 John Wiley & Sons, Ltd. Published 2016 by John Wiley & Sons, Ltd.

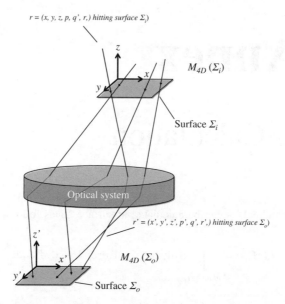

Figure 4-I.1   The *étendue* of a bundle M_{4D}, evaluated at two different surfaces Σ_i and Σ_o (plane on this occasion) remains the same after propagation through an optical system

Reference

1. Greivenkamp, J.E., Field Guide to Geometrical Optics (2003) DOI: 10.1117/3.547461.

4-II Annex

2D Treatment of Rotational and Linear 3D Optical Systems

These symmetries impose a constraint on the optical equations that reduces the number of unknowns in the optical system. For instance, if a linear symmetry is applied along the z-axis, r is an invariant for each ray and the system can then be studied in the x-y plane, as Miñano pointed out [1], by simply substituting the refractive index $n(x, y)$ by a fictitious value $n^*(x,y) = (n^2(x, y) - r^2)^{1/2}$, or, relative to vacuum:

$$n_r^*(x, y) = \sqrt{\frac{n^2(x, y) - r^2}{1 - r^2}}, \qquad (4\text{-II}.1)$$

Notice that the trajectories of rays would remain straight lines insofar as $n(x, y)$ is homogeneous, since $n^*(x, y)$ will be homogeneous as well (although different for each r). If the optical system consists of mirrors only, i.e., $n(x, y) = 1$, the relative apparent refractive index will be $n_r^* = 1$. This fact, along with the homogeneity of n^*, implies that the projections of 3D rays on a plane $z = constant$ match 2D trajectories; thus, the study of these kinds of devices in 2D is sufficient.

In rotational devices, the most suitable system of coordinates is the cylindrical one, with variables (ρ, θ, z, g, h, r) and the symmetry condition implies $h = $ constant. The value $h = n \cdot S \cdot \sin\gamma$, with S being the shortest distance between the ray and the optical axis, and γ the angle between the ray and the axis, is conserved in this kind of system, and is called the *skew invariant* or, simply, the *skewness*. Accordingly, these systems can be studied in a so-called *meridian plane*, where $h = constant$. The rays with $h = 0$ are called meridian rays, and rays with $h \neq 0$ are referred to as *skew rays*. The trajectories of the latter in 2D are not rectilinear.

Reference

1. Miñano, J.C. (1985) Aspectos relativos a la síntesis de concentradores para fuentes solares extensas, Tesis doctoral, E.T.S.I. Telecomunicación, Madrid.

Handbook of Concentrator Photovoltaic Technology, First Edition. Edited by Carlos Algora and Ignacio Rey-Stolle.
© 2016 John Wiley & Sons, Ltd. Published 2016 by John Wiley & Sons, Ltd.

4-III Annex

Design of the XR Concentrator

The design procedure of the free-form XR is common to all 3D SMS devices. It consists of the simultaneous calculation of points that, at the end of the process, in the form of cloud points, will define the free form surfaces connecting the input and output edge-ray bundles. According to the *edge ray theorem*, if the process succeeds, such surfaces will connect rays inside the boundaries defined by edge rays at the *phase space* as well. The 3D SMS design method often starts with the calculation of a first couple of SMS chains ('*seed chain*', one at each surface, in this case of one at the POE mirror and another at the SOE lens, as shown in Figure 4-III.1), comprising a set of 3D points whose locations and normals are such that input edge rays ($\delta M1_i$ and $\delta M2_i$, in the example of the figure these are rays hitting the POE with angles $+\alpha$ and $-\alpha$ with respect to sun direction, matching the acceptance angle, and comprised in plane $x = 0$) and output edge rays ($\delta M1_o$ and $\delta M2_o$, in this example rays coming from the edges of the intersection of cell plane with plane $x = 0$) passing through these points are connected, observing Fermat's Principle. The *seed* SMS *chain* starts at two points (P_i and P_o in Figure 4-III.1) that can be selected by designer. The selection of these points fixes the design optical path length and is critical since they will eventually define the overall shape of the optical train and important features such as aspect ratio and the possibility of hiding the receivers under the next mirrors, to maximize light collection area.

Once the *seed chains* are available, their points act as initial points for the calculation of a set of SMS chains that typically run more or less transversally, both to the left and to the right of the original ones and therefore, can be considered as 'ribs' (Figure 4-III.2) hanging from the first chain. There is a rib starting at each original SMS chains point, and these ribs comprise the majority of points defining the final free-form surfaces. The SMS ribs are transversal to the first chain because they are calculated using a different set of input and output design edge bundles, transversal to the original ones as well. The input edge rays are $\delta M3_i$ and $\delta M4_i$, comprising rays hitting the POE with angles $+\alpha$ and $-\alpha$ with respect to sun direction, and included in plane $y = 0$, while the output edge rays are $\delta M3_o$ and $\delta M4_o$, i.e. rays coming from the edges of the intersection of cell plane with plane $y = 0$.

The final solution consists of a couple of NURBS (*Non-Uniform Rational B-Spline*) surfaces (one for the POE mirror, another for the SOE lens) calculated from the SMS cloudpoints. The

Handbook of Concentrator Photovoltaic Technology, First Edition. Edited by Carlos Algora and Ignacio Rey-Stolle.
© 2016 John Wiley & Sons, Ltd. Published 2016 by John Wiley & Sons, Ltd.

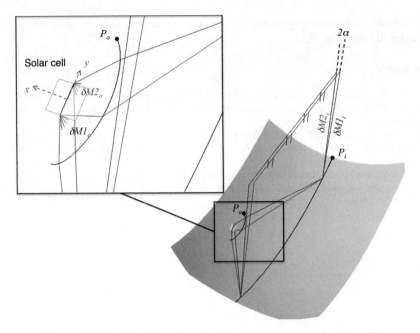

Figure 4-III.1   Calculation of seed chain in the canonical free-form XR

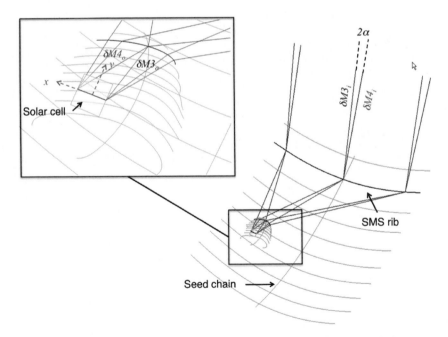

Figure 4-III.2   Calculation of SMS ribs, leaning on the SMS seed chain

XR has a simpler rotationally symmetric version [1] whose cross section can be designed in 2D and Köhler 3D versions [2,3].

References

1. Hernández, M., Benítez, P., Miñano, J. C. *et al.* (2007) The XR nonimaging photovoltaic concentrator. *Proceedings SPIE* vol. 6670, Nonimaging Optics and Efficient Illumination Systems IV, p. 667005 DOI:10.1117/12.736897; http://dx.doi.org/10.1117/12.736897.
2. Benítez, P. and Miñano, J.C. (2008) High-performance Kohler concentrators with uniform irradiance on solar cell. *Proceedings SPIE* vol. 7059, Nonimaging Optics and Efficient Illumination Systems V, p. 705908 DOI:10.1117/12.794927.
3. Hernandez, M., Benítez, P., Mendes-Lopes, J. *et al.* (2012) 1000x shadow-free mirror-based Köhler concentrator. *Proceedings SPIE* vol. 8468, High and Low Concentrator Systems for Solar Electric Applications VII, p. 84680C. DOI:10.1117/12.931657.

5

Temperature Effects on CPV Solar Cells, Optics and Modules

Iván García, Marta Victoria, and Ignacio Antón

Instituto de Energía Solar, Universidad Politécnica de Madrid, Spain

5.1 Introduction

Temperature affects the operation, performance and reliability of CPV systems. In the field, temperature fluctuates as humidity, wind and irradiance change. The effects of temperature occur at all levels of the CPV system: solar cell, receiver, optics, module, etc. and must be predicted in order to minimize as much as possible their negative impact on the energy output of the system. In this chapter we present a detailed analysis of the effect of temperature on CPV solar cells, optics and modules since they are the most affected parts of the CPV system. Section 5.2 is devoted to the analysis at the solar cell level and section 5.3 deals with the concentrator optics and module. Concerning the solar cell, a didactic approach is followed in order to elucidate the characteristics of the temperature sensitivity of the solar cell bandgap and electrical parameters (short circuit current, open circuit voltage, fill factor and efficiency). Theoretical calculations for ideal solar cells are presented aiming to analyze the qualitative trends of the temperature coefficients with temperature and concentration, and to explain the weight of each electrical parameter on the overall temperature dependence of the solar cell performance. Lastly in section 5.2, a review of experimental data of real multijunction solar cells is presented and discussed. In section 5.3 the focus is put on the effect of temperature on concentrating optics, in particular the behavior of silicone on glass (SOG) Fresnel lenses is described in detail. Then, experimental data on the thermal sensitivity of electrical parameters at module level is shown. Finally, a comprehensive analysis on the alternative designs to improve heat dissipation is presented in two steps: firstly, the different architectures for the stack that thermally connect the solar cell to the module chassis are described; secondly, the strategies followed to transport heat from the module to the ambient are reviewed.

Handbook of Concentrator Photovoltaic Technology, First Edition. Edited by Carlos Algora and Ignacio Rey-Stolle.
© 2016 John Wiley & Sons, Ltd. Published 2016 by John Wiley & Sons, Ltd.

5.2　Effects of Temperature on CPV Solar Cells

The sensitivity to temperature of semiconductors has been exploited to invent devices capable of sensing the temperature and creating a readable signal, such as NTC (negative temperature coefficient) thermistors or diode-based temperature sensors. However, in most of cases, the performance sensitivity of a semiconductor device to the temperature is generally an undesired effect. In fact, a number of different strategies have been developed to mitigate it. These include techniques to control the temperature in the device, such as the well-known heat dissipation solutions, and also techniques to cancel out the effect of the temperature on the internal operation of the device, such as the paired transistor architectures used in microelectronic circuits.

A solar cell is a semiconductor device and, thus, its performance is inevitably affected by temperature. The efficiency of a solar cell describes the percentage of the input light power that is converted into output electrical power. A fraction of the input power that is not converted into electricity is lost by optical reflection or transmission, and the rest is transformed into heat in the cell by carrier thermalization and ohmic losses. This heat causes an increase of the temperature in the cell, which is higher as the light intensity is increased. Temperature control strategies are actually used to remove as much of the heat generated in the solar cell as possible (see section 5.3). Although both active and passive cooling strategies are possible, only the latter is energy- and cost-effective for the case of CPV modules, where optics and cells are pre-assembled in a factory to form indivisible units referred to as modules as defined in Chapter 7. The commonly used passive solutions are cheaper, but the temperature of the solar cell can rise a few tens of °C during operation (see Table 5.6).

Therefore, efforts have been devoted to understand the influence of temperature on the performance of the solar cell and obtain trends that can be used to predict it for a given temperature. This way the optimum solar cell design that maximizes the power output for variable temperature conditions during operation can be found.

The main material property governing the dependency of a solar cell performance with temperature is its bandgap. Variations in the bandgap due to fluctuations in the temperature affect the current and voltage delivered by the solar cell and, ultimately, its efficiency of power conversion. In section 5.2.1 the dependence on temperature of the solar cell bandgap and in 5.2.2 of the electrical merit figures (J_{sc}, V_{oc}, FF and efficiency) are examined.

5.2.1　Dependence of the Bandgap on Temperature

The temperature sensitivity of the solar cell performance is primarily an effect of the variation of the bandgap of its active layers with temperature. The bandgap of the majority of semiconductor materials decreases as temperature increases. Put simply, this effect is originated by the lower energy potential 'seen' by the electrons as the semiconductor lattice expands due to an increased temperature. A number of publications can be found in the literature in which the temperature dependence of the bandgap in semiconductors is theoretically analyzed by accounting for the interaction of the electron with the vibrating semiconductor lattice using different approaches [1,2]. However, this kind of analysis falls beyond the scope of this chapter. A convenient way, for our purposes, to express the variation of the bandgap with temperature is to use the well-known Varshni's semi-empirical equation [3]:

$$E_g(T) = E_{g_0} - \frac{aT^2}{T+b} \tag{5.1}$$

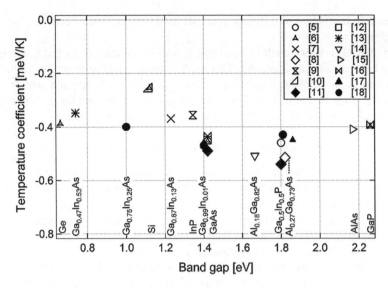

Figure 5.1 Measured bandgap temperature coefficient (*a*) of different materials, plotted against their bandgap. The filled symbols correspond to cases where the temperature coefficient is obtained by measuring the EQE of solar cells made with these materials. The empty symbols correspond to results obtained using a variety of methods including photoluminescence, spectroscopic ellipsometry, spectral absorption and transmission, photoreflectance, etc

where E_g [eV] is the bandgap, T [K] the absolute temperature and E_{g0} [eV], a [eV·K^{-1}] and b [K] are fitting parameters for each material[1].

Improvements to this expression, that allow obtaining a better fit of the E_g vs. T data for a wider range of temperatures (particularly for very low temperatures approaching 0 K), can be found in the literature (see for example [4]). However, for the temperature ranges that are dealt with in concentrator solar cells, the Varshni equation fits the experimental data to a good accuracy. Moreover, for the materials commonly employed in concentrator solar cells, the *a* and *b* parameters are such that, for the range of practical temperatures around 300 K, the $E_g(T)$ can be accurately approximated by a linear function. This means that a linear coefficient–which is *a* in Varshni Eq. (5.1), since when b <300 K, $T^2/(T+b) \approx$ T–can be used, in our context, to describe the variation of the bandgap with temperature in these materials.

A number of measurement results for the bandgap temperature coefficient of the semi-conductor materials used in concentrator cells can be found in the literature. Different measurement methods are used in these works, including photoluminescence, spectroscopic ellipsometry, spectral absorption and transmission, photoreflectance, etc. (see references in legend of Figure 5.1 [5–18]). In some recent works [11,17,18], the temperature coefficient is obtained by measuring the external quantum efficiency (EQE) vs. temperature of actual solar cells and determining the variation of the cut-off energy. In Figure 5.1, a compilation of

---

[1] We use here '*a*' and '*b*' in Varshni equation instead of the most common symbols 'α' and 'β' in order not to confuse them with the 'α' and 'β' used in section 5.3 for the short circuit current and open circuit voltage temperature coefficients.

the temperature coefficient of different materials used in concentrator cells is plotted against their bandgap.

An important observation on this graph is the rather low variation of the bandgap temperature coefficient between different materials. This implies that the relative sensitivity of the bandgap to temperature in high bandgap materials is lower than in low bandgap materials. A consequence of this is a lower sensitivity of the solar cell power conversion efficiency to temperature in higher bandgap materials, as will be shown in the following sections.

5.2.2 Dependence of the Solar Cell Parameters on Temperature

In this section, the influence of temperature on the short circuit current density (J_{sc}), open circuit voltage (V_{oc}), fill factor (FF) and efficiency (η) of single and multijunction solar cells is theoretically analyzed and the effect of the bandgap on the sensitivity of all these parameters to temperature is studied. For a detailed explanation of these parameters the reader is referred to Chapter 2. We focus on the temperature effects at one-sun ($1\times$) operation for didactic purposes while in section 5.2.3 the effect of concentration is studied.

The theoretical trends presented in this section are obtained for ideal, optically thick cells (i.e., EQE is 1 above the bandgap energy) working in the radiative limit, and Eq. (5.2) is used to determine the solar cell performance parameters. The general trends obtained will serve to illustrate the effect of temperature. In section 5.2.4, the measurement of real solar cells is tackled.

$$J = J_{sc} - J_0 \exp\left(\frac{qV}{kT}\right)$$ (5.2)

The derivative with respect to temperature is used to quantify the absolute sensitivity of the electrical parameters to temperature. In general, this magnitude is assumed to be constant with temperature but this approximation is not always accurate as will be shown in the following sections, particularly for the J_{sc} in multijunction solar cells. The derivative divided by the magnitude of the parameter at a nominal temperature (usually 25 °C), called the 'temperature coefficient', is also frequently used, to represent the relative variation of the parameter with temperature. In this section we use both: the temperature derivative when we want to ascertain its linearity with respect to temperature, and the temperature coefficient when comparing the sensitivity to temperature of different solar cell materials, parameters or working conditions (concentration). For didactic purposes, the notation used in this section for graph labels and in the text is the explicit derivative or relative derivative (temperature coefficient). In section 5.3, the commonly used symbols α and β are used to represent the temperature coefficient of the J_{sc} and V_{oc}, respectively.

5.2.2.1 Short Circuit Current Density, J_{sc}

5.2.2.1.1 Single Junction Solar Cells

The short circuit current density (J_{sc}) of a solar cell is, briefly, the result of a certain amount of photons being absorbed, converted into charge carriers and collected in the p-n junction that composes the solar cell, at 0 volts bias. The number of photons absorbed and converted into carriers depends on the spectral reflectance, the absorbance of the solar cell, and on the spectral composition of the impinging light. The fraction of these photogenerated carriers that are collected depends on the electronic properties of the solar cell, mainly the diffusion length and the recombination at the interfaces of the semiconductor structure including the perimeter.

These three solar cell properties (spectral reflectance and internal absorbance, and electronic properties) configure the spectral response (SR) and external quantum efficiency (EQE) of the solar cell that, in conjunction with the solar spectrum, determine the J_{sc} of the cell. The EQE is affected by the temperature mainly by a shift in the cutoff energy, due to the modification of the bandgap of the material that forms the p-n junction. The effect of the temperature on the reflectance (through the optical properties of the cell) and electronic properties of the cells do not produce a significant modification of the EQE, for the range of temperatures of operation of a concentrator solar cell. This is experimentally illustrated in the measurement results available in the literature for III-V concentrator cells (for example [11,18–20]).

The shape of the solar spectrum influences the dependence of J_{sc} on temperature. The direct solar spectrum (see Chapter 1), which is the one relevant for concentrator solar cells, exhibits an irregular shape, with pronounced dips caused by the atmospheric absorption and scattering. As the cutoff energy of the cell EQE changes with temperature, a different portion of this irregularly-shaped solar spectrum is absorbed in the cell. In a single-junction cell, as the temperature and its bandgap increases, the J_{sc} may increase also, or may remain constant, depending on the shape of the spectrum around the energy corresponding to the bandgap of the cell. This is illustrated in Figure 5.2, where the calculated relative percentage of variation of the J_{sc} in an ideal single-junction cell (EQE unity for energies equal or above the bandgap, and 0 below it), with respect to the J_{sc} at 25 °C, is plotted against the temperature and wavelength corresponding to its bandgap. An average bandgap temperature coefficient of –0.45 meV/K is used in the calculations for all bandgap energies (see Figure 5.1). The dashed vertical lines indicate the wavelength corresponding to some relevant materials for concentrator multi-junction solar cells. A clearly different behavior of their J_{sc} as temperature changes is observed. For example, the J_{sc} of GaAs increases at a faster rate than the J_{sc} of the 0.7 eV GaInAs cell (used as the bottom subcell in 4-junction solar cells [21]) as temperature is raised. In the same figure, the G173 direct solar spectrum is plotted below to clarify the reason of this behavior. While the bandgap of an ideal GaAs cell moves with temperature around a region away from any of the dips in the spectrum, this is not the case for the ideal 0.7 eV GaInAs cell.

The derivative of the J_{sc} with respect to temperature is shown in Figure 5.3 for some typical materials indicated in the legend. These values, calculated for ideal solar cells, are reasonably close to the values measured in experimental cells [17], except in the case of the Ge subcell. This is mostly due to the fact that the square-QE idealization used here is not as good for the indirect Ge material as for the other direct III-V materials (see QE figures in [17]). On the other hand, this graph suggests that the derivative of the J_{sc} at one temperature, used as a unique 'temperature coefficient' for J_{sc}, is not always representative of the overall sensitivity of J_{sc} to temperature in the range of temperatures of interest. For, convenience, however, a constant temperature coefficient is used in most practical cases.

5.2.2.1.2 Multijunction Solar Cells

In a multijunction solar cell, the J_{sc} in each subcell depends on its bandgap and on the bandgaps of the subcells on top, which depend all on temperature. Figure 5.4 shows the experimental EQE of a GaInP/GaAs/GaInAs triple-junction solar cell for 25 °C and 100 °C that exemplifies this. The bandgap change with temperature in each subcell determines how much light is absorbed in that subcell and how much light is transmitted to the underlying subcells.

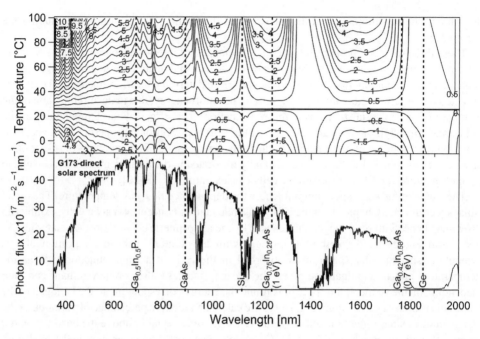

Figure 5.2 Top: iso-contours of the variation, in %, of the J_{sc} in a single-junction solar cell as temperature changes along the Y axis, using as reference the J_{sc} at 25 °C. The cells are assumed to have an ideal spectral response where only the cutoff energy changes with temperature. The bandgap temperature coefficient used is −0.45 mV/K for all bandgap energies. Bottom: the G173 direct solar spectrum used in these calculations, whose shape determines the variations in J_{sc} shown on the top graph. The dashed lines indicate some of the most relevant bandgaps used in state of the art CPV solar cells

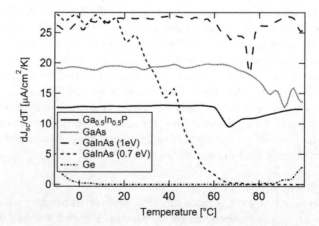

Figure 5.3 Derivative of the J_{sc} with respect to temperature for some typical multijunction solar cell materials indicated in the legend

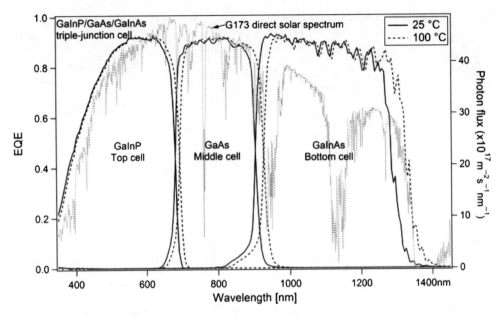

Figure 5.4 Experimental EQE of a GaInP/GaAs/GaInAs triple-junction cell, for two different temperatures, that illustrates the effect of the temperature on the shape of the EQE of each subcell

As a consequence, while increasing the temperature gives always rise to an increased or constant J_{sc} in a single-junction cell (as shown in Figure 5.2), the J_{sc} in a subcell (other than the top cell) in a multijunction solar cell can also decrease as the temperature increases, if the increased photon absorption in the upper subcells is not compensated by a higher absorption in the subcell under examination. Illustrative examples are shown in Figure 5.5, for ideal GaInP/GaAs/GaInAs and GaInP/GaAs/Ge triple-junction solar cells. All the subcells in each of these multijunction cells are current matched at 25 °C for the G173 direct solar spectrum, except for the bottom Ge subcell in the GaInP/GaAs/Ge triple-junction cell.

The Ge bottom cell in the ideal GaInP/GaAs/Ge cell is a clear example of J_{sc} decreasing as the temperature increases, as a result of a higher absorption in the middle cell and a negligible increase in the absorption in the bottom cell for the energies around their band edges.

The J_{sc} of the current limiting subcell determines the J_{sc} of the multijunction solar cell. Therefore, the overall dependence of the multijunction solar cell J_{sc} on temperature is determined by the limiting subcell at each temperature. This can give rise to abrupt changes in the slope of the J_{sc} derivative vs. temperature of the multijunction cell, at the temperatures where the limiting subcell changes, as can be observed in Figure 5.5.

According to these theoretical results, using a linear temperature coefficient to quantify the dependence of J_{sc} on temperature in a multijunction solar cell is, in general, not accurate. However, they are normally used. In fact, the J_{sc} vs. T data available in the literature for this kind of triple-junction cells do not show abrupt changes in the slope of J_{sc} such as in Figure 5.5 [18,22]. This can be due to multiple factors, related to the cell characteristics or the measurement conditions that affect the actual current matching of the cell. Nevertheless, the use

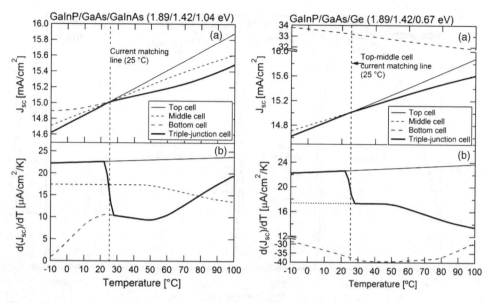

Figure 5.5 J_{sc} parameters of ideal triple-junction solar cells with the bandgap combinations indicated, for the G173 direct solar spectrum and using −0.45 meV/K as the temperature coefficient of the bandgap for all the materials: (a) J_{sc}; (b) dJ_{sc}/dT

of a linear temperature coefficient for the J_{sc} in a multijunction cell cannot be assumed to be appropriate in all cases [23].

5.2.2.2 Open Circuit Voltage, V_{oc}

5.2.2.2.1 Single-Junction Solar Cells
As stated in Chapter 2, the V_{oc} of a solar cell is commonly expressed as:

$$V_{OC} = \frac{nkT}{q} \ln\left(\frac{J_{SC}}{J_0}\right) \qquad (5.3)$$

where n is 1 or 2, if the recombination in the cell at the operating point is dominated by kT-(neutral regions) or $2kT$-(depletion regions or perimeter) components, respectively; J_{sc} is the short circuit current density of the cell at the solar spectrum used and J_0 is the saturation current density for the case considered ($n = 1$ or $n = 2$). We are going to focus on the case of $n = 1$, as the most representative for the current levels and characteristics of the solar cells used in CPV. Extended analysis including the case when the operation of the cell is dominated by the recombination in the depletion region can be found in the literature [20].

The derivative of V_{oc} with temperature is:

$$\frac{\mathrm{d}V_{OC}}{\mathrm{d}T} = \frac{V_{OC}}{T} + \frac{nkT}{q}\left(\frac{1}{J_{sc}}\frac{\mathrm{d}J_{sc}}{\mathrm{d}T} - \frac{1}{J_0}\frac{\mathrm{d}J_0}{\mathrm{d}T}\right) \qquad (5.4)$$

The derivative of J_{sc} is relatively small (see Figure 5.5) and can be neglected. The variation of J_0 with temperature of non-ideal solar cells is modeled using the minority carrier diffusion equations. In some works, the dependence of the minority carrier properties on temperature is taken into account using a constant γ [22] that can be used as an empirical fitting parameter for experimental data [24]. In other works, the dependence of the minority carrier properties on temperature is neglected [20,25] to derive the expression for J_0. In the following, we calculate J_0 and V_{oc} for ideal solar cells working in the radiative limit (see Chapter 3):

$$J_0 = const \; E_g^2 \cdot T \cdot e^{-qE_g/kT} \tag{5.5}$$

$$\frac{dV_{oc}}{dT} \approx \frac{1}{T}\left[V_{oc} - \frac{kT}{q} - \frac{E_g}{q} + \frac{T}{q}\frac{dE_g}{dT}\right] = \frac{V_{oc} - E_g/q}{T} - \frac{k}{q} + \frac{1}{q}\frac{dE_g}{dT} \tag{5.6}$$

where the absorptivity of the cell is approximated to be 1 and 0 for photon energies above and below the bandgap respectively, and the relatively small variation of J_{sc} with temperature has been neglected. This expression is similar to the one obtained using the diffusion equations [25], differing only in a small $2kT/q$ term.

This derivative is not constant over temperature. However, its variation with temperature is normally small over the temperature range of interest in concentrator solar cells, as will be shown later. Therefore, a linear temperature coefficient can be used to quantify the sensitivity of the V_{oc} to temperature. Besides, from Eq. (5.6) it can be derived that the variations of the V_{oc} with temperature do not depend on the solar spectrum, in contrast with the case of the J_{sc}. Strictly speaking, the V_{oc} depends on the J_{sc} (see Eq. (5.4)) which varies with temperature and with the solar spectrum. But these variations do not propagate to the V_{oc}, given their small value and the logarithmic dependence of V_{oc} on J_{sc}.

From Eq. (5.6) it can be concluded that, since the bandgap of the cell has a negative temperature coefficient (see section 5.2.1) and E_g/q is always higher than V_{oc}, the temperature coefficient of the V_{oc} is always negative. Another important characteristic derived from Eq. (5.6) is that, for a given temperature and bandgap, the temperature coefficient of the V_{oc} depends on the actual V_{oc} of the cell: the higher the V_{oc}, the lower its temperature coefficient. This means that, for the same material, higher quality solar cells have lower temperature sensitivity. As an example, in Figure 5.6, the V_{oc} vs. temperature of experimental GaAs cells with different qualities and V_{oc} is shown. All the cells have the same structure but their V_{oc} is different as a result of a different material quality and photon recycling obtained using different back reflectors [26]. The temperature derivative and temperature coefficient of the V_{oc} for each cell is indicated, showing a lower value for the best quality (higher V_{oc}) cells.

The sensitivity of the V_{oc} of a solar cell to the temperature depends on its bandgap too. However, the direction of this dependency is not evident from Eqs. (5.4) and (5.6) since different bandgap materials always produce different V_{oc} solar cells, and the evolution of the $(V_{oc} - E_g/q)$ term is not obvious. It is well known, though, that the difference between E_g/q and V_{oc} is slightly higher for higher bandgap materials [27]. Moreover, dE_g/dT is roughly constant for all bandgaps, as shown in section 5.2.1. Therefore, the temperature derivative of the V_{oc} is, in theory, higher for higher bandgap materials. However, the temperature coefficient of the V_{oc} is lower for higher bandgap materials, since the V_{oc} grows faster with E_g than its derivative. To illustrate this, in Figure 5.7, the contours of the derivative of the V_{oc} with respect to temperature,

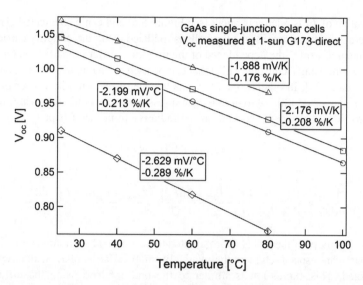

Figure 5.6 V_{oc} vs. temperature for different quality experimental GaAs cells. The temperature coefficients obtained by linear fit of these curves show higher values for lower quality (lower V_{oc}) cells. The empty triangles data correspond to a cell with enhanced photon recycling, that operates close to the radiative limit and shows the highest V_{oc} and the lowest V_{oc} temperature coefficient

calculated in the radiative limit, are plotted against the temperature and the bandgap wavelength. In the inset, the derivative or temperature coefficient at 25 °C in its absolute and relative versions is plotted against the bandgap wavelength.

It can be observed that the higher the bandgap energy (lower bandgap wavelength), the higher the V_{oc} derivative. The contours appear as almost vertical lines, meaning that the derivative of the V_{oc} with respect to temperature is constant over temperature, as commented before, supporting the use a 'constant temperature coefficient' to describe the sensitivity of V_{oc} to temperature as a good approximation. From the inset of Figure 5.7 we can draw one of the main conclusions about the temperature sensitivity of V_{oc}: the higher the bandgap of the material, the higher the magnitude of the temperature derivative but the lower the temperature coefficient of the V_{oc}. Since the weight of the sensitivity of the V_{oc} to temperature in the overall sensitivity to temperature of the performance of a solar cell is the largest, the performance of higher bandgap solar cells is less affected by temperature, as will be demonstrated in section 5.2.4.

5.2.2.2.2 Multijunction Solar Cells

The V_{oc} of a multijunction solar cell is the sum of the V_{oc} of each subcell. Therefore, the derivative of the V_{oc} in a multijunction solar cell is the sum of the derivatives of each of its subcells. As explained in the previous sections, the influence of variations of the J_{sc} with temperature on the V_{oc} derivative is negligible, for the typical materials and temperature ranges in CPV. As a consequence, linearity over temperature is assumed and linear temperature coefficients can be (and are) used, in a good approximation, to quantify the sensitivity of V_{oc} to temperature in multijunction solar cells as well as in single-junction solar cells.

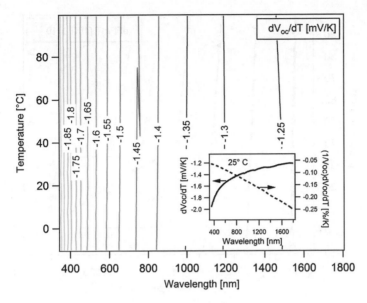

Figure 5.7 Contour plot of the V_{oc} derivative with respect to temperature, plotted against the temperature and the bandgap wavelength. The inset shows the derivative and the relative derivative, for 25 °C, plotted against the bandgap wavelength. All calculations are done for the radiative limit at 1-sun under the G173 direct solar spectrum

5.2.2.3 Fill Factor, *FF*

5.2.2.3.1 *Single-Junction Solar Cells*

The fill factor is a figure of merit of the performance of solar cells that does not convey a direct physical meaning as the J_{sc} and V_{oc} do. It is just a convenient parameter that relates the maximum electrical power generated by the cell to the $J_{sc} \cdot V_{oc}$ product. It is also a useful way to graphically assess the performance of a cell by examining the shape of its I-V curve under illumination: the higher the *FF*, the more 'square' is this I-V curve. The power losses caused by parasitic resistance components in the cell give rise to a lower *FF* that can be relatively easily identified as a less 'square' shape of this I-V curve.

However, not only the parasitic resistance components influence the fill factor. It is easy to recognize that the fill factor is also related to the V_{oc}: the higher the V_{oc}, the higher the fill factor. An empirical expression was obtained for this dependency [28] that fits the *FF* calculated for ideal single-junction cells with an excellent accuracy:

$$FF = \frac{q\dfrac{V_{oc}}{kT} - \ln\left(q\dfrac{V_{oc}}{kT} + 0.72\right)}{q\dfrac{V_{oc}}{kT} + 1} \qquad (5.7)$$

As explained in previous sections, the V_{oc} depends on the bandgap, and thus it follows that the fill factor depends on the bandgap too: lower bandgap cells have a lower fill factor. Moreover, since bandgap and V_{oc} depend on temperature, the fill factor depends on temperature

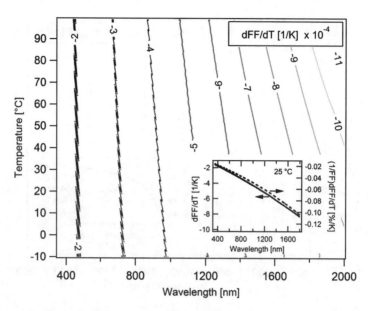

Figure 5.8 Contour plot of the *FF* derivative with respect to temperature, plotted against the temperature and the bandgap wavelength. The inset shows the temperature derivative and the temperature coefficient, for 25 °C, plotted against the bandgap wavelength. All calculations are done for the radiative limit at 1-sun under G173 direct solar spectrum

as well. In Figure 5.8, the contours of the derivative of the fill factor, at 1-sun G173 direct solar spectrum, of ideal single-junction cells, is plotted against the bandgap and temperature. Note that the same contour plot can be obtained using the derivative of eq. (5.7) or numerically constructing the I-V curves of the cell in the radiative limit and then calculating the *FF*. The sensitivity of the *FF* to temperature is negative, as could be expected from the dependency of *FF* on the V_{oc}. Besides, both the temperature derivative and temperature coefficient of the *FF* are higher for lower bandgaps, as shown in the inset of Figure 5.8.

As commented before, the fill factor depends on the parasitic resistance components of the solar cell too. In typical, good quality III-V concentrator cells the parallel resistance is high enough so it does not affect the performance of the cell at concentration operation. On the contrary, the series resistance does generally influence, in many cases drastically, the performance of the solar cell working under concentrated light. While in extreme cases the J_{sc} and V_{oc} can also be affected, the *FF* is the merit figure mostly influenced by the series resistance. Therefore, an effect of the series resistance on the sensitivity of the fill factor to temperature can be expected. As shown in [25], a lumped series resistance used to model the overall effect of the different resistance components in the cell, added to the ideal solar cell does not affect the temperature derivative of the *FF* to temperature. However, since this series resistance gives rise to a lower *FF*, the temperature coefficient of the *FF* increases with a higher series resistance. In practice, the series resistance in a concentrator solar cell has a distributed nature that cannot be modeled as a lumped component [29] (see Chapter 2). Moreover, depending on the size of the cell, the irradiation profile and the heat extraction technique used (see section 5.3.3), the distribution of temperatures across the solar cell during concentration operation may not be homogeneous, giving rise to an even more intricate influence of the series resistance on the sensitivity of the *FF* to temperature.

5.2.2.3.2 Multijunction Solar Cells

In a multijunction solar cell, the *FF* depends strongly on the current matching between subcells (see Chapter 2). Therefore, the subcells J_{sc} and, consequently, the solar spectrum, influence the sensitivity of the *FF* to temperature. The same two ideal triple-junction solar cell cases analyzed in section 5.2.2.1.2, are used here to illustrate this point. The calculated *FF* and its temperature derivative are shown in Figure 5.9.

The *FF* derivative fluctuates sharply around the temperature where the current limiting subcell changes. In a dual-junction solar cell, the *FF* derivative follows the same trend as the derivative of the limiting subcell, as shown in [25] for a GaInP/GaAs cell (see in Figure 5.4 what subcell is limiting in each case depicted in Figure 5.8). In the case of the GaInP/GaAs/Ge triple-junction cell (Figure 5.9, graph on the right), whose Ge bottom cell is highly current mismatched with respect to the other two subcells, a similar behavior as in a GaInP/GaAs dual-junction cell is observed. In the case of the GaInP/GaAs/GaInAs cell, the derivative of the *FF* is more complicated, since the three subcells are current matched at 25 °C, and the J_{sc} of all of them influence the *FF*. Comparing Figure 5.5 and Figure 5.9, note the similarity in the shape of *FF* derivative and the J_{sc} derivative, that illustrates this point.

One important conclusion that can be extracted from this analysis is that, similarly as in the case of the J_{sc}, the temperature derivative of the *FF* is, in most cases, not constant over

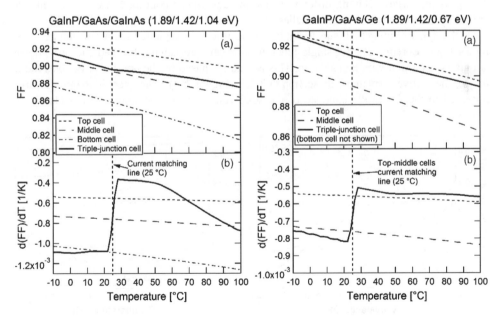

Figure 5.9 *FF* parameters of ideal triple-junction solar cells with the bandgap combinations indicated, for the G173 direct solar spectrum and using −0.45 meV/K as the temperature coefficient of the bandgap for all the materials: (a) *FF*; (b) d*FF*/d*T*. All the subcells in the GaInP/GaAs/GaInAs triple-junction cell (left) are current matched at 25 °C, while only the top and middle subcells are current matched in the GaInP/GaAs/Ge triple-junction cell (right)

temperature and, therefore, its temperature sensitivity cannot be described with a single temperature coefficient. Another important conclusion is that, the temperature coefficient of the *FF* in typical multijunction solar cells is lower than the temperature coefficient of the J_{sc} and V_{oc} and, therefore, its influence on the temperature derivative of the efficiency is the lowest.

5.2.2.4 Efficiency, η

5.2.2.4.1 Single-Junction Solar Cells

The efficiency of a solar cell is proportional to the V_{oc} J_{sc} *FF* product and, therefore, the sensitivity to temperature of V_{oc}, J_{sc} and *FF* determines the sensitivity of the efficiency to temperature. The temperature coefficient of the efficiency is the sum of the temperature coefficients of V_{oc}, J_{sc} and *FF*.

$$\frac{1}{\eta}\frac{d\eta}{dT} = \frac{1}{V_{oc}}\frac{dV_{oc}}{dT} + \frac{1}{FF}\frac{dFF}{dT} + \frac{1}{J_{sc}}\frac{dJ_{sc}}{dT} \tag{5.8}$$

In Figure 5.10-left, the calculated temperature coefficients of J_{sc}, V_{oc}, *FF* and efficiency (η) of ideal single-junction cells under 1-sun G173 direct solar spectrum are plotted against their bandgap wavelength. The sign of each parameter's relative derivative and their weight on the efficiency relative derivative is made apparent in this graph. The irregular shape of the derivative of the J_{sc} calculated for the G173 direct solar spectrum (see section 5.2.2.1) propagates to the derivative of the efficiency. The weight of the temperature coefficient of the V_{oc} is the largest. Having different signs, the aggregated contribution of the J_{sc} and FF temperature coefficients is much smaller.

On the right of Figure 5.10, the derivative of the efficiency is plotted against the temperature, for a set of most representative single-junction solar cells. As can be observed, the derivative is not always constant with temperature. This is mainly caused by the non-linearity observed in the J_{sc} derivative explained in section 5.2.2.1 (Figure 5.3). The increased non linearity observed in the *FF* as the bandgap of the solar cells decreases (Figure 5.8) has also some

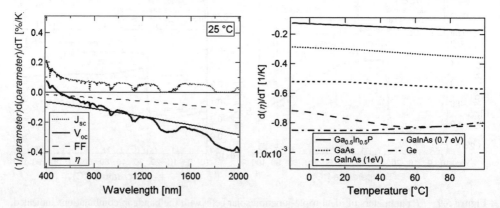

Figure 5.10 Left: calculated temperature coefficient of J_{sc}, V_{oc}, *FF* and efficiency of ideal single junction solar cells at 25 °C, plotted against the bandgap wavelength. Right: derivative of the efficiency of the most significant single junction solar cells used in multijunction concentrator solar cells, plotted against temperature. All calculations are done for the radiative limit at 1-sun under G173 direct solar spectrum

influence. However, the non-linearities observed in Figure 5.10 are relatively weak, and, therefore, using a constant temperature coefficient for the efficiency can be considered accurate enough in most cases.

5.2.2.4.2 *Multijunction Solar Cells*

The sensitivity of the efficiency to temperature is contributed to by the J_{sc}, V_{oc} and *FF* the same way in multijunction as in single-junction solar cells. In the previous sections the complex sensitivity of J_{sc} and *FF* to temperature in multijunction solar cells is shown. The same triple-junction solar cell case-examples are again used in this section to illustrate how the sensitivities of J_{sc}, V_{oc} and *FF* propagate to the sensitivity of the efficiency to temperature. The results are shown in Figure 5.11.

As can be observed, the variation of the efficiency with temperature is smoother than could be expected from the irregularities in the J_{sc} and *FF* temperature derivatives shown in Figure 5.5 and Figure 5.9. In fact, the opposite direction in the change of these derivatives with temperature contributes to smooth the efficiency temperature derivative [25], to the point that no abrupt changes are observed around the current matching transition point (25 °C). However, in the temperature range shown, the efficiency derivative in these ideal triple-junction solar cells examples is not constant over temperature, an effect more pronounced in the case of the GaInP/GaAs/Ge solar cell. This indicates that using a constant temperature coefficient for the efficiency of multijunction solar cells has to be taken with caution.

In this section, we have examined the influence of temperature on the merit figures of single and multijunction solar cells, focusing on qualitative trends computed using ideal solar cells.

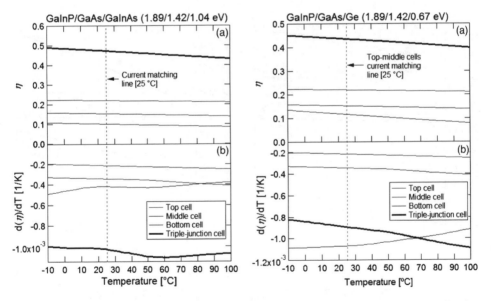

Figure 5.11 Efficiency parameters of ideal triple-junction solar cells with the bandgap combinations indicated, for the G173 direct solar spectrum at 1-sun and using −0.45 meV/K as the temperature coefficient of the bandgap for all the materials: (a) η; (b) efficiency temperature derivative, dη/dT. All the subcells in the triple-junction shown on the left are current matched at 25 °C, while only the top and middle subcells are current matched in the triple-junction cell shown on the right

In section 5.2.4, a set of representative measurement results among those published in the literature is analyzed.

5.2.3 Influence of Concentration on the Sensitivity to Temperature

As concentration (C) increases, the solar cell parameters (J_{sc}, V_{oc}, FF, η) change and, therefore, their sensitivity to temperature could be expected to change too. The J_{sc} is normally assumed to increase linearly with concentration, and this is a good assumption in the majority of concentrator solar cell cases (see Chapter 2). It is easy to recognize, then, that the relative variation of J_{sc} with temperature is constant with concentration. The sensitivity of the V_{oc} to temperature, which can be quantified by a constant temperature coefficient as shown in section 5.2.2.2, dramatically changes with concentration: the temperature derivative goes down logarithmically with concentration, as can be easily deduced from Eq. (5.6). The dependence of the FF on the V_{oc}, shown in section 5.2.2.3, leads to a similar reasoning for the influence of concentration on the FF sensitivity to temperature. From Eq. (5.8), it follows that the rate of change with concentration of the temperature coefficient of the efficiency is the sum of the rates of V_{oc} and FF.

The evolutions of the temperature coefficient with concentration of V_{oc}, FF and efficiency are illustrated in Figure 5.12–left. Four ideal single-junction solar cells, whose bandgap wavelengths are shown in the legend, are used.

Since the relative increase of V_{oc} with concentration is higher in lower bandgap solar cells, their temperature coefficient of the V_{oc} decreases faster (in absolute value) with concentration, as can be seen in Figure 5.12–left. This effect propagates to the temperature coefficients of the

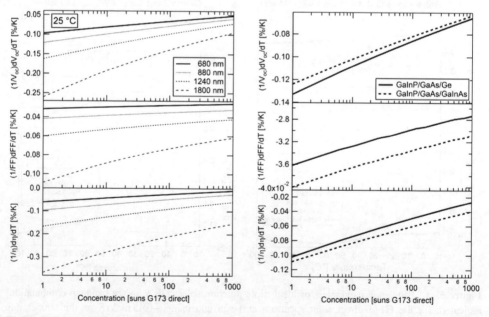

Figure 5.12 Temperature coefficient of V_{oc}, FF and η, at 25 °C, plotted against concentration for single-junction solar cells whose bandgap wavelength is indicated in the legend (left) and the triple-junction solar cells analyzed in previous sections (right)

FF and efficiency. The same discussion applies to multijunction solar cells. As shown in the previous sections, current matching affects the overall sensitivity to temperature of multi-junction solar cells. However, in principle, concentration does not affect current matching and, therefore, the sensitivity to temperature of single and multijunction solar cells is influenced in the same way by concentration. In Figure 5.12–right, the evolution with concentration, at 25 °C, of the temperature coefficient of V_{oc}, *FF* and efficiency of the two ideal triple-junction solar cells analyzed in the previous sections is plotted. Having a higher V_{oc}, the GaInP/GaAs/GaInAs triple-junction solar cell exhibits a lower temperature coefficient of the V_{oc}, but it decreases more slowly with concentration than in the case of the GaInP/GaAs/Ge cell. The Ge-based triple-junction cell exhibits a higher *FF* (due to its current mismatch at 25 °C), enough to produce a lower efficiency temperature coefficient that decreases noticeably faster with concentration.

We have assumed above that the J_{sc} varies linearly, and its temperature coefficient is constant, with concentration. In practice, for higher concentration systems, the cell temperature is higher too, as determined by the heat transfer characteristics of the CPV module. If for higher concentration the current limiting junction in the multijunction solar cell changes because the temperature changes too, the evolution of the temperature coefficients with concentration would be much more complex than shown in Figure 5.12, for the reasons explained in the previous sections.

As a main conclusion, the trends obtained in this section for ideal solar cells show decreasing sensitivity to temperature of the solar cell electrical parameters as the concentration increases. In next section, this trend is also shown in real solar cells.

5.2.4 *Experimental Measurements on Real Solar Cells*

The measurement of the temperature coefficients in multijunction solar cells is of central importance for predicting the performance of real concentrator systems and for being able to estimate the energy output. Therefore, it has been the focus of many studies such as [22,24,30]. In this sense, the multijunction solar cell design, which is usually optimized for operation under standard conditions (25 °C), may require modifications if the optimization of the solar cell performance at operation temperatures is pursued. For example, it is well known that the optimum bandgap combination in a multijunction cell is different for different operation temperatures, and has been shown experimentally for example for a GaInP/GaAs/GaInAs triple-junction cell in [19].

Important as it is, the measurement of the temperature sensitivity of multijunction solar cells is not straightforward. As shown in the previous sections, the I-V curve and the sensitivity to temperature of J_{sc} and *FF* strongly depend on the subcell current matching conditions. Therefore, the exact current balance between subcells under the reference spectrum must be reproduced during the I-V curve measurement at different temperatures. The measurement technique is not complicated conceptually (see Chapter 10): isotype cells are used to calibrate the light source output so that the light spectral composition produces, at each temperature, the same J_{sc} in the subcells of the multijunction cell as under the reference spectrum. If the available isotype cells are such that they perfectly match the spectral response and sensitivity to temperature of the subcells in the multijunction cell, the measurement over temperature is not more complicated that the standard measurement at 25 °C. Otherwise, as frequently happens, a spectral correction factor must be applied, that requires the knowledge of the sensitivity of the spectral response of subcells and isotype cells, and of the spectrum of the light source. In addition to this complication that affects the general measurement procedure, other equally important technical hitches occur and are detailed in section 5.3 of this chapter.

Outdoor measurements of concentrator systems involve elements such as optics, tracking errors, spectrum variations, etc, that can contribute to a higher measurement complexity and uncertainty [30]. Even in the ideal case of perfectly characterized optics and no tracking errors, the variation in the spectrum with atmospheric conditions is unavoidable and can affect significantly the temperature sensitivity of the multijunction solar cell.

It is noteworthy that, although this effect has been generally ignored, in good quality multijunction solar cells a significant photon coupling between subcells can occur [31], particularly when current mismatch exists and one or more subcells are forward biased beyond the maximum power point at a voltage where radiative recombination is high. This effect introduces non-linearities in the evolution of J_{sc} with concentration, and in the current matching conditions that cannot be tracked by using isotype cells and that affect the sensitivity of the cell to temperature. Given the fact that multijunction cells are usually designed to be current matched at the reference spectrum, photon coupling may not have a significant effect during well calibrated indoor measurements. However, in outdoor measurements the current matching is dictated by the spectrum of the light impinging the solar cell, which is determined by the solar spectrum available during the measurement and the optics.

Among all solar cell performance parameters, the sensitivity of V_{oc} to temperature is the most studied and reported in the literature, given the fact that it dominates the overall sensitivity temperature of the efficiency of single and multijunction solar cells. Fortunately, the V_{oc} is much less affected by current matching. Moreover, errors in the measured J_{sc} in each subcell at different temperatures propagate only logarithmically to the V_{oc}. In Figure 5.13, the measured V_{oc} temperature coefficient of a set of different triple-junction solar cell technologies is plotted against concentration. Most of the data correspond to $Ga_{0.50}In_{0.50}P/Ga_{0.99}In_{0.01}As/Ge$ triple-junction solar cells, which is the most mature technology to date. The cells of this technology

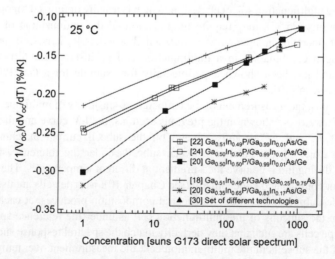

Figure 5.13 Measured V_{oc} temperature coefficient at 25 °C of a set of triple-junction solar cells (see references in legend), plotted against the concentration

measured in [22] and [24] show very similar temperature coefficients, and an almost identical trend with respect to concentration. For the same kind of cell, the result in [20] shows a significantly different temperature coefficient and it changes at a higher rate with concentration. As for other technologies, the cells measured in [18] and [20] show a higher and lower temperature coefficient, respectively, due to the higher and lower bandgap and V_{oc} in these cells. Despite the difficulty in analyzing the irregular slopes shown by these curves, slight differences can be noticed, with a higher slope for lower V_{oc} cells, in qualitative agreement with the trends shown in Figure 5.12 for ideal cells. Note, however, how the quantitative results obtained for ideal cells working in the radiative limit (Figure 5.12) differ significantly from these experimental results. Finally, the relative temperature coefficient of the V_{oc} at 550 suns, for a set of different (undisclosed) technologies from [30] agrees well with the results obtained by other authors for the $Ga_{0.50}In_{0.50}P/Ga_{0.99}In_{0.01}As/Ge$ and $Ga_{0.50}In_{0.50}P/GaAs/Ga_{0.25}In_{0.75}As$ cases.

Notably scarcer is the data available for temperature coefficients of the J_{sc} in multijunction solar cells. In fact, as commented above, the accurate measurement of this parameter is not straightforward, given its strong dependency on the subcell current matching conditions, which depend on the spectral composition and solar cell bandgaps. For completeness and to provide exemplary figures for this temperature coefficient, a summary of some data that can be found in the literature is presented in Table 5.1. Observe that there is a significant dispersion in the coefficients, even for cells of the same bandgap combination, although it is difficult to know the exact conditions used during the measurements.

As shown in section 5.2.3 for ideal solar cells, the relative sensitivity of efficiency to temperature decreases with concentration. In Figure 5.14, the temperature coefficient of the efficiency measured on triple-junction solar cells is plotted against the concentration, and shows the same trend predicted theoretically. The reasons for the discrepancies observed between solar cells of the same technology are difficult to be ascertained, but a different error in the measurement method and temperature used may be playing an important role. It is noteworthy the close agreement between the data published in [20] and the manufacturers [34], [35] and [36] for GaInP/GaInAs/Ge triple-junction solar cells at 500 suns.

From a practical perspective, an important conclusion of this section is the beneficial effect of the concentration on the temperature sensitivity of the performance of solar cells, predicted theoretically in previous sections and demonstrated experimentally with the data shown in this section. As shown in Figure 5.14, measured temperature coefficients of the efficiency are around three times lower at usual operation concentrations (500–1000 suns) than at 1 sun, for the triple-junction solar cells of the most mature technology available at present.

Table 5.1 Experimental temperature coefficients of the J_{sc} in triple-junction solar cells

Cell type	J_{sc} temperature coefficient [%/K] at 25 °C	Reference
$Ga_{0.49}In_{0.51}P/Ga_{0.99}In_{0.01}As/Ge$ 'C1MJ'	0.07	[32]
$Ga_{0.49}In_{0.51}P/Ga_{0.99}In_{0.01}As/Ge$ 'C2MJ'	0.09	[32]
$Ga_{0.49}In_{0.51}P/Ga_{0.99}In_{0.01}As/Ge$	0.0107	[33]
$Ga_{0.35}In_{0.65}P/Ga_{0.83}In_{0.17}As/Ge$	0.0547	[33]
$Ga_{0.49}In_{0.51}P/Ga_{0.99}In_{0.01}As/Ga_{0.25}In_{0.75}As$	0.056	[18]

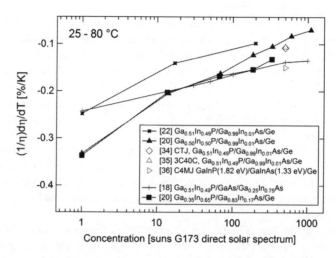

Figure 5.14 Measured temperature coefficient of the efficiency plotted against concentration, for a set of triple-junction solar cells. The data of three commercial triple-junction solar cells, obtained from the datasheets provided by the manufacturers, are included (empty symbols)

5.2.5 Summary of Temperature Effects in CPV Multijunction Solar Cells

In section 5.2 we have studied the effect of temperature on the bandgap and electrical parameters (J_{sc}, V_{oc}, FF and efficiency) of solar cells. The evolution of these parameters with temperature has been calculated for ideal solar cells working in the radiative limit, and temperature coefficients have been obtained. The effect of concentration on the temperature coefficients has been also calculated. The magnitude of the temperature coefficients obtained in this study can differ significantly from the values obtained for real solar cells, but the trends obtained are similar and some important conclusions are:

1. The temperature coefficient of the bandgap is very similar for most of the relevant semiconductor materials used in concentrator solar cells (Figure 5.1).
2. The evolution of the short circuit current density (J_{sc}) with temperature depends on the solar spectrum and the position of the solar cell bandgap in the spectrum. The temperature derivative and, therefore, the temperature coefficient of the J_{sc} is not always constant with temperature, in both single and multijunction solar cells (Figures 5.2, 5.3, 5.5). In multijunction solar cells, the current matching conditions play an important role in this effect.
3. The temperature derivative of the open circuit voltage (V_{oc}) is similar for most materials as a result of 1) and, therefore, the temperature coefficient is lower for higher bandgap materials. It is constant over temperature for both single and multijunction solar cells (Figure 5.7).
4. The temperature coefficient of the fill factor (FF) is closely related to the V_{oc}, and is rather constant for single-junction solar cells (Figure 5.8). In multijunction solar cells, where the FF depends strongly on the current matching between subcells, sharp changes in the FF temperature coefficient can happen if the current matching conditions vary with temperature (Figure 5.9).
5. The temperature coefficient of the solar cell efficiency (η) is the aggregate of the temperature coefficients of J_{sc}, V_{oc} and FF, but the weight of the temperature coefficient of the V_{oc} is the largest. Thus, the temperature coefficient of η is reasonably constant over temperature (Figure 5.10).
6. The temperature coefficient of the V_{oc}, FF and η decreases with concentration (Figure 5.12).

In section 5.2.4 we presented a compilation of some experimental results available in the literature about temperature coefficients in real multijunction cells. Since it has the largest weight in the performance of the solar cell, the temperature coefficient of the V_{oc} is the most studied and reported. In Table 5.2 a summary of the theoretical and measured values for some relevant single and multijunction solar cells, at 1-sun and under concentration is presented. The values for silicon solar cells are also shown for comparison.

Some important points can be outlined from this table. First, the significant difference between the theoretical temperature coefficients calculated for ideal solar cells working in the radiative limit and the temperature coefficients measured on actual solar cells. Observe that for the case of the GaAs solar cells with different qualities, as the photon recycling and quality of the cell increases (see Figure 5.6 also), its temperature coefficient approaches the theoretical value. Second, the reduction in the temperature coefficient with concentration is observed in both theoretical results and measurements, and is an important effect to counteract the usual cell temperature increase when going to high concentrations. Lastly, in the light of the theoretical analysis carried out in this section and the data shown in Table 5.2, a brief comment on the temperature coefficient of III-V vs. silicon solar cells is warranted. It is traditionally said that silicon solar cells have a higher sensitivity to temperature than III-V solar cells. Concerning single-junction solar cells, this assertion is not sensible, because of the intrinsic dependency of the temperature sensitivity on the material bandgap, which varies in wide range for III-V compounds. The quality of the solar cell plays also an important role, as shown in this section and as demonstrated in Table 5.2: some III-V single-junction solar cells exhibit a higher temperature coefficient than 'advanced' silicon solar cells. For a global comparison of both

Table 5.2 V_{oc} temperature coefficient at 25 °C for III-V solar cells, comparing the calculated values for ideal solar cells in the radiative limit and measured values that can be found in the literature (see section 5.2.4). The values for silicon solar cells are also shown for comparison

Cell	Calculated V_{oc} coefficient, ideal cells (this book) [%/K]		Measured V_{oc} coefficient (as reported in Refs.) [%/K]		
	1 sun	concentration	1 sun	concentration	Refs
Si (standard)	−0.162	−0.087 (250×)	~−0.35	—	[37]
Si (advanced)	−0.162	−0.087 (250×)	−0.258	−0.172 (250×)	[38]
GaAs (standard)	−0.121	−0.065 (500×)	−0.289	—	This book, Figure 5.6
GaAs (high photon recycling)	−0.121	−0.065 (500×)	−0.177	—	This book, Figure 5.6
GaInP	−0.096	−0.056 (500×)	~−0.163	—	[18]
$Ga_{0.25}In_{0.75}As$ (1 eV)	−0.162	−0.079 (500×)	~−0.343	—	[18]
$Ga_{0.49}In_{0.51}P/$ $Ga_{0.99}In_{0.01}As/Ge$	−0.133	−0.071 (500×)	−0.245	−0.151 (200×)	[22]
	−0.133	−0.071 (500×)	−0.250	−0.141 (500×)	[24]
	−0.133	−0.071 (500×)	−0.302	−0.127 (500×)	[20]
$Ga_{0.49}In_{0.51}P/$ $Ga_{0.99}In_{0.01}As/$ $Ga_{0.25}In_{0.75}As$	−0.124	−0.069 (500×)	−0.215	−0.118 (500×)	[18]

technologies, III-V multijunction solar cells must be used. In this case, the III-V technology exhibits a lower sensitivity to temperature, and even more for high voltage designs (such as $Ga_{0.49}In_{0.51}P/Ga_{0.99}In_{0.01}As/Ga_{0.25}In_{0.75}As$) or for work under concentration. However, it has to be acknowledged that the significant leap forward in the voltage of advanced silicon solar cells has impressively narrowed the gap between temperature sensitivities of both technologies.

5.3 Temperature Effects and Thermal Management in CPV Optics and Modules

5.3.1 Temperature Effects on CPV Optics and Modules

Once we have reviewed the temperature effects on concentrator solar cells, it is worth to devote some effort studying how temperature influences the entire module. In the first part of this section, the study of thermal sensitivity of optical concentrators is undertaken, focusing on optics using Fresnel lenses as primary optical element (POE) which, owing to its inherent chromatic aberration (see Chapter 4), show much higher temperature sensitivity compared to reflective POE. In the second part, the effects of temperature on both the solar cell and the optics are studied to understand how the electrical parameters of the entire module are modified by temperature.

Since, as shown in the previous section, the solar cell efficiency decreases with temperature, module thermal management approaches should be designed to maintain the solar cell temperature within a certain operating range. Heat transmission from the solar cell to the surrounding air comprises, at least, two steps: a) heat extraction from the cell to the module chassis, which is mainly dominated by conduction, and b) heat transport from the chassis to the air, by convection. Several options exist to improve the efficiency of both thermal connections. The final part of this section deals with this issue and summarizes the alternatives proposed and implemented by different researchers and manufactures.

5.3.1.1 Temperature Dependence of the Optics

One of the most common configurations for CPV modules consists of using a Fresnel lens as POE to concentrate light over a multijunction (MJ) solar cell. In some cases, but not always, a secondary optical element (SOE) is added to the system with one or more of the following objectives: to increase the geometric concentration, to widen the acceptance angle, to improve the tolerance to assembly errors, to smooth the irradiance distribution over the cell and to protect the solar cell from degradation. Currently, two main techniques are being used by the industry to manufacture Fresnel lenses. The first and most popular is the silicone-on-glass (SOG) hybrid lenses [39]. The injection-molded SOG process features a rigid glass substrate to which a Fresnel lens structure composed of optical silicone rubber is directly molded. A mold is pressed against the glass, the uncured silicone is injected and the silicone is allowed to cure at elevated temperatures. This process has proven to be cost effective and highly scalable, allowing the production of large parquets of lenses. The main advantages of SOG Fresnel lenses are related to the materials that they are made of. Silicone allows high precision of the profile, creating sharper corners and reducing inactive areas between facets caused by draft angles. Additionally, the glass provides its inherent rigidity, flatness, resistance to scratch and well-proven long-term reliability. The similitude between the refractive indices of both materials allows their optical coupling without significant losses. The second manufacturing process currently used by some companies is based on molding polymethyl methacrylate (PMMA) lenses and it is presented as an alternative capable of obtaining high efficiency optics while reducing the cost. PMMA parquets composed of several lenses can be manufactured by hot embossing or injection molding [40] (see Chapter 4 for further details on these concepts and processes).

The performance of Fresnel lenses varies with temperature due to two different effects. In the first place, the refractive indices of the materials that compose the lenses vary with temperature. Lenses manufactured by any of the techniques described above suffer from this effect. However, those made of SOG are much more sensitive since the coefficient of variation of the refractive index with temperature is approximately three times larger for silicone than for PMMA (Figure 5.15). In the second place, temperature variations cause the materials to expand or shrink. Manufacturing temperature (i.e. curing temperature of silicone) and operating temperature are different, so the groove of the lenses are physically constrained, that might cause significant deformation in the facets slopes and therefore, deviations in the refracted rays. This effect is more noticeable in SOG Fresnel lenses due to the large mismatch between the coefficient of thermal expansion (CTE) of glass and silicone.

Figure 5.16 shows the deformation predicted by finite elements (FE) modeling of the outermost groove of a SOG Fresnel lens with f-number equal to 1.6 (calculated using the lens diagonal) when the assumed silicone curing temperature is 60 °C and the operating lens temperature is 35 °C, that is, the silicone has experienced a temperature variation of −25 °C. Since temperature effects are more significant on SOG Fresnel lenses, and because this architecture is the most common nowadays, the rest of this section focuses on them.

We shall start reviewing the effects that temperature variations have on the irradiance distribution cast by a Fresnel lens on a solar cell, which have been studied by several authors. Rumyantsev and co-workers [44] measured the size of the spot cast by the lens at different focal distances and temperatures using the solar simulator developed at Ioffe Institute. The authors even built a magnified model of a silicone groove over a glass substrate and heated it up to

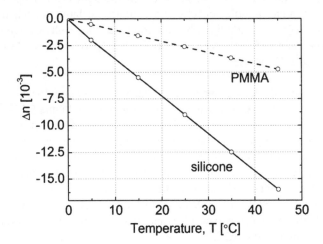

Figure 5.15 Refractive index variation vs. temperature for materials typically used to manufacture Fresnel lenses. Data for silicone comes from reference [41] and for PMMA from [42,43]

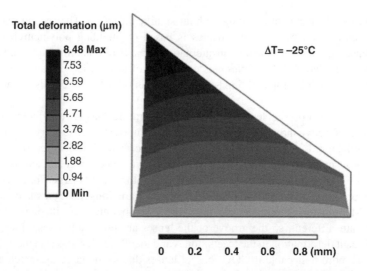

Figure 5.16 Finite elements (FE) modeling of the outermost groove of a SOG Fresnel lens when it experiences a temperature variation of −25 °C. A two-dimensional model for the FE is assumed. To simulate the fact that glass CTE is two orders of magnitude smaller than that of silicone, a fixed support is assumed in the base (displacement is equal to zero along the base). The continuous line represents the initial geometry of the groove. Deformations have been magnified by 13× so they can be easily appreciated since real scale deformations would not be distinguishable from the original structure in this picture

observe the deformation on the facet slope. Hornung and co-workers [45,46] analyzed the size of the spot for different temperatures and significant wavelengths illuminating the Fresnel lens with collimated light emitted by LEDs and used those results to check the validity of a FE model of the lens. Askins *et al.* [47] obtained the variations on the spot size for different temperatures and focal distances by photographing the spot using a CCD camera and adequate band-pass filters to discriminate between the spot created by the wavelengths converted by the top and middle subcells in a commercial lattice matched triple-junction GaInP/InGaAs/Ge solar cell. Additionally, they compared the results with those predicted by FE modeling. The main conclusions emerged from the experimental works of those authors can be summarized as follows:

1. Increasing the operating lens temperature lengthens the optimum focal distance, considering it as the focal point where the size of the spot is minimum (Figure 5.17).The temperature dependence of silicone refractive index is the main cause for this effect; it is inherent to the material and, consequently, unavoidable. In practice, if the Fresnel lens illuminates a bare cell, this means that as the temperature increases, the 'blue' light (converted by the top subcell) focuses while the 'red' light (converted by the middle subcell) defocuses. The middle subcell may limit the current of the MJ solar cell decreasing the system efficiency. If the system includes a SOE the irradiance distribution at its entrance varies with the lens temperature. In that case, the ratio of 'blue' and 'red' light over the cell strongly depends on the particular design of the SOE.
2. The spot size minimizes when the lens is operating at the same temperature as the silicone has been cured because at that particular temperature the silicone is in a stress-free state and

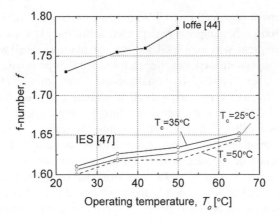

Figure 5.17 An increment on the operating temperature T_o, causes a decrease on the refractive index and consequently an increment on the optimum f-number f, of a SOG Fresnel lens [44,47]. f-number is defined as the ratio between the lens focal distance F and its diagonal D ($f = F/D$). T_c stands for the silicone curing temperature

the facets slopes are closest to the design profile (Figure 5.18) [45,47]. At this point the significant question that arises is: what is the range of temperatures reached by a Fresnel lens when the CPV module is operating outdoors along the year? The answer to this question depends, of course, on the location of the CPV system and on the module thermal management, but a delimitation of that temperature range is worth doing before optimizing a Fresnel lens design. The design temperature, should be representative of outdoor operation, and ideally selected as the curing temperature, but the latter is also constrained by the cost and yield of the manufacturing process.

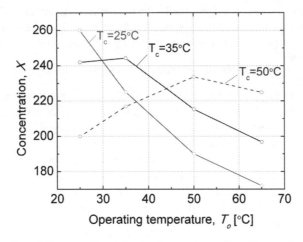

Figure 5.18 The size of the spot is minimum (i.e. concentration increases) when the operating temperature (T_o) of a SOG Fresnel lens coincides with the curing temperature (T_c) of the silicone [47]. Concentration is defined as the ratio between the areas of the lens and the projected spot

3. In general, the current generated by a CPV module comprising a SOG Fresnel lens results from the interaction of the temperature dependencies of the MJ solar cell and the Fresnel lens. As it has been previously explained, the bandgap of each subcell within a MJ solar cell decreases as the temperature increases. Hence, the spectral response of a MJ solar cell varies with temperature (see Figure 5.4). Simultaneously, the amount of light that can be transformed into electricity by each subcell is modified as the primary optics change with temperature varying the irradiance distribution at the entrance of the SOE, or directly over the bare solar cell. Interaction between this two effects[2] determines the current matching among the subcells and therefore, the short-circuit current of the MJ solar cell. Fill factor, and efficiency of the cell, will be influenced not only by the variation on the photogenerated current but also by the fact that changes of the refractive index and deformation on the facets slopes modify the irradiance distribution profile over the cell. As a final conclusion, it can be added that the total effect that temperature variations have on the distribution of light over the cell strongly depends on the system tolerance, i.e. with its acceptance angle. Those systems with a high tolerance design (either because they use a low geometric concentration or because they include a highly tolerant SOE), will be less sensitive to thermal effects than others with a 'tight' optical design.

5.3.2 Thermal Coefficients of CPV Modules

5.3.2.1 Short Circuit Current, I_{sc}

As stated before, sensitivities to temperature of the main electrical parameters of a CPV module (I_{sc}, V_{oc}, FF and efficiency η) result from the interaction of the optical system and the solar cell. In order to determine the temperature coefficients of a CPV module, measurements should be performed under stable conditions where the irradiance level and the spectral content remain as constant as possible. Otherwise, the influence of these aspects on the electrical parameters of the module will blur the temperature effect, which will be more difficult or impossible to ascertain.

For a CPV module including MJ solar cells, the temperature dependence of the short-circuit current of the module is influenced by the spectral content of the light reaching the solar cell and the spectral response (SR) of the cell at a certain temperature. The temperature coefficient $\alpha_i = \dfrac{1}{J_{sc}^{subcell\ i}} \dfrac{\mathrm{d}J_{sc}^{subcell\ i}}{\mathrm{d}T}$ of the subcell i limiting the current will determine the evolution of the short-circuit current of the module. It should be noticed here that, if at a certain temperature, the current limitation condition changes from one subcell to another, the dependency of the module short-circuit current on temperature can experience a sharp variation [48] (as shown for bare multijunction solar cells in section 5.2.2.1 and Figure 5.5).

Peharz and co-workers [49] measured the electrical parameters for Soitec modules indoors at several operating temperatures using a CPV solar simulator and heating up the modules with infrared bulbs. They characterized two different module configurations. Both comprise $Ga_{0.50}In_{0.50}P/Ga_{0.99}In_{0.01}As/Ge$ triple-junction solar cells and SOG Fresnel lenses with a

---

[2] Furthermore, changes on the spectral distribution of irradiance reaching the aperture of the optics, as the air mass and atmospheric conditions change along the day and the year, will also add to temperature effects to determine the current matching of the multijunction solar cell.

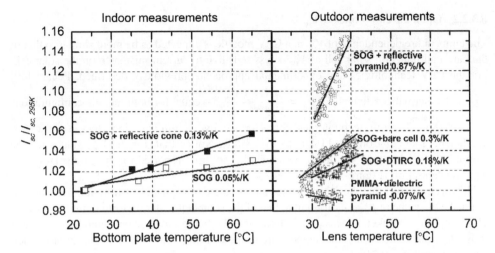

Figure 5.19 Left: Indoor experimental determination of the relative short-circuit current for CPV modules including reflective SOE and only consisting of a SOG Fresnel lens. Reproduced from [49]. Right: Outdoor experimental determination of relative short-circuit current for four different CPV module technologies. Adapted from [50]

geometrical concentration of 385×. One of them includes reflective cones as SOEs whereas the other includes bare solar cells. Both modules consisted of five parallel strings with 30 solar cells connected in series each. They found that the short-circuit current sensitivity to temperature could be as different as 0.13%/K for modules including reflective SOE and 0.05%/K for modules only comprising SOG Fresnel lenses (Figure 5.19).

This result indicates that the temperature coefficient α of a CPV module is significantly impacted by the optical configuration. The experiments reported by Domínguez and Besson [50] also support this idea. These researchers followed a different approach: they measured four different CPV module technologies outdoors and filtered the data to remove periods with significant irradiance variations or spectral distribution far from the reference. The study included CPV modules with high-concentration ratios (300–1000×). Three of the modules included a SOG Fresnel lens and either a reflective pyramid, a dielectric totally internally reflecting concentrator (DTIRC) or a bare solar cell. The fourth technology used a PMMA Fresnel lens as POE and a dielectric pyramid as SOE.

The technologies including SOG Fresnel lenses with reflective pyramids and with bare solar cells showed the largest sensitivity to lens temperature. In particular, for the module including reflective pyramids, the sensitivity of the module current was considerably higher than the thermal coefficient of the bare solar cell (see Table 5.1 for experimental values of the temperature coefficient of triple-junction cells). The module with a DTIRC showed a larger tolerance to SOG primary lens temperature variations while the module with a PMMA primary lens showed no significant sensitivity (within measurement uncertainty).

Finally, the Soitec CX-S530 module data sheet includes two different values for the α coefficient: 0.20%/K for $T_{amb} < 21\,°C$ and $-0.20\%/K$ for $T_{amb} > 21\,°C$ [51]. This further demonstrates the great influence of the optics in the temperature sensitivity of the module short-circuit current.

5.3.2.2 Open Circuit Voltage, V_{oc}

The temperature dependence of V_{oc} in a CPV module is expected to be more similar to that of the bare multijunction solar cell as V_{oc} is less sensitive to current mismatch on the solar cell. For the CPV technologies and experimental approaches previously described, Peharz [49] reported a thermal coefficient $\beta_{module} = \dfrac{1}{V_{oc}^{module}} \dfrac{dV_{oc}^{module}}{dT}$ equal for both module configurations (-0.18%/K). The four technologies characterized by Domínguez and Besson showed β coefficients ranging from -0.11%/K to -0.18%/K. Notice that, for the V_{oc}, values for the thermal coefficient at the module level are similar to those reported for the bare solar cell at high concentration (Figure 5.13).

However, the fact that the optics creates a non-uniform irradiance distribution over the cell can make the value of β_{module} different to that of the solar cell under uniform illumination β_{cell}. This effect must be taken into account if the solar cell is being used as a temperature sensor, as it is usually done to determine the temperature of the cell illuminated by the concentrator when operating outdoors [52].

One experimental approach to determine the temperature dependence of the open-circuit voltage of a solar cell under non-uniform illumination can be described as follows. A flash solar simulator is used to illuminate a solar cell covered by a mask and whose temperature is controlled by a thermal plate. If the mask reproduces the irradiance distribution caused by a real optical system, β_{module} can be considered equal to the value of β_{cell} determined using this method. Thereafter, this information can be employed to determine the cell temperature when it is illuminated by that particular CPV system.

Different sensitivities of β_{cell} under non uniform irradiance distribution have been reported for silicon and III-V solar cells. For large area (4×10 cm) silicon solar cells designed for low concentration, Antón et al. [53] found that the effect of non-uniform irradiation must be taken into account to obtain a value of β_{module} that agrees with experimental data and that allows the accurate prediction of the solar cell temperature illuminated by the Euclides linear concentrator. On the other hand, for 3 mm in diameter round GaAs single-junction concentrator cells, Andreev et al. [54] reported almost indistinguishable β_{cell} under uniform and non-uniform irradiance distribution over the cell. Although in the first case the higher temperature sensitivity of silicon together with the large size of the solar cell causes differences in the β_{cell} parameter, in the second case, for smaller III-V solar cells, non-uniform irradiance distributions are not expected to show noticeable influence on β_{cell}.

5.3.2.3 Fill Factor, FF

There is a lack of experimental data about the temperature dependence of FF at the module level. Peharz and co-workers [49] reported a temperature coefficient of -0.16%/K for both module configurations described before. In addition, they measured a module comprising the same solar cells and optics (without SOE) but including two parallel connected strings each of them comprising 24 series-connected solar cells. For this module, the FF temperature coefficient was -0.12%/K.

Furthermore, Andreev and co-workers [54], using bare solar cells covered with masks, measured an increased sensitivity of the FF with temperature when the non-uniformity of the irradiance profile is increased. Consequently, the temperature sensitivity of the efficiency reported was also higher for the non-uniformity case.

So far, it has been implicitly assumed that the temperature effect is the same on every elementary unit of the array that constitutes the module. Nevertheless, the temperature distribution among the cells in a module may vary depending, for example, on the module size, the velocity and direction of wind, etc. In general, cells located at the edges and corners of the module are expected to be at a lower temperature than inner cells [55]. The temperature variations among the solar cells in a module cause a non-uniform distribution of the current and voltages between each solar cell which, for most of the cases, is negligible. The effect of these non-uniformities will depend, of course, on the module connection layout. For a Soitec module, where the greatest difference in temperature is 15 K, researchers at Fraunhofer-ISE [55,56] estimated the reduction in power output due to temperature differences between cells to be less than the reduction in power output due to statistical distribution differences in photocurrent among the solar cells.

5.3.2.4 Efficiency, η

Finally, the temperature dependence of the efficiency of a CPV module will be directly related to that of the solar cell but also influenced by the aspects mentioned above: the spectral and spatial variations of the irradiance caused by the optical system. Consequently, the relationship between the efficiency of a CPV module and the lens operating temperature (which in turn depends on the ambient temperature) is not necessarily linear, but a more complex dependency can be found. For example, Figure 5.20 reproduces the efficiency of several modules as a function of the lens operating temperature reported in references [57,58]. For most systems, the efficiency peaks at a certain temperature value and decreases at both sides. Consequently, when designing a CPV system not only the efficiency at the reference temperature must be maximized but tolerance to temperature effects has to be ensured. This tolerance must be wide enough to include the entire range of operating temperatures reached by the module outdoors in order to maximize the energy harvesting along the year.

Hornung and co-workers [59] estimated the losses due to thermal effect on PMMA and SOG Fresnel lens in the harvested energy along the year for six different locations. These authors

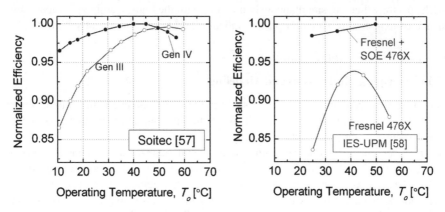

Figure 5.20 Efficiencies reported for the different generations of the Soitec module comprising a SOG Fresnel lens [57] (left) and for a 476× CPV module with two different configurations [58] (right). Reported efficiencies were normalized to the maximum value to show their sensitivity to the Fresnel lens operating temperature T_o. Figure adapted from [57] and [58]

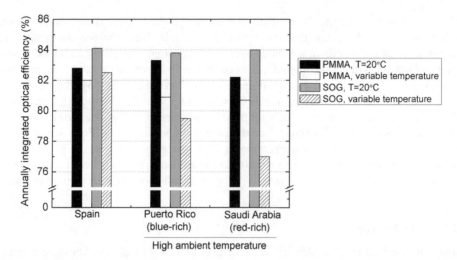

Figure 5.21 Optical efficiency of a system comprising a PMMA or SOG Fresnel lens integrated along the year at three different locations for variable lens temperature and for constant 20 °C lens temperature. Adapted from [59]

used a model which included the temperature and spectra variation along the year and the optics and MJ solar cell dependence with temperature. They estimated the integrated optical efficiency of the optics for variable Fresnel lens temperature and compared it with the calculated value if temperature is kept constant and equal to 20 °C along the year. The optical efficiency defined by the authors includes losses due to the sun shape and a 0.1° tracking tolerance but excludes the spectral sensitivity of MJ solar cells.

Figure 5.21 reproduces the main results from those simulations. For close to market locations a decrease in optical efficiency around one to two percentage points compared to optics constantly at 20 °C was estimated. For the two particular locations with extremely blue-rich and red-rich spectra which are simultaneously the ones with the highest ambient temperatures, the influence of temperature variation on the optical efficiency is significantly higher.

5.3.3 Heat Extraction Strategies

5.3.3.1 Principles, Mechanisms and Modeling

The efficiency of a certain heat extraction strategy for a CPV module is strongly dependent on its particular configuration (see Chapter 7). According to thermal management, concentrators can be classified in three different families.

The first one includes modules comprising several point-focus elementary units, either refractive or reflective. Several modules are placed in an array which is mounted on a tracker. In this case, the solar cell can make use of an area roughly equal to that of its corresponding primary optics to extract heat which allows an effective passive dissipation. If the optics is refractive or reflective comprising two stages, the cell is located at the module rear wall and dissipative fins or the entire rear wall can be used to dissipate heat by convection. If the optics is single-stage reflective, the solar cell is placed at the module front wall and the area of the dissipative elements must be reduced to avoid significant shadowing losses.

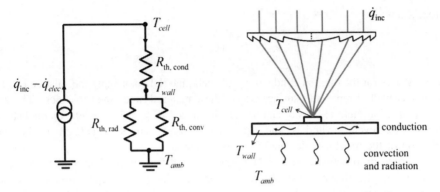

Figure 5.22 Simplest equivalent circuit representing the thermal behavior of a CPV module with passive dissipation to the atmosphere

The second type of concentrators includes linear assemblies, the most common configuration being a parabolic through concentrator. The linear concentrator allows less dissipative area for each cell than the point-focus configuration but, as linear concentrators attain lower concentration ratios, passive dissipation is also a valid heat extraction strategy. Nevertheless, more dissipative fins would probably be necessary in this case.

The third family of concentrators includes assemblies where a single large optics, most probably a parabolic dish, is used to concentrate light onto a densely packed receiver. This configuration presents greater problems for cooling than the two previous families because each solar cell has only its own rear side available for heat sinking. This condition entails the necessity of active dissipation. Nevertheless, this Handbook is mainly devoted to point-focus systems so, although we consider below the three kinds of systems, we will pay more attention to the point-focus ones.

The thermal behavior of a CPV module can be represented and modeled by its equivalent thermal circuit. The heat flow and temperature performs in the same way as current and voltage do in electrical circuits. Thermal resistors are equivalent to ohmic resistors.

Figure 5.22 represents a simplified scheme of the main heat extraction mechanisms in a CPV module. More complex analysis using the same equivalent circuit approach can be found elsewhere [60,61]. Only a fraction of the total incident energy flux $\dot{q}_{inc}$, concentrated over the solar cell is transformed into electricity $\dot{q}_{elec}$, while the rest $\dot{q}_{inc} - \dot{q}_{elec}$ must be dissipated to the atmosphere. The heat flow is transported first by conduction, next by convection and radiation so steady-state conditions lead to:

$$\dot{q}_{inc} - \dot{q}_{ele} = \dot{q}_{cond} = \dot{q}_{conv} + \dot{q}_{rad} \tag{5.9}$$

In order to allow the heat to be extracted to the atmosphere, the heat flow goes from the solar cell at a certain temperature T_{cell}, to the module rear wall at a temperature T_{wall}. The heat transport mechanism is conduction and its efficiency is determined by the thicknesses and thermal conductivities of the materials underneath the solar cell. Although a more detailed description of the different architectures proposed for the thermal stack is carried out below, the basic heat flow $\dot{q}_{cond}$, in the one-dimensional approach is described by

Fourier's law:

$$\dot{q}_{cond} = \frac{\kappa A}{l}(T_{cell} - T_{wall})$$ (5.10)

where κ stands for the thermal conductivity of each material connecting the solar cell and the module rear wall, A represents surface area of these materials and l their thickness. The value of the conduction thermal resistor can be calculated as $R_{th,cond} = l/\kappa A$. If the thermal stack is composed of several materials, the equivalent resistor can be calculated as the series connection of the resistors calculated for every material.

The thermal drop between the module rear wall and the surrounding ambient at T_{amb} is determined by the thermal resistors $R_{th,conv}$ and $R_{th,rad}$, which represent the incapacity (resistance) to transport heat by convection and radiation. The power dissipated to the atmosphere by convection in the rear wall can be estimated by using Eq. (5.11) where h stands for the heat transfer coefficient and A_w for the wet area, or area in contact with the fluid extracting the heat, i.e. the air surrounding the module in passive dissipation systems.

$$\dot{q}_{conv} = hA_w(T_{wall} - T_{amb})$$ (5.11)

Table 5.3 summarizes some typical values for h. If natural convection is not enough to maintain the cell temperature below a certain value, using forced convection, by for example circulating water underneath the solar cell, will increase the efficiency of this mechanism by two orders of magnitude. However, pumping a fluid consumes some power and increases the cost and complexity of the system, restraining commercial CPV modules from using this solution. Additionally, as pointed out by some authors, not only the wind velocity influences the efficiency of heat dissipation but also the direction of the wind has a significant impact. The efficiency of convection is significantly improved when the wind blows from the lateral side of the module rather than from the backside [62,63].

Another way to enhance this mechanism consists in increasing the area A_w in contact with the ambient by, for example, adding a heat sink with dissipative fins. Their design must be carried out carefully to optimize their performance. Since this metallic material is one of the main module cost drivers, the total volume of the heat sink must be minimized. On the one hand, the wet area must be as large as possible leading to very thin elements. On the other hand, a certain thickness for the dissipative fins is required to allow good heat conduction across them to keep the fin efficiency high. If conduction along the fins is deficient, their tips will be too cold, and since the temperature drop $T_{wall} - T_{amb}$ directly impacts on $\dot{q}_{conv}$ (see Eq. (5.11)) the efficiency of this mechanism will deteriorate. A well described theoretical and practical method to optimize dissipative fins can be found in [64].

Table 5.3 Typical values of heat transfer coefficients, h, at 300 K (units: $Wm^{-2}K^{-1}$). Reproduced from: [65]

Air		Water	
Natural convection	5	Laminar mode	380
Wind at 1 ms^{-1}	10	Intermediate mode	1270
Wind at 2 ms^{-1}	15	Turbulent mode	3683

In addition to convection, radiation is another dissipative mechanism that promotes heat extraction. The power dissipated to the atmosphere by radiation in the rear wall $\dot{q}_{rad}$, can be estimated using the Stefan-Boltzmann law:

$$\dot{q}_{rad} = \varepsilon \sigma A_r (T_{wall}^4 - T_{amb}^4) \tag{5.12}$$

where $\sigma = 5.67 \cdot 10^{-8}\,\mathrm{Wm^{-2}\,K^{-4}}$ stands for the Stefan-Boltzmann constant, ε represents the emissivity of the surface of the rear wall, and A_r stands for the radiating energy area. Equation 5.12 is commonly linearized to obtain:

$$\dot{q}_{rad} = 4\varepsilon \sigma A_r T_{amb}^3 (T_{wall} - T_{amb}) \tag{5.13}$$

so an equivalent thermal resistor $R_{th,rad} = 1/4\varepsilon \sigma A_r T_{amb}^3$ can be estimated.

Material emissivity can be improved by adding a thermal radiation layer or paint to the module heat-sink. Nishioka et al. [66] have studied this effect and reported a cell temperature decrease up to $10\,°C$ when an uncoated aluminum rear wall is taken as benchmark. For the wall temperature usually reached by CPV modules with metal housing, about 25 to 50 degrees above ambient temperature, convection is far more efficient than radiation. In those cases, the power dissipated by radiation is usually neglected or an increased heat transfer coefficient is used in Eq. (5.11) which accounts for radiation. Using the electrical circuits analogy, it can be said that, when the value of the radiation thermal resistor $R_{th,rad}$ is significantly higher than that of the convection resistor $R_{th,conv}$, the first one is considered to be infinite and only $R_{th,conv}$ is taken into account in the calculations.

The very simple equivalent circuit analyzed here only accounts for the main heat transfer path but depending on the module configuration, there could be other significant heat transport mechanisms that must be taken into account. For example, the approximation shown here neglects the heat transmitted to the air inside the module by convection or radiation from the solar cell and subsequently transported to the atmosphere by the module front side and lateral walls. Some authors estimated that these thermal paths can transfer between a quarter [61] and a third [67] of the total heat flow.

5.3.3.2 Solar Cell Packaging for Efficient Heat Transfer

Once we have summarized the basic equations describing the heat transfer, let's analyze common architectures and materials used in the CPV industry to promote heat conduction from the solar cell to the module rear wall. The stack of materials underneath the solar cell seeks two objectives simultaneously. On the one hand, the thermal conductivity must be as high as feasible to keep the temperature drop $T_{cell} - T_{wall}$ as small as possible. On the other hand, the electrical insulation between the different cells must be high enough to avoid short circuits. If the rear wall or the dissipative elements are metallic, this implies insulating the solar cell from the housing. In fact, the electrical insulation test described by the IEC62108 standard [68] requires that the module withstands a voltage equal to 1000 V plus twice the maximum system voltage without dielectric breakdown or significant current leakages. Table 5.4 summarizes the thermal conductivity of materials commonly involved in the thermal stack of CPV modules. The coefficient of thermal expansion (CTE) of those materials is also included. If materials with large CTE are coupled, the tensile or compressive thermal stress caused by temperature

Table 5.4 Thermal conductivities and coefficient of thermal expansion (CTE) of the main materials involved in the thermal management of a CPV module

	Material	Thermal conductivity [Wm^{-1}K^{-1}]		Coefficient of thermal expansion [10^{-6} K^{-1}]		Dielectric strength [kV·mm^{-1}]	
Electrical	Silicon	130	[69]	2.6	[69]		
conductors	Germanium	58	[69]	5.8	[69]		
	Gallium Arsenide	55	[69]	6.5	[69]		
	Copper	401	[69]	17	[69]		
	Aluminum	201	[65]	23	[69]		
	Solder (Sn:Pb:Ag)	50	[65]	25–35	[69]		
Electrical	AlN	180	[69]	4.5	[69]	14	[70]
insulators	BeO	280	[69]	7	[69]	13.8	[70]
	Al$_2$O$_3$	37	[65]	7.1	[69]	13.4	[70]
	alumina filled epoxy	1.3	[71]	40	[71]	112	[71]
	thermal adhesive	1.59	[72]			20	[72]
	thermal pad	3.5^3	[73]			19	[73]

variations along the day may induce degradation or failure mechanism that reduces drastically the module lifetime. In addition, the dielectric strength of the materials classified as insulators is also indicated in Table 5.4.

The final configuration of the thermal stack underneath the solar cell will result as a trade-off where thermal conductivity, electrical insulation and cost are the main aspects to be considered. Figure 5.23 represents some of the architectures selected by CPV module designers described hereafter.

Direct bonded copper (DBC) is fabricated by bonding copper sheets on the top and bottom areas of a ceramic insulation material (AlN of Al$_2$O$_3$) by formation of eutectic melt over 1080 °C. The contact layout structure is etched into the top side copper surface, the solar cell is attached on it and the connection to the contacts on the top of the cell is done by wire bonding or soldering tabs. Either eutectic or epoxy die attach are used to bond the solar cell. Vacuum must be applied during the bonding to remove air trapped between the cell and the copper, since the increased thermal resistance resulting from the presence of voids can give rise to catastrophic thermal runaway (see Chapter 9). The insulation materials in the middle of the DCB sandwich, either AlN of Al$_2$O$_3$, have excellent combined properties as electrical insulator and thermal conductor being the high cost of DBC its only drawback. DBC substrates are usually designed to minimize their area around the solar cell in order to reduce its cost impact on the module. A proper design (wide margins and round corners) prevents electrical arcs in the insulation test and also during operation between the two copper layers or between the upper copper layer and the metallic wall of the module. The lower copper layer must be thermally attached to the module rear wall by using a thermal silicone or epoxy, a thermal pad or simply thermal grease. If the module includes a heat sink, it should also be thermally coupled to the rear wall. A possible solution consists on connecting directly the DBC to the heat sink and making a hole in

---

^{3} Although thermal conductivity of pads is higher than alumina filled epoxi or adhesives, it should be noticed here that pads have a minimum thickness of 0.5 mm.

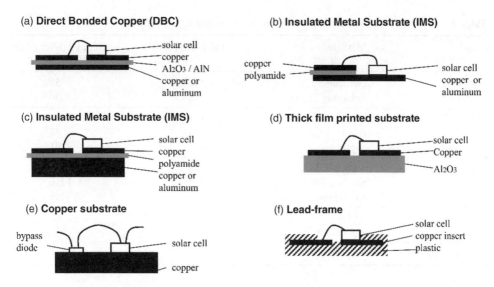

Figure 5.23 Schematics of different material stacks under the solar cell

the rear wall where this package is assembled. Nevertheless, this solution implies a significant increment on the length of the joints that must be sealed in the module.

Insulated metal substrate (IMS) is an alternative material consisting of a sandwich made up of two metallic layers and an insulation material. The metallic layers are commonly copper for the top, compatible with manufacturing process of printed circuit board, and aluminum for the bottom. For the insulation layer, polymers such as polyamides or alumina filled epoxies are used. The solar cell is attached to the top copper layer, as in DBC, or directly to the bottom copper layer, although this last configuration requires other means to provide electrical insulation between the cell and the heat-sink or rear module wall.

Thick film printed substrates constitute other alternative. They use an insulation ceramic as basic material for the substrate which is glued directly to the heat sink. The tracks on the top side of the ceramic substrate are made of copper and are applied by screen printing.

Other effective solution consists in using only a copper substrate where the solar cell is soldered. The top of the cell is connected to the top of the bypass diode using heavy wire bonding and it serves as negative contact avoiding the use of a sandwich structure or an expensive insulation material. This solution is probably limited to small solar cells where connecting the top of the solar cell and diode by heavy wire bonding is an adequate technique. Nevertheless, it should be remarked here that this configuration requires adding an extra layer to electrically insulate the solar cells and the module chassis. If the module rear wall is made of a dielectric material, such as glass, this problem is solved [55].

Finally, the possibility of manufacturing a lead-frame by injecting plastic around a copper insert has also been proposed by some designers. This is an effective but expensive solution so, similarly to DBC, lead-frames should be designed to have an area only slightly larger than the solar cell.

Using data from Table 5.5, a preliminary analysis can be performed using a one dimensional model and assuming uniform temperature across each material. Additionally, finite element (FE) modeling can be applied to analyze the temperature distribution on three-dimensional

Table 5.5 Properties of the materials considered in FE simulations of Figures 5.24–5.27

	Material	Thickness [mm]	Thermal conductivity [$Wm^{-1}K^{-1}$]
Solar cell	Germanium	0.15	58
	Soldering Sn:Ag	0.2	50
DBC	copper	0.3	401
	Al_2O_3	0.63	37
	copper	0.3	401
IMS	copper	0.3	401
	alumina filled epoxy	0.075	1.3
	aluminium	2	201
Thick film	Al_2O_3	0.63	37
	aluminium	2	201
Copper substrate	copper	2	401

structures. Hereafter, the results predicted by three sets of FE simulations are depicted to compare the different thermal stacks and to illustrate some meaningful concepts. The FE models include only the stack beneath the cell. To simulate the fact that the stack is thermally connected to some dissipative elements, the surface at the bottom of the substrate is assumed to dissipate heat by natural convection. Since that surface is significantly smaller than the surface of the dissipative area of the module –commonly the flat rear wall of the module which can optionally include fins– a modified convection coefficient h^* is assumed:

$$h^* = 2h \frac{A_{lens}}{A_{substrate}} = 744 \text{ W/mK} \tag{5.14}$$

where $h = 10$ W/mK represents the convection coefficient for a wind velocity 1 ms^{-1}; A_{lens} and $A_{substrate}$ are the aperture area of the lens and the substrate, respectively; the factor 2 comes from the fact that the wet area of the dissipative elements has been assumed to be twice the area of the lens and ambient temperature is 25 °C.

In the first set of FE simulations four different architectures are compared: DBC, IMS, thick film printed substrate and copper substrate. Figure 5.24 shows the temperature distribution when a 5×5 mm solar cell with 40% efficiency is under a 1000× uniform illumination, that is, thermal power on the solar cell is 60 W/cm^2. Thermal conductivities and thicknesses assumed in the simulations, and summarized in Table 5.5, are based on designs currently employed by the industry.

The last configuration depicted in Figure 5.24 illustrates the concept of the heat spreader, i.e., connecting the cell directly to a material with high thermal conductivity that mostly spreads the heat horizontally before conducting it downwards. The effectiveness of a copper heat spreader depends on the size of the solar cell. For example, Figure 5.25 shows three solar cells with different sizes (1×1, 5×5 and 10×10 mm) attached to a copper substrate with an area proportional to the solar cell area. The power over the solar cell is also proportional to its area while the thickness of the copper substrate is kept equal to 2 mm.

The third set of FE models is intended to study the influence of the position of the different materials composing the stack. Besides the copper or an equivalent electrical conductor, the

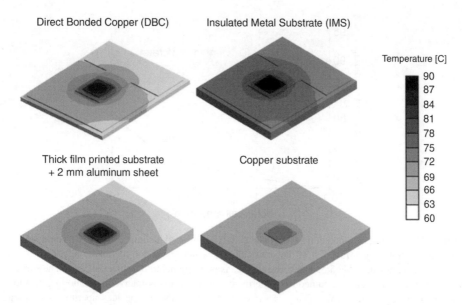

Figure 5.24 Temperature distribution (°C) estimated by finite element (FE) modeling for a 1000× concentration over a 5 × 5 mm solar cell and different thermal stacks. Thicknesses and thermal conductivities of the materials are shown in Table 5.5. Notice that the solution based on a copper substrate does not include electrical insulation. Such insulation must be provided by some additional element or, alternatively, a dielectric material such as glass must be employed for the module rear wall. In the configuration based on the thick film printed substrate, the metal housing where it is attached contributes to the smoothing of the temperature across the Al_2O_3 layer. Hence, a 2 mm thick aluminum sheet has been added to better simulate real life conditions

stack should include an electrically insulation layer. Figure 5.26 depicts two equivalent cases where a 1 × 1 mm solar cell at 1000× is simulated. In the first case, a 75 µm thick insulation layer of alumina-filled epoxy is added underneath the solar cell. In the second case, the cell is directly soldered to the copper and the insulation layer is beneath the copper. This last case

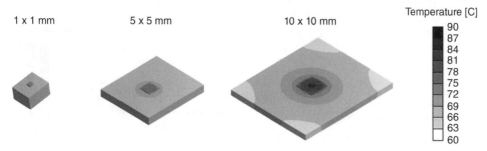

Figure 5.25 Temperature distributions (°C) estimated by finite element (FE) modeling for a 1000× concentration over solar cells with several dimensions (1 × 1, 5 × 5, 10 × 10 mm). The cells are soldered on a 2 mm thick copper substrate with an area proportional to the cell area. The surface at the bottom of every substrate is assumed to be thermally connected to dissipative fins removing heat to the atmosphere by convection. An ambient temperature of 25 °C is assumed

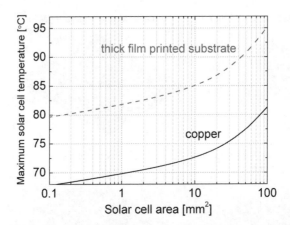

Figure 5.26 Maximum temperature at the solar cell (°C) estimated by finite element (FE) modeling for a 1000× concentration as a function of the solar cell size. Thicknesses and thermal conductivities of the materials are shown in Table 5.5. The substrate area is proportional to the solar cell area. The surface at the rear of every substrate is assumed to be thermally connected to dissipative fins removing heat to the atmosphere by convection. Wet area is assumed to be equivalent to the lens aperture. An ambient temperature of 25 °C °C is assumed

is representative of a module configuration like Soitec's [55] where the rear wall made of glass acts as electrical insulator. As Figure 5.27 shows, the insulation between the copper and the cell prevents the heat spreading across the copper increasing the cell temperature 15 °C above the second configuration.

Although in the previous simulations some details have been omitted –as tracks on the copper in some of them–, the results obtained provide valuable insights for the design of thermal stacks. Additionally, some authors [57,67,74] have extended the FE modeling to the whole CPV module which provides a good understanding about the temperature differences across the module and the strategies to reduce them.

Before concluding this chapter a general review of the heat extraction strategies designed for a great variety of concentrator configurations is presented. Following the classification by

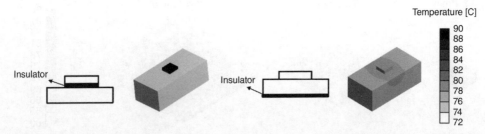

Figure 5.27 Temperature distributions (°C) estimated by finite element (FE) modeling for a 1000× concentration on a 1 × 1 mm solar cell. For the plot on the left hand side a 75 μm thick alumina filled epoxy layer is added directly underneath the solar cell. For the plot on the right hand side, the solar cell is directly attached to the copper and a 75 μm thick alumina filled epoxy layer is added underneath the 2 mm thick copper. An ambient temperature of 25 °C is assumed

Royne and coworkers [75], they are organized considering the kind of concentrator that they have been proposed for: modules comprising an array of point-focus elementary units, assemblies with linear geometries and assemblies including a densely packed array. Table 5.6 at the end of this chapter summarizes several heat extraction strategies and the cell to ambient temperature drop reported by them. It is difficult to establish a maximum operating cell temperature. As shown previously, the cell efficiency decreases as the temperature increases, so there is a certain temperature where the cost of the dissipative element is compensated by the efficiency gain. However, the optimum temperature in the trade-off resulting from the previous argument is usually too high, i.e., it causes too large thermal cycles as the solar cell temperature varies between operation and dark conditions (at night or when the module is shadowed by clouds). Well-designed modules maintain the cell temperature below 85 °C under extreme ambient conditions (low wind velocity and ambient temperature around 40 °C) but the final decision on the heat extraction mechanism is a matter not only of its efficiency but also of its cost and long-term reliability. Nevertheless, the comparisons stated in the next pages and Table 5.6 may provide a good background for the reader.

5.3.3.3 Modules Comprising an Array of Point-Focus Elementary Units

As a rule of thumb, passive dissipation works good enough to maintain solar cells within a CPV module below a safety temperature if the thermal resistance between the cell and the dissipative elements is low enough and if the design includes sufficient area to exchange heat with the surrounding environment. For classic CPV modules comprising refractive optics, an area roughly equivalent to the primary optics is available for heat dissipation. The classic approach consists in equipping the system with dissipative fins. An alternative approach is based on conducting the heat efficiently across the module rear wall which is thick enough to approach the isothermal case, without any additional heat sink [76]. This approach requires a good heat spreading underneath the cell that enlarges the area in which conduction between the cell substrate and the module rear wall is taking place. FLATCON modules use a thick copper substrate significantly larger than the solar cell glued to a glass rear wall. The glass has a low thermal conductivity leading to large temperature differences in the module but it is inherently an electrical insulator avoiding the need of any additional insulating material [60,76]. Additionally, as glass emissivity is higher than that of aluminum, for low wind velocities, radiation becomes as efficient as convection [67]. An alternative solution, proposed by Araki and co-workers (Daido Steel), consists in a rear wall made of aluminum. The aluminum is printed with epoxy filled with conductive agents to enhance heat conduction from the cell while maintaining electrical insulation [77,78]. It should be noticed that the efficiency of the heat spreader approach decreases as the solar cell size increases and, probably, using a heat sink with dissipative fins becomes mandatory for solar cells larger than 5×5 mm. Conversely, for small solar cells the rear wall is enough to maintain the cell at low temperature. This is the approach followed by designs including solar cells smaller than 1 mm^2, as Semprius' module [79].

When a CPV module is based on mirrors instead of lenses the heat extraction becomes a more difficult problem as the heat sinks cast shadow on the primary optics (mirror) leading to unbearable losses. De Nardis and co-workers [80] presented a concentrator based on a single reflective stage where a thin aluminum dissipative fins is optimized to extract the heat from the solar cell while minimizing the shadow cast to the primary optics. Plesniak and co-workers [81,82] presented a reflective concentrator based in a mirror working off-axis to avoid

Table 5.6 Summary of heat extraction strategies and the cell to ambient temperature drop

			Reference		Description	C [suns]	$\Delta T_{c\text{-}a}$ [°C]
Point focus (single cell)	Passive cooling – Refractive optics		Edenburn [92]	T	Linear fins surrounding each solar cell, no temperature limit is assumed to avoid cell degradation	175×	73
			Araki [77]	E	Heat spreader, aluminum sheet printed with filled epoxy	400×	21
			FLATCON Jaus [60] Wiesenfarth [67]	FE	Copper heat spreader + glass rear wall	500×	45
			Beach and White [83]	E	Copper heat pipe with copper fins, working fluid: water or acetone	732×	35
	Passive cooling – Reflective optics		Miñano [100]	T	Aluminum heat spreader	—	—
			BOEING Plesniak [81] Anderson [82]	E	Cooper heat pipe with water + aluminum fins	809×	40
			BECAR De Nardis [80]	FE	Thin aluminum fins	1300×	—
			FLUIDREFLEX Victoria [74,84,85]	E	Dielectric optical fluid that fills the module	636×	37
Linear geometries	Passive cooling		Edenburn [92]	—	V-type geometry with heat sink that reduces shadow	20–40×	—
			EUCLIDES Luque [88]	E	Thin aluminum finned heat sink	32×	35
			Skyline [89]	—	Aluminum finned heat sink	14×	—
			Arquimedes [90]	E	Parabolic and thinned heat sink obtained by a single aluminum extrusion	10×	10
			Akbarzadeh [91]		Copper heat pipe	20×	46
	Active cooling		Edenburn	—	Circulating water in a square channel	—	—
			O'Leary [93]	T	Working fluid: water and ethylenglycol mixture	—	—
			Chenlo [94]	E	Water through a galvanized steel pipe	24×	15
			CHAPS. Smeltink [95]	E	PV-thermal hybrid system. Water with anti-freeze additives	38×	35
Dense array	Active cooling		C.L.Tilford [96]	E	Water circulating underneath a monolithic module	250×	31
			Verlinden [97,98]	E	Water circulating underneath a dense-array receiver	340×	38
			Vicenzi [99]	—	Micro channels machined on silicon wafers	120×	—

T: Theoretical work; E: experimental work; FE: finite element simulation.

shadowing losses. In this design, the cell is connected to a copper heat pipe with aluminum fins and water as heat transfer fluid which maintains the temperature drop between the solar cell and the ambient below 40 °C. Beach and White [83] reported an increased efficiency of the heat pipe when the water is replaced by acetone as a working fluid inside the copper heat pipe.

Finally, the FluidReflex concept consists in using a single reflective stage and filling the module volume with a fluid dielectric that improves optical performance while simultaneously enhances the thermal management in the module [74,84,85]. Heat is transported efficiently by conduction and convection in the fluid from the solar cell to both, rear and front, module walls where it is dissipated to the atmosphere. Fluids candidate to be a part of a CPV system must surpass several requirements related to their thermal conductivity, boiling temperature, durability and –if they are used as optical elements– transmittance. References [86,87] provide extensive reviews on the main characteristics and durability of candidate materials.

5.3.3.4 Assemblies with Linear Geometry

Prominent among the systems with linear geometries is the EUCLIDES concentrator composed of a parabolic through mirror focusing on a linear array of high efficiency silicon solar cells [88]. A 480 kW$_p$ power plant using this technology was installed in 1998 in Tenerife where mirrors concentrate irradiance 32× on high-efficiency silicon solar cells that were cooled down by natural convection using extremely thin aluminum dissipative fins. Passive dissipation is also the most common configuration among linear concentrators, since natural convection is efficient enough to cool down the solar cells thanks to the low/middle concentration ratios attained by systems of this kind [89]. For example, the 10× Archimedes concentrator comprises a piece obtained by a single aluminum extrusion which includes the parabolic mirror on one side and the substrate for the solar cell and dissipative fins on the other [90].

Heat pipes were also proposed for linear geometries as a way of enhancing the efficiency of natural convection [91], although their higher cost has restrained its use from commercial systems. Finally, active cooling by circulating a fluid underneath the solar cells was proposed for linear geometries based on mirrors [92,93] or Fresnel lenses [94]. Of particular interest is the CHAPS concentrator developed at the Australian National University which is a hybrid PV-thermal parabolic through where concentration is 37×. The working fluid is water with anti-freeze and anti-corrosive additives. The authors reported a total efficiency higher than 60% although the thermal component of that figure is significantly higher than that of electricity and it is not correct to add both components to calculate the output [95].

5.3.3.5 Assemblies Including a Densely Packed Receiver

Heat dissipation is a higher demanding task in concentrators comprising densely packed arrays of solar cells; in fact, active cooling becomes mandatory for this kind of concentrators. In the early nineties, the increase in efficiency of back point-contact silicon solar cells led to the development of concentrators based on parabolic dishes illuminating densely packed arrays of solar cells. A monolithic module on a silicon wafer was manufactured fully interconnected and cooled by water circulating underneath the module. The cooling system was designed to minimize the total power loss, that is, the sum of the pumping power of the coolant fluid and the loss of the electrical output power due to increased temperature [96,97]. A slightly modified design was implemented by the company Solar Systems using point-contact solar cells densely packed in a receiver illuminated by a parabolic dish. The active dissipation was also based on water circulating underneath the solar cell array. The cooling circuit was part of the supporting structure of the photovoltaic receiver thus reducing the shadowing over the mirror [98].

Another proposal consists in using a silicon wafer where microchannels have been machined and water circulates directly underneath the cell. In this way, the cooling function is integrated in the cell manufacturing process [99]. Other designs based also on microchannels, as well as an study on the applicability of heat extraction strategies developed for other industries to CPV, can be found in reference [75].

Glossary

List of Acronyms

Acronym	Description
AM	Air mass
AM1.5d	Reference spectrum for direct radiation
DCB	Direct bonded copper
EQE	External quantum efficiency
IMS	Insulated metal substrate
MJ	Multijunction
POE	Primary optical element
SOE	Secondary optical element
SOG	Silicone on glass
SR	Spectral Response

List of Symbols

Typical units given in square brackets. If no units are given, variable is dimensionless.

Symbol	Description [Units]
A	Surface area [mm^2, cm^2 or m^2]
A_r	Radiating area [m^2]
A_w	Wet area [m^2]
E_g	Semiconductor bandgap [eV]
FF	Fill factor [%]
h	Heat transfer coefficient [Wm^{-2}K^{-1}]
I_{sc}	Short circuit current [A]
J_{sc}	Short circuit current density [A/cm^2]
l	Thickness [m]
$R_{th,cond}$	Conduction thermal resistor [KW^{-1}]
$R_{th,conv}$	Convection thermal resistor [KW^{-1}]
$R_{th,rad}$	Radiation thermal resistor [KW^{-1}]
T	Temperature [K or °C]
T_{cell}	Solar cell temperature [K or °C]
T_{wall}	Temperature on the module rear wall [K or °C]
T_{amb}	Ambient temperature [K or °C]
V_{oc}	Open circuit voltage [V]
α	Short-circuit current temperature coefficient [%/K]

β	Open-circuit voltage temperature coefficient [%/K]
ε	Emissivity
κ	Thermal conductivity [$\mathrm{Wm^{-1}K^{-1}}$]
η	Efficiency [%]
σ	Stefan-Boltzmann constant ($\sigma = 5.67 \cdot 10^{-8}\mathrm{Wm^{-2}K^{-4}}$)

References

1. Fan, H.Y. (1951) Temperature dependence of the energy gap in semiconductors. *Physical Review*, **82**(6) 900–905.
2. Allen, P.B. and Heine, V. (1976) Theory of the temperature dependence of electronic band structures, *Journal of Physics C Solid State Phys.* **9**(12), 2305–2312.
3. Varshni, Y.P. (1967) Temperature dependence of the energy gap in semiconductors. *Physica*, **34**(1), 149–154.
4. O'Donnell, K.P. and Chen, X. (1991) Temperature dependence of semiconductor band gaps. *Applied Physics Letters*, **58**(25), 2924–2926.
5. Lu, S.C., Wu, M.C., Lee, C.Y. and Yang, Y.C. (1991) Temperature dependence of photoluminescence from Mg-doped $\mathrm{In_{0.5}Ga_{0.5}P}$ grown by liquid-phase epitaxy. *Journal of Applied Physics*, **70**(4), 2309–2312.
6. Thurmond, C.D. (1975) The standard thermodynamic functions for the formation of electrons and holes in Ge, Si, GaAs, and GaP. *Journal of the Electrochemical Society*, **122**(8), 1133–1141.
7. Levinshtein, M.M., Rumyantsev, S.L. and Shur, M. (1996) *Handbook Series on Semiconductor Parameters. 2. Ternary and Quaternary III-V Compounds*, World Scientific Publishing Company, London.
8. Logothetidis, S., Cardona, M. and Garriga, M. (1991) Temperature dependence of the dielectric function and the interband critical-point parameters of $\mathrm{Al_xGa_{1-x}As}$. *Physical Review B*, **43**(14), 11950–11965.
9. Levinshtein, M.M.E., Rumyantsev, S.L. and Shur, M. (1996) *Handbook Series on Semiconductor Parameters. 1. Si, Ge, C (diamond), GaAs, GaP, GaSb, InAs, InP, InSb*. World Scientific Publishing Company.
10. Alex, V., Finkbeiner, S. and Weber, J. (1996) Temperature dependence of the indirect energy gap in crystalline silicon. *Journal of Applied Physics*, **79**(9), 6943–6946.
11. Thériault, O., Wheeldon, J., Walker, F.A. *et al.* (2011) Temperature-dependent quantum efficiency of quantum dot enhanced multi-junction solar cells. *AIP Conference Proceedings*, **1407**, 50–53.
12. Blakemore, J.S. (1982) Semiconducting and other major properties of gallium arsenide. *Journal of Applied Physics*, **53**(10), R123–R181.
13. Zielinski, E., Schweizer, H., Streubel, K., Eisele, H. and Weimann, G. (1986) Excitonic transitions and exciton damping processes in InGaAs/InP. *Journal of Applied Physics*, **59**(6), 2196–2204.
14. Shen, H., Pan, S.H., Hang, Z. *et al.* (1988) Photoreflectance of GaAs and $\mathrm{Ga_{0.82}Al_{0.18}As}$ at elevated temperatures up to 600 °C. *Applied Physics Letters*, **53**(12), 1080–1082.
15. Vurgaftman, I., Meyer, J.R. and Ram-Mohan, L.R. (2001) Band parameters for III–V compound semiconductors and their alloys. *Applied Physics Letters*, **89**(11), 5815–5875.
16. Panish, M.B. and Casey, H.C., (2003) Temperature dependence of the energy gap in GaAs and GaP, *Journal of Applied Physics*, **40**(1), 163–167.
17. Aiken, D., Stan, M., Murray, C. *et al.* (2002) Temperature dependent spectral response measurements for III-V multi-junction solar cells, in *Conference Record of the Twenty-Ninth IEEE Photovoltaic Specialists Conference*, 2002, 828–831.
18. Steiner, M.A., Geisz, J.F., Friedman, D.J. *et al.* (2011) Temperature-dependent measurements of an inverted metamorphic multijunction (IMM) solar cell, in *37th IEEE Photovoltaic Specialists Conference (PVSC) 2011*, 002527–002532.
19. Geisz, J.F., Duda, A., France, R.M. *et al.* (2012) Optimization of 3-junction inverted metamorphic solar cells for high-temperature and high-concentration operation. *AIP Conference Proceedings*, **1477**, 44–48.
20. Siefer, G. and Bett, A.W. (2012) Analysis of temperature coefficients for III–V multi-junction concentrator cells. *Progress in Photovoltaic Research and Applications*, **22**, 515–524.
21. Friedman, D.J., Geisz, J.F., Norman, A.G., Wanlass, M.W. and Kurtz, S.R. (2006) 0.7-eV GaInAs Junction for a GaInP/GaAs/GaInAs(1eV)/GaInAs(0.7eV) Four-Junction Solar Cell. *Conference Record of the 4th IEEE World Conference on Photovoltaic Energy Conversion*, pp. 598–602.

22. Nishioka, K., Takamoto, T., Agui, T. *et al.* (2006) Annual output estimation of concentrator photovoltaic systems using high-efficiency InGaP/InGaAs/Ge triple-junction solar cells based on experimental solar cell's characteristics and field-test meteorological data. *Solar Energy Materials Solar Cells*, **90**(1), 57–67.

23. Antón, I., Martínez, M., Rubio, F. *et al.* (2012) Power rating of CPV systems based on spectrally corrected DNI. *AIP Conference Proceedings*, **1477**, 331–335.

24. Kinsey, G.S., Hebert, P., Barbour, K.E. *et al.* (2008) Concentrator multijunction solar cell characteristics under variable intensity and temperature. *Progress in Photovoltaic Research and Applications*, **16**(6), 503–508.

25. Friedman, D.J. (1996) Modelling of tandem cell temperature coefficients, in *Conference Record of the Twenty Fifth IEEE Photovoltaic Specialists Conference, 1996*, 89–92.

26. Steiner, M.A., Geisz, J.F., García, I. *et al.* (2013) Optical enhancement of the open-circuit voltage in high quality GaAs solar cells. *Journal of Applied Physics*, **113**(12), 123109.

27. King, R.R., Bhusari, D., Boca, A. *et al.* (2011) Band gap-voltage offset and energy production in next-generation multijunction solar cells, *Progress in Photovoltaic Research and Applications*, **19**(7), 797–812.

28. Green, M.A. (1982) *Solar Cells: Operating Principles, Technology, and System Applications*, Prentice-Hall.

29. Garcia, I., Algora, C., Rey-Stolle, I. and Galiana, B. (2008) Study of non-uniform light profiles on high concentration III-V solar cells using quasi-3D distributed models, in *33rd IEEE Photovoltaic Specialists Conference, 2008, PVSC '08*, 1–6.

30. Bagienski, W., Kinsey, G.S. Liu, M., Nayak, A. and Garboushian, V. (2012) Open circuit voltage temperature coefficients vs. concentration: Theory, indoor measurements, and outdoor measurements, in *AIP Conference Proceedings*, **1477**, 148–151.

31. Steiner M.A. and Geisz, J.F. (2012) Non-linear luminescent coupling in series-connected multijunction solar cells. *Applied Physics Letters*, **100**(25), 251106–251105.

32. Kinsey G.S. and Edmondson, K.M. (2009) Spectral response and energy output of concentrator multijunction solar cells, *Progress in Photovoltaic Research and Applications*, **17**(5), 279–288.

33. Siefer, G., Abbot, R., Baur, C., Schlegl, T. and Bett, A.W. (2005) Determination of the temperature coefficients of various III-V solar cells, in *Proceedings 20th European Photovoltaic Solar Energy Conference*, Barcelona, Spain, pp. 495–498.

34. EMCORE Corporation [Online]. Available at: http://www.emcore.com/ [Accessed: January 2014].

35. AZUR Space [Online]. Available at: http://www.azurspace.com/ [Accessed: January 2014].

36. Spectrolab [Online]. Available at: http://www.spectrolab.com/concentrator.htm [Accessed: April 2015].

37. Ponce-Alcántara, S., Connolly, J.P., Sánchez, G. *et al.* (2014) A statistical analysis of the temperature coefficients of industrial silicon solar cells. *Energy Procedia*, **55**, 578–588.

38. Yoon, S. and Garboushian, V. (1994) Reduced temperature dependence of high-concentration photovoltaic solar cell open-circuit voltage (V_{oc}) at high concentration levels, in *Conference record of the 24th IEEE Photovoltaic Specialists Conference/IEEE First World Conference on Photovoltaic Energy Conversion*, Waikoloa, Hawaii, USA, **2**, 1500–1504.

39. Lorenzo, E. and Sala, G. (1979) Hybrid silicone-glass Fresnel lens as concentrator for photovoltaic applications, in *Proceedings 2nd European Photovoltaic Solar Energy Conference*, Berlin, 536–539.

40. Luce, T. and Cohen, J. (2010) The path to volume production for CPV optics, in *35th IEEE Photovoltaic Specialists Conference*, 487–492.

41. Schult, T., Neubauer, M., Bassler, Y., Nitz, P. and Gombert, A. (2009) Temperature dependence of Fresnel lenses for concentrating photovoltaics, in *2nd International Workshop on Concentrating Photovoltaic Optics and Power*, Darmstadt, DOI./10.1155/2014/539891.

42. Weber, M. J. (2003) *Handbook of Optical Materials*, CRC Press.

43. Cariou, J.M., Dugas, J., Martin, L. and Michel, P. (1986) Refractive-index variations with temperature of PMMA and polycarbonate, *Applied Optics*, **25**(3), 334–336.

44. Rumyantsev, V.D., Davidyuk, N.Y., Ionova, E.A. *et al.* (2010) Thermal regimes of fresnel lenses and cells in 'all-glass' HCPV modules. *AIP Conference Proceedings*, Freiburg, Germany, **1277**, 89–92.

45. Hornung, T., Bachmaier, A., Nitz, P. *et al.* (2010) Temperature dependent measurement and simulation of Fresnel lenses for concentrating photovoltaics. *AIP Conference Proceedings*, Freiburg, Germany, **1277**, 85–88.

46. Hornung, T., Bachmaier, A., Nitz, P. and Gombert, A. (2010) Temperature and wavelength dependent measurement and simulation of Fresnel lenses for concentrating photovoltaics. *Proceeding of SPIE* **7725**, 77250A.

47. Askins, S., Victoria, M., Herrero, R. *et al.* (2011) Effects of temperature on hybrid lens performance. *AIP Conference Proceedings*, **1407**, 57–60.

48. Antón, I., Martínez, M., Rubio, F., *et al.* (2012) Power rating of CPV systems based on spectrally corrected DNI. *AIP Conference Proceedings*, **1477**(1), 331.

49. Peharz, G., Ferrer Rodríguez, J.P., Siefer, G. and Bett, A.W. (2011) Investigations on the temperature dependence of CPV modules equipped with triple-junction solar cells. *Progress in Photovoltaic Research and Applications*, **19**(1), 54–60.

50. Domínguez, C. and Besson, P. (2014) On the sensitivity of 4 different CPV module technologies to relevant ambient and operation conditions. *AIP Conference Proceedings*, **1616**, 308–312.

51. Technical Data Sheet – Soitec CX-S530-II CPV System 29.4 kWp. 2015-04-21. URL:http://www.soitec.com/pdf/CX-S530-II_Technical_Data_Sheet.pdf. Accessed: 2015-04-21. (Archived by WebCite® at http://www.webcitation.org/6Xx7E7m8B).

52. Muller, M., Deline, C., Marion, B., Kurtz, S. and Bosco, N. (2011) Determining outdoor CPV cell temperature. *AIP Conference Proceedings – 7th International Conference on Concentrating Photovoltaic Systems*, **1407**, pp. 331–335.

53. Antón, I., Sala, G. and Pachón, D. (2001) Correction of the V_{oc} vs. temperature dependence under non-uniform concentrated illumination, in *17th European Photovoltaic Solar Energy Conference and Exhibition*, Munich, 156–159.

54. Andreev, V., Grilikhes, V., Rumyantsev, V., Timoshina, N. and Shvarts, M. (2003) Effect of nonuniform light intensity distribution on temperature coefficients of concentrator solar cells, in *Proceedings of 3rd World Conference on Photovoltaic Energy Conversion*, **1**, 881–884.

55. Jaus, J., Hue, R., Wiesenfarth, M., Peharz, G. and Bett, A.W. (2008) Thermal management in a passively cooled concentrator photovoltaic module, in *Proceedings of the 23rd EUPVSEC, Valencia*, 832–836.

56. Steiner, M., Siefer, G. and Bett, A.W. (2012) An investigation of solar cell interconnection schemes within CPV modules using a validated temperature-dependent SPICE network model. *Progress in Photovoltaic Research and Applications*, DOI 101002pip2284.

57. Van Riesen, S., Gombert, A., Gerster, E. *et al.* (2011) Concentrix Solar's progress in developing highly efficient modules. *AIP Conference Proceedings*, **1407**(1), 235–238.

58. Askins, S., Victoria, M., Herrero, R. *et al.* (2012) Optimization of tolerant optical systems for silicone on glass concentrators, in *8th International Conference on Concentrating Photovoltaic Systems*, Toledo, Spain.

59. Hornung, T., Steiner, M. and Nitz, P. (2012) Estimation of the influence of Fresnel lens temperature on energy generation of a concentrator photovoltaic system, *Solar Energy Materials Solar Cells*, **99**, 333–338.

60. Jaus, J., Hue, R., Wiesenfarth, M. Peharz, G. and Bett, A.W. (2008) Thermal management in a passively cooled concentrator photovoltaic module, in *Proceedings of the 23rd European Photovoltaic Solar Energy Conference and Exhibition*, Valencia, 2008, pp. 832–836.

61. Martínez, M., Antón, I. and Sala, G. (2007) Prediction of PV concentrators energy production: influence of wind in the cooling mechanisms, first steps, in *Proceedings of the ICSC-4*, El Escorial, Spain.

62. Castro, M., Domínguez, C., Nuñez, R. *et al.* (2013) Detailed effects of wind on the field performance of a 50KW CPV demonstration plant, in *AIP Conference Proceedings - 9th International Conference on Concentrating Photovoltaic Systems*, **1556**, pp. 256–260, Japan, 2013.

63. Sueto, T., Yano, H., Shibata, N. *et al.* (2013) Cooling effect of wind direction on CPV module, in *Proceedings CPV-9*, Japan, 2013.

64. Çengel, Y.A. (2003) *Heat Transfer: A Practical Approach*, McGraw-Hill.

65. Sala, G. (1989) *Solar Cells and Optics for Photovoltaic Concentration*, A. Hilger.

66. Nishioka, K., Ota, Y., Tamura, K. and Araki, K. (2013) Heat reduction of concentrator photovoltaic module using high radiation coating. *Surface Coatings Technology*, **215**, 472–475.

67. Wiesenfarth, M., Gamisch, S., Kraus, H. and Bett, A.W. (2013) Investigation on 3-dimensional temperature distribution in a FLATCON-type CPV module. *AIP Conference Proceedings*, **1556**, 189–194.

68. IEC 62108 Concentrator Photovoltaic (CPV) Modules and Assemblies – Design Qualification and Type Approval.

69. Martinelli G. and Stefancich, M. (2007) Solar cell cooling, in *Concentrator Photovoltaics*, (eds A. Luque and V. M. Andreev), Springer Series in Optical Science, pp. 133–149.

70. Lide, D.R. (2004) *Handbook of Chemistry and Physics*, CRC Press, Boca Raton, Florida.

71. Bergquist Thermal Substrates 2013-05-21. URL:http://www.bergquistcompany.com/thermal_substrates/dielectrics/summary.htm. (Archived by WebCite® at http://www.webcitation.org/6GmZ56boM).

72. Thermal adhesive Dow Corning SE 4486. 2013-05-21. URL:http://www.dowcorning.com(Archived by WebCite® at http://www.webcitation.org/6GmmxYRzR).

73. Press Release 19/12/2012: Soitec opens its solar manufacturing facility in San Diego. URL:http://www.soitec. com/en/news/press-releases/soitec-opens-solar-manufacturing-facility-san-diego-locally-produce-cpv-modules-for-us-renewable-energy-market-1143/ (Archived by WebCite® at http://www.webcitation.org/6LvjWv5t2).

74. Victoria, M., Chiappori, E., Askins, S. *et al.* (2010) Finite elements model for thermal analysis of FluidReflex CPV system, in *Proceedings of the 25th EUPVSECand and 5th WPEC*, Valencia, 2010, pp. 902–905.

75. Royne, A., Dey, C.J. and Mills, D.R. (2005) Cooling of photovoltaic cells under concentrated illumination: a critical review. *Solar Energy Materials Solar Cells*, **86**(4), 451–483.

76. Rumyantsev, V.D., Andreev, V.M., Bett, A.W. *et al.* (2000) Progress in development of all-glass terrestrial concentrator modules based on composite Fresnel lenses and III-V solar cells, in *Conference Record of the 28th IEEE Photovoltaic Specialists Conference*, 2000, 1169–1172.

77. Araki, K., Uozumi, H. and Yamaguchi, M. (2002) A simple passive cooling structure and its heat analysis for 500 times concentrator PV module, in *Conference Record of the 29th IEEE Photovoltaic Specialists Conference*, 2002, 1568–1571.

78. Araki, K., Kondo, M., Uozumi, H. and Yamaguchi, M. (2003) Development of a robust and high efficiency concentrator receiver, in *Proceedings of 3rd World Conference on Photovoltaic Energy Conversion*, 2003, **1**, 630–633.

79. Ghosal, K., Lilly, D., Gabriel, J. *et al.* (2012) Performance of a micro-cell based transfer printed HCPV system in the South Eastern US, *AIP Conference Proceedings*, **1477**(1), 327–330.

80. De Nardis, D. (2012) A single reflection approach to HCPV: Very high concentration ratio and wide acceptance angles using low cost materials, *AIP Conference Proceedings*, **1477**(1), 94–97.

81. Plesniak, A., Jones, R., Schwartz, J. *et al.* (2009) High performance concentrating photovoltaic module designs for utility scale power generation, in *34th IEEE Photovoltaic Specialists Conference*, 2009, 2231–2236.

82. Anderson, W.G., Dussinger, P.M., Sarraf, D.B. and Tamanna, S. (2008) Heat pipe cooling of concentrating photovoltaic cells, in *33rd IEEE Photovoltaic Specialists Conference*, 2008, 1–6.

83. Beach, R. T. and White, R. M. (1981) Heat pipe for passive cooling of concentrator solar cells, in *15th Photovoltaic Specialists Conference*, 1981, **1**, 75–80.

84. Victoria, M., Domínguez, C., Askins, S., Antón, I. and Sala, G. (2013) Experimental analysis of a photovoltaic concentrator based on a single reflective stage immersed in an optical fluid. *Progress in Photovoltaic Research and Applications*, DOI 101002pip2381.

85. Victoria, M., Askins, S., Domínguez, C., Antón, I. and Sala, G. (2012) Outdoor performance of fluid dielectric CPV modules. *AIP Conference Proceedings*, **1477**(1), 208–212.

86. Vivar M. and Everett, V. (2012) A review of optical and thermal transfer fluids used for optical adaptation or beam-splitting in concentrating solar systems. *Progress in Photovoltaic Research and Applications*, DOI 101002pip2307.

87. Victoria, M., Askins, S., Domínguez, C., Antón, I. and Sala, G. (2013) Durability of dielectric fluids for concentrating photovoltaic systems. *Solar Energy Materials Solar Cells*, **113**, 31–36.

88. Luque, A., Sala, G., Arboiro, J.C. *et al.* (1997) Some results of the EUCLIDES photovoltaic concentrator prototype. *Progress in Photovoltaics Research and Applications*, **5**(3), 195–212.

89. Finot, M., MacDonald, B. and Lance, T. (2012) Validation of energy harvest modeling for X14 system. *AIP Conference Proceedings*, **1477**(1), 336–339.

90. Klotz, F.H., Mohring, H.D., Cruel, C. *et al.* (2000) Photovoltaic V-through concentrator system with gravitational tracking (ARCHIMEDES), in *16th European Photovoltaic Solar Energy Conference and Exhibition*, 2000, 2229–2232.

91. Akbarzadeh A. and Wadowski, T. (1996) Heat pipe-based cooling systems for photovoltaic cells under concentrated solar radiation. *Applied Thermal Engineering* **16**(1), 81–87.

92. Edenburn, M.W. (1980) Active and Passive Cooling for Concentrating Photovoltaic Arrays, in *14th IEEE Photovoltaic Specialist Conference*, San Diego, USA, 771–776.

93. Oleary M.J. and Clements, L.D. (1980) Thermal-electric performance analysis for actively cooled, concentrating photovoltaic systems. *Solar Energy*, **25**, 401–406.

94. Chenlo F. and Cid, M. (1987) A linear concentrator photovoltaic module: analysis of non-uniform illumination and temperature effects on efficiency. *Solar Cells*, **20**(1), 27–39.

95. Smeltink J.F.H. and Blakers, A.W. (2006) 40 kW PV Thermal Roof Mounted Concentrator System, in *Conference Record of the 4th World Conference on Photovoltaic Energy Conversion*, 2006, **1**, 636–639.

96. Tilford, C.L., Sinton, R.A., Swanson, R.M., Crane, R.A. and Verlinden, P. (1993) Development of a 10 kW reflective dish PV system, in *Conference Record of the 23rd IEEE Photovoltaic Specialists Conference*, 1993, 1222–1227.
97. Verlinden, P., Sinton, R.A., Swanson, R.M. and Crane, R.A. (1991) Single-wafer integrated 140 W silicon concentrator module, in *Conference Record of the 22nd IEEE Photovoltaic Specialists Conference*, 739–743.
98. Verlinden, P., Terao, A., Smith, D.D. *et al.* (2002) Will we have a 20% efficient (PCT) photovoltaic system?, in *17th European Photovoltaic Solar Energy Conference and Exhibition*, 2002.
99. Vicenzi, D., Stefancich, M., Bizzi, F. *et al.* (2003) Micromachined silicon heat exchanger for water cooling of concentrator solar cells, in *Conference Record ISES*, Gothenburg, Sweden, 2003.
100. Miñano, J.C., Gonzalez, J.C. and Zanesco, I. (1994) Flat high concentration devices, in *24th IEEE Photovoltaic Specialists Conference*, 1994, **1**, 1123–1126.

6

CPV Tracking and Trackers

Ignacio Luque-Heredia,[1] Pedro Magalhães,[2] and Matthew Muller[3]

[1]*BSQ Solar, Spain*
[2]*Versol Solar, United States*
[3]*National Renewable Energy Laboratory, United States*

6.1 Introduction

Most CPV systems use only direct solar radiation, and they must therefore permanently track the Sun's apparent daytime motion, and hence incorporate an automatic sun-tracking structure able to mount and position the concentrator optics in such a way that direct sunlight is always focused on the cells. This sun tracker is basically composed of a structure presenting a sunlight collecting surface on which to attach concentrator modules or systems. Such a collecting surface – sometimes also referred to as *aperture surface* or *tracker table* – is in some way coupled to a one- or two-axis mechanical drive, and also to a sun tracking control system which operates over the drive axes, and maintains an optimum aim of the collecting surface towards the sun.

Static mounts are only feasible today for low concentration factors below 5×; in the long term static concentrators with higher ratios making use of luminescence and photonic crystals might appear; however all these topics are beyond the scope of this chapter.

Line-focus reflective concentrators, such as troughs, only require one-axis tracking to maintain the PV receiver along the focus line. However, due to the daily variations in the sun's elevation, sunlight incidence on the tracker's aperture is usually somewhat oblique, thus reducing the intercepted energy and causing the sun's image to move up and down within the focal axis thus producing further losses whenever it surpasses the receiver's ends. Line-focus refractive concentrators, for example those based on linear Fresnel lenses, experience severe optical aberrations when light incidence is not normal, thus requiring two-axis sun tracking. The same happens to most of the point–focus concentration concepts that have been developed, except for some low–concentration factor devices with sufficient acceptance angle to admit the variations in the sun's altitude.

Almost all PV concentrators that are already commercialized or currently under development use two-axis tracking and a so-called pedestal tracker, usually with azimuth–elevation

Handbook of Concentrator Photovoltaic Technology, First Edition. Edited by Carlos Algora and Ignacio Rey-Stolle.
© 2016 John Wiley & Sons, Ltd. Published 2016 by John Wiley & Sons, Ltd.

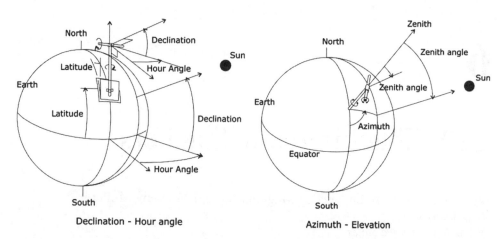

Figure 6.1   The two most common sun tracking axes geometries used in solar trackers, declination- hour angle or also called tilt-roll trackers (left) and azimuth-elevation, usually implemented with pedestal trackers (right)

axes. Tilt-roll trackers operating on the declination–hour angle axes (Figure 6.1) are also used, though are less common.

With regard to sun tracking control, most of the early systems consisted of analog sun pointing sensors based on the shadowing or illumination differences of a couple or quad of PV cells, integrated in an automatic closed loop with the tracker's driving motors. The advent of cheap microcontrollers motivated the appearance of sun tracking control systems requiring no sun sensing, and based only on the digital computation of precise analytic sun ephemeris equations. To date the need for an efficient and reliable sun tracking control in CPV applications has driven the state of the art towards a blend of these two initial approaches, producing hybrid strategies that integrate both sun alignment error feedback and ephemeris based positioning.

This chapter begins with an explanation in Section 6.2 of the functional requirements of a CPV sun tracker and how design specifications of a CPV tracker are derived from these. Section 6.3 presents a taxonomy of trackers describing the most common tracking architectures that can be found on the market, based on the number of axes, their relative position, and the foundation and placing of tracking drives. Section 6.4 deals with the structural issues related to tracker design, mainly related to structural flexure and its impact on the system's acceptance angle. In Section 6.5 sun tracking control is analyzed, firstly by describing the state of the art and its development background, to follow with a more detailed description of the most effective tracking control strategy developed to date, namely the autocalibrated sun tracking control. Then, section 6.6 will further complete this analysis from a quality assurance perspective, by explaining about sun tracking accuracy measurement and providing a practical example. Section 6.7 discusses tracker manufacturing and tracker field works, and section 6.8 presents a survey of different types of tracker designs obtained from different manufacturers. Finally, Section 6.9 deals with IEC62817, the technical standard that is presently being developed for CPV sun trackers.

6.2   Requirements and Specifications

Strictly speaking, the main commitment to be fulfilled by a CPV sun tracker is to permanently align the pointing axis of the supported concentrator module array with the local sun vector, in this way producing maximum power output. As we will see throughout this chapter there are several error sources to take into account and therefore, some off-tracking tolerance is required. Usually this tolerance, or minimum tracking accuracy required, is characterized by means of the acceptance angle of the concentration system, usually defined as the off-tracking angle at which power output drops below 90% (see Chapters 4 and 7).

As presented in Figure 6.2, the reasons for the decrease of sun tracking performance can be classified into two main types: (i) those purely related to the precise pointing of the tracker to the sun and (ii) those causing shrinkage of the overall acceptance angle of the concentrator system thus indirectly increasing the tracking accuracy required. Among those related to the tracking accuracy, these are basically on the one hand, the exactness of the sun's positional coordinates generated by the control system, expressed in terms of rotation angles of the tracking axes, either by sun ephemeris based computations or derived from the feedback of sun–pointing sensor readings, or a combination of both, and which in any case is affected by numerous error sources. The other factor determining tracking accuracy is the precision with which the tracker can be positioned at these dictated orientations, i.e. the positioning resolution of the tracking drive and its control system; this essentially depends on the performance of tracking speed control and on the mechanical backlash introduced by the drive's gearings.

Regarding acceptance angle losses caused by the tracking system, these are due to the accuracy which can be attained in the mounting and alignment of the concentrator modules atop of the tracker. Such accuracy is basically, in the first instance, a design problem having to do with the fixtures provided for this purpose, their accurate assembly and the regulation means provided for in-field fine tuning, but also with the mounting protocols devised to carry out these tasks. Also resulting in acceptance angle cuts is the stiffness designed into the tracker, which is to say the deformations allowed in the different elements of its structure under service conditions.

In a first iteration, characterization of the service conditions for a CPV tracker basically consist in determining the CPV array payload (modules' weight) and fixing the maximum wind load, i.e. wind speed, to be withstood during sun tracking operation with no effect on the concentrator's power output (MSWS: Maximum Service Wind Speed). To ensure service

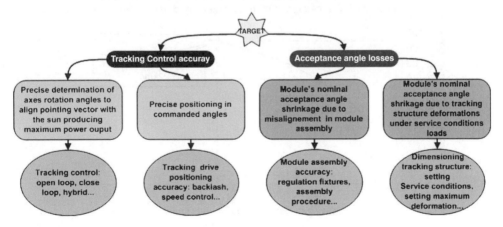

Figure 6.2 Factors conditioning sun tracking performance related either to the tracking control accuracy or the acceptance angle losses

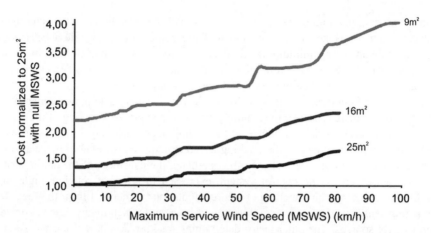

Figure 6.3 Tracker cost *vs.* maximum service wind speed for a specific tracker design dimensioned for three different aperture surfaces and 1000 units/yr. productions

under these maximum permanent and variable loads, a maximum flexure deformation should be specified for the tracker structure.

As for the variable loads, the higher the maximum wind speed to be resisted maintaining productive operation, the heavier and more expensive the tracking structure required to maintain deformation below the threshold required for accurate tracking. This behavior can be seen in Figure 6.3 for a $9\,m^2$ CPV pedestal tracker, in which normalized cost vs. maximum service wind speed for a maximum $0.1°$ flexure is plotted [1].

A cost–effective approach is to determine this maximum wind load from the cross–correlation between wind speed and direct radiation, in the location or set of locations in which the trackers are planned to be marketed and installed. A case example of this type of analysis is presented in Figure 6.4, worked out with one year of continuous wind speed and

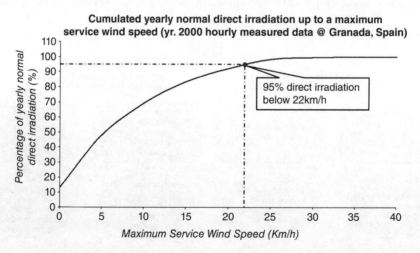

Figure 6.4 Wind speed *vs.* yearly DNI correlation for the determination of optimum service conditions applied to Granada

direct normal irradiation hourly data (assuming a two–axis tracker), for the Spanish city of Granada. Above this threshold, stiffness specifications do not have to be met and, consequently, service is not guaranteed and the tracker should switch to some low wind profile stow position to decrease stress and increase operative lifetime. As a general rule of thumb, a 47 km/ h maximum wind load both windward or leeward to the modules in any of the sun tracking orientations of the tracker can be considered a reasonably conservative value. It has been proven that this value comprises a minimum of 95% of the direct radiation measured by the 26 weather stations of the SOLMET network distributed over the contiguous United States [2].

We therefore infer how structural dimensioning of a tracking design can be cost optimized for a given location, an option which may be worthwhile when building large scale CPV plants provided there are some tuning possibilities in the tracking structure design and manufacturing, and also if it can be done without compromising the cost effectiveness of the supply chain.

Regarding the CPV module array payload, and again having in mind the targets to be placed on a possible CPV module design team in order to decrease tracker cost, it would be a worthwhile effort to decrease the module's weight as much as possible. In this respect the weights of present module technologies range from 40–140 gW^{-1}. But not only this, module sizing also proves to be an important issue, and an optimum is to be sought in which the size of modules does not require redundant framing from the tracker just to hold them, but on the other hand in which module stiffness doesn't develop into an excess of self weight.

The main structural design variable related to the said service conditions, namely, the maximum allowed structural bending measured in the aperture surface, intends to place a bound on the losses caused by the tracker's flexure on the acceptance angle of its CPV modules. Determination of the value for this maximum possible bending depends on the acceptance angle of the particular CPV module technology object of the design, bearing in mind that the finally obtained tracking accuracy is to be comprised within the overall CPV array acceptance angle. The higher the tracking accuracy and the wider the nominal acceptance angle of the modules, the bigger the deformation tolerance which can be set for the structure, and consequently the lower its cost.

Apart from sheer tracking performance, downtime (i.e. the availability ratio) is the other main concern with a CPV tracker (see the section 'Reliability of Plants' in Chapter 9). Here the mechanical part of the tracker is usually free of suspicion, provided the pertinent structural codes are respected in its design. Moreover, it should be taken into account that when recurring to off-the-shelf drive gearings, these are subject to very mild operation conditions in CPV trackers – one axis turn per day – as compared to their usual market applications in machine tools, cranes, etc. Instead, most of the reported problems arise in the electrical and electronic parts which, first of all, are to be designed to operate reliably in outdoor conditions, but also comply with a set of electromagnetic compatibility (EMC) and electrical safety standards suitably chosen, thus anticipating common field problems such as power spikes or surges. When considerable amounts of software are involved, as happens with today's tracking control systems integrating microprocessors, it is not just a matter of a reliable and well protected electronic design but also of a redundant code, immune to hangs and able to gracefully recover from power outages or sags.

6.3 Basic Taxonomy of CPV Trackers

As outlined above, tracking structures will either have one or two axes. When having two axes, the only necessary condition to allow them to position the collection surface towards any

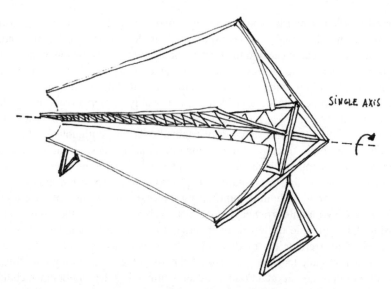

Figure 6.5 Single axis, parabolic trough tracker typically used for low and medium concentrations (<100×). Source: Joana Magalhães, VERSOLSOLAR, 2015. Reproduced with permission of Joana Magalhães

possible orientation, is that the axes are at right angles. In this chapter we refer to the axis connected to the foundation as the primary axis, regardless of its orientation with respect to the Earth's surface. In the case of two axis trackers, we define the secondary axis as the one that is connected to the primary axis. There are sun tracking assemblies in which one can find more than two rotating axes, but this is usually done through a single primary axis and several parallel secondary axes, an architecture that we here name the multi-secondary axis.

There are many tracking axes assemblies; however, we will here present those that have more commonly been proposed for CPV tracking structures. They are here only presented conceptually and in section 6.8 some real implementation cases are shown.

Regarding one axis trackers, the N-S oriented horizontal axis has been the most used in CPV applications, with reflective optics such as in the parabolic trough case depicted in Figure 6.5. Here the single primary axis is parallel to the ground following the North-South direction. There are also other one axis concepts being manufactured nowadays such as the polar axis tracker, with its axis parallel to the Earth's rotation axis (thus tilted to the local latitude), and also E-W oriented horizontal axis trackers, but these have only been used for conventional flat plate PV modules and not for CPV.

One of the most common two-axis approaches for CPV trackers is the pedestal or single-pole azimuth-elevation (Az.-El.) configuration, shown in Figure 6.6. In these trackers the primary axis is vertical to the ground and therefore coincident with an azimuth axis. The secondary axis is orthogonal to the primary axis and will then provide elevation angles in the azimuth orientation of the collecting surface.

These trackers are normally mounted onto a single pole, which, most times, causes the tracker to have a single anchoring point to the ground. This results in a concentrated foundation right around the anchoring point. In addition, the frame of the collecting surface is usually supported by another tube, sometimes called the torque tube, along the line of the secondary or elevation axis. The length of both tubes, the pole mount and the torque tube, is not very

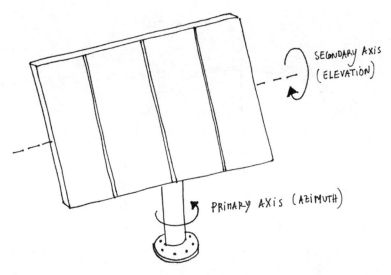

Figure 6.6 Single Pole Az.-El. tracker. Source: Joana Magalhães, VERSOLSOLAR, 2015. Reproduced with permission of Joana Magalhães

different, the secondary axis (El.) typically being the longest. A unique characteristic of these trackers is tracker-to-tracker independency. This has pros and cons that are further discussed in section 6.7

The tilt-roll tracker, shown in Figure 6.7, has its primary axis horizontal, parallel to the ground. The secondary axis is orthogonal to the primary axis. The primary axis can be E-W oriented in which case its rotation angles correspond to the Earth's declination angle and those of the secondary axis at right angles correspond to the hour-angle. In their mechanical realization, in contrast to the pedestal trackers, the tube along the direction of the primary axis is much longer

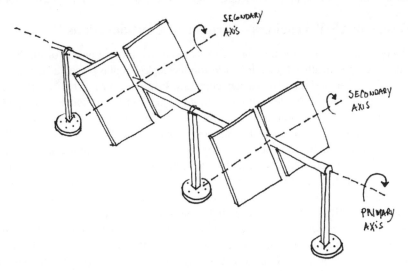

Figure 6.7 Tilt and roll tracker. Source: Joana Magalhães, VERSOLSOLAR, 2015. Reproduced with permission of Joana Magalhães

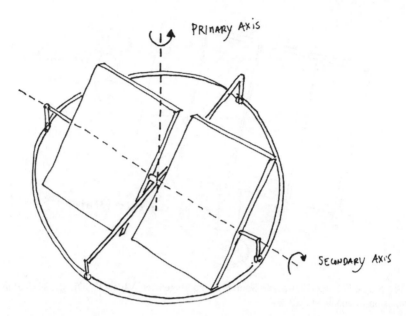

Figure 6.8 Carousel tracker. Source: Joana Magalhães, VERSOLSOLAR, 2015. Reproduced with permission of Joana Magalhães

than that of the secondary axis (20–30 times). In fact, usually the primary axis holds several secondary axes at the same time, being an example of multisecondary axes configuration.

The carousel tracker (Figure 6.8), as its name indicates, is a round structure that spins in azimuth around the primary axis that sits in the center of its circle and then has several elevation axes connected to the outer circumference, parallel to each other. It is therefore also an Az.-El. configuration with the primary axis normal to the ground. However, unlike the pedestal tracker there is a multiple point ground support, usually ending in wheels that run around a circular rail. This approach therefore, requires an extended foundation providing a leveled surface for the rail.

6.4 Design of CPV Trackers – Structural Considerations

In this section, an overview of how the stiffness issue was tackled in one of the first tracker designs that was specifically targeted for CPV volume deployment is presented [3]. It was a two axis pedestal tracker designed by tracker company Inspira, with a 30 m² aperture surface, customized for the CPV modules with 1,000× concentration factor being developed by the Spanish PV manufacturer Isofoton (Figure 6.9). The modules had a nominal acceptance angle of ±0.6°, determined through indoor lab measurements using collimated laser light. Subtracting the approximately 0.26° subtended half angle of the sun, this meant that a 0.34° minimum tracking accuracy was required. As we will see in section 6.6, a feasible value for the minimum tracking accuracy is 0.1° (i.e. 95% probability that the off-track angle remains below 0.1°) so acceptance angle overall loss on the array must not surpass 0.24°. Introducing some overestimation, to allow for extra acceptance angle losses due to some degree of error in module leveling atop of the tracker's aperture surface, a maximum 0.1° bending was the starting point set for the tracking structure design. This means that this is the maximum allowed turn induced by structural flexure for any vector normal to the aperture surface (i.e. the tracker table or

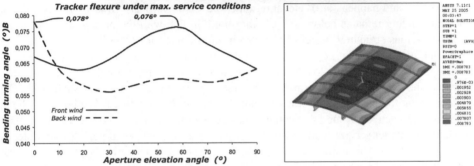

Figure 6.9 CPV pedestal tracker designed and produced by Inspira was subject to flexure analysis (top). Maximum bending turning angle in aperture surface when subject to maximum service conditions (43 km/ h wind speed) as a function of aperture elevation function (bottom left). Sample finite elements analysis of flexure bending (bottom right)

collecting surface) when the tracker is subject to maximum service conditions (maximum operating wind speed set in this case to 43 km/h, blowing from either the front or the back of the tracker) at any aperture elevation angle.

First in the design of the metal structure forming the tracker's aperture is choosing what we can loosely call its *topology*. Here only the lengths of metal beams and the connections among them and with the drive block are decided, seeking the optimization of different aspects such as transportation, field installation, mounting of CPV modules, etc. Once the tracker frame is settled, this is to be sized playing with the precise form of the structural beams. For example, using I-beams, angles and channels, tubes, etc. if directly opting for off-the-shelf construction standards, or using instead others requiring more processing such as trussed parts, but in all cases taking into consideration their stiffness to weight ratios and their manufacturing costs.

It is when getting to this point that the stiffness constraints start to rule over the design, and precise finite elements (FE) analyses are to be carried out over the complete tracker structure, when subject to the specified maximum service loads (CPV modules payload and maximum operative wind loads). When this was done for the referred tracker of Figure 6.9, a solution based on standard structural beams was obtained, which resulted in the least tracker's self weight, and according to FE simulations, did not surpass the 0.1° bending at any aperture elevation. In the case of this pedestal tracker, the design was separately considered in three segments: (i) aperture frame; (ii) pedestal and drive block and (iii) foundation. Firstly, a certain percentage of that total 0.1° maximum flexure was allocated to each segment, taking into account that, while bending in the aperture will usually result in overall acceptance angle shrinkage, bending in the pedestal or the foundation works as an overall pointing vector turn, which as will be seen in the sun tracking control section 6.5, can eventually be characterized and handled by a tracking controller. In the case of the tracker's foundation, meeting its flexure quota requires a standard geotechnical analysis of the ground where it will be installed, in order to choose the best suitable solution. Quite obviously, in a pedestal tracker with a rectangular aperture surface, maximum bending at whichever elevation will occur at its corners. Final results for this design can be seen in Figure 6.9, where maximum bending when maximum service wind load comes frontally is 0.076° and occurs at 57° aperture elevation. When this same wind speed is received in the aperture's rear face, this maximum bending is slightly bigger, 0.078° and happens at 0° elevation (aperture frame vertical). In any case, maximum structural bending remains below the 0.1° threshold. In line with the present analysis, other flexure conscious structural designs have also been presented for tilt-roll trackers [4].

Once the structure sizing is optimized below this maximum bending threshold, next step is to estimate the acceptance angle losses induced by structural flexure, using the bending rotation values of the set of vectors normal to the aperture, obtained in the FE simulation. An approximation to this problem was attempted through a geometrical method in which, assuming that each CPV module mounted on the aperture could be considered to remain non-deformed under service loads, a single normal vector is considered per CPV module, as seen in Figure 6.10. This normal vector is taken as the pointing vector of the module, i.e. that vector that when aligned with the local sun vector produces the module's maximum power output. The acceptance angle for each module is characterized by the cone drawn by the vectors at this acceptance angle from the pointing vector, which is then the cone's axis. For the sake of simplicity, power is assumed to drop down to zero outside the acceptance angle cone, and also a worst case scenario is assumed in terms of acceptance angle losses, in which all modules are supposed to be connected in series. Thus, the set of tracker orientations producing nominal power output for a certain aperture elevation angle is taken, as the set of vectors pertaining to the acceptance angle cones of all the modules, i.e. their intersection. The acceptance angle at this aperture elevation can then be defined as the maximum cone contained in this intersection of cones, and the axis of this cone is taken as the overall concentrator pointing vector.

The problem of determining this overall acceptance angle cone can be better viewed and solved, if the pointing vectors and their respective acceptance angle cones are projected in the plane. This more precisely means the projection of the intersection of pointing vectors and cones with a unit radius sphere whose center coincides with the origin of all the pointing vectors. In this way, as can be seen in Figure 6.11, every module pointing vector is transformed into a point in the plane, having as Cartesian coordinates its direction cosines with respect to the plane reference axes; and acceptance cones are transformed into ellipses. The flexure turning angle of a certain pointing vector will be small and therefore its projected coordinates appear very close to the reference system origin, which represents the pointing vector of the concentrator system if the tracker was ideally rigid and non-deformable, and the distance of each pointing vector to the origin is its bending rotation angle. For these pointing vector points located close to the origin, their corresponding acceptance angle ellipses can be approximated by a circle, centered in the

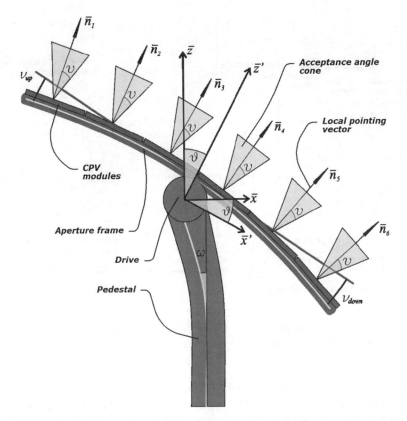

Figure 6.10 Cross section of pedestal tracker subject to flexure at a certain aperture elevation (zenith angle θ). The local pointing vector to each module $\bar{n}_i$ and its acceptance angle (υ) cone within the aperture's local reference system ($\bar{x}'$, $\bar{y}'$, $\bar{z}'$). In the figure ω represents the pedestal bending angle and ν_{down} and ν_{up} the bending angles of the upper and lower ends of the aperture frame

pointing vector coordinates. On the other hand, high concentration CPV modules usually have small acceptance angles (in the sub-degree range) and, in this case, the radius of the projected circle representing the acceptance angle cone equals the acceptance angle itself.

Therefore, after this projection, we can reformulate the problem of obtaining the maximum cone contained in the intersection of module acceptance angle cones, as that of determining the maximum incircle to the intersection of acceptance angle circles in the plane. The center of this incircle —the *incenter*— will then represent the projection of the overall concentrator pointing vector. It can be proven that finding this maximum incircle is equivalent to determining the minimum enclosing circle (MEC) containing all the pointing vector points, where the center of this MEC coincides with the incenter of the maximum incircle. Also the radius of the maximum incircle, i.e. the overall acceptance angle, equals the acceptance angle of a single module minus the radius of the obtained MEC, which in this way represents the acceptance angle loss due to flexure. MEC determination for a set of points in the plane is a classical computational geometry problem first stated by Sylvester in 1857, for whose solution the most efficient algorithm to date due to Welzl, – able to achieve $0(n)$ linear running time –, was implemented [5].

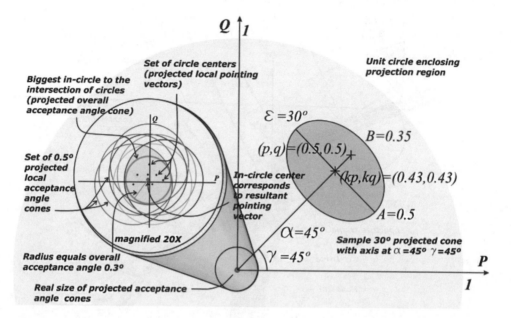

Figure 6.11 Schematic of the plane projection of the local pointing vectors of the modules and associated acceptance angle cones, for the determination of the worst case for complete array pointing vector and acceptance angle

Applying the procedure stated above to the FE simulations obtained from the pedestal tracker of our case example, produced the plot of acceptance angle loss as a function of aperture elevation for both front and back maximum service wind speeds shown in Figure 6.12. Also in this figure, the MECs for aperture elevation angles taken every 10° from 0° to 90° are shown along with the centers of each MEC, which shows how the overall pointing vector also moves due to flexure. In

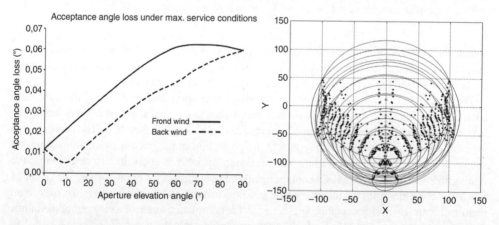

Figure 6.12 Estimation of a worst case acceptance angle loss in the 30 m² pedestal tracker of Figure 6.9 as a function of aperture elevation and subject to maximum service conditions (left). Pointing vectors and MECs at different elevations with maximum service wind speed windward and leeward to module's active surface (right)

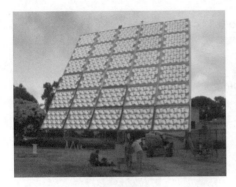 

Figure 6.13 Examples of trackers designed and manufactured by Inspira for different CPV module technologies subject to flexure constraints. Tracker with 48 m^2 aperture surface and 0.3° maximum flexure for Solfocus CPV (left). Tracker with 36 m^2 aperture surface and 0.2° maximum flexure for Concentrix CPV (right)

this case, the local pointing vectors used at each elevation are only those corresponding to the modules mounted along the aperture perimeter, which are those suffering the largest bending. As can be seen from the acceptance angle loss graph, the maximum acceptance angle loss is 0.063° and it occurs with maximum service front wind at 70° of aperture elevation. So having started with 0.1° maximum bending threshold under maximum service conditions, and then obtaining an optimum design with 0.076° maximum flexure, the calculations finally result in a maximum acceptance angle loss of 0.063°. Getting back at this point to our module's nominal acceptance angle, and the expected tracking accuracy, suggests we could try to further relax the bending threshold and go for a second optimization iteration to further lighten the structure and reduce its cost. It is to be pointed out that the aperture elevation angles, producing, on the one hand, maximum bending of local pointing vectors and, on the other hand, maximum acceptance angle loss, do not necessarily coincide because, as said, the turning angle computed is also affected by the pedestal and other global components which equally affect all aperture pointing vectors, and therefore do not contribute to acceptance angle losses.

Several other trackers were designed and manufactured by Inspira following this flexure constrained approach, being representative samples the ones built for Concentrix (today Soitec) and Solfocus, with maximum flexure 0.2° and 0.3°, respectively (Figure 6.13).

The cost reduction achievable in the tracking structure through the increase of the flexure threshold and hence, weight reduction, has been plotted for a particular CPV tracker design. This was resized for each different threshold [3] and its overall weight decreased following a potential law (Figure 6.14). It must be said that this behavior is to some extent design specific, and other design approaches to the tracker structure, that for example do not rely exclusively on standard construction parts or seek a reduction in projecting lengths, may behave differently. However, this analysis does provide a good example of how the reduction of the flexure constraint reaches a limit, that we can call the *flexure floor*, beyond which it is ultimate structural strength under the maximum loads specified by standard building codes, the more stringent condition determining the sizing of the tracking structure. In the present example, this flexure floor occurs between 0.3° and 0.4°. Obviously, this flexure floor will also depend on the maximum service wind speed and the set of maximum loads, due to snow or wind, specified by the construction codes ruling in the region for which the tracker design is intended.

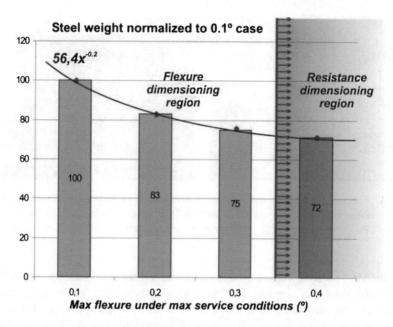

Figure 6.14   Steel weight for a CPV tracker design when varying its maximum flexure constraint normalized to the 0.1° most restrictive case

This design and sizing process for the structural part of the tracker is presented here in the sequential and iterative way in which it happened in the early designs by Inspira, which on the other hand, appears as the most intelligible when compared to present design methods which have evolved afterwards, more automated and parallel. For example, present design method at BSQ Solar, sketched in Figure 6.15, have further gained accuracy by introducing additional

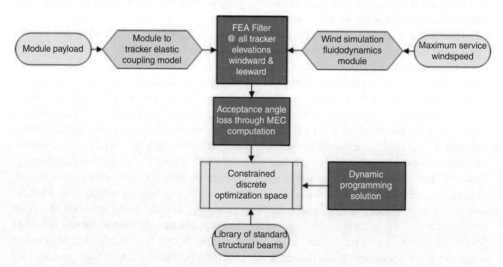

Figure 6.15   Tracker structural dimensioning process carried out at BSQ Solar

modeling to simulate the elastic coupling of quasi-rigid modules and the aperture frame as well as fluido-dynamics modeling to simulate wind loads. This method also merges FEA structural bending simulation and acceptance angle loss estimation through the MEC method, in a single process that enables obtaining the optimum tracker topology and sizing, in a constrained discrete optimization space built from a library of standard structural beams, through the use of dynamic programming methods devised for the so called Knapsack-kind problems [6].

However, further refinements of this design methods are still being developed such as the very significant of not considering overall tracker weight as the only merit function to minimize, but also take into account materials, manufacturing, and installation costs, which –even whilst being much more case dependent– will provide, the last fine tuning to the optimization when the data is available. On the other hand, the MEC method that has proven to be computationally simple in most practical cases, requires further linkage with the electric behavior of the CPV array so that the interconnection of modules, and of the cells within them, can also be taken into account and optimized [7].

6.5 Sun Tracking Control

6.5.1 Background

Early sun tracking controllers were developed following the classical control systems closed-loop approach by integrating a sun sensor able to provide pointing error signals, one per tracking axes, which generate motor correction movements [8,9]. This sun sensor is essentially constituted by a pair of photodiodes and some sort of shading device which casts a different shade on these photodiodes, therefore generating different photocurrents, whenever it is not aligned with the local sun vector (Figure 6.16.a). Added to this, the photodiodes can be mounted on tilted planes in order to increase the photocurrent sensitivity through cosine effect (Figure 6.16.b) and, very commonly in CPV applications, the shading device is presented as a collimating tube which prevents diffuse irradiation from entering the sensor and masking a precise measurement of the sun alignment position (Figure 6.16.c).

Even if this closed-loop approach can be very cheap and simple to implement, it has already gathered significant field experience to unveil some recurrent problems affecting its

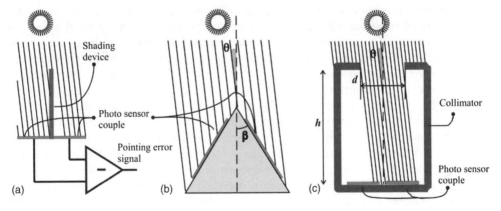

Figure 6.16 Schematics of the shade balancing principle of sun pointing sensors (a); tilted mount of photo sensors to increase sensitivity (b) and precise sun pointing by means of a collimator (c)

reliability [10], mostly caused by drifts in the analog electronics involved, and the requirement of cleanliness. This imposes the requirement of frequent maintenance which is usually affordable in research centers, where cared by attentive technical personnel, but is not feasible for the control of remote large scale tracker fields. Furthermore, closed loop controllers have proved not to perform well under less than ideal illumination conditions. For example, due to the fact that the irradiance within a sensor's acceptance angle is averaged, there is a funny phenomenon by which the bright reflection of a nearby cloud can cause tracking errors in the one degree range even when the sun is visible. This to the extent that, when the sun is hidden, closed loop controllers have been reported to track bright clouds drifting away from the sun. This simple closed loop controller is also by itself unfit to manage non-tracking and stowing situations. Due to their limited acceptance angle, the reacquisition of the sun after overcast periods is usually time consuming and inefficient, if not complemented with auxiliary control electronics. Finally, in high accuracy applications, a fundamental handicap arises provided they are to be aligned with the peak power output of the CPV array under control which, being a difficult operator requirement in itself, may even not suffice in big aperture trackers where, as seen in the past section, structural flexure varying with the tracker orientation impede a stable alignment. Nevertheless, these sun pointing sensors remain a fast pathway to CPV compliant sun tracking control. Highly integrated versions of these devices have been developed [11], and they remain an auxiliary constituent of the tracking control unit in several CPV systems.

Also brought up in the early days of sun tracking, as an alternative to sun pointing sensor controllers, the possibility of digital computing of sun ephemeris and converting this output into tracking axes turning angles, gave way to, again using the control theory term, open-loop controllers which required no feedback of sun position measurements. These were able, in principle, to keep on tracking no matter the degree of clearness of the sky, and easily programmed the management of non-tracking situations such as night or emergency stowage e.g. when subject to high winds. However, a precise timing source is to be provided to feed the computation of the ephemeris equations, and for implementations seeking sub-degree accuracy, some sensing device able to measure axes turning angles is required. Heliostat fields in solar thermal, such as in DOE's precursory Solar One plant (10 MW, yr. 1981) were the first to implement open-loop tracking, soon followed by also a grand CPV forerunner such as ARCO's Carissa Plains plant (6 MW, yr. 1985). At that time, being digital computers still expensive, this first open-loop demonstrations were carried out in a centralized way where a single computer continuously calculated turning angles for all trackers in the plant and transmitted them using a field data network. The advent of cheap microprocessors and embedded electronic systems enabled the development of specific open loop tracking controllers, at affordable unit prices which enabled the autonomous control of every tracker in a plant. Autonomous tracking control is not only inherently more reliable due to its distributed approach, but also because it gets rid of a complex and expensive field communication system, which due to its broad coverage, was frequently reported vulnerable to, for example, ground loop currents. Although first patents and publications proposing these specific open loop controllers can be traced back to the 1980s [12], it is the SolarTrak™ controller developed in the early 1990s by Sandia Labs' Alexander Maish the first serious and well documented effort done in this direction.

However, an open-loop controller, even if operating on the very precise sun ephemeris equations available to date, is affected, once connected in the field to its corresponding concentrator, by a set of error sources which can highly degrade its final tracking accuracy well below its ephemeris' nominal value to the point of even missing its specifications. Among these error sources the most significant ones have a deterministic nature. They result from a defective characterization of the concentrator system by the controller, and operate over the transform employed to convert sun ephemeris coordinates, usually in the Az.-El. horizontal topocentric format used in solar applications, into tracking axis turns. Tolerances of the manufacturing, assembly and installation processes of a concentrator system will produce some deviations with respect to specifications and therefore, to the assumptions made in the sun coordinates to axes turning angles transform. Drift in the internal timing required for the computation of the sun ephemeris, is the other major error source which is to be restrained. Second order error sources, and also to some extent predictable, such as gravitational bending in wide aperture trackers, the effect of mismatch in multi-secondary axis trackers, or even ephemeris inaccuracies, due to the effect of local atmospheric refraction, might have to be considered as well. Feedback of the tracking errors caused by the referred sources is to be somehow integrated in the control strategy in order to suppress them. This open loop core strategy blended with a feeding backloop is sometimes referred as the hybrid approach.

There are basically two types of hybrid sun tracking controllers: the model based calibrated approach and the model free predictive approach. The calibrated approach relies in a mathematical error model, able to characterize the set of systematic error sources responsible of degrading tracking accuracy below that provided by the core sun ephemeris equations. After a full clear day session obtaining tracking error measurements, these are used to fit the model parameters. Provided the acquisition of error measurements is a time consuming task, some degree of automation in this process is required when used in large tracker fields, in order to permit the simultaneous set up of them all and avoid the need of personnel to carry out this task. After the calibration session, the error model tuned with these best fit parameters will be used as the transform converting the sun coordinates supplied by the sun ephemeris to tracker's axes turning angles, and thus will, in principle, operate from then on, on a purely open-loop basis with no further requirement of tracking error feedback.

Automatic calibration routines are commonly featured in electronic instrumentation products. Very similar approaches to this type of hybrid sun tracking control are the ones commonly found for the calibration of the pointing control of many telescopes in scientific observatories worldwide, such as happens with the commonly used TPoint software [13]. Among the early developers of this technique is Nobel Laureate Arno Penzias, the discoverer of the cosmic microwave background radiation. When Penzias first joined Bell Laboratories, he was put on the pointing committee of an antenna built to communicate with the Telstar satellite. Aiming errors occurred because the steel antenna bent under gravity, wind load, and temperature changes, neither were the antenna's gears perfect nor its was its foundation perfectly horizontal. Penzias' solution was to calibrate it using an error model fitted by pointing to a known and precisely located radio galaxy [14].

On the other hand, the predictive approach to hybrid sun tracking avoids getting into any error modeling and its subsequent fitting. It intends, instead, to avoid initial assumptions regarding the tracking errors that will be encountered, thus seeking a general purpose conception able to cope with any sort of tracking errors at whichever tracker design. However, this will require the integration of permanent tracking error surveillance implying, once again, as in the case of a calibration session, some scanning scheme to determine correct sun positions. So, in this case, corrections to sun positions provided by computed ephemeris will result from an estimation based on some set of past tracking error measurements and estimations, and for this purpose, the wide mathematical toolbox for time series forecasting is at hand. The more

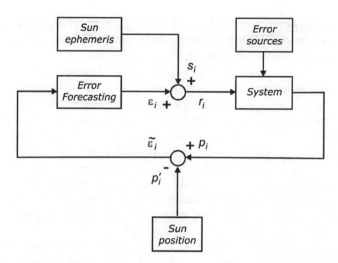

Figure 6.17 Error model free hybrid sun tracking control relying on error scanning and iterative forecasting

general form of this approach is presented in Figure 6.17 where the starting point is the computation of the sun ephemeris to provide a first set of sun coordinates. As Figure 6.17 shows these ephemeris have to be corrected, due to whichever error sources or simply because the type of sun orientation coordinates employed are not matched with the real tracking axes employed, such as for example would happen if providing horizontal Az.-El. coordinates to a two axis tilt-roll tracker. Some scanning scheme is used by the tracking axes to obtain some precise sun pointing and the correct axes turning angles from which to obtain tracking error measurements, which are then to enter the box labeled error forecasting which produces corrections to be added to the next 'raw' sun ephemeris coordinates.

First implementation of this tracking control approach was that developed by Inspira for the EUCLIDES™ CPV technology called *EPS-Tenerife* (Figure 6.18). Its error correction estimates were computed using one of the most simple and widely used time series forecasting methods: the exponential smoothing, however, in this case with an adaptive scheme for the variation of its parameter. The required gathering of tracking error measurements to feed the estimator is highly simplified whenever EUCLIDES was a single axis linear trough and accurate sun pointing measurements at each time could be acquired by exploring with back and forth turns [15,16]. Other model free hybrid approaches have been proposed such as the one using a discrete version of a classical proportional–integral (PI) controller as the error forecasting method [17]. Correction estimation makes sense when precise sun pointing is a costly task, such as can happen when this is obtained through power output maximization, so that in this case prediction will to some extent reduce scanning time and increase mean tracking accuracy. However, as said, some present concentrator tracking controllers work on a two stage basis, firstly coarse aiming based on sun ephemeris computed coordinates followed by fine pointing using a sun sensor. Leaving aside the discussed reliability of a sun pointing sensor, and provided it is always kept well aligned with maximum power output, this is a feasible method moreover when pointing a sensor is simpler than seeking for maximum power orientation, and can be classified with the hybrid model free approaches as the simplest case involving no forecasting at all.

Figure 6.18 View of the string of 14 *EPS –Tenerife* tracking controllers designed and produced by Inspira for the EUCLIDES 480 kW$_p$ CPV plant in Tenerife

6.5.2 The Autocalibrated Sun Tracking Control Unit

Inspira's *EPS-Tenerife* tracking control unit made the correction estimates dependent on the tracking angle of the EUCLIDES single axis tracker. These estimates were kept in a memory-stored look-up table, one per each one-degree tracking sector, being permanently updated based on the above referred adaptive forecasting process. Even if these forecast estimates together with the tracking errors measured for their generation happen to vary continuously during the year, this variation is basically seasonal because it is mostly caused by the above referred systematic characterization errors. We can attempt to model these systematic errors and correct them from a start, thus making the scheme of permanent scanning movements unnecessary, and therefore reducing motor fatigue and increasing tracking accuracy. This will be even more advantageous in the case of two axis trackers which require more complex scanning routines, which will further subtract from the accurate tracking operating time.

Later on, an error model was also developed by Inspira, which was termed 'calibration transform', that assumed the tracker's axes and their reference orientations have the same reference system as the horizontal Az.-El. coordinate system used by the ephemeris. This means that the axis connected to the foundation, defined as the primary axis, is pointing to the local zenith with its reference orientation pointing south, and the secondary axis, the one which is linked to the primary, always remains at right angles with it, and has its reference orientation pointing the horizon. In other words, the ideal Az.-El. pedestal tracker.

The error model is based on a six parameter kernel characterizing the departure of the real tracker to be controlled from that which ideally assumed:

- Primary axis azimuth (φ) and zenith angle (θ): These two parameters are the azimuth and zenith angle coordinates determining the real orientation of the primary axis, in which, regardless of the axes configuration, is always defined as the axis which is fixed to the ground. This is mainly an installation error due to the imprecise foundation of the tracker.

- Primary axis offset (β): This parameter determines the location of the reference orientation of the primary axis. Reference orientation is usually determined by a specific sensor, or the index mark when working with incremental optical encoders, directly installed in the primary axis. Misplacement of this sensor during manufacturing or its incorrect alignment at installation may cause this error. When $\varphi = \theta = 0$, β simply becomes the angular offset to the south.

These first three parameters are in the referred order, the nutation, precession, and spin Euler angles which relate any two reference systems with a common origin, and only these will be required if a tracker has only one axis, the primary axis, such as in present polar, azimuthal, or E-W or N-S horizontal axis trackers. When a secondary axis is attached to the primary, three more parameters are required and the pointing vector is defined as that one which is oriented by the joint action of the two tracking axis, and if aligned with the sun vector produces the maximum power output of the concentrator array:

- Non-orthogonality of axes (λ): This parameter takes the value of the difference to the right angle between the primary and secondary axis. This mainly being a manufacturing error source, a non-zero value for this angle implies the two axis tracker is no longer ideal, and a cone of orientations around the primary axis will remain out of reach.
- Pointing vector axial tilt (δ): The pointing vector is assumed normal to the secondary axis and contained in the horizontal plane when this axis rotation is zero. The axial tilt of the pointing vector is the difference angle to a plane normal to the secondary axis. This error can have its origin in the defective assembly of the tracker's aperture frame, but also in the misalignment of the concentrator optics.
- Secondary axis offset (η): The secondary axis offset accounts not only for the difference angle between the plane normal to the primary axis and the reference orientation of the secondary axis, but also for the difference angle between this reference orientation and the plane containing the pointing vector and the secondary axis, i.e. a radial tilt which is the second value characterizing pointing vector departure from assumptions. This error therefore derives from both the misplacement of the secondary axis reference sensor or, again, the improper assembly of aperture frame or optics.

These six parameters appear in a $\mathfrak{R}^2 \rightarrow \mathfrak{R}^2$ function, consisting in the composition of five partial transforms, which convert the ephemeris horizontal coordinates into pairs of angular rotations for both tracking axes. For single axis trackers, only the first three parameters enter into play and it is the primary axis turning angle the valid output. The behavior of this calibration function can be visualized through the usual grid transform representation in complex variable analysis (Figure 6.19).

The fact that the assumed reference system for the tracker under control, is that of an ideal Az.-El. pedestal tracker is just a convention, and the model is able to correct horizontal ephemeris coordinates to any one or two axes configuration, including others frequently used such as for example the tilt-roll assembly (ideally $\varphi = \theta = \pi/2$). In order to maintain the generality of the model, no simplifying assumptions are made regarding the transform parameters, which otherwise would restrain its application range.

As has been said, the parameters characterizing a specific tracker and its in-field installation are to be fitted to a set of tracking error observations and, due to the non-linear nature of the

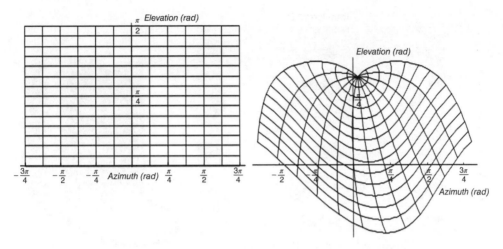

Figure 6.19 Calibration transform applied to a rectangular grid in the Az.-El. coordinates plane, into the two tracking axes turning angles plane for parameter values $\theta = 30°, \lambda = 20°, \delta = -20°$

model, it is to be by means of numerical optimization techniques. The target was to integrate this numerical procedure in a low cost embedded system, and therefore programming efficiency was a must, as well as the good conditioning of the maximum likelihood estimation (MLE) function. Least squares (LS) is the MLE chosen, which, even if there are other more robust estimators less sensitive to outlier measurements and fat tailed distributions, is by far the one presenting the most effective non-linear minimization techniques. The existence of local minima obliges to resort to the global optimization toolbox, and finally a clustered multi-start algorithm with Levenberg-Marquardt (LM) [18,19] based local searching is implemented. These local minima sometimes depend on the day of the year on which the tracking error measurements are made. For example, in N-S oriented single axis trackers particularly strong local minima appear when calibrating close to the equinoxes.

Even if first prototypes relied on the manual collection of tracking errors, this proved to be a tiresome and error prone task which had to be automated in order to prevent outliers from creeping in and thus to increase the accuracy of the corrected ephemeris. An automatic error collection scheme was developed, in which direct search of the alignment of the CPV array pointing vector and the local sun vector, in order to obtain each tracking error measurement, required the maximization of the concentrator's power output. However, in the first instance and in order to avoid interaction with the inverter's MPPT, or to be obliged to dissipate a high power, an approximately equivalent variable such as the CPV array's short circuit current was employed as feedback signal. Sun precise alignment proceeds in three stages. First a coarse approach by maximizing the irradiance in a PV cell mounted parallel to the concentrator's aperture. Beyond this point, search proceeds blindly scanning by means of spiral search of the Koopman kind [20] until the sun enters the concentrator's acceptance angle and then, a two dimensional short circuit current maximization is to be carried out [21] whose complexity largely depends on whether the power output vs. off-track angle function exhibits rotational symmetry or not [22].

The above described capacities were implemented in an electronic embedded system, based on an 8 bit microprocessor, along with the required chipset and sensors to carry out the

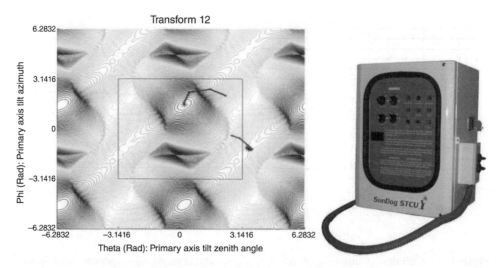

Figure 6.20 Levenberg-Marquardt local searches in the least squares function used to fit a simulated error model transform using for visualization purposes only its two first parameters (φ, θ) (left). SunDog STCU® (right)

described algorithms and perform the analog measurements, and also to provide motor driving capacities [23,24] (Figure 6.20). Remarkable hardware elements further enhancing performance are the temperature compensation circuit devised to restrain drifts in the quartz oscillator responsible for internal timing, as well as its encoder decoding and interpolation subsystem which increases the axis turn measurement accuracy. Named *SunDog STCU®* – always follows his master the Sun – it was supplied with SunDog Monitor, a Windows application to run in a locally or remotely connected PC as a virtual user interface. It also integrated an interchangeable modem for PSTN, RF, Ethernet or GSM/GPRS Internet connectivity, which enabled e-mail reporting and web based control and monitoring. Intended for operation in harsh environments, it was tested in electronic certification labs attaining CE labelling covering the corresponding EMC and electrical safety standards, and also successfully passed climatic tests (temperature cycling, humid and high salinity ambient, water and dust tightness etc.).

This tracking controller designed by Inspira should rightly be considered a landmark, in the development of sun tracking control systems specifically designed for CPV Systems, as it was the first that fully automated the referred model based calibrated approach, being in this way able to attain, as we will see below, very high tracking accuracies. However, further refinements could still be introduced in this type of controllers, such as for example extensions in the calibration model to account for second order error sources such as flexure effects on the pointing vector position, which are usually specific to the tracker concept employed. Another relevant improvement could come from further adjustments in internal clock drifts. In this respect, even if Internet connectivity or GPS might provide atomic time synchronicity and this may be cost efficiently implemented in networked CPV tracker fields, the availability of high accuracy ephemerides provides an immediate and autonomous alternative to precise time-keeping. A time drift parameter can be included in the calibration model in such a way that it can be fitted with a tracking error set either, jointly with the rest of the parameters or individually within periodic time adjustment procedures. Finally, the tracking error set in calibrated controllers is not necessarily to be obtained from sun position measurements, as in principle any other light source with precise analytic kinematics, and with enough emitting power to extract a measurable output from the concentrator, will suffice. In this respect, the full moon proves to be a good candidate, as far as it will enable night calibrations not interfering with concentrator's daily production and even more, it will also permit these to be done with a maximum power point bias but at much lower power levels than nominal thus highly easing its handling in terms of switching and biasing hardware. As is well known, the moon's 0.49° apparent diameter is very similar to that of the sun, while its irradiance is six orders of magnitude smaller, so its photogenerated current is still within reach of cheap current sensing devices. On the other hand, the full moon irradiance is three orders of magnitude above that of the most brilliant planets and stars so it will be easily distinguishable by the concentrator when searching the night sky.

6.6 Sun Tracking Accuracy

6.6.1 The Tracking Accuracy Sensor

In the PV field there are few past experiences on which to base the development of a sensor, able to measure the incidence angle of direct sun radiation with respect to some built-in axis [25]. However, fairly accurate devices of this kind can be found in the aerospace sector, which based on CCD and CMOS arrays contribute to satellite attitude control [26]. These usually feature hemispheric acceptance angles which preclude the extraction of higher accuracies from their very high resolution image sensors, nonetheless attaining the 0.05–0.01° range. They are produced at very high costs due to their required compliance with spacecraft specifications, and usually on a custom made basis without an open commercial intention, so even if they could serve our means, some CPV related companies have recently found it worthwhile to develop a specific sensor.

This was again the case of Inspira, which to our knowledge produced the first tracking accuracy sensor (TAS) for CPV applications, which they refer to as the *SunSpear* [27]. The position-sensitive diode (PSD) sensor that was chosen for this TAS has no discrete elements such as in CCDs, and provides continuous data of a light spot on its surface by making use of the surface resistance of a planar PIN photodiode. Due to its analog nature, these sensors feature excellent position resolution in the micron range and very high speed; moreover, they detect the 'center of gravity' position of the light spot and have proven a very high reliability. As usual, placing a light collimator on top of this sensor, will produce the required light spot, and enable the measurement of the angle of the incoming light beam with respect to sensor's axis, where the acceptance angle and also the angular resolution of this measurement is basically determined by the height of the collimator's pinhole over the sensor's surface. With this SunSpear design, resolutions in the 1/1000 of a degree range at a ±1° acceptance angle can be achieved (Figure 6.21).

Along with the off-track angle measurement system, the TAS was completed by a custom made electronic system providing signal conditioning to the PSD's output, AD conversion and driving a RS-232 serial output. It is this electronic system that hosts the PSD sensor in its PCB and is contained in a watertight enclosure also providing a fixture point for the machined collimator tube.

Confirming the requirement for a rigorous control of tracking accuracy for the development and manufacturing of CPV products, once this industry is growing in maturity, other tracking accuracy measurement sensors have later been developed following the trail of the Sunspear that based in similar design concepts, are at last truly commercial products such as that of the Silicon Valley company *GreenMountain* [28], or that of the Freiburg based *BlackPhoton* [29].

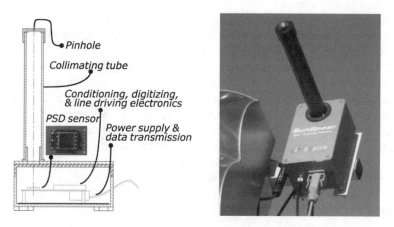

Figure 6.21   The *SunSpear* Tracking Accuracy Sensor

6.6.2 The Monitoring System

In the case of the Inspira SunSpear TAS, the sampled position data generated is sent through a serial link to a PC which is to run a monitoring software application able to store it but also to display it in real time, convert sunspot Cartesian coordinates to off-tracking angles, generate plots and statistics of selected tracking periods, and to periodically e-mail tracking data and reports (Figure 6.22).

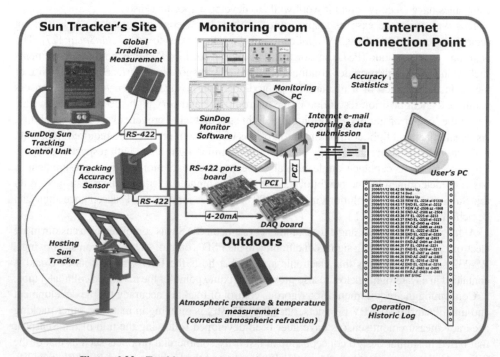

Figure 6.22   Tracking accuracy monitoring system developed by Inspira

The first application given by Inspira to this tracking error monitor was in the assessment of the tracking accuracy of whatever hybrid tracking routines, which no matter if based on an error model or on error forecasting schemes, all have in common the requirement of obtaining power output feedback, or some other equivalent of this variable, from the CPV array, in order to get precise sun position measurements. The method used to test these strategies without having to mount a CPV array and its power output measurement electronics, relies on using the TAS as a virtual power output, i.e. perfect alignment of the TAS built in axis with the solar vector is assumed to represent peak power output orientation. Aside from the simplicity, an added advantage of this setup is that, in principle, it allows for very precise measurements at a faster rate than real power output maximization requiring the involvement of the MPPT stage. In this way, this method exposes the specific weaknesses of a certain tracking strategy, when almost not affected by errors in the sun position measurements whether these are used to feed the calibration model fitting or in error forecasting schemes.

Inspira gets into the specific case of testing an autocalibrated *SunDog* controller, in such a way that once these calibration measurements are completed, some selection of them are fed into the error model Levenberg-Marquardt LS fitting routine, whose basic parameters regarding e.g. the clustered multi-start optimization routine or the stopping criteria, can also be fixed within the software running in the monitoring PC. Depending on the selection of error observations, some variation in the best fit parameters obtained can appear. In this respect, this software application also offers the possibility to periodically change the parameters being used by a connected *SunDog* unit for sun ephemeris correction, in order to jointly obtain the tracking accuracy statistics of an assortment of varying sets of parameters and thus help to estimate those measurement schemes obtaining the best performance. In addition, the monitoring software application version implementing these tracking accuracy assessment tools when using the TAS as virtual power output for calibration, features additional resources to further restrict tracking error margins, by integrating the connection to an outdoors thermometer and a barometer. Ambient temperature and atmospheric pressure readings are employed in the computation of atmospheric refraction corrections to the sun ephemeris elevation coordinate, based in Bennet's model [30]. Measured values of these corrections have been reported to amount up to an average $0.6°$ for near the horizon elevations [31], and as will be seen in next section, they do have a measurable impact in a tracking accuracy monitoring campaign. These corrections are applied to computed sun ephemeris both in the error model fitting stage and also afterwards during real-time tracking, in which the corrections are periodically transmitted to the *SunDog* controller. Finally, and also contributing to tracking accuracy enhancement, the monitoring software can use top quality arc second precise ephemeris during the error model fitting data, thanks to a built-in direct connection to the USNO MICA software [32].

The second and more general application of the tracking error monitoring system consists in directly measuring the real tracking accuracy of a concentrator system. This system will have to be initially calibrated against the peak power output of the CPV array, i.e. recording the sunspot coordinates on the PSD surface when maximum power is delivered and taking these as the TAS' reference system origin when converting its readings to off-track angle. Provided the effect of the tracker's self-weight and its CPV array payload upon the bending of the concentrator's structure varies with its orientation, the calibration is to be carried out at a set of different positions. Then, the reference points to use in off-track angle conversion at each orientation will be interpolated from them [33]. This second function is independent of the tracking control means employed. However, during calibration, it requires combined automatic readings both from the TAS and array's power in order to precisely locate the sunspot coordinates on the PSD when power output is maximum.

6.6.3 Accuracy Assessment: Example of the Autocalibrated Tracking Strategy

Following the procedure advanced above, we present here an example of long term monitoring with the tracking error monitor applied to the assessment of the tracking accuracy performance of a *SunDog* STCU, when controlling a small aperture (4 m^2) laboratory sun tracker. The calibration was based on an error measurement session carried out on 31/01/2006 that included 368 points. Four different model based hybrid routines were set to compete:

- Case no. 1: The six core parameters of the autocalibrated SunDog STCU error model were fitted directly in this controller's processor using all the error measurements. This case represents the normal SunDog performance.
- Case no. 2: Same as case no.1 but the best fit values for the six parameters were calculated by the monitoring SW application running in a PC, with its added float point accuracy, and also using the superior performance of the MICA ephemeris when compared to the analytic ones computed by SunDog (0.03° mean accuracy taking MICA as reference). This case works with more precise fit than that being attained by the LM embedded implementation in SunDog. However as in case no.1, once fitting is completed, tracking control solely relies on SunDog and its less accurate built-in ephemeris. It represents an operative alternative in which SunDog units operating a CPV plant are networked and send their tracking error measurement sets for fitting to a more powerful central computer.
- Case no. 3: Same as case no. 2 but temperature and atmospheric pressure measurements are activated, and atmospheric refraction correction in the ephemeris' elevation is integrated both during the tracking error acquisition session and also later providing the SunDog STCU real time measurement pairs every 30 s for it to internally calculate the corrections. This case is in principle the one which should present the highest performance and is useful to assess the possibility of integrating a thermometer and barometer in the SunDog STCU hardware.
- Case no. 4: A simple two-parameter linear model is used to fit the error measurement set essentially obtaining the mean offsets in both tracking axes. This is the most straightforward tracking error model which requires no numerical fitting algorithm, and in the same way that the nominal SunDog performance of case no.1, it should rank below the enhanced case no.3. This case is to serve as a low performance benchmark to rate the benefit of using the SunDog non-linear model along with its fitting procedures.

The monitoring ran uninterruptedly till 30/04/2006. Every day the tracking control of the lab tracker was assumed by a different case, no.1 to no.4 above, following a fixed sequence. Every TAS sample includes along with the Cartesian coordinates of the sunspot on the PSD surface, the incidence light level. This incidence light level measurement is to be above a certain threshold to ensure the minimum required resolution, and in order to accept a monitoring daily session, 90% of its samples were to have its light level above this threshold. This basically means that only full clear sky days were considered for the analysis in order to compare the performance of the different cases on identical grounds. This means that, at the end of the monitoring period, some cases had more valid days than others, but nevertheless, all had enough to draw some interesting conclusions. For every day the tracking accuracy statistics were calculated: mean, standard deviation, and the daily probabilities of accuracies below 0.1° and 0.05°. Probability density and distribution functions were plotted, and also the probability density of the sunspot point over the PSD surface was obtained.

In Figure 6.23, daily probability density plots of the collimated sunspot over the PSD surface, are included for all four cases, both for the first valid day and the last one comprised in the monitoring period. The 0.1° and 0.2° tracking accuracy rings are displayed in these plots. The purpose of presenting plots for the two ends of the monitoring period is to see how tracking accuracy may drift with time. In Figure 6.24 (above) a tracking accuracy statistics sample is shown. This is representative of the best ratings obtained with the case no.3 setup in this monitoring period, showing a mean daily tracking error of 0.05°, with Std. Dev. 0.02°, and having accuracy better than 0.1° with 98% probability and better than 0.05° with 50%. Also shown is the sunspot trace over the PSD's surface from which all statistics are obtained. Also, on Figure 6.24 (below), the daily plot of the tracking error is presented and it can be seen this may also prove to be useful to detect positioning resolution defects in a certain tracking drive, such as occurs in the two error peaks appearing symmetrically with respect to solar noon, and which have to do with a momentary unleashing of the drive's backlash at that precise elevation in which push and pull loads equate in the tracker's aperture.

Figure 6.25 plots the evolution of daily mean tracking error for the four cases during the monitoring period. The most obvious result, as can be also inferred from density plots in Figure 6.23, is the superiority of the complete six parameter error model (cases no.1–3) over the simplified mean offsets model benchmark (case no.4), moreover when the latter drifts apart during the monitoring period and, as can be seen in its last recorded density plot of 11/04/2006, it is finally incapable of entering the 0.1° ring with its mean accuracy rising over 0.2°. Besides, a more subtle drift is also found in the tracking errors of cases no.1 and no.2 which disregard the atmospheric refraction effect. Even if case no.3 accuracy was already slightly better than cases no.1 or no.2 from a start, just after the calibration day, case no.3 maintains the same ratings all over the reported monitoring period while cases no.1 and no.2 suffer a slight decrease in accuracy clearly noticeable when entering the second monitoring month. The main reason for this effect is to be found in the fact that calibration is done in winter when sun elevations are lower and therefore, the atmospheric refraction correction is relatively more important during the day. In cases no.1 or no.2 the atmospheric refraction effect is erroneously taken for an elevation axis offset effect and absorbed by the corresponding error model parameter (η secondary axis offset above). This defective identification is exposed when approaching summer and low elevations requiring atmospheric refraction correction become relatively less important. Mean tracking error slightly increases as seen encircled in Figure 6.25 and regarding the probability distribution first 93–95% probabilities for errors below 0.1° drop down to 83–85%. Also remarkable is the fact that cases no.1 and no.2 have a similar behavior and no significant advantage seems to derive from using the PC's higher accuracy float point arithmetic for the model parameter fitting, or the very accurate MICA ephemeris which based on the interpolation of tabulated numerically integrated solutions of the equations of celestial motion, are not feasible for a low cost embedded integration.

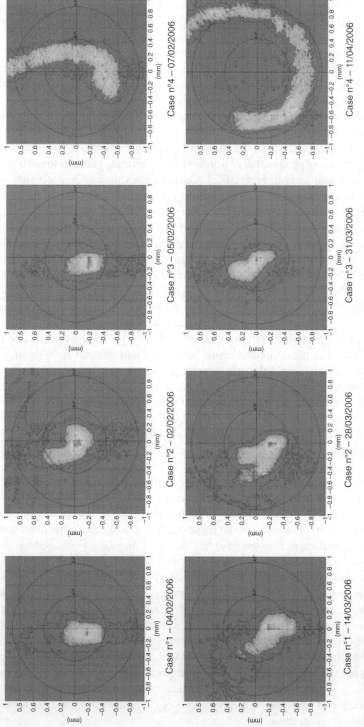

Figure 6.23 Daily probability density of the sunspot over the PSD surface for the first and last monitoring day in each of the four calibration cases. The 0.1° and 0.2° off-tracking circles are represented. It can be seen how mean value approaches zero with increasing precision of the calibrated ephemeris, while the standard deviation directly related to the tracker's positioning resolution

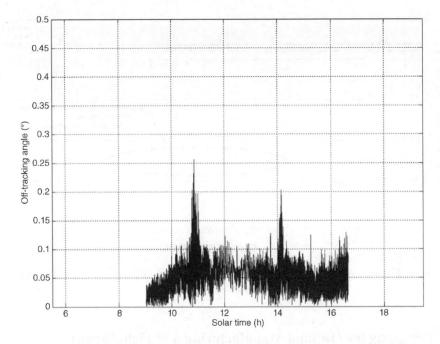

Mean = 0.050788 / Std. Dev = 0.023195 / 50.0675%<0.05° / 98.0908%<0.1°

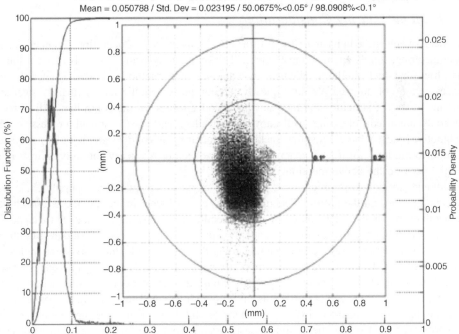

Figure 6.24 Off-tracking angle during a day remains below 0.1° excepting two symmetric points with respect to noon at which aperture elevation has equal push and pull forces thus releasing backlash (above). A typical daily off-track angle probability density and distribution functions with 0.05° mean and 0.02° standard deviation, and superimposed the sunspot trace over the PSD (below)

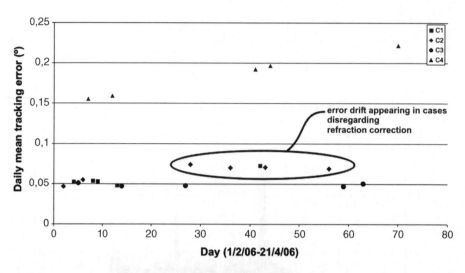

Figure 6.25 Daily mean tracking error for the fully clear days during the three months monitoring period for the four cases (C1–C4) considered in the main text

6.7 Designing for Optimal Manufacturing and Field Works

6.7.1 Manufacturing Considerations

We here switch the emphasis of the tracker development to the 'operations competencies': manufacturing, supply chain and installation. Tracker manufacturing has evolved according to industries that had similar challenges by adopting their methods. Tracker materials and processes are not unique. We are typically talking about metallic alloys (mainly steel and aluminum) for the structure and mechanisms and electronics for the controllers. All of those materials are used following well-known, controlled, and vastly spread processes. The automotive industry largely inspires tracker manufacturing as volumes go up. Fully automated manufacturing lines for tracker manufacturing have been developed where robots, instead of people, perform the above-mentioned processes.

Although specialists in process and production engineering, quality and supply chain will execute the work on the ground, design decisions are still what influences more those tasks. Rather than an extensive description of all the specific operational tasks that need to happen in the production floor, we will here focus attention on the design decisions that constrain them, from a systems engineering point of view. During the design cycle, there are some trade-offs that need to be made with respect to system performance. The tracker greatly influences several variables, among which we would like to highlight the already mentioned system accuracy and acceptance angle. At the end of the day, as any commodity related engineering challenge, this is no more than a cost consideration. The most successful design is the one that solely achieves the necessary objectives by the minimum cost possible. Over-specifying results does not seem to make sense because it either pushes the cost up or throws unnecessary functions that will not be valued. Poetic design considerations that make the product 'cooler' or 'look better' are important in industries where the final consumer has direct interaction with product and appreciates and does not mind paying more for features. Examples are the consumer electronics or clothing industries, where look, touch and feel are very important. However, the solar industry is about selling electricity to the grid. There are no records of better looking electrons. Also, nice colored trackers with many 'cool' functions do not necessarily maximize kWh or, even more important, optimize LCOE (Levelized Cost of Electricity, see Chapter 14), which is the metric that relates investment with revenue. All of this to say that every single feature put in a tracker is better to have a real purpose and better to be performing a fully utilized function. Otherwise it is probably superfluous.

While making that evaluation, the trade-offs are always between design specifications and how to manage the tolerances budget and where to allocate them throughout the lifecycle phase: design, manufacturing or field installation. As the tracker lifecycle evolves through these three stages there is less and less flexibility to change and more and more people involved. The impact of any process or change at the end of the lifecycle is always much bigger than at the beginning. It is commonly accepted that design needs to be the most tolerant possible, making life easy at the factory and even easier in the field. As we progress in the lifecycle it is more about doing than questioning, so the great challenge of the designer is to collaboratively make all the trade-offs and reduce complexity up front and then document all the actions very well so, people downstream can execute as smoothly as possible.

In the ideal tracker, the design is tolerable enough not to require special manufacturing processes, machines or very tight tolerances. These requirements just increase complexity and hence cost. Special processes and machines can be mitigated by specific investment, but this only makes sense with high volume demand. Designing tracker components that can only be materialized with very expensive tooling, such as by casting or stamping, often is equivalent to putting oneself in a corner, depending on others to get high volumes so the design decision makes sense. It is therefore important to design all components with scalability in mind. This means the design needs to be compatible with simple manufacturing processes that do not require major investment, but also compatible with high volume techniques for the moment the business grows and demand increases significantly.

A lost bet – unfortunately often seen in the solar industry – was major multi-million investments up-front, which never had a sustained pipeline that paid for them. Although the concentration solar industry adopted the automotive methods, its volumes are still not comparable. Its demand intermittency so far resulted in many companies that invested enormities up-front bleeding money until their extinction.

So in sum and as a general rule, it pays off more to accept the biggest challenges in the design phase in order to minimize complexity, labor and cost in the factory and in the field so that the curve effort vs. lifecycle phase results descendent. A quick example very relevant to CPV trackers to illustrate this point is the challenge of minimizing the unevenness of the final array of CPV modules on top of the tracker's aperture surface. In the early times of the industry, all manufacturers of reference would have a step of panel leveling in the field. This was to be done on an individual panel basis, which quickly proved to be one of the most time-consuming activities in the field, driving the overall CPV balance of systems (BOS) cost up. Leveling in the field violates the rule of 'least progressive effort' suggested above. Throughout the times, when it has been mandatory to reduce costs for the CPV industry to survive, this was corrected and engineering departments have come up with more tolerant approaches that would either have enough buffer not to require any leveling or, if that proved not cost-effective due to excessive performance degradation, leveling would occur at the factory and not in the field.

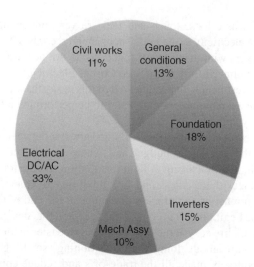

Figure 6.26 BOS cost distribution in January 2012 for a CPV plant. Source: Adapted from SolVida Energy Group, Inc

6.7.2 Field Works Considerations

Regarding field works, oftentimes referred as BOS, we can consider the following categories:

- *civil works:* include site survey and preparation, trenching and fencing;
- *foundation:* includes procurement and installation costs;
- *mechanical assembly:* includes assembling the trackers and populating them with CPV modules;
- *inverters:* includes procurement cost;
- *electrical DC and AC:* includes stringing, grounding, combiner boxes and connections to inverters and transformers;
- *general conditions:* includes plant engineering, permitting, project administration and management, commissioning, site security and insurance.

The pie chart in Figure 6.26 shows an example of the distribution of costs per categories above in the case of double axes trackers like the ones used for CPV courtesy *of SolVida Energy Group* (data for January 2012).

As can be seen, even more important than the magnitude of each category is to individualize the optimization opportunity in each of them and, for the sake of this chapter, how much can be influenced by the tracker design decisions. It is fair to say the tracker design affects every single category, although some categories more than others. Let us try to cover them all in order of potential influence by the tracker design.

From the least to the most influenced by tracker design, we can sort the categories as follows:

1. inverters
2. general conditions
3. civil works

4. electrical DC and AC
5. foundation
6. mechanical assembly.

Let us analyze the influence of tracker design decisions on the above categories:

1. *Inverters:* inverters are typically dissociated from tracker design. They are directly related to the PV power installed. Several different approaches have been used in real plants, from big centralized inverters who handle thousands of panels to micro inverters on a panel basis, despite tracker architecture (see Chapter 8). The one design link may occur related to the maximum voltage of the array hosted by the tracker, which is related to the number of modules connected in series which determines to some extent the size of this array.
2. *General conditions:* only slightly affected, in the case the tracker design allows for the Engineering, Procurement and Construction (EPC) contractor to spend less time at the job site per MW_p installed. This may happen when touch points in the field are minimized. This can be achieved with consolidation at the factory. However, permitting and other project bureaucratic fees are not affected.
3. *Civil works:* site survey and fencing are not altered, but site preparation and trenching are greatly affected by tracker design and size. Carousel and tilt-roll trackers require much more stringent site grading requirements than single-pole Az.-El. trackers, which can be installed almost in any orographical condition. Tracker size will affect trenching. The larger the tracker the fewer trenches will be needed.
4. *Electrical DC and AC:* here the main consideration is tracker size. Smaller trackers penalize this category as DC wiring is more dispersed. In a limit case, very small trackers will necessarily call for much higher DC electrical costs, as the strings will not be confined to a single tracker.
5. *Foundation:* this is something that offers plenty of options and where people have been very creative. From foundation-less trackers to screw piles, driven piers or the old fashioned concrete slabs, there are many variables that allow for a serious optimization exercise. Also, tracker size highly affects foundations. In opposition to the previous wiring category, larger trackers drive the cost up.
6. *Mechanical assembly:* this category is fully determined by the design decisions made during tracker development. A serious trade-off exercise between pre-assembly at the factory, packaging and transportation costs, tracker bill of materials (BOM) costs, handling and associated equipment is mandatory to achieve the optimum solution.

We noted above there are some variables that pull optimization in different directions. For instance, tracker size lowers wiring cost, but increases foundation and mechanical assembly costs.

More than looking at the above pie chart and immediately try to tackle all categories separately; it is imperative to perform a true optimization exercise where all variables are taken into consideration. In this way, we can list all variables available and firstly map how they qualitatively affect the categories above. Then, we should cross that result with the opportunity available in each category, and sort all the projects based on opportunity cost. That will help to form the design and engineering teams according to a scale of effort, resources, time, risk and payoff.

Table 6.1 Savings opportunity for each BOS category due to tracker design improvements

Element	Absolute cost	Cost Reduction (%)	Success Probability (%)	Savings Opportunity
Electrical DC/AC	0.41	30	70	0.09
Foundation	0.22	60	80	0.11
Inverters	0.19	5	90	0.01
General Conditions	0.16	10	80	0.01
Civil Works	0.14	30	90	0.04
Mechanical Assy.	0.12	70	80	0.07

Let us imagine a scenario in accordance with the pie chart shown in Figure 6.26, list all BOS categories and sort them by cost as in Table 6.1. Note the costs are normalized to the total cost and just for the sake of exemplifying the method in discussion.

Each category has different optimization opportunities. In this section we are interested in how tracker design affects those categories. Consistently with the analysis above we can score each category with respect to what we think may be achieved in a cost reduction effort and its success probability, factoring risk, difficulty and dependencies. These scores are just examples and appear in the two middle columns of the table.

To finally assess savings opportunity, we should multiply the absolute cost by both probabilities. Figure 6.27 compares absolute cost and savings opportunities for each BOS category.

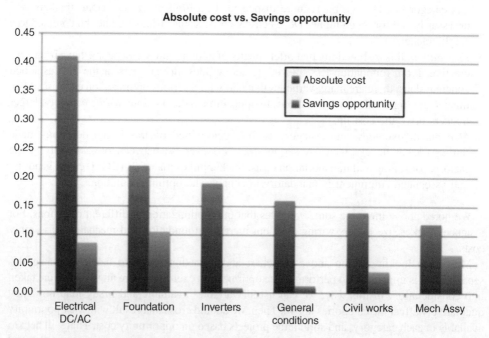

Figure 6.27 Absolute cost and cost saving opportunity due to tracker design improvements for each BOS cost category. Dark bars correspond to absolute cost while the light ones to savings opportunity

Table 6.2 Value of tracker design improvements per BOS category

Element	Absolute cost	Cost reduction (%)	Success probability (%)	Savings opportunity	Value
Foundation	0.22	60	80	0.11	12
Electrical DC/AC	0.41	30	70	0.09	10
Mechanical Assy.	0.12	70	80	0.07	8
Civil Works	0.14	30	90	0.04	4
General Conditions	0.16	10	80	0.01	2
Inverters	0.19	5	90	0.01	1

Note that categories are sorted by absolute cost (dark bars), but the biggest savings opportunities do not necessarily follow the same order.

In the real world, the action would be to organize projects and teams to go after savings organized according to opportunity. Accordingly, Table 6.2 illustrates the subjective 'value' of possible projects and suggests that for any unit of 'effort' put into an inverter project we would have to dedicate 12 units to a foundation project, 10 to a project tackling electrical connections and so forth. It is up to the reader to give a real meaning to 'effort'. Possibilities are resources, time, money or a combination of all.

6.8 Description and Performance of Current Tracker Approaches

There are several tracker designs currently operating in CPV power plants. The general tracking axes architectures used are the ones described in section 6.3:

6.8.1 Parabolic Trough

These trackers are used in systems with lower to medium concentration factors (see Figure 6.28 and Figure 6.29). Their objective was to lower PV cost using the least disruptive and probably

Figure 6.28 EUCLIDES® CPV System – designed by the Solar Energy Institute of the Technical University of Madrid (IES-UPM) and Inspira, SL, and commercialized by BP Solar Ltd, all at Madrid, Spain

Figure 6.29   X14 CPV System - designed and commercialized by Skyline Solar, Mountain View, CA, USA

less complex way. Hence these trackers are used in systems that use c-Si cells, instead of very sophisticated and more expensive multi-junction cells.

The benefit of this approach was to take advantage of existent and well-established PV technology, trying to push its boundaries instead of coming up with a whole new CPV paradigm by recreating every single component of the system. The tracker and controller inherently present fewer challenges due to the single-axis nature of the approach.

A disadvantage associated with lower concentration is that single axis CPV is heavily dependent on the flat PV industry trends. In fact, since c-Si cost came down so much, it is really difficult to justify the extra cost for the tracker; controller and all optical components, when compared with energy yield augmentation.

6.8.2   Single-Pole Az.-El. Trackers

This is the most common tracker architecture of the industry and the one adopted by the current manufacturers of reference with more MW_p installed in the field. Its basic axes configuration is an Az.-El. one. The most unique characteristic of these trackers is that the whole group of CPV modules is mounted onto a single pole. Hence, there is one individual foundation per tracker in a fairly small ground area. In soils with an average hardness, this could allow for a more favorable foundation per MW_p ratio. This has also the big advantage of allowing for installation in almost all orographies, since there is no tight site-grading requirement. It also enables for less stringent installation tolerances in the field. All poles are individually installed and then a compact group of mechanisms is handled on a pole-by-pole basis.

On the other hand, each individual tracker is usually supplied with its controller, set of Az. and El. motors and corresponding mechanisms, which probably make the ratio of these components per kW_p less attractive cost-wise and probably reliability-wise, as there are more components prone to failure per kW_p. However, having more components reduces the severity of any possible failure, as fewer CPV panels would be affected. The only way to mitigate these

Figure 6.30 SF1100S CPV system – designed, manufactured and commercialized by SolFocus Inc., San Jose, CA, USA

high fixed costs per kW_p is to increase tracker size significantly, which, however, also increases steel utilization and penalizes the ratio (kg of steel)/kW_p.

The differences among the systems from different manufacturers that selected the single-pole Az.-El. architecture are based on tracker size, types of motors, driving mechanisms and way of supporting and leveling the matrix of CPV panels. Tracker sizes in the field vary from 25 to 250 m^2. Figure 6.30 to Figure 6.33 show some pictures and companies that adopted this architecture.

Figure 6.31 CX-S530-II CPV system – commercialized by Soitec, Freiburg, Germany, with Concentrix™ technology

Figure 6.32 7700 CPV Solar Power Generator – designed and commercialized by Amonix, Seal Beach, CA, USA. Source: Amonix. Reproduced with permission of Arzon

6.8.3 Tilt-Roll Trackers

The motivation of this approach is to lower the fixed costs per kW$_p$ without increasing tracker size. Its axes configuration usually follows the declination-hour angle type. They are usually built following a multisecondary approach aggregating many smaller groups of modules along a common primary axis, which indeed allows for having fewer controllers, motors and mechanisms per kW$_p$. The structure has a low visual profile keeping the modules close to the ground. This

Figure 6.33 13.44 kW$_p$ HPCV systems jointly developed by Daido Steel, Co. Ltd., Nagoya, Japan and BSQ Solar and Solar Energy Institute of the Technical University of Madrid for the NGCPV Euro-Japanese project

Figure 6.34 Suncore tilt-roll tracker and DDM-1090-X CPV modules – Designed, manufactured and commercialized by Suncore Photovoltaic Technology Co., Ltd., Huainan, Anhui, China. Source: Suncore US, 2015. Reproduced with permission of Suncore US

helps the ratio weight of steel to kW_p, when compared with single-pole trackers that have to grow in size a lot to keep the same ratio. However, the count of steel members goes up due to the numerous linkages necessary to connect the small groups of panels in a row.

Tilt-roll trackers also require many foundations per tracker. Fans of this architecture claim these foundations have much lower penetration than the one used by single-pole trackers. Although it is true loads are more evenly distributed, dependency on soil conditions is a dangerous thing to have as the technology provider does not have control over that parameter, which can bring unpleasant surprises.

Another disadvantage of a distributed foundation strategy is a tighter site grade requirement and the absolute need for additional leveling of the vertical posts in the field, in order to keep the primary axis parallel to the ground over a long extension of land. This seems to violate the law of 'least progressive effort' mentioned in section 6.7.1, as complexity has been displaced to the field.

Reliability-wise, fewer components per kW_p lower failure rates, which is a good thing. However, even though the probability of occurrence is lower, the severity of a failure is much higher as greater number of generators would be affected.

Examples of tracker companies that installed CPV tilt-roll trackers in the field are Suncore (see Figure 6.34) and GreenVolts.

6.8.4 Carrousel Trackers

Carrousel trackers appear with the exact same motivations as tilt-roll trackers: decrease the ratio fixed components per kW_p without significantly increasing steel utilization (kg/kW_p). This is inherent to architectures that have the primary axis holding several secondary axes and carrousels are not an exception. This is achieved because the structure is supported more evenly with several load-discharging points instead of only one, as in the single-pole architecture. This potential yields a lighter structure per kW_p.

Compared with tilt-roll, there is not such a strong dependency on soil, which is a very nice advantage.

On the other hand, this is the architecture that is tighter on grading requirements. Carrousels need an even surface to place their wheels, which calls for a big flat ground area or, even worse, additional concrete to shape a rolling track.

Figure 6.35 TT PTK-130 PRECISION – developed by Titan Tracker, Torrijos, Toledo, Spain. Example of a single secondary axis carrousel. Source: Titan Tracker, Spain, 2015. Reproduced with permission of Titan Tracker

Technologically, carrousels bring something new to the table. That is the ability of easily measuring the Az. angle, which ultimately eases the tracker controller and requires less resolution in the Az. position feedback sensor. This happens when parts of the structure are far away from the primary axis, which allows for a small angle to be very noticeable at the perimeter of the carousel.

Finally, this architecture is the one with more unique components, like wheels, track, and possibly an anti-sliding break for when there is ice on the track (see Figure 6.35).

6.8.5 Variations to Main Architectures in the Field

Some other companies operating in the CPV market opted for designing their system based on one of the technologies described previously, in spite of introducing some new ideas to the traditional designs. A photo guide is presented below for these CPV trackers that do not quite make separate architectures, but present some noticeable differentiation elements.

Solar Systems' tracker gets its inspiration from a regular single-pole Az.-El. tracker, being practically identical to those in all components, except the frame that holds the CPV panels. In this case, there are no panels and the whole system is a gigantic power unit focusing light in space where the generator was placed (Figure 6.36).

Figure 6.36 Solar Systems, Victoria, Australia. Source: Solar Systems, 2015. Reproduced with permission of Solar Systems

Figure 6.37 Morgan Solar, Toronto, Ontario, Canada. As in the carrousels, multipoint ground support is attained through aggregation of pedestal trackers. Source: Morgan Solar. Reproduced with permission of Morgan Solar

Morgan Solar's *Sun Simba*™ CPV system also has a tracker whose architecture was based on the common Az.-El. approach. The difference here was on the elements that connect the mechanism to the ground. Instead of a pedestal or tripod and a foundation, a grid of steel beams connects all trackers among each other (Figure 6.37). This allows for a foundation-less solution, which seems clever, as foundation costs are not negligible, as we saw in section 6.7.2. The drawback is that this approach only works in a flat site, which requires a serious effort of grading.

Energy Innovations' Sunflower CPV system was inspired in the tilt-roll architecture but integrating Az.-El. axes where the Az. axes are mechanically linked and driven by a single motor (see Figure 6.38). Its differentiator was that the whole group of panels could be installed manually by operators on foot in the field, preventing the use of expensive heavy equipment, in a tentative to lower the cost of the mechanical assembly chunk of the cost. As we saw in Figure 6.26, this kind of tracker presents a fairly low savings opportunity to justify an innovation that solely addresses it.

Among the many developments of Spanish manufacturer Isofotón in the CPV field, was the sun tracker approach presented in Figure 6.39. It is essentially a pedestal tracker with a single

Figure 6.38 Energy Innovations, Los Angeles, CA, USA

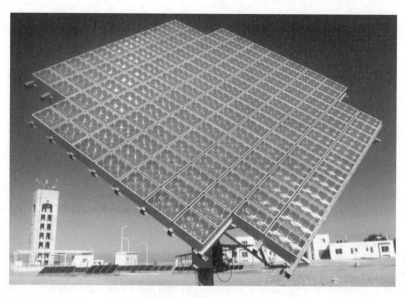

Figure 6.39 ICST-70 tracker and Gen CPV modules – designed and commercialized by Isofoton, Malaga, Spain

pint foundation but using an hour angle–declination two axes architecture instead of the more common Az.-El. The main difference here was the use of linear actuators for both axes, avoiding the use of the usual slewing drive in the azimuth axis.

6.9 International Standards for Solar Trackers

Solar trackers were in existence for many years prior to the development of associated international standards. When the concentrating photovoltaic (CPV) industry was in its infancy leading members of that industry formed Working Group 7 (WG7) of the International Electrotechnical Commission (IEC) to create standards for the CPV industry (see Chapter 9 for further details about CPV standards). While the group's initial focus was to publish a design qualification standard for CPV modules, WG7 was quick to realize that a perfectly designed CPV module would fail to produce electricity without a carefully designed solar tracker. In some instances at this point in history, standard flat-plate photovoltaic PV modules were already making use of solar trackers but the requirements for these trackers were not as rigorous as the CPV industry knew they would need. Solar trackers were being used to enhance the performance of PV but in the events of poor tracking or failure to track, the PV would still produce significant power. Alternatively, CPV requires tracking to produce significant power and high accuracy tracking is often necessary to achieve optimum performance.

Due to the increased requirements for CPV trackers, WG7 formed a tracker subgroup in 2007. The initial tracker subgroup came to a consensus that WG7 should aim to produce an IEC design qualification for trackers. IEC standards are drafted through international consensus and completely through voluntary participation. Accordingly, it can often take many years to go from forming an idea for a new standard to the final publication of that standard. Due to the challenges associated with this process and resources available within WG7, it was decided that a first goal would be to draft and publish a technical specification for solar trackers and then work to build a design qualification standard from this technical specification.

A technical specification provides terminology and technical guidance to an industry but has no qualification tests and is binding in no way. For this reason, it is easier to achieve consensus on technical specifications and publication can often be achieved in a shorter time period. In May of 2012, the work of the tracker subgroup resulted in the publication of IEC TS 62727 Ed.1 *Photovoltaic Systems-Specifications of Solar Trackers*. This specification primarily provides a consistent set of definitions and terminology for discussing and comparing trackers, a suggested specification sheet for manufacturers of trackers, a procedure to follow for measuring tracking accuracy, and a statistical means of reporting tracking accuracy. In the interim while TS 62727 was proceeding through the formal publications procedure, WG7 and the tracker subgroup were working towards the preplanned design qualification standard for solar trackers. In early 2012 WG7 submitted to IEC a new work item proposal and a draft document titled '*Photovoltaic Systems-Design Qualification for Solar Trackers*'. In 2013 the first draft document was numbered IEC 62817 which would then work its way through the formal IEC publication process till it was finally published in August 2014.

The design qualification standard includes all the text associated with TS 62727 as well as testing and pass fail criteria appropriate for solar trackers. TS 62727 expired in 2014 and IEC 62817 became the only one document that provides both definitions/terminology and design test criteria. It is useful to quote the *object* directly from IEC 62817 as the purpose of a design qualification standard is often misunderstood. The object of IEC 62817 is as follows:

1. 'This document ensures the user of the said tracker that parameters reported in the specification sheet were measured by consistent and accepted industry procedures. This provides the customer with a sound basis for comparing and selecting a tracker appropriate to their specific needs. This standard provides industry-wide definitions and parameters for solar trackers. Each vendor can design, build, and specify the functionality and accuracy with uniform definition. This allows consistency in specifying the requirements for purchasing, comparing the products from different vendors, and verifying the quality of the products.
2. 'The tests with pass/fail criteria are engineered with the purpose of separating tracker designs that are likely to have early failures from those designs that are sound and suitable for use as specified by the manufacturer. Mechanical and environmental testing in this standard is designed to gauge the tracker's ability to perform under varying operating conditions as well as to survive extreme conditions. Mechanical testing is not intended to certify structural and foundational designs, as this type of certification is specific to local jurisdictions, soil types, and other local requirements.'

The above *object* makes clear the intended use IEC 62817 as well as uses for which it is not intended. In other words, this document was written to help distinguish quality tracker designs from poor designs but in no way does it imply a design is *reliable* or that it meets the codes of a given local jurisdiction. In many cases PV and CPV modules are being expected to perform under outdoor operating conditions for 20 or more years. One might expect the same of solar trackers but it should be clear that the ability to consistently perform over the lifetime of a product is *reliability* and not design qualification (see Chapter 9 for clarifying both concepts). A quality design has the ability to perform the intended functions within the intended environment over the short term (for trackers the short term is not fully established but might be considered the first few years of life). It should be further noted that many tracker manufacturers are designing trackers with the intent to replace certain parts and provide maintenance at given intervals. This is an engineering design choice and IEC 62817 is written in a way to allow testing of trackers with an array of design choices without bias towards one design or another.

It is clear that IEC 62817 is intended to distinguish quality designs from poor designs and the following tests are included specifically for this purpose:

- tracking accuracy;
- functional validation tests (verify basic functions, stow, tracking limits, etc.);
- basic performance metrics such as energy usage, time to stow, etc.;
- mechanical testing;
- drive train pointing repeatability;
- deflection under static load;
- torsional stiffness, drive torque, backlash;
- moment testing under extreme wind loads;
- accelerated environmental testing applied to a moving drive train;
- 40 temperature cycles from $-20\,°C$ to $55\,°C$ (dust exposure during test);
- 10 humidity freeze cycles;
- freeze/spray;
- accelerated mechanical testing (3650 cycles or $\sim$10 years following sun);
- qualification testing for tracker electronic equipment similar to IEC 62093 [34];
- functional testing, surge immunity, shipping vibration, shock test, IK class (mechanical impact), IP class (dust ingress), 1000 hours damp heat, UV test, 200 thermal cycles, 10 humidity freeze cycles, and robust terminal test.

While the above tests are intended to test the quality of the design, all the tests should be viewed through the IEC consensus based development process. The tracker subgroup that drafted these test procedures was comprised mainly volunteers from the CPV industry, the tracker industry, and international labs associated with CPV work. While every effort was made to develop meaningful tests these tests were often not developed through experimental research. Tracker and CPV companies have made efforts to implement the tests through the drafting stage and to provide feedback to improve the document but ultimately IEC 62817 will be further tested after it is published. The community will provide feedback with time and changes will be made in later editions to improve the value of the standard with time. It should also be understood that the consensus from the community results in trade-offs between test rigor and other variables such as time to completion and test cost. A simple example of this is measurement of tracking accuracy. It would be ideal to measure tracking accuracy over a long period of time that covers more than one season but the desire is always to have tests that can be completed quickly. Again, experience will be the eventual judge of each test in IEC 62817.

WG7 as well as the CPV community has periodically discussed the need to go beyond design qualification and to develop a reliability standard for solar trackers. The primary argument for such a standard has been due to challenges associated with financing projects that use solar trackers. The financial community is tasked with placing risk and therefore cost on each aspect of a proposed project. Trackers are perceived as an increased risk over fixed mounting configurations and therefore can be more expensive to finance. If the project owner can document tracker reliability over the course of 20–30 years, costs can be lowered. Although the usefulness of reliability standard is understood it is unclear if WG7 will move to develop one. Currently there are very little data pertaining to performance and reliability of solar trackers over the long term, namely, 10–30 years (see section 'Reliability of Plants' in Chapter 9). When such data becomes available there still must be a research effort to develop reliability tests that correlate with failure mechanisms that are seen in the field.

References

1. Luque-Heredia, I., Martin, C., Mananes, M.T. *et al.* (2003) A subdegree precision sun tracker for 1000X micro-concentrator modules. Conference Record of the 3rd World Conference on Photovoltaic Energy Conversion, Osaka, 857–860.
2. Randall, D.E., Grandjean, N.R. (1982) Correlation of Insolation and Wind Data for SOLMET Stations SAND82-0094, Sandia National Laboratories.
3. Luque-Heredia, I., Quéméré, G., Magalhães, P.H., *et al.* (2006) Modelling structural flexure effects in CPV sun trackers. Conference Record of the 21st European Photovoltaic Solar Energy Conference, Dresden, 2179–2183.
4. Lin, C.K., and Fang, J.Y. (2013) Analysis of Structural deformation and concentrator misalignment in a roll-tilt solar tracker. *AIP Conference Proceedings*, **1556**, 210–213.
5. Welzl, E. (1991) Smallest enclosing disks (balls and ellipsoids), in *New Results and New Trends in Computer Science, Lecture Notes in Computer Science*, (ed. Hermann Maurer) **555**, 359–370.
6. Luque-Heredia, I., Chiappori, E. and Laurent, O. (2009) CPV Tracking system: Performance issues, specifications and design for CPV applications. Conference Record of the 2nd Concentrated Photovoltaic Summit, Toledo, 2009.
7. Kim, S.K., Kang, S.M. and Winston, R. (2014) Tracking control of high concentration photovoltaic systems for minimizing power losses *Progress in Photovoltaics*, **22** (9), 1001–1009.
8. McCay, T.E. (1975) Sun Tracking Control Apparatus US Patent no. 3 917 942, Nov. 1975, available at: http://patft.uspto.gov/netacgi/nph-Parser?Sect2=PTO1&Sect2=HITOFF&p=1&u=/netahtml/PTO/search-bool.html&r=1&f=G&l=50&d=PALL&RefSrch=yes&Query=PN/3917942
9. Luque, A., Sala, G., Alonso, A. *et al.* (1978) Project of the Ramón Areces Concentrated Photovoltaic Power Station. Conference Record of the 13th IEEE PV Specialist Conference, Washington DC 1139–1146.
10. Maish, A.B., (1991) The SolarTrak Solar Array Tracking Controller, SAND90-1471, Sandia National Laboratories.
11. Quero Reboul, J.M., García Franquelo, L. and Guerrero, A. (2001) Light source position microsensor. Conference Record of the 2001 IEEE International Symposium on Circuits and Systems, Sidney, 648–651.
12. Weslow, J.H. and Rodrian, J.A. (1980) Solar Tracker US Patent nº 4 215 410, available at: http://patft.uspto.gov/netacgi/nph-Parser?Sect1=PTO1&Sect2=HITOFF&d=PALL&p=1&u=%2Fnetahtml%2FPTO%2Fsrchnum.htm&r=1&f=G&l=50&s1=4,215,410.PN.&OS=PN/4,215,410&RS=PN/4,215,410 (July 1980).
13. Software Bisque, T-Point Telescope Pointing Analysis Softtware, http://www.bisque.com/sc/media/p/29538.aspx (August 28, 2014).
14. Bernstein, J., (1984) *Three Degrees Above Zero: Bell Labs in the Information Age*. Charles Scribner and Sons, New York, 213–233.
15. Arboiro, J.C. and Sala, G. (1997) Self-learning tracking: a new control strategy for PV concentrators, *Progress in Photovoltaics*, **5**, 213–226.
16. Luque-Heredia, I., Moreno, J.M., Quéméré, G. *et al.* (2005) CPV sun tracking at Inspira. Conference Record of the 3rd International Conference on Solar Electric Concentrators for the Production of Electricity or Hydrogen, Scottsdale, 2005.
17. Luque-Heredia, I., Gordillo, F. and Rodríguez, F.A. (2004) PI Based hybrid sun tracking algorithm for photovoltaic concentration. Conference Record of the 19th European Photovoltaic Solar Energy Conference, Paris, 2383–2387.
18. Marquardt, D.W. (1963) An algorithm for least-squares estimation of non-linear parameters. *Journal of the Society of Industrial and Applied Mathematics*, **11** (2), 431–441.
19. Törn, A. and Viitanen, S. (1992) Topographical global optimization, in *Recent Advances in Global Optimization* (eds C.A. Floudas and P. M. Pardalos), Princeton University Press, Princeton, 384–398.
20. Koopman, B.O. (1980) *Search and Screening: General Principles with Historical Applications*, Pergamon Press, New York, 213–227.
21. Brent, R.P. (1973) *Algorithms for Minimization without Derivatives*, Prentice-Hall, Englewood Cliffs, New Jersey.

22. Quémeré, G, Cervantes, R. Moreno, J.M. and Luque-Heredia, I. (2008) Automation of the calibration process in the SunDog® STCU Conference Record of the 23rd European Photovoltaic Solar Energy Conference and Exhibition, Valencia, 887–890.

23. Luque-Heredia, I., Moreno, J.M., Quémeré, G., Cervantes, R. and Magalhães, P.H. (2005) SunDog STCU TM A Generic Sun Tracking Unit for Concentration Technologies. Conference Record of the 20th European Photovoltaic Solar Energy Conference, Barcelona, 2047–2050.

24. Luque-Heredia I., Moreno, J.M., Quémeré, G., Cervantes, R. and Magalhães, P.H. (2008) Equipo y procedimiento de control de seguimiento solar con autocalibración para concentradores fotovoltaicos. Spanish Patent no. 2273576, Mar. 2008.

25. Galbraith, G. (1988) Development and Evaluation of a Tracking Error Monitor for Solar Trackers, SAND88-7025, Sandia National Laboratories.

26. Zabiyakin, A.S., Prasolov, V.O., Baklanov, A.I. et al. (1999) Sun sensor orientation and navigation systems of the spacecraft. Proceedings – SPIE the International Society for Optical Engineering, 3901, 106–111.

27. Luque-Heredia, I., Cervantes, R. and Quémeré, G.A. (2006) Sun tracking error monitor for photovoltaic concentrators. 4th World Conference on Photovoltaic Energy Conversion, Hawaii, 706–709.

28. Davis, M., Lawler, J., Coyle, J. et. al. (2008) Machine vision as a method for characterizing solar tracking performance. Conference Record of the 33rd IEEE Photovoltaics Specialist Conference, San Diego, pp. 1–6.

29. Missbach, T. and Jaus, J.A. (2012) A new sensor for measuring tracking accuracy, tracker vibration, and structural deflection. AIP Conference Proceedings, 1477, 262–265.

30. Bennett, G.G. (1982) The calculation of atmospheric refraction in marine navigation. Journal of the Institute of Navigation, 35, 255–259.

31. Schaefer, B.E. and Liller, W. (1990) Refraction near the horizon. Publications of the Astronomical Society of the Pacific, 102, 796–805.

32. U.S. Naval Observatory (1998) Multiyear Interactive Computer Almanac 1990–2005, Version 1.5, Willmann-Bell, Richmond.

33. Cervantes, R., Quémeré, G. and Luque-Heredia, I. (2008) SunsSpear calibration against array power output for tracking accuracy monitoring in solar concentrators. Conference Record of the 23rd European Photovoltaic Solar Energy Conference and Exhibition, Valencia, 890–894.

34. International Electrotechnical Commission (2005) IEC 62093 Balance-of-system components for photovoltaic systems – Design qualification natural environments. IEC, Geneva.

7

CPV Modules

Stephen Askins and Gabriel Sala Pano
Instituto de Energía Solar, Universidad Politécnica de Madrid, Spain

7.1 Introduction

In preceding chapters of this handbook, the optics (Chapter 4) and the high efficiency solar cells (Chapter 2) used in concentrator photovoltaics have been discussed in detail. But, at some point, these two elements must be physically integrated to form a concentrator photovoltaic device. If the micro-concentrator architecture is used; where many small solar cells, a few centimeters square or less, are each combined with many individual optical systems, whose size is on the order of 1500 square centimeters or less; then the optics and the solar cells are usually pre-integrated in a factory to form an assembly referred to as a *module*, whose size is on the order of a few meters square or less, and which will be later integrated into larger assemblies in the factory or in the field and finally mounted on a tracker (see Chapter 6).

In this chapter, we intend to provide a global view of the process involved in the design of the module and its components, describe the various trade-offs and decisions that a system designer faces, and propose some simplified models and rules of thumb for system designers at the very beginning of the process. The chapter is organized as follows; first we will define the module, as used in this chapter, and describe its primary components. Then we will discuss the major steps in its design, following the order in which the design decisions must be taken, and providing equations to be used in the early stages of module design and optimization. Finally, we will touch on some discussion of the relevant IEC standards for rating and qualification of modules. At the end of the chapter (in Annexes 7-I to 7-IV), several case studies illustrate how companies (Abengoa Solar, Daido Steel, Soitec and Suncore) have faced the design and manufacturing of their modules.

7.2 What is a CPV Module?

The technical principal of concentrator photovoltaics (CPV) can seem, at first, to be simple. One must place a converging optical element, either a mirror or lens, in the path of the near-parallel

---

Handbook of Concentrator Photovoltaic Technology, First Edition. Edited by Carlos Algora and Ignacio Rey-Stolle.
© 2016 John Wiley & Sons, Ltd. Published 2016 by John Wiley & Sons, Ltd.

rays of light produced by the sun, and then locate a solar cell, preferably with as high efficiency as possible, in the focus of this optic. However, as will become clear in Chapter 14, concentration technology has only one goal: cost competitive production of solar electricity. The objective of CPV, then, is not only a laboratory demonstration, but the manufacture of a product that can be economically manufactured and can withstand up to 30 years or more of outdoor exposure without a failure or significant performance degradation (see Chapter 9).

Therefore, the overall level of concentration produced by the CPV module, that is density of luminous power that is focused onto the solar cells, should never be chosen with the criteria 'as much as possible', but should always be the minimum required to reach the cost objectives that an economic analysis shows are necessary. To operate at any level higher than this is to cause unnecessary stress on the solar cell and to will result in the over-dimensioning of other elements of the system design. In fact the concentration of sunlight, while providing a possible path to economic generation of solar electricity, also causes a series of complications in a photovoltaic device:

- The solar cells themselves must be constructed in such a way to minimize series resistance, R_S, in order to avoid Joule losses, which are proportional to I^2Rs.
- The electrical connections between cells must also be of sufficiently low resistance and robustness in order to avoid resistive losses.
- As in other PV technologies, the electricity produced tends to be of low voltage and high current, the opposite of that which is desired for the transportation of this electricity. In CPV systems, the effective area for current generation is that of the lens, and so currents may be higher than typical flat plate modules. In order to boost voltage and limit current, many solar cells are connected in series, which can lead to losses if each solar cell does not generate the same current as its neighbors.
- Any luminous flux incident on the cells that is not converted into electricity, which is the majority even for the highest efficiency cells, is instead converted into heat. The very concentration that allows us to minimize solar cell area also concentrates this heat into a very small area. A path of low thermal resistance must be provided between the solar cell and the ambient, both to protect the components from dangerously high temperatures as well as to limit voltage losses (see Chapter 5). Thus, the materials which come into contact with the solar cell must very good conductors of heat. This means that materials commonly used in conventional flat plate photovoltaics, such as glasses and plastics, are not adequate for CPV. The solar cells cannot be laminated between such materials to provide environmental protection.

It can therefore be seen that any CPV technology necessarily consists of the following elements:

1. High efficiency solar cells with low series resistance and capable of operating at high concentration;
2. a convergent optical system capable of focusing parallel rays of light to a sufficiently high concentration level, while allowing reasonable tolerance of alignment and of assembly;
3. a mechanical structure that can hold the last two elements at the correct distance, with a reasonably high accuracy;
4. the ability to connect such cells in series, which additionally requires ensuring that each cell produces very similar current;

5. a thermal pathway that can remove the concentrated heat produced on the cell such that the cell and its substrate are kept at tolerable temperature levels;
6. dielectric materials that keep such cells isolated electrically from the outside world even at high voltage levels, without impeding heat removal; and finally
7. a wall or enclosure that keeps the sensitive portions of the system, including optics and solar cells, protected from dust, water, wind, insects, and any contaminants that would otherwise cause system degradation over time in the field.

Any device which is able to cover the aforementioned seven functions can be called a CPV module, and is the subject of this chapter.

7.3 Definition, Functions, and Structure of a CPV Module

In section 7.2 we have offered a possible definition for a concentrator photovoltaic module. However, for the purposes of this chapter, we must be even more specific. As we have already mentioned, we focus on the high concentration, micro-concentrator architecture. In these systems, the optics and the solar cells are generally integrated into an assembly referred to as a *module*, which will be later integrated into larger assemblies, or *systems*, either in the factory or in the field. IEC standard 62108 [1] defines a CPV module as 'A group of receivers, optics, and other related components, such as interconnection and mounting, that accepts un-concentrated sunlight' and goes on to state that 'All above components are usually prefabricated as one unit, and the focus point is not field-adjustable.' This is the definition we will adhere to in this chapter.

There are many CPV systems that do not include any assembly that fits the above definition, notably single-axis low concentration systems[1] like the Euclides or Skyline systems [2,7,8], high concentration 'dish' systems such as those of Solar Systems, or the newer REhnu [9,10], or heliostat/fixed tower systems, like that proposed by RayGen. In these systems, relatively large area optics (multiple meters square) are integrated with assemblies of multiple solar cells in the field. These prefabricated assemblies of solar cells are sometimes also referred to as 'modules;' however this chapter discusses only modules according to the IEC 62108, and therefore excludes all CPV architectures other than micro-concentrator. It should be noted that at the time of this writing, both the installed base and the forward momentum of CPV development is primarily in high concentration micro-concentrators, most likely because their modularity allows a fuller leveraging of mass-production with its attendant cost benefits, much as the modularity of flat-plate solar has surely been influential in the precipitous decline of crystalline silicon-based photovoltaics in recent years.

In the flat-plate crystalline silicon solar industry, a 'common' module design has emerged from standard cell sizes (which are in turn are related to standard wafer sizes), required voltages, and a simplified set of module functions (protection of the cell from the ambient while ensuring light capture). On the contrary, the wider design space in CPV, even when limiting our discussion to the micro-concentrator modules, has resulted in much more widely varying module design.

---

[1] The principles and experiences of low (2× to 15×) and medium (15× to 150×) concentration systems have been widely studied, developed and demonstrated in the 1990s, and the reader should turn to [2] for an in depth discussion of their engineering. SunPower, an early innovator in back-contact solar cells for concentration [3] has returned to the low-concentration market with a 7× product [4], which uses a modification of the same interdigitated back contact solar cells [5] used in their flat plate offerings. This product appears to be especially successful in China, where the company has recently completed a 300 MW_p receiver manufacturing facility, and claims a 3 GW pipeline [6].

If there is one aspect of CPV design that is becoming standardized, it is the choice of the point-focus Fresnel lens as the primary optic. For obvious reasons of material use, the Fresnel lens (see Chapter 4) is the only practical refractive primary optic to be considered: a normal plano-convex lens would either require a very high f-number,[2] or use far too much material.[3] The only exception to the use of Fresnel lenses by a serious commercial CPV provider has been SolFocus, a venture-backed CPV systems firm which installed nearly 15MW$_p$ [12,13] of its planar Cassegrain reflector type system [14] from 2007 until 2013, when they were forced to declare bankruptcy for lack of additional funding. This design ameliorated the primary disadvantage of the mirror-based modules: the shadowing caused by placing the receivers on the sun side of the modules.[4] However, doing so requires the addition of an intermediate optic and the attendant loss in optical efficiency. The two extant CPV systems providers with 'large' installed capacity: SunCore with around 100 MW in the Gobi Desert [12], Soitec/Concentrix, with a cumulative 75 MW installed to date [15], use silicone-on-glass [16] Fresnel lenses (see the corresponding annexes at the end of this chapter). This is because, while mirrors offer the promise of better performance due to the of chromatic aberration (theoretically they can achieve a better concentration–acceptance product as defined in Section 7.6.1 and described in Chapter 4), a Fresnel design offers simplicity. The module becomes an empty box, with the lenses, which are manufactured of-a-piece as parquets the size of the entire module on one side, and a backplane with the mounted cells on the other. This greatly reduces both the number of components and the number of precise assembly operations involved in module production. For this reason, and for the sake of simplifying the discussion, we will focus exclusively on Fresnel-based CPV modules.

However, even with the design space whittled down to Fresnel micro-concentrators we still see a wide variety of module sizes and shapes. This is because the system designer must first choose the concentration ratio and the cell size (since the cells are diced from wafers, theoretically any cell size can be chosen, however it should be noted that cell providers are moving towards only a few standard cell sizes). This choice impacts both the performance and cost of many aspects of the module design, as well as its balance of systems and the installation, but as is common in engineering, there are many competing trade-offs, and the optimum choice is not clear. Although we may not be able to define once and for all the optimum concentration ratio and cell size, we hope that this chapter will at least give the system designer a good understanding of the advantages and disadvantages of the available choices.

7.3.1 Functions of a CPV Module

A CPV module is a complex engineered product with many components. We could divide these components into active and non-active components, where the former consist basically of the optics and the solar cells, which are amply covered by Chapters 2 and 4 of this handbook and the latter consists of all other components which form the module and the focus of this

---

[2] The f-number of a condensing optic is its focal distance divided by its diameter.
[3] The module design of Semprius Inc. [11] is a notable exception. This is enabled by their very small cell size, which allows them to use a relatively large f-number without making the module unduly thick.
[4] The intermediate optics, however, did shadow approximately 4% of the entrance aperture which is a loss of optical efficiency that would be considered intolerable in Fresnel-based designs.

chapter. It could be said that the active components have only one function: to generate solar electricity economically. The function of the 'non-active' components is to support them in that objective, specifically by carrying out the following roles:

- opto-mechanical: to support the optics such that they focus light onto the solar cells;
- electrical: to electrically connect each cell and allow for external connection as well as provide the proper electrical insulation where needed;
- thermal: to provide a conduit for heat to be extracted from the cell to the ambient;
- environmental: to protect the optics and the cell from the elements while also preventing high positive or negative internal pressures due changes in internal temperature.

Each of these functions will be addressed in the following sections. However, first we propose some general definitions, as well as a taxonomy, or bill of materials, for a generalized CPV module that is applicable to the extant module designs at the time of writing, so that we have a list of accepted terms to use in later sections.

7.3.2 General Terms and Definitions

Optics-cell unit: A CPV module as defined above is composed of many repeating concentrating units, that is concentrating optics paired with receivers, although these units are not generally fabricated as a sub assembly, but rather come into being when the module itself is assembled. We use the term optics-cell unit to refer to this minimum unit of concentration that is then repeated a number of times in given module.

Lens size categories: In later discussion we will see that for a given chosen concentration ratio, 1) we must first choose a cell size and 2) this choice of cell size then controls our lens size. We will also note that there are considerable differences in the thermal management, electrical interconnection, and assembly techniques used in large- vs. small-lensed systems. Therefore, for the purposes of this chapter, we will use to use a lens size of $100 \, cm^2$, corresponding to cells just over $3 \times 3 \, mm$ in size used at 1000×, as an arbitrary dividing line between large and small-lensed system. In the following pages, we will explain that large-lensed modules will require dedicated heat sinks, will generally contain only serial connections, and may have sufficiently wide assembly tolerances that allow receivers to be located by mechanical features of the module backplane, whereas small-lens systems use parallel-serial connections to reduce voltage levels, do not require additional dedicated heat sinks but can rely on a thin metal backplane for dissipation, and require high tolerance positioning of receivers using pick and place. Within the large-lensed systems there are also apparent differences between the smaller lenses in this group, which we may refer to as 'medium-lensed systems' and the largest, in terms of appropriate heat sink approach and chassis manufacturing techniques. However, the reader will understand that this separation of all CPV modules into discrete groups is a merely a simplification, and that the module requirements for any given optical design must be evaluated individually.

7.3.3 Structure of a CPV Module

We propose the breakdown and terminology for the components of a basic CPV module outlined in this section. The elements are illustrated using three example module designs (Figures 7.2–7.4).

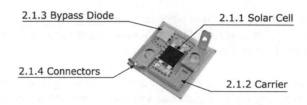

Figure 7.1 Typical cell-on-carrier assembly design for medium- to large- lens module designs [17] (Reprinted with permission of Azur Space)

1. *Window subsystem*:
    1.1 *Primary optical element* (POE): The POE redirects the incoming parallel light rays from the sun and to a focal point, where the receiver is placed. If the POE is a lens, then they usually also form the front face of the enclosure.
2. *Receiver subsystem*: we use the term *receiver* to indicate the solar cell and all associated components that are pre-assembled to it before integration the module, including any secondary optics, and possibly a heat sink.
    2.1 *Cell-on-carrier* (CoC): in most extant CPV systems the solar cell is first bonded to a small circuit board or carrier. This carrier allows for easy electrical connection to the cell, acts as a heat spreading layer for thermal dissipation, and allows for a bypass diode to be mounted in parallel to the cell (Figure 7.1).
        2.1.1 *Solar cell*: clearly the most important component of the receiver (and perhaps the entire CPV module) is the solar cell. We understand it to mean a single die as removed from a wafer, ready for electrical connection.
        2.1.2 *Cell carrier*: the circuit board, generally composed of alternating layers of electrically conductive and insulating material, with a very simple circuit in the topmost conductive layer to differentiate the surface on which the solar cell is to be mounted (and is therefore electrically connected to the back surface of the cell) from that which will be connected to the topside metallization via wire or tab bonding.
        2.1.3 *Bypass diode*: in almost all CPV designs, a diode is placed in parallel with the solar cell to provide a current bypass, as described in Section 7.7.2.
        2.1.4 *Connectors*: in large-lens systems, small connectors are often soldered onto the carrier for easy (manual) attachment of cell-to-cell interconnections. In small-lens systems this is not usually the case, and cell-to-cell interconnections are accomplished by automated means, such as heavy wire bonding.
    2.2 *Secondary optical element* (SOE): By placing a small optical element on the cell itself, we can increase the effective aperture area of the receiver. This allows us to operate at a higher overall concentration ratio than a Fresnel lens alone would allow us to provide. There are two main classes of secondary optics in use: reflective (hollow inverted pyramids of thin reflective sheet with a reflective coating) or glass/dielectric (composed of a transparent material that is usually molded or ground and polished glass, but could also be silicone, and which use a refraction or a combination of and total internal reflection).
3. *Thermal subsystem*: the following components will be part of the receiver subsystem or the mechanical subsystem, depending on the lens size.

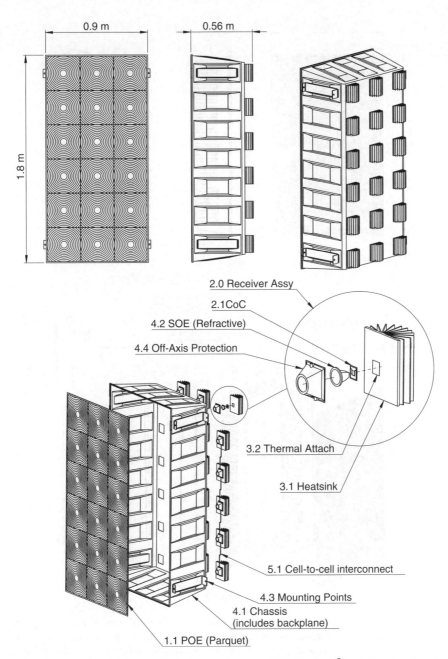

Figure 7.2 Example module design for large-lensed system ($300 \times 300 \, \text{mm}^2$ lens) showing the principal components. The module design uses a parquet POE, structured sheet-metal chassis, and a complex receiver assembly including not only the CoC and SOE, but individual heat sink and SOE support/off axis protection. The module is shown to possess upper and lower mounting points, so that when the module are mounted together in a sub-array frame, rigidity can be provided by the modules themselves (see section 7.5.2.3)

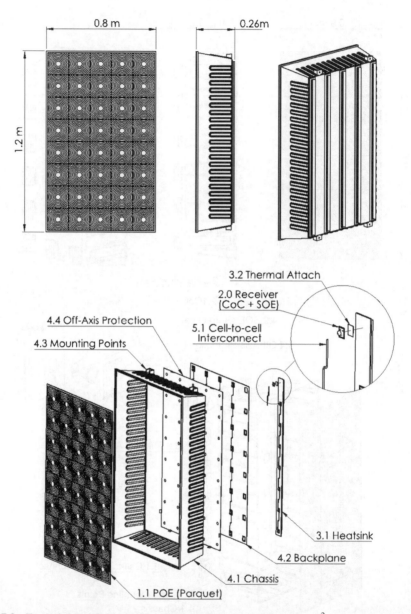

Figure 7.3 Example module design for medium-lens size ($150 \times 150\,\text{mm}^2$ lens) showing principle components. Example module design uses a parquet POE, plastic injection molded chassis with separate metal backplane, a linear heat sink that acts as a stiffener and simple receiver composed of a CoC and an over-molded dome SOE

3.1 *Heat sink/spreader*: depending on the size of cell selected, the backplane alone, based on its thickness and amount of area exposed to the ambient, may not allow sufficient thermal transfer from the cells to keep their temperature in operation at an acceptable value. In the largest cell designs (7×7 mm and up), acceptable temperatures can only be

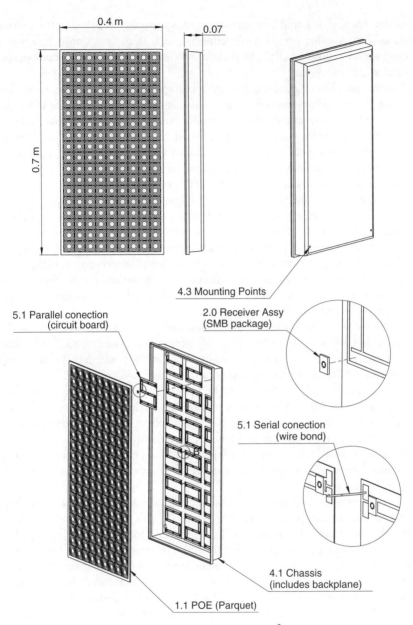

Figure 7.4 Example module design for small-lens size (40×40 mm^2 lens) showing principle components. Example module design uses a parquet POE, single-piece stamped metal chassis/backplane, simple receiver package (no SOE) that is surface mounted on a multi-cell circuit board providing parallel connections

reached with the use of large, individual finned heat sinks, which should be thermally attached directly to the cell carrier to reduce the number of transitions in keep the thermal stack, and can be considered a component of the receiver. For intermediate cell sizes, another option is an additional thick aluminum or copper plate bonded to the carrier that acts as initial heat spreader, and is thermally attached to the interior of the backplane. Another option for intermediate cell sizes is a finned heat sink that is common to multiple cells, and therefore should be considered a component of the mechanical subsystem.

    3.2 *Thermal attachment*: depending on the topology, as discussed above, either the carrier or the receiver subsystem is fixed to a heat sink (which may simply be the back plane). This can be done mechanically or with adhesive, but in either case usually a material is placed at the interface to enhance the thermal transfer between the two. Depending on the properties of the cell carrier, this interface may also need to act as a dielectric barrier.

4. *Mechanical subsystem*: the 'balance of system' of the module.

    4.1 *Chassis*: the chassis is the mechanical part that ensures a protective envelope around the light path between the window and the cell, and provides a structure onto which the rest of the components are mounted. It may be itself composed of a single piece or multiple piece parts. Common materials are thin steel or aluminum sheet, fiber-reinforced polymers, and in some module designs, glass. Its depth usually fixes the distance between the lens and the receiver (such that the receiver is in the focus of the optics). Its primary function is opto-mechanical support and environmental protection, although in metal designs it may also be relied on as a thermal dissipater, especially if it is of a piece with the backplane.

    4.2 *Backplane*: The rear surface of the module onto which the receivers are mounted. In almost all CPV designs the backplane is metallic (aluminum or steel) even when other portions of the chassis, e.g. the side walls, are of another material. This is in part to help with thermal dissipation, and in part because off-axis rays could melt plastic parts in this area. In designs with small cells, the backplane may be the only thermal dissipater. In designs with finned heat sinks, the CoCs are affixed directly to the heat sink, which are then mounted to the back surface of the backplane, with the active portions of the receiver allowed to pass through a hole in the backplane.

    4.3 *Mounting points*: mounting points are generally fixed to the chassis for later mounting to an array frame. These should be analyzed using finite element analysis (FEA) for strength and rigidity.

    4.4 *Off-axis protection*: the system designer must always anticipate that the tracking system may be shut off unexpectedly at any moment. Then, as the sun continues to travel across the sky, the 'spot' of concentrated light produced by the primary optics will start to travel away from the receiver aperture and will then concentrate onto nearby areas. Either both the receiver and the cell interconnection must be prepared to withstand high irradiances, or an additional component, often a thin metal sheet with holes aligned over the receiver apertures, must be included in the module design.

    4.5 *Vent*: the module vents allow the module to 'breathe' according as the density of the enclosed air changes due to temperature, while preventing the ingress of dust, liquid water, insects, and other contaminants.

5. *Electrical subsystem*

    5.1 *Cell-to-cell interconnection*: the cells are connected together in series or parallel according the requirements of each design. In large-lens systems, small pieces of insulated copper cable are usually attached to connectors on the receiver by hand. In small-lens systems, heavy wire bonding may be used to robotically connect each receiver to the next.

5.2 *Module-to-module interconnection*: either cables of length sufficient to reach to the next module ('pigtails') pass through the chassis wall via a cable gland, the ends of which are provided with solar connectors, or panel type solar connectors are mounted directly to the chassis. For large-lens systems, commonly the module to module cable (which carries the same current as any of the cell-to-cell cables), simply passes through a cable gland to the first and last receivers respectively, where they attach to the connector on the cell carrier as any other cell-to-cell cable. In small-lens systems, the module-to-module cable may be connected to terminals from which various parallel connections to the strings of cells are joined.

5.3 *Bypass/blocking diodes*: in large-lens systems, the only bypass diode required is the individual cell bypass diode. In small-lens systems with parallel and series connections, blocking or bypass diodes will be required in the module-level circuit, depending if the configuration is series-parallel or parallel series. See Section 7.7.

7.4 Design Process and Prototyping Stages

In this section we describe our recommended process for the design of the architecture and inactive components CPV module, including prototype stages, for a CPV module using an array of point-focus Fresnel lenses. The following sections of this chapter describe each step of the design process, including equations for initial sizing of components. We also describe the phases in which experimental testing should be performed to confirm simulations results and to inform further design. This process is illustrated in Figure 7.5 and has the following steps:

1. *Architecture selection*: the first stage involves making the big, global decisions that will affect every other aspect of the design. These decisions should ideally be made using a cost model optimization at every step. The four steps are:

   1.1 As described in the next section, the design process must begin with two critical decisions: the concentration ratio, and the solar cell size to be used.

   1.2 The optical architecture must be chosen, specifically the question of whether to include an SOE and what type to use. Designers can turn to Section 7.6 to estimate the effect this choice will have on acceptance angle and required tolerances.

   1.3 Based on our cell-size and geometrical concentration ratio, we can select a thermal architecture based on the guidelines of Section 7.8, which is required for mechanical design.

   1.4 Assuming that the *f*-number is fixed, we know our lens size and module depth. We must choose an number of lenses, and their arrangement, to calculate our total module size, in order to begin preliminary mechanical design

2. *Prototype CoC purchase*: once we know our cell size, we should purchase an off-the-shelf pre-assembled CoC (if one exists for our size) or otherwise procure high-efficiency multi-junction solar cells mounted on a carrier, that we can use for optics testing and initial module prototypes. These need not fulfill all our requirements for our final CoC in terms of cost, reliability, etc.

3. *Initial design*: the initial design steps can proceed in parallel, but they are listed here in order of priority.

   3.1 *Optical design*: both the primary and the secondary optical elements are designed using geometric optics or more advanced methods. The designs are transferred to optical

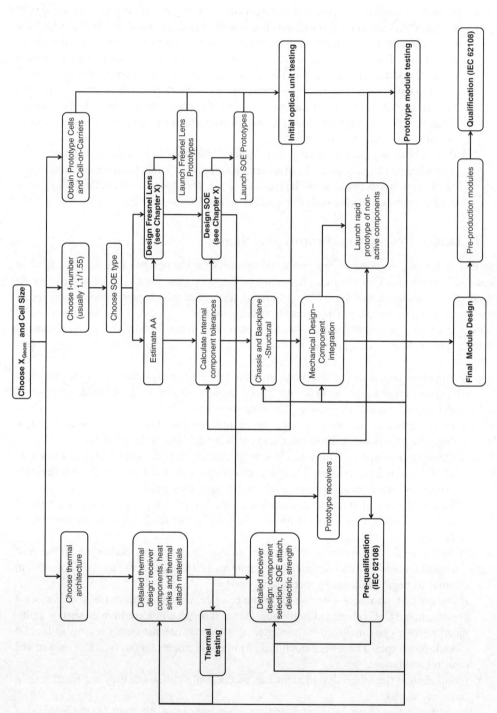

Figure 7.5 Design process for CPV modules

simulations package for analysis and design improvement. Finally, the electronic design files are sent out for prototyping (see Chapter 4).

3.2 *Thermal design*: we must now move beyond the simplifications of Section 7.8 and perform a detailed thermal analysis and optimization of our thermal solution using finite element analysis.

3.3 *Chassis/mechanical design*: the 'nuts and bolts' of designing a CPV module is the same as that of any other engineered product. If results from Step 4 (see below) are not yet available, the simplifications of Section 7.6 should be more than sufficient for setting mechanical tolerances and deformation limits. We should consider a derivative of the design that is amenable to rapid prototyping.

3.4 *Receiver design*: this is a critical area for reliability and cost, as well as supply chain management (we do not need to co-locate receiver assembly and module assembly, and there may be advantages to not doing so) However, we can move forward with prototyping our optics-cell unit and even initial modules without our own custom designed receivers, and instead rely on outsourced CoCs and a relatively 'non-industrial' receiver for the initial stages.

4. *Optics-cell unit testing*: for the first testing stage we need only a single prototype POE and SOE, and CoC prototype. These components should be mounted on a test bench that allows their relative movement of the component using opto-mechanical actuators. (Figure 7.6). At the IES-UPM, we find using a solar simulator [18] to be of great use for these evaluations, since we can achieve repeatable illumination conditions as we make small changes in the optics configuration. If the test is performed outdoors, great care must be taken that the conditions do not change unduly during measurements. Measurement sets should be taken very quickly, near solar noon on an exceedingly clear day, and with a temperature controlled platform for the cell, if possible. The evaluations to be made, in decreasing order of priority, are:

4.1 *Initial performance evaluation*: measurement of electrical efficiency and acceptance angle. If the CoC used for evaluation has also been measured under concentration with no optics (i.e. with a cell flasher) we can also estimate optical efficiency.

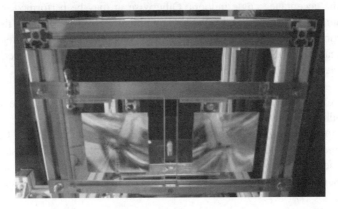

Figure 7.6 Two prototype lenses mounted in an open test bench for initial optics-cell unit evaluation. In these tests, the left hand lens is being used with a prototype CoC and SOE for the evaluations described in Step 4, while the right hand lens is being evaluated using the imaging technique of [20]

4.2 *Lens to receiver distance optimization*: it has been shown to be very difficult to predict the optimum distance between the POE and receiver aperture. Additionally, certain types of POEs have relatively high temperature coefficients with focal distance. We recommend directly measuring electrical efficiency versus receiver distance, if possible as a function of lens temperature. This will also provide an opto-mechanical tolerance value for the receiver distance

4.3 *Determination of spectral effect of optics*: we can use the methods of [19] to find how well matched the sub-cells are at the reference spectrum.

4.4 *Additional tolerances:* with this kind of test bench, it is trivial to additionally extract tolerances for lens-to-cell alignment and lens rotation.

5. *Thermal testing*: when we have an initial design for our thermal stackup (from Step 3.2), we recommend attempting to evaluate its thermal resistance through a simple laboratory experiment. For instance, a solar cell can be forward biased to produce heat, and temperatures measured on either side. This will prevent investments in the next step with inadequate thermal designs.

6. *Module rapid prototyping and test*: assuming the performance measured in step 4.1 meets expectations, we can feed the focal distance from step 4.2 into our mechanical design. Once step 3.2 is complete, we should move to making prototype modules. We mention the thermal design specifically, because we believe that this is the element of the full module design that is most important to test at this stage, because typically we find that early prototypes do not meet expectations of cell temperature. See Chapter 12 of this handbook for detailed descriptions of these tests. The evaluations to be completed, in decreasing order of priority, are:

6.1 *Outdoor performance testing*: the focus is on estimating electrical efficiency in operation and spectral dependence. The cell temperature in operation should be estimated, using, for example, the methods of [21].

6.2 *Indoor performance testing*: can also be useful to understand performance. Comparing the V_{oc} measured indoors (at known temperature) to that during the outdoor measurements can be useful for the estimation of operational cell temperature.

7. *Acceptance angle*: if indoor testing is to be performed, then it may be most convenient to measure the acceptance angle indoors. Otherwise, it may be performed outdoors.

8. *IEC prequalification*: although we list IEC qualification of the module as the last step, we would recommend prequalifying certain module components early and often. A laboratory knowledgeable about this standard can recommend which tests to perform on each component. Candidate components are:

   - optics
   - receivers with cells, especially the die attach and the SOE attachment
   - any adhesive joint
   - any 'novel' materials choices (those not used in competing certified CPV modules).

9. *Module final design*: based on the feedback from our initial evaluation stages, we move into final design, design-for-manufacturing, adhesive selection, and the incredibly long laundry list of tasks common to any engineered product, and which is beyond the scope of this chapter.

10. *Pre-production*: we recommend attempting to fabricate on the order of 5 to 50 kW as soon as possible, in order to have enough products in the sun to learn about early failures, energy harvested over long time periods, etc.

11. *Qualification*: the final step of the design process is to qualify the CPV module to IEC 62108. That is, until work begins on generation two.

7.5   Concentration Ratio and Cell Size

The first step in designing a CPV module is to make two key decisions: the concentration ratio to be used, and the size of the solar cells onto which this concentrated light is to be focused. It is only after these key aspects of the CPV architecture are defined that we can proceed to design the optics themselves, choose a solar cell supplier and design the solar cell carrier, and design the module itself, including the chassis, the heat sinks, as well as all the other components.

7.5.1   Concentration Ratio

We define geometric concentration ratio as:

$$C_g = \frac{A_{lens}}{A_{cell}} \tag{7.1}$$

In the authors' opinion, as expressed in the introduction, the only reasonable way to choose a geometric concentration ratio for a given concentrator system, at least initially, is to choose the minimum concentration ratio that will reach the economic goals of the module. Since this goal is actually the economic production of solar electricity, the decision should really start with a *levelized cost of energy* (LCOE) calculation [22] in order to determine which is the installed cost of the system, in euros, dollars, etc. per Watt-peak that will produce solar electricity at an economical price (see Chapter 14).

Current estimates of CPV system capital cost are between 1.75 to 2.75 $/W$_p$ [23,24].. A relatively recent estimate of installed cost is near 3.8 $/W$_p$ for the Amonix system in Alamosa [25, p. 5]. Spectrolab currently mentions on their website an advisory pricing for volume purchase of its largest cells of $ 0.2/W when used at 1000× [26], which agrees well the calculations of Chapter 14. It seems clear that at $C_g = 1000$ the cell is no more than 10% of 10% of the total cost of a CPV plant (see Chapter 14), so increasing it further may only serve to increase the total cost. For instance, imagine we have a 1000× design that costs 2 $/W$_P$, installed, and uses Spectrolab cells, purchased for the price listed above, which represents 10% of our total cost. We could increase to 1500× and pay less for cells, but the savings for cells in the final installed product would be only 3.3%, which would be offset by either decreased performance or more expensive optics and tracking, or both.

7.5.2   Cell Size Selection

Of the many decisions to be made when designing a CPV system with a given concentration level, the most important is the cell sized to be used. There are a number of trade-offs to be considered, moving from smaller to larger cells makes some aspects of the CPV module (and the balance of systems) easier and cheaper, while making other aspects more difficult and expensive.

At the time of this writing, a wide variety of cell sizes and module thicknesses and sizes can be found on the market, each resulting from the individual expertise, expectations, prejudices, and informed or uninformed decisions and guesses of their designers and manufacturers in search of an optimum.

7.5.2.1 Cell Size, Lens Size, and f-Number

Fresnel-based CPV modules are, in general, rectangular boxes that may appear to be hermetically sealed, but in fact almost never are, for reasons which will be discussed in Section 7.9. The height or thickness of this box is defined exclusively by the focal length of the lens. Efficient Fresnel lenses must have an f-number between 1 and 2. Values below this will suffer from high chromatic aberration, and Fresnel losses, while values above this start to lose concentration due to long path lengths. Since, for reasons of tiling, the lenses are generally square in shape, they actually have f-numbers that are higher when considering the dimension of the lens across a meridian plane that is parallel to its edges and lower than when considering the dimension from corner to corner. For clarity in this chapter, we will use refer to both f_{side} and f_{diag} to distinguish these two values for f-number:

$$f_{side} = \frac{w}{F} = \sqrt{2} f_{diag}, \tag{7.2}$$

where w is the width (square dimension) of the lens, and F is the nominal focal distance of the system. The most common choice for f_{side}/f_{diag} is 1.55/1.1, since below $f = 1.1$ its lens efficiency curve starts to drop precipitously. For the following examples we will use $f_{side} = 1.5$ to simplify the math.

Let us imagine a module whose lens is 30 cm on a side, the module is necessarily 45 cm deep, which could appear to be excessive for material use, handling, and shipping concerns. On the other hand, for very small lenses of only 5 cm on a side, then the module is a svelte 7.5 cm, which many observers would select as a better choice. However, the designer cannot simply choose a lens size, but rather must calculate it from the previously chosen cell size and concentration ratio. So, if we wish to achieve a concentration ratio of, for example 900×, the lens dimension will fixed at 30 times that of the cell by Eq. (7.1), and our only liberty is to choose a cell size. If we choose small, 1 mm² cells, then our lenses are 3×3 cm², while if we use 1 cm² cells the lens will instead be 30 cm on a side, with module thickness increasing from 45 mm to 45 cm.

7.5.2.2 Impact on Module Construction, Assembly, and Shipping Costs

After these simple examples, it may seem that the obvious choice is to choose as small a cell as possible in order to obtain the cost savings in chassis materials and transport that come with thin modules. The cost savings due to reduced material use in thin modules is almost trivial to calculate. Based on recent discussions with industry, the shipping aspect may be the most compelling of these savings. Consider the dimensions of the example module of Figure 7.2 versus that of Figure 7.4. The volume occupied by the former is eight times that of the latter for a given area (and, therefore, capacity). That means that every MW_p of power plant installed using the large module design will require the transport of eight times as many standard shipping containers, eight times as many truck deliveries to the plant site, etc. However, we cannot ignore that the number of cells required increases as the square of decrease in cell size, in the last example each 1 cm² cell, would have to be replaced with 100×1 mm² cells, which implies 100 times the number of manipulations, soldering operations, and complementary components such as connectors, bypass diodes, etc. Small cells must be placed with greater precision beneath lenses which are themselves fabricated to tighter tolerances. The equipment required may be more expensive, and the time required to fabricate each module may be longer, all of which implies greater costs that may be difficult to estimate until detailed cost models are developed.

7.5.2.3 Impact on Tracker Structural Rigidity

There are other aspects in the cell size/module thickness trade-off that must be considered before this all-important decision is made. The modules will eventually be mounted on a solar tracker. The primary consideration when sizing the structural elements of the array frame (that is, the flat structure that is turned to face the sun) is deformation (bending) due to wind loading, which must be sufficiently small that modules are still within their angle of acceptance with respect to the sun at the highest wind speed at which the system is to operate. We can assume that this load is a bluff drag which is dependent entirely on the area tracked. That is, the thickness or even the weight of modules is not important in determining the rigidity required of the structure, and a thin 4.5 cm module will require just as much steel in the tracker to keep it pointed at the sun as a 45 cm thick module.

We consider then that taller modules inherently have a greater moment of inertia and therefore are more mechanically rigid. This is, first of all, useful for the mechanical design of the module itself, that is larger modules can be built with relatively less material while maintaining deformation due to wind and their own weight within acceptable limits. This inertia can also, with intelligent design of the module-tracker interface, result in savings to the tracker structure itself. If a large-lens module is designed to act as a structural element of the array frame, as opposed to a simple dead weight, then the additional material required for construction of modules themselves are at least partially compensated by the reduction in the material in the structural elements of the tracker, as shown in Figure 7.7

Figure 7.7 Examples of large-lens CPV systems that use their focal distance to their advantage. The Suncore system uses a *tilt-roll* style tracker, with four of their approximately 350 W modules on each array frame. Because the modules are over 0.5 m deep, they are extremely rigid, as evidenced by the fact that the outer supports reach only just past the module mid-point (left) [Photo courtesy of Folium Energy]. The 'mega-module' design from Amonix/Guascor (now Arzon Solar). In this photo of an early system design, it can be seen how the tracker frame itself only extends over a small section of the length of the mega-modules (right)

7.5.2.4 Impact on Thermal Design

On the other hand, the thermal aspect of module design in general tends to disincentivize large-lens designs. The amount of thermal power that must be dissipated from individual cells increases by the square of its size, and the area required to dissipate this heat to the ambient increases apace. Therefore, larger cells need increasingly thicker and heavier heat spreaders, generally fabricated from extruded aluminum, in order to first 'de-concentrate' this heat load prior to its convection to the outside (see Chapter 5). The act of first dividing up this thermal load by concentrating smaller lenses onto smaller cells is a much more effective way of spreading the heat than conduction in aluminum could possibly be. So, smaller cell modules, (cell sizes of 3×3 mm or less) do not generally need any dedicated heat spreader, but rather can be mounted directly on a relatively thin aluminum sheet. In section 7.8.2 we will provide a simplified model so the thermal factor can be taken into account for cost modeling at the time of cell-size selection.

7.5.2.5 Impact on Cell and Receiver Costs

Finally, turning to the economics of the cells and receivers, it is generally assumed that larger cells are less expensive per watt (at a given C_g) than smaller cells. First, for smaller cells the area taken up on a wafer by non-power-producing features; such as front metallization patterns, dicing street widths, and mesa isolation areas; take up a larger proportion of the overall wafer size, which may lead to a 15% cost advantage for 1 cm^2 cells over 1 mm^2 cells [27]. Added to this are the costs of die-singulation, testing, and other post-wafer processing steps that can be expected to scale with the number of devices. However, recall that cells are only 10% of our total cost (when operating at 1000×), so these cost variations may not be the deciding factor in final cell size decision. Finally, it is important to note that the marketplace has decided on a maximum cell size to be used in high-concentration CPV, that of 1 cm^2. Beyond this level, the series resistance in the cell may become important, and wafer yields begin to drop. Even more importantly, the thermal problem becomes next to impossible to resolve at high concentration and cell sizes larger than 1 cm^2.

7.5.3 Module Size and Length

The cell size trade off may be the most important, but it is not the only decision that must be made regarding module dimensions. Additionally, the module area (exterior length and breadth) must be chosen. Many manufacturers choose to limit module dimensions at a size that can be comfortably handled by 1 or 2 persons. Other manufacturers choose to renounce non-automated handling and produce even bigger modules so as to reduce the number of operations that must be performed in the field. For instance, the latest generation module from Soitec measures 3.7 m by 2.4 m and produces over 2 kW in operating conditions (Figure 7.8). Another common approach is to assemble a number of relatively small sized modules into larger units in the factory, where their alignment is ensured (Figure 7.9). Module voltage must also be taken into account, which should be limited to the levels discussed in section 7.7. It should also be noted that the cell size chosen has an impact here as well: large-lens modules can be more easily built in larger sizes for a given allowable deformation due to their inherent rigidity.

Figure 7.8 A Soitec CPV system with a single CX-M500 module highlighted (darkened). Two workers washing the array show the scale of the module. Note that this was an early installation, hand washing is not normally required in an industrial Soitec plant. Source: © Robert Berkman, 2013

7.5.4 Market Survey

Although we can't provide a foolproof optimum cell size and concentration ratio to use in a new CPV design, we can present a summary of the values selected by the most prominent CPV systems providers at the time of the writing. We limit ourselves to produces that have a reasonably commercial product, and who have fabricated and installed at least $50\,kW_p$. These are listed in Table 7.1.

We find it interesting to graph these commercial CPV proposals on the Cartesian space of cell size and geometric concentration ratio (see Figure 7.10). We have also added data from a snapshot of the industry in 2008 taken from [28]. In fact there are very few overlapping players between the two data-sets, those that are represented in both datasets, but with new designs in 2014 are connected by dots. With a few exceptions we observe a general movement by the industry towards higher concentrations.

Figure 7.9 Six Heliotrop modules are pre-mounted in a sub-array frame in the factory to reduce operations in the field. Source: Heliotrop, 2014. Reproduced with permission of Heliotrop

Table 7.1 A list of CPV systems providers, showing the chosen cell size and concentration ratio, and how much product they have installed. Cell size refers to the width along edge of the active area of the solar cell (the cell is assumed to be square). [12,15,29–39]

Company	Country	SOE type	C_g	Cell size (mm)	Installed capacity (MW$_p$)
Abengoa Solar, S.A.	Spain	Glass	1300	7.5	1.5
Arima EcoEnergy Technologies Corp.	Taiwan	N/A	476	10	3.4
Arzon Solar LLC (formerly Amonix)	US	Reflective	900	8	40
Daido Steel Co. Ltd.	Japan	Glass	820	7	1[5]
Heliotrop SAS	France	Reflective	1024	10	0.05
MagPower	Portugal	N/A	800	10	7.2 (31)[6]
Semprius Inc.	US	Glass	1111	0.6	0.4(0.6)
Soitec (acquired Concentrix Solar)	France	None	500	2.2	75 (20)[7]
Sumitomo Electrical Industries, Ltd	Japan	None	278	3	0.122
Suncore Photovoltaic Technology Co.	China	Reflective	1090	10	140

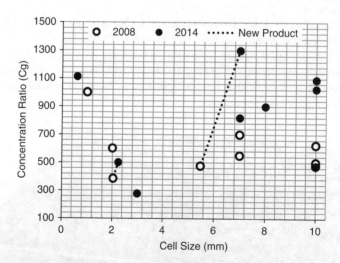

Figure 7.10 Map of prominent CPV providers in the C_g, cell size space in 2008 and today. The dotted lines indicate a new generation of product from the same supplier

[5] Includes installed capacity of licensees.
[6] Currently under installation or near-term pipeline.
[7] Currently under installation or near-term pipeline.

More importantly, we see a dearth of systems in the middle cell size range (those that measure 4–6 mm on a side). We believe that this is a difficult part of the CPV design space due to the cost of the carrier. At this size, the thermal load is still too high to use less expensive carrier materials, and compared to larger cell size, many substrates are needed. A cell carrier using, for instance, direct bonded copper for a 5 mm side cell may cost almost as much as that for a 10 mm, but four times as many are required. As receiver and carrier costs are at least as large a proportion of overall cost as the cells, this is an important economic driver towards the largest cell sizes. The smallest cell sizes that use radically simplified receiver designs and microelectronic techniques may find a different optimum.

7.6 Opto-Mechanics of CPV Modules

The most obvious role of the CPV module is to support the active components: the cell and the lens. This section attempts to address the question of how accurately it must do so. What assembly precision is required for each component and dimension? How much deformation can we allow these components, such that module continues to function at or near its optimum power?

7.6.1 Acceptance Angle

The acceptance angle, θ_Y, describes how far a given concentrator can be rotated from its optimal pointing vector[8] (where the optical axis aligned to the vector pointing at the sun) before its efficiency falls below a certain threshold, Y, of its maximum value:

$$Y = \frac{\eta(\theta_Y)}{\eta(0)} \tag{7.3}$$

where η refers to the electrical efficiency of the optics-cell unit. The most commonly used threshold at which this angle is reported is 90%, which we will refer to as the 90% acceptance angle, θ_{90}, however this is useful only as a generalized performance parameter: in reality we should never allow the module to operate at a 10% power loss. We can consider the acceptance angle where at least 98% of the power is produced, θ_{98}, to be a more realistic description of the allowable angular tolerance.

Conservation of *étendue* (see Chapter 4), and the fact that the angle of incidence of the light arriving on the cell can be no greater than 90° leads to the following relation between the ideal acceptance angle (maximum achievable for 100% power capture), C_g, and the index of refraction in which the cell is submerged, n [40]:

$$C_g = n^2 \cdot \sin\left(\theta_{ideal,100\%}\right) \tag{7.4}$$

In practice, no concentrator is ideal, and the acceptance angle achieved by real concentrator modules is much lower. Solving (7.4) for n, we can formulate what is commonly called the

---

[8] We define the 'optimal pointing vector' as the center of the angular region for which the electrical efficiency is greater than $Yh(0)$, so in practice to determine we must measure the electrical performance of the concentrator while varying its incidence angle from less than $-\theta_Y$ to greater than $+\theta_Y$ to determine both this vector as well as the angular acceptance around it.

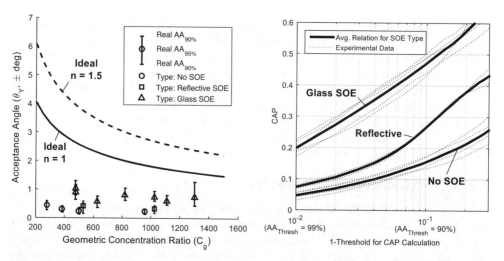

Figure 7.11 A summary of concentration-acceptance characteristics of various industrial CPV modules measured at the IES-UPM. Acceptance angle plotted versus concentration for a range of threshold values, with indication of secondary optic type (left). When we plot the data in terms of CAP versus threshold levels (right), we can see how the three secondary types define three families of curves. The average for each secondary type is shown as a thick line

concentration-acceptance product (CAP), which is equal to n for ideal concentrators and equal to a smaller value for real concentrators [41]:

$$\sqrt{C_g}\sin\left(\theta_{ideal,100\%}\right) = n \qquad \Rightarrow \qquad CAP_Y = \sqrt{C_g}\sin(\theta_Y) \qquad (7.5)$$

That is, the CAP is a measure of the concentrator's ideality. Note that the CAP value is calculated using a real (measured or simulated) acceptance angle, and is therefore linked implicitly to a threshold value used to define the acceptance angle in the first place. While in literature it is most common to use $\theta_{90\%}$ to calculate CAP, any threshold may be used.

In Figure 7.11 (left), we show actual acceptance angles for a number of commercial and prototype CPV modules, as measured by the authors. These measurements have been made on different CPV modules, from different manufacturers, at the IES-UPM, using the Helios 3198 Solar Simulator for CPV, between the years of 2009 to 2014. The figure shows the acceptance angle for each module over the range of threshold values from 90% to 98%. Only in-house data is presented, because in such detailed information of acceptance angle versus threshold is not available generally available in the literature. All measurements were of full modules consisting of multiple optics-cell units, measured by the direct method (repeated I-V curve measurements for different alignment angles.) For the purposes of this discussion, it would have been more useful to use the acceptance data of optics-cell units, but this data was not available for a wide array of technologies. Due to misalignments and assembly errors in the modules, the θ_Y of a full module is not the same as that of a single unit. In fact, in modules with series only connections, which represent the majority of the modules in Figure 7.11, the acceptance angle of the parameter P_{max} (maximum power point of the I-V curve) will be less than that of the single unit, while that of the of the module I_{sc} will usually be greater. Therefore, the values in Figure 7.11 represent the average of the angular transmission curves of these two parameters.

For each module we have calculated the CAP for a range of threshold. In Figure 7.11 (right), these values are plotted against the allowed power drop, (1-Y) on a log scale. We see then that the relationship is highly dependent on kind of SOE used, and in fact we can calculate average CAP-vs-threshold curves for each SOE type. It is reasonable to expect that any new CPV point-focus Fresnel design will fall on or near these curves. Therefore, we propose that during the initial phase of design and optimization, that assuming we have selected a concentration ratio and an SOE type, Figure 7.11 (right) is sufficient to derive an initial estimate of acceptance angle at a given threshold level.

7.6.2 Acceptance Angle Budget

We assume now that the value for the acceptance angle for 98% power loss, or if necessary, for a lower threshold, of our chosen CPV technology has been determined, either by an initial estimate using the data in the last section, by optical simulation, or most accurately, by experimental measurements on prototype optics and receivers in the sun or a solar simulator. This value of θ_{98} is the amount that a single optics-cell unit can be rotated away from its optimal direction before power produced drops below 98%. But in reality we will have many such optics-cell units assembled into modules with attendant assembly errors, and then we will have many such modules assembled planar to each other, within a certain tolerance, onto an array frame that is only flat and rigid to a certain degree, all of which is tracked to the sun with controllers and actuators that are themselves not perfect. The result is that we are lucky if we can keep the system operating at 98% performance during normal operation.

We posit that we can linearly divide the angular acceptance 'budget' provided by the characteristics of our optics-cell unit, amongst the different contributors to error. This is equivalent to assuming that the power reduction of our entire CPV system is the same as the power reduction on a single unit which is at the tolerance limit. Given that most of the electrical connections between cells are series, this is a reasonable assumption.

First we must identify the major assembly and operational tolerances where this acceptance budget can be spent. They can be divided into two groups; internal, or those that arise from assembly errors and deformations inside the module; and external; those that arise from the module's installation on a tracker. There are additional important assembly tolerances to be taken into account within the receiver itself, but these are not accounted for within our acceptance angle budget, but rather result in slight maximum power losses to the optics-cell unit. Accordingly:

- *External tolerance budget*
    - $\delta\theta_{tracker}$, tracking error (controller and actuators): the amount of misalignment error the tracker allows with the vector representing the sun according to its own calibration during normal operation, including the error due to calibration of the tracker, that is the error between the vector that the tracker is attempting to align to the sun and the vector which would produce the maximum power from the array.
    - $\delta\theta_{array}$, array frame structural deflection and flatness: the amount of module alignment due to bending of the structure under the weight of the modules and the maximum wind load under which the array is intended to function correctly.
    - $\delta\theta_{module}$, module-to-module alignment error: the angle between the optimal pointing of the two modules who are least aligned on the tracker, due to the mounting.

- *Internal tolerance budget*
  - $\delta\theta_x$, cell-to-lens misalignment: this is a misalignment of the optimal pointing vector of each optics-cell unit due to misalignment of the optical axis of each lens with each receiver in the direction perpendicular to that axis. We define the optical axis to be the z-axis, so this represents movements in the x–y plane.[9]
  - $\delta\theta_z$, *receiver distance errors:* the acceptance angle budget is for optics-cell units who lens-to-cell distance is at the optimum, as determined empirically or by simulation. It must be assumed that both mechanical tolerances and bending of the backplane will cause real optics-cell units to have errors in the distance between the lens and the receiver (in the z-direction, or parallel to the optical axis). We can approximate the losses caused by these focal distance errors as acceptance angle errors in order to incorporate them in our tolerance budget.

Using an arithmetical tolerance stack-up, we can then see that, if we want to ensure that produced power doesn't drop more than $(1-Y)$, the sum of the magnitude of all of these tolerances must be equal to θ_Y:

$$\theta_Y = \underbrace{\left(\delta\theta_{tracker} + \delta\theta_{array} + \delta\theta_{module}\right)}_{External} + \underbrace{(\delta\theta_x + \delta\theta_z)}_{Internal} \tag{7.6}$$

7.6.3 External Tolerances

Many CPV trackers are quoted to have an 'accuracy' of $\pm0.1°$. This figure usually refers only to $\delta\theta_{tracker}$, leaving out the errors associated with the deformation of the array frame. It is also up to the tracker customer to ensure that this value includes all aspects of $\delta\theta_{tracker}$, and not just software and hardware error. Most importantly, the tracker should provide an adequate calibration method that ensures that the variation of $\pm0.1°$ is centered on the optimal vector for the array of modules as mounted (that is, when power is maximized), and not some other direction.

In general it should be expected that the other external tolerances will have to be on the same order, a tenth of degree. That is, asking our tracker structure for deformations of less than $\pm0.1°$, will probably be too expensive, and aligning all of the modules on a tracker of 50 or 100 m^2 to less than $\pm0.1°$ is also probably not reasonable. All told, we can expect, as an initial rule of thumb that the external tolerances will contribute at least $\pm0.3°$ to our overall tolerance budget.

Turning back to the last section, we see that only the most tolerant systems have values for θ_{98} of greater than $\pm0.3°$. In these cases, it is clear that in order to make room in our tolerance budget for the internal tolerances, we must lower our threshold of allowable power loss. That is, we must make a conscious decision to increase our tolerance to energy generation losses in the field. As always, the trade-off is economic. We must ensure that the costs of making our system more tolerant (for instance, by adding a secondary optic, or changing from a cheaper reflective secondary optic, to a more expensive molded glass secondary optic) are offset by an increase in energy harvested.

---

[9] We will refer to these misalignments as in the 'x' direction, but they refer to any movement in the x–y plane.

7.6.4 Internal Tolerances

Once we subtract the external tolerances from our angular acceptance budget, we can divide the remainder among the internal positional tolerances. However, in this case we are interested in applying our acceptance angle budget to concepts that are related, but not equivalent, to rotations of the optics-cell unit. We can find equivalency factors, K, between the way power is lost as we rotate a single solar cell unit by θ, and when we translate the lens and the receiver with respect to each other by an amount δ. These factors can be used to determine the correct tolerance setting for each translation (as a proportion of the lens focal distance, F) for a given acceptance angle allocation in radians:

$$\frac{\delta}{F} = K \cdot \theta \tag{7.7}$$

The values for K for the two internal tolerances are shown in Table 7.2. Their derivations are shown in sections 7.6.4.1 and 7.6.4.2.

7.6.4.1 Cell-to-Lens Misalignment

In a CPV module square lenses are generally tiled in a square window assembly, at a certain spacing, with a certain tolerance, δx_1. If the window assembly is manufactured of a piece (i.e. as a parquet of lenses) we can expect this spacing to be very precise. At the same time, the individual receivers must be mechanically assembled to the backplane, and in this step the tolerance, δx_2, is surely greater, and finally, these two relatively large parts must be assembled together with the chassis between them, in a final assembly step, usually the least accurate step of the three, δx_3. Each of these individual tolerances can be summed and related to the allotted portion of the overall angular acceptance:

$$\delta x = \delta x_1 + \delta x_2 + \delta x_3 = f(\delta \theta_x) \tag{7.8}$$

For a certain displacement, δx, we consider that the *spot*[10] of light, formed by lens on the aperture of the receiver, which may be either an SOE or the cell itself, will be displaced with respect to the center of that aperture in a way that is not unlike how it is displaced when the

Table 7.2 Summary of tolerance equivalency factors

Translation direction	K
X and Y	1
Z– (towards lens)	3.5
Z+ (away from lens)	4

---

[10] We use the colloquial term 'spot' to indicate the area which is illuminated by the lens with concentrated light. In reality, the definition of the extent of this area is complex (since the light contained in it is non-uniform) and usually is defined as the area into which a certain high percentage of all the rays leaving the lens intersect the focal plane, or of all energy incident on the focal plane. However, for the purposes of the very simplified analysis of this section, we imagine an ideal and uniform illuminated zone, that is circular when the lens and the focal plane are parallel.

entire optics cell unit is rotated by $\delta\theta_x$. If we consider first the ray passing through the exact center of the lens, where the lens acts like a window (no refraction, or more accurately, two equal refractions as the ray enters and leaves the window assembly) we can see, trivially, that:

$$\sin(\delta\theta_x) = \frac{\delta x_{center}}{F} \tag{7.9}$$

and for small angles:

$$\frac{\delta x_{center}}{F} \approx \delta\theta_x \tag{7.10}$$

This is equivalent to saying that for every 0.001 radians ($\sim$0.057°) of acceptance angle, we can allow a tolerance on the lens-to-cell alignment equal to 0.1% of the focal distance. However, for any other part of the lens, we should consider whether the effect of refraction, as well as the increased distance that the ray travels from the point of refraction to the focal point, will affect this relation. We are also well aware that while in the case of a lens-to-cell misalignment the spot is displaced but does not change its characteristics (either shape or uniformity of illumination) while in the case of a rotation the optics-cell unit, the spot both translates, and changes shape (becoming larger and elongated). However, since we assume we already know, or can estimate, the $\delta\theta_x$ for a given power loss, we are therefore confident that if we focus only on relating δx and $\delta\theta_x$, the neglected effects of the change in spot characteristics will serve only to make our tolerances more conservative.

In Figure 7.12 we show an aligned and non-aligned meridian ray passing through a point that is distance x from the center of the lens, and located in the plane perpendicular to the axis of rotation. The ray-angle from that point to the focus is φ, and the facet angle, α, of the Fresnel lens at that point can be found from the index of refraction, n, and Snell's law:

$$n \sin\alpha = \sin(\alpha + \varphi) \Rightarrow \tan\alpha = \frac{\sin\varphi}{n - \cos\varphi} \tag{7.11}$$

Equation (7.11) is the Fresnel lens design equation. We can extend this to the case where the incoming ray is misaligned by $\delta\theta_{lens}$, defined as a rotation around the y-axis, and further, lets restrict ourselves, for the moment, to the meridian plane ($y = 0$) such that the this angular misalignment is in the same plane as φ and α. There are now two refractions to take into

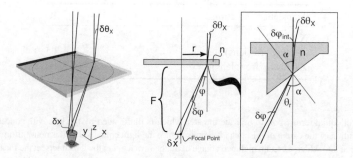

Figure 7.12 Equivalence between $\delta\theta_{lens}$ and δx_{lens}

account, at the entrance, forming an interior angle $\delta\varphi_{int}$ and a second as the ray exits through the Fresnel facet (Figure 7.12):

$$\sin\delta\theta_{lens} = n\sin(\delta\varphi_{int}) \tag{7.12}$$

$$n\sin(\alpha + \delta\varphi_{int}) = \sin(\alpha + \varphi + \delta\varphi) \tag{7.13}$$

Assuming the values of $\delta\theta$ and $\delta\phi$ are small, such that $\sin\delta = \delta$ and $\cos\delta = 1$, and taking (7.11) into account, we can rearrange (7.12) and (7.13) to find an approximately linear relationship between the radial and tangential components of $\delta\theta_{lens}$ and $\delta\phi$:

$$\delta\varphi_x \approx \frac{1}{n}(\cos\varphi_x - \tan\alpha_x \sin\varphi_x)^{-1}\delta\varphi_{int} = \frac{n - \cos\varphi_x}{n\cos\varphi_x - 1}\delta\theta_x \tag{7.14}$$

The x subscripts here indicate that these angles lie in the x–z plane (ie. the meridian plane). This angular divergence causes a small movement, δx, of the ray in the focal plane, which can be found by simple trigonometry, for any point on the lens, (x,y):

$$\frac{\delta x}{F}(x, y) = \tan[\varphi_x + \delta\varphi_x(x, y)] - \tan\varphi_x \tag{7.15}$$

Note that we find it convenient to express this as a fraction of the focal distance of the system. Again, we can use a small angle approximation to obtain a linear proportionality between δx and $\delta\varphi$:

$$\frac{\delta x}{F}(x, y) \approx (1 + \tan\varphi_x)\delta\varphi_x(x, y) = (1 + \tan^2\varphi_x)\delta\varphi_x(x, y) \tag{7.16}$$

Finally, we can combine (7.14) and (7.16) to find a single linear relation, $k_{\delta x\delta\theta}$, which is valid for all points such that $y = 0$:

$$\frac{\delta x}{F}(x, 0) \approx \left(\frac{n - \cos\varphi_x}{n\cos\varphi_x - 1}\right)(1 + \tan^2\varphi_x)\delta\theta_{lens} = k_{\delta x\delta\theta}\delta\theta_x \tag{7.17}$$

When $x = \varphi = 0$, we see that the equation collapses to the one-to-one relation of (7.10), while as φ approaches normal Fresnel lens rim angles, the slope $k_{\delta x\delta\theta}$ increases to values of around 1.5. That is, as we tilt the optics-cell unit, we see the central ray will move linearly according to the sine law, the meridian rays farther from the center will move by increasing amounts creating a 'smear' of light, which may be familiar to anyone that has visually aligned a concentrator system. However, from the stand point of our analysis of the tolerance of the cell to lens alignment, this 'smear' in fact allows for greater receiver moment. That is, if we know that a 0.001 radian rotation is equivalent to 98% power, even considering this light smearing, then we can reasonably expect at least 98% power for a simple lateral receiver movement of 0.1% of the focal distance, if not more.

We would like to find a single value, $K_{\delta x\delta\theta}$, that is representative of our entire lens, and therefore acts as an appropriate multiplier for our tolerance budget. First, we assume a lens of unit size, such that $F = f_{side}$ and the ray angle at any point is $\tan(\varphi) = r/f_{side}$. The simplest option would to ignore the *skew* rays, or those that is off the meridian plan and the average value of the expression (7.17) over the period $-0.5 < x < 0.5$, using the integral as follows, where we have expressed the trigonometric functions in terms of x and f_{side}:

$$K_{\delta x \delta \theta} = \int_{x=-\frac{1}{2}}^{\frac{1}{2}} k_{\delta x \delta \theta} dx = \int_{x=-\frac{1}{2}}^{\frac{1}{2}} \left(\frac{n - \dfrac{x}{(x^2 + f_{side}^2)}}{n \dfrac{x}{(x^2 + f_{side}^2)} - 1} \right) \left(1 + \frac{x^2}{f_{side}^2} \right) dx \qquad (7.18)$$

In order to ensure that the errors involved in ignoring all of the lens except for the central strip on are not too large, we also find an exact solution. For a given f-number we can successively find the average δx over the lens for increasing values of $\delta \theta$ via equations (7.11)–(7.13) and (7.15). Note that to account for skew angles, we must divide the angle $\delta \theta_{int}$ into radial and tangential components, apply Eq. (7.13) only to the radial component, and then use trigonometry to find the x-component of the resulting angle. If we do not consider angles greater than 1.5 degrees, we continue to see a very linear relationship, allowing us to calculate values of $k_{\delta x \delta \theta}$ via regression, as shown in Figure 7.13 (left). In Figure 7.13 (right) we show the resultant factors for a range of f-numbers.

First, we see that the simplified treatment, taking into account only the meridian rays, is sufficiently close to the exact solution, so (7.18) can be used to find the value of $K_{\delta x \delta \theta}$ for arbitrary n and f. Although we have performed this calculation for a range of f-numbers, as mentioned before, almost all Fresnel-lens based CPV systems use a value very close to $f_{side} = 1.55$. Figure 7.13 shows us that the value of $K_{\delta x \delta \theta}$ is about 1.1 at this point. This would indicate that for every 0.001 radians of acceptance angle budget dedicated to misalignment of cell and lens, we can allow a lateral movement of a little over 0.1% of the focal distance. Keep in mind that these relations are to be used only in early design stages or for trade-off studies. If prototype lenses and receivers are available, then their lateral tolerance should simply be measured, as discussed in Section 7.4, step 4).

In Figure 7.14 we plot the results of such a measurement for two CPV technologies. The receiver of a single-unit concentrator system, mounted on a 3-axis opto-mechanical stage, was translated in two directions (parallel with the cell edges) and many I-V curves were recorded using the IES-UPM solar simulator, and efficiency as percentage of the maximum recorded was calculated. Also, the same concentrator was measured for various angles of misalignment with

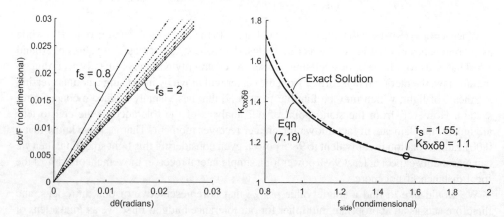

Figure 7.13 Numerical solution of the relationship between $\delta \theta$ and δx for different values of f_{side}, assuming $n = 1.5$ (left). Comparison of the numerical solution to Eq. (7.18) (right)

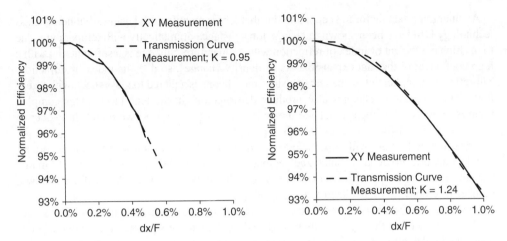

Figure 7.14 Comparison of measured lens-to-cell tolerance data compared to single-unit transmission curve data for two systems. The angle of the transmission curve data (in radians) has been scaled by an experimentally determined constant so that it matches to the tolerance data. For a system with a lower acceptance angle (left), the value of $K_{\delta x \delta \theta}$ found experimentally is 0.95, while for the system with a higher acceptance angle, the value found is 1.24

the collimated light beam (by rotating the unit precisely), also in two directions, and the normalized efficiency calculated as well. For both sets of data, δx and $d\theta$, the two vectors, x and y, were averaged together to reduce noise. Finally, we found the factor $K_{\delta x \delta \theta}$ that produced a good match between the two curves.

Unsurprisingly, the real world data do not match exactly the conclusions of our simplistic analysis. We note that in one example, with a system that had a relatively low acceptance angle, the value for $K_{\delta x \delta \theta}$ was even slightly less than one. Therefore, we recommend for tolerance purposes to use a multiplier of $K_{\delta x \delta \theta} = 1$. In other words, the previous three pages may be safely ignored, and equation (7.10) can be applied.

7.6.4.2 Receiver Distance Tolerance

We also want to derive a linear factor $K_{\delta z \delta \theta}$ for changes in lens to receiver distance. As we move away from the nominal focal distance of a Fresnel lens, the light that arrives at the receiver undergoes a number of changes, all of which affect the performance of a multi-junction solar cell (see Chapter 2) as follows:

- The concentration level (irradiance) decreases.
- Due to chromatic aberration, the overall spectrum (integrated over the aperture of the receiver) shifts in one direction or the other: blue-shift for movements toward the lens, and red-shift for movements away from it [42]. This shift is asymmetric, as will be explained below.
- The spatial uniformity of light over the cell will decrease, causing a decrease in performance due to series resistance losses [43].
- Depending on the type of SOE, the spatial uniformity of spectrum will also decrease, possibly causing a situations where part of the cell are top-limited and other parts are middle-limited, which can cause a decrease in fill factor [44].

Another important factor to keep in mind is that lenses manufactured using silicone-on-glass technology [16] have been shown to change focal distance dramatically with temperature, due to the high coefficient of index of refraction with temperature. Silicone is a soft material with a high coefficient of thermal expansion. As its density changes, so does its index of refraction. This effect is complicated by the fact that since the silicone is adhered to glass sheet, with a very low CTE, temperature changes also produce deformations in the lens facets. This complex issue is, however, completely outside the scope of this chapter, and we direct the reader to references [45–47] for discussion on this topic.

Our simplified model will only include the irradiance decrease as it is the most important, and the only effect that can be analyzed geometrically, but we should keep in mind that in this respect the model is optimistic, and safety factors should be considered. We will also show a number of empirically calculated values for $K_{dzd\theta}$.

The decrease in irradiance over the receiver as it moves away from its optimal focal position, is due to the fact that the 'spot' of light produced by the lens gets progressively larger. In our simplistic model, we assume that both the spot and aperture of the receiver are circular, and that the aperture radius $r_{receiver}$ is larger than the nominal spot radius, r_{spot}, as shown in Figure 7.15. The spot size increase, δr_{spot} is related geometrically to the rim angle, φ_R:

$$\tan \phi_R = \frac{\delta r_{spot}}{\delta z} \tag{7.19}$$

In this simplistic model, a change in size of the spot will initially have no effect on the electrical efficiency of the concentrator, but when $(r_{spot} + \delta r_{spot}) > r_{rec}$, the efficiency will be decreased proportionally to the ratio of areas:

$$\frac{\eta(\delta z)}{\eta(0)} = \frac{r_{rec}^2}{\left(r_{spot} + \delta r_{spot}\right)^2} = Y \tag{7.20}$$

We note that $\eta(\delta z)/\eta(0)$ is equivalent to the threshold value, Y, that we are using to define our tolerances. Now we would like to relate δr_{spot} to $\delta\theta_{focus}$. In Figure 7.15, we show a simplified version of our spot translated by a distance δx (which we know we can relate to $\delta\theta$ using the method of the last section). We observe that the movement required for the efficiency loss to reach our chosen threshold Y is that where the fraction of the spot area remaining inside the receiver is Y. This area could be calculated with a rather long equation [48], but it is simpler just

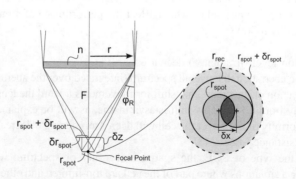

Figure 7.15 Simplified model of the effect of changing the focal distance

to set our threshold, for the moment, at $Y = 1$, such that the inner circle just touches the outer circle. Then it is evident that:

$$r_{rec} = r_{spot} + \delta x \tag{7.21}$$

Now we can combine Eq. (7.19) and (7.20) to (7.21), giving:

$$\delta z = \frac{1}{\sqrt{Y}\tan \varphi_R}\left(\delta x + \underbrace{\left(1 - \sqrt{Y}\right)}_{\text{very close to } 0} r_{spot} \right) \tag{7.22}$$

Since $\sqrt{Y} \approx 1$, we can assume that the y-intercept goes to zero, and therefore we do not need to explicitly calculate the spot size, nor explicitly assume a threshold value. Finally, we combine this expression with that of (7.17) (after first dividing both sides by F) to obtain:

$$\frac{\delta z}{F} = \frac{n - \cos \varphi_R}{\tan \varphi_R(n \cos \varphi_R - 1)}\left(1 + \tan^2 \varphi_R\right)\delta\theta_Z = k_{\delta F \delta\theta} \cdot \delta\theta_Z \tag{7.23}$$

In this case we do not want to take the average over the surface of the lens, but around the perimeter, because we are assuming that it is always the rim, or outermost rays, that are those that are related to the growth of the edge of the spot. If we assume the lens is circular, then $\delta z/F$ is constant ($K_{\delta z \delta\theta} = k_{\delta z \delta\theta}$), but we could also take into account that our lens is square by taking the average along on edge of the perimeter, from the point $(0, \frac{1}{2})$ to $(\frac{1}{2}, \frac{1}{2})$:

$$K_{\delta z \delta\theta} = \frac{1}{\sqrt{2}} \int_{r=0}^{1/\sqrt{2}} \frac{n - \dfrac{r}{\sqrt{r^2 + f_{side}^2}}}{\dfrac{r}{f_{side}}\left(n\dfrac{r}{\sqrt{r^2 + f_{side}^2}} - 1\right)}\left(1 + \dfrac{r^2}{f_{side}^2}\right)dr. \tag{7.24}$$

The result of this integral is shown in Figure 7.16. For the case of the standard f-number of $f_{side} = 1.55$, we predict that we can allow a multiplier of more than four times, that is for every 0.001 radians of tolerance budget dedicated to the lens-to-receiver distance, we could allow a movement of up to 0.4% of the overall focal distance. This is an important point when considering module design and assembly techniques: we can allow about 4 times more positional error in the Z-direction than in the X- or Y-directions.

This follows from our simplistic analysis, where the principle driver is the change in the area of intersection of a fairly oblique cone, but we should keep in mind that we ignored three of the four effects of focal distance on cell behavior, so a safety factor should be applied. To gain more insight, we will examine real values of $K_{\delta F \delta\theta}$ measured for different prototype concentrator optical system at the IES-UPM. In all cases, the efficiency versus lens-to-cell distance has been measured on an open single optics-cell unit prototype for a given CPV design. Once the optimal focal distance was determined, we also measured the angular transmission curve of that same unit at the optical distance. As before, we find a value of $K_{\delta z \delta\theta}$ such that these two curves match, by minimizing the root-sum-square error between them.

However, the shape of the efficiency versus lens-to-receiver distance in HCPV systems using multi-junction cells is innately non-symmetric. That is, the efficiency drops off faster as we move the cell closer to the lens than when we move it farther away. This is primarily due to

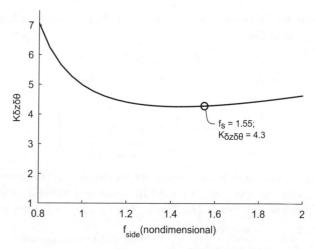

Figure 7.16 The relationship between $\delta\theta_{lens}$ and δz for different values of f_{side}, assuming $n = 1.5$

the current-matching within the multi-junction cell and the shape of the quantum efficiency curves. The photocurrents of the top and middle subcells of the multi-junction cell are usually matched the nominal receiver distance. When we move the receiver farther away, due to chromatic aberration, we will experience a *red-shift*, that is the overall spectrum on the cell will become redder than the reference spectrum, with lowest wavelength rays being lost first. Looking at Figure 7.17 (left) it is clear that the top subcell, since it has very little efficiency for the highest energy photons, will not necessarily see an immediate decrease in current. The

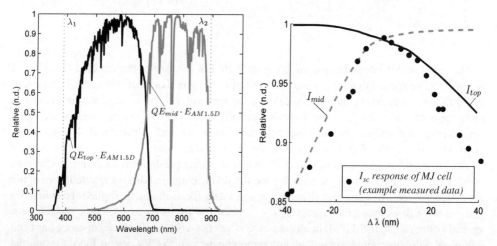

Figure 7.17 Solar-spectrum-weighted quantum efficiencies of the top and middle sub-cells. (left) If we integrate of these curves over a moving window to estimate sub-cell photocurrents during a focal scan, we can produce an estimated photocurrent (I_{top} and I_{mid}) (right). Due to low response of the top cell to UV, a red shift (equivalent to moving the cell farther from the lens) causes approximately half as much top current loss as an equivalent blue shift causes in middle current. The dots in the right-hand figure indicate measured data of the relative change in photocurrent of an example system during a focal distance scan, showing how the shape is similar (note, x-axis for this data is scaled arbitrarily)

middle cell, on the other hand, is not as tolerant the losses in the longest-wavelengths that occur when the receiver is moved towards the cell, because these losses affect the abrupt cut-off of its QE curve at the bandgap. We can emulate this by simply looking at the effect on the subcell photocurrents of a moving window that is first centered such that the subcells are current matched and then displaced by an amount $\Delta\lambda$ (Figure 7.17 right). The practical effect is that the designer, should be more worried about, and specify tighter tolerance for, receiver movements towards the lens.

With this in mind, we turn to experimental results for three different systems in Figure 7.18. Due to the non-symmetry discussed above, separate values of $K_{\delta F\delta\theta}$ have been found for the regions $\delta F/F < 0$ and $\delta F/F > 0$. The experimentally determined values of $K_{\delta F\delta\theta}(+)$ and $K_{\delta F\delta\theta}(-)$, for each of the systems are shown in the figure. We note that while the exponential

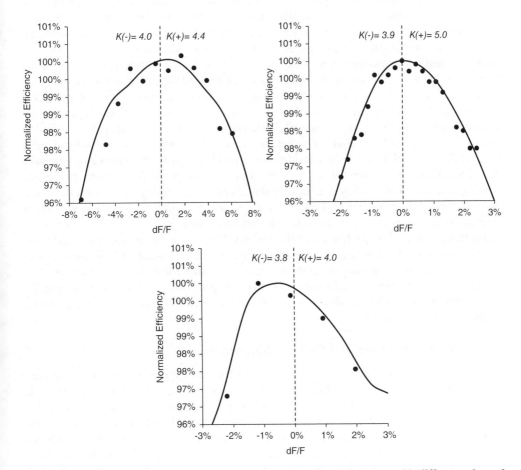

Figure 7.18 Relationship between $\delta F/F$ and $\delta\theta$ for three real world systems, with different values of CAP. Measurements of power versus receiver distance are shown with dots, the transmission curve measurements are shown as lines. Due to the systematic non-symmetry of the power loss, the scaling factor applied to the angle is lower for negative values of $\delta F/F$ (closer to the lens), then for positive values of $\delta F/F$ (farther from the lens)

multipliers in the positive direction are very similar to our theoretical value of 4.3, the values on the negative side are lower. We recommend that the highest values used in system design are 3.5 for the negative side and 4.0 for the positive side, although a prudent designer may reduce these numbers even further. At any rate, once initial optical prototypes are obtained, optics-cell testing should be used to experimentally determine the curve of efficiency versus δz.

7.7 Electrical Design

This section focuses on electrical aspects in the design of a CPV module, while discussion of thermal issues is in Section 7.8. But in reality the designer must consider both thermal and electrical characteristics in parallel because they interfere with each other. The electrical design of a CPV module has four main aspects.

The first aspect of electrical design is the circuit design of connections between the cells comprising a module, which are usually either series-only (for large-lens systems, with relatively few cells) or a combination of series and parallel. The second aspect is the connection of the cell to the carrier. Common methods and materials widely used in integrated circuit packaging and power electronics have been adopted by the CPV industry, such as die attach adhesion or soldering for the mechanical fixing of the cell and the electrical connection of the back surface contact (usually the positive side of the cell) to a substrate (the cell carrier); and wire or tab bonding for the electrical connection of the front (negative) electrodes and interconnection to other components. The cell carrier is a key element, not only for the electrical connection of the cells but also for a good thermal performance, as it was discussed in Chapter 5. The third aspect to be considered is insulation, that is, the absence of undesired electrical connection between the active electrical parts and other metallic elements such as the housing and heat-sinks, which is critical to the reliability and safety of the module. These dielectric materials must be placed between the cells and the heatsink, directly in the path of thermal extraction, and so they require high electrical resistance but low thermal conductivity, a characteristic found only in a few materials, most notably ceramics such as alumina, and in the epoxies and silicones that are filled with alumina powder. The last facet is performance optimization, not only focusing on ohmic losses but also other components such us protecting diodes which can improve reliability but decrease performance; and the series/parallel connection strategy which, combined with optical misalignments would impact in the angular tolerance of the system [49].

7.7.1 Module Voltages and Dielectric Strength

The lattice-matched GaInP/Ga(In)As/Ge triple-junction cells typically used in commercial HCPV at the time of writing operate at a little more three volts, although this may increase slightly in coming years as a fourth or even a fifth junction will be commercially available (see Chapter 2). In general, the number of cells in series is limited by the system voltage, or the highest voltage that will be reached between any two points in the circuit, considering the number of modules that will be connected in series in the field, and the ground, to prevent the perforation of the electrical isolation. CPV modules must be in compliance with international and national safety standards applicable to electrical installations, which require safety tests such as high voltage test (also called the dielectric voltage withstand test, or hi-pot test), insulation resistance test, ground continuity test and others. The conditions required for the electrical insulation depend on the module class defined according to classes for sources and circuits defined within IEC 61140 [50] A specific safety standard for CPV modules is currently being developed [51] whose scope is to describe the fundamental construction and testing requirements for CPV modules and assemblies in order to provide safe electrical and mechanical operation during their expected lifetime as well as to define the basic requirements for various application classes of concentrator photovoltaic modules and assemblies

A minimum requirement for the electrical insulation is imposed in the section 10.4 of the CPV qualification standard, IEC 62108 [1], which requires that the module be able to withstand twice the system voltage it is to be used in plus 1000 V (that is, a module that is to be used in 800 V systems must be able to withstand 2600 V). This corresponds to the minimum condition required in the dielectric voltage withstand test (or dry hi-pot test) for a class B module (while more demanding conditions would be required to meet the standard for Class A modules). Nevertheless, the most demanding requirement is typically the so-called wet insulation tests, which evaluates the insulation of the CPV module under wet operating conditions which can cause corrosion, ground fault or a safety hazard. The minimum requirement for insulation resistance values, 500 V is given by IEC 62108, section 10.5.

This challenge of implementing a sufficient dielectric solution in CPV is complicated by the fact that any material used as an electric isolator must also be a reasonably good thermal conductor, and materials showing these two opposing properties are not common nature. Usually ceramics are used to meet this requirement, either in sheet form, or in resins or adhesives containing ceramic powder.

For all these reasons, system voltages in CPV do not generally exceed 600 V, and module voltages do not generally exceed around 150 V. Though there is no theoretical reason that module voltages could not reach the maximum allowed by the electrical isolation, this would mean that modules would have to wire in parallel in the field, or individually connected to the inverter equipment, either of which is more complex. For large-lens modules, the small number of lenses that fit into a few square meters means that normal module voltages are considerably less than 150 V. For modules using smaller cells and lenses, a series parallel connection is commonly used to ensure the voltage does not surpass these levels.

7.7.2 Series Connections and Bypass Diodes

Regardless of the size of the cell and primary optics, a considerable number of series connections of cells within the module is necessary to reach suitable operating voltages. However, very different kinds of electrical connections are required depending of the area of the primary optics, which in practice determines the current generated by an optics-cell unit, from the wire bonding techniques used in the semiconductor or microelectronics industry for the case of small-lens systems to heavy connectors and wires for the case of large-cell systems generating tens of amperes.

An indispensable element to protect cells in operation is bypass diodes, which are connected in parallel but with opposite polarity to a cell or string of cells connected in series. Under even illumination, each solar cell will be forward biased and therefore the bypass diode will be reverse biased and effectively be an open circuit. But under uneven illumination causing mismatch in the generated currents between several cells connected in series, the diode is activated and allows the current excess generated by better cells or strings to bypass the cell, thereby reverse biasing the poorest illuminated cell or string connected to the diode. In conventional flat PV modules, the common practice is to place a bypass diode per string of cells, typically one diode per 20 cells as maximum, in order to prevent hot spot formation in partially shadowed conditions. However, CPV modules require many more of bypass diodes, typically one per cell, not only to prevent hot spots but also to minimize power losses considering that a certain degree of uneven illumination caused by optics and misalignments is almost unavoidable and inherent to the technology and not only caused by external factors like shadows. For the same reason, to reduce power losses, Schottky diodes are preferable to junction diodes, because of their lower conduction voltage.

Unlike conventional PV modules where bypass diodes are placed in junction boxes, in CPV modules these diodes are integrated on the cell carrier along with the solar cell [52]. In some designs, instead of an encapsulated diode, a bare die is attached to the same copper substrate than the cell and used for the interconnection between receivers by wire bonding.

7.7.3 Parallel Connections and Blocking Diodes

As mentioned above, in modules with relatively small lenses and cells, series-parallel connections are used to prevent voltages from increasing beyond the range which would cause the requirements for electrical insulation between cell and chassis to become prohibitive. However, the existence of these parallel connections causes additional problems, as well as a choice. We refer to the order in which the connection is made. Either groups of cells are first connected in parallel to increase the current, and then these groups are connected in series to increase voltage, or cells are first series-connected, and various of these series connected strings are connected in parallel to a common bus so that they all operated at the voltage of the module itself while summing their current.

The first choice is attractive because it effectively allows a number of cells to operate in conjunction with no series-induced losses due to current non-uniformities. (V_{mpp} is much more consistent across different cells in operation than I_{mpp}), whereas the second option is more similar to a number of small modules connected in parallel. However, there are two disadvantages to the parallel then series solution. The first is that the number of inter-cell connections is effectively doubled, although in some very small cell designs, if the inter-connections are achieved not with cabling but a PCB-style substrate or lead frame, this may be less problematic. The other problem is that while parallel-connected cells behave better than series-connected cells for reasonable small current variations, the problem of shading must still be considered. If a single lens is shaded, and the majority of the concentrated light is removed from the cell, the voltage produced by that cell will be greatly reduced. It this situation, the shaded cell will immediately be placed in a forward-bias situation condition by its neighbors, which continue to produce their nominal V_{mpp}. High quality III-V solar cells should be able to withstand a forward bias conditions at current levels on the order of their I_{sc} for some period of time, and in fact must do so to pass relevant qualification certifications [1], however, if only a single cell is shadowed, and that cell is connected to, for instance, nine others in parallel, than that cell may theoretically have to absorb up to nine times its nominal I_{sc}, a situation which is clearly dangerous. In addition, this mechanism has positive feedback, the injected current in the cell forward biased heats the cell, thus reducing its voltage, which leads to an increase in the injected current, starting a self-feeding process called thermal runaway if a certain threshold current is reached. In [53], Steiner et. al. conclude that by limiting the number or cells in parallel, thermal runaway can be prevented. The other solution would be to place a blocking diode in series with each cell, which would prevent reverse current flow. Surprisingly, the module designs known to the authors that use this connection type, usually modules composed of very small ($\sim$1 mm^2 or less) cells and lenses, have opted not to protect the cells from this situation via diodes. This is most likely because, on the one hand, the voltage loss due to the requirement to keep open an equal number of blocking diodes as there are cells would be an important limitation to module performance, as well as the fact that the simple cost of this many diodes, and the associated assembly operations may be prohibitive. It remains to be seen whether the issue of forward-biasing will cause lifetime problems in these modules.

The alternative solution, that of connecting in series and then in parallel, simplifies the problems listed above, and is currently more prevalent in manufacturers who employ cells on the order of a few square millimeters with lenses on the order of a few centimeters. For one, the number of cell-to-cell interconnections is very similar to that of a classical series connected module, with the addition of the parallel bus. In fact, using clever connection schemes it is possible to have multiple strings in parallel all of which start and stop at the same point, such that the bus is in essence the module output terminals. And while, if a single string is shaded sufficiently, the other strings could cause reverse current flow through that string, it is much more reasonable to add a few blocking diodes, one per string, to the module, although it should be mentioned that some prominent CPV manufacturers choose not to do so, instead relying on intelligent connection layout that ensure that it would be physically unlikely that only a single string receive shadowing.

7.8 Thermal Design

Conventional flat-plate PV modules can dissipate the portion of the absorbed luminous power that is not converted into electricity using both convection as well as radiation using both their upper and lower surfaces. In other words, we could assume that we have two square meters of thermal dissipation for every one square meter of solar collector. In flat-plate modules, generally the cells are heated to about 30°C above ambient, under the Standard Reference Environment of 800 W/m^2, 20°C ambient, and 1 m/s wind speed [54].

On the other hand, concentrator solar cells receive almost the same incident power, but their dissipation area is up to 1000 times smaller. If flat-plate thermal solutions were employed in CPV, the solar cells would reach a steady state temperature of 400 or 500°C, which is the point where radiation could become the dominant means of heat extraction. Of course in the real world, the solar cells or the surrounding components would either melt or incinerate long before such temperatures were achieved. To prevent CPV solar cells from immediately disappearing in a puff of smoke, the module designer must design an efficient thermal pathway that conducts and spreads the heat from the small area of the cell to relatively larger areas, such that heat can be exchanged with the exterior at tolerable cell temperatures (see Chapter 5).

In a CPV module, the front and the back of the module are not equivalent in terms of heat dissipation. Normally, the cell is mounted on, and thermally connected to, a metal backplane or heat sink which redistributes the heat and dissipates it through the back side of the module to the exterior air. But the only thermal connection between the cell and the module front surface is the air trapped inside the module. Although some heat is transferred out the front surface via this air, to a first-order analysis we can disregard this thermal pathway as insignificant, and say simply that a CPV module has only its back surface available for heat dissipation.

7.8.1 Target Cell Temperature

To choose a thermal strategy for a given cell size and concentration ratio (thermal load), we must first choose a desired cell operating temperature. Two criteria inform this decision:

1. The cell should be kept beneath a level that is thought or known to produce a loss of lifetime in the cell, the die-attach connecting it to the solder, or surrounding critical components. At the least, it should not exceed the maximum recommended operating temperature provided by the cell manufacturer.
2. The cell temperature should then further be reduced until further reduction would imply an incremental cost greater than the value of the additional harvested energy that would result from such a reduction. This trade-off may be calculated more in terms of energy rather than cost if it is simpler to do so, using embodied energy of the components used.

The first criterion is still very much up for debate. Manufacturers often specify a maximum temperature of 110°C, although while Azur Space states in no uncertain terms that 'the cell junction shall not exceed [this] maximum operation temperature' [55] while Spectrolab is content to simply specify a 'recommended operating temperature' [56]. At any rate 110°C, is clearly an upper limit of the operating temperature we should allow in our module, and while it is unlikely that the cell will sustain significant damage if its temperature exceeds this limit for short periods, it is well known that temperature is an acceleration factor for many failure modes, and in fact temperatures in the range of 120°C and 130°C are used for accelerated aging of CPV cells and CoCs (see Chapter 9).

While high temperatures alone affect lifetime, recent research has shown that an additional 'killer' is the very fast temperature swings that the cell is subjected to when, on a clear, high DNI day, a cloud passes in front of the sun. In an instant, the DNI can fall from 1000 W/m2 to almost zero. Given the right conditions (multiple, fast moving low clouds) many of these variations can occur per day. In Madrid, for example, we have calculated that about 1500 of these cycles occur per year [57] (see Figure 7.19 for an exemplary day). The solar cell, which has negligible thermal mass, is only connected to the greater thermal mass of the module itself

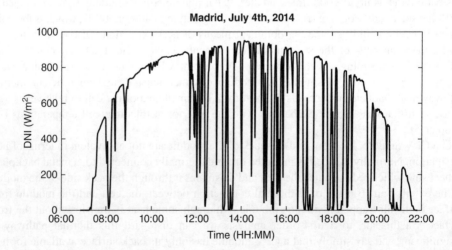

Figure 7.19 Example of a day in Madrid with relatively high irradiance and 'deep-cloud' cycles

through a resistive thermal circuit (see later in this section, and Chapter 5). Therefore the temperature differential between the cell and any part of the module with significant thermal mass can in the range of 10 to 20°C or even higher. When the DNI, and therefore the thermal load causing this temperature differential, is removed, the cell cools extremely rapidly by this amount. This can cause mechanical stress to both the cell and the die-attach materials, due to differing coefficients of thermal expansion between the materials in the cell-on-carrier stack, and these repeated expansions and contractions produce a clear risk of fatigue failure, most likely in the die attach. Both Bosco and co-workers al [58]., as well as Espinet and co-workers [59] have studied the effect of these thermal cycles. Additionally, Herrero and co-workers have proposed a promising approach for qualifying cell-on-carriers and receivers (including SOEs) against this thermal fatigue using accelerated light cycling [57].

Returning to our two criteria for setting desired cell temperature, we can see that it may be reasonable to in fact specify two target temperatures, at different sets of ambient conditions, as follows:

1. Worst case: Highest expected air temperature, no electrical generation ($\eta_{cell} = 0$, corresponding to the module in an open or short circuit condition), and additionally a maximum reasonable albedo value in the surface underneath the rear surface of the module. This case is used to make sure we do not supersede the maximum operation temperature of any component, including the cell.
2. Operation: Standard operating conditions of 20°C air temperature, a 2 m/s wind in the most favorable direction, and the module operating at a maximum power condition, with $\eta_{cell} = 40\%$. This condition is used for the purposes of the economic analysis: to answer the question of whether the costs of increasing the thermal performance above and beyond those stipulated by the worst case will yield positive or negative returns in terms of the value of the additional energy harvested due to the increase in operating voltage.

7.8.2 Simplified Thermal Model

Chapter 5 goes in great detail into the thermal analysis of the cell carrier, as well as material sections, and possible thermal strategies. In this section, we will elaborate some simple thermal relations, the aim of which is to support the initial choice of thermal strategy, that is, whether or not dedicated heat-sinks are required, and to support the economic trade-off study described above. As before, this discussion limits itself to CPV modules using an array of point-focus Fresnel lenses, beneath which the receivers are arrayed in a matching matrix on the module backplane.

For this discussion, we will use a very simplified model of the thermal dissipation system, consisting of a quasi-isothermic plate, of area A_p and thickness w_p, into which is flowing P_{in} of thermal flux. This plate represents the module backplane, and no additional heat sink is added, as the purpose of the exercise is precisely to determine whether a heat-sink-less configuration will provide sufficient cooling with a very simple calculation. We assume that 85% of the luminous flux incoming on the lens is transferred to the cell, as this is a typical upper bound for CPV optical efficiencies. Of those 15% losses, perhaps 10% is returned to the atmosphere by the Fresnel reflections at the entrance and exit surface of the primary optics, or completely scattered because it passes through the lens in a non-active portion of the Fresnel structure (that is, the draft angle). Of the other 5%, a small portion is absorbed in the optics, and larger portion is not sufficiently well concentrated to enter the receiver aperture, either because of scattering and dispersion on the primary optics surfaces, or simply because the receiver aperture is not large enough to capture all the rays. We can assume the surfaces around the receiver are typically metallic with a fairly high reflectance, so we will assume all rays that do not enter the receiver aperture are retro-reflected back out of the module, as shown in Figure 7.20. These rays are what give on-sun CPV modules a 'bright' appearance when viewed from an off-axis direction. In conclusion, the overall percentage of luminous power that is absorbed uniformly by any surface of the module other than the solar cell itself, for any reasonably well-designed CPV module is insignificant, and will be completely neglected in the following treatment.

As is well known, the portion of this power that must be evacuated under normal operation is that which is not converted into electricity. If the cell has a conversion efficiency of η_{Cell}, and an optical efficiency of η_{op}, then total thermal load on the cell is:

$$P_{in} = \underbrace{(B)}_{\sim 0.1^{W}/_{cm^2}} \cdot \underbrace{(\eta_{op})}_{\sim 85\%} \cdot C_g A_{cell}(1 - \eta_{cell}) \tag{7.25}$$

where B is the direct irradiance and A_{cell} the area of the solar cell. We must keep in mind that all components must be able to withstand the thermal condition when the module is aligned to the sun, but not producing power, that is, when η_{Cell} falls to zero.

Conservation of energy implies that once we have reached a steady-state temperature, all of this incident power will be transferred to the environment, either through radiation or convection:

$$P_{in} = P_{radiation} + P_{convection} \tag{7.26}$$

For an isothermal plate of area A_p, and T_p, we can use equations of Stefan-Boltzmann and Newton for $P_{radiation}$ and $P_{convection}$ respectively:

$$P_{convection} = A_p h\left(T_p - T_{amb}\right) \tag{7.27}$$

$$P_{radiation} = A_p \epsilon \sigma\left(T_p^{\,4} - T_{amb}^4\right) \tag{7.28}$$

where h is the coefficient of convection, ε is the emissivity of the surface, and σ is the Stefan-Boltzmann constant. As a thought experiment, let's place our solar cell in outer space, where it can depend only on radiation to evacuate the heat. Let's say additionally that the irradiance on the cell is 100 W/cm^2, the cell is 1 cm^2 in size, and the cell is mounted on a carrier that provides it with 9 cm^2 of area ($A_{carrier}$). Even if we assume a relatively high emissivity, for instance, $\varepsilon = 0.5$, the cell will come to an equilibrium at temperatures approaching 1000°C:

$$T_p \approx \left(\frac{P_{in}}{A_{carrier} \cdot \epsilon \sigma}\right)^{1/4} = \left(\frac{100\frac{W}{cm^2} \cdot 1\ cm^2}{(2 \cdot 9\ cm^2)(0.5)\left(5.67 \cdot 10^{-12}\frac{W}{K^4 cm^4}\right)}\right)^{1/4} = 1183\ K = 910°C$$

$$\tag{7.29}$$

This simple calculation only serves to indicate that convection plays an important role in heat transfer for CPV, and that we need to provide each cell with a surface area for heat transfer that is much larger than itself, that is, a heat-sink (which may or may not simply be the backplane of the module). This also serves to underline the importance of the thermal conduction between the cell and this heatsink, which usually flows through some type of thermal attachment material. This material must not only provide a high thermal conductance, but a high dielectric strength, and be very reliable over the lifetime of the module, since any reduction in its performance can quickly lead to the thermal death of the cell and surrounding components.

7.8.2.1 Radiative Heat Transfer: How to Neglect It

Since the heat sink of a CPV module in general attains only modest temperatures (<100°C) we already know that our primary heat transport mechanism to the ambient will be convection. However, we would like to know if radiation will have any significant effect at all. For a simple box-like module, as shown in Figure 7.20, the window assembly and the backplane are of the same size, and therefore we assume that the plate in our simple model has the same area as the lens, and therefore can also be related to the area of the cell through the geometric concentration ratio:

$$A_p = A_{lens} = C_g \times A_{cell} \tag{7.30}$$

The lens materials used are transparent to the far infrared radiation corresponding to the emissions of our relatively low-temperature plate, so we can further assume that the plate radiates equally on both sides. We also will assume for the moment that the plate is entirely isothermal. We can find the radiated power form the plate as a proportion of the incoming radiation from the sun, or the 'radiation efficiency':

$$\eta_{rad} = \frac{2C_g A_{cell} \epsilon \sigma \left(T_p^4 - T_{amb}^4 \right)}{C_g A_{cell} \eta_{Op} B} = \frac{2\epsilon \sigma \left(T_p^4 - T_{amb}^4 \right)}{\eta_{op} B} \tag{7.31}$$

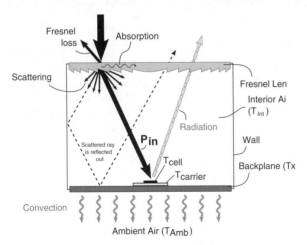

Figure 7.20 Diagram of flux (luminous and thermal) as input and outputs from the solar cell in a typical CPV module. Note that both scattered rays and radiation produced by the solar cell are not assumed to contribute to heating of the solar cell's surroundings

As it is likely that we are using bare aluminum plate, the emissivity will be low: we can assume a value of approximately 0.1. In Figure 7.21, we show the evaluation of Eq. (7.31) over a range of reasonable values of T_P. We can see that this theoretical calculation gives us a proportion that is small, but not insignificant. The proportion is high enough that, were we to be able to increase the emissivity somewhat, for instance with special coatings, than we may be able to improve performance, as long as these coatings do not affect thermal transfer through convection, which is still carrying the majority of the load in cooling our cells.

From Figure 7.21 we can see that over the most common operational ranges of heat sink temperature and DNI, we will be operating with a radiation efficiency of between 5% and 10%. In the following discussion, we will assume a value of 10%. Therefore, to a first order, we can use $\eta_{rad} = 90\%$ to reduce the thermal load on the cell and take radiation into account implicitly. The effective thermal load on the cell is found by the irradiance, the lens area, and the optical, electrical, and radiative efficiencies:

$$P_{in} = A_{lens}B\eta_{op}\eta_{rad}(1 - \eta_{cell})$$ (7.32)

7.8.2.2 Convection: Ignoring the Interior Surface of the Module

We now turn to the most important heat transport mode for CPV: convection between the module backplane and the air. As shown in Figure 7.22, there are two convective pathways from the cell to the ambient. The first, is out of the back surface to the relatively cool outside air, where additionally any wind or breeze will increase the convective constant and increase thermal transfer. Heat will also convect from the interior surface of the back plane to the air trapped inside the module, which is stagnant and at a much higher temperature. There, internal natural convection will transfer the heat through the air to the chassis walls and the front window assembly. This second pathway is obviously much less efficient, and also has the disadvantage of being more difficult to estimate through simple equations, and in fact can only be calculated exactly for a given module design using

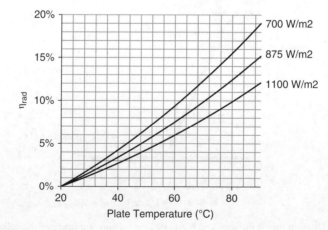

Figure 7.21   Radiation efficiency (percentage of incoming radiation that is re-radiated) for several values of DNI

a fairly complex fluid dynamics simulation. Therefore, we will make some approximations, in order to account for this thermal pathway in the following discussion.

First, we will make an estimate of the lens temperature during operation. In our experience, the difference in temperature between the front surface of the module and the ambient is about one-third that of the back surface of the module. If we assume that the temperature of the air in the interior of the module is at an intermediate temperature between the backplane and the lens, that is the average of the two, then, we can find the ratio of the temperature drops on either side of the backplane as follows:

$$\frac{\Delta T^p_{lens}}{\Delta T^p_{amb}} \frac{(T_{lens} - T_{amb})}{(T_p - T_{amb})} = \frac{1}{3} \Rightarrow T_{lens} = \tfrac{1}{3} T_p + \tfrac{2}{3} T_{amb}$$

$$T_{int} = \frac{\tfrac{1}{3} T_p + \tfrac{2}{3} T_{amb} + T_p}{2} = \tfrac{2}{3} T_p + \tfrac{1}{3} T_{amb} \qquad (7.33)$$

$$\frac{\Delta T^p_{int}}{\Delta T^p_{amb}} = \frac{T_p - \tfrac{2}{3} T_p + \tfrac{1}{3} T_{amb}}{T_p - T_{amb}} = \frac{1}{3}$$

where the definition of each temperature variable is given in Figure 7.22. Since convection, as seen in Eq. (7.27), is proportional both to ΔT and to area, the resultant temperature is the same whether we consider the interior surface to be convecting through a smaller temperature difference or through a smaller area. So instead of assuming that our plate convects equally through two surfaces of area A_{lens} we can assume that there are $n_s = 4/3$ surfaces available for convection.

7.8.2.3 Convection: Calculating the Basic Parameters of the Thermal Solution

The basic mechanism of dissipation away from the receiver, which acts as a heat source, consists of lateral conduction of the heat through the plate, while at the same time the heat is transferred to the ambient through convection. The latter is a complex phenomenon, whereby air molecules, in constant motion, make intermittent contact with a surface, robbing them of small amounts of heat, and then transferring that heat away from the surface either through their own physical motion or through collisions with other air molecules. The efficacy of these complex fluid dynamics is usually summarized in a single constant, the coefficient of convection, h, used in Eq. (7.27).

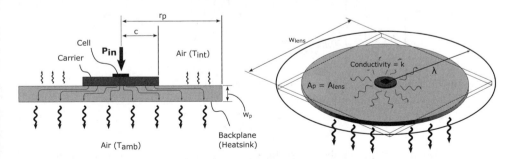

Figure 7.22 Simplified thermal model

In order to simplify the analysis, we will take the plate to be circular, but of the same area as the lens above it, as shown in Figure 7.22. This flux flows into the plate uniformly from a much smaller area in the center, representing the point at which the cell-on-carrier is attached to the heat sink, and is conducted through the material, which has a coefficient of thermal conduction, k.

We know that we can no longer consider our plate to be isothermal, because convection is robbing the plate of heat as it conducts away from the source. This field of thermodynamics is called heat transfer in extended surfaces, and most applicable to the fins used in heat sinks. In order to resolve the steady state temperature profile for any given shape of fin, we must resolve a second order differential equation representing the heat flow balance in any infinitesimal element of the plate. For any real three dimensional fin, this leads to fairly complex solutions using Bessel functions, as originally proposed by Gardner [60] and well summarized by Kraus and coworkers [61]. In our case, we are approximating the area of the backplane around a single cell as a circular fin with thickness w_p, an outer radius r_p, and an inner radius r_c that represents the size of the cell carrier. We assume that the temperature is held constant at a fixed temperature over the area where it is attached. The ratio between the area of lens and carrier can be seen as a 'thermal concentration' factor, C_{th}. Common values for carrier size are between 10 and 30 times that of the cell, so for a system with $C_g = 1000\times$, we may typically have C_{th} between $40\times$ and $100\times$:

$$\left. \begin{array}{l} \pi r_p^2 = A_{lens} \\ \pi r_c^2 = A_{carrier} \end{array} \right\} r_p = r_c \sqrt{C_{th}} \tag{7.34}$$

In the following discussion we will use $\Theta_p(r)$ to indicate the excess temperature of the fin compared to ambient at a given radius from the center. As mentioned above, the temperature of the plate for radii $r < r_c$, is fixed at the initial temperature $\Theta_p(0)$, which we would like to solve for. For larger radii, and assuming dissipation at the edge of this radial fin is insignificant, the temperature profile is given by and expression composed of modified Bessel functions of first and second class, and orders zero and one: I_0, K_0, I_1, K_1 (see [61, p. 27]). We simplify later analysis by defining this piecewise profile in terms of the normalized radius that varies from 0 to 1, defined as $r_n = r/r_p$, such that:

$$\Theta_p(r_n) = \Theta_p(0) \cdot \Theta_p'(r_n)$$

$$\Theta_p'(r_n) = \begin{cases} 1, & \text{if } r_n < \sqrt{C_{th}} \\ \dfrac{I_0\left(\dfrac{r_n}{\lambda_n}\right) \cdot K_1\left(\dfrac{1}{\lambda_n}\right) + I_1\left(\dfrac{1}{\lambda_n}\right) \cdot K_0\left(\dfrac{r_n}{\lambda_n}\right)}{I_0\left(\dfrac{1}{\sqrt{C_{th}} \cdot \lambda_n}\right) \cdot K_1\left(\dfrac{1}{\lambda_n}\right) + I_1\left(\dfrac{1}{\lambda_n}\right) \cdot K_0\left(\dfrac{1}{\sqrt{C_{th}} \cdot \lambda_n}\right)}, & \text{if } r_n > \sqrt{C_{th}} \end{cases} \tag{7.35}$$

where the *normalized dissipation length*, λ_n, is found by:

$$\lambda_n = \sqrt{\frac{k w_p}{n_s h}} \cdot \frac{1}{r_p} \tag{7.36}$$

In most references the number of surfaces, n_s, $= 2$ because the disc is assumed to be transferring heat equally on both sides. However, in our case, we are assuming that the

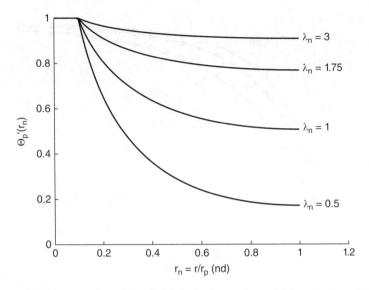

Figure 7.23 Evaluation of Eq. (7.35) for various values of λ/r_p with $C_{th} = 100\times$

interior surface is only capable of transferring one-third as much heat as the exterior surface, for any given temperature and area, so we will use $n_s = 4/3$ (see previous section).[11] Also note that in many treatments, the exponential is written in terms of the *fin performance parameter*, $m = 1/\lambda$, but we find using lambda to be more useful, since it functions as a length constant. For reasonable values of the conduction and convection coefficients, it can be shown that the dissipation length is many times r_p. This is extremely important, as it means that the backplane is more or less isothermal, and much of the area is used for convection. In Figure 7.23 we evaluate Eq. (7.35) for different values of λ_n, assuming $C_{th} = 100\times$.

Now that we know the temperature profile in the plate, we can find heat flowing out of it via convection. First, we consider that differential heat flux that is convected away through the ring shaped area at a given r_n. is given by Newton's equation:

$$dP_{out}(r_n) = (4/3) \cdot h \cdot \Theta_p(r_n) \cdot dA(r_p r_n)$$
$$dP_{out}(r_n) = (4/3) \cdot 2\pi h \cdot \Theta_p(0) \cdot r_p^2 \cdot r_n \cdot \Theta_p'(r_n) \cdot dr_n$$

(7.37)

where the 4/3 term is used to adjust the total plate area to account for convection to the interior, as mentioned earlier. The total heat flux out of the plate is then found by the integral of Eq. (7.37). We find that we can separate the result into a portion that is equivalent to the convective heat transfer of an isothermal plate at temperature $\Theta_p(0)$, and a fraction that

---

[11] Adjusting the dissipation length for reduced convection on the interior face in only part of the adjustment. We still must adjust the effective area over which the convection occurs by 4/3 as well.

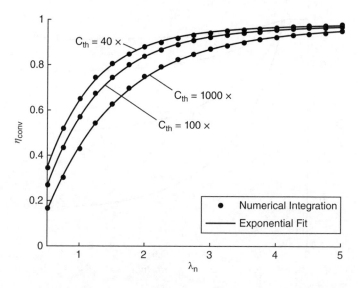

Figure 7.24 Convection efficiency versus normalized dissipation length as found by numerical integration

reduces this convective performance due to the fact that portions of the plate are at lower temperatures, a value that we will refer to as the *convection efficiency*, η_{Conv}.[12]

$$P_{out} = \overbrace{\left[\tfrac{4}{3} \cdot \pi r_p^2 \cdot h \cdot \Theta_p(0) \right]}^{Q_{out};\text{iso-thermal plate}} \cdot \overbrace{\left[2 \int_0^1 r_n \cdot \Theta_p'(r_n) \cdot dr_n \right]}^{\eta_{Conv}} \tag{7.38}$$

$$P_{out} = \tfrac{4}{3} \cdot \eta_{conv} \cdot A_{lens} \cdot h \cdot \Theta_p(0)$$

The convection efficiency is most easily calculated via numerical integration of Eq. (7.35). In Figure 7.24 we show the value of this integration for a range of λ_n and for two values of C_{th}. Additionally, we show that a two-term exponential fit allows a smooth function between λ_n and η_{conv} to be derived to sufficient accuracy. The coefficients of this fit are listed in Table 7.3 for C_{th} of 100× and 1000×. Furthermore, we can derive a relationship between these coefficients and C_{th}, thereby allowing the derivation of η_{Conv} for an arbitrary system.

Now, with a known value of η_{Conv}, we can use conservation of energy to impose $P_{out} = P_{in}$, that is, set Eq. (7.38) equal to Eq. (7.32), and solve for the temperature $\Theta(0)$, which, if it was not yet clear, is the point of this analysis:

$$\Theta_p(0) = B \frac{\eta_{op} \eta_{rad}(1 - \eta_{cell})}{\tfrac{4}{3} h \cdot \eta_{conv}} \tag{7.39}$$

Note that this temperature is the point at the interface between the receiver and the back plane. The cell carrier consists of multiple layers of materials, including electrically insulating materials

---

[12] Equivalently the product of $\Theta(0)$ and η_{Conv} is equal to the area-weighted average temperature on the plate, and then we can consider that the expression of eq. (7.38) is the calculation of the heat flow due to convection of a quasi-thermal plate at that temperature.

Table 7.3 Summary of fitting parameters of η_{Conv} (from numerical integration) to λ_n, and the fits of those coefficients to C_{th}. Over the space $\lambda_n = [1, 5]$; $C_{th} = [10\ 400]$ this combination of fits returns the numerically calculated value for convection efficiency with an error of less than 1.5%

	Fit of numerical integration: $\eta_{conv} = Ae^{(B\lambda_n)} + Ce^{(D\lambda_n)}$			
C_{th}	A	B	C	D
40×	9.55E-01	4.51E-03	−1.19E+00	−1.33E+00
100×	9.48E-01	4.96E-03	−1.22E+00	−1.16E+00
1000×	9.41E-01	4.92E-03	−1.23E+00	−8.88E-01
	Fit of Coefficients A–D: $Coef = Ee^{(F\sqrt{C_{th}})} + Ge^{(H\sqrt{C_{th}})}$			
E	6.48E-02	5.27E-03	−1.22E+00	−1.71E+00
F	−2.94E-01	−1.71E-03	2.94E-04	−3.40E-01
G	9.46E-01	−6.34E-03	1.26E+00	−1.21E+00
F	−1.89E-04	−3.45E-01	−5.83E-01	−1.03E-02

that are not especially good thermal conductors, and we cannot ignore the important temperature drop between the cell, T_{cell}, and the back surface of carrier, $T_p(0)$. This thermal drop is most easily calculated using relatively simple finite element analysis, which is beyond the scope of this discussion. Therefore, we will turn to Chapter 5 of this handbook where simulation results are provided for the most common kinds of cell carrier materials (direct bonded copper, DBC, and insulated metal substrate, IMS) for a 25 m² cell, as well as a comparison of different cell sizes (1 mm², 25 mm², and 100 mm²); see Chapter 5, Figure 5.24 and Fig. 5.25. In Table 7.4, we summarize the results of the first simulation. The values for T_{cell} and T_p are the average temperatures across the cell surface and the back surface of the receiver, respectively. For both carrier types we calculate an effective thermal resistance for the carrier, $R_{carrier}$. This simulation was performed assuming $DNI = 1000$ W/m², $C_g = 1000×$, $\eta_{cell} = 40\%$ and $\eta_{op} = 1$. The size of the carrier was fixed approximately 27 times that of the cell, such that $C_{th} \approx 40$. An important note about this simulation is that it assumes that the backplane does not aid at all in initial heat spreading over the carrier area, which is not always true. These results are specific to that configuration, so these assumptions may need to be reevaluated for higher thermal

Table 7.4 Summary of finite element simulation of common cell carrier configuration [Source: Chapter 5, Figure 5.24 of this book]

	DBC	IMS
A_{cell} (mm²)	25	25
$A_{carrier}$ (mm²)	672	672
P_{in} (W)	15	15
T_{cell} (°C)	82	76
T_p (°C)	65	65
$\Delta T_{carrier}$ (°C)	17	11
$R_{carrier} = \Delta T_{carrier}/P_{in}$ $\left(\frac{°C}{W}\right)$	0.29	0.18

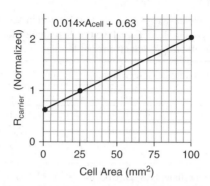

Figure 7.25 Relationship for the effective thermal resistance of the carrier with the cell area

concentration ratios. See Chapter 5 for more details, including the materials assumed and their physical constants.

We can conclude that a value of $R_{carrier} = 0.25°C/W$ is reasonable for cells and carriers of this size: that is, A_{cell} and $A_{carrier} = 27 \times A_{cell}$. Note that we have used a parameter, R, which does not take into account the area through which the heat is conducted. The carrier is performing both conduction and heat spreading functions, so such an area is not well defined, and we also see that even if we control for area, for instance by calculating an equivalent coefficient of conductivity, this constant continues to change for different cell/carrier areas. In the second simulation, a third type of receiver, a simple copper plate with a thin alumina layer below it, was investigated for three cell sizes. The resultant variation in $R_{carrier}$, is seen to be linear with cell area. In Figure 7.25 we show the value of $R_{carrier}$, normalized to the value at 25 mm^2 versus cell area. Applying a linear regression to these points, and using the resistance value of $R_{carrier} = 0.25°C/W$ for 25 mm^2, we can finally write a simplified equation for the ΔT between the cell and the heatsink, with P_{in} that of Eq (7.39):

$$\Delta T_{p(0)}^{cell} = \left(0.36 \, \tfrac{\text{K}}{\text{Wmm}^2} A_{cell} + 0.16 \, \tfrac{\text{K}}{\text{W}} \right) P_{in} \tag{7.40}$$

Now, we can calculate the differential between the cell temperature and the ambient (Θ_{cell}) as the sum of Eqs. (7.39) and (7.40):

$$\Theta_{cell} = \Delta T_p^{cell} + \Theta_p \tag{7.41}$$

We will evaluate this expression at for a variety of lens sizes and backplane thicknesses, and we hope to provide clarity on initial thermal decision. The thermal concentration ratio was fixed at 40× in order to match the simulation results of Chapter 5, and a concentration ration of 1000× was used to calculate the cell area required for Eq. (7.40). However, the lens area is the primary driver for the thermal load, and we think these results can be reasonably applied as an initial estimate of CPV systems of a given lens size, regardless of either concentration ratio.

We find it convenient to plot iso-thickness plots defined by the lens size and the cell temperature attained. This is shown in Figure 7.26 for a *DNI* of 1000 W/m^2; a relatively low coefficient of convection, $h = 10$ W/m^{2}K, which corresponds to a light wind; the conduction coefficient of aluminum, 200 W/mK. Optical efficiency is assumed to be 85%, and radiation efficiency is assumed to be 90%. The calculation is carried out for $\eta_{cell} = 0\%$ (module in open or

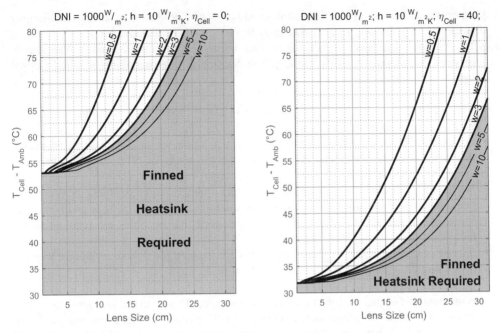

Figure 7.26 Iso-thickness curves for different geometric concentrations (left to right) and with and without electrical power extraction (top to bottom) for a CPV model with a smooth aluminum backplane as the only means of convection. Conditions are: 1000 W/m² and no wind ($h = 10$ W/mK). $\eta_{Op} = 85\%$ and $\eta_{Rad} = 90\%$

short circuit) and $\eta_{cell} = 40\%$ (operating at P_{max} with current cells). The curves represent the estimated cell temperature resulting from using different backplane thicknesses. The y-axis represents the excess cell temperature over ambient. In other words, we must assume that the cell temperature is 20°C higher under standard operating conditions and up to 40°C higher in the summer in many climates where CPV is cost-competitive. Keeping this in mind, it is clear, we require $\Delta T_{cell} < 70°C$, and preferably closer to 50°C if not below.

The minimum on the left-hand of each graph is the predicted temperature excess if the backplane was completely isothermal, for instance if it were a large solar cell encased in a box with a glass front but no concentration. We observe that the temperature swing between operation and V_{oc} can be in the range of 20°C even for the smallest lenses, and for large lenses can grow to 30° or 40°C. This is important for two reasons: this should be kept in mind during operating of CPV power plants, whenever the modules are not connected to a load, the trackers should move them off-sun, to prevent unnecessary thermal stress on the cells and CoCs.[13] Second, that we should carefully choose the two target temperatures as discussed earlier.

The behavior shown in Figure 7.26 is that of a CPV module whose only thermal dissipation is the backplane itself. Beyond a certain thickness, it would become more economical to add a finned heatsink. We would stipulate that any more than 3 mm for a monolithic aluminum backplane is thicker than desirable, and have therefore indicated the area to the right of the

---

[13] Off-axis rays can be damaging as well if we redirect the trackers only a few degrees off-axis. The trackers should be stowed or at least 10 degrees away from the sun.

3 mm profile as such. However this cutoff should obviously be chosen by the system designer after careful consideration of heat sink manufacturing vs bulk material costs.

These graphs already allow us to draw some interesting conclusions, and we can start to understand the importance of the lens/cell size decision in determining the required thermal strategy:

- Very small lenses (10 cm or less on a side) will always operate at low temperatures, even when mounted on a simple aluminum sheet of 1 mm thickness or less.
- For intermediate lenses, in the range of 10–15 cm on a side, we must increase the backplane thickness, or allow the temperature to rise into the range of 50°C above ambient (in operation), or both. However, finned heat sink is not required.
- For lenses larger than 15 cm, finned heat sinks are probably required, unless the concentration is low. We should also consider an intermediate thermal design, using an additional heat spreader between the CoC and the backplane, such that no holes are required in the backplane.
- For lenses larger than 25 cm, finned heat sinks are required for all concentrations; the only open question is the heat sink design (i.e. how many fins).

In the final part of this section, we will make an attempt to answer the last question, that of how many fins to use. We will assume that fins conduct heat away from the source in parallel to the back sheet, and that they all have a temperature drop as calculated by (7.39). We will assume that the area of each fin is a fixed percentage of the area of our modeled plate (ie. of the area of the backplane dedicated to each cell). As an example, we will assume that each fin has a radius that is 50% that of the plate representing the backplane. So the value of r_p to use to calculate $\Theta(0)$ is:

$$r_{p,fin} = \tfrac{1}{2}\sqrt{A_{cell}C_g} \tag{7.42}$$

Also, we will use $n_s = 1$ instead of 4/3. In reality the fins transfer heat to the air from both sides, but by only assuming one surface, we don't have to account for the fact the fins are not radially symmetric, but a semicircle. Finally, we will use a simplified thermal circuit model to add in the effect of the new fins by calculating thermal resistance values for the backplane and for all the fins, and then summing their inverses:

$$R_p = \frac{\Theta_P(0)}{P_{in}}; \; R_{fin} = \frac{\Theta_{fin}(0)}{P_{in}}; \; \frac{1}{R_{eq}} = \frac{1}{R_p} + \frac{N_f}{R_{fin}}; \; \Theta_{eq} = R_{eq}P_{in} \tag{7.43}$$

This Θ_{eq} is then summed to T_p^{cell} as before. This allows us to determine the temperature attained for the same simulation above but adding number of fins, N_f. The results of this calculation are shown in Figure 7.27. In these plots we have held the backplane thickness at 1mm and added a number of fins, also of thickness $w = 1$ mm.

Of course, this is only a crude approximation, not a substitute for finned heat sink design. Once the final system architecture is decided upon, the desired thermal resistance can be determined, and an optimum heat sink can be designed taking into account the many intricacies of their operation.

In summary, in section 7.8 we have presented a simple method for predicting cell temperatures for a variety of cell sizes, conditions, backplane and heatsink configurations.

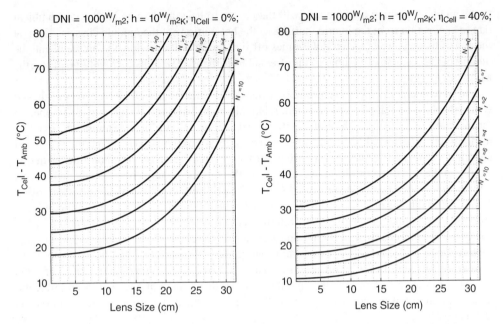

Figure 7.27 The system analyzed in Figure 7.26 with the addition of finned heatsinks. All thickness are held at $w_p = 1$ mm

We do not claim that the results of these calculations will be exceedingly accurate. However, the fact that they are built from first principles means that their application be sufficient to provide the right behavior for cost modeling and early-stage design decisions.

7.9 Venting Considerations

There is no technological reason why CPV modules couldn't be manufactured with a completely hermetic seal. After all, there are hermetic enclosures and bottles beyond counting in the world. However, the temperature variation of the gasses contained within the module – usually air – from night to day would cause such an enclosure to undergo significant stress. If the volume were kept constant, a temperature swing from 10°C to 50°C, which is reasonable in desert operation, would represent a pressure increase of 0.15 bar, or 150 N of force over the upper and lower interior surfaces of a 1 m² module. Unless the module was manufactured to be extremely rigid, this force would deform the optics and the plane of the cells and prevent their correct operation. Therefore, module manufactures opt to allow their modules to breathe through venting apertures. Such vents must allow the easy passage of air, but cannot allow the passage of liquid water, dust, insects, or other contaminants. One common solution is to use a waterproof but breathable fabric like Gore-Tex™ [62] However, it should be noted that no breathable solution prevents the ingress of water vapor, and solutions that use a labyrinthine passage to prevent the possibility of liquid water or wind-blown dust may be sufficient.

It is also important to consider the schedule for module breathing: in general the module will breathe out whenever the air inside heats up, and breathe in whenever this air cools. In general, in the morning when the module first comes on sun, the trapped module air, which will be equal to the ambient morning temperature, will then begin to heat up to some level above ambient due to convection from the interior surface of the hot portions of the module chassis. Then, at the end of the day, when the module comes off sun, the air will cool down to ambient, and the module will breathe in a volume of air of equal to perhaps 15% of its own volume, using the assumptions above.

The air temperature in the late afternoon when the module inhales is still relatively high, and so the absolute humidity of this air is still relatively high. This air is then relatively trapped in the module and cooled with the ambient through the night, until the outside temperatures reach its minimum. These are ideal conditions for a condensation event to occur, and the water will generally condense on the coldest surface. If the modules are mounted on a tracker that is stowed in the horizontal position overnight with the lens facing upwards, then the lenses will radiate their heat to the night sky, and it will be the cold surface onto which the condensation forms. Although this process should be modeled in detail for each individual module, it is the authors' opinion that in general, and excluding active systems, such as the injection of dried air, condensation events in CPV modules should be considered unavoidable, and designs should focus on forcing their evaporation as quickly as possible. This leads not only to increasing the overall ventilation aperture area, but also consideration of the location of these vents to promote a certain flow through the module, due to natural convection, to lower the relative humidity of the air inside the module and allow the condensation to re-evaporate as fast as possible.

7.10   Manufacturing Processes for CPV Modules

In recent sections we have covered, first, how to define our module architecture, and second, how to set specifications for the four essential functions of a CPV module. In this section, we would like to touch briefly on how these requirements might be realized. How is a CPV module made? What fabrication processes may be used?

Of course there is no single way to fabricate a CPV module, and as a relatively young technology, novel materials and techniques are still being proposed [63], but at least we would like to present some of the most applicable technologies, and compare which cell sizes they are most applicable for.

7.10.1   Chassis and Backplane Fabrication

We first must choose how to fabricate the chassis and the backplane. Leaving aside for the moment the rather specialized choice of a glass backplane in Soitec's Concentrix technology (Figure 7.28), which will be discussed further in the case study dedicated to that company, and is also discussed in detail in [64], we can say that we will almost certainly be using either an all-metal design, or a combination of metal for the backplane and fiber-reinforced plastic for the chassis walls. The metal used is usually aluminum, but for cost reasons steel should be considered if separate aluminum heat sinks are to be used, i.e. in a large-lens design. If we consider the range of module depths (that is, the lens size) we can see that two types of all-metal designs are most adequate for the smallest and the largest cells, while the hybrid metal-plastic design can be considered most adequate for intermediate sized cells.

The choice of techniques is dependent, not only on the module depth, but on the volume of modules to be manufactured per year, since different processes have different tooling costs and therefore become economic at different volume levels. We show the suitability of each of the

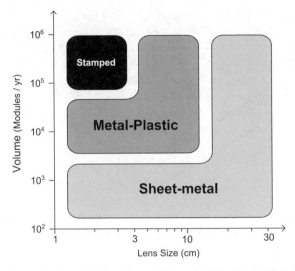

Figure 7.28 General guidelines for the commonly used CPV chassis fabrication techniques

three most common chassis fabrication techniques, in approximate terms, for varying lens sizes and productions volumes in Figure 7.29. The example module designs at the beginning of the chapter (Figures 7.2–7.4) each use one of these three options.

7.10.1.1 Single-Piece Stamped

For modules on the order of 100 mm deep or less we can consider stamping the entire chassis, including the backplane, out of a single piece of sheet aluminum. This is the only chassis

Figure 7.29 Upper (left) and lower (right) views of an old Soitec 'all-glass module', wherein the back of the module is a second sheet of glass, which was originally developed by a collaboration between the Ioffe and Fraunhoffer ISE Institutes [65]. The main advantages of this concept are that glass is a cheap and rigid material, that it eliminates concerns of module deformation due to different coefficients of thermal expansion in the front of the back plane, and that glass is an electric insulator, so there is no need to add a dielectric layer to the carrier or between the receiver and the backplane. Note that due to the low coefficient of thermal conduction, the cell carrier is very large with respect to the cell and acts as a heat spreader

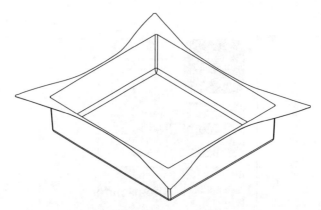

Figure 7.30 Predicted shape of a chassis drawn from a rectangular sheet after stamping and before trimming. Based on a simulation by Teknia R&D [67]. Note the wasted material at the corners

fabrication option that does not require further assembly steps, and it therefore eliminates the need for sealing at the seams during chassis assembly. When combined with appropriate structuring on the back surface, sufficient rigidity can be achieved without the need for additional stiffeners. This method implies a significant investment in tooling, but then produces the lowest per part process cost, because of high cycle times and absence of any further processing.

Stamping, or deep-drawing, as it is called when the depth of the feature is on the order of its width, can achieve depths and aspect ratios much greater than those we would recommend using it for in CPV: just look at your kitchen sink. The downside is that for progressively deeper-drawn pans, the ideal blank approaches a circle of the same surface area as the part, and even so leaves extra material at the corners (Figure 7.30); see [66, p. 240] for further details. This will result in a large amount of waste material, which at high volumes will be a significant contributor to overall cost.

This method was notably used for the final generation SolFocus product. Their design has a relatively large POE (1 cm^2 cell, 625× concentration) but the Cassegrain-type system reduced the profile significantly to around 100 mm [68], so that stamping could be used. The back of the one-piece chassis was formed into an 'egg-crate' shape (into which fit their parabolic primary mirrors) with connecting troughs that added stiffness (see Figure 7.31), thereby minimizing interior volume (which in theory should reduce the venting/condensation problem). The small-

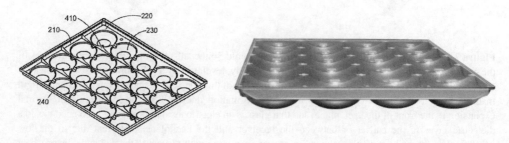

Figure 7.31 SolFocus's single-piece stamped backplane design. Left figure taken from Young et al, 2011 [69] and right one from the author's records

lens example of Figure 7.4 is intended to illustrate a module whose chassis was fabricated with this method.

7.10.1.2 Hybrid Metal-Plastic

For modules up to around 250 mm in depth, and for certain volume levels in smaller modules, using plastic injection molding for the chassis walls may be the best choice. This process offers material costs of the same order as aluminum but without material waste. The tooling costs are moderate, making it a good technology choice for pre-production and low-production manu-facturing, as well as a candidate for volume manufacturing at certain intermediate cell sizes. Although plastic-injection molding is often thought of as a process for small consumer goods, very large parts, including automotive bumpers and even large garbage bins are injection molded plastics. A good choice are fiber reinforced thermoplastics, including reinforced polyethylene or polybutylene terephtalate (PET or PBT), such as those developed by DuPont specifically for the PV industry [70]. These composites are used widely in the automotive industry as for parts with demanding mechanical requirements, along with a variety of reinforced thermosets, formed with the *reaction injection molding* process, which may be more adequate for the largest parts [71,72, p. 343]. The full range of polymer combinations, molding processes, and method for adding the reinforcing fibers goes beyond the scope of this chapter, but suffice it to say that the automobile industry research has resulted in a number of high strength, low cost injectable plastics that are ripe for use by the CPV industry.

The addition of fibers increases the tensile strength of these polymers to the range of 150 MPa, about half of the value of Aluminum 6061-T6. The value for flexural modulus is about a third of that of aluminum. Both of these parameters increase linearly with fiber content (see Figure 7.32). And while the overall size limitations of the plastic injection molding process will not limit their use for CPV chassis, the cost of the mold, the cycle time, and the availability of a molding machine with sufficient tonnage will probably limit the application for molding the chassis of a piece to relatively small modules (<0.5 m^2). Of course, each side of the chassis could be made individually and assembled, but the addition of the assembly steps may negate to some extent the advantage of the use of plastics. Surprisingly few CPV systems providers have opted to use plastics. The Taiwanese supplier Arima EcoEnergy, uses injection molded

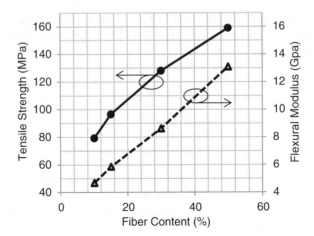

Figure 7.32 Mechanical properties of a reinforced plastic material (Schuladur) vs. fiber content. Source: Askins et al 2012 [73]

Figure 7.33 The module designed by the IES for Martifer Solar [74] uses a two-part plastic chassis. The upper and lower halves are identical. The narrow Abengoa/Sol3G module on the far right uses a similar design. Source: Askins et al 2012 [73]

side panels that are assembled using adhesives and fasteners. The demonstration CPV module developed in collaboration between IES-UPM and Martifer Solar [74], used an innovative plastic housing, which was divided into two identical parts (front and back). The required depth of mold was thereby halved, reducing both tooling costs as well as relaxing the required tonnage of the injection molding press (Figure 7.33). The medium-lens example of Figure 7.3 is intended to illustrate a module whose chassis was fabricated with this method.

7.10.1.3 Sheet Metal Fabrication

The most conventional of options is to use sheet metal for all five sides of the box, similar to how outdoor weatherproof electrical enclosures are made. Either all five sides can be separate sheets, or a bending step can be incorporated so that less than five panels are required. Assembly methods can range from fasteners to welding, depending on the volume to be produced.

A simple assembly method for producing a low number of prototype modules to the optical, electrical, and thermal performance of the design might be to assemble a frame from aluminum extrusions (beams) with plain aluminum panels fastened to them, and using sufficient sealant to close the (many) gaps, as shown in Figure 7.34. Although such an approach will clearly yield a module that is heavier per unit area than a more industrial approach, it may be a good option for producing the module rapid-prototypes of Section 7.4, Step 6, where the focus is on opto-electrical and thermal performance, although in this case it should be taken into account that the thermal design closely mimics the eventual production module designs.

As we move up the scale of production volumes, we enter into the realm of bent sheet metal. Ideally sheet metal should be structured, that is embossed to form protrusions out of the plane,

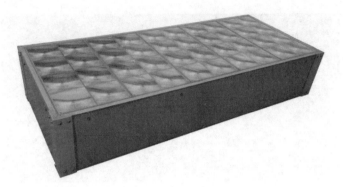

Figure 7.34   The Everphoton module uses the extrusion-plus-panel technique, even for production modules. This module type was used for a 1 MW$_p$ installation completed in 2009, the largest in Asia at that time

to enhance its rigidity and allow the amount of material to be used to be minimized. However, such embossing processes require a significant investment in tooling, so a minimum volume is required. For early designs that would not reach sufficient volume, steel stiffeners can be added.

There are various ways to fix bent sheet metal to itself to form a box without an underlying frame. First, rivets are cheap, can be installed manually or automatically and may be reasonable for low and medium volumes. A riveted structure should be sealed with structural silicone. For low volumes, a sheet metal enclosure may be welded manually, although given the cost and time required, the advantage may be mostly aesthetic (that is for the prototype module to appear closer to a planned high-volume version). At high volumes, welding can be performed using automatic welding stations (Figure 7.35), similar to those used in automotive assembly, however for medium volumes the cost of these stations may be prohibitive and welding may not be a good option. The large-lens example of Figure 7.2 is intended to illustrate a module whose chassis was fabricated using automated bending and welding.

The French startup, Heliotrop, who uses a large lens design (module thickness >0.5 m), has alighted on another interesting option that is promising for inexpensive assembly and scalability from medium to high volume. The chassis is composed of five separate thin aluminum sheets, all of which are embossed for rigidity, but since the depth of the embossing is low, tooling costs are reasonable. Using a robot, silicone sealant is first applied in the seams and the panels are brought together with an automated jig (Figure 7.36). Before opening the jig, the metal panels are joined using automated clinching. This is a process where a small tool and die are used to cold form small depressions such that the two sheets interlock and form a cold-formed joint with no rivet or additional fastener (Figure 7.37). Because the panels are also joined with structural silicone, these clinch points do not have to take all of the required force, but function primarily to hold the panels in place while the silicone cures. As soon as the clinching operations are complete, the completed chassis is removed from the jig and proceeds to the next module assembly step.

7.10.2   Heat Sink Fabrication

To the author's knowledge, the only fabrication method applicable to the manufacture of heat sinks for CPV module is aluminum extrusion. The tooling costs are reasonable so even at moderate volumes custom heat sinks should be considered for those modules that require them. This fabrication process limits us to designs with a constant section in one plane. It also means

Figure 7.35 Chassis being mounted in a fixture for automated welding at the Suncore Photovoltaic Ltd. factory in China (top). The welding occurs out of the frame to the right. Note that the main portion of the chassis, which measures about $1\,m \times 1.6\,m$, is bent only in the long direction, minimizing both wasted material and the length of the weld seams. (bottom) Close up of back of module, showing weld line and structuring features. Also, note the large heatsink required (Suncore uses a $1000\,cm^2$ lens). Source: Suncore, 2015. Reproduced with permission of Suncore

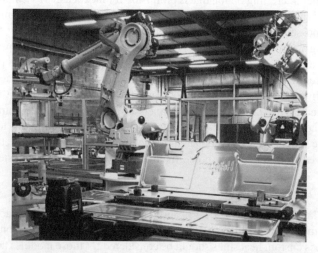

Figure 7.36 Heliotrop chassis assembly station, mounting the panels prior to assembly. Source: Heliotrop, 2014. Reproduced with permission of Heliotrop

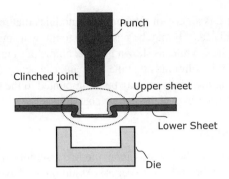

Figure 7.37 Diagram of the clinching process

we need to consider the mounting orientation of our modules, so that the fins are vertical when the modules are on the tracker, to promote convection rather than block it.

The addition of a heat sink also affects the module design in other ways. Once we have invested in a heat sink, we would like to make the best possible use of it, and that means minimizing the temperature difference between the cell and the point at which the heat flows into heat sink. To keep the resistance of the thermal stack low, the carrier should be mounted directly on the heat sink, which means that the backplane must be provided with holes at the carrier locations.

For the largest-lens modules, which have the highest thermal load and require many fins, we will have to be very careful to optimize our use of material in order to keep costs down. Therefore, we will find that we must have one heat sink per cell as shown in Figure 7.35 (bottom) and Figure 7.38 (left). This changes the assembly process flow. The heat sink becomes part of the receiver, and the carrier to receiver assembly may be carried out in another part of the production line, or even in another facility than the module assembly. In these large lens modules, there is no flat backplane assembly of cells; generally the chassis is assembled, and then the receivers are installed through the back, as shown in the example of Figure 7.2.

Figure 7.38 Two approaches for finned heat sinks. Back side of a Heliotrop module, showing individual heat sinks. Note also the embossed structures on all panels to increase stiffness (left) Source: Heliotrop, 2014. Reproduced with permission of Heliotrop. Back side of the Martifer/IES-UPM module, showing how the heat sinks extend in the long direction of the module, providing stiffness and reducing part count (right)

On the other hand, there is a class of medium-lens modules that require heat sinks but one with just 4 or 5 fins. In this case, it may be reasonable to allow a single heat sink to continue along all of the length of the module, as shown in the example of Figure 7.3 and in Figure 7.38 (right). The advantage of this solution is that the heat sink can act as a stiffener for the module assembly. In such a design, the receivers may be pre-mounted to the heat sink on a linear line before being assembled to the backplane and/or chassis.

7.10.3 Module Assembly

In this section we will say a few short words about module assembly, despite the fact automated production line design is not the authors' expertise. Assuming we have an assembled chassis, a backplane, a parquet of primary lenses, and a plurality of receivers, we have two main assembly operations to perform: the mounting of the receivers in an appropriate matrix on the backplane, and the assembly of the backplane and the window assembly to the chassis.

Receiver assembly is another of the many areas where large -and small- lens assemblies are quite different. There are basically two possibilities for assembling the receivers. In the first, precise physical features (punched holes, for instance) are provided in the backplane or chassis such that when the receiver is installed manually or with a simple automated system, the location tolerance is known. In the second, a high precision pick and place machine adapted from the electronics industry is used (Figure 7.39). In this latter procedure, in general receivers are only attached to the backplane using an adhesive or similar method. Considering that we might only have 0.1° or 0.2° of acceptance angle to spare on receiver positioning, we know from Section 7.6.4.1 that the tolerance on our lens to receiver alignment is in the range of 0.2% to 0.4% of our focal distance to spare on lens to receiver alignment. If we assume that the cutoff for mechanical location is 0.5 mm, we can easily calculate that such mechanical positioning is only feasible for lenses of greater than 100 mm on a side. For lenses smaller than 100 mm, we must use automated pick and place machines, both to achieve placement accuracies in the tens to hundreds of microns, as well as to be able to place the hundreds of cell in each module quickly.

Figure 7.39   Soitec's backplate manufacturing line in San Diego, California. With a small-lens system, Soitec relies on automated pick and place and wire bonding equipment from the electronic industry. Source: Eduardo Contreras/UT San Diego, 2015. http://www.utsandiego.com/photos/galleries/2013/feb/ 14/soitec-solar-manufacturing/ Reproduced with permission of ZUMA

Generally for the large-lens modules we can again rely on mechanical alignment. If the backplane is separate from the chassis, then it is good practice to use exterior tooling as the alignment surface for both the lens plate and the backplane (the location of the chassis is not as important). For small-lens modules a good option may be to use active alignment for the lens plate to backplane alignment. In an active system, a measurement system is attached to the module outputs during alignment, and a collimated light source (it can be of very low intensity) is shone through the lens plate. Simple passive resistors can be used to ensure that the array of cells is at or near its maximum power point voltage for the irradiance level of the alignment lighting system. Then, simple current measurements suffice to find the lens plate to receiver plate position (and rotation) that maximizes the module efficiency. Such a feedback loop is simple to implement with industrial robotics and has the advantage that it reduces errors in this assembly step, and therefore allows more tolerance in the earlier receiver assembly step.

7.11 Standards Applicable to CPV Modules

A major issue for the development of the CPV industry is the lack of specific standard normative to assess quality, safety, performance and service life of this technology. The development of such normative is carried out by the International Electrotechnical Commission (IEC), whose Technical Committee 82 (TC82) is devoted to the preparation of standards for systems of photovoltaic conversion of solar energy into electrical energy and for all the elements in the entire photovoltaic energy system. Within this committee, the Working Group 7 (WG7) is charged with the development of international standards for photovoltaic concentrators in the areas of safety, photoelectric performance and environmental reliability tests (see Chapter 9). At the time of the writing only a few of specific normative for CPV modules is in force but several others are being developed and will be approved in the forthcoming years. Table 7.5 collects all them, reflecting the current status of each.

For the design of a CPV module, the reference document is IEC 62108. This standard is not intended to determine the reliability or life cycle of a CPV module, but only minimum requirements for the design qualification and type approval of concentrator photovoltaic (CPV) modules and assemblies suitable for long-term operation. Nevertheless, several of the test sequences defined in this standard are quite demanding for the module design and subparts such as cells, receivers and optics. Chapter 9 presents the accelerating ageing test carried out on modules and describes with more detail the current and forthcoming standards for CPV.

Table 7.5 A list of CPV of IEC standards, both published and in process of being drafted

Standard	Title	Description
IEC 62108 Ed.1.0 Dec 2007 Ed.2.0 Draft	Concentrator Photovoltaic (CPV) Modules and Assemblies - Design Qualification and Type Approval	Determines the electrical, mechanical, and thermal characteristics of the CPV modules and assemblies and to show, as far as possible within reasonable constraints of cost and time, that the CPV modules and assemblies are capable of withstanding prolonged exposure.

(continued)

Table 7.5 (*Continued*)

Standard	Title	Description
IEC TS 62108 - 9 Ed. 1.0 Draft	Retest guidelines	Sets forth a uniform approach to maintain the certification of products that have, or will, undergo modification from the articles originally certified.
IEC 62670	Concentrator Photovoltaic (CPV) Modules and Assemblies Performance Testing	
IEC 62670 - 1 Ed.1.0 Sep 2013	Part 1 – Standard Conditions	Defines standard conditions for measuring the power produced by a CPV module so that power ratings noted on data sheets and nameplates will have a standard basis. Two sets of conditions are included to characterize a) operating conditions that represent on-sun performance relative to commonly measured meteorological conditions (CSOC) and, b) test conditions that represent performance when the module is isothermal as measured with a flash or fast shutter (CSTC).
IEC 62670 - 2 Ed.1.0 Draft	Part 2 – Energy Measurement	Specifies minimum requirements for determining the energy output and Performance Ratio for CPV modules, arrays, assemblies and power plants using an on-sun, measurement-based method.
IEC 62670 - 3 Ed.1.0 Draft	Part 3 – Performance Measurements and Power Rating	Defines measurement procedures and instrumentation for determining concentrator photovoltaic performance under both sets of standard conditions.
IEC 60904-9-1[14] Ed. 1.0 Draft	Collimated Beam Solar Simulator Performance Requirements	Adds additional requirements and measurement procedures to existing IEC 60904-9 so that collimated beam solar simulators that are applicable to CPV measurements may be classified.
IEC 62688 Ed.1.0 Draft	Concentrator Photovoltaic (CPV) Module And Assembly Safety Qualification	Describes the fundamental construction and testing requirements for Concentrator Photovoltaic (CPV) modules and assemblies in order to provide safe electrical and mechanical operation during their expected lifetime. Specific topics are provided to assess the prevention of electrical shock, fire hazards, and personal injury due to mechanical and environmental stresses.
IEC TS 62727 Ed.1.0 May 2012		This technical specification provides guidelines for the parameters to be specified

---

[14] Expected to be number for standard at the time of writing, but not yet officially assigned by the IEC.

Table 7.5   (*Continued*)

Standard	Title	Description
	Photovoltaic systems – Specifications for solar trackers	for solar trackers for photovoltaic systems and provides recommendations for measurement techniques. No attempt is made to determine pass/fail criteria for trackers.
IEC 62817 Ed.1.0 Aug 2014	Solar trackers for photovoltaic systems – Design qualification	Defines test procedures for both key components and for the complete tracker system. In some cases, test procedures describe methods to measure and/or calculate parameters to be reported in the defined tracker specification sheet. In other cases, the test procedure results in a pass/fail criterion.
IEC 62787 Ed. 1.0 Draft	Concentrator photovoltaic (CPV) solar cells and cell-on-carrier (COC) assemblies - Reliability qualification	Specifies the minimum requirements for the qualification of concentrator photovoltaic (CPV) cells and Cell on Carrier (CoC) designs for incorporation into CPV receivers, modules, and systems. The test is designed to demonstrate that cell or CoC components are suitable for typical assembly processes, and when properly assembled, are capable of passing IEC 62108.
IEC TS 62789 Ed 1.0 Jul 2014	Specification of concentrator cell description	Define the information communicated on a concentrator cell data sheet in order to standard.ze the way that data is reported and to serve as a reference for the power rating document.
IEC 62108 Ed.1.0 Dec 2007 Ed.2.0 Draft	Concentrator Photovoltaic (CPV) Modules and Assemblies - Design Qualification and Type Approval	To determine the electrical, mechanical, and thermal characteristics of the CPV modules and assemblies and to show, as far as possible within reasonable constraints of cost and time, that the CPV modules and assemblies are capable of withstanding prolonged exposure.

Glossary

α	angle of a single facet of a Fresnel lens
$A_{carrier}$	area of the cell carrier
A_{cell}	area of the solar cell
A_{lens}	area of the lens
A_p	area of the plate,[15]
B	Direct normal irradiance (DNI)

---

[15] The 'plate' representing the portion of the backplane dedicated to heat extraction of a single optics-cell unit

CAP_Y	Concentration-acceptance product calculated for θ_Y
C_g	Geometric concentration ratio
C_{th}	Thermal concentration ratio, ratio between the areas of the lens and the cell carrier
δ	Mechanical assembly tolerance between the lens and the receivers
δ_x	Translation errors in the x–y plane.
δz	Translation errors in the z-axis (focal direction)
$\delta\theta_{xxx}$	Portion of the acceptance angle budget dedicated to a given physical tolerance 'xxx'. For example, $\delta\theta_x$ is the portion dedicated to misalignments of the lens and receiver in the x and y direction
$\delta\varphi$	Deviation of the ray angle
$\delta\varphi_{int}$	Deviation of the ray angle (after first refraction, before second refraction)
ΔT_2^1	Delta temperature between elements 1 and 2.
ε	Emissivity
F	Focal distance of the lens
f_{diag}	f-number calculated using diagonal dimension of the lens
f_{side}	f-number calculated using lateral dimension of the lens
h	Coefficient of convection
η	Conversion efficiency of the optics-cell unit
η_{Cell}	Electrical efficiency of the solar cell
η_{conv}	Efficiency of convection (used to account for convection heat transfer of a non-isothermal plate compared to an isothermal plate)
η_{rad}	Efficiency of radiation (used to account for radiative heat transfer)
I_0, I_1, K_0, K_1	Modified Bessel functions
φ	Ray angle from any point on a Fresnel lens to the focus
φ_R	Rim angle (ray angle at the edge of the lens)
φ_x, α_x	The components of these angles that lie in the x–z (meridian) plane
K	Equivalency factor between d and q. Subscripts may be used to indicate which equivalency is evaluated, e.g. $K_{dxd\theta}$
k[16]	Equivalency factor between d and q for only one point on the lens K represents the area weighted average of k across the entire lens.
k[17]	Coefficient of conductivity
λ	Dissipation length
λ_n	Dissipation length normalized to r_p
n	Index of refraction
Pin	The overall luminous flux on the solar cell
P_{out}	Total thermal heat transport out of the cell, must be equivalent to P_{in}
QE	(External) quantum efficiency
$\theta_{ideal,100\%}$	Ideal acceptance angle for 100% of ray captured, according to conservation of *étendue*
θ_Y	Acceptance angle at a given threshold, for instance θ_{90} is the 90% acceptance angle)
Θc	Excess temperature (over ambient) of the cell
$\Theta p(r)$	Excess temperature (over ambient) at a given location on the plate[18]
r	Radial position on the plate
r_c	Equivalent radius of the cell carrier, calculated from the carrier area
r_n	Radial position on the plate normalized to plate radius

---

[16] When used in an optical context.

[17] When used in a thermal context

[18] The 'plate' representing the portion of the backplane dedicated to heat extraction of a single optics-cell unit

r_p	Equivalent radius of the plate, calculated from the lens area
r_{rec}	Receiver aperture radius for purposes of focal tolerance calculation.
r_{spot}	Radius of 'spot' of light formed by the lens.
σ	Stefan-Boltzmann constant
T_{amb}	Ambient temperature of the air surrounding the module
T_{cell}	Temperature of the cell
T_{int}	Temperature of the air internal to the module
T_p	Temperature of plate representing the portion of the backplane dedicated to heat extraction of a single optics-cell unit
w_{lens}	Width of lens
Y	Threshold used for calculation

References

1. International Electrotechnical Commission (2007) Concentrator photovoltaic (CPV) modules and assemblies - Design qualification and type approval. IEC 62108 ed1.0.
2. Antón, I. and Sala, G. (2007) The EUCLIDES Concentrator, in *Concentrator Photovoltaics* (eds. Luque, A.L., and Viacheslav, A.), Springer Berlin Heidelberg, pp. 279–299.
3. Swanson, R.M., Beckwith, S.K., Crane, R.A., *et al.* (1984) Point-contact silicon solar cells. *IEEE Transactions Electron Devices* **31**(5), 661–664.
4. SunPower Datasheet: Sunpower C7 Tracker, available at: http://us.sunpower.com/sites/sunpower/files/media-library/data-sheets/ds-sunpower-c7-tracker-datasheet.pdf (accessed: Jan. 31, 2015, archived: http://www.webcitation.org/6dBAMvPlh).
5. Mulligan, W.P., Cudzinovic, M.J., Pass, T., *et al.* (2008) Solar cell and method of manufacture, US7339110 B1.
6. SunPower secures 3GW pipeline in China with its C7 LCPV technology, *PV - Tech*, available at: http://www.pv-tech.org/news/sunpower_secures_3gw_pipeline_in_china_with_its_c7_lcpv_technology (2014) (accessed: Jan. 31, 2015 archived: http://www.webcitation.org/6dFlT8RwA).
7. Vivar, M., Antón, I., Pachón, D., and Sala, G. (2012) Third-generation EUCLIDES concentrator results. *Progress in Photovoltaics: Research and Applications* **20**(3), 356–371.
8. MacDonald, B., Finot, M., Heiken, B., *et al.* (2009) High gain solar photovoltaics. *SPIE 7407 High Low Concentrator Systems Solar Electricity Applications IV*, **7407**, 740708–740708–7.
9. Solar Systems *About Solar Systems*, Brochure, [Online] http://solarsystems.com.au/getmedia/9a237c4e-dbed-4e72-ba42-0707baf5c08e/AboutSolarSystems_Flyer.pdf (accessed: Dec. 12, 2014, archived: http://www.webcitation.org/6dFo0m99T).
10. Angel, R., Connors, T., Davison, W., *et al.* (2011) Development and on-sun performance of dish-based HCPV, *7TH Int. Conference Concentrator Photovoltaic Systems CPV-7* 1407, 34–37.
11. Ghosal, K., Lilly, D., Gabriel, J., *et al.* (2014) Semprius field results and progress in system development. *IEEE Journal of Photovoltaics* **4**(2), 703–708.
12. CPV Consortium Project Library, [Online] http://cpvconsortium.org/projects (accessed: Dec. 13, 2014, archived: http://www.webcitation.org/6dFoIDbAf).
13. Heras, J. (2014) Personal communication.
14. Gordon, J.M., and Feuermann, D. (2005) Optical performance at the thermodynamic limit with tailored imaging designs. *Applied Optics* **44**(12), 2327–2331.
15. Soitec CPV Installations, [Soitec Brochure online] http://www.soitec.com/pdf/Soitec_CPV_Installations_V11.0_EN_October_2014.pdf (archived: http://www.webcitation.org/6dFoh89AS).
16. Lorenzo, E., and Sala, G. (1979) Hybrid silicone-glass Fresnel lens as concentrator for photovoltaic applications. *Proceedings 2nd European Photovoltaics Solar Energy Conference*, 536–539.
17. Enhanced Fresnel Assembly - EFA, Type: 3C44A – with 5.5 × 5.5mm^2 CPV TJ Solar Cell. Azur Space Solar Power GMBH [Online] http://www.azurspace.com/images/products/DB_4360-00-00_3C44_AzurDesign_EFA_55x55_2015-04-02.pdf (accessed: Apr. 27, 2015, archived: http://www.webcitation.org/6dFpj8ljX).
18. Domínguez, C., Antón, I., and Sala, G. (2008) Solar simulator for concentrator photovoltaic systems. *Optical Express* **16**(19), 14894–14901.

19. Domínguez, C., Antón, I., Sala, G., and Askins, S. (2013) Current-matching estimation for multijunction cells within a CPV module by means of component cells. *Progress in Photovoltaics: Research and Applications*, **21**(7), 1478–1488.

20. Herrero, R., Victoria, M., Domínguez, C., *et al.* (2012) Concentration photovoltaic optical system irradiance distribution measurements and its effect on multi-junction solar cells. *Progress in Photovoltaics: Research and Applications* **20**(4), 423–430.

21. Núñez, R., Antón, I., Askins, S., *et al.* (2014) Characterization of CPV arrays based on differences on their thermal resistances, *AIP Conference Proceedings* **1616**, 144–148.

22. Short, W., Packey, D.J., and Holt, T. (1995) A Manual for the Economic Evaluation of Energy Efficiency and Renewable Energy Technologies, NREL/TP-462-5173.

23. Jo, J.H., Waszak, R., and Shawgo, M. (2014) Feasibility of concentrated photovoltaic systems (CPV) in various United States geographic locations. *Energy Technology Policy* **1**(1), 84–90.

24. C. Kost, J.N. Mayer, J. Thomsen, *et al.* (2014) Levelized cost of electricity: PV and CPV in comparison to other technologies, *Proceedings 29th EUPVSEC*, 4086–4090.

25. Bolinger, M. (2013) Utility-Scale Solar 2012: An empirical analysis of project cost, performance, and pricing trends in the United States, (archived: Utility-Scale Solar 2012).

26. Spectrolab Concentrator Photovoltaics (cpv): Frequently asked questions. [Online] http://www.spectrolab.com/faqs-terrestrial.htm (accessed: Apr. 29, 2015, archived: http://www.webcitation.org/6dFqZqv5y).

27. Aiken, D., Clevenger, B., Newman, F., *et al.* (2008) The first ten megawatts of III-V multi-junction concentrator solar cell production, 33rd IEEE Photovoltaics Specilists Conference 2008 PVSC 08, 1–4.

28. Zubi, G., Bernal-Agustín, J.L., and Fracastoro, G.V. (2009) High concentration photovoltaic systems applying III–V cells. *Renewable and Sustainable Energy Review* **13**(9), 2645–2652.

29. PV Insider (2012) CPV World Map 2012. [Online] http://www.pv-insider.com/cpv/documents/CPVWorldMap2012.PDF (accessed: Feb. 2, 2015, archived: http://www.webcitation.org/6dFqecYhx).

30. Antonio De Dios, Abengoa Solar (2015) personal communication.

31. Arima Eco Energy Technologies Corporation [Online] http://www.arimaeco.com/module.html (accessed: Dec. 15, 2014, archived: http://www.webcitation.org/6dFqinPJj).

32. Arzon Solar 8700 CPV Solar Power Generator [Online] http://arzonsolar.com/amonix-8700-solar-power-generator/ (accessed: Dec. 16, 2014, archived: http://www.webcitation.org/6dFqqTwJm).

33. Kenji Araki, Daido Steel (2015) personal communication.

34. Thomas Bouzanquet, Heliotrop (2014) Personal communication.

35. Jaime Silva (2014) Concentrated PhotoVoltaic Technology - CPV - MAGPOWER.

36. MAGPOWER: Power Plants - Installations [Online] http://www.magpower.eu/X/pagina.cgi?pagina_id=50. (accessed: Apr. 30, 2015, archived: http://www.webcitation.org/6dFrBYW5U).

37. Scott Burroughs, Semprius Inc. (2015) personal communication.

38. Yoshiya Abiko, Sumitomo Electrical Industries (2015) personal communication.

39. Suncore Photovoltaics, DDM-1090X Concentrator - Datasheet.

40. Benítez, P., and Miñano, J.C. (2003) Concentrator optics for the next-generation photovoltaics, in *Next Generation Photovoltaics: High Efficiency Through Full Spectrum Utilization*, Taylor & Francis.

41. Benítez, P., Miñano, J.C., Zamora, P., *et al.* (2010) High performance Fresnel-based photovoltaic concentrator. *Optics Express* **18**(S1), A25–A40.

42. Victoria, M., Herrero, R., Domínguez, C., *et al.* (2013) Characterization of the spatial distribution of irradiance and spectrum in concentrating photovoltaic systems and their effect on multi-junction solar cells. *Progress in Photovoltaics: Research and Applications* **21**(3), 308–318.

43. Victoria, M., Herrero, R., Askins, S., *et al.* (2010) Indoor characterization of non uniformity light distribution due to concentration optics and its effects on solar cell performance. 25th European Photovoltaics and Solar Energy Conference, 5th World Conference Photovoltaic Energy Conversion, 143–146.

44. Askins, S. (2014) Selected Effects of Chromatic Aberration on Triple-Junction Cell Performance.

45. Rumyantsev, V.D., Davidyuk, N.Y., Ionova, E.A., *et al.* (2010) Thermal regimes of Fresnel lenses and cells in 'all-glass' HCPV modules, *AIP Conference Proceedings* **1277**, 89–92.

46. Hornung, T., Bachmaier, A., Nitz, P., *et al.* (2010) Temperature dependent measurement and simulation of fresnel lenses for concentrating photovoltaics, *AIP Conference Proceedings*, 85–88.

47. Askins, S., Victoria, M., Herrero, R., *et al.* (2011) Effects of temperature on hybrid lens performance. *AIP Conference Proceedings* **1407**, 57.

48. Weisstein, E.W. 'Circle-Circle Intersection', MathWorld- Wolfram web resource.

49. Antón, I., and Sala, G. (2005) Losses caused by dispersion of optical parameters and misalignments in PV concentrators. *Progress in Photovoltaics: Research and Applications* **13**(4), 341–352.

50. International Electrotechnical Commission (2001) Protection against electric shock - Common aspects for installation and equipment, IEC 61140 ed3.0.

51. International Electrotechnical Commission Concentrator photovoltaic (CPV) module and assembly safety qualification, IEC 62688 draft.

52. Foresi, J.S., Yang, L., Blumenfeld, P., *et al.* (2010) EMCORE receivers for CPV system development, 2010 35th IEEE Photovolt. Spec. Conf. PVSC, 000209–000212.

53. Steiner, M., Siefer, G., and Bett, A.W. (2013) Modeling the thermal runaway effect in CPV modules, *AIP Conference Proceedings*, **1556**, 230–233.

54. Alonso García, M.C., and Balenzategui, J.L. (2004) Estimation of photovoltaic module yearly temperature and performance based on nominal operation cell temperature calculations. *Renew. Energy* **29**(12), 1997–2010.

55. Azur Space (2014) Concentrator Triple Junction Solar Cell, 3C42 – $10 \times 10mm^2$ Datasheet.

56. Spectrolab (2011) C4MJ Metamorphic Fourth Generation CPV Technology Datasheet.

57. Herrero Martin, R., Askins, S., Victoria Pérez, M., *et al.* (2012) Thermal effects and other interesting issues with CPV lenses.

58. Bosco, N., Silverman, T.J., and Kurtz, S. (2011) Modeling thermal fatigue in CPV cell assemblies. *IEEE Journal of Photovoltaics* **1**(2), 242–247.

59. Espinet-González, P., Algora, C., Núñez, N., *et al.* (2013) Evaluation of the reliability of commercial concentrator triple-junction solar cells by means of accelerated life tests (ALT), *AIP Conference Proceedings* **1556**, 222–225.

60. Gardner, K.A. (1945) Efficiency of extended surfaces. *Transactions ASME* **67**(1), 621–631.

61. Kraus, A.D., Aziz, A., and Welty, J. (2000) Convection with simplified constraints, in *Extended Surface Heat Transfer*, John Wiley & Sons, Inc., pp. 1–58.

62. Ellis, S. (2014) Impact of environmental hazards on internal soiling within concentrator photovoltaic (CPV) modules, *AIP Conference Proceedings* **1616**, 246–249.

63. Victoria, M., Domínguez, C., Askins, S. *et al.* (2013) Experimental analysis of a photovoltaic concentrator based on a single reflective stage immersed in an optical fluid. *Progress in Photovoltaics Research and Applications*.

64. Bett, A.W. and Lerchenmüller, H. (2007) The FLATCON System from Concentrix Solar, in *Concentrator Photovoltaics* (eds. Luque, A.L., and Viacheslav, A.), Springer, Berlin Heidelberg, pp. 301–319.

65. Rumyantsev, V.D., Andreev, V.M., Bett, A.W., *et al.* (2000) Progress in development of all-glass terrestrial concentrator modules based on composite Fresnel lenses and III-V solar cells, Conference Record Twenty-Eighth IEEE Photovolt. Specialists Conference 2000, 1169–1172.

66. William F. Hosford, R.M.C. (2011) *Metal Forming: Mechanics and Metallurgy*, 4th edn, Cambridge University Press.

67. Orpez, Antonio, Teknia Research & Development Personal communication

68. SolFocus, SF-CPV-205 High Efficiency Concentrator PV Panel Datasheet.

69. Young, P., Horne, S.J., Conley, G.D., and Vasquez, M. (2011) Solar concentrator backpan, US patent US7928316 B2.

70. Osborne, M. (2010) DuPont develops 'Rynite' PET grades for CPV, CSP and PV applications, *PV-Techorg*, [Online] http://www.pv-tech.org/product_reviews/new_product_dupont_develops_rynite_pet_grades_for_cpv_and_pv_applications. (archived: http://www.webcitation.org/6dFtKMNgP).

71. Helps, I.G. (2001) *Plastics in European Cars, 2000–2008*, iSmithers Rapra Publishing.

72. Dym, J.B. (1987) *Injection Molds and Molding: A Practical Manual*, Springer Science & Business Media.

73. Matweb *Schuladur A GF Property Data*. [Online] http://www.matweb.com/search/QuickText.aspx?SearchText=SCHULADUR%20gf. (Archived: http://www.webcitation.org/6ej1RqMwU).

74. Askins, S., Victoria, M., Herrero, R., *et al.* (2012) Optimization of tolerant optical systems for silicone on glass concentrators. *8th International Conference on Concentrator Photovoltaic Systems,CPV-8* Toledo.

7-I Annex

Abengoa's CPV Modules and Systems

José A. Pérez, Sebastián Caparrós, Justo Albarrán, and Antonio de Dios
Abengoa Research, Spain

7-I.1 Abengoa

Abengoa is an international company that applies innovative technology solutions for sustainability in the energy and environment sectors, generating electricity from renewable resources, converting biomass into biofuels and producing drinking water from sea water. Abengoa is the world's leading company in solar thermal electricity (STE) in terms of capacity installed with 1603 MW in commercial operation and 680 MW under construction or in pre-construction phase as of April 2015. Its experienced solar project team develops and applies proprietary solar thermal electricity and PV solar energy technologies to foster sustainable development and energy independence. Abengoa continuously improves in product development, manufacturing and installation through rigorous R&D. It has become a pioneer in the construction of commercial solar plants through technological advances and financial investments.

7-I.2 CPV Systems Development Principles in Abengoa

Leveraging the unique experience acquired with the development, operation and maintenance of CSP plants, Abengoa has developed a CPV technology targeted for utility scale applications in high radiation regions. The key principles considered by Abengoa in the system development will be discussed in the following.

7-I.2.1 CPV as a System

There is a general tendency in the PV market to give excessive relevance to the cost of the technology, typically expressed in €/W_p, \$/W_p, etc. of the module, tracker or system. Although,

Handbook of Concentrator Photovoltaic Technology, First Edition. Edited by Carlos Algora and Ignacio Rey-Stolle.
© 2016 John Wiley & Sons, Ltd. Published 2016 by John Wiley & Sons, Ltd.

as a comparison of different systems with the same technology, the direct cost of the equipment can be a good indicator, it can be misleading to compare two systems profoundly different such as CPV and conventional PV. Even within CPV, there are significantly different approaches which cannot be compared based on the cost of the module or system alone.

The main metric to evaluate the feasibility of a technology for energy generation in utility scale applications is the *levelized cost of energy* (LCOE). The best technology for a specific location will be the one that leads to the lowest LCOE, expressed in €/kWh, $/kWh, etc. (see Chapter 14). Consequently, in the development of CPV technology the LCOE needs to be competitive with conventional PV technologies to achieve a significant market share in locations with high radiation. In order to succeed, the CPV technology has to be developed considering the entire system instead of focusing on individual components of the technology such as the module or tracker. The module and tracker are parts of a complete system and therefore cannot be optimized independently. Furthermore, all the cost implications from the rest of the elements of a CPV plant, as well as the installation, operation and maintenance costs have to be carefully considered. Foremost it is key to understanding the performance of the technology in operating conditions and the energy generation of the plant throughout its lifetime.

7-I.2.2 A High Efficiency, High Concentration and High Acceptance Angle Module

One of the main differences of CPV when compared to other PV technologies is the high efficiency and the potential to continue improving both the cell and the module with new advances. CPV requires precise tracking structures. Taking advantage of the high efficiency potential of the technology, module and system costs can be reduced. The high efficiency optimizes the utilization of the trackers and minimizes the operation and maintenance (O&M) costs.

On the other hand, a high concentration design can minimize the cell and receiver costs, allowing for the highest efficiency materials to be used. Additionally, the number of parts in a module (receivers, solar cells, etc.) may be reduced by means of using high concentration, which can in turn reduce the assembly costs.

The acceptance angle of the module is of great importance when the complete system is taken into consideration. The acceptance angle increases the tolerances in the module factory assembly and in the tracker installation in the field which reduces system installation costs. O&M costs are also reduced, as the installation does not require constant adjustment of the modules for optimum operation.

The combination of high efficiency, high concentration and high acceptance angle is a challenge in the design of cost competitive CPV modules. An outstanding optical design together with the utilization of both high efficiency optics and cells are needed. Additionally, in order to keep the high performance in operating conditions and to benefit from the excellent temperature coefficient of III-V multijunction solar cells, the thermal management in the module has to be carefully considered and optimized.

7-I.2.3 Reliable Tracking

As the modules produce energy only when they are accurately aligned with the sun, reliable and accurate trackers are essential for the operation of CPV systems. The acceptance angle for CPV modules developed by Abengoa is in the range of $\pm 1°$, which implies that the tracking accuracy has to be well below that value.

The acceptance angle considered in the design of the tracker has to take into account the tolerances in the assembly and module alignment. Additionally, modules working within the range of the acceptance angle would produce almost 10% less than those aligned on axis and could increment the mismatch in the array, producing significant yield losses.

There are two factors to be considered when analyzing the tracking accuracy in a CPV system: the accuracy of the tracking control system and the structural deformations. Both factors can be affected by the design and installation of the tracker and they change with the operating conditions. Abengoa usually sets the specification of maximum tracking error for the design of the tracker as half of the acceptance angle. The maximum tracker error is calculated considering both the tracking accuracy and structural deformations in the worst operating conditions.

Trackers have to operate according to specifications with a high availability during the entire life of the plant, which typically ranges between 20 and 30 years. Otherwise both the energy yield and the O&M costs would be affected, increasing the LCOE of the CPV plant. A minimum 99.5% tracker availability – i.e. downtime resulting from tracker failure or tracker control system malfunction is limited to 0.5% – is considered by Abengoa in the design of CPV systems.

Designing a cost competitive tracking system which meets both the requirements of high accuracy and high availability is feasible but remains a significant challenge. As stated previously, both the module and tracker have to be considered together in the design and optimization process in order to meet the cost and performance targets of CPV systems.

7-I.2.4 Demonstrated Performance Before Commercialization

Every new technology in a market such as the energy generation needs to be well proven before deploying it in volume. Utility scale plants involve high initial investments and financing a project requires a demonstrated bankability of the technology. Rushing to commercialization before the performance and reliability is validated can lead to unexpected results or failures in the field which would generate distrust in the technology.

Abengoa has built the Solúcar Platform which is the largest concentrating solar installation in Europe. The platform combines solar thermal and photovoltaic technologies to supply a total of 183 MW_p. In addition to the on-site commercial plants, the platform features extended research and development facilities, where a broad range of new technologies are assessed through demonstration plants. The validation in demonstration plants provides the confidence in the market to facilitate the bankability of the technology.

7-I.3 Abengoa's CPV Technology

The CPV technology of Abengoa includes modules and both electro-mechanical and hydraulic trackers. Different generations have been developed progressively improving the cost competitiveness required for volume industrialization of the technology. As CPV is still in an early development stage and there is a high potential to continue improving the cell and module efficiency, the technology will continue evolving rapidly in upcoming years.

7-I.3.1 M300S Module

M300S is the third generation of CPV modules from Abengoa (see Figure 7-I.1). The same principles are considered in the design of previous generations, although M300S uses a larger

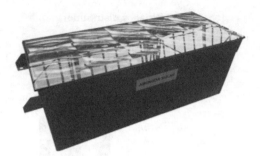

Figure 7-I.1   Abengoa M300S Module (see details in Table 7-I.1)

solar cell ($7.5 \times 7.5\,\text{mm}^2$) and a higher geometrical concentration (1330×). The module features silicon on glass primary optics and a round to square glass secondary optics. The optics design, based on proprietary IP, provides an acceptance angle of $\pm 1°$, an exceptional result considering the high concentration of the design.

Each M300S module contains 10 multijunction cells which are interconnected in series. As of today, lattice-matched GaInP/GaAs/Ge triple junction solar cells are used but this module allows the integration of new solar cell architectures. In the evolution from the initial generation to the M300S, the cell size and concentration have been increased, reducing significantly the number of parts in the module (see Table 7-I.1). As the cell operation temperature target is the same than in the previous generation, this approach increases the heat dissipation requirements for each individual cell. Therefore, M300S technology requires a more efficient heat dissipation design. Considering the module configuration and distribution of the heat sources in the module, only passive cooling can be considered as an economically feasible option. The heat sinks are bigger and more expensive than the ones previously used; however, the impact in $€/W_p$ is insignificant because the output power per solar cell is significantly increased.

Although smaller solar cells could achieve a higher efficiency, the assembly costs would be significantly higher and could require important investments for industrialization, only feasible with very high volume manufacturing. Taking into account that before full scale commercialization smaller demonstration and pre-commercial plants need to be demonstrated, Abengoa's approach has been to develop a solution which requires a moderate investment for industrialization and can be cost competitive not only with high volume manufacturing but also with medium scale volumes.

At the time of writing (March 2015), the M300S is in a qualification and certification phase. M300S average production efficiency will be above 30% in the initial demonstration systems. With new improvements in the cell structures and optics efficiency there is a clear roadmap with the target of achieving 40% module efficiency. Optical and cell efficiency above 83% and 48% respectively are needed to achieve such target.

The previous generation, M300 module, is IEC certified and being produced in a pilot line for demonstration purposes. The M300 module utilizes $5.5 \times 5.5\,\text{mm}^2$ triple-junction solar cells with a 1000× geometrical concentration. The M300 module is not cost competitive in the current PV market scenario, consequently Abengoa will not pursue its volume industrialization and will focus on the new generation. However, as both generations are based in the same proprietary IP and design principles, all the experience acquired with the M300 systems in the

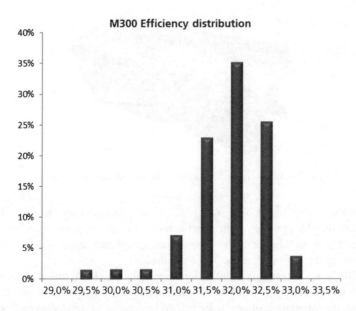

Figure 7-I.2 Production DC efficiency distribution. 400 kW M300 modules produced in 2014

field will be an excellent starting point for the demonstration of the technology and future optimization. Figure 7-I.2 shows the DC efficiency distribution of 400 kW produced with M300 modules for a demonstration plant.

Before the development of the M300 module, Abengoa worked on a design using also $5.5 \times 5.5 \, mm^2$ triple-junction solar cells with a 475× geometrical concentration (Gen 1). Table 7-I.1 summarizes the evolution of the main design parameters of Abengoa's CPV modules. In principle the objective of the different changes in the evolution is mainly driven by the module costs. Two main changes can be identified in the evolution of the parameters: Increase the geometrical concentration from 475× to 1300× and increase the cell active area from $5.5 \times 5.5 \, mm^2$ to $7.5 \times 7.5 \, mm^2$. A significant cost reduction is achieved at the cell level thanks to the higher concentration. Additionally, the combination of both changes translates into a reduction of the number of parts per module area, minimizing the module assembly costs. The changes in the dimensions of the module are driven by the optical requirements for the higher concentration designs and by the optimization of the module enclosure and primary lens costs.

Table 7-I.1 Evolution of Abengoa's CPV module generations

	Gen 1	M300	M300S
Geometric concentration	475×	1000×	1300×
Cell active area in mm^2	5.5 × 5.5	5.5 × 5.5	7.5 × 7.5
Number of cells per module	10	42	10
Module active aperture in m^2	0.14	1.27	0.72
Module dimensions in m^3	0.15 × 1.25 × 0.35	1.04 × 1.22 × 0.42	1.4 × 0.575 × 0.5

Figure 7-I.3 On the left, hydraulic tracker ($140\,m^2$). On the right, electro-mechanic tracker ($65\,m^2$)

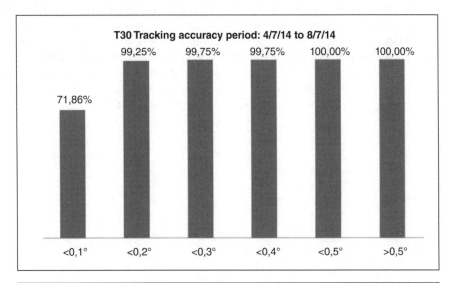

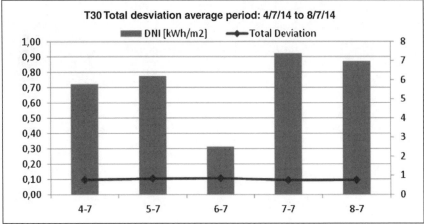

Figure 7-I.4 Tracking accuracy measured during five consecutive days in a prototype tracker

7-I.3.2 CPV Trackers

Abengoa has developed two different lines of CPV trackers: electro-mechanic and hydraulic. The hydraulic mechanism is typically more cost effective because larger size trackers can be deployed. However, electro-mechanic trackers can have a wider movement range and offer advantages when smaller size trackers or using concrete-less foundations are required. The electro-mechanic tracker can have an aperture of up to $90\,m^2$, while the hydraulic can reach $180\,m^2$ (see Figure 7-I.3) The trackers designed in Abengoa are optimized considering the complete system, taking into account the specific size, weight, acceptance angle and gravity center of the modules. The optimum solution for a particular project would be selected considering the location, site requirements, type of soil and plant size.

As for the tracking algorithm, both solutions can use an open or closed loop control, or a combination of both. Although the closed loop algorithm provides a higher accuracy, it is important in a tracker that the open loop works also within specifications. Figure 7-I.4 shows the tracking accuracy measured during five consecutive days in a prototype.

At the end of 2014, a new certification for CPV trackers has been developed by the IEC, where Abengoa took part as a member of the working group. Since Abengoa has deployed more than 1.2 GW of tracking structures for STE and PV, Abengoa has an extensive experience and know how in this field. A thorough validation to guarantee the tracking accuracy and reliability of every new design is completed before commercialization.

7-II Annex

CPV Modules and Systems from Daido Steel

Kenji Araki

Daido Steel Co., Ltd, Japan
Present address: Toyota Technological Institute, Japan

7-II.1 Introduction

Daido Steel is one of the pioneers in CPV technologies using III-V multijunction cells and high concentration advanced optics, including the development of a 28 % 200 W module in 2003 [1,2]. This module is still generating power in Tsuyama National College of Technology in Japan after 12 years and is possibly the oldest HCPV module using III-V cells. The R and D of CPV in Japan started in 1998 by joint research with Toyota Technological Institute on the development of CPV-Thermal system using low concentration Si cells [3]. In 2000, R and D of HCPV system using III-V multijunction cell started with Sharp and Daido Metal with a project granted by NEDO. The design policy, such as dome-shaped Fresnel lens for expanding tolerance and heat removal without relying heat sinks, has been included in Daido developments since then. After the success of the 28% 200 W module in 2003, the R and D target moved to higher concentration and establishing technologies for local assembly [4].

In this annex, a brief explanation of Daido's CPV technology is presented as well as how its system is integrated to meet the specific purposes and constraints in a CPV power plant.

7-II.2 Design Philosophy of Daido's CPV Module and System

Unlike flat-plate PV technology, the CPV manufacturing technology is an integration of plastic, metal and machine technologies. Although the system itself contains high-tech and sophisticated components, most of the manufacturing may be done by the local workforce (see Figure 7-II.1) . This is particularly true for the tracker and the system assembly step. Module assembly and lens manufacturing can be also accomplished onsite. The barrier for local assembly may be high but the key fact is that upstream products can be transported easily and

Handbook of Concentrator Photovoltaic Technology, First Edition. Edited by Carlos Algora and Ignacio Rey-Stolle.
© 2016 John Wiley & Sons, Ltd. Published 2016 by John Wiley & Sons, Ltd.

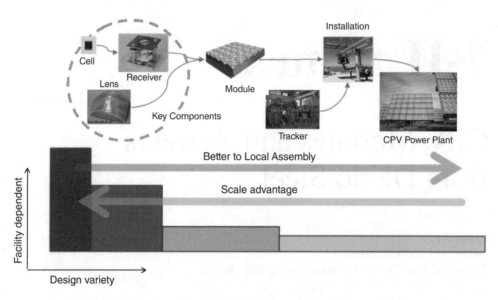

Figure 7-II.1   Value chain of CPV components and system integration

have more chances of cost reduction by volume mass production (see Figure 7-II.1). Another advantage of local assembly is that technologies used by downstream CPV products can be assumed by local commodity industries.

Another aspect that needs to be considered is logistics. Typical CPV modules are bulky and have fragile optics. This particular structure increases both packaging cost and transportation cost. Local assembly of modules and local manufacturing of other downstream products offers the advantage of saving transportation costs.

The cost reduction strategy commonly accepted in PV industry, relying on simple scale-expansion, is not always effective. A practical and low-risk alternative approach is to use local assembly. CPV is the only PV technology that can achieve low-cost power generation not relying on scale merit [4].

The CPV technology of Daido Steel has been developed under this philosophy. All the technology complexity is packed into the upstream components like receivers and lenses so that downstream technologies can be accomplished by local manufacturers using hand tools and commonly-used industry tools. In other words, both lens and receiver are designed to be tolerant to the assembly errors associated with manual or semi-manual assembly. Accordingly, the Daido module has a wide acceptance angle so that many kinds of trackers, regardless of their high or moderate precision, can properly illuminate the concentrator cells without relying on special tuning and time-consuming field alignment procedures. Therefore, Daido systems operate without the need for frequent and periodic cleaning and alignment calibration in the field.

Another key subject is reliability. Daido Steel's philosophy is the control of key components such as receiver and lens without relying on external environmental control devices and components such as an air conditioning unit and external breathing filters. Receivers are designed and tested against strong UV irradiation [5], harsh environment [6], cycles of heat stress [7] and lightning and external surges [8,9]. The high reliability at the component level was realized mainly by strong sealing and careful selection of materials and structure. The philosophy that reliability is secured by upstream components is similar to the computer industry; analogously, circuit engineers do not have to worry about semiconductor failure physics caused by the ambient because of reliable IC packaging.

Unlike other modules that are designed to operate in dry areas, Daido Steel's CPV modules were designed to survive in damp environments:

- First, all the materials in the module enclosure – including receiver components and solar cells – were tested in a wet environment, with a risk of water condensation. Some materials commonly used in CPV receivers could not pass our tests and were replaced with more robust alternatives.
- Second, breathing holes were optimized to avoid water dew on the backside of the Fresnel lens. The main idea is to enhance air ventilation while prohibiting water intrusion, like traditional houses in East Asia.

In the following section, optics, that is one of the most distinctive features of Daido's CPV module, is briefly introduced.

7-II.3   Optics in Dado Steel's CPV

Figure 7-II.2 illustrates the basic configuration of the CPV module from Daido Steel. The enclosure is made from aluminum alloy by *monocoque* frame without heat sinks or fins for reducing weight. This latter fact is important to keep high tracking accuracy which can be easily lost by deflection of the tracker rack. The sunlight is concentrated by the dome-shaped Fresnel lens and collected by an inverted pyramid kaleidoscope SOE named the homogenizer. The cell is placed under the homogenizer surrounded by strong sealing resin.

The homogenizer does many things. Its basic function is to homogenize the concentrated light in both intensity and color. The performance of the concentrator solar cell is enhanced by improving the illumination uniformity. The uniform distribution of illumination is also important to achieve a homogeneous temperature distribution on the solar cell. The kaleidoscope effect re-shapes the circular concentrated beam into a square that matches the cell aperture. The inverted pyramid structure expands aperture of the solar cell and thus enlarges the assembly tolerance. Off-axis protection is also enhanced by the shading effect around the periphery of the cell by the expanded base of the pyramid and shifting the focal point above the cell level to the base of the pyramid.

The dome shaped Fresnel lens has been optimized as a low loss collector. A key break-through in the development of the lens was the successful injection molding as in automobile

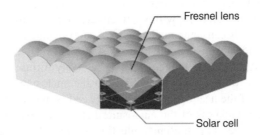

Fresnel lens

Solar cell

Figure 7-II.2   Structure of CPV module of Daido Steel

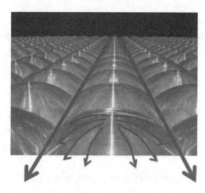

Figure 7-II.3   Self-washing effect favored by the arrays of dome-shaped Fresnel lenses

lamps. This process allows a rapid and inexpensive means for manufacturing high quality lenses that are durable enough for use in a concentrator system.

The main reason to use the dome shape is the minimization of chromatic aberration and the minimization of the focal distance. The shorter focal distances associated with the dome structure allow low profile module enclosures as well as enlarge tracking tolerance. Thus, no field alignment is necessary. This enables high and stable performance even if the module is installed on lower precision or flattering trackers. Besides, surface reflectance loss is minimized. It has also a structural advantage: the dome shape structure is able to support 1.5 m depth snowfall.

It is known that the soiling of optics is one of the main reasons of a lower performance in the field. The resemblance of our module to traditional roofs in East Asia − using curved roof tiles for a proper channeling of water flow by rain − implies that a dome-shaped lens surface helps wash out dust even with a small quantity of water by light rain. It is also useful to save water for cleaning. This is because a clear water flow channel is allowed and thus the dust dragged from the lens surface does not flow into other lenses (see Figure 7-II.3).

7-II.4   Heat Removal and Other Module Technologies in Daido Steel

The size of the lens of Daido Steel modules is 200×200 mm for 820× concentration and 165×165 mm for 550× concentration. The cell aperture size is 7×7 mm so that the receiver is commonly used in both 550× and 820× types. The selection of the lens size is important because it is closely related to heat removal from the cell. In the module of Daido Steel, the heat of the concentrator cell is exchanged by the enclosure itself without the use of heat sinks or other heat removal devices. The basic design is shown in Figure 7-II.4. It is true that the heat is concentrated on the cell but it will be effectively delivered to the environment just as happens in flat-plate PV, because the total area for heat exchange is the same. There are two conditions to achieve this. One is that the heat resistance between the cell and the enclosure is sufficiently small. Another one is that heat is effectively diffused in the enclosure. The latter condition depends on the heat conductivity of the enclosure (made of aluminum) as well as the magnitude of the heat concentrated to the cell, which is proportional to the lens size. Given the limited thickness of the enclosure that is constrained by the weight of the module, the reasonable size of the sunlight collection is 200×200 mm as a maximum, that is, 820× concentration when using a 7×7 mm cell. With these sizes, the assembly tolerance can

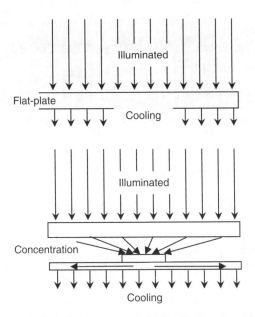

Figure 7-II.4   Heat removal design of the module of Daido Steel compared to that of flat-plate PV modules

be reasonably wide and is within the level of accuracy attainable of hand tools and standard tolerance of metal working.

7-II.5   Performance of the a CPV System of Daido Steel

This is a case study of implementation of a CPV system in a landfill area near the Nagoya international airport, Japan (see Figure 7-II.5) and constructed in 2008. The electricity

Figure 7-II.5   30 kW (5 kW× 6) CPV system for energy supply to a sewage center

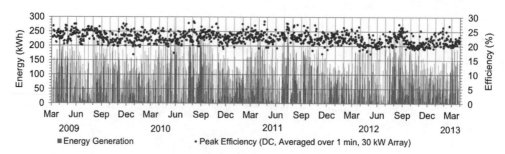

Figure 7-II.6   Energy generation record of 30 kW CPV power plant to the local sewage center

generated by this power plant is delivered to a local sewage center that operates the entire sewerage treatment of the Nagoya International airport and urban area of Tokoname city (population 55 000). It was expected that the energy generated by this CPV plant covered 10 % of the electricity demand in a clear sky day.

Usually, CPV power plants are constructed in dry and inland areas to receive better direct solar resources and to avoid degradation from moisture and salt. However, the challenge of this power plant was to show the feasibility of CPV in an undeveloped coastal area with much chance of receiving chloride damage as well as strong wind.

Figure 7-II.6   indicates the energy generation record to the local sewage center. The performance ratio in six years of operation was 0.824, which was sufficiently good and contributed to a significant CO_2 emission reduction in the International Airport. In April 2013, the energy supply to the sewage center is terminated due to the expiry of contracts. The generated energy has been sold to a local utility company by the FIT since then.

Nowadays, the efficiency of the 820× standard module of Daido is 28%. The cost of the module is about €1.1/W_p for the manual assembling line with workforce in Japan. It is expected to reach to €0.67/W_p by the introduction of an advanced optical design and the increase of geometrical concentration ratio up to 1000×.

References

1. K. Araki, M. Yamaguchi, *et al.* (2003) A 28 % efficient, 200 Wp, 400× concentrator module and its packaging technologies. Solar Concentrator Conference 10–14 November, Alice Springs, Australia.
2. K. Araki, M. Yamaguchi, *et al.* (2004) Achievement of 28 % efficient and 200 Wp concentrator module and the technological roadmap toward realization of more than 31 % efficient modules. *Technical Digest of PVSEC-14* **12**(1), 427–428.
3. K. Araki, M. Yamaguchi *et al.* (2000) Energy flow examination on PV generator system using cold mirror. *Proceedings 16th EUPVSEC.*
4. K. Araki, (2007) 500× to 1000× – R and D and Market Strategy of Daido Steel. *Proceedings SCC2007.*
5. K. Araki, M. Yamaguchi, *et al.* (2003) Material study for the solar module under high concentration UV exposure. *Technical Digest of the 3rd World Conference on Photovoltaic Energy Conversion.*
6. K. Araki, M. Yamaguchi *et al.* (2004) Development of a new 550X concentrator module with 3J cells – performance and reliability. *Proceedings 31st IEEE PVSC.*
7. K. Araki, H. Nagai and K. Tamura (2012) Fatigue failure of concentrator III-V solar cells - Does forward bias current injection really kill III-V CPV cells? *AIP Conference Proceedings,* **1477**, 281.
8. K. Tamura, K. Araki, I. Kumagaiand H. Nagai, (2011) Lightning test for concentrator photovoltaic system. *Proceedings 37th PVSC.*
9. K. Araki, H. Nagai and K. Hobo, (2011) Mortality of III-V Multi-junction solar cells of CPV systems in the real field. *Proceedings 25th EUPVSEC.*

7-III Annex

Soitec CPV Modules and Systems

Francisca Rubio, Sven T. Wanka, and Andreas Gombert
Soitec, Germany

7-III.1 Introduction

Soitec Solar GmbH started 2005 as a spin off from the Fraunhofer Institute for Solar Energy Systems ISE and developed the Concentrix™-Technologies to industrial mass production in order to gain a significant share of the PV market. The three key drivers which had to be addressed by the development are increasing efficiency, cost reduction and long-term stability. The largest project size realized with this technology to date is 44 MW_p DC with CX-S530 systems in Touwsrivier, South Africa [1]. In this Annex the path leading to the final product and the planned innovations will be described.

7-III.2 The Principles of Concentrix™ Module Technology

Firstly, the main characteristic of the Concentrix™ module technology is its simple, symmetric design. Two parallel glass plates, one for the optics and one carrying the solar cells, are fixed by a metal frame (see Figure 7-III.1). The first objective of Concentrix technology was to create a simple and low cost technology; therefore, no secondary optical element was used. This means fewer components and thus a reduced number of process steps, and fewer failures during manufacturing and operation. This first objective provided guidance for the following decisions. Without secondary optics the geometric concentration cannot be extremely high, therefore Concentrix used a concentration of 500×. The second consequence is the size of the cell. In order to reduce the series resistance losses of the solar cells due to an inhomogeneous illumination, the cell should be small. A small cell has the added advantage of better thermal behavior compared with a larger cell size.

The cooling of the module is passive and the heat distribution is realized by plain metal heat spreaders which are glued on to the bottom glass plate. The geometric layout enables the heat, which is generated locally on the cell as a result of the concentrated energy which is not

---

Handbook of Concentrator Photovoltaic Technology, First Edition. Edited by Carlos Algora and Ignacio Rey-Stolle.
© 2016 John Wiley & Sons, Ltd. Published 2016 by John Wiley & Sons, Ltd.

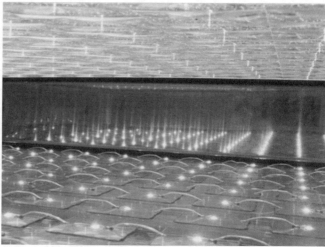

Figure 7-III.1 Principal design of Concentrix™ technology with SOG (Silicon on Glass) primary lens and planar heat sink on glass receiver (left). A look into a real module between the two glass plates (right)

converted into electricity, to be distributed over the full available area and transferred to the environment via convection and radiation.

The reason for using a glass-glass module design was to have a receiver plate substrate with a small coefficient of thermal expansion (CTE) and to have the same CTE on the lens plate and on the receiver plate, but this approach is also perfect for ensuring electrical insulation. The thermal radiation behavior of the glass is quite good, and additionally this glass-glass approach offers the possibility to use a standard industrial solution from the double glazing industry with respect to the assembly and sealing technology.

7-III.3   How Size Matters

An intrinsic difference to flat PV modules is the need for CPV modules to have a certain distance between the optics and the cells. The CX-M500 module described here is designed with a gross height of 102 mm. The distribution of the heat is much easier and cheaper if numerous, smaller local heat injections have to be distributed. On the other hand, optimized wafer usage and kerf loss reduction or throughput in the manufacturing process favors seeking a global optimum. Another aspect to consider is the capability at different manufacturing process steps to place cells, receivers and lens plates with an accuracy range from 10 to 100 μm (see Figure 7-III.2). In order to solve this optimization problem, a deep understanding of all relevant impact factors is required.

Over the last five generations of Concentrix™ modules, the single lens and heat spreader sizes were modified only moderately. However, the size of the module changed more drastically from $0.27\,m^2$ to $8.77\,m^2$ and this enabled a cost reduction over the full value chain due to reduction of cables and connectors as well as a higher integration in the fully automated manufacturing line and easier installation. The first prototype power plant of CX-S530 equipped with CX-M500 modules has been connected to the grid since 2013. Already on

Figure 7-III.2 Placement and electrical interconnection is performed by fully automated machines with very high precision

the module level the full operating voltage of around 600 V is available. The latest variant of this module type has a nominal power of 2650 W_p and an efficiency of 33.8% (as of 2015) under concentrator standard test conditions (CSTC).

7-III.4 Reliability and Long Term Stability

The importance of high efficiency paired with low cost is very obvious and this is the main market focus. However, to achieve economic success, the system reliability is at least as important as the efficiency and cost. Field data to prove the track record of the Concentrix™ technology is available for more than five years. This data shows that up to now there has been no degradation in module performance [2]. Soitec's modules are certified according to IEC 62108 and UL 8703 in accordance to the industry standard, as well as extended tests with an overall duration of up to three times the duration defined by IEC [3].

This high level of robustness is reached by minimizing the potential negative impact of humidity inside the module. Each CX-M500 leaving the manufacturing plant is tested to ensure that there are no leaks at the silicone seam connecting the glass plates to the frame and also that there are no gaps at the joints where the aluminum frame structure has been welded together. During operation, the air inside the module can be exchanged by an air drying unit (ADU) and additionally, the germanium substrate of the solar cells is covered by a plasma layer applied in the receiver plate line in order to prevent potential corrosion. The long term stability of the module further benefits from the use of the well-known metal bonding process which is used to connect 2400 solar cells and the protecting bypass diodes within the CX-M500 module.

Finally, resistance against UV, sand or other negative environmental impacts is achieved as the materials exposed to the outside world are mainly glass and aluminum and some silicone and plastic connectors. The parts required to maintain a high efficiency, such as the solar cells and silicone on glass (SoG) lenses, are well protected inside the module.

7-III.5   Efficiency

As was the case in the past, increasing efficiencies for CPV modules will be based on the ongoing development of multijunction solar cells using the potential offered by semiconductor technology. In the solar cell efficiency tables for a reasonable sized solar energy converter the trend is heading towards efficiencies in excess of 40% (see Chapter 2). This is backed by solar cell developments moving to four junctions with the perspective to exceed the 50% efficiency landmark.

Higher efficiency cells provide a better conversion over the full sun spectrum which means that the matching of the different junctions together with the optical design has to be considered. Optical and electrical simulations are used when new types of solar cells are integrated into the module. A further requirement for the cells is that they have to operate under locally high concentration of up to few thousand suns as is the case with the point focusing of the CX-M500. Several cell structures including lattice matched or metamorphic from different vendors, three or four junction cells have been tested and integrated successfully into Concentrix™ technology. This shows the flexibility of this module design which has the potential to benefit from further cell developments.

A variety of different optical designs have been implemented in the past, however today the dominant design is based on Fresnel lenses using the SoG technology. For the CX-M500 an optimized point focusing design is used which can be manufactured at high volume with a narrow standard distribution of the optical efficiency of each individual lens.

7-III.6   Tracking System

As with all CPV technologies, the Concentrix™ technology requires a tracking system in order to generate electricity, because CPV only uses the DNI radiation. The accuracy of the tracking is given by the acceptance angle of the module's optics. As the acceptance angle is quite small (see Figure 7-III.3) it is necessary that the tracking system is very accurate.

One of the most difficult choices of the tracking system is the size. Large systems have a higher cost for the foundations but a lower cabling and trenching cost. In comparison, smaller tracking systems require small, less expensive foundations but the cost of cables and trenches is higher.

In any case, the system should be stiff and resistant enough to maintain the weight of the CX-M500 modules (210 kg each), but at the same time, it should be very accurate to track the sun properly.

In the first system generations [4], Soitec chose small systems with DC motors. For the power plant in Touwsrivier (South Africa), Soitec has developed, together with suppliers, a larger tracker with a tracking area of around $110 \, m^2$. The tracker consists of a galvanized steel structure and a dual axis slew drive with AC motors. This tracker size was selected as the aim was to optimize the foundation costs, installation time and the cabling. The evolution of the tracker size in Soitec's technology portfolio is presented in Figure 7-III.4.

To assure the correct movement of the tracker, the drive is fitted with an encoder to give feedback of the tracker position. Additionally, there are limit switches for the elevation and azimuth axes which define the allowable movement range of the tracker.

Of course, the mechanical system could not work without a control unit, which controls the movement of the tracker. Soitec uses a customized tracker control unit (TCU) with a self-developed tracking algorithm. The hardware of the TCU is developed together with electronic companies depending of the type of tracker and even the location of the manufacturing.

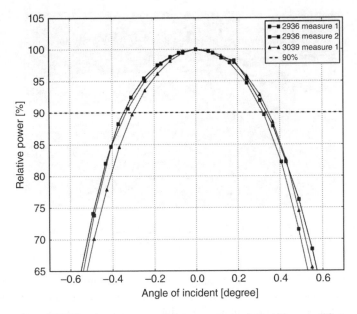

Figure 7-III.3 Acceptance angle measurements in outdoor conditions

The algorithm of Soitec is unique and includes three different types of tracking methods:

- Astro tracking gives the theoretical movement of the tracker. This method provides the basis of the tracking and it should be calibrated during the first sunny days after the installation.
- Sun sensor tracking. This tracking method is mainly used during the commissioning, in the first days of installation, during calibration and in failure cases.
- Power tracking is the optimum tracking method implemented by Soitec, where the power produced by the modules is measured during every tracker movement and the algorithm calculates the best position to maximize the output power of the system. This algorithm was initially developed by Fraunhofer ISE.

The hardware of the TCU includes a control board, a power measurement board as well as all of the required protection and fuses. The TCU distributes power to the motors, and receives the input signals from the encoders, limit switches and sun sensor.

Figure 7-III.4 Evolution of the size of the tracker from 35 m², 70 m² and 110 m² of Soitec CPV system

Figure 7-III.5   Drive mechanism and cable guidance on a tracker system

A unique component of Soitec's technology is the air drying unit (ADU). This component is used to avoid any condensation inside the modules as it supplies dry air to the modules on a daily basis. The first generation contains a ventilator, a heater, desiccant, valves and control electronics. This unit is connected by tubing to all modules and safely avoids condensation inside the modules in all weather conditions. For future developments, Soitec is studying other technologies to perform this function.

Equally as important as all these individual components is the system integration of the complete system (see Figure 7-III.5). System integration refers to brackets, cables, cable management, fuses, grounding, air tubing, the sun sensor, and meteorological sensors, in addition to consideration of the interfaces between the system components. If some of these small components are not well defined either electrically and/or mechanically the system cannot function properly. For example, as the tracker is a large moving system and all the cables and tubing are also subject to frequent movement it is essential that they are well protected so that any damage or incorrect tracker operation can be avoided.

The system integration components are defined to be installed during the assembly of the tracker table while the final cabling connection will be carried out after the mounting of the tracker table on the mast. Currently the efficiency of the complete CPV system is around 29.5%

In summary, together with suppliers Soitec develops the modules, ADU, tracker structure, drive mechanism, tracker control unit and the system integration components.

7-III.7   Future Improvements

The use of very high efficiency cells, like the *smart cell* [5] is one of the most important objectives of Soitec to improve the technology. The implementation of a secondary optical element to achieve higher concentration and be more independent of the ambient temperature is an attractive next step but this must go along with a new thermal management development due to the use of increased concentration. The first tests of this approach with a demonstration sub-module have resulted in an outstanding module efficiency of 39% [6].

On the system side, the most important challenge is to further reduce the cost of the whole system, including structure and tracking mechanism but also to reduce the power plant costs such as the foundations, installation and cabling. Therefore, Soitec has built up a multi-disciplinary team to address and optimize all of the variables which have an impact on the total cost of ownership of the CPV system. New tracker concepts, installation procedures, foundation designs and cabling optimization are currently being studied.

References

1. Gerstmaier, T., Zech, T., Röttger, M., Braun, C. and Gombert, A. (2015) Large-scale and long-term CPV power plant field results. *11th International Conference on Concentrator Photovoltaic Systems (CPV-11)*, Aix-les-Bains, France, AIP Conference Proceedings **1679**, pp. 030002-1–030002-8.
2. T. Zech, T. Gerstmaier, M. Röttger, *et al.* (2014) Return of experience from 5 years of field data: long term performance reliability of Soitec's CPV technology. *Proceedings of the 29th European PV Solar Energy Conference and Exhibition*, Sep 22th–26th, Amsterdam, Netherlands.
3. A. Gombert and F. Rubio (2013) The importance of manufacturing processes and their control for the reliability of CPV systems. 9th International Conference on Concentrating Photovoltaic Systems CPV-9, Miyazaki, Japan, *American Institute of Physics Conference Proceedings* **1556**, 279–283.
4. A. Gombert, T. Gerstmaier, M. Gomez, I. Heile, M. Röttger, and J. Wüllner, (2011) Field experience of Concentrix Solar's CPV systems in different climatic conditions. *American Institute of Physics Conference Proceedings* **1407**, p. 327.
5. T.N. Tibbits, P. Beutel, M. Grave, *et al.* (2014) New Efficiency Frontiers with Wafer-Bonded Multi-Junction Solar Cells. *Proceedings of EU-PVSEC Conference*, Amsterdam, September 2014.
6. S. van Riesen, S. Wanka, M. Neubauer, *et al.* (2015) New module design with 4-junction solar cells for high efficiencies. *Proceedings of the CPV11 Conference*, Aix les Bains, France, April 2015. *American Institute of Physics Conference Proceedings* **1679**, 100006-1–100006-8. http://dx.doi.org/10.1063/1.4931553.

7-IV Annex

Suncore Photovoltaics' CPV Modules

James Foresi

Suncore Photovoltaics, Inc., Albuquerque, United States

7-IV.1 Introduction

Suncore's DDM 1090X Module (Figure 7-IV.1) is a high-concentration, Fresnel lens-based design incorporating high-efficiency, triple-junction solar cells. The DDM 1090X CPV module had its origins within Emcore Solar Power, a division of Emcore Corporation with the mission to design, develop and deploy utility-scale CPV power plants. Suncore, a joint venture between Emcore and SanAn Optoelectronics, was formed in 2010 to ramp Emcore's Generation 3 (Gen 3) design to volume production and to pursue opportunities for deployment in China. To date over 110 MW of the DDM 1090X have been deployed. This section reviews the development history of the DDM 1090X, from its inception at Emcore through its volume production ramp at Suncore. The development includes basic design considerations for cost optimization, design verification and reliability assessments, as well as manufacturing and deployment considerations.

7-IV.2 Gen 3 Module

Suncore's DDM 1090X module is based, in large extent, on Emcore's Gen 3 module design. The following sections describe the development of that design and the engineering trade-offs that were considered. Much of the design work centered on developing an appropriate cost model that allowed for trade-offs. Once the initial design was selected, two development iterations, namely, Baseline and Alpha, were built and evaluated. These iterations allowed for fine-tuning of the design and optimizing the assembly approach for volume production. The initial design, Baseline and Alpha are reviewed below.

Handbook of Concentrator Photovoltaic Technology, First Edition. Edited by Carlos Algora and Ignacio Rey-Stolle.
© 2016 John Wiley & Sons, Ltd. Published 2016 by John Wiley & Sons, Ltd.

Figure 7-IV.1 Front and back views of the Suncore DDM 1090X module

7-IV.2.1 Initial Design

When the Gen 3 program began in July of 2008, the cost target for the module and tracking system combined was set at $1.50/Watt (this is a DC-uninstalled cost, not including system assembly, inverters, etc.). This was expected to be highly competitive with silicon flat panel, which in 2008 had an average selling price of $3.16/W with expectations going to $2.22/W in 2010 (module only). It was this cost advantage that lead Emcore, as well as many others, to invest in CPV development. The design space for Gen 3 was initially unlimited. Multiple designs were considered including reflective (dish) systems with dense or semi-dense grid receivers, refractive (Fresnel lens) solutions, guided wave solutions and others. Initial cost assessments indicated that any of these approaches could meet the cost target with varying levels of risk or ultimate system complexity. Based on Emcore's prior experience with their Generation 1 and 2 prototypes, the decision was made to continue work with Fresnel lens based designs.

To optimize the final configuration of the refractive design, a cost model was generated that allowed for costs to be assessed as a function of concentration and lens size (with cell size varying accordingly). Some basic assumptions drove the cost model. In particular, the receiver package for the system would use direct-bonded-copper (DBC) substrates (Figure 7-IV.2), which had been proven to meet thermal performance and reliability requirements. Additionally, silicone-on-glass (SOG) technology was selected over acrylic for the primary optical element. SOG was preferred over acrylic for reliability reasons; in particular acrylic suffered from performance issues after damp heat or humidity freeze stresses. Another key design consideration for utility-scale installations was to have the option for final module assembly to be done on-site to reduce shipping costs and incentivize local economic development near the power

Figure 7-IV.2 Direct-bonded-copper (DBC) receiver design with box connectors, bypass diode and triple junction solar cell

plants. Based on these considerations and others, the optimization chart shown in Figure 7-IV.3 was developed. The chart incorporates only the receiver and lens costs, however the conclusions were the same for the complete cost model. Costs in the chart are based on volume quotes from 2008 and should not be compared to modern CPV system costs.

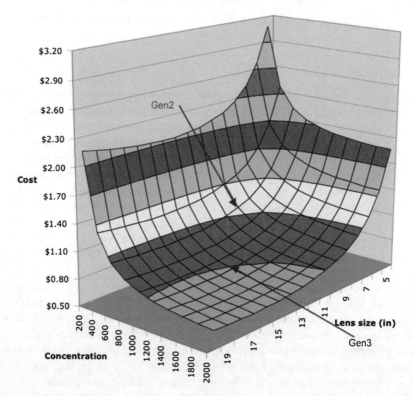

Figure 7-IV.3 Trade-off of costs (dc uninstalled $/W) for the optics and packaged cell (exclusive of module and structure costs) that result from variations in lens size & concentration. The arrows show the actual design points for the Gen 2 and Gen 3 systems

The conclusion reached from the optimization of the refractive system was that the ideal solution would produce as many Watts as possible from a single receiver. This is a result of fixed costs in the system that arise for many of the system components. For example: the size of the DBC (and thus the cost) does not scale with the solar cell size due to fixed area required for the bypass diode and connectors, the bus bars on the solar cells are of fixed size and become a larger fraction of the solar cell as the solar cell size decreases, the labor associated with receiver assembly does not scale with the solar cell size. The only inputs constraining the concentration, and thus the power produced by a single receiver, was a maximum current set to 12.5 A to minimize grid finger losses and a requirement that thermal management be entirely passive. The final design point selected was a 13″ square Fresnel lens operating at 1090× concentration. Figure 7-IV.3 shows this design point (labeled 'Gen 3') compared to the Gen 2 design point. Note that this design optimum is wholly attributed to the cost model assumptions. Other CPV designers, for example, made the assumption that the entire process for assembly would be automated. Taking this assumption as a starting point, would lead to very different, but entirely valid, conclusions and product designs.

Because the cost model assumed on-site final assembly for the module, the shipping cost was not a driver to keep the module thin. The only negative impact of a thicker module is the incremental material cost of taller sidewalls, and this incremental cost was not high enough to drive the module design to a thinner configuration.

The optical design that resulted from the Gen 3 cost optimization is shown in Figure 7-IV.4. The lens size used is $1090\,\text{cm}^2$ with a $1\,\text{cm}^2$ solar cell for a geometric concentration of 1090×. The optical design included a glass secondary optic that is attached directly to the solar cell with an optically transparent silicone. The glass secondary operated as a 2× concentrator, which allowed us to reduce the lens-to-cell distance, improve flux uniformity and still maintain good optical efficiency. A 'tertiary' optical element is mounted above the glass secondary. This is a reflective kaleidoscope design that helped to increase the acceptance angle for the system as

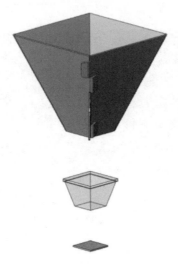

Figure 7-IV.4   The optical design (at the receiver) for the Gen 3 Baseline design. The solar cell (bottom), glass secondary (middle) and reflective tertiary (top) are shown

well as protect module components from off-axis beam damage. The optical design achieved an acceptance angle of approximately ± 0.7°. Thermal management for the modules was accomplished using a finned heat sink design. The total thermal resistance for the design was set at 0.8°C/W under standard test conditions. The operating temperature of the solar cell was approximately 40 °C above ambient under max power point conditions. The module included 15 receivers, each generating approximately 30 W, for a total of 450 W.

7-IV.2.2   Gen 3 Baseline

The first completed Gen 3 system was dubbed 'Baseline'. It was constructed at Emcore's location in Albuquerque, NM and consisted of 24 modules (two strings) on a tilt/roll tracker (see Figure 7-IV.5). The Baseline system was used to evaluate the operation of the modules at 1090× concentration (the Gen 1 and Gen 2 designs were both 500×), the performance of the tracker in real-life conditions, and to identify any near-term reliability concerns from on-sun operation. The Baseline system ran for approximately six months and several areas were identified for improvement. In particular, the DBC-to-heatsink attach was found to vary over time and was re-engineered.

7-IV.2.3   Gen 3 Alpha

The Gen 3 Alpha system was a cost-reduced and improved version of the Baseline system. Modifications of the Baseline design were made to improve the long-term performance and reliability of the DBC-heat sink attach, to reduce the cost of many of the components that were used and to address issues identified from internal stress testing. The Alpha system ran for approximately two years providing a wealth of valuable data. The year-over-year performance is shown in Figure 7-IV.1 below. The system showed less than 1% per year degradation which was not unexpected. Seasonal variations, due to the SOG lens temperature dependence (which has been well-documented), are evident in the performance of the system. To better understand

Figure 7-IV.5 The Gen 3 Baseline system

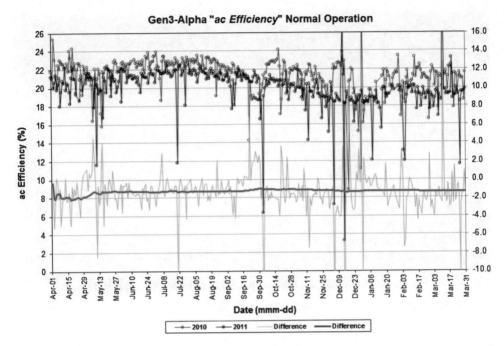

Figure 7-IV.6 Year-over-year performance for the Gen 3 Alpha system during its two years of operation. The light 'difference' line indicates the instantaneous difference while the dark 'difference' line is a running average

if the SOG lenses were contributing to the year-over-year loss, lenses were removed from the Alpha modules and attached to new modules. The results showed no degradation in lens performance. Ultimately, the loss mechanism for the Alpha system was identified and corrected for the DDM 1090X.

7-IV.3   DDM 1090X Module

While the Gen 3 module was originally designed to be assembled on or near the site of deployment, the involvement with San'an Optoelectronics changed Suncore's market paradigm in two significant ways. First, through San'an, Suncore had significant opportunities for deploying utility-scale solar power plants in western China, which did not require ocean transport. Second, to improve the manufacturability of the module, Suncore implemented a stamped (or Deep Drawn – the DD in DDM) module housing. The stamped version was more mechanically robust, but did not allow for stacking of the module housing. The combination of these two factors made the original deployment plan of on-site assembly obsolete. All manufacturing for the utility-scale solar power plants in China was done in Suncore's factory in Huainan, China, and special shipping arrangements were made to transport the modules from the factory to the install sites. In addition to changing the module housing design, Suncore developed a fully automated receiver manufacturing line and automated lens alignment and attachment capabilities. The production line in Huainan is capable of producing 250 MW of DDM 1090X modules annually when operating at full capacity.

Figure 7-IV.7   Suncore's 50 MW CPV power plant in Golmud, Anhui Province, China

7-IV.4   Current Status and Future Projections

Suncore's DDM 1090X module is the result of a long history of Fresnel lens-based CPV design work at Emcore and Suncore. The resulting module achieved 28.5% efficiency at Standard Test Conditions with a 38.5% efficient solar cell. The module was certified to IEC 62108 by a third party and in addition was UL-listed and CE marked. Over 110 MW of this module design have been deployed near Golmud, China (Figure 7-IV.7 shows the 50 MW field). The original paradigm of on-site assembly for the product was discarded in favor of centralized manufacturing. This decision was successful in getting large scale fields in China deployed and supported the ramping of volume production. The design met or exceeded all of the cost targets for the product and was competitive with silicon flat plate systems at the time of deployment. Suncore continues to reduce the cost of their CPV modules through evolutionary design changes and supply chain management. In addition, the roadmap for higher concentrator solar cell efficiency keeps CPV competitive in a PV landscape that continues to drive to lower costs.

8

CPV Power Plants

María Martínez,[1] Daniel Sánchez,[1] Francisca Rubio,[2]
Eduardo F. Fernández,[3] Florencia Almonacid,[3]
Norman Abela,[2] Tobias Zech,[2] and Tobias Gerstmaier[2]
[1]*Instituto de Sistemas Fotovoltaicos de Concentración –ISFOC, Spain*
[2]*Soitec, Germany*
[3]*Centro de Estudios Avanzados en Energía y Medio Ambiente,
Universidad de Jaén, Spain*

8.1 Introduction

The ultimate objective of CPV technology is to build large scale power plants to supply clean energy to the population with competitive costs. In order to achieve this goal, efficient, reliable and cost-effective CPV power plants have to be engineered. This chapter reviews the most important aspects to consider when building a CPV power plant, with the goal of maximizing its energy output and reducing the costs of installation, operation and maintenance.

First, in section 8.2, we describe a general procedure to be followed for the construction of a CPV plant. Second, we focus on some special issues to take into account for the construction, namely, the selection of the inverter (section 8.3), the distribution of the trackers (section 8.4) and some considerations of the environmental impact and the possible dual use of the land of CPV plants (section 8.5). Next, we define the basics for plant monitoring (section 8.6), to subsequently address operation and maintenance procedures (section 8.7). Following this, two key aspects in any power plant, namely the power rating and the modeling of energy produced by a CPV plant, are discussed in detail in sections 8.8 and 8.9, respectively. Finally, a list of symbols and abbreviations commonly used in the chapter is included for easier reference. In addition, the chapter is complemented with some ancillary material included in a set of annexes. In Annex 8-I the main software tools that can be used for the modeling, design and/or analysis of a CPV plant are listed and briefly described. In the following annexes, several case studies illustrate how ISFOC and Soitec have managed the design and construction of their

Handbook of Concentrator Photovoltaic Technology, First Edition. Edited by Carlos Algora and Ignacio Rey-Stolle.
© 2016 John Wiley & Sons, Ltd. Published 2016 by John Wiley & Sons, Ltd.

multi-Megawatt CPV plants. The selection criterion for the case studies was power plants greater than 1 MW_p.

8.2 Construction of CPV Plants

The realization of a CPV plant can be divided in four phases. The first phase involves the preliminary works, with an in-depth analysis of all potential sites in order to confirm which site is the most viable to construct a CPV plant. The second phase involves the creation of a first draft detailed engineering document based on the basic engineering requirements for the CPV system being considered. The third phase allows preliminary costing to be made based on the draft detailed engineering. The final version of the detailed engineering and costing is then done until an economically viable CPV power plant design is achieved for the chosen installation site. Once the detailed engineering is completed, the fourth phase may be initiated which covers the construction works themselves from the civil works to the fine tuning of the CPV systems and the startup and commissioning of the complete installation. Additionally and in parallel, financing should be started to establish the economic resources of the project.

8.2.1 Preliminary Works

The first step to build a CPV power plant is the analysis of a potential new site. Before even starting any detailed engineering it is necessary to analyze the technical parameters of the site.

8.2.1.1 Meteorological Data

In the following, the most important meteorological data that needs to be analyzed to characterize the site are described:

- Estimation of the DNI to analyze whether the site is suitable for CPV or not. As a rule of thumb, suitable sites for CPV are locations with DNI higher than 2000 kWh/m^2 and excellent locations are those with DNI higher than 2300 kWh/m^2 and ratio between DNI and GNI is also a key point to understand the performance of a CPV power plant. The minimum recommended ratio DNI/GNI should be larger than 0.85.

  DNI ground measurements are rarely available for proposed sites; therefore, data within a 100 km distance from the site is acceptable only in the very first analysis. For a detailed study it is mandatory to have ground measurements at least for a period of 3–6 months and use the data in order to calibrate satellite derived data which are on several locations available on a 3 km grid. DNI data should be acquired from different sources to minimize uncertainty. These meteorological data can be obtained from different databases like *Meteonorm*, *PVGIS*, *SolarGIS*, *SolarAnywhere*, NASA, etc. (see Annex 8-I for further details about these software packages).

  The air quality is very important too, because too much humidity, dust or any other pollution will decrease the DNI incident on the CPV modules.

  Conducting on-site measurements for at least six months and using this data to enhance the quality of satellite derived data, is offered by independent companies and is strongly recommended to reduce the uncertainty.

- Temperature: CPV is more suitable for hot regions because is less sensitive to the heat than the traditional PV technology [1,2] (see Chapter 5). The technology could be installed in cold regions too, but maybe some kind of optics should be adapted to the particular temperature range for the region, since they are typically optimized for high temperatures [2–4] (see Chapter 4). The temperature is also very important to define the specifications of the *balance-of-system* (BOS) components (i.e. control unit, inverter, electrical cabinets, etc.). The typical range of operation temperatures is between −20°C to +55°C.

- Wind: wind data are extremely important to define the loads and the characteristics of the tracker and foundations. CPV trackers are designed to withstand a certain operating wind speed. When the wind exceeds this level, most of the trackers move into a stow or protection position, where the system is out of focus and no energy is generated. If the limiting wind speed is reached often, then a reduction of the performance ratio is observed. For this reason, it is preferred that CPV power plant sites are located where the wind speed does not frequently exceed the operational wind speed limit of the tracker. On the other hand, a low and constant wind is beneficial for the plants due to the cooling effect which increases the module output.

  Wind information can be obtained from national or private weather services. If no local measurements are available, it can also be acceptable to use data collected from a meteorological station (measuring at a height of 10 m, at least), located within a distance of no more than 100 km from the plant site and with similar surrounding terrain. This information is very critical and for a detailed analysis is mandatory to take data from an anemometer during some months. There are many companies who offer historical wind data like *Meteonorm*, *3TIER*, etc. (see Annex 8-I for further details about these software packages).

  Not only the average wind data are important, but the evolution of wind over time is also important to size the drive system. Typical wind speed values to start moving to stow position are between 12 m/s and 16 m/s depending on the tracker (see Chapter 6). Therefore, a detailed analysis of the specific location should be done to calculate the energy loss due to the time spent at stow position. Typically, the maximum wind speed up to which the drive can operate is around 20 m/s to 24 m/s, which allows the tracker to move to a stow position within some minutes. Most the trackers are prepared to resist maximum winds of 40 m/s in stow position.

- Rain: the rain data are important to determine the requirement for cleaning [5].

8.2.1.2 Geographical and Technical Data

In this section, the most important geographical and technical data from the site that need to be analyzed are described:

- Altitude: the altitude of the proposed CPV plant site must be considered. All components which are used within the CPV plant must be rated for the specific altitude.
- Topography, seismic grade, ground conditions (inclination, rocks, mountains, shadowing profile, corrosion, etc.).
- Latitude: if the site is in the tropics, the system should be adapted to track the sun during the whole year because the sun will change from South to North very quickly. For example, polar axis trackers are suitable for latitudes within the tropics.
- Special environmental needs (flood level, animals, etc.).
- Status of grid connection, kVA rating and voltage of the nearest existing grid connection point.

8.2.1.3 Socio-Political Situation

Next, the most important socio-political data from the site that need to be analyzed are described:

- region, country: applicable laws for renewable electricity generation and grid codes;
- local content requirements;
- permitting requirements;
- site owner (public, private, etc), jurisdiction, utility company, permitting agencies;
- expected PPA (power purchase agreement) or FIT (feed-in tariff) value.

When all this data is analyzed and the results show that there is no factor that impedes the construction of a CPV power plant at this site, then a basic study should be carried out to determine the typical performance of the plant and the resulting revenue of the energy generated by the CPV plant.

8.2.2 Basic Engineering Study

A basic engineering study should be made in order to present the engineering constraints of the CPV system when designing the power plant. The resulting basic engineering document is then used to create the detailed engineering document which is the basis of all cost calculation for the CPV project. The following steps describe the process of the basic engineering study.

1. Shadowing study: the latitude, type of tracker (movement, size, shape, etc.), landscape and electrical connection between cells and modules are inputs to calculate the losses of the CPV power plant due to shading. Section 8.4 describes in detail a procedure for optimizing the distribution of the trackers in a CPV plant. Typically, *PVsyst* software is used to determine the losses due to shading of the CPV systems. It can also be used to have a first estimate of the energy output. With this purpose, in addition to *PVsyst* data, some other information should be simulated such as: topography (change in the terrain slope in the power plant plot), external shading (e.g. mountains, trees, buildings.), and latitude requirements (loses due to restrictions in the movement of the tracker). Some other useful tools used for the calculation of the energy production are summarized in Annex 8-I.
2. Using *PVsyst*, iterative studies are made to determine the most cost-effective distance between trackers. The difference in energy output due to shading between trackers is compared with the cost of additional cable length and trenches. Based on these results, a layout of the power plant is then made. Section 8.4 describes in detail a procedure for optimizing the distribution of the trackers in a CPV plant.
3. Power plant layout: A preliminary power plant layout (see Figure 8.1) showing all the elements of the installation is created; this includes all AC and/or DC for power supply and power output cabling, the selection of inverters, combiner boxes, transformers, communications wiring, other tubing and wiring (e.g. air drying unit, active refrigeration unit, etc.) when needed.

The use of string or central inverters has a significant influence on the use of DC or AC cables [6]. The cost of the inverter is lower in the central inverter configuration, but the cost of the cabling and the losses may be higher. Section 8.3 makes a comparison of different inverter configurations. A good design will determine the optimum combination. The use of a

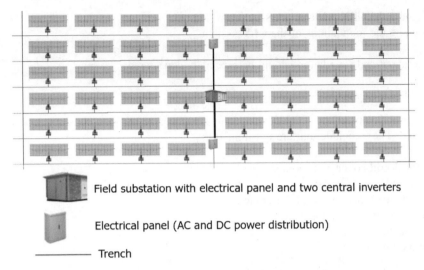

Field substation with electrical panel and two central inverters

Electrical panel (AC and DC power distribution)

——— Trench

Figure 8.1 Schematic layout of a CPV power plant

DC combiner box (with connection and protection fuses) is needed in the case a central inverter is used.

The power output of the tracker influences the dimensions of the cables. For a fixed power plant size, the power output of the tracker determines the configuration of the DC cables. A greater quantity of low power trackers is required to attain a certain power plant power output, or alternatively this could be done with a smaller number of CPV systems with a higher power output. The DC cables used in high output power CPV systems are larger as they need to carry a higher DC output.

An estimation of the foundation cost can also be done in this phase.

8.2.3 Detailed Engineering

The *detailed engineering document* is the most important document of the project. All project calculations are present in this document together with drawings and diagrams of how the plant should be built and assembled. Foundation drawings as calculated from the geological study are included. Details and drawings of the trenches and actual layout of the plant together with a shading analysis for the best layout should be also available.

The detailed engineering document includes, but is not limited to the following:

- designing the layout of the CPV power plant, including the position of each CPV power plant component;
- dimensioning and selection of the power and communication cables, including the cable trenches;
- designing the electrical grid and communication connection to the CPV power plant;
- designing the power and communication distribution of the CPV power plant, including the electrical panels and substations;
- dimensioning and final designing of the backup power system of the CPV power plant;

- designing the CPV power plant security system;
- designing the foundations of the CPV systems.

Final layout and foundation should be especially considered:

8.2.3.1 Final Layout

For the final layout of the CPV power plant, a number of relevant constraints need to be considered. These include, but are not limited to:

- electrical grid connection points;
- streets and access roads;
- water bodies, such as rivers, lakes etc.;
- inclination of the terrain;
- infrastructure facilities;
- conservation areas;
- ecological compensation areas;
- topography and soils data;
- local emissions requirements for generator exhaust;
- an environmental impact assessment (EIA) is also an essential report that should be made before any layout can be finalized; this determines the allowable height of trackers that can be used, and specifies any areas that may not be used to build upon due to the presence of some indigenous plants or animals.

8.2.3.2 Foundations

The cost of the foundation for a CPV tracker at a particular site depends on the type of tracker used, the properties of the terrain and the foundation technology. A geological study should be performed in order to determine the terrain properties and suggest possible foundation technologies to be used for a particular CPV tracker and wind profile at the site.

The geological survey (Figure 8.2) depicts an overview of how the land looks beneath the surface by giving details of the soil formations and granulation together with an indication of the humidity level. A geological survey employs techniques from the traditional walk-over survey studying outcrops and landforms, to intrusive methods, such as hand auguring and machine driven boreholes, to the use of geophysical techniques and remote sensing methods, such as aerial photography and satellite imagery.

The main important loads and factors to take into account when designing the foundations are:

- The static loads of the tracker including the weight of the CPV modules.
- The wind load: The wind is very important to size not only the tracker and drive, but also the foundation of the whole system. A compromise between the performance and the wind trigger to stow position should be agreed to achieve the best performance ratio of the plant at the minimum cost. Once the wind data is studied, the output torques of the drive could be calculated. Recently, a new standard IEC 62817 ed1.0. 'Photovoltaic systems - Design qualification of solar trackers' [7] that defines qualification procedures was approved and published by the IEC Committee. The first analysis to this standard has already been done [8].

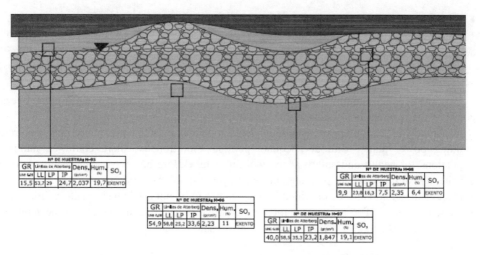

Figure 8.2 Geotechnical study example. Source: Soitec. Reproduced with permission of Soitec

- The snow load and seismic grade (depending of the location).
- The position in the power plant (e.g. a tracker installed on the boundary of the power plant would need a larger foundation than one installed inside within the boundaries of the plant due to the shielding effect of the external trackers which cause a reduction in wind speed).

Various foundation types exist that can be used according to the different geotechnical reports, for example:

- Block foundations are essentially made of a block of concrete and the tracker mast is either embedded in the block or bolted on via a flange interface.
- Pile foundations are typically used for embedded masts. Various technologies exist for pile foundations. The most economical is the pile driven foundation where no concrete is used and the mast is essentially vibrated into the ground. The mast may also be dropped down a wide borehole and the spacing between mast and the ground is filled with concrete and rebar.
- Steel root: In this case a three dimensional steel structure is used which is dropped in a hole and covered up with the same material previously dug out from the ground. For all non-concrete foundations, the corrosive effects of the soil with the metal mast should be considered.

8.2.3.3 Power Plant Calculation

Once the detailed engineering document is finished the power plant cost can be calculated. The main costs should include:

- Product BOM (*bill of materials*): modules, tracker structure, drive, control unit, cabling, system components.
- Project BOM: It includes all the cost of the power plant components as:
  - BOS cost: inverter, transformers, field power distribution and supply panel (circuit breakers, fuses, surge protection, etc.).
  - SCADA and communication system.

- Other costs:
  - assembly cost
  - installation cost
  - logistics cost.

Once the overall costing is analyzed and the power plant is optimized to be as cost effective as possible, the contract is signed and the construction phase can begin.

Depending on the country of installation, legislation may exist that prevents companies from building power plants without adequate permits. Additionally, there could be national regulations for health and safety at the working site. Before any work can be initiated on the site, a health and safety plan should be presented to all the workers on site

Documents should also be provided for the heavy equipment used on site such as fork lifters and cherry pickers. The equipment must be certified that it is compliant with the required regional standards and has been tested to confirm safe operation.

8.2.4 Construction Phase

Before the construction phase can proceed, all the equipment should be on site and the terrain should be prepared. Therefore the following works are part of the preparation phase for the project:

- Logistic and supply chain: some of the components of the power plant could have a very long lead time. Parts such as trackers, drives and inverters could require more than four months from the time of order placement to the delivery on site. Therefore, all the equipment should be ordered a long time before the scheduled beginning of the construction. The transport costs can be quite significant for the large components (especially trackers), therefore the design should be optimized to reduce the transport cost as much as possible.
- Mobilization of the workforce.
- Site preparation: Any tall weeds or large rocks are removed from the site and any large bumps in the land which would interfere with the power plant construction or general operation are flattened out. Again the tracker design will have an influence in how the terrain should be prepared (small or large size, ring or pylon) and the type of foundation too.
- Tools and construction jigs (Figure 8.3) must also be purchased during the preparation phase.

8.2.4.1 Civil Works

Civil works involve heavy construction works. During this process foundations are created for the trackers and trenches and manholes for cables (Figure 8.4) are prepared and completed.

Grounding cable should be laid in the trenches before they are filled in together with all cables (AC, DC and communication cables), air tubing and other required installations. Internal roads and foundations for field distribution boards should also be prepared.

8.2.4.2 Tracker Assembly

Typically, the tracker should be assembled close to the field or on the field, due to the high cost of transporting, for example, a completely assembled tracker. For small trackers, however, it may be feasible to pre-assemble the tracker to reduce onsite installation work.

Figure 8.3 Construction jig for assembling the modules on the tracker table at ground level

The tracker assembly could be done directly on the field (Figure 8.5) or in a central assembly hall (Figure 8.6). This decision should be taken depending on the size of the power plant and the cost of the equipment on site (very dependent on the location of the power plant). The central assembly allows for work in a controlled environment, simple logistics, high level quality control, but it requires an initial investment for the central building. The field assembly does not need this initial investment and it is less susceptible to bottleneck issues which could occur in a centralized assembly area.

The tracker construction must pass various quality checks before it can proceed to the next process of assembly step. This ensures a good performance when the tracker is commissioned.

Figure 8.4 Trenches for AC cables

Figure 8.5 Field assembly

8.2.4.3 Module Mounting on the Tracker Table

The modules can be mounted on the tracker table either during the tracker assembly or once the tracker is installed. The pre-mounting allows a much easier construction, because workers can work at ground level (Figure 8.7). In this phase, the final alignment or pre-alignment of the modules can be done to facilitate the final commissioning. Also in this step, the module cabling could be connected.

8.2.4.4 Tracker Erection

Once the tracker table is assembled with or without modules, it can be installed on the mast (Figure 8.8), which is typically already fixed to the ground.

Figure 8.6 Central assembly hall

Figure 8.7 Modules mounted on the tracker table during the tracker assembly

8.2.4.5 Cabling of the Tracker and System Integration

After the tracker is completely installed with the modules, the cabling that is lying in the trenches can be connected to the control unit and all the other electrical equipment (string inverters, combiner box, etc.).

8.2.4.6 Module Alignment (Pre-Alignment vs. Field Alignment)

When the CPV system installation is complete a realignment of the modules could be done to improve the performance of the CPV System. This step depends mainly on the acceptance

Figure 8.8 Tracker erection

angle of the module and the methodology used during the assembly and pre-alignment of the modules.

Various techniques exist for the alignment process ranging from visual alignment of the sun's focus on the cell, to current measurement of individual modules. Alignment with a laser or string can also be used.

8.2.4.7 Commissioning

After the completion of the steps commented in the preceding sections the plant is ready for operation. At this point, the commissioning of the CPV plant should be carried out to ensure that it will operate as expected. Commissioning a CPV plant means making all the necessary revisions and tests to ensure that the installation fulfills the design requirements presented in the contract and project. In brief, this process needs to verify that:

- The plant has been constructed following the specifications of the Contract and Detailed Engineering Project.
- All the documentation requested by the local government has been provided.
- The plant operates as expected in regular operating conditions and in maintenance mode.
- The plant is of adequate quality both in components used as well as in workmanship.
- The plant is safely constructed and does not pose a danger to the equipment or to animals and humans.
- The power installed agrees with the contract.
- The energy generated will meet the targets set in the warranties.

All the works listed above can be grouped in two main groups:

1. The inspection of a CPV plant covers all the works to be done to verify that the plant fulfils with all the documentation and technical issues defined in the contract and project, according always to the local regulation. A detailed description of the activities to be carried out in this phase is given below.
2. The rating of a CPV plant covers the determination of the power in the standard conditions defined in the contract and the estimation of the energy losses to ensure that it will conform the warranties. The power rating problem is one of the key challenges in CPV technology and will be treated extensively in section 8.8.

A key part of the commissioning process is a thorough inspection of the CPV plant, which, in essence, comprises the verification of the documentation as well as all the elements of the installation to ensure its quality and reliability.

First, to legalize the installation, all the documentation of the plant has to be revised according to the local government regulation. Different countries may require different documentation for obtaining the final permissions for the connection to the grid [9], such as:

- list of components, serial number
- warranty certificates of all components
- warranty certificates for construction
- maintenance book

- maintenance manual
- operation manual
- record of health and safety incidents
- 'As Built' documentation
- 'Finished Works' certificate
- 'Low Voltage Form' certificate.

Secondly, all the elements of the installation are verified, looking for any faults or hidden defects of the construction that may occur [10]. The objective of this visual inspection is to ensure the quality and reliability of the installation, specifically if they fulfill with health and safety requirements and with the ruling normative and technical specifications (UNE normative, local regulation . . .). In general, the elements to verify are: the tracker (the structure itself, the screws, nuts and bolts . . .), the CPV modules, the inverters, the cabling (DC, AC and communications), the DC and AC protections, the grounding connections and the grid connection.

The defects detected can be classified into different groups according to its severity: faults that affect directly the operation and to be solved before commissioning the plant (health and safety requirements), faults to be solved during the first months and faults with follow-up recommendation that surely will affect the warranty period and maintenance contracts.

Finally, the monitoring system in charge of controlling the CPV plant – which will be extensively discussed in section 8.6 – is tested. The communication between the monitoring system and the trackers is checked and the operation of the plant is verified. That means that all the trackers track the sun in automatic mode as in normal operation and that all of them can be operated in manual mode for special issues or emergencies, like stow position, maintenance or cleaning mode. Also, the data acquisition of all the components is tested, like trackers, inverters, meteorological sensors, etc. ensuring that the monitoring system manages as expected all the incidences and alarms coming from the different elements.

In summary, the objective of a CPV plant is to generate income and it this in this last commissioning step, before beginning the actual operation, where the manufacturer should verify the installation to ensure that it will operate as designed and that it will generate, the expected incomes for the final customer.

8.3 CPV Inverters: Configurations and Sizing

The inverter is the device used to convert the DC power provided by a solar generator into AC power with the required features and quality demanded by the electrical grid. To achieve this, the inverter should perform two main functions: switching and filtering. By means of the switching, the DC signal is converted into a square AC signal. These circuits are also called synthesizers. The second stage is the filtering in which a sine wave is obtained. For doing the filtering, power filters made up with inductors and capacitors are used.

As commented above, an inverter is a device for converting the DC power into AC power but it performs also a number of additional tasks. In brief, an inverter for a PV system connected to the grid should perform the following functions:

- convert DC energy into AC energy;
- get the maximum power from the PV generator every instant;

- monitor operating variables of the generator and electrical grid;
- ensure the proper operation of the system, avoiding problems in the electrical grid or cause security problems for users.

For a proper operation the inverter should have the following operating features:

- high efficiency
- low power consumption
- tracking the maximum power point
- protection
- isolation
- not islanding
- automatic connection/disconnection
- low total harmonic distortion (THD)
- reliable operation for long periods of time (20 years).

8.3.1 Types of Configurations

There are different kinds of configurations between the inverter and the generator in a CPV plant. The configurations can be classified in three types:

1. module inverter
2. tracker inverter
3. central inverter.

8.3.1.1 Module Inverter

In this case, there is one inverter for each CPV module, as shown in Figure 8.9. This combination is called AC module. The inverter power is in the range of 100 W, although it depends on the module to which the inverter is connected. In this kind of installations the DC part is simplified while the AC part is more complex.

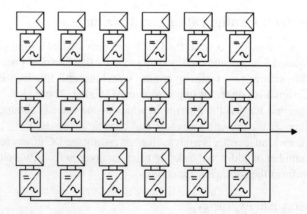

Figure 8.9 CPV plant with module inverter configuration (i.e. with AC modules)

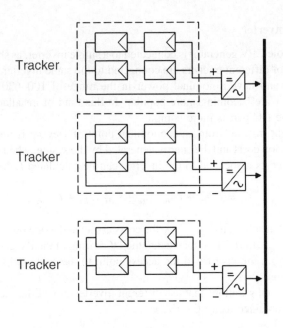

Figure 8.10   CPV plant with tracker inverter configuration

8.3.1.2 Tracker Inverter

In this case, there is one inverter for each solar tracker, as shown in Figure 8.10. These inverters have a nominal power in the range of 5–20 kW.

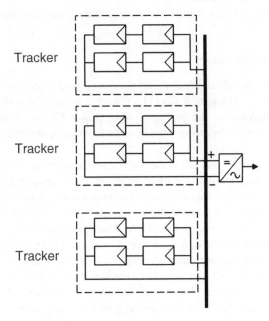

Figure 8.11   CPV plant with central inverter configuration

8.3.1.3 Central Inverter

In this case, the whole CPV generator is connected to a single inverter, as shown in Figure 8.11. Hence, generators of different trackers are connected to the same inverter. These inverters are used in big plants and have a nominal power in the range of 100–630 kW; although MW inverters and higher are becoming more popular. In this kind of installations the AC part is simplified while the DC part is more complex.

To select the right inverter from an economical point of view, it is necessary to take into account the installation costs and the energy generated by the system when this inverter is used.

The installation costs associated to the inverter can be calculated following Eq. (8.1):

$$C_I = C_{INV} + C_{DC\_wire} + C_{AC\_wire} + C_{others}, \tag{8.1}$$

where C_I is the total installation costs; C_{INV} the costs associated to the purchase and installation of the inverter; C_{DC_wire} the costs associated to the DC wiring of the PV generator; C_{AC_wire} the AC wiring costs (from the inverter to the access point to the grid) and C_{others} represents all remaining installation costs.

To compare two configurations that use different inverters, the difference between the costs of both of them is obtained using Eq. (8.2):

$$\Delta C_I = (C_{INV\_1} + C_{DC\_wire\_1} + C_{AC\_wire\_1}) - (C_{INV\_2} + C_{DC\_wire\_2} + C_{AC\_wire\_2}), \tag{8.2}$$

In this case, C_{DC_wire} and C_{AC_wire} depend on the configuration while C_{other} remains constant for both of them.

Finally, it is necessary to estimate the energy difference obtained with both configuration and quantify the cost of this difference ΔC_E.

Thus, to select the right inverter, if ΔC_I is lower than ΔC_E then configuration 1 will be better than configuration 2.

8.3.2 Sizing of the Inverter

The inverter efficiency mainly depends on its nominal power (P_{INV}) and the power produced by the generator ($P_{DC,oper}$) (Figure 8.12) [11]. Because of this, the optimum nominal power of an inverter will depend on the generator peak power ($P_{DC,STC}$) and the configuration between both. So, it is necessary to size the inverter in order to maximize the energy yield of a CPV system.

There are several methods that allow the CPV generator power or energy to be predicted based on parameters such as direct normal irradiance, cell temperature or spectrum [12]. However, there are other losses that should be taken into account to attain an adequate modeling of the DC part of a CPV system. Each configuration studied above has different kinds of losses. Based on simulations, measurements over CPV generators and different works that address this issue [13–17], it is possible to define a range of losses depending on the configuration used that the current models do not taken into account (Table 8.1). It is important to remark that the losses shown in Table 8.1 should be only considered as representative trends for each configuration. Therefore, they should be carefully analyzed for each CPV system under study because their impact could be different depending on the configuration, and size of the CPV system and power plant under study.

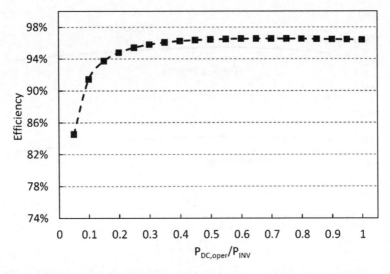

Figure 8.12 Efficiency curve of inverter versus the ratio of its input power ($P_{DC,oper}$, i.e. the power produced by the generator) to its nominal power (P_{INV})

Taking into account the losses shown in Table 8.1, it is possible to simulate the performance of the DC part for each configuration used [18]. In order to simulate the whole system, taking also into account the AC part, it is necessary to model the inverter performance based on the output provided by the generator [19]. Once the DC and AC parts are modeled, the optimum size of the inverter is calculated to obtain the maximum annual energy yield (Y_F). A definition and the significance of the energy yield will be discussed in section 8.8.

Figure 8.13 shows the annual energy yield (Y_F) simulated for a CPV system, located at Jaén (Spain), with the three possible inverter configurations versus the ratio of the inverter nominal power to the generator peak power ($P_{INV}/P_{DC,STC}$). The annual energy yield of each configuration has been obtained with the losses shown in Table 8.1 and the procedure described in [20]. As can be seen from the analysis of each configuration, the maximum value of Y_F is obtained for an inverter size 0.9 times the generator peak power. This proportion holds for the three configurations under study despite that the level of losses is different for each of them. Also, as can be observed, the module inverter configuration is the

Table 8.1 Percentage of losses depending on the configuration used

Losses	Module inverter	Tracker inverter	Central inverter
Mismatch	0 %	1.5 %	3 %
Dust and dirt	2 %	2 %	2 %
Ohmic losses	0 %	0.5 %	1.5 %
Losses in the tracking of the inverter maximum power point	3 %	3 %	3 %
Misalignment	1 %	1.5 %	2 %
TOTAL	6 %	9 %	12 %

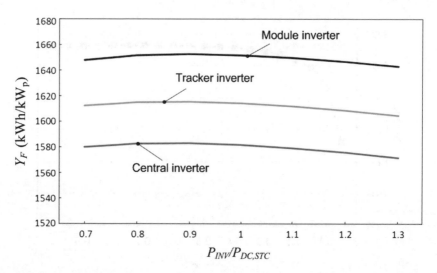

Figure 8.13   Annual energy yield (Y_F) versus the ratio of the inverter nominal power to the generator peak power ($P_{nominal}/P_{DC,STC}$) for each inverter configuration

one with lower losses. This configuration reaches 4% more Y_F than the central inverter configuration, which is the one that has higher losses. These data are consistent with those obtained by other authors [21].

8.4   Optimized Distribution of Trackers

In CPV systems modules must be always perpendicular to solar rays, so it is needed to follow the sun trajectory over time. Therefore, the tracker is directly related to the optical train used since the tracker pointing accuracy needed is going to depend on the acceptance angle of the optics.

Although there are several kinds of trackers [22], the ones used in high CPV technology are two-axis trackers. These trackers allow the modules to be always pointing towards the sun so that concentrator lenses can focus the radiation on the small solar cell area. This kind of tracking system is more complex than linear tracking or static systems, from a mechanical point of view, but it presents the advantage of maximizing the solar radiation capture along the daily trajectory.

In a CPV power plant there is always some moments of the day where some trackers are shaded by others; mainly in the sunrise and sunset (Figure 8.14). This phenomenon is the so called tracker *inter-shading effect*. Therefore, to reduce the shading losses and increase the energy yield, it is necessary to increase the distance between trackers, as it is shown in Figure 8.15, where typical evolution of shading losses and yearly energy produced based on land occupation are shown. Hence, the land occupation by trackers is also increased [23,24] and, consequently, the costs of the civil works and installation are increased. An ideal plant layout should solve this energy yield vs. cost trade-off in an optimum manner. Therefore, the first step is to quantify the effects of shadows on the energy yield of the system in order to obtain an adequate system design and performance analysis. However, the studies on shading in CPV plants are scarce because of the novelty of this technology [25–28].

Figure 8.14 Example of tracker inter-shading during the sunrise in a real CPV plant in operation located at ISFOC (Puertollano, Spain)

8.4.1 State of the Art

Currently there are few methods for calculating the losses due to tracker inter-shading in CPV technology. The main conclusions of the literature available on this issue are as follows:

- There are several methods for calculating tracker inter -shading losses for flat panel PV technology [29–31].
- In [32] authors provided some examples to calculate the tracker inter -shading losses for a specific CPV installation.

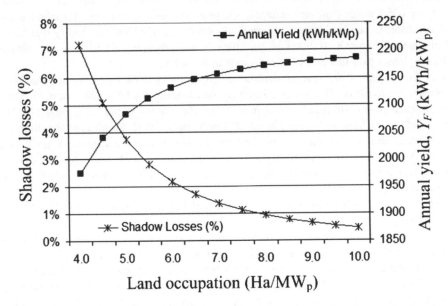

Figure 8.15 Typical evolution of shading losses and yearly energy produced versus land occupation

- In [25] authors propose several methods for calculating the auto-shading losses in CPV although the procedure followed is not given.
- In [28] authors present the performance of high CPV receivers under partial-shading and propose a method for the calculation of the maximum power output of concentration revivers under partial shading on the primary lens.
- In [27] authors propose a complete frame for modelling concentrating system. The proposed model takes into account the tracker inter-shading effects.

Also, there are several commercial software packages that implement these calculations. Some of these programs can calculate the shadows from the global solar irradiance and from direct solar irradiance, so it is possible to use them for CPV technology. See Annex 8-I for further details on these software packages.

8.4.2 Procedure for Optimizing the Distribution of Trackers

8.4.2.1 General Vision of the Process

The University of Jaén has participated in several studies related to CPV trackers [33,34]. This has allowed the development of a new and detailed method for calculating the energy losses due to tracker inter-shading. This procedure leads to the determination of some key aspects in CPV systems design such as the best location of solar trackers in a given field or the optimum size and shape of the tracker table of solar trackers.

To optimize the distribution of trackers in order to minimize the losses due to shadows, wires and land occupation, the following procedure is used:

Step 1. Estimating the shading that a tracker casts on another tracker. The percentage of shaded area that a tracker causes on other trackers as a function of their size and relative distance is calculated for every instant in the year. The subsequent annual irradiance losses are estimated taking into account the yearly distribution of DNI and the shading that happens during a year. Figure 8.20 shows an example of a butterfly graph detailing the percentage of shading losses as a function of the distance to a tracker.

Step 2. Estimating of the annual irradiance losses on a tracker due to the shadows of surrounding trackers. The losses caused by surrounding trackers that cast a shadow on the tracker under study are integrated from the values obtained in step 1. These surrounding trackers are those placed at South, East, West, Southwest and Southeast. Figure 8.22 shows an example of percentage of losses versus the distance of trackers and Figure 8.21 sketches the tracker disposition assumed in this study: the tracker under study would be the one in light grey (in the center) and surrounding trackers would be those in dark grey.

Steps 1 and 2 will be explained in more detail below in sections 8.4.2.2 and 8.4.2.3, respectively.

Step 3. Cost of the energy losses due to shadows during the system life-cycle. From the irradiation values obtained in step 2 and using approximate methods (i.e. performance ratio method), the annual energy losses by tracker inter-shading are calculated. Considering the cost of the energy in €/kWh (in this study, a feed-in-tariff of 0.44€/kWh was assumed) and the system life-cycle it is possible to obtain the economic cost of the energy losses due to shadows during the system life. Figure 8.16 shows an example of these costs versus the

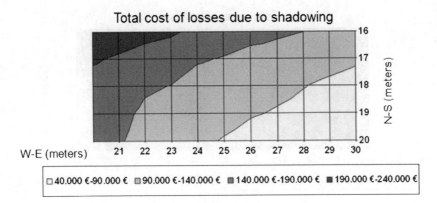

Figure 8.16 Grey-scale map of costs due to inter-tracker shading in a system of 100 kW located in Jaén (Spain) during its life-cycle as function of the distance between trackers

distance between trackers. It can be observed that costs due to inter-shading are inversely proportional to the distance.

Step 4. Cost of the installation, cabling and land and energy losses in cabling. In this step of the process, costs directly proportional to the distance between trackers are calculated. Mainly, they are the installation and land costs. Costs of the energy losses due to wires are also taken into account since they will increase with the distance among trackers. As in the previous case, the cost of the energy in €/kWh and the system life-cycle have been considered, but also the land cost, the cable cost and the civil work cost, etc. Figure 8.17 shows an example of these costs as function of the distance between trackers.

Step 5. Final balance. The final balance is obtained as the sum of the previous costs obtained in step 3 and step 4. Figure 8.18 shows a graphic example of these final costs. It can be observed that an optimum spacing for the trackers can be found.

Next, step 1 and step 2 of the described procedure are going to be explained in more detail.

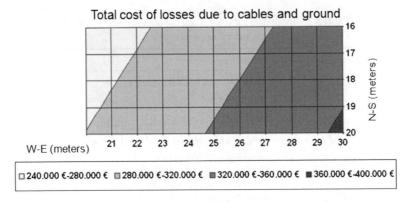

Figure 8.17 Grey-scale map of costs due to cables and the land in a system of 100 kW located in Jaén (Spain) during its life-cycle as function of the distance between trackers

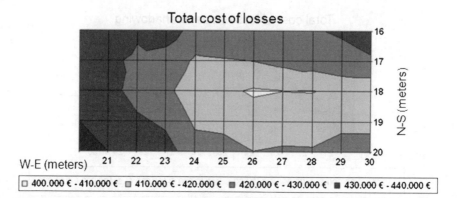

Figure 8.18 Grey-scale map of the total costs in a system of 100 kW located in Jaen (Spain) during its life-cycle as function of the N–S and E–W distance between trackers. It can be observed that the optimal spacing between trackers is of 18 m N–S and 26 m E–W

8.4.2.2 Calculating Shadows Due to Inter-Tracking Shading

In step 1 of the procedure detailed in section 8.4.2.1, the annual losses of direct solar irradiance in a tracker due to the shadow produced by another tracker are calculated as function of the distance and the orientation between both trackers. The variables used in the calculations are presented in Table 8.2 and a scheme of the calculation can be found in Figure 8.19.

The procedure can be summarized as follows:

1. Estimation of the shading surface S_{SH} (m^2) for every instant. This is done with the help of Figure 8.19 and using Eq. (8.3), Eq. (8.4) and Eq. (8.5):

$$P_Y = \frac{A_Y}{\cos(\theta_S)} - d_S, \tag{8.3}$$

$$P_X = A_X - d_S \cdot \sin(\psi_S), \tag{8.4}$$

$$S_{SH} = P_Y \cdot P_X, \tag{8.5}$$

Table 8.2 Input data for calculating the annual losses of DNI in a tracker due to the shading of another tracker

Parameter	Definition
DNI	Direct normal solar irradiance
θ_S	Zenith angle
Ψ_S	Azimuth angle
A_X, A_Y	Length and width of the tracker table
P_X, P_Y	Length and width of the shaded surface
A	Tracker table surface ($S_G = A_X \cdot A_Y$)
S_{SH}	Shaded surface ($S_{SH} = P_X \cdot P_Y$)
d_S	Distance between trackers
DNI_{SH}	Solar irradiance losses
E_{DNI}	Yearly direct normal irradiation
H_{SH}	Yearly solar radiation losses from shadowing

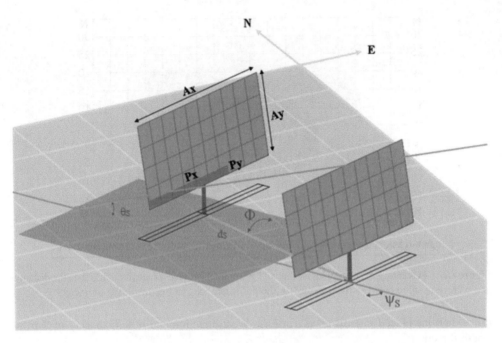

Figure 8.19 Shading cast by one tracker on another

2. Estimation of the solar irradiance losses for every instant (DNI_{SH} in kW/m^2). It is considered that these losses are proportional to the generator shaded surface:

$$DNI_{SH} = DNI \cdot \frac{S_{SH}}{A}, \qquad (8.6)$$

3. Estimation of the percentage of annual solar radiation losses (H_{SH} in kWh/m^2 or in %) in a tracker caused by another tracker (Eq. (8.7), Eq. (8.8) and Eq. (8.9)):

$$E_{DNI} = \sum_{year} DNI \cdot \Delta T, \qquad (8.7)$$

$$H_{SH} = \sum_{year} DNI_{SH} \cdot \Delta T, \qquad (8.8)$$

$$H_{SH}[\%] = \frac{H_{SH}}{E_{DNI}} \times 100, \qquad (8.9)$$

with E_{DNI} being the yearly direct normal irradiation (in kWh/m^2).

The value obtained from the previous procedure is the percentage of solar direct irradiance losses on the generator plane due to shadows produced by another tracker. This calculation allows the so called 'butterfly graphs' to be obtained (Figure 8.20). These graphs plot the annual solar direct irradiance losses that a tracker placed in the coordinates (0, 0) produces

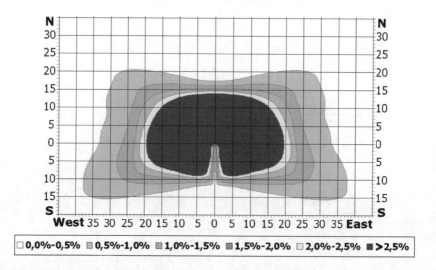

Figure 8.20 Percentage of shading losses that a 12 × 9 m tracker causes on another, based on the distance and relative position between them. Units for the horizontal axis and vertical axis are meters

in another one placed in a distance x in East–West direction and in a distance y in North–South direction.

8.4.2.3 Calculating Irradiance Losses from Inter-Tracking Shading

In step 2 of the procedure detailed in section 8.4.2.1, the yearly solar irradiance losses in a tracker due to shadows produced by surrounding trackers are calculated. These losses are calculated considering the actual distribution of the trackers, being the basic tool for this the butterfly graphs described in the latter section.

Following the proposed methodology it is possible to know the yearly percentage of DNI lost in a tracker due to shadows H_{SH} (%) caused not by one but by all surrounding trackers, based on the N–S and E–W distance among them. Consider the rectangular mesh distribution of trackers of Figure 8.21. There, E–W distance represents the separation between trackers in the same row whilst N–S distance represents the separation between adjacent rows. As shown by Figure 8.21, in a rectangular distribution each tracker plots a butterfly graph on the tracker under study (the central tracker in Figure 8.21). If the losses of all generators are summed up, the grayscale map in Figure 8.22 is obtained. In essence, Figure 8.22 summarizes the percentage of DNI lost in the tracker under study (i.e. central tracker in Figure 8.21) as a function of its distance to surrounding trackers.

8.5 Considerations of Environmental Impact and Dual Use of the Land

The installation of a solar plant has often an impact in the environment because it typically involves removal of the topsoil, land leveling and other actions such as compaction or gravel addition. A detailed review of the environmental impact of CPV technology including the whole life-cycle assessment is given in Chapter 13. In brief, it can be summarized that in CPV plants this impact is lower because, due to the high efficiency of the systems, the ground

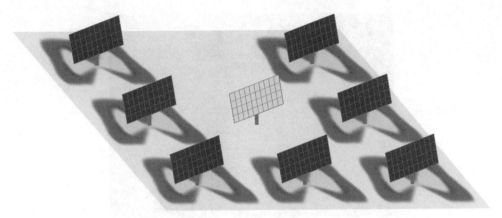

Figure 8.21 Base configuration considered to assess the shadowing losses in a tracker due to surrounding trackers (details in the text)

coverage ratio is better than in other technologies. That essentially means that, to install a power plant the use of the terrain is lower, meaning that there is more unused space between systems. This empty space between systems could be used for a dual use, as animal breeding or agriculture.

The re-vegetation of a proportion of the land occupied by the photovoltaic plant is, in some locations, mandatory, in an effort to reduce the environmental impact of the installation. This re-vegetation could be done in the surroundings of the solar plant, but it could also be carried out inside the power plant land. This plantation could fulfill the double objective of regenerating the used land, but also maintaining the ground (avoiding erosion) and keeping a low level of dust thanks to the vegetation, which will reduce the soiling on the modules.

The vegetation can be used only for this objective, but also as profitable agriculture which some economic benefits.

A study carried out in ISFOC between 2009 and 2011 [35], shows that the best method to regenerate a power plant installation site is the use of native species in the edges of the power

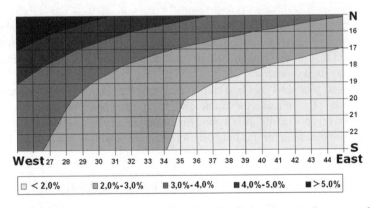

Figure 8.22 DNI losses (in %) due to the shading in a 12 × 9 m tracker cast by surrounding trackers, based on the distance and position among them

Figure 8.23 Plants growing next to a CPV power plant at ISFOC (Puertollano, Spain)

Figure 8.24 Vegetation growing in the sun and shadow in the middle of a CPV power plant at ISFOC (Puertollano, Spain)

plant, but also in the middle of the power plant plot, under the shading of the solar modules. The results show no difference between the plants on the sun (Figure 8.23) or on the shading (Figure 8.24), and that means that agriculture could be used in a CPV power plant in the same way that it can be used outside.

One of the most typical dual uses is to take animals to graze (Figure 8.25). That has the double advantage of feeding the animals (for example sheep or horses) and maintains the land in good condition (excessively tall plants are not good for the systems, because they can create shading). Not all animals are desirable for this dual use; for example, goats are not wanted, because they could eat the cables.

8.6 CPV Plant Monitoring and Production Data Analysis

The monitoring system of a CPV plant has a twofold objective: on one hand it is the system in charge of registering all the operating parameters and on the other hand it is the tool to be used

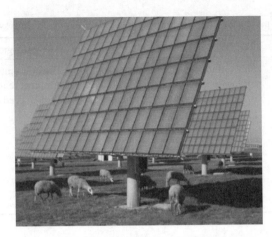

Figure 8.25 Sheep grazing in a CPV power plant at ISFOC (Puertollano, Spain)

for the operation of the plant. This twofold objective makes the monitoring system the cornerstone of the O and M as is described in section 8.7: it is the tool used for controlling the plant and, in addition, for providing the operating raw data to be analyzed according to the maintenance strategy. The monitoring system contributes to decreasing the time of detection of system failures and the consequent loss of energy production, which directly will suppose a reduction in the profits.

The general philosophy for the monitoring system of a CPV plant [36] is that a SCADA (Supervisory Control And Data Acquisition) system gathers all the information coming from the different elements that compose the plant. In addition, this SCADA permits the remote operation of the complete plant.

8.6.1 Monitoring System: Registering the Operating Parameters

The monitoring system registers a lot of information from all the elements that compose the CPV plant (modules, inverters, trackers, energy meters, meteorological stations . . .). With this purpose, the monitoring system of a CPV plant is based in the same rules that traditional PV plants being perhaps, the only feature to remark, that apart from all the electrical data related with energy generation it has to register some parameters of the solar trackers.

Several reference documents that establish monitoring guidelines and analysis of the electrical performance of PV systems exist in the literature. They recommend procedures for the monitoring of energy-related PV system characteristics such as irradiance, array output and power conditioner input and output. Some examples are:

- The international standard IEC 61724 'Photovoltaic system performance monitoring – Guidelines for measurement, data exchange and analysis' [37].
- The guidelines of the European Joint Research (JRC), Report EUR 16338 'Guidelines for the Assessment of Photovoltaic Plants. Document A: Photovoltaic System Monitoring' [38].

From the point of view of the authors, the minimum variables that should be registered by the monitoring system that offer to the operator the complete vision and control of the CPV plant

Table 8.3 Advisable parameters that should be registered by the monitoring system of a CPV plant

Inverter data	Tracker data	Meteorological station data
DC input voltage (V)	Elevation angle (current and target)	Direct normal irradiance (W/m^2)
DC input current (A)	Elevation pulses (current and target)	Ambient temperature ($^\circ$C)
DC input power (W)	Azimuth angle (current and target)	Wind, speed (m/s) and direction ($^\circ$)
AC output voltage (V)	Azimuth pulses (current and target)	
AC output current (A)	STATUS	
AC output power (W)	ALARMS	
Frequency (Hz)		
Power factor		
STATUS		
ALARMS		

are described in Table 8.3, grouped by the element of the installation that is commonly used for obtaining it [36]. All the electrical parameters related with the energy generation are registered by the inverters and energy meters, and the control unit of the trackers registers the position of the solar tracker. For both elements, inverters and trackers, it is also very useful to obtain the status and alarms that will indicate the operator if there is any malfunction or incidence occurring in the CPV plant. Finally, it will be always convenient to have a meteorological station at the site to measure the most critical ambient conditions during the operation of the plant, but it is always mandatory to have an anemometer (one or more distributed over the plant) to measure the wind speed because it is necessary to place the trackers in stow position in the presence of high wind gusts because of safety requirements.

Apart from the advisable parameters described in Table 8.3 there are additional parameters (spectral measurements, module temperature, tracker alignment etc.) that will offer some added-value when analyzing the performance of the CPV plant.

All inverter manufacturers offer a tool for monitoring the electrical performance of the installation that could be enough to be used as a low-cost monitoring system in traditional flat plate PV plants [39]. The results obtained are that the electrical measurements performed by the inverter are usually accurate enough for this purpose. Anyhow, for CPV it is strongly recommended to implement a SCADA that gathers the information of all the components, as the trackers are a crucial element, including then the control of these elements in a centralized platform.

8.6.2 Monitoring System: Controlling a CPV Plant

For CPV plants the monitoring system is a crucial element because to ensure the quality of the performance and maximize the incomes of the installation it is necessary to accurately operate the solar trackers of the plant. This issue is not critical for traditional flat plate PV installations, because the use of solar trackers is not usually implemented, only in some cases for increasing the generation all along the day.

The monitoring system offers the operator a general view of the status of the CPV plant, displaying the energy generated by the concentrators and the ambient conditions and it will indicate also the status of all the inverters and trackers and if there is any incidence in any of the elements and what is the origin of it. The picture in Figure 8.26 shows a screenshot of a monitoring system that is implemented for one of the CPV plants operated at ISFOC facilities.

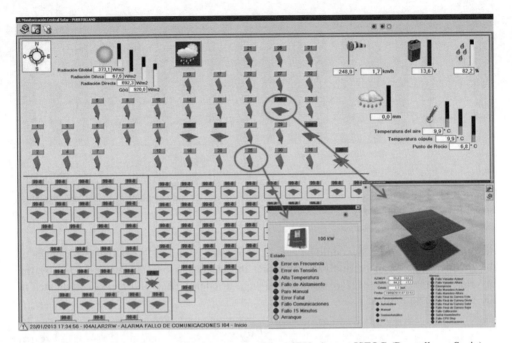

Figure 8.26 Picture of the monitoring system of one CPV plant at ISFOC (Puertollano, Spain)

As has been said before, this tool centralizes the operation of all the solar trackers which allows the operator to carry out remotely any manual operation over the solar trackers. Typical operations include placing them in stow position if necessary or in maintenance mode, for repair or cleaning. Also, it is important to remark that it permits to move several trackers at the same time what saves time when making any maintenance over the plant.

8.6.3 Analysis of Production Data

The production data of a CPV plant is analyzed during the operation to control its performance. All the data to analyze are obtained from the monitoring system. The same standard, IEC 61724 'Photovoltaic system performance monitoring – Guidelines for measurement, data exchange and analysis' [37], that establishes general guidelines for the monitoring system also establishes the guidelines for the analysis of the performance of PV systems. Different kinds of analysis can be carried out with this data for the different elements of the installation.

First, it is advisable to check the quality of the raw data before any data processing is executed. The objective is to ensure the consistency of the data, looking for gaps or incoherent data to identify obvious anomalies during the monitoring. A reasonable set of limits should be defined for each parameter according to its known characteristics and the characteristics of the CPV plant and the environment.

Once the consistency of the data is checked various parameters related to the system's energy balance and performance may be calculated. Standard IEC 61724 [37] defines all the derived parameters that are of interest for PV installations.

For example, the AC data provided by the inverter can be used to analyze the quality of energy injected to the grid and thus ensure that the installation is not introducing any

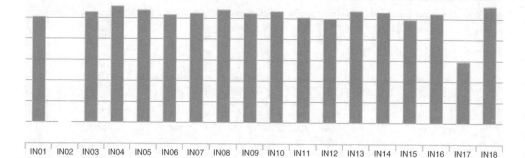

IN01 IN02 IN03 IN04 IN05 IN06 IN07 IN08 IN09 IN10 IN11 IN12 IN13 IN14 IN15 IN16 IN17 IN18

Figure 8.27 Daily energy generation for each inverter in a CPV power plant at ISFOC (Puertollano, Spain) for a representative day

distortion into the grid, what could be penalized in some countries. Also, the performance of the inverters (η_{inv}) can be analyzed by using the input and output data provided by the inverter in terms of power ($P_{AC,oper}$ and $P_{DC,oper}$) or energy (E_{AC} and E_{DC}), as indicated in the following equations:

$$\eta_{inv} = \frac{P_{AC,oper}}{P_{DC,oper}}, \tag{8.10}$$

$$\eta_{inv} = \frac{E_{AC}}{E_{DC}}, \tag{8.11}$$

Finally, the most interesting data can be obtained from the analysis of the systems production and even more when there is a meteorological station in the field. The efficiency and the performance ratio (a key quality metric of the installation that will be defined in section 8.8) of the complete installation and of the concentrator systems can be independently calculated.

The comparison of the production between the concentrators can be used to detect anomalies or incidences in the operation of the CPV plant. Graphs like the ones presented in Figure 8.27 and Figure 8.28 are a simple representation of the analysis carried out by the monitoring system. Figure 8.27 shows the daily generation of each inverter for one CPV plant installed at ISFOC project in Puertollano, Spain. The graph shows that one inverter is non-operating during the complete day and another one has generated less energy than the average. By analyzing the hourly generation presented in Figure 8.28 it can be observed how that inverter was stopped during some hours without generation. More advanced analysis can be performed on this data, as the ones presented to illustrate the concept of advanced plant operation focused on maintenance in section 8.7.

Finally, the next two graphs show an example analysis of the performance ratio and system energy losses as described in the standard, IEC 61724 'Photovoltaic system performance monitoring – Guidelines for measurement, data exchange and analysis' [42]. Figure 8.29 shows the performance ratio calculated each month for a CPV plant installed at ISFOC project in Puertollano, Spain. This calculation is done using Eq. (8.13), Eq. (8.14) and Eq. (8.15) described in section 8.8. Figure 8.30 shows the associated calculation of the energy losses for

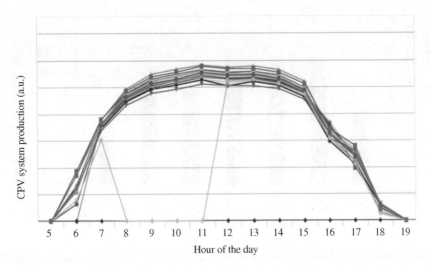

Figure 8.28 Hourly energy generation for each inverter in a CPV power plant at ISFOC (Puertollano, Spain) for a representative day

the same CPV plant, where represents the *capture losses* (L_C) and the *BOS losses* (L_{BOS}). These two concepts, to be defined in the following, summarize the system losses.

- The concept of capture losses (L_C) represents the losses associated to the operation of the CPV generator. This loss groups the effects of temperature higher than the standard, mismatch between modules, tracking incidences, wiring losses, shadowing, maximum power point (MPP) tracking error etc.
- The BOS or rest of the system losses (L_{BOS}) represent the losses associated to the DC–AC conversion of the inverter.

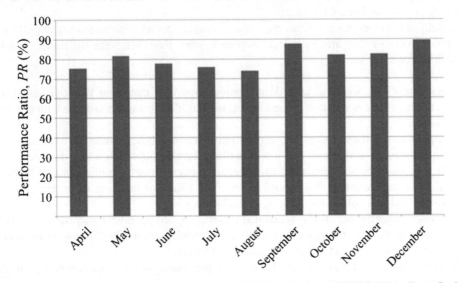

Figure 8.29 Monthly performance ratio (*PR*) of a CPV power plant at ISFOC (Puertollano, Spain)

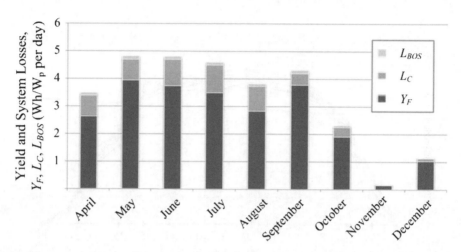

Figure 8.30 Monthly final yield (Y_F), capture losses (L_C) and BOS losses (L_{BOS}) of a CPV power plant at ISFOC (Puertollano, Spain). For a precise definition and calculation of this terms refer to section 8.8

8.7 Operation and Maintenance

The operation and maintenance (O and M) of a CPV plant is similar to what it is done in other industries. In this respect, a CPV plant could be considered as a 'production system' whose 'last product' is the energy generated from the 'raw material' that is the solar radiation. In the manufacturing industry, with the aim of decreasing the final cost of a product and thus increasing the benefit, the manufacturing process is controlled by O and M activities, which are the efforts of conserving the normal functioning of the process. Therefore, with this hypothesis a CPV plant can take advantage of the progress made in the industry during the past years. Accordingly; definitions and procedures can be adapted and O and M tools developed in the industry can be used in a CPV plant.

Usually, the operation and maintenance of a process are defined simultaneously in the O and M strategy because they are dependent upon each other. However, some differences can be established between them:

- Operation: all the technical and administrative actions for managing and supervising the functioning of an element or process to ensure that it performs the task for which it was designed.
- Maintenance: all the technical and administrative actions whose objective is to conserve an element or process to ensure that it will perform the task for which it was designed.
    In general, in the maintenance of any industrial process, three types of actions are defined:
- Corrective maintenance: this is applied once the failure or breaking of an element has occurred and it is solved by repairing or replacing it.
- Preventive maintenance: this is scheduled and it is applied periodically trying to avoid the failure or breaking of the element. Usually the schedule is established following the specifications of the manufacturer and/or existing standards.
- Predictive maintenance: modifications over the preventive maintenance resulting from the experience and the analysis of the components.

All the preventive and predictive maintenance tasks should be described in the *Maintenance Manual* of the plant, together with the typical tasks of corrective maintenance given by the experience of the manufacturer.

In the industry, usually, nobody discusses the usefulness of controlling a process but the maintenance has always been the origin of controversy. Maintaining a process supposes a cost which is not easily understood; but, in the long term, it has been demonstrated that when the maintenance strategy is not optimized, frequently it will incur a higher cost because of failures in the elements. Generally, investing in maintenance will reduce the long-term cost of the process, because operating costs are reduced and the lifecycle of the element increases.

8.7.1 Operation

8.7.1.1 Plant Operation Basics: Monitoring System and SCADA

The main tool for the operation of CPV plants is the monitoring system that has been described in section 8.6. This system is responsible for registering the operating parameters and controlling the CPV plant.

The difference existing in the operation of a CPV plant as compared to a PV plant is the requirement of high accuracy tracking. While in PV the operation tasks are practically non-existing and the main responsibility of the monitoring system is the registration of the operating parameters, for CPV it is mandatory to ensure that all concentrator systems in the plant are tracking the sun perfectly. This daily operation of the CPV plant is carried out automatically by the SCADA of the monitoring system, which is in charge of controlling the status and position of all trackers. The SCADA is usually also responsible for some safety actuations, like placing all the trackers in stow position in case of high winds, power cut off etc.

The operator of the CPV plant can verify the status of the plant with the monitoring system. The operator communicates any incident detected in the plant (alarms and failure status) to the maintenance manager in order to solve it and minimize, as much as possible, the energy losses associated to the incidence. Moreover, incidents that are resulting in lower performance than the expected can be generated if an advanced tool is implemented in the monitoring system that permits the automatic comparison of the energy production with ambient conditions. Finally, all of these incidents could be managed also by the SCADA system, which could notify them directly to the maintenance manager.

8.7.1.2 Advanced Plant Operation Focused on Maintenance

One step further in CPV plant operation could be the introduction of *statistical process control* (SPC) to govern the plant in order to maintain energy production at its maximum potential. SPC is the application of statistical methods to the monitoring and control of a process to ensure that it operates at its full potential to produce a conforming product. In [40] an approach to the application of this tool in CPV plants is presented, where the 'process to control' is the CPV Plant and the 'conforming product' is the maximum energy yield. This work shows how controlling the energy generation of each concentrator, compared with the total energy generated by the complete plant, can result in the early detection of possible failures before they actually occur. In this way, incidents are minimized and, consequently, so are costly interruptions of the production of energy. The procedure followed is the calculation of the full production potential for each concentrator and subsequently associated control limits are

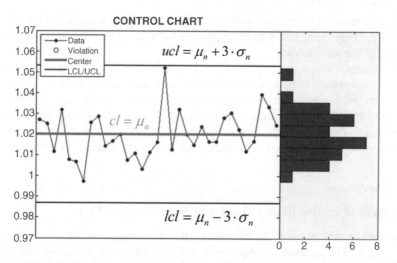

Figure 8.31   SPC techniques applied to CPV. Control chart calculation

defined. Therefore, each concentrator is monitored daily using a control chart, as the one shown in Figure 8.31. The daily production of the concentrator system is normalized using the average value of the production of all the systems in the CPV plant to generate the sample to analyze. Then, the average value (μ_n) and standard deviation (σ_n) of this sample is calculated to generate the control chart.

The control chart of Figure 8.31 represents in dots at the left the evolution of the daily production of a CPV system and at the right a histogram representing the percentage of occurrence of each value. The center line (*cl*) represents the estimated average value for the system production; the lower line (*lcl*), represents the lower control limit estimated for the system production and the line at the top, the upper control limit (*ucl*), represents the high limit estimated for the system production.

The area between the lower and upper control limits (*lcl* and *ucl*) Figure 8.31 is defined like the *safe operation area* (SOA) where the concentrator is operating within its specifications, at its full potential between the limits. The control limits can be used to raise alarms to indicate that the energy production is out of control. In this case, a corrective maintenance action will be necessary. Making a detailed analysis of Figure 8.31, two new limits can be established inside the SOA: the lower alarm limit (*lal*) and upper alarm limit (*ual*). These new limits are inside the SOA and the area between them and their respective control limits is the *guaranteed operational area*. When the system operates in this last area, an early action can be undertaken in order to correct the performance before the system actually fails. Figure 8.32 shows an example of this technique applied to a concentrator. As can be observed, the performance is presenting a decreasing trend in daily production until it reaches the lower limit where a corrective action is required. Applying the lower alarm limit, this malfunction could be solved before.

Therefore, if these kinds of techniques are applied in the daily operation of a CPV plant, the unavailability of the installation will be surely reduced, the energy generation could be maximized, and the cost of maintenance would decrease as well because of the early detection of failures.

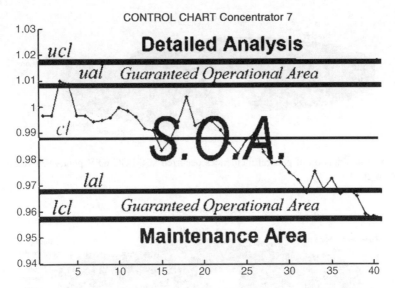

Figure 8.32 SPC techniques applied to CPV: relation between operation and maintenance

8.7.2 Maintenance

In CPV plants, a failure can be defined as any issue affecting any element of the installation that reduces the energy generation. This does not imply that the affected element is broken; it may only need to be adjusted.

As has been said before, three different types of maintenance can be carried out in a CPV plant. In the following, the most common tasks included in each type are presented.

8.7.2.1 Corrective Maintenance

Corrective maintenance includes all the corrective actions that have to be carried out occasionally in a CPV plant to ensure the maximum production of electricity.

The most common tasks of corrective maintenance for flat plate PV are carried out over the inverter and the modules. Nevertheless, if trackers are present in the installation, they usually become the most critical elements. For CPV, there is not too much literature available covering this issue because of the novelty of the technology. The O and M experience in CPV power plants is scarce and manufacturers are not very prone to disclosing data about their own experience, because in some cases most of the incidences are still part of their learning phase. However, some results on this topic have been published by the team at ISFOC, that are the result of their experience in the operation of CPV plants at Puertollano (Spain) for more than five years [40,41]. This section will be based upon these data.

First of all, it has to be taken into account that ISFOC CPV plants were among the first prototypes installed for many CPV manufacturers, so probably these numbers are the result of an early learning phase. Having said this, it is important to anticipate that the results are not alarming, though really point at the key element to attain reliability: the trackers.

The results obtained define that the most vulnerable element of the installation, registering most of the incidences, are the trackers. Therefore, most of the corrective maintenance tasks will

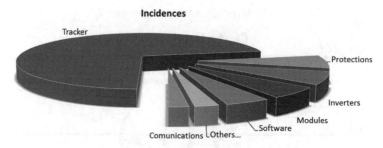

Figure 8.33 Classification of general incidences occurred at ISFOC CPV plants (Puertollano, Spain) from 2008 to 2013

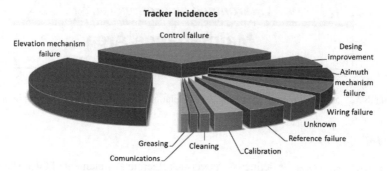

Figure 8.34 Classification of tracker incidences occurred at ISFOC CPV plants (Puertollano, Spain) from 2008 to 2013

be carried out on the different elements of the tracker. Figure 8.33 shows the classification of the incidences registered at ISFOC CPV plants during their operation from 2008 to 2013 according to the element affected. Figure 8.34 breaks down the tracker incidences according to the element that was causing it.

Therefore, analyzing these incidences, it can be observed that most of the corrective maintenance tasks were carried out over the trackers, being the elevation drive (including motors) and the control unit the issues responsible for roughly two thirds of the incidences. See also Chapter 9 for further details about reliability of plants.

8.7.2.2 Preventive Maintenance

Preventive maintenance includes all the preventive actions that have to be carried out periodically in a CPV plant to ensure the maximum production of electricity.

From time to time, most of the elements of the installation must be inspected to ensure that they fulfill the task for what they were designed. Usually, the manufacturer of the CPV plant should establish the best schedule for the preventive maintenance in the maintenance handbook. Table 8.4 collects a typical schedule for the preventive tasks to be carried out in a CPV plant. The tasks described in the table are only an example as the elements and the best way to maintaining them will depend on the technology. It is also important to remark that to reduce the energy losses of the installation it could be reasonable to schedule the tasks according to the meteorological forecast to avoid as much as possible unavailability of the plant in sunny days.

Table 8.4 Typical schedule for preventive maintenance in a CPV plant

Element	Task	Regularity
CPV module	Wiring and ground connection verification	1 year
	Fixings verification	1 year
Tracker	Tighten structure nuts and bolts	3 years
	Tracker structure inspection (deformations and rusting)	1 year
	Inspection of the foundation and surrounding terrains	1 year
	Limit switches inspection	1 year
	Wiring verification	1 year
	Motors and actuators inspection	1 year
	Azimuth and elevation drives greasing	6 months
	Verification of the motor consumption	6 months
Tracker control unit	Backup batteries replacement	5 years
	Solar sensor inspection and adjustment if necessary	6 months
	Backup batteries verification	6 months
Inverter	DC and AC wiring connectors verification	6 months
	Heat-sink inspection/thermography	1 year
Electrical cabinets	Tighten protections connections	5 years
	Thermography	1 year
	Electrical protections inspection	6 months
Communications	Communication cabinet inspection	6 months
Trenches	Manhole inspection	6 months
Transformer	Inspection	1 year

Finally, the cleaning of the CPV plant, to avoid the effect of soiling in energy production, can be also defined as a preventive maintenance. As before, there are not many references about this topic in the literature [5,42,43]. The first conclusion that is obtained from these studies is that the effect of soiling on the performance of CPV modules is strongly influenced by the location and also by the technology. Different locations have been analyzed in the studies and also different technologies (different concentration ratios, types of optics, sizes of lenses etc.). The results obtained for energy losses due to soiling span a wide range with numbers from 6% to 26%. Hence, these numbers seem to suggest that it is always necessary to clean. However, a more detailed analysis is needed to assess the real impact of cleaning in the economic balance of a CPV plant. In [5], the time elapsed until a stable level of soiling was established on the module surface was analyzed, together with the potential incomes of the installation and the cost of cleaning. The conclusion obtained for that location (continental climate without pollution) and high CPV systems was that the energy losses because of soiling were around 7% and this value was reached in around a week. The subsequent economic analysis carried out, revealed that it was only cost-effective to clean the systems during the summer. This value was also validated analyzing the effect of rain on the energy losses; it was determined that after several days of strong rain the energy of the installation increased around 5%.

Therefore, as soiling effects are technology- and site-specific and difficult to generalize, each particular installation should be analyzed in detail to maximize the incomes derived from cleaning and taking advantage of the cleaning effect of the rain.

8.7.2.3 Predictive Maintenance

Predictive maintenance includes all the modifications made in the preventive maintenance strategy because of the analysis carried out on the corrective actions implemented in the installation.

This means that some of the corrective actions carried out in the field are caused by a less than optimum preventive maintenance strategy or even are a failure in the design of the installation. Therefore, the early introduction of changes in the preventive maintenance handbook – or even the preventive change of unsound elements in the installation – could, in the long term, become beneficial for the installation.

8.8 Power Rating of a CPV Plant

Nowadays there is no official standard or procedure for rating or determining the nominal power of a CPV plant. Therefore, each installation made up to now has defined its own procedure that is usually inspired by what it is done in the flat plate PV industry.

Basically, the rating of a CPV plant consists in the determination of the nominal power, what means the power of the CPV plant at some fixed operating conditions (standard reporting conditions). In the following different approaches to solve that problem are presented and the range of applicability is discussed.

8.8.1 ISFOC Approach

From the point of view of the authors, the best way to rate a CPV plant is to determine the nominal power of the plant in DC followed by an estimation of the energy losses of the complete installation. Firstly, the DC power of the plant is obtained from the measurements of I-V curves of a representative population of concentrators. Secondly, the measurements should be translated from the operating conditions to the standard reporting conditions defined in the contract. Finally, for the whole installation, the energy generated is monitored during a period of time. This monitoring will allow estimation of the different losses present in the installation.

Several publications [44–49] cover the procedure followed by ISFOC for rating the CPV plants installed in their facilities. From 2007 on, several CPV plants of up to 2.4 MW of different technologies have been installed in ISFOC facilities. As there was no standard available, ISFOC developed its own standard taking into account the hypothesis described before and based in the Shockley model for the solar cells. This procedure – described in detail in Annex 8-II – is as follows:

- Basically, the power of some concentrator systems of the plant is measured at operating conditions and then translated to the standard reporting conditions specified in the contract, namely, DIN of 850 W/m^2 and 60°C cell temperature. This yields the DC power of the plant. The equations used in the translation are based in the Shockley model adapted to triple-junction cells (see Chapter 2) and the equivalent cell temperature is calculated via heat-sink temperature assuming knowledge of the thermal resistance between the cell and the heat-sink.
- Then, during the acceptance period, the energy generated by the complete plant is monitored and compared with the theoretical energy calculated from the DC power (using the same translation equations). The result from this comparison is a correction factor that represents

all the 'energy losses' of the plant, like spectrum, tracking, mismatch, shading etc. Finally the AC power of the plant is calculated by applying this correction factor to the DC power.

This procedure has been proven with different technologies [48,49] demonstrating its usefulness for rating a CPV plant and its effectiveness for estimating the normal energy losses of the plant and its production. Moreover, ISFOC experience has demonstrated that the procedure for rating a CPV plant needs to be practical, easy and fast to implement in the field – where the surrounding conditions cannot be controlled while doing the measurements – and applicable in any location and time of the year, for example to finish the commissioning of a CPV plant prior to its operation as fast as possible. With this purpose, the IEC committee is working in developing new standards that could be used for rating CPV plants.

8.8.2 International ASTM Standards

Up to now, the only standard that deals with the determination of the nominal power of a CPV plant, system and/or module is the ASTM E2527 09 [50] 'Standard Test Method for Electrical Performance of Concentrator Terrestrial Photovoltaic Modules and Systems under Natural Sunlight.'

The goal of this standard is to provide an accepted procedure for testing and reporting the electrical performance of a CPV module or system. Depending on whether the inverter input or output is monitored the rating is defined in DC or AC. As the DNI is measured using a normal incidence pyrheliometer, it can be only applied to CPV systems with a geometric concentration ratio greater than 5×. In addition, it can be only applied to concentration systems using passive cooling where the cell temperature can be related to air temperature.

The standard concentrator reporting conditions set in this standard for rating the maximum power of the concentration system are:

- direct normal irradiance of 850 W/m^2
- ambient temperature of 20°C
- wind speed of 4 m/s.

The maximum power (P_{oper}) of the concentration system is measured together with the operating conditions, – direct normal irradiance (DNI), ambient temperature (T_a) and wind speed (w_s) – over a range of irradiance and air temperature values. A multiple linear regression of the power as a function of the operating conditions (using Eq. (8.12)) is performed to compute the regression coefficients (a_i). For best results, the data points to be used in the multiple linear regression should be distributed around the standard concentrator reporting conditions:

$$P = E \cdot (a_1 + a_2 \cdot DNI + a_3 \cdot T_a + a_4 \cdot w_s), \tag{8.12}$$

Then, the maximum power of the concentrator is extrapolated to the standard concentrator reporting conditions using the same multiple linear regression (Eq. (8.12)).

This test method assumes that the regression equation accurately predicts the concentrator performance as a function of total irradiance with a fixed spectral irradiance, wind speed, and air temperature. The spectral distribution will be seasonal and site specific because of the optical air mass, water vapor, aerosols, and other meteorological variables. Moreover, it is already known that the dependence with temperature for CPV systems has a double contribution in the electrical performance (see Chapter 5): the cell electrical output depends on cell temperature and the optical focusing characteristics depend on the temperature of the optics, which also affects the cell electrical outputs [2–4]. This issue is not taken into account with this approach and perhaps it could result in a nonlinear dependence of the performance with temperature because of different weights of contributions of these two effects.

All these issues, which were not taken into account in the first version of the standard, are currently being addressed in the new versions under development by the IEC committee.

8.8.3 International IEC Standards

At this time, the TC82/WG7 IEC committee, in charge of developing international standards for photovoltaic concentrators and receivers, is working in the development of several standards for CPV. One of its priorities is the IEC 62670 'Photovoltaic concentrators (CPV) - Performance Testing' that covers the methodology of the performance tests that should be carried out over a CPV module, concentrator or plant for determining its power. This standard is divided in several parts as the objective is to cover different purposes.

IEC 62670-1:2013 [51] Part 1 'Standard Conditions' was approved and published in 2013. It defines standard conditions for assessing the power produced by CPV systems and their subcomponents. The goal is to define a set of conditions so that power ratings noted on datasheets, nameplates or contracts will have a standard basis independently of the technology and manufacturer. Because of the special features related with the measuring procedures that can be applied to this technology, two sets of conditions are included to characterize a CPV system based on the environment:

- Concentrator Standard Test Conditions (CSTC). These conditions are used when the characterization is carried out under controlled conditions and the objective is to obtain the performance when the system is in a readily reproducible environment – typically in the laboratory.
- Concentrator Standard Operating Conditions (CSOC). These conditions are used when the characterization is carried out under operating conditions – typically outdoors – and the objective is to obtain the on-sun performance relative to commonly measured meteorological conditions.

The definition of both standard conditions is summarized in Table 8.5. While CSTC are really useful for measuring CPV modules indoors, in a solar simulator for controlling the

Table 8.5 Definition of Concentrator Standard Test Conditions (CSTC) and Concentrator Standard Operating Conditions (CSOC)

Parameter	CSTC	CSOC
Irradiance (direct normal)	$1000 \, W/m^2$	$900 \, W/m^2$
Temperature	25°C cell temperature	20°C ambient temperature
Wind speed	—	2 m/s
Spectrum	Direct normal AM1.5 spectral irradiance distribution consistent with conditions described in IEC 60904-3 [52]	

manufacturing line, CSOC are applicable outdoors and could be used for rating a CPV plant.

IEC 62670-2 *draft P4* [53] Part 2 'Energy Measurement' is currently a draft version. It specifies the minimum requirements for determining the energy output and performance ratio for CPV modules, assemblies and power plants using an on-sun measurement based method. The purpose is to define testing methods to establish a standard energy measurement for CPV systems and to establish the minimum reporting information. This standard describes the measurements that should be done over a CPV power plant, establishes the minimum requirements of the equipment to be used and defines the processing of the measurement data and the calculations to be carried out. The target is to calculate the performance ratio of the installation for evaluating its efficiency and reliability in terms of energy generation.

The *performance ratio* is a measure of the quality of the CPV plant that is independent of the location and describes the relationship between the actual and theoretical energy outputs of the plant. The performance ratio (PR) is calculated as the ratio of the final yield (Y_F) to the reference yield (Y_R) of the installation using Eq. (8.13):

$$PR = \frac{Y_F}{Y_R}, \tag{8.13}$$

The reference yield (Y_R) represents the number of hours during which the DNI needs to be 1000 W/m^2 (CSTC DNI defined in IEC 62670-1) in order to contribute the same direct irradiation (E_{DNI}) received by the system over a certain period of time (typically a year):

$$Y_R = \frac{E_{DNI}}{1000W/m^2}, \tag{8.14}$$

The final yield (Y_F) represents the number of hours that the system should operate at CSTC (i.e. producing the output power P_{CSTC}) or at CSOC (i.e. producing the output power P_{CSOC}) to equal its monitored net energy (E) over a certain period of time (typically a year):

$$Y_F = \frac{E}{P_{CSTC}} \quad \text{or} \quad Y_F = \frac{E}{P_{CSOC}}, \tag{8.15}$$

Depending where the energy is being measured, DC or AC, and thus the point at which final yield is being calculated different losses of the installation can be estimated:

- The capture losses (L_C) represent the losses associated to the CPV generator itself. This loss groups the effects of temperature higher than the standard, mismatch between modules, tracking incidences, wiring losses, shadowing, MPP tracking error etc. It is calculated as the difference between the reference yield (Y_R) and the final yield in DC when the energy measured is before the inverter $(Y_{F,DC})$:

$$L_C = Y_R - Y_{F,DC}, \tag{8.16}$$

- The BOS or rest of the system losses (L_{BOS}) represent the losses associated to the DC–AC conversion of the inverter. This parameter is calculated as the difference between the DC final yield ($Y_{F,DC}$) and the AC Final Yield ($Y_{F,AC}$):

$$L_{BOS} = Y_{F,DC} - Y_{F,AC}, \tag{8.17}$$

When measuring the energy, all the parasitic terms should be included, that means all the energy consumed by different elements of the installation that are essential for its operation, such as the energy consumed by the trackers, the control units, the electrical equipment (inverters, transformers, etc.), the drying or cooling systems where relevant. It is important to remark that the power plant should be completely finished and in standard operation to apply this standard and that all maintenance performed over the plant during the period should be reported and taken out the calculation.

IEC 62670-2 *draft P4* - Part 2 does not establish a period for making the measurements but as it only normalizes the measurements with the DNI, the desirable period would be a whole year in order to avoid seasonal influences in the result. This standard could be applied during a short period of time to rate a CPV plant during its commissioning, in order to make a first approximation of the energy losses of the installation. However, its intended use is to serve as a method for measuring the energy produced during the warranty period as a long-term verification of the installation.

IEC 62670-3 *draft C02* [54] Part 3 'Performance Measurements and Power Rating' is currently a draft version at an advanced stage in the IEC committee because of its importance to determine nominal power of CPV modules or systems. It defines measurement procedures and instrumentation for determining concentrator photovoltaic performance at CSOC and CSTC, including power ratings.

This standard is more focused on the determination of the power of CPV modules and assemblies in CSTC and CSOC but some of its sections can be used for determining the power of a CPV plant. In general, it describes the procedure for measuring CPV modules indoor using a solar simulator and outdoors under real sunlight as well as the procedure for making the translation of outdoor measurements to CSOC and even to CSTC, together with data filtering requirements for this translation. So the outdoor measurement procedure can be applied directly to concentrators and the translations can be used if the manufacturer has previously rated its technology following the complete standard for the determination of the coefficients to be used in the translations.

This standard deals with the spectral influence by measuring the spectral distribution of the direct normal irradiance and establishing some limitations to this variable during the measurement period. For measuring the spectral distribution, a spectroradiometer and/or component reference cells can be used, mounted always inside collimated tubes for measuring only the direct normal irradiance. The component reference cells are defined as devices that are analogous to the individual junctions of the solar cells mounted in the CPV module. The standard establishes these component cells as the individual junctions of a conventional triple-junction lattice matched GaInP/GaInAs/Ge solar cell (bandgaps of 1.9 eV, 1.4 eV and 0.7 eV) but also other component reference cells that are specifically matched to the module to be measured can be used. The variable used for characterizing the spectral distribution is the *spectral matching ratio* (SMR) that represents the ratio between two of the sub-bands of the spectrum that are absorbed by each two junctions. Eqs. (8.18), (8.19) and (8.20) are used for the calculation, considering three sub-bands of direct normal irradiance. I_{top}, I_{mid} and I_{bot} are defined as the short-circuit current generated by each component cell, being the values with the superscript *STC* the ones obtained in the calibration under the IEC60904-3 [52] direct spectrum:

$$SMR1 = \frac{\frac{I_{top}}{I_{mid}}}{\frac{I_{top}^{STC}}{I_{mid}^{STC}}}, \tag{8.18}$$

$$SMR2 = \frac{\frac{I_{top}}{I_{bot}}}{\frac{I_{top}^{STC}}{I_{bot}^{STC}}}, \tag{8.19}$$

$$SMR3 = \frac{\frac{I_{mid}}{I_{bot}}}{\frac{I_{mid}^{STC}}{I_{bot}^{STC}}}, \tag{8.20}$$

Therefore, to ensure that the spectral distribution during the rating corresponds to the one defined in the CSOC and the CSTC, the SMR of the junctions that are limiting the current generation of the solar cell (usually the top and the middle junctions) should be around SMR = 1.

The data filtering requirements are listed below:

- Measurements of at least three different days with at least five separate *SMR1* crossings.
- The CPV module or concentrator should be thermally stable (*DNI* fluctuation smaller than 40% during the previous 30 minutes and lower than 10% during the previous 10 minutes, no changes greater than 5°C in the ambient temperature during the previous 30 minutes).
- *DNI* between 700 W/m² and 1100 W/m².
- Ambient temperature between 0°C and 40°C. Special care must be taken to ensure that there are no changes in the optical focusing in this range of temperature.
- Five minute average wind speed between 0.5 m/s and 5 m/s.
- *SMR1*, *SMR2* and *SMR3* should be within 2.5% of unity. If only two junctions are limiting the current generation of the solar cell the range of the *SMR* is ±1% for those two component cells.

The CSOC power determination can be calculated using Eq. (8.21) to Eq. (8.27) where: P_{CSOC} and η_{CSOC} denote power and efficiency at CSOC. A is the aperture area of the CPV module or system, measured from inside edge to inside edge of the frame. $\eta_{CSOC,i}$ denotes the maximum efficiency of an individual *I-V* measurement (*i* from *N*) after correcting to CSOC while $\eta_{I\_V,i}$ is the maximum power point efficiency of the I-V measurement before correction. $MPP_{I\_V,i}$, $V_{oc,i}$ and $I_{sc,i}$ are the maximum power point, open circuit voltage and short-circuit current, respectively, of an individual *I-V* curve. $f_{V,CSOC,i}$ is an intensity correction factor for the voltage and f_{DNI} is a temperature correction factor based on DNI. T_a is the ambient temperature and $T_{c,i}$ is the calculated average cell temperature inside the modules during the I-V measurement. δ is the efficiency temperature coefficient and β is the open circuit voltage temperature coefficient. These parameters should be obtained in the characterization of the CPV module. N_S is the total number of cells in series. n is the diode ideality factor for cell (assumed three for triple junction cells, four for four junctions . . .), k is the Boltzmann's constant $(1.38066 \cdot 10^{-23} \text{ J/K})$ and q is the elementary charge $(1.60218 \cdot 10^{-19} \text{ C})$. T_r is the reference cell temperature 298.15 K. Finally, V_{ocr} and I_{scr} are the open circuit voltage and short-circuit current, respectively, at the reference cell temperature and 900 W/m^2:

$$P_{CSOC} = 900 \cdot \eta_{CSOC} \cdot A, \tag{8.21}$$

$$\eta_{CSOC} = \frac{\left(\sum_i \eta_{CSOC,i}\right)}{N}, \tag{8.22}$$

$$\eta_{CSOC,i} = f_{V,CSOC,i} \cdot \left[\eta_{I\_V,i} - \delta \cdot \left\{(T_{a,i} - 20) + f_{DNI} \cdot (DNI_i - 900)\right\}\right], \tag{8.23}$$

$$\eta_{I\_V,i} = \frac{MPP_{I\_V,i}}{DNI_i \cdot A}, \tag{8.24}$$

$$f_{V,CSOC,i} = 1 + \frac{N_S \cdot n \cdot k \cdot T_{c,i}}{q \cdot V_{OC,i}} \cdot \ln\left(\frac{900}{DNI_i}\right), \tag{8.25}$$

$$f_{DNI} = \frac{\left(\sum_i \frac{(T_{c,i} - T_{a,i})}{DNI_i}\right)}{N}, \tag{8.26}$$

$$T_{c,i} = \frac{V_{OC,i} - V_{OCr} + \beta \cdot T_r}{N_S \cdot \left[\frac{n \cdot k}{q}\right] \cdot \ln\left(\frac{I_{SC,i}}{I_{SCr}}\right) + \beta}, \tag{8.27}$$

The CSTC power determination can be calculated using Eq. (8.28) to Eq. (8.31) and some of the previous equations. All the parameters have been defined before for CSOC; they now have analogous meaning just changing CSOC with CSTC:

$$P_{CSTC} = 900 \cdot \eta_{CSTC} \cdot A, \tag{8.28}$$

$$\eta_{CSTC} = \frac{\left(\sum_i \eta_{CSTC,i}\right)}{N}, \tag{8.29}$$

$$\eta_{CSTC,i} = f_{V,CSTC,i} \cdot \left[\eta_{I\_V,i} - \delta \cdot (T_{c,i} - T_r)\right], \tag{8.30}$$

$$f_{V,CSTC,i} = 1 - \frac{N_S \cdot n \cdot k \cdot T_{c,i}}{q \cdot V_{OC,i}} \cdot \ln\left(\frac{DNI_i}{1000}\right), \tag{8.31}$$

Therefore, the process described above can be applied for rating a concentrator with outdoor measurements both at CSOC and CSTC. Probably, to apply it in the commissioning of a CPV plant could be not so practical and fast, mostly because of the restrictions imposed to SMR, (i.e. measurements of at least three different days with at least five separate *SMR1* crossings, per concentrator) that would take the period needed to make the DC measurements to much longer than the desirable.

In summary, several works have been carried out by the CPV community to identify the best procedure to rate CPV systems [55–58]. In general, the procedures that are based on regression models are performing worse than procedures based in the physical model of the solar cells which include DNI and temperature corrections. Moreover, the results show that the only way to minimize the uncertainty is to increase the restrictions to environmental conditions during valid measurements or to introduce advanced instrumentation so the restrictions can be relaxed and thus increase the available days for making the measuring works. Therefore, relaxing the restrictions and reducing the number of measurements per concentrator as well as introducing advanced equipment could be an opportunity to apply this standard to rate CPV plants.

Finally, the IEC committee has just approved IEC62817 'Photovoltaic systems – design qualification of solar trackers' [7]. The objective of this standard is to qualify the design of a solar tracker by testing one prototype or its components. From this standard section 7 describes how to characterize the tracker accuracy. This section could also be used to verify the tracker accuracy during the commissioning of a CPV plant.

8.9 Modeling the Energy Production of CPV Power Plants

8.9.1 Basic Models

Standard IEC 62670-2 [53] provides a very simple expression to calculate the electrical energy generated by a CPV power plant over a certain period. In fact, an analogous expression has been successfully used by the flat plate PV community for years. The energy produced (E) in a given reference period can be calculated as:

$$E = \frac{P_{CSTC}}{DNI_{CSTC}} \cdot E_{DNI} \cdot PR, \tag{8.32}$$

where P_{CSTC} is the rated power of the system (kW_p) at concentrator standard testing conditions (CSTC); DNI_{CSTC} is the reference irradiance for CSTC that equals 1 kW/m^2; E_{DNI} is the average direct irradiation (kWh/m^2) received by the system in that reference period; and PR is the performance ratio of the plant.

It should be made clear that E_{DNI} is an irradiation or, in other words, a solar *energy* areal density expressed in kWh/m^2. E_{DNI} is calculated as the integral of the DNI –which is an irradiance or, in other words, a solar *power* areal density expressed in kW/m^2– received by the plant over the reference period of time, typically a year.

The performance ratio (PR) was described in section 8.8 and its calculation is done using Eq. (8.13), Eq. (8.14) and Eq. (8.15). Unfortunately the PR is neither the same for different observation periods nor at different locations. In fact, it depends on every detail of the time series of multiple environmental conditions, primarily, irradiance, ambient temperature, wind speed and solar spectrum as well as on the state of the entire electrical chain. The PR can be decomposed into the different loss mechanisms which reduce the conversion efficiency compared to its nominal value at CSTC. A reasonable approach is to multiply all the different losses factors (L_i) where i labels the different loss mechanisms:

$$PR = (100\% - L_0) \cdot (100\% - L_1) \cdot \ldots (100\% - L_i), \tag{8.33}$$

Typical loss factors that can be included in Eq. (8.33) to model the *PR* are: soiling; shading; tracking, misalignment, lens temperature, tracker stow, spectrum related, solar cell temperature, module mismatch, DC to AC conversion, unavailability etc.

The value of the different loss factors (L_i) have to be estimated by developing specific models (typically semi-empirical models attending the specific nature of the loss) and finely tuned, in a second stage, with the help of extensive simulations. In the following sections, the most critical loss mechanisms are discussed in depth in order to help develop such models. Regarding the simulations, they are typically done with hourly average values; the effect of the usage of time averages for simulations is discussed in general in [59] and specifically for CPV in the next sections.

The most popular modeling tools for PV and CPV power plants include *PVsyst* software originating from the University of Geneva and now developed by *PVsyst* S.A., the *System Advisor Model (SAM)* software by NREL and the *PV_LIB* toolbox managed by the Photovoltaic Performance Modeling Collaborative (PVPMC), an industry and Sandia National Laboratories collaborative to improve photovoltaic performance modeling. All these tools will be discussed in Annex 8-I.

The use of Eq. (8.32) to model the energy produced by CPV plants has some hidden difficulties. While the power at CSTC is defined by the installed module capacity, both, the DNI values and the simulation of the performance ratio, yield a large range of ambiguity. For CPV energy modeling, as for modeling in general, the quality of the input data can hardly be overestimated. Therefore, in the next section we discuss how the creation of a high quality dataset should proceed. Secondly, in succeeding sections, all loss mechanisms relevant for CPV are reviewed in order to establish the model for the performance ratio.

8.9.2 Input Data and Quality Checks

For the model building and validation phase, the quality of the meteorological data and the electrical data (i.e. measured DC/AC output) that is used to validate the model are equally important. Once the model is built and validated, most attention should be paid to the quality of the DNI dataset (see Chapter 1 for further details), as the electrical energy output strongly depends on the DNI. Ambient temperature and wind speed only have second order influence on the modeled AC power. The uncertainty of the full data set should be closely inspected for statistical and systematic components. Histograms are helpful for plausibility checks and identification of outliers. Scatter plots, especially of DC or AC power against DNI are also a helpful tool to detect compromised parts of the data set. The described data quality checks usually result in a truncation of the dataset. As typically DC or AC power is compared to the DNI, it is important to filter the dataset symmetrically in order not to compromise the overall results [60]. For example, periods of missing DNI should also be removed from the AC power time series, and vice versa. It is also important to document each applied filter and resulting truncation and estimate the impact on the overall result, as required by IEC 62670-2 [53]. Statistical uncertainty results from limited sample size and thus can be reduced by increasing the number of sensors (e.g. several pyrheliometers distributed over the plant) and the period covered with data. Systematic uncertainty is determined by the accuracy and the maintenance of the measurement instruments. It can be reduced by regular cleaning and verification of the alignment of the pyrheliometer and sensor recalibration as required by the manufacturer. Measures to reduce both uncertainties are desirable but may be opposed to each other, e.g. ensuring pyrheliometer cleanliness for long periods may be difficult.

8.9.3 Loss Mechanisms

This section is focused on the extension of models for non-concentrating photovoltaics to consider the particularities of CPV.

8.9.3.1 Shading and Soiling

Partial or full shading of CPV modules leads to a reduction of the electrical power output as discussed in section 8.4.2. The ratio of the shaded to the overall module area is called 'geometrical shading loss.' Modules that are non-uniformly shaded show different voltage-current characteristics, depending on the size and shape of the shaded area and the internal wiring of cells and by-pass diodes. When such modules are connected to a single electrical maximum power point (MPP) tracker, the observed shading loss is higher than the pure geometrical shading loss, because of the additional electrical mismatch loss. This effect is considered in *PVsyst* by the empirically determined parameter 'fraction for electrical effect' [61]. Shading may be caused by distant objects like hills ('far shading') and mutual shading of the trackers ('near shading') which are the predominant obstacles at typical large-scale CPV plants. As for PV, far shading can be modeled using a horizon line [62]. The calculation of the near shading loss requires reasonable simplifications – as those described in section 8.4.2.2 –, as the shape of the shades and their impact on the electrical characteristics can become very complex [63]. The modeling of the near shading loss typically consists of three steps: Firstly, a three-dimensional model of the plant or a representative section thereof is built. Secondly, the geometrical shading is calculated by surface projection and intersection processing as explained by Capdevila *et al.* [64]. Thirdly, the before mentioned empirically determined factor is applied in order to consider the additional loss due to electrical mismatch. Capdevila *et al.* show that the electrical mismatch loss due to shading may be modeled by addition of piecewise-linear approximations of the I-V curves of the CPV modules. They also showed that using 5-minute instead of hourly data further increases the accuracy of the shading model. For a Mediterranean site, as in Figure 8.22, it is expected to have shading losses between 1% and 6% over the entire year, depending on the spacing between trackers.

Soiling loss is caused by deposition of dirt particles on the module surface. M. Winter *et al.* [65] showed that soiling loss can have a strong impact on CPV performance, but may be easily reduced by small rain events. A. Kimber *et al.* [66] empirically derived a model based on a daily soiling rate, a cleaning threshold (i.e. minimum amount of rainfall to achieve a cleaning effect) and a grace period (i.e. number of days with negligible soiling loss after a cleaning event). In a first approximation, soiling loss may be modeled as a constant loss of a few percentage points. More detailed models require careful consideration of site and season dependent effects. Typical values for the soiling loss range from 3% up to 7% [5] for continental climates without high levels of pollution.

8.9.3.2 Misalignment Losses: Acceptance Angle and Tracker Pointing Error

The usage of concentrating optical elements entails a limitation of the acceptance angle under which incoming sunlight is directed onto the photovoltaic cells. IEC 62670-3 [54] contains a method for characterizing the misalignment sensitivity of CPV modules, by the determination of the 'acceptance angle curve', as the one shown in Figure 8.35 in the middle. An acceptance angle curve can be determined by maintaining tracking in one axis and permitting the tracker to

Figure 8.35 Left: Normalized frequency of tracking error for low and high wind speed according to IEC62817; Middle: Acceptance angle curve determined according to IEC 62670-3 (draft); Right: Example values for frequency of low and high wind speed bin

misalign in the other axis, while *I-V* sweeps are performed for at least ten increments of misalignment (see Chapter 12 for more details on the procedure).

IEC 62817 [7] describes how the pointing error of solar trackers can be determined using a pointing error measurement tool (see Chapter 6). Data over a minimum of five days is measured and divided into a high and low wind speed bin based on a 4 m/s threshold (Figure 8.35 right). An example of the distributions is shown in Figure 8.35 at the left.

The energy loss due to misalignment of a specific CPV module mounted on a specific CPV tracker can be estimated by combining the results of both procedures, following Eq. (8.34):

$$
\begin{aligned}
L_{mis} = freq_{low-wind} \cdot &\sum \left[freq_{low-wind}(Et_i) \cdot eff(Et_i) \right] \\
+ freq_{high-wind} \cdot &\sum \left[freq_{high-wind}(Et_j) \cdot eff(Et_j) \right]
\end{aligned}
\tag{8.34}
$$

where L_{mis} are the losses associated to the misalignment; *freq* represents the frequency of occurrence of each bin of wind and *eff* the efficiency obtained from the acceptance angle; being for both variables E_t the tracking error value.

Typical average values for the misalignment losses are between 1% and 2%.

8.9.3.3 Temperature Dependence of the Optics

Optical elements used in CPV modules are often composed of different materials (such is the case of silicone-on-glass lenses). Their efficiency is temperature dependent due to the temperature dependency of the refractive index and different coefficients of thermal expansion of the materials. The reader may refer to Chapter 5 for a detailed description of temperature effects pertaining both to the optics and solar cells. Lenses are often optimized for a certain temperature and show losses when working below and above this temperature due to internal misalignment, as shown by Kurtz and co-workers [67]. Depending on the design of the module, a first order approximation of the losses due to non-optimal lens temperature ($L_{lens,temperature}$), as implemented in PVsyst's *Utilization Factor* (see Annex8-I), may be possible with expression:

$$
L_{lens,temperature} = \mu \cdot (T_{lens} - 35\,^\circ C),
\tag{8.35}
$$

where T_{lens} is the lens temperature. In (8.35), it is frequently assumed $\mu = -0.5$ %/K for $T_{lens} < 35°C$ and $\mu = 0.3$ %/K for $T_{lens} \geq 35°C$.

In addition; lens temperature (T_{lens}) may be estimated from ambient temperature (T_a) and wind speed (w_s) by a linear regression, like Eq. (8.36):

$$T_{lens} = T_a + A + B \cdot w_s, \tag{8.36}$$

with example values of A = 12°C and B = −1.5°C/(m/s). Such a model typically leads to annual losses factors between 2% and 6% strongly depending on the deviation of the site's temperature from the optimum temperature of the optics.

Steiner *et al.* [68] used the finite element technique and ray tracing to account for the temperature sensitivity of the optics in their *YieldOpt* model. Faiman *et al.* [69] described an empirical outdoor-based method for recording heat-up and cool-down lens temperature measurements, using a small two-axis tracker encapsulated in an air-conditioned and quickly removable housing.

8.9.3.4 Tracker in Stow Position

CPV modules mounted on two-axis trackers present significant obstacles to wind particularly when the tracker table is in the upright position. Cost-efficient trackers can withstand only certain wind speed values in tracking mode and need to be moved to 'stow position', i.e. tracked plane parallel to ground, from a certain wind speed on. The loss ($L_{wind,stow}$) can be estimated as:

$$L_{wind,stow} = \begin{cases} 0\% & for \quad w_s < w_{s,threshold} \\ 100\% & for \quad w_s \geq w_{s,threshold} \end{cases}, \tag{8.37}$$

Note that the uncertainty of the estimated loss may be high if hourly data is used, so the temporal resolution should be at least 15 min., better 5 min. or ideally 1 minute. Complex quantities, like the maximum wind speed over 10 seconds within the hour, are another option to reach a higher temporal resolution. While they require a high sampling rate during data collection in the first place too, they can fit into a dataset of hourly values and hence can be processed more easily. Improved modeling also considers the behavior of the tracker in detail. The time required for moving from tracking mode to stow position and vice versa, as well as hysteresis periods used by SCADA to avoid toggling between the modes, should be considered. However, often the available data does not justify such high level of detail since either the temporal resolution is too low or the dataset is too short in order to be representative for rare events which is what high wind gusts are [70].

Typical wind stow losses, hence, strongly depend not only on the site's wind conditions, but also on the level of detail in its modeling. Therefore, values can easily reach between 0 and 10% for the same power plant project, while an acceptable wind stow loss is supposed to be below 3%.

8.9.3.5 Spectrum

Multijunction cells are inherently sensitive to the sunlight spectrum due to the serial interconnection of the subcells. The overall current is limited by the subcell which produces the lowest current. Depending on the sunlight spectrum and the spectral transmission of the optics, different subcells may limit the overall current over time. As measurements of the DNI spectrum are rarely available, simple models use geometrical, pressure corrected air mass (*AM*) as a surrogate.

A piecewise linear model using *AM* as the independent variable is used by *PVsyst* [71] and leads to annual losses of about 2% to 6%. Strobach and co-workers [72] described a parabolic model. Steiner *et al.* [73] use measurements of a multifilter rotating shadowband radiometer to estimate key atmospheric parameters and model the DNI spectrum with SMARTS to calculate the individual subcell currents.

8.9.3.6 Solar Cell Temperature

Temperature effects and thermal coefficients are described in detail in Chapter 5. IEC 62670-3 [54] describes how CPV module temperature coefficients for efficiency, open circuit voltage (V_{oc}) and short-circuit current (I_{sc}) can be determined by outdoor measurements. As the temperature of CPV cells if often non-homogeneously distributed and direct access to cells may be difficult, cell temperature can be estimated using the measured V_{oc}. This requires the V_{oc} temperature coefficient and V_{oc} at a reference conditions (e.g. 25°C and 900 W/m^2) which can be obtained following the procedures described in IEC 62670-3 [54]. For modeling, *PVsyst* and NREL *SAM* contain two different temperature models which compute the solar cell temperature from ambient temperature, wind speed and irradiance [74,75].

On an annual basis, cell temperature losses between 5% and 7% can be expected.

8.9.3.7 Module I-V Characteristics

The electrical behavior of a CPV module can be modelled with the same two approaches which are known from flat-plate PV. Therefore, the two approaches for modelling the *I-V* curve of a CPV module are:

- To use the equations derived from the so called 'equivalent circuit models' that represent the full current-voltage characteristics of a PV or CPV module. The characteristic points of the I-V curve (I_{sc}, V_{oc}, I_{mpp}, V_{mpp}) are obtained by solving the equations. Such models are implemented for example in *PVsyst* (one-diode model based on Duffie *et al.* [76]), in NREL *SAM* (five parameter single-diode equivalent circuit model [77]) and were used in the Spice network model by Steiner *et al.* [68].
- To use the so called 'point-value models' that focus on the characteristic points of the *I-V* curve (I_{sc}, V_{oc}, I_{mpp}, V_{mpp}) and do not try to model the full I-V curve. Such models include the Sandia PV *Array Performance Model* [75] and the *Simple Efficiency Module Model*, both optionally available in NREL *SAM*.

The single diode model may be extended by an irradiance dependence of the shunt resistance [78]. This irradiance dependence of the shunt leads to typical annual losses of 1% to 4%.

There are more complex equivalent circuit models for multijunction solar cells [79], as described in detail in Chapter 2. However, highly complex models need to be justified by high quality of the power plant data which is often not the case. Such models are more suitable for high quality data acquired from single or few modules monitoring; e.g. during power rating measurements.

8.9.3.8 Module Mismatch

Mismatch loss is caused by not perfectly identical *I-V* characteristics of the modules that are connected to a common MPP tracker ('module array'). The voltage and current differences of the *I-V* curves can be result from manufacturing and installation variations (e.g. solar cell quality, alignment of the modules on the trackers) and different DC cable lengths and cross-sections.

Mismatch loss can be regarded as the gap between two scenarios. In the no mismatch scenario, each module would be equipped with its own MPP tracker and therefore operated exactly at its MPP current and voltage. In this ideal scenario the total power of the system would be the sum of the maximum powers of each individual module, i.e. the denominator of Eq. (8.38). In the typical utility scale scenario, hundreds of modules form an array and they are connected to a common MPP tracker, what it is represented like the numerator of of Eq. (8.38). The losses then can be estimated with the next expression:

$$L_{missmatch} = 1 - \left(\frac{P_{mpp_{module,array}}}{\sum P_{mpp_{single,module}}} \right), \tag{8.38}$$

Studies about the module mismatch show losses ranging from 1% to 3%, which are mainly depending on the number of modules, their size and the interconnections between them.

8.9.3.9 DC to AC Conversion and Auxiliary Consumption

The DC to AC conversion of a CPV plant is usually carried out by off-the-shelf inverters which are not specifically designed for CPV applications. Inverter behavior can usually be modeled in the same way as for flat-plate PV with one exception. Due to clouds, DNI may change almost instantaneously whilst, in contrast, the capability of inverters to follow steep ramps may be limited. Hence, the DNI during ramp-up periods may not be fully converted. As shown in Figure 8.36, the technical constraints of the inverter may be characterized by a DNI or DC power threshold (1) which triggers the wake-up of the inverter. After the wake-up time (2) has passed, the inverter starts ramping up the DC power until the full DC power is reached, once the ramp-up time (3) has been covered. The resulting 'switch-on loss' can be modeled by means of

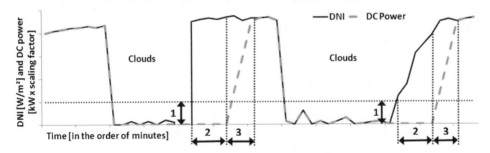

Figure 8.36 Modeling of inverter switch-on losses. Actual DNI (black solid line) on a cloudy day. The DC power of the inverter (dashed grey line) cannot follow the steep DNI ramps due to technical constraints of the inverter. (1) indicates the inverter specific DNI or DC power threshold which triggers the wake-up phase of the inverter. (2) indicates the wake-up time before the inverter starts ramping up the DC power. (3) indicates the ramp up time which the inverter needs to reach full DC power output

an effective DNI to describe the inverter's behavior. Due to a large number of different inverter types and parameters, it is hard to provide a value for the expected loss due to switch-on. Annual losses between 2% and 4% might be both realistic and acceptable for the 'switch-on loss'.

The DC to AC conversion itself is not different from non-concentrating PV, so existing inverter models such as part load curves from *PVsyst*, *SAM*, the CEC database or the Sandia inverter model [80] can be used. Inverter efficiency curve typically result in annual losses of 2% to 4%. AC voltage transformation and medium and high voltage line losses are treated in the same way as for non-concentrating PV.

CPV power plants usually show a higher loss due to parasitic consumption than non-concentrating plants, as tracker motors, control boards and possibly active cooling and air drying system consume electrical power. IEC 62670-2 [53] contains procedures to measure the power consumption of such devices. Self-consumed energy can be estimated by considering the operating time of the different devices. The energy loss depends on the size of the plant and its detailed engineering. Energy consuming factors like the tracker motors scale with the size of the plant, while, for instance, the energy requirements of the SCADA system are less dominant in a larger power plant compared to a small installation. Typical losses can be between 1% and 3%.

Glossary

List of acronyms

Acronym	Description
AC	Alternating current
AM	Air mass
ASCII	American Standard Code for Information Interchange
BOM	Bill of materials
BOS	Balance of system
BSRN	Baseline Surface Radiation Network
CEC	California Energy Commission
CFD	Computational fluid dynamics
CPV	Concentrator photovoltaics
CSOC	Concentrator Standard Operating Conditions as in IEC 62670-1:2013 [51]
CSTC	Concentrator Standard Test Conditions defined in IEC 62670-1:2013 [51]
DC	Direct current
DNI	Direct normal irradiance
EPC	Engineering, procurement and construction
EPW	Energy Plus Weather data
FIT	Feed-in tariff
GEBA	Global Energy Balance Archive
GNI	Global normal irradiance
IEC	International Electrotechnical Commission
IEC TC82/WG7	IEC Technical Committee 82/Working Group 7 (Concentrator Modules)
JRC	Joint Research Centre
lcl/ucl	Lower and upper control limit
lal/ual	Lower and upper alarm limit
MPP	Maximum power point
MPPT	Maximum power point tracking
NASA	National Aeronautics and Space Administration

NREL	National Renewable Energy Laboratory
O and M	Operation and maintenance
PPA	Power purchase agreement
PV	Photovoltaic
PVGIS	Photovoltaic geographical information system
PVPMC	PV Performance Modeling Collaborative
RMS	Root mean square
SAM	System advisor model
SCADA	Supervisory control and data acquisition
SDK	Software development kit
SMARTS	Simple model of the atmospheric radiative transfer of sunshine
SOA	Safe operational area
SPC	Statistical process control
SSE	Surface meteorology and solar energy
THD	Total harmonic distortion
TMY	Typical meteorological year
TRNSYS	Transient system simulation tool
TRY	German test reference years
UV	Ultraviolet radiation
WMO/OMM	World Meteorological Organization

List of symbols

Typical units given in square brackets. If no units are given, variable is dimensionless.

Symbol	Units	Description
A	[m^2]	Tracker table surface ($A = A_X \cdot A_Y$)
A_X, A_Y	[m]	Length and width of the tracker table
C_I	[€]	Total installation costs associated to the inverter selection
C_{INV}	[€]	Costs due to the purchase and installation of the inverter
C_{DC_wire}	[€]	Costs associated to the DC wiring of the PV generator
C_{AC_wire}	[€]	Costs associated to the AC wiring costs (from the inverter to the connection to the grid point)
C_{others}	[€]	Installation costs associated to the rest of the installation
DNI	[W/m^2]	Direct Normal Solar Irradiance
DNI_{CST}	[W/m^2]	Reference DNI level at CSTC ($= 1000 \, \text{W/m}^2$)
DNI_{SH}	[W/m^2]	Solar irradiance losses
d_S	[m]	Distance among trackers
eff	[%]	Efficiency
E	[Wh]	Energy generation in the period under analysis (DC or AC depending on the analysis)
E_{DNI}	[Wh/m^2]	Direct Normal Solar Irradiation (solar energy aerial density accumulated over the period under analysis)
$E_{DNI,y}$	[Wh/m^2]	Yearly direct normal solar irradiation
H	[Wh/m^2]	Annual solar radiation
H_{SH}	[Wh/m^2]	Annual solar radiation losses
I_{sc}	[A]	Short circuit current

I_{top}	[A]	Short-circuit current generated by each component cell (top, middle and bottom) in a triple-junction GaInP/GaInAs/Ge multijunction. An asterisk superscript, indicates measurement taken at reference conditions
I_{mid}	[A]	
I_{bot}	[A]	
k	[J/K]	Boltzmann's constant
L_{BOS}	[H]	Rest of the system or BOS losses
L_C	[H]	Capture losses
L_i	[%]	Loss factor (subscript i referring to the different loss mechanisms)
n		Diode ideality factor
N_S		Total number of cells connected in series
P_{CSOC}	[W]	Concentrator power in CSOC (DC or AC depending on the analysis)
P_{CSTC}	[W]	Concentrator power in CSTC (DC or AC depending on the analysis)
$P_{DC,STC}$	[W]	Peak power – concentrator power in standard test conditions (CSTC or CSOC)
$P_{DC,oper}$	[W]	Inverter input power – DC power produced in operating conditions
P_{INV}	[W]	Inverter nominal power
PR	[%]	Performance ratio
P_X, P_Y	[m]	Length and width of the shading surface
q	[C]	Elementary charge
SMR		Spectral matching ratio
S_{SH}	[m^2]	Shading surface ($S_{SH} = P_X \cdot P_Y$)
T_c	[°C]	Calculated average cell temperature
T_a	[°C]	Ambient temperature
T_{lens}	[°C]	Lens temperature
V_{oc}	[V]	Open circuit voltage
w_s	[m/s]	Wind speed
Y_F	[hours]	Final yield (DC or AC depending on the analysis)
Y_R	[hours]	Reference Yield
β	[V/K]	Open circuit voltage temperature coefficient
δ	[1/K]	Efficiency temperature coefficient
η_{CSOC}	[%]	Efficiency at CSOC
η_{CSTC}	[%]	Efficiency at CSTC
η_{inv}	[%]	Inverter efficiency
θ_S	[°]	Zenith angle
μ	[%/K]	Cell temperature coefficient
Ψ_S	[°]	Azimuth angle

References

1. F. Rubio, M. Martínez, A. Hipólito, A. Martín, P. Banda. Status of CPV Technology. *Proceedings 25th EU PVSEC/ WCPEC-5*, Valencia, pp. 1008–1011. 2010.
2. G. Peharz, J.P. Ferrer Rodriguez, G. Siefer, A.W. Bett. Investigations on temperature dependence of CPV modules equipped with triple-junction cells. *Progress in Photovoltaics: Research and Applications.* **19** (1), 54–60. 2011.
3. S. Askins, M. Victoria, R. Herrero, C. Domínguez, I. Antón, G. Sala. Effects of Temperature on Hybrid Lens Performance. Proceedings CPV-7, Las Vegas, 2011. *AIP Conference Proceedings* **1407**, 57. 2011.
4. T. Hornung, A. Bachmaier, P. Nitz, A. Gombert. Temperature Dependent Measurement and Simulation of Fresnel Lenses for Concentrating Photovoltaics. Proceedings CPV-6, Freiburg, 2010. *AIP Conference Proceedings,* **1277**, 85. 2010.

5. D. Sanchez, P. Trujillo, M. Martinez, J. P. Ferrer, F. Rubio. CPV performance versus soiling effects: Cleaning policies. Proceedings CPV-8, Toledo, 2012. *AIP Conference Proceedings* **1477**, 348. 2012.
6. C. Alamillo, O. de la Rubia, E. Gil, *et al*. Analysis of inverter configuration on CPV plants. *Proceedings 25*[th] *EU PVSEC/WCPEC-5*, Valencia, pp. 4733–4736. 2010.
7. IEC 62817:2014 ed1.0. "Photovoltaic systems - Design qualification of solar trackers"
8. W. Aipperspach, S. Bambrook, F. Rubio. Comparison of wind tunnel testing and CFD wind load results for a CPV tracker. *Proceedings 29*[th] *EU PVSEC*, Amsterdam, pp. Xxxxxx. 2014.
9. O. De la Rubia, D. Sánchez, Mª L. García, *et al*. Acceptance procedure applied to ISFOC's CPV plants. *Proceedings ICSC–5*, Palm Desert, pp. Xxxxxx. 2008.
10. E. Gil, D. Sánchez, O. de la Rubia, *et al*. Field technical inspection of CPV power plants. *Proceedings 25th EU PVSEC/WCPEC-5*, Valencia, pp. 4483–4486. 2010.
11. B. Bletterie, R. Bründlinger, G. Lauss. On the characterisation of PV inverters' efficiency-introduction to the concept of achievable efficiency. *Progress in Photovoltaics: Research and Applications*, **19** (4), 423–435. 2011.
12. P. Rodrigo, E.F. Fernández, F. Almonacid, P.J. Pérez-Higueras. Models for the electrical characterization of high concentration photovoltaic cells and modules: A review. *Renewable and Sustainable Energy Reviews*, **26**, 752–760. 2013.
13. Y.S. Kim, S.-M. Kang, R. Winston. Tracking control of high-concentration photovoltaic systems for minimizing power losses. *Progress in Photovoltaics: Research and Applications*, **22** (9), 1001–1009. 2014.
14. M. Liu, G. S. Kinsey, W. Bagienski, A. Nayak, V. Garboushian. Measurements of mismatch loss in CPV modules. Proceedings CPV-8, Toledo. *AIP Conference Proceedings* **1477**, 165–167. 2012.
15. M. Mendelsohn, T. Lowder, B. Canavan. Utility-scale concentrating solar power and photovoltaics projects: a technology and market overview. NREL/TP-6A20-51137, April 2012.
16. W. Nishikawa, E. Green, and S. Crowley. Energy production of CPV power plants at ISFOC. *Proceedings ICSC–5*, Palm Desert. 2008.
17. Concentrator Photovoltaic (CPV) workshop. Understanding the technology and related implications for scaled deployment. *Solar Power International*, 11, Dallas, Texas. 2011.
18. E.F. Fernández, F. Almonacid., P. Rodrigo, P. Pérez-Higueras. Model for the prediction of the maximum power of a high concentrator photovoltaic module. *Solar Energy*, **97**, 12–18. 2013.
19. M. Jantsch, H. Schmidt, J. Schmid. Results on the concerted action on power conditioning and control. *Proceedings 11*[th] *EU PVSEC*, Montreux, pp. 1589–1592. 1992.
20. Eduardo F., Fernández, P., Pérez-Higueras, F. *et al*. Model for estimating the energy yield of a high concentrator photovoltaic system. *Energy*, in press. 2015.
21. Y.S. Kim, R. Winston. Power conversion in concentrating photovoltaic systems: central, string, and micro-inverters. *Progress in Photovoltaics: Research and Applications*, **22** (9), 984–992. 2014.
22. E. Muñoz, P.G. Vidal, G. Nofuentes *et al*. CPV standardization: An overview. *Renewable and Sustainable Energy Reviews*, **14** (1), 518–523. 2010.
23. L. Narvarte, E. Lorenzo. Tracking and ground cover ratio. *Progress in Photovoltaics: Research and Applications*, **16** (8), 703–714. 2008.
24. V. Fthenakis, H. Chul Kim. Land use and electricity generation: A life-cycle analysis. *Renewable and Sustainable Energy Reviews*, **13** (6–7), 1465–1474. 2009.
25. K. Araki, I. Kumaga, H. Nagai. Theory and experimental proof of shading loss of multi-trackers CPV system. *Proceedings 25*[th] *EU PVSEC/WCPEC-5*, Valencia. 114–117. 2010.
26. K. Araki. Two interactive and practical methods for optimization of tracker allocation in a given land. Proceedings CPV-8, Toledo, 2012. *AIP Conference Proceedings* **1477**, 244. 2012.
27. Y.S. Kim, S.M. Kang, R. Winston. Modeling of a concentrating photovoltaic system for optimum land use. *Progress in Photovoltaics: Research and Applications*, **21** (2), 240–249. 2013.
28. P. Rodrigo, E.F. Fernández, F. Almonacid, P.J. Pérez-Higueras. Outdoor measurement of high concentration photovoltaic receivers operating with partial shading on the primary optics. *Energy*, **61**, 583–588. 2013.
29. J. Monedero, F. Dobon, A. Lugo, *et al*. Minimizing energy shadow losses for large PV plants. *Proceedings WCPEC-3*, Osaka, vol. **2**, pp. 2043–2045. 2003.
30. O. Perpiñan, E. Lorenzo and M. A. Castro. On the calculation of energy produced by a PV grid-connected system. *Progress in Photovoltaics: Research and Applications*, **15** (3), 265–274. 2007.
31. M. Garcia, J.A. Vera, L. Marroyo, E. Lorenzo, M. Perez. Solar-tracking PV Plants in Navarra: A 10 MW Assessment. *Progress in Photovoltaics: Research and Applications*, **17** (5), pp. 337–346. 2009.

32. K. Araki. Design of 30 kW CPV power plant optimized to Japanese environment. *Proceedings ICSC–5*, Palm Desert. 2008.

33. P.J. Perez, G. Almonacid, P.G. Vidal. Estimation of shading losses in multi-trackers PVsystems. *Proceedings 22ⁿᵈ EU PVSEC*, Milano, pp. 2295–2298. 2007.

34. P.J. Pérez, G. Almonacid, J. Aguilera, *et al.* Multitrackers systems calculation of losses due to selfshadowing. *Proceedings ICSC–4*, El Escorial. 2007.

35. C. Perez-de-los-Reyes, M. Sánchez, J. A. Amorós, *et al.* Revegetation in solar photovoltaic farms in mediterranean areas. *Fresenius Environmental Bulletin*, **22** (12a), 3680–3688. 2013.

36. M. L. García, G. Calvo-Parra, A. Hipólito, J. L. Pachón. Monitoring, communications and data processing of CPV plants. Proceedings CPV-6, Freiburg, 2010. *AIP Conference Proceedings*, **1277**, 273. 2010.

37. International Electrotechnical Commission. Photovoltaic system performance monitoring - Guidelines for measurement, data exchange and analysis, 61724:1998 ed1.0. 1998.

38. G. Blaesser, D. Munro. Guidelines for the assessment of photovoltaic plants document A: photovoltaic system monitoring. Report EUR16338EN, JRC. 1993.

39. L. Fanni, M. Giussani, M. Marzoli, M. Nikolaeva-Dimitrova. How accurate is a commercial monitoring system for photovoltaic plant? *Progress in Photovoltaics: Research and Applications*, **22** (8), 910–922. 2014.

40. D. Sánchez, M. Martínez, E. Gil, F. Rubio, J.L. Pachón, P. Banda. First experiences of ISFOC in the maintenance of CPV plants. Proceedings CPV-6, Freiburg, 2010. *AIP Conference Proceedings*, **1277**, 248. 2010.

41. F. Rubio, M. Martínez, D. Sánchez, R. Aranda and P. Banda. Two years operating CPV plants: Analysis and results at ISFOC. Proceedings CPV-7, Las Vegas, 2011. *AIP Conference Proceedings* **1407**, 323. 2011.

42. M. Vivar, R. Herrero, I. Antón, *et al.* Effect of soiling in CPV systems. *Solar Energy*, **84**, 1327–1335. 2010.

43. K.W. Stone, V. Garboushian, D. Dutra H. Hayden. Four years of operation of the AMONIX high concentration photovoltaic system at Arizona public service utility. *Proceedings ASME 2004 International Solar Energy Conference*, Portland OR, USA. ISEC2004-65014, pp. 441–448. 2004.

44. F. Rubio, M. Martinez, P. Banda. Concentrator photovoltaic field installations, in *Solar Cells and their Applications*, 2nd edn, (eds L. Fraas and L. Partain), John Wiley & Sons, Inc., Hoboken, NJ, USA, chapter 17. DOI: 10.1002/9780470636886.

45. M. Martinez, O. de la Rubia, F. Rubio, P. Banda. Concentration photovoltaics. *Comprehensive Renewable Energy, Volume 1: Photovoltaic Solar Energy*, pp. 745–765. 2012. DOI: 10.1016/B978-0-08-087872-0.00141-4.

46. F. Rubio, M. Martinez, R. Coronado, J.L. Pachón, P. Banda. Deploying CPV power plants – ISFOC Experiences. *Proceedings 33ʳᵈ IEEE PVSC*, San Diego. 2008.

47. M. Martínez, O. de la Rubia, D. Sánchez, *et al.* Concentrator photovoltaics connected to the grid and systems rating. *Proceedings 23ʳᵈ EU PVSEC*, Valencia, pp. 146–150. 2008.

48. M. Martínez, D. Sánchez, F. Rubio, J.L. Pachón, P. Banda. CPV systems rating, results and lessons learned at ISFOC. *Proceedings ICSC–5*, Palm Desert. 2008.

49. M. Martínez, D. Sánchez, J. Perea, F. Rubio, P. Banda. ISFOC demonstration plants: Rating and production data analysis. *Proceedings 24ᵗʰ EU PVSEC*, Hamburg, pp. 159–164. 2009.

50. ASTM E2527 – 09 Standard Test Method for Electrical Performance of Concentrator Terrestrial Photovoltaic Modules and Systems Under Natural Sunlight. ASTM International, West Conshohocken, PA, USA. 2009.

51. IEC 62670-1:2013 ed1.0 Photovoltaic concentrators (CPV) - Performance testing - Part 1: Standard conditions. 2013.

52. IEC 60904-3:2008 ed2.0 Photovoltaic devices – Part 3: Measurement principles for terrestrial photovoltaic (PV) solar devices with reference spectral irradiance data. 2008.

53. IEC 62670-2 *draft P4* Photovoltaic concentrators (CPV) – Performance testing – Part 2: Energy Measurement.

54. IEC 62670-3 *draft C02* Photovoltaic concentrators (CPV) – Performance testing – Part 3: Performance Measurements and Power Rating.

55. M. Muller, B. Marion, J. Rodriguez, S. Kurtz Muller. Minimizing variation in outdoor CPV power ratings. Proceedings CPV-7, Las Vegas. *AIP Conference Proceedings*, **1407**, 336. 2011.

56. P. Trujillo, J. P. Ferrer, M. Martínez, F. Rubio. Influence of ambient conditions in the power rating of CPV modules. *Proceedings CPV-8*, Toledo. *AIP Conference Proceedings*, **1477**, 135. 2012.

57. M. Muller, S. Kurtz, J. Rodriguez. Procedural considerations for CPV outdoor power ratings per IEC 62670. Proceedings CPV-9, Miyazaki. *AIP Conference Proceedings*, **1556**, 125. 2013.

58. I. Antón, M. Martínez, F. Rubio, *et al.* Power rating of CPV systems based on spectrally corrected DNI. Proceedings CPV-8, Toledo. *AIP Conference Proceedings*, **1477**, 331. 2012.

59. C.W. Hansen, J.S. Stein, D. Riley. Effect of time scale on analysis of PV system performance. SANDIA REPORT, SAND2012-1099, Sandia National Laboratories, Albuquerque. 2012.

60. D.C. Jordan, S.R. Kurtz. The dark horse of evaluating long-term field performance – data filtering. *IEEE Journal of Photovoltaics*, **4** (1), 317–323. 2014. DOI: 10.1109/JPHOTOV.2013.2282741.

61. A. Mermoud. PVsyst help section: Partition in module strings. Retrieved from http://files.pvsyst.com/help/near_shadings_partition.htm (23 Feb. 2015).

62. A. Mermoud. PVsyst help section: Importing an horizon profile. Retrieved from http://http://files.pvsyst.com/help/horizon_import.htm (26 Feb. 2015).

63. A. Mermoud, "PVsyst help section: Shading factor", Retrieved from http://files.pvsyst.com/help/near_shadings_factor.htm (18 Feb. 2015).

64. H. Capdevila, A. Marola, M. Herrerías Azcué. High resolution shading model and performance simulation of suntracking photovoltaic systems. *Proceedings of 42nd ASES Annual Conference*, Baltimore. 2013.

65. M. Winter, G. Flynn, J. Foresi, *et al.* Reliability testing of high-concentration PV module and soiling issues. Presentation given at NREL Photovoltaic Module Reliability Workshop. 2011.

66. A. Kimber, L. Mitchell, S. Nogradi, H. Wenger. The effect of soiling on large grid-connected photovoltaic systems in California and the Southwest Region of the United States. *Conference Record of the 2006 IEEE 4th World Conference on Photovoltaic Energy Conversion*. 2006. DOI: 10.1109/WCPEC.2006.279690

67. S. Kurtz, M. Muller, D. Jordan, *et al.* Key parameters in determining energy generated by CPV modules. *Progress in Photovoltaics: Research and Applications*. 2014. DOI: 10.1002/pip.2544

68. M. Steiner, G. Siefer, T. Hornung, G. Peharz, A. W. Bett. YieldOpt, a model to predict the power output and energy yield for concentrating photovoltaic modules. *Progress in Photovoltaics: Research and Applications*. 2014. DOI: 10.1002/pip.2458

69. D. Faiman, V. Melnichak, D. Bokobza, S. Kabalo. A sun-tracking environmental chamber for the outdoor quantification of CPV modules. *AIP Conference Proceedings*, **1616**, 106. 2014. DOI: 0.1063/1.4897039

70. P. Friederichs, M. Göber, S. Bentzien, A. Lenz, R. Krampitz. A probabilistic analysis of wind gusts using extreme value statistics. *Meteoroloische Zeitschrift*, **18** (6), 615–629. 2009.

71. T. Gerstmaier, M. Gomez, A. Gombert, A. Mermoud, T. Lejeune. Validation of the PVSyst performance model for the concentrix CPV technology. *AIP Conference Proceedings*, **1407**, 366. 2011. DOI: 10.1063/1.3658363

72. E. Strobach, D. Faiman, S. Kabalo, *et al.* Modeling a grid-connected concentrator photovoltaic system. *Progress in Photovoltaics: Research and Applications*, 2014. DOI: 10.1002/pip.2467

73. M. Steiner, G. Siefer, T. Hornung, G. Peharz, A. W. Bett. YieldOpt, a model to predict the power output and energy yield for concentrating photovoltaic modules. *Progress in Photovoltaics: Research and Applications*, 2014. DOI: 10.1002/pip.2458

74. A. Mermoud. PVsyst help section: array thermal losses. Retrieved from http://files.pvsyst.com/help/thermal_loss.htm (26 Feb. 2015).

75. D.L. King, W.E. Boyson, J.A. Kratochvil. Photovoltaic Array Performance Model. Sandia Report SAND2004-3535, Sandia National Laboratories, Albuquerque. 2004.

76. J.A. Duffie, W.A. Beckman. *Solar Engineering of Thermal Processes*, 4th edn, John Wiley & Sons, Ltd. 2013.

77. P. Gilman. SAM Photovoltaic Model (pvsamv1) Technical Reference. NREL Draft Report. 2014.

78. A. Mermoud, T. Lejeune. Performance assessment of a simulation model for PV modules of any available technology. *Proceedings of the 25th European Photovoltaic Solar Energy Conference and Exhibition* (EU PVSEC), Valencia, Spain, 6–10 September. 2010.

79. P. Trujillo, J.P. Ferrer Rodríguez, *et al.* Analysis and comparison between CPV indoor/outdoor characterisation results. *Proceedings of the 26th European Photovoltaic Solar Energy Conference and Exhibition*, Hamburg, p. 664–669. 2011. DOI: 10.4229/26thEUPVSEC2011-1DV.4.14

80. D.L. King, S. Gonzalez, G.M. Galbraith, W.E. Boyson. Performance Model for Grid-Connected Photovoltaic Inverters. Sandia Report SAND2007-5036, Sandia National Laboratories, Albuquerque. 2007.

8-I Annex

Software Tools for CPV Plant Design and Analysis

In this Annex, the most important software tools that have been referenced throughout Chapter 8 are listed and described.

8-I.1 Meteonorm

Meteonorm [1] is one of the most popular tools for assessing the solar resource. It is a comprehensive meteorological reference, based on more than 25 years of experience in the development of meteorological databases for energy applications. It is developed by Meteotest and the first version of *Meteonorm* was created in the early 1980s thanks to intensive research activities in collaboration with universities and the industry. Since then, it has been continuously developed.

Meteonorm gives access to a catalogue of meteorological data for solar applications at any desired location in the world. Currently, it has 8325 weather stations installed all around the world (1600 in Europe, 2625 in North America, 800 in Central and South America, 1900 in Asia including Russia, 600 in Africa and 800 in Australia and Oceania); 1325 of which include irradiation measurements.

Most of the data are taken from the GEBA (Global Energy Balance Archive), from the World Meteorological Organization (WMO/OMM). The periods 1961–1990 and 2000–2009 are available for temperature, humidity, wind speed and precipitation; the periods 1981–1990 and 1991–2010 are available for solar radiation. Monthly climatological (long term) means are available for the following parameters: global radiation, ambient air temperature, humidity, precipitation, days with precipitation, wind speed, wind direction and sunshine duration. The station data are supplemented by surface data from five geostationary satellites; being available on a global grid with a horizontal resolution of 8 km (3 km in Europe and Northern Africa). Usually, measurement data can only be used in the vicinity of a weather station; elsewhere the data have to be interpolated between different stations. The sophisticated interpolation models implemented in *Meteonorm* allow a reliable calculation of solar radiation, temperature and additional parameters at any site in the world. Moreover, monthly or hourly radiation and

Handbook of Concentrator Photovoltaic Technology, First Edition. Edited by Carlos Algora and Ignacio Rey-Stolle.
© 2016 John Wiley & Sons, Ltd. Published 2016 by John Wiley & Sons, Ltd.

temperature data from the user can be imported into *Meteonorm* and, subsequently, *Meteonorm* models and procedures can be applied to this data.

From the monthly values *Meteonorm* calculates hourly values of all parameters using a stochastic model. The resulting time series correspond to 'typical years' used for system design. Additionally, the following parameters are derived: solar azimuth and elevation, global, diffuse and beam (direct normal) radiation as well as radiation on inclined planes, longwave radiation, luminance, UVA/UVB, erythemal radiation, precipitation, driving rain and humidity parameters (dewpoint, relative humidity, mixing ratio, psychrometric temperature). Additionally, a stochastic model for producing minute-periodicity values of radiation parameters is also available.

Finally, 36 different export formats are available, covering most of the established in solar energy simulation software, including *TMY2* and *TMY3*, *EPW*, *TRNSYS* and *TRY*, *POLYSUN*, *PVSOL*, *PVSyst*. All export formats are available for hourly and monthly values.

Meteonorm software is not freeware, so the acquisition of a license is needed.

8-I.2 PVGIS

PVGIS (Photovoltaic Geographical Information System) [2] is a tool for research, demonstration and political support for the geographic resource assessment of solar energy in the context of the management for distributed power generation. It provides a map-based inventory of solar energy resource and assessment of the electricity generation from photovoltaic systems in Europe, Africa, and South-West Asia. It is a part of the SOLAREC action that contributes to the implementation of renewable energy in the European Union as a sustainable and long-term energy supply. The SOLAREC action was promoted European Commission through the Joint Research Centre (JRC), in particular its Renewable Energy Unit.

The interactive maps are web applications to browse and query GIS databases of solar radiation and other climatic parameters in two regions: Europe-Africa-Mediterranean Basin and South-West Asia. A location can be chosen either by browsing/zooming and clicking on a map, entering an address, or by directly setting latitude/longitude values.

- The solar irradiation data utility provides monthly and yearly averages of global irradiation at horizontal and inclined surfaces and tracking systems, as well as other climatic and PV-related data (Linke turbidity, diffuse to global irradiation ratio and optimum inclination angle of the surface). For the chosen location, these parameters are displayed in a separate window as a table and graphs in a report or it can be exported as text file or pdf format. An estimate of the deficit in yearly horizontal irradiation due to terrain shadowing is also provided.
- With the solar irradiance data utility, for a selected PV technology with a module inclination and orientation, daily profiles of clear-sky and real-sky irradiances for a chosen month can be obtained. The results are displayed in a separate window and can be exported as text file and pdf format. The daily variance is estimated by a standalone calculator, running on a server. The calculator takes into account also the shadowing by local terrain features.
- The solar electricity utility calculates the monthly and yearly potential electricity generation of a PV configuration with the defined module inclination and orientation.
- Finally, the PV potential (in built-up areas) utility (only available for Europe) summarizes the statistical characteristics of global irradiation and PV potential in regions that were

calculated for the horizontal, vertical and optimally inclined surfaces. The most widespread grid-connected PV technology, installed within the existing building infrastructure, was been considered for this calculation.

Additionally, *PVGIS* offers animations that are free for public use [3] of daily variation of sun shadows determined by terrain features (mountains), daily variation of global horizontal irradiance and monthly variation of global irradiation and other parameters.

Finally, several solar radiation and photovoltaic electricity potential and regional maps for Europe and Africa are available and free for public use [4,5]. The maps represent yearly sum of global irradiation on horizontal and optimally inclined surface. For most of the regions, the data represent the average of the period 1998–2011. The same color legend represents also potential yearly solar electricity generated by a 1 kW$_p$ system with photovoltaic modules mounted at an optimum inclination and assuming a system performance ratio of 0.75.

PVGIS software is freeware, so no license is needed.

8-I.3 SolarGIS

SolarGIS [6] is a geographical information system designed to meet the needs of the solar energy industry. It integrates solar resource and meteorological data with tools for planning and performance monitoring of solar energy systems.

The *SolarGIS* database is a high resolution database. Its model runs 24 hours a day and processes data from four geo-stationary satellites, which cover almost Earth's entire surface. The data are calculated using proprietary algorithms that process the satellite imagery and atmospheric and geographical inputs. The database consists of the following parameters:

- Solar and PV data: global horizontal irradiation, direct normal irradiation, diffuse horizontal irradiation, global tilted/in-plane irradiation for fixed and sun-tracking surfaces, optimum angle for PV modules on fixed mounted construction and PV electricity yield. The solar radiation database is derived from satellite data and atmospheric parameters. The original spatial resolution of ~3–5 km (depending on latitude) is disaggregated to ~90 m for the studied site or region using high resolution digital elevation model SRTM-3. The data are compared with high-quality ground measurements. Global and Diffuse horizontal and Direct Normal irradiation data available include: 1994 to 2010 for Europe, Middle East and Africa (PRIME region); 1999 to 2011 for most of Asia (IODC region); 1999 to 2013 for America (GEOS-EAST region) and 2007 to 2013 for Australia, Japan, Korea and other Pacific countries (MTSAT region).
- Meteorological data: air temperature at 2 m, relative humidity, wind speed and wind direction at 10 m. Meteorological data are provided by third party organizations, which operate numerical models of the atmosphere.
- Geographic data: terrain, landscape, population. These data parameters are very useful in the site prospection stage, are available globally.

SolarGIS offers different applications to process the information in the database:

- *iMaps*. Offers an interactive map that provides solar radiation and meteorological information with very high detail. It offers Global Horizontal Irradiation and Direct Normal

Irradiation maps at spatial resolution of 90 m, including additional parameters like diffuse horizontal irradiation, temperature, landscape, terrain, and population data.

- *pvPlanner.* Offers an accurate PV electricity energy calculator. It helps in the site prospection of solar energy systems and supports the comparison from various PV technology options.
- *pvSpot.* Allows reliable and independent performance monitoring of any PV power plant. It permits the comparison of the expected electricity production with the actual power production.
- *climData.* The database itself for the solar and meteorological data. The available time resolution includes: 1) *Historical time-series* - fully harmonized dataset (flagged and with gaps filled); 2) *Recent time-series* – Data for a continuous 12-month period; 3) *Typical Meteorological Year (TMY)*, constructed by SolarGIS concept and Forecast data – Forecast horizon of up to 72 hours. Time-series data are available with minimal time resolution 15- or 30-minute time-step or as aggregated values (hourly, daily, monthly, yearly sums).

SolarGIS software is not freeware, so the acquisition of a license is needed.

8-I.4 SSE-NASA

SSE (Surface meteorology and Solar Energy) software [7] is a renewable energy resource web site sponsored by NASA's Applied Science Program in the Science Mission Directorate and developed by POWER (Prediction of Worldwide Energy Resource Project). It offers over 200 satellite-derived meteorology and solar energy parameters in monthly resolution averaged from 22 years of data.

The meteorological data are on a 1 degree longitude by 1 degree latitude equal-angle grid covering the entire globe (64 800 regions). It offers a great group of meteorological data for different solar energy purposes, the most important and related to this chapter are:

- Solar irradiation parameters: insolation on horizontal surface global solar radiation (in average and percent difference minimum and maximum), diffuse radiation on horizontal surface, direct normal radiation, insolation at 3-hourly intervals, insolation clearness index, insolation normalized clearness index, clear sky insolation (cloud coverage < 10%), clear sky insolation clearness index (cloud coverage < 10%) and clear sky insolation normalized clearness index. Support information: top-of-atmosphere insolation and surface albedo.
- Solar Geometry: solar noon, daylight hours, daylight average of hourly cosine solar zenith angles, cosine solar zenith angle at mid-time between sunrise and solar noon, declination, sunset hour angle, maximum solar angle relative to the horizon, hourly solar angles relative to the horizon and hourly solar azimuth angles.
- Parameters for sizing the battery or other energy-storage systems: minimum available insolation as % of average values over consecutive-day period (1, 3, 7, 14, and 21 days), horizontal surface deficits below expected average values over consecutive-day period (1, 3, 7, 14, and 21 days) and equivalent number of no-sun days over consecutive-day period (1, 3, 7, 14, and 21 days).
- Cloud information: daylight cloud amount.
- Meteorology: air temperature, daily temperature range, dew/frost point temperature, wind speed at 50 m, percent of time for ranges of wind speed at 50 m, wind direction at 50 m,

Table 8-I.1 Regression analysis of SSE versus BSRN monthly averaged values for the time period July 1983 through June 2006

Parameter	Region	Bias (%)	RMS (%)
Horizontal Isolation	Global	−.01	10.25
	60° Poleward	−1.18	34.37
	60° Equatorward	0.29	8.71
Horizontal Diffuse Radiation	Global	7.49	29.34
	60° Poleward	11.29	54.14
	60° Equatorward	6.86	22.78
Direct Normal Radiation	Global	−4.06	22.73
	60° Poleward	−15.66	33.12
	60° Equatorward	2.40	20.93

wind speeds may be adjusted for different terrain by selecting from 17 vegetation types, relative humidity, humidity ratio, atmospheric pressure, total column precipitable water and precipitation.

The levels of uncertainty for different solar and meteorological parameters are analyzed through comparisons with ground site data on a global basis. Radiation parameters were compared with data from the Baseline Surface Radiation Network (BSRN), see Table 8-I.1. SSE-NASA software is freeware, so no license is needed.

8-I.5   Solar Anywhere

SolarAnywhere [8] is a web based service that provides hourly estimates of solar irradiance for locations within the continental US and Hawaii. Irradiance estimates are provided for years 1998 through the current day and seven day forecasts. The common uses of *SolarAnywhere* data are:

- identify optimal solar sites: locate the highest producing site for a residential, commercial, or utility-scale system;
- monitor solar power production: compare measured to simulated production without expensive on-site weather stations;
- forecast solar power production: manage system to provide reliable, low cost power and facilitate efficient market trading or meet PPA requirements;
- administer incentive programs: compare expected and actual solar production to identify sub-optimal installations or fraud;
- financial analysis: review payback periods, net-metering effects, capital investment returns, etc.

SolarAnywhere generates irradiance estimates using NOAA GOES visible satellite images. The hourly satellite images are processed using the most updated algorithms developed and maintained by the State University of New York at Albany (SUNY-Albany). The algorithm extracts cloud indices from the satellite's visible channel using a self-calibrating feedback process that is capable of adjusting for arbitrary ground surfaces. The cloud indices are used to modulate physically-based radiative transfer models describing localized clear sky climatology. More information on the Perez model can be found in [9–11] and on the SUNY Atmospheric Sciences Research Center website [12].

SolarAnywhere software is not freeware, so the acquisition of a license is needed.

8-I.6   3TIER

3TIER [13] brings an integrated suite of renewable energy assessment, forecasting, asset optimization, and measurement solutions oriented to utility scale solar (and wind) projects. The Solar Prospecting Tools are a powerful, web-based application with a simple point and click map for easily accessing renewable resource data. Both annual and monthly mean values, of global horizontal irradiance, direct normal irradiance and diffuse horizontal irradiance are available from a 3 km resolution dataset based on over a decade of satellite data.

3TIER software is not freeware, so the acquisition of a license is needed.

8-I.7   PVsyst

PVsyst [14,15], is one of the most popular tools for the modelling of PV and CPV power plants. It permits the design and development of any kind of photovoltaic installation at any place in the world. Originally, it was developed by the University of Geneva and is now developed by the company *PVsyst* S.A.

PVsyst is a PC software package for the study, sizing and data analysis of complete PV systems. It deals with grid-connected, stand-alone, pumping and DC-grid (public transport) PV systems, and includes extensive meteorological and PV systems components databases, as well as general solar energy tools. For the electrical model of the PV and CPV module the user can choose between the one-diode model by Duffie *et al.* [16] and Sandia's Array Performance Model [17]. Specifics features for CPV, such as the spectral sensitivity and the temperature dependence, are modeled by means of a de-rating factor on DNI referred to as 'Utilization Factor' [18]. *PVsyst* contains several additional models, for example for shading, cable, inverter, and transformer losses. It offers three levels of PV system study, roughly corresponding to the different stages in the development of real project:

- Preliminary design: this is the pre-sizing step of a project. In this mode the system yield evaluations are performed very quickly in monthly values, using only a very few general system characteristics or parameters, without specifying actual system components. A rough estimation of the system cost is also available.
- Project design: it aims to perform a thorough system design using detailed hourly simulations. In a first step, it assists in the design of the CPV array. In a second step, the user can specify more detailed parameters and analyze fine effects like thermal behaviour, wiring, module quality, mismatch and incidence angle losses, horizon (far shading), or partial shadings of near objects on the array, and so on. Results include several simulation variables, which may be displayed in monthly, daily or hourly values. The 'loss diagram' (Figure 8-I.1) is particularly useful to identify the weaknesses of the system design. A detailed economic evaluation can be performed using real component prices, any additional costs and investment conditions.

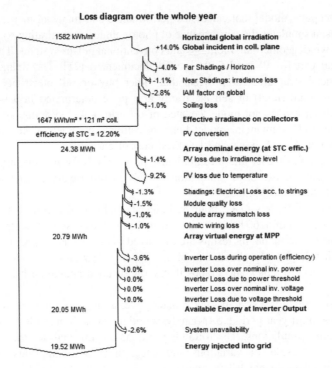

Loss diagram over the whole year

1582 kWh/m²	Horizontal global irradiation
+14.0%	Global incident in coll. plane
-4.0%	Far Shadings / Horizon
-1.1%	Near Shadings: irradiance loss
-2.8%	IAM factor on global
-1.0%	Soiling loss
1647 kWh/m² * 121 m² coll.	**Effective irradiance on collectors**
efficiency at STC = 12.20%	PV conversion
24.38 MWh	**Array nominal energy (at STC effic.)**
-1.4%	PV loss due to irradiance level
-9.2%	PV loss due to temperature
-1.3%	Shadings: Electrical Loss acc. to strings
-1.5%	Module quality loss
-1.0%	Module array mismatch loss
-1.0%	Ohmic wiring loss
20.79 MWh	**Array virtual energy at MPP**
-3.6%	Inverter Loss during operation (efficiency)
0.0%	Inverter Loss over nominal inv. power
0.0%	Inverter Loss due to power threshold
0.0%	Inverter Loss over nominal inv. voltage
0.0%	Inverter Loss due to voltage threshold
20.05 MWh	**Available Energy at Inverter Output**
-2.6%	System unavailability
19.52 MWh	**Energy injected into grid**

Figure 8-I.1 Loss diagram produced by *PVsyst* software for the case of a flat plate PV installation.

- Measured data analysis: when a PV system is running and carefully monitored, the tool permits the import of measured data, to display tables and graphs of the actual performances, and to perform close comparisons with the simulated variables.

PVsyst software is not freeware, so the acquisition of a license is needed.

8-I.8 SAM

System Advisor Model (*SAM*) [19,20] is one of the most popular tools for PV and CPV power plant modelling. It combines electrical with financial performance modeling of renewable energy projects. It was originally called the Solar Advisor Model and it was first developed by the National Renewable Energy Laboratory (NREL), Golden (CO) in collaboration with Sandia National Laboratories, Albuquerque (NM) in 2005 for internal use by the U.S. Department of Energy's Solar Energy Technologies Program. NREL released the first public version in August 2007 as Version 1, making it possible for solar energy professionals to analyze photovoltaic systems and concentrating solar power parabolic trough systems in the same modeling platform using consistent financial assumptions. Between 2007 and 2013, two new versions were released each year, adding new technologies and financing options. In 2010, the name changed to System Advisor Model to reflect the addition of non-solar technologies. Beginning in 2014, NREL releases one new version of the software each year, with periodic maintenance updates as needed.

SAM is a computer model that calculates performance and financial metrics of renewable energy systems; it simulates the performance of photovoltaic, concentrating solar power, solar water heating, wind, geothermal, biomass, and conventional power systems. The functionality of the electrical part for PV and CPV is well documented [21]. The financial model can represent financial structures for projects that either buy or sell electricity at retail rates (residential and commercial) or sell electricity at a price determined in a power purchase agreement (utility); it does not model isolated or off-grid power systems, and systems with electricity storage. This financial model helps to estimate commercial key figures such as internal rate of return, cash flows and levelized cost of energy.

SAM offers a graphical user interface but it also provides a software development kit (*SDK*) that makes possible to use *SAM* simulation models in proprietary applications written in C/C++, Java, Python and Matlab.

SAM performance models make hour-by-hour calculations of a power system's electric output, generating a set of 8760 hourly values that represent the system's electricity production over a single year. The modeling results are displayed in tables and graphs, ranging from the metrics table that displays levelized cost of energy, first year annual production, and other single-value metrics, to tables and graphs that show detailed annual cash flows and hourly performance data.

In addition to simulating a system's performance over a single year and calculating a project cash flow over a multi-year period, *SAM* analysis options make it possible to conduct studies involving multiple simulations. The following options are for analyses that investigate the impacts on model results of variations and uncertainty in assumptions about weather, performance, cost, and financial parameters:

- Parametric analysis: assign multiple values to input variables to create graphs and tables showing the value of output metrics for each value of the input variable. Useful for optimization and exploring relationships between input variables and results.
- Sensitivity analysis: create tornado graphs by specifying a range of values for input variables as a percentage of a base value.
- Statistical analysis: create histograms showing the sensitivity of output metrics to variations in input values.
- Probability of exceedance analysis (P50/P90): for locations with weather data available for many years, calculate the probability that the system's total annual output will exceed a certain value.

SAM software is freeware, so no license is needed.

8-I.9 PV_LIB Toolbox

PV_LIB toolbox [22] is managed by the PV Performance Modeling Collaborative (PVPMC), an industry and National Laboratory collaboration to improve photovoltaic performance modeling. Sandia National Laboratories is facilitating a collaborative group of photovoltaic professionals (PVPMC) interested in improving the accuracy and technical rigor of PV performance models and analyses. Such models are used to evaluate current performance (performance ratio) and determine the future value of PV generation projects (expressed as the predicted energy yield) and, by extension, influence how PV projects and technologies are perceived by the financial community in terms of investment risk. The *PV_LIB toolbox* consists of a collection of algorithms and tools implemented in Matlab and Python, for example atmospheric and irradiance translation functions, modules and inverter models and data handling functions. As the toolbox does not contain a graphical user interface it mainly targets engineers and modelers with programming skills. Full transparency of the modeling process is achieved as the source code is disclosed. Users are encouraged to extend the algorithms and add new software functions, which can be easily shared with the community through a Github repository [23].

References

1. http://Meteonorm.com
2. http://re.jrc.ec.europa.eu/PVGIS
3. Súri, M., Huld, T.A., Dunlop, E.D. PVGIS: a web-based solar radiation database for the calculation of PV potential in Europe. *International Journal of Sustainable Energy*, **24**(2), 55–67. 2005.
4. Súri, M., Huld, T.A., Dunlop, E.D. Ossenbrink, H.A. Potential of solar electricity generation in the European Union member states and candidate countries. *Solar Energy*, **81**, 1295–1305 2007. http://re.jrc.ec.europa.eu/pvgis/
5. Huld, T., Müller, R., Gambardella, A. A new solar radiation database for estimating PV performance in Europe and Africa. *Solar Energy*, **86**, 1803–1815. 2012.
6. http://SolarGIS.info/
7. https://eosweb.larc.nasa.gov/cgi-bin/sse/sse
8. https://www.SolarAnywhere.com
9. Perez, R., Ineichen, P., Moore, K. *et al.* A new operational satellite-to-irradiance model. *Solar Energy*, **73**(5), 307–317. 2002.
10. Perez, R., Ineichen, P., Kmiecik, M. *et al.* Producing satellite-derived irradiances in complex arid terrain. *Solar Energy*, **77**(4), 363–370. 2004.
11. Perez, R., Schlemmer, J., Renne, D. *et al.* Validation of the SUNY satellite model in a Meteosat environment. Proceedings ASES Annual Conference, Buffalo, New York. 2009.
12. http://asrc.albany.edu/people/faculty/perez/directory/ResourceAssessment.html
13. http://www.3TIER.com/en/
14. http://www.PVsyst.com/en/
15. A. Meremoud, B. Wittmer. PVsyst user's manual. January 2014.
16. J. A. Duffie, W. A. Beckman. *Solar Engineering of Thermal Processes*, 4th edn, Wiley, 2013.
17. D. L. King, W. E. Boyson, J. A. Kratochvil. Photovoltaic Array Performance Model. Sandia Report SAND2004-3535, Sandia National Laboratories, Albuquerque NM, USA. 2004.
18. T. Gerstmaier, S. van Riesen, A. Gombert, *et al.* Software modeling of FLATCON CPV systems. AIP Conference Proceedings, 1277, 183. 2010. DOI: 10.1063/1.3509185
19. https://sam.nrel.gov/
20. N. Blair, A. P. Dobos, J. Freeman, *et al.* System Advisor Model, SAM 2014. 1.14: general description. Technical Report NREL/TP-6A20-61019, National Renewable Energy Laboratory, February 2014.
21. P. Gilman, SAM Photovoltaic Model (pvsamv1) technical reference. NREL Draft Report. National Renewable Energy Laboratory, 2014.
22. https://pvpmc.sandia.gov
23. https://github.com/Sandia-Labs/PVLIB_Python

8-II Annex

CPV Power Plants at ISFOC

María Martínez,[1] Daniel Sánchez,[1] Óscar de la Rubia,[1] and Francisca Rubio[2]

[1]*Instituto de Sistemas Fotovoltaicos de Concentración (ISFOC), Spain*
[2]*Soitec, Germany*

8-II.1 Introduction

The Institute of Concentration Photovoltaic Systems (ISFOC) was established in 2006 in Puertollano (Spain), following a R&D program promoted by the regional Government of Castilla la Mancha and the Solar Energy Institute of the Technical University of Madrid (IES-UPM) [1].

ISFOC is a R&D center focused on CPV with the objective of fostering its industrialization. With this purpose, the aim of the ISFOC project was to install CPV power plants (with a minimum geometric concentration ratio of 200×) connected to the grid from a range of available and developing technologies and in various locations. The objective was to test CPV technologies and assist the manufacturers in the development of the product; moreover, the information obtained during the whole cycle of the project has demonstrated to be very valuable for helping in the development of specific standards for CPV.

ISFOC LA NAVA was the first installation connected to the grid by ISFOC. It has $800\,kW_p$ nominal power with three different CPV technologies. It was connected to the grid in September 2008 and has been in operation since then; offering very valuable information about CPV field performance. Therefore, most of the information presented in this case study was obtained from this installation that has provided a wealth of knowledge about CPV operation and performance.

8-II.2 Choice of Sites and Technologies

At the beginning of the ISFOC project, the objective was to install five different technologies in the five different provinces of Castilla la Mancha. Nevertheless, it was very soon detected that the permitting and paperwork required to carry out five installations in five different sites was too challenging and long. So, finally, only three installations, located in two sites, were

Handbook of Concentrator Photovoltaic Technology, First Edition. Edited by Carlos Algora and Ignacio Rey-Stolle.
© 2016 John Wiley & Sons, Ltd. Published 2016 by John Wiley & Sons, Ltd.

completed: one in Almoguera (Guadalajara) and two in Puertollano (Ciudad Real); this latter site being also where ISFOC headquarters are located.

Because ISFOC is a State-owned company, the choice of the technologies to be installed should be open and transparent, so two different public Call for Tenders were launched in 2006 and 2007, respectively. For the choice of the technologies ISFOC was supported by an International Scientific Advisory Committee who judged the proposals received. The criteria of choice were based on a scientific analysis of the technologies, including the capacity of manufacturing a reliable product in time, the expected price and the potential for decreasing costs with the experience obtained.

Furthermore, because of the novelty of CPV technology, no long-term energy guarantee was introduced in the contract, but some specific milestones were defined to ensure the reliability of the product prior to making the final commissioning.

1. First, related to CPV modules, when the first Call for Tenders was launched, there was no official standard to qualify their design. However, the IEC Committee was working in the development of IEC62108 *Concentrator photovoltaic (CPV) modules and assemblies – Design qualification and type approval* [2] that was finally approved in 2007. Therefore, the first milestone to overcome by all the technologies selected was to obtain the qualification of their product following this standard. With this pre-qualification, the technologies installed at ISFOC were the first in the world to be tested under IEC 62108 [3,4]. As a result of this, a first outcome from ISFOC was to prove the benefits of this standard, because practically all manufacturers made improvements in their design to comply with the IEC tests. In general, the most critical and useful tests were the damp heat test to ensure the electrical insulation and the thermal cycling for testing the heat dissipation capacity. Finally, after the experience obtained, ISFOC advises that perhaps some new tests should be included like dust deposition and sand abrasion or salty-fog to ensure that this technology can withstand in those types of climates that could be potential CPV installation areas like deserts or coastal regions.

2. The second technical milestone was the characterization of a prototype in order to test its operation and performance. Not only the nominal power of the prototype was characterized to ensure it was within the limits provided by the manufacturer but also a verification of the operation of the prototype was carried out. This verification included: the automatic tracking function and its influence on energy production, the manual mode including the end of run detection, operation after power cut-off and wind stow mode. All the technologies installed in the ISFOC project overcame successfully the prototype verification prior to field installation and very useful information was obtained for the manufacturers in order to optimize the proper operation of the power plants.

In the first Call for Tenders in 2006, six companies participated, but only three were chosen by the International Scientific Advisory Committee: 1) *Concentrix Solar* (Germany), now *Soitec Solar* (France); 2) *Isofotón* (Spain); and 3) *Solfocus* (USA). In the second Call for Tenders in 2007, another four CPV companies were selected for the second phase: 1) *Arima Eco* (Taiwan); 2) *Emcore* (USA); now *Suncore* (China); 3) *Concentración Solar la Mancha*, later *Renovalia CPV* (Spain); and 4) *Sol 3G*, now *Abengoa Solar* (Spain).

Additionally, ISFOC installed new prototypes from several CPV companies in their facilities like *Daido Steel* (Japan), *Semprius* (USA) and *Isofotón* (Spain). With all these systems, ISFOC can analyze both their long-term performance and the behavior of the newer prototypes with higher efficiency.

8-II.3 Permitting and Basic Engineering Study

The reason to choose the site of La Nava (Puertollano, Ciudad Real) for the first installation was that the regional Government donated the plot to ISFOC for the installation of the $800\,kW_p$ plant and the building of its headquarters in Puertollano. Even with this step already clarified, the rest of paperwork, permitting, etc. was a long and difficult process.

The basic engineering documents were done by the ISFOC team and the shading study was carried out by the University of Jaen (Spain) [5,6] to make the distribution as *fair* as possible for the three different manufacturers. Chapter 8 describes the procedure developed by the University of Jaen for making the shading study.

Figure 8-II.1 shows at the left the distribution proposed by the University of Jaen after the shading analysis and at the right the actual distribution. The proposed layout was modified by the manufacturers and ISFOC to optimize the distribution of each power plant and improve the accessibility to the power plant with the main roads showed in the plan.

As a standard reference, to minimize the shading losses, the trackers were distributed along a north-south axis, with an inter-tracker spacing enough to avoid any shade during two hours before and after solar midday on the day when the sun is at the lower elevation in the horizon.

8-II.4 Engineering

The installation of ISFOC La Nava consists of eight photovoltaic plants of $100\,kW_p$ each totally independent but connected to a common transformer substation for the evacuation of the energy. The reason for designing several $100\,kW_p$ plants was strictly administrative since in Spain by those days the largest subsidy for PV electricity was granted for power stations of $100\,kW_p$ or less inverter nominal power but finally, because of technical and administrative reasons, the legalization of the installation was carried out as a unique power plant of $800\,kW_p$ with only one transformer substation instead of eight installations of $100\,kW_p$. The distribution between the manufacturers is: $200\,kW_p$ for each, *Concentrix Solar* and *SolFocus*, and $400\,kW_p$ for *Isofotón*.

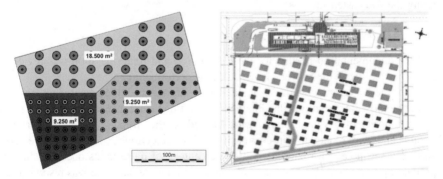

Figure 8-II.1 ISFOC La Nava power plant layout. Left: distribution proposed by the University of Jaen after the shadowing study. Right: actual distribution

Each $100\,kW_p$ has available an energy meter for the assessment of the energy production and there is one meteorological station at the site that completes the available data to make detailed studies of the performance of the plant, studying the relationship between the energy production and the available solar resource and ambient conditions.

The installation of La Nava was one of the first CPV power plants connected to the grid and was built with the objective of testing different CPV technologies as well as different electrical configurations. The three electrical configurations available in ISFOC La Nava CPV plant are:

- 3-phase centralized inverter installed in a building;
- 3-phase inverter per tracker installed at the pole;
- 1-phase inverter per tracker installed at the pole.

Chapter 8 discussed the different configuration options for inverters and thus here only the main results obtained in the analysis carried out at ISFOC installation are presented [7].

The use of distributed inverters reduces the DC cabling between modules and strings, because usually the inverter is installed in the pole of the solar tracker, so the greater part of the electrical installation is AC wiring. In the case of using a centralized inverter – installed inside a building –, the greater part of the electrical installation is in DC, from the concentrators to the inverters, and the AC cabling is reduced to only the connection to the substation and/or to the utility grid.

The AC wiring efficiency is very close to 100% in both cases whilst the DC cable losses are significant in centralized inverters topologies but not in distributed inverters. The higher losses for any configuration are due to a bad inverter performance under real operating conditions. Finally, also, the mismatch between concentrators introduces losses in the case of using centralized inverters so it is important to analyze in detail the most suitable configuration for each installation and technology.

8-II.5 Commissioning and Rating

The procedure followed for the commissioning and acceptance of ISFOC's CPV plants is as described in Chapter 8, but here a detailed description of the main results and more challenging works is given. The objective was to verify that the installations met the requirements and specifications defined in the contracts and to ensure that they would operate as expected, generating the forecasted incomes.

8-II.5.1 Inspection

First, all the documentation needed for the legalization and final operation was reviewed to guarantee the quality of the power plants and their performance.

In a second step, a technical verification of all the components of the power plant was carried out. Next, the most important results obtained during this visual inspection [8,9] are summarized. The majority of the faults found were located on the structure of the tracker and the tracker table and were related to anti-corrosion metallic protections: minor rusts were detected as well as tracker table deformations. A high percentage of these faults occurred during the transportation process and the construction works. Another element that was responsible of a lot of faults was the drive mechanisms, especially the elevation drive. In the cases where faults were found in the wiring and protections, most of them are related to risk warnings stickers and mixed communication wires with power wire. Finally, the foundation was the element where fewer defects were found. It is important to remark that all the defects detected could be solved before continuing with the commissioning.

Finally, also the monitoring system in charge of controlling the plant was tested [10], the SCADA software was evaluated and the communications within all the components were tested and also the data acquisition system was checked.

8-II.5.2 Rating

The last step of the commissioning and acceptance procedure was the determination of the nominal power of the power plant. In Chapter 8 different approaches were discussed. At of 2007, when ISFOC was launched, there was not a consensus on CPV rating procedures so the ISFOC project defined its own approach in the Call for Tenders. Such CPV rating procedure will be commented on below and has been widely discussed in the literature [3,4,11–14].

First, the standard test conditions used in the ISFOC project have to be clarified. Nowadays, the standard IEC 62670-1:2013 [15]. Part 1 *Photovoltaic concentrators (CPV) – Performance Testing – Standard Conditions* establishes the standard conditions for assessing the power produced by CPV systems and their subcomponents, namely, CSTC and CSOC (see Chapter 8 for details). But in 2007 this standard did not exist, and the standard conditions were defined in the Call for Tenders as follows:

- Direct solar irradiance of 850 W/m².
- Equivalent operating cell temperature of 60°C.

This means that the CPV installations made under ISFOC project should be of $100 \, kW_p$ measured under these particular conditions.

The objective of ISFOC's rating method is to determine the nominal power at those standard test conditions of each CPV power plant in the point at which connection to the grid occurs. This means not only determining the nominal power in DC generated by the CPV system but including the average energy losses of the complete installation because of its normal operation. Therefore, the rating procedure used by ISFOC is divided in two steps:

- DC rating. The determination of the DC power generated by the power plant obtained from the I-V curves measured at some operating conditions and translated to standard test conditions.
- AC rating. The determination of the AC power of the complete plant by the correction of the previous value with the real losses measured during the acceptance period.

8-II.5.2.1 DC Rating

The objective of this first step is to determine the potential of conversion of the CPV technology installed. The procedure defines that a set of CPV systems in the plant is randomly selected for their characterization, being finally the DC power of the complete plant the average value obtained from the systems measured. In general, for most of the cases, ISFOC performed the DC rating of all the CPV systems in each plant to increase the wealth of data available, following its R&D aim, and also because that data will be the base of future performance and degradation analyses.

The characterization of each CPV system is carried out by measuring the I-V curve during at least 30 minutes to ensure thermal stabilization. The procedure establishes that the DNI should be monitored during the measurements with two independent pyrheliometers mounted on two independent trackers. Also the heat-sink temperature is measured on several modules with thermal probes installed at the back plate of the module, right behind the cells. Finally, the procedure establishes that the ambient temperature and the wind speed and direction are monitored to complete the characterization with some ambient conditions, though these variables are not included in the translation calculations.

To ensure stable and repetitive conditions during the measurements, some ambient requirements were defined in the procedure. The DNI should be higher than $700 \, W/m^2$, guaranteeing clear sky conditions with no clouds around the sun when making the measurements. Also, the wind speed while making the measurements should be lower than 3.3 m/s, to ensure thermal stabilization.

Then, from all the measurements, a period of five minutes is selected for making the translation using Eqs. (8-II.1) to (8-II.3), which were previously published in the Call for Tenders. This translation method is based on the Shockley model adapted to multi-junction (J being the number of junctions) solar cells and being the equivalent operating cell temperature calculated by measuring the heat sink temperature:

$$T_{cell} = T_{h-s} + DNI \cdot R_{th,sys}, \tag{8-II.1}$$

$$I_{STC} = I \cdot \frac{DNI_{STC}}{DNI}, \tag{8-II.2}$$

$$V_{STC} = V + N \cdot \frac{0.0257 \cdot \left(T_{cell,STC} - T_{cell}\right)}{297} \cdot \ln \left(\frac{\prod\limits_{j}^{J} \left(I_{Lj} - I\right)}{\prod\limits_{j}^{J} I_{Lj}} \right),$$

$$+ \left[N \cdot \left(\sum\limits_{j}^{J} E_{gj} \right) - V_{OC} \right] \cdot \left(1 - \frac{T_{cell,STC}}{T_{cell}} \right) \tag{8-II.3}$$

Equation (8-II.1) is used to calculate the equivalent cell temperature while Eqs. (8-II.2) and (8-II.3) to make the translation of the current and voltage of the I-V curves measured.

Table 8-II.1 collects the meaning of all the symbols used in the equations. The parameter internal thermal resistance ($R_{th,sys}$) is provided by the manufacturer for the complete system. It is defined like the thermal drop between the cell and the module heat-sink but using DNI as input instead of the heat flow. The equivalent short-circuit current for each junction (I_{Lj}) is calculated by using the short-circuit current measured and applying a constant factor of proportionality assumed for the photo-current generated by each junction.

So, finally the DC power at standard test conditions of a CPV system is the average value obtained of the translation of the measurements of the five-minute period selected.

8-II.5.2.2 AC Rating

The objective was to estimate what the energy losses of the plant during normal operation would be. Therefore, the final power value – i.e. AC power in standard test conditions –, allows

Table 8-II.1 Meaning of the symbols used in eqs. (8.II-1) to (8.II-3)

Parameter	Units	Definition
DNI	W/m^2	Measured direct normal irradiance at operating conditions
DNI_{STC}	W/m^2	Direct normal irradiance at standard test conditions (850 W/m^2)
I	A	Measured current
I_{STC}	A	Current at standard test conditions
V	V	Measured voltage
V_{STC}	V	Voltage at standard test conditions
T_{cell}	K	Calculated equivalent cell temperature under operating conditions
$T_{cell,STC}$	K	Equivalent cell temperature under standard test conditions (60°C)
T_{h-s}	K	Measured heat-sink temperature in operating conditions
$R_{th,sys}$	K/(W/m^2)	Internal thermal resistance
N		Number of cells connected in series in the system
I_{Lj}	A	Measured equivalent short-circuit current for each junction at operating conditions
E_{gj}	eV	Band gap energy for each junction
J		Number of junctions of the multijunction solar cells
V_{OC}	V	Measured open-circuit voltage at operating conditions

the calculation of the real net energy fed to the grid that directly generates the incomes of the installation. This AC rating is carried out by assessing the energy generated by the plant during the acceptance period.

The procedure defines that the AC power in standard test conditions ($P_{AC,plant}$) is calculated as defined in Eq. (8-II.4). The DC power value ($P_{DC,plant}$) obtained in the DC rating is corrected with the inverter efficiency (η_{inv}) and a correction factor (F) that represents the real performance of the plant:

$$P_{AC,plant} = P_{DC,plant} \cdot \eta_{inv} \cdot F, \qquad (8\text{-}II.4)$$

The result of the assessment of the energy generation during the acceptance period is the correction factor (F) that it is calculated like the comparison between the real energy generated by the plant and theoretical energy that should have generated; see Eq. (8-II.5). During the acceptance period, the real energy generated by the plant is registered together with the operating conditions (DNI and heat-sink temperature for several systems). The theoretical energy is calculated as the integral, over the acceptance period, of the theoretical power of the plant in operating conditions; this power is obtained from the DC power at standard test conditions translated to the registered operating conditions using again Eqs. (8-II.1) to (8-II.3):

$$F = \frac{E_{AC}}{E_{AC,th}} = \frac{E_{AC}}{\int P_{AC,th}dt} = \frac{E_{AC}}{\eta_{inv}\int P_{DC,th}dt}, \qquad (8\text{-}II.5)$$

The factor ($1\text{-}F$) represents the actual losses of the plant, namely, shadowing, wiring, mismatch between modules and systems, inverter actual performance, tracking losses along the day, actual performance of the systems in varying ambient conditions, any malfunction occurring during the acceptance period. However, this factor doesn't differentiate the source of the losses, so it cannot be used to identify the influence of each loss in the energy generation.

Particularly, during the acceptance of ISFOC plants, some limitations were established in the Call for Tenders for making the final selection of the acceptance period. The reason is that the objective was only to analyze the performance of the CPV plants when they are operating near the optimum conditions:

- DNI should be higher than 600 W/m². All the data collected with low DNI were rejected, like, cloudy days and hours near dawn and dusk. Usually these periods are when spectrum is farther from reference conditions and causes a big influence in the performance of the concentrators.
- Wind speed lower than 3.3 m/s. Fast changes in the thermal behavior of the systems were also rejected.
- Shading periods. Because the area for the installation was defined by ISFOC and not by the suppliers no shading losses were introduced in the power determination.
- Any malfunction. Because ISFOC power plants were demo power plants, no malfunctions were taken into account during the acceptance. In general, the malfunction of the power plant should be recorded as unavailability.

Therefore, for ISFOC power plants, considering the abovementioned limitations, $(1\text{-}F)$ represents only the losses due to wiring, mismatch between modules and systems, inverter actual performance, tracking losses along the day, and actual performance of the systems in varying ambient conditions.

8-II.5.2.3 Rating Results

ISFOC made the rating of all the power plants installed in the project. The results have been previously presented [12,13,14] and here only the most important issues obtained are highlighted and some graphs updated with the latest results.

First, during the DC rating of each 100 kW$_p$ the power of all the systems were evaluated to verify that they met the requirements; the DC power of the concentrators should reach at least the 90% of the expected value. Figure 8-II.2 is a representation of the results obtained in the DC rating of a complete power plant. The nominal power in standard test conditions obtained for

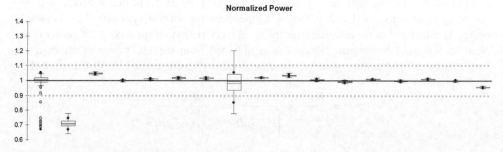

Figure 8-II.2   100 kW$_p$ ISFOC power plant DC rating results. Source: Martinez, 2008. Reproduced with permission of Elsevier

each concentrator is represented like a normalized value of the expected value (blue line is the expected value), yellow lines are the $\pm 10\%$ variation permitted in the value. Finally, also the dispersion obtained in the calculation of each concentrator is plotted.

This graph permits the analysis of the rating process, both the application of the procedure in the field and the results obtained. In the particular case shown in the graph there are a couple of anomalous data. The first one is a system with a nominal power lower than the 90% expected; the process was reviewed and it was found out that during the measurements the concentrator was not tracking correctly and the modules were not accurately aligned to the sun. The second one is a system whose associated dispersion is very high, around a 20%; once the results were checked it was observed that the DNI was not stable during the measurements. Both measurements were repeated in optimum conditions for calculating the final DC power of the plant.

Figure 8-II.3 shows a Pareto chart representing the distribution of the nominal power obtained for the systems characterized by ISFOC at the end of 2010, accounting for more than $1MW_p$ distributed in the three installations. The nominal power has been normalized with the expected value stated by the manufacturer for comparing the different technologies.

The most important conclusion obtained is that for 80% of the concentrators, the DC nominal power obtained is higher than the 95% of the expected value and that for more than 50% is higher than the expected value. This means that in general terms the installations were performing as expected and the DC rating was passed by all the manufacturers.

Finally, also the AC rating results were analyzed. The key parameter for calculating the AC power of the plants is the correction factor that represents the performance of the plant in

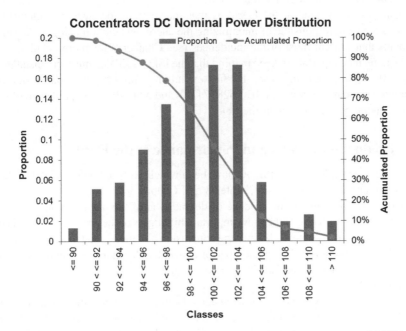

Figure 8-II.3 Distribution of the DC nominal power obtained for the concentrators (1 MW$_p$) adapted from [14] with updated values. Source: Adapted from Martinez, 2009 [14]

normal operation. The average value obtained for all the plants accepted in AC at the end of 2010 was of 95.8%, being the AC/DC ratio of 93%; what means that the energy losses were just 7%.

It is important to remark that these were the first large scale projects for most of the manufacturers involved, being the manufacturing process in low level of automation and without an exhaustive manufacturing process control, what makes more significant the results: This demonstrated that CPV was ready for its industrialization and how ISFOC project pushed it through the lessons learned during the whole cycle.

8-II.6   Power Plant Monitoring and Operation and Maintenance

For the power plant monitoring, each manufacturer developed their own SCADA system to control the plant, and to carry out the data acquisition as well. However, for the in-depth analysis of the data and for the exploitation of the results, ISFOC created a proprietary system called *GoCPV* (Gestion y Optimización de CPV) [16] (www.gocpv.net). This system central-izes the data of all the installations, including all CPV systems and meteorological stations at each site, what enables an easy access to production data and ambient variables facilitating the comparison of the different installations.

The operation and maintenance (O&M) is carried out by ISFOC team and the results have been presented in different conferences [16,17] and in Chapter 8 and Chapter 9. Therefore, no further results are presented here; and refer to those chapters for further details.

With the analysis that ISFOC carries out for the O&M, not only O&M issues are solved but also paves the way for a deeper analysis in order to improve designs in key elements, solving as well the root cause of the malfunctions issues.

The main result obtained at the O&M is that trackers are the 'weakest' element of the installation. They are responsible of maximizing the energy generation being at the same time the elements that present most of the incidences (see Chapter 9 for more details).

Furthermore, it is important to keep in mind that the ISFOC CPV plants were installed during 2007 and 2008, being, in many cases, the first evolution of prototypes. The new qualification and validation standards of trackers (IEC 62817 [18]) and other BOS components will also help reduce the operational problems in the field.

8-II.7   Production Results and Performance of the Plant

The results and lesson learned in ISFOC have been widely diffused before [3,4,13,14,16] and some of them are also presented in Chapter 8 and Chapter 9.

The good performance and the absence of degradation of the mature technologies for the long term performance is one of the most important highlights of the project, the details of these results can be found in Chapter 9. Mainly the conclusion obtained is that there is no visible sign of strong degradation in the CPV plants' performance, the diminution observed in the energy generation is up to now of around a 0.7% per year, which is a similar value to those obtained for PV modules [19].

On the other side, very important lessons have been learned since the installation of these first CPV power plants. In particular, a deeper understanding of all the elements involved in the life cycle of CPV systems (shown in Figure 8-II.4) has been a key outcome. The meaning of this figure is that to build a commercial CPV power plant it is not only the design that is

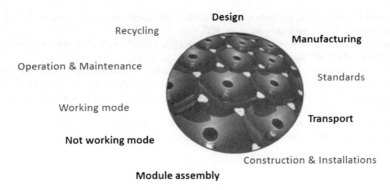

Design

Recycling

Manufacturing

Operation & Maintenance

Standards

Working mode

Transport

Not working mode

Construction & Installations

Module assembly

Figure 8-II.4 Life cycle of the CPV systems as one of the first lessons learned from ISFOC. Source: Martinez, 2008. Reproduced with permission of Elsevier

important, but also the manufacturing, the qualification following the prescribed standards, the construction and installation, and the Operation and Maintenance strategy, etc.

References

1. F. Rubio, P. Banda, J.L. Pachón and O. Hofmann. Establishment of the Institute of Concentration Photovoltaics Systems – ISFOC. Proceedings of 22nd PVSEC, Milan. 2007.
2. International Electrotechnical Commission. IEC 62108 ed1.0 Concentrator photovoltaic (CPV) modules and assemblies - Design qualification and type approval. IEC, Geneva.
3. F. Rubio, M. Martinez, P. Banda. Concentrator photovoltaic field installations, in *Solar Cells and their Applications*, Second edn, (eds L. Fraas and L. Partain), John Wiley and Sons, Inc., Hoboken, NJ. 2010. USA. DOI: 10.1002/9780470636886.ch17
4. F. Rubio, M. Martinez, R. Coronado, J.L. Pachón, P. Banda. Deploying CPV power plants – ISFOC experiences. Proceedings 33rd IEEE PVSC, San Diego. 2008.
5. P. J. Perez, G. Almonacid, P.G. Vidal. Estimation of shading losses in multi-trackers PV systems. Proceedings 22nd EU PVSEC: pp. 2295–2298, Milano, Italy. 2007.
6. P. J. Pérez, G. Almonacid, J. Aguilera *et al.* Multitracker systems calculation of losses due to selfshadowing. Proceedings ICSC–4, El Escorial, Spain. 2007.
7. C. Alamillo, O. de la Rubia, E. Gil, *et al.* Analysis of inverter configuration on CPV plants. Proceedings 25th EU PVSEC/WCPEC-5, pp. 4733–4736, Valencia, Spain. 2010.
8. O. De la Rubia, D. Sánchez, M. L. García, *et al.* Acceptance procedure applied to ISFOC's CPV plants. Proceedings of the 5th International Conference on Concentrating Photovoltaic Systems ICSC-5, Palm Desert, USA. 2008.
9. E. Gil, D. Sánchez, O. de la Rubia, *et al.* Field technical inspection of CPV power plants. *Proceedings 25th EU PVSEC/WCPEC-5*, pp. 4483–4486, Valencia, 2010.
10. M. L. García, G. Calvo-Parra, A. Hipólito, J. L. Pachón. Monitoring, communications and data processing of CPV plants. Proceedings CPV-6, Freiburg, 2010. *AIP Conference Proceedings*, 1277, 273. 2010.
11. M. Martinez, O. de la Rubia, F. Rubio, P. Banda. Concentration photovoltaics. *Comprehensive Renewable Energy,* 1: Photovoltaic Solar Energy, pp. 745–765, 2012. DOI: 10.1016/B978-0-08-087872-0.00141-4
12. M. Martínez, O. de la Rubia, D. Sánchez, *et al.* Concentrator photovoltaics connected to the grid and systems rating. *Proceedings 23rd EU PVSEC*, pp. 146–150, Valencia, Spain. 2008.
13. M. Martínez, D. Sánchez, F. Rubio, J.L. Pachón, P. Banda. CPV systems rating, results and lessons learned at ISFOC. Proceedings ICSC–5, Palm Desert, USA. 2008.
14. M. Martínez, D. Sánchez, J. Perea, F. Rubio, P. Banda. ISFOC demonstration plants: Rating and production data analysis. *Proceedings 24th EU PVSEC*, pp. 159–164, Hamburg, Germany. 2009.

15. International Electrotechnical Commission. IEC 62670-1:2013 ed1.0 Photovoltaic concentrators (CPV) - Performance testing - Part 1: Standard conditions. IEC, Geneva.
16. M. Martínez, F. Rubio, G. Sala, *et al*. CPV plants data analysis. ISFOC And NACIR projects results. Proceedings CPV-8, Toledo, 2012. *AIP Conference Proceedings*, 1477, 323. 2012.
17. D. Sánchez, M. Martínez, E. Gil, F. Rubio, J.L. Pachón, P. Banda. First experiences of ISFOC in the maintenance of CPV plants. Proceedings CPV-6, Freiburg, 2010. *AIP Conference Proceedings*, 1277, 248. 2010.
18. International Electrotechnical Commission. IEC 62817:2014 ed1.0 Photovoltaic systems - Design qualification of solar trackers. IEC, Geneva.
19. E. D. Dunlop. Lifetime performance of crystalline silicon PV modules. *Proceedings of the 3rd World Conference on Photovoltaic Energy Conversion*, Osaka Japan, pp. 2927–2930. 2003.

8-III Annex

Soitec Power Plants

Andreas Gombert, Norman Abela, Tobias Gerstmeier, Shelley Bambrook, and Francisca Rubio
Soitec, Germany

8-III.1 Introduction

Soitec Solar GmbH started 2005 as a spin off from the Fraunhofer Institute for Solar Energy Systems ISE and developed the Concentrix™-Technologies to industrial mass production in order to gain a significant share of the PV market. The largest project size realized with this technology to date is 44 MW_p DC with CX-S530 systems in Touwsrivier (see Figure 8-III.1), South Africa [1].

8-III.2 Description of a CPV Power Plant

A CPV power plant with Concentrix Technology consists essentially of a group of CPV Systems cabled together in an electrical building block. Depending on the rating of the power plant, a number of building blocks are installed together and through inverters and transformers, the voltage and frequency of the power plant is configured to match the grid parameters at the grid connection point.

The basic element of the power plant is therefore the CPV system. Various systems exist that utilize the Concentrix Technology. An example of one such system: the CX-S530-II is shown in Figure 8-III.2.

8-III.3 The Site

The main criteria for selecting a suitable region for a CPV installation should be, of course, a high DNI level (higher than 2300 kWh/m^2/year) and a DNI/GNI ratio higher than 0.85, as specified in Chapter 8. But this is not always the case and there are numerous other reasons for developing a project on a specific site.

Political or legal factors are among the main reasons to choose one region. Specific feed-in tariffs for CPV or favorable PPAs (Power Purchase Agreements) for new technologies are some examples.

Handbook of Concentrator Photovoltaic Technology, First Edition. Edited by Carlos Algora and Ignacio Rey-Stolle.
© 2016 John Wiley & Sons, Ltd. Published 2016 by John Wiley & Sons, Ltd.

Figure 8-III.1 Touwsrivier Power Plant, South Africa

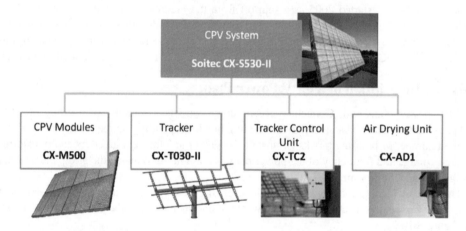

Figure 8-III.2 Example of the components of a Soitec CPV System

Other motivating factors could be economic, research, new development, new market strategy, or electrification of remote areas. In any case, once the main region is chosen, the project development should follow the same processes as described in Chapter 8, with a meteorological data study including DNI, wind, and rain, and an analysis of geographical and technical data, for successful execution and operation.

8-III.4   The System

The development of the standard CPV system is completed before the deployment of a power plant. However, following the choice of the site, it is possible that some adaptations are

Figure 8-III.3   Two systems developed by Soitec. On the left there is a small system with $26\,m^2$ and a height of less than 4.5 m and on the right a large system with $108\,m^2$ and a height of 8 m

required to meet the local regulations, or even a new system may need to be developed. This may be necessary, for example, in regions with certain height restrictions. Therefore, Soitec has systems of different sizes as shown in Figure 8-III.3, to be able to install in as many regions as possible.

Other adaptations are almost always needed due to difference in the legislation (e.g. electrical safety, cable sizing), national building standards (e.g. Eurocode, US building code) and especially due to site dependent requirements (flood level, earthquake area, humidity or corrosion level, etc.).

The labor cost and the availability of heavy installation equipment in the specific region could also strongly influence the type of system chosen. For example, expensive labor cost and readily available heavy equipment in USA is favorable for large systems. Low labor cost and/or unavailability of heavy equipment as in China, for example, could lead to a preference for small systems.

8-III.5   Layout of the Systems

Essentially the power plant is made up of a number of standard building blocks, which are repeatedly arranged within the site boundaries. The repetition of standardized building blocks, if possible, is highly recommended to reduce cost and to ease the installation. The optimum layout of the building block can be determined using *PVsyst* software (see Annex 8-I) once the site and the CPV system is selected.

The first step in specifying the layout is to identify the optimal distances between the CPV systems in north–south, and east–west directions, in order to minimize losses due to shading. Shading losses grow with decreasing distance between the CPV Systems. Consequently, the layout is determined by a trade-off between land-use, cost of land and energy production yield loss due to shading.

The standard building block as shown in Figure 8.1 of Chapter 8 is determined by the *PVsyst* analysis. This includes the size of the inverter according to given AC/DC ratios and cutout parameters. The layout consists of a number of CPV Systems arranged in rows and columns according to a calculated distance between trackers.

The power generated by the CPV systems is collected in DC combiner boxes via DC cables and is then fed to a central inverter. The tracker however also requires power to operate the azimuth and elevation motors. In the CX-S530-II system, the motors are supplied by three-phase AC. A tracker control unit (TCU) fixed to the mast of the tracker is connected to a single phase AC supply from a field power distribution board on the field and transforms the power to a three-phase supply using a variable frequency drive (VFD). The TCUs are also connected in a computer network to exchange data and enable remote control.

Therefore, the trenches shown in the above diagram would contain not only the DC, AC and Ethernet cables described above but also air pipes utilized for the air drying units.

The power converted by the central inverter shown in the diagram above is typically of a voltage that is non-standard in order to prevent parallel connections of inverters. This voltage is transformed to a suitable medium voltage utilizing a transformer. The output of a standard block is typically 11 kV or 22 kV.

A number of standard building blocks are connected together depending on the rating of the power plant through ring main units and are finally connected to a power transformer to be transform the electricity generated to the grid connection voltage.

8-III.6 Engineering Choices in Power Plant Design

8-III.6.1 Inverters

Basically, PV inverters are suitable for use with Soitec CPV technology and both architectures (centralized or decentralized) can be used. Two particular features related to CPV must be taken into account:

1. *AC disconnection delay*
   The inverter should not disconnect the grid if the DC current drops to zero for several minutes (e.g. a cloud passing over the power plant). If the inverter disconnects from the grid, it will take 1–3 minutes after the recovery of DC power to re-connect to the grid (due to initialization tests, DC insulation measurements, grid synchronization). This will have a negative impact on the production of the power plant and will overstress the AC contactor of the inverter which is normally designed for one open/close cycle per day. This CPV specific issue can easily be solved by adjusting an inverter parameter or by updating the firmware.
2. *Inverter MPPT algorithm accuracy*
   The IV curve of the Soitec CPV modules is more 'rectangular' compared to standard PV modules (crystalline or thin films). The fill factor is approximately 0.80 (standard PV is about 0.75). This high fill factor requires a high MPPT voltage accuracy ($\sim$1 V) in order to detect the real maximum power point. In other words, the inverter MPPT algorithm must be suitable for modules with high fill factor.

Once the type of inverter technology is chosen, the inverter size should be selected based on the DNI level and the temperature of the region. During the design phase, the inverter configuration must be modeled with a dynamic simulation software (*PVsyst*) in order to define an optimized inverter configuration for each power plant. The results of the dynamic simulation will confirm the DC/AC ratio (DC CSTC power/AC inverter power). To ensure an optimized inverter design, the losses above the nominal power, also called clipping losses, should remain below 0%.

The choice of an inverter in a power plant is nowadays more focused on the conformity to the grid codes of the country where the power plant will be installed. As grid codes become more stringent, inverters are being developed that can handle most parameters including the control of active power on demand, providing frequency control and also reactive power control.

8-III.6.2 Cabling

The dimensioning of the cable is one of the most critical items in the design of the power plant. This is not only because the dimension (cross-section and length) of the cable determines the overall cost of the electrical block but also since minimizing the cross section or increasing the length of the cable leads to higher voltage drops and hence power loss. Therefore a balance should be reached where the value of the lost energy is less than the actual capital cost of the cable over the lifetime of the power plant. The material of the cables selected also plays an important role in the cost of the electrical block with Aluminum currently being favored due to its lower cost. Aluminum cables however have a higher maintenance cost since they require more frequent periodic tightening of the terminations.

This discussion can be taken a step further in the determination of choosing between a central inverter or a string inverter. The major problem here is the sizing of the cables in order to have the least voltage mismatch and therefore a higher throughput of the energy. Generally a central inverter is very suitable for CPV plants, but only if the generators provide almost the same DC voltage and if the DC cabling is sized to have a low voltage drop.

8-III.6.3 Power Transformers and Protection

The power transformer is defined according to the requirements of the inverters selected (rated power, rated voltages, with or without neutral connection, simple or double secondary winding, $U_k\%$).

The power transformer must be compliant with the local standards (cast resin or oil cooled, temperature monitoring and protection, pressure monitoring and protection).

An adequate protection unit is selected according to the requirements of the utility and should allow for discrimination and selectivity between faults on the power plant side and the grid side.

8-III.6.4 Back-up Power Supply

The function of the backup power supply is to provide over 15 minutes of sufficient electrical energy during an electrical power cut, to move all the CPV systems into the safety (stow) position. All CPV systems within the CPV power plant must move towards the safety position at the same time.

It is very important to reduce the power consumption required by the CPV systems to go to stow position, because the size of the back-up power supply will depend on this and the cost could be very high.

A safety position controller has been developed by Soitec to ensure that during a strong wind event or a power cut, the back-up power supply will switch on immediately and move the CPV systems to stow position. This controller brings an additional safety factor for the customer and certification body.

8-III.6.5 Foundations

The optimal design and layout of the foundation is highly dependent on the soil conditions defined in the site geo-technical report. Foundation type and calculations are to be finalized by a structural engineer, using findings from a local project land study and/or geo-technical report.

The legislation and national standards such as Eurocode and the US building code have a large influence on the design of the foundation. Foundation specifications are developed during the detailed engineering of the CPV power plant.

Foundations can be one of the more expensive elements of a power plant. One of the approaches to try to reduce the cost is the use of pile-driven foundations or, for trackers placed in the middle of a power plant, lower winds loads could be used for the calculation of the foundation design.

8-III.7 Construction Phase

A utility scale installation process provides a cost-effective method for CPV system assembly on the field. The utility scale installation process commences with the arrival of materials to the construction site and ends when a CPV system is completely assembled and connected to the distribution cabling.

The organization of the logistic and supply chain flows is essential to the success of the project. Any delay in one of the key components or a malfunction in the plan of the income could heavily impact the whole project. For example, the mast should arrive before the other components and the warehouse should be sized depending on the logistic flow and the cost.

The incoming inspection and quality controls of the components are mandatory to ensure the quality of the final systems.

The installation is typically done with different assembly stations. At each assembly station jigs and assembly tooling are used to facilitate the works. The jigs and tooling should be identified and manufactured, prior to the start of the construction phase.

These stations could be placed in the power plant installation area or in a centralized area as shown in Figure 8-III.4 which facilitates the logistics and the quality controls. The pre-alignment of the modules could be done in one of these pre-assembly stations.

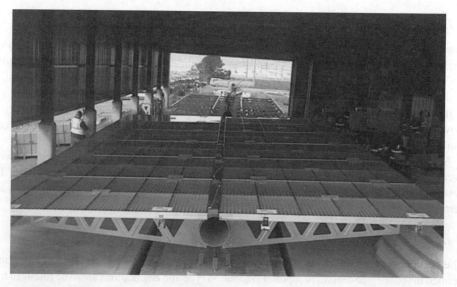

Figure 8-III.4 Example of a centralized tracker table assembly in an assembly hall

At least two of the assembly processes can be done only in the final installation area; one is the foundation of the mast as first step and the other is the erection of the tracker table on the mast as the final assembly step.

Once the system is assembled, all the cabling and electrical connections of the systems should be finished.

It is very important that at every step of the assembly and/or installation process, some quality checks are carried out, to ensure that the assembly was done according to the requirements. The quality checks are identified during the development of the product.

8-III.8   Commissioning and Acceptance Phase

Ideally, in a large power plant, the commissioning of the installed system should be done immediately after a building block has been finished. This would avoid any major problem, because the first equipment will be tested immediately after the installation.

The first steps of the commissioning are the TCU firmware and parameter flashing, module alignment (if needed) and tracker calibration. The power data of every system could be measured with the TCU and these could be read with the SCADA system (therefore it is very important that the Ethernet connection is available from the first step).

Once every system has been commissioned and is working properly, all the DC section tests and AC tests can be performed.

With the different power plant sections commissioned and accepted, the production ratio test could be started and the performance yield can be calculated. The long term analysis of the very first Concentrix™ power plants and the first large scale data of the Touwsrivier power plant show very good results [2].

References

1. Gerstmaier, T., Zech, T., Röttger, M., Braun, C. and Gombert, A. (2015) Large-scale and long-term CPV power plant field results. 11th International Conference on Concentrator Photovoltaic Systems (CPV-11), Aix-les-Bains, France. AIP Conference Proceedings, **1679**, 030002-1–030002-8.
2. T. Gerstmaier *et al.*, Large-scale and long-term CPV Power Plant Field Results. Paper presented at the 11th International Conference on Concentrator Photovoltaic Systems (CPV-11), Aix-les-Bains, France, 2015.

9

Reliability

Carlos Algora,[1] Pilar Espinet-Gonzalez,[5] Manuel Vázquez,[1] Nick Bosco,[2] David Miller,[2] Sarah Kurtz,[2] Francisca Rubio,[3] and Robert McConnell[4]

[1]*Instituto de Energía Solar, Universidad Politécnica de Madrid, Spain*
[2]*National Renewable Energy Laboratory, United States*
[3]*Instituto de Sistemas Fotovoltaicos de Concentración (ISFOC), Spain. Present address: Soitec, Germany*
[4]*Amonix, United States. Present address: CPVSTAR, United States*
[5]*California Institute of Technology, United States*

9.1 Introduction

Photovoltaic systems require a high initial investment cost that only can be paid off if the life of the system is large enough. Silicon module manufacturers are aware of this issue and they have increased the warranty time of the photovoltaic modules to around thirty years. Therefore, silicon flat plate PV systems have been deployed widely over the past three decades and operate using proven technology with annual degradation around 0.5% or lower.

In the case of CPV, open questions about its reliability remain because, in many cases, CPV systems are still in the prototyping and testing phases. Reliability data collection has only begun since 2005 running into commercial development, with the exception of a few multiple-hundred kilowatt scale plants. Therefore CPV lacks a historical record of proven longevity and reliability. Furthermore, reliability is key in reducing CPV cost as is shown in Chapter 14 (Cost Analysis).

Accordingly, this chapter describes in detail the accumulated knowledge on CPV reliability. Firstly, the fundamentals of reliability and qualification are shown. After that, the reliability of solar cells, modules (including optics) and plants is widely described. The chapter ends with a summary of the CPV qualification and reliability standards.

9.2 Fundamentals of Reliability

Reliability engineering is the discipline of ensuring that a product will work properly during a specified period of time. Therefore the aim of reliability engineering is to delay the failures and

Handbook of Concentrator Photovoltaic Technology, First Edition. Edited by Carlos Algora and Ignacio Rey-Stolle.
© 2016 John Wiley & Sons, Ltd. Published 2016 by John Wiley & Sons, Ltd.

then to maximize the life of the product. In this section a brief introduction of fundamentals of reliability is done. A much deeper knowledge of the foundations of reliability engineering can be found in [1–3]. As it is not possible to assure that a product will work properly at a given instant in time, it is necessary to use statistics and define the *probability of working without failure*.

Therefore, reliability is defined as the probability that a product performs its intended function without failure under specified conditions for a given period of time. This definition contains four important aspects (probability, intended function, given period of time and specified conditions) that are explained as follows:

- As reliability is quantified by a *probability*, it is necessary to use the probability theory and statistics in order to analyze the product reliability.
- The *intended function* must be specified unambiguously in order to know if a product works properly or has failed.
- *Given period of time*. It is assumed that at initial stage ($t = 0$) the product works and therefore reliability is 1 (as will be shown in the next section, the reliability function $R(t = 0) = 1$) and the probability to perform the intended function decreases with time.
- *Specified conditions*. Reliability depends on some specified electrical and/or environmental conditions. The higher the electrical and/or environmental stress the lower the reliability is. Therefore it is necessary to specify clearly the work conditions when reliability analysis is done.

9.2.1 Reliability Functions

Once the concept of reliability has been presented it is necessary to define the main reliability-related functions, namely:

- failure probability density function, $f(t)$
- cumulated failure probability density function, $F(t)$
- reliability function, $R(t)$
- failure rate function, $\lambda(t)$.

The failure probability density function is the probability density function associated with the random variable *time to failure*. The larger the value of $f(t)$ the higher the number of failures which will occur in a small interval of time around t. As $f(t)$ is a failure probability density function and all the failures occur in a time greater than 0 then $f(t)$ must fulfill the following equation:

$$\int_0^\infty f(t) = 1 \tag{9.1}$$

The *cumulated failure probability density function* is also called *unreliability* and it is understood as the population fraction that has failed at an instant of time, t. Hence, $F(t)$ is the probability of failure until an instant of time t and being the cumulated probability density function of $f(t)$ it is expressed as:

$$F(t) = \int_0^t f(t) \tag{9.2}$$

where $F(t)$ is an increasing function being $F(0) = 0$ and $F(\infty) = 1$.

The reliability function, $R(t)$ is interpreted as the population fraction that has not failed at an instant of time t. $R(t)$ is the probability of success, which is complementary to $F(t)$:

$$R(t) = \int_t^\infty f(t) = 1 - F(t) \tag{9.3}$$

where $R(t)$ is a decreasing function being $R(0) = 1$ and $R(\infty) = 0$.

The failure rate, $\lambda(t)$, measures the rate of change in the probability that a surviving product will fail in the next small interval of time. In other words, the failure rate function indicates the frequency with which surviving products will fail. Accordingly, $\lambda(t)$ has units of failures per unit time (e.g. failures/hour). It can be calculated as:

$$\lambda(t) = \frac{f(t)}{R(t)} \tag{9.4}$$

The failure rate function is very useful for maintenance purposes because it gives the failure probability but only of the surviving products. In addition, the failure rate function gives information about the periods in the life of a product:

- *Early failure period*, if the product has a decreasing failure rate. This period of time typically lasts several weeks or few months. In this period, the failures due manufacturing defects appear. Failures in this period constitute the so called *infant mortality*.
- *Random failure period*, if the product has a constant failure rate. This period of time starts after the early failure period and, in many cases, spans most of the life of the product.
- *Wear-out failure period*, if the product has an increasing failure rate. In this period the materials wear out and degradation failures occur at an increasing rate.

As shown in Figure 9.1, the shape of the failure rate function versus time with these three periods of life has a bathtub shape so it is called the *bathtub curve*.

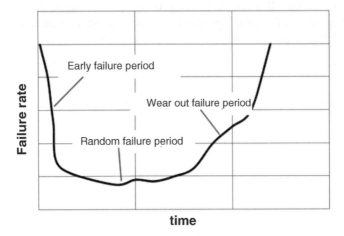

Figure 9.1 Failure rate bathtub curve

One important time parameter in the life of a product is its *MTTF* (mean time to failure). From the statistics point of view, *MTTF* is the mean time at which failures occur and mathematically can be defined as the expected value of the failure probability density function:

$$MTTF = \int_0^\infty t \cdot f(t) \mathrm{d}t \qquad (9.5)$$

MTTF is intensively used in reliability but it is necessary to consider that the reliability at *MTTF* – i.e. $R(t = MTTF)$ – is not large enough to design the life of a product with this time. For instance, if failures follow a normal distribution, the reliability at *MTTF* time is $R(MTTF) = 0.5$ which essentially means that at *MTTF* half of the population has failed. Therefore, *MTTF* is a time parameter that gives us information about when failures will occur but it must not be considered for design purposes.

9.2.2 Statistical Distribution Functions

Once the different reliability functions have been defined it is necessary to know the main statistical distributions used in reliability. Below, such statistical distributions – i.e. exponential, normal and Weibull – will be described. The exponential distribution is widely used in electronic devices working at low stress level and it fits the reliability of devices with a constant failure rate, namely, in the random failure period of the bathtub curve of Figure 9.1. On the other hand, the normal or Gaussian distribution is used in devices working under a high stress level and it fits the reliability of devices with increasing failure rate, namely, in the wear-out failure period of the bathtub curve of Figure 9.1. Finally, the Weibull distribution is much more flexible and by using conveniently its parameters, is able to fit the reliability of devices in the three periods of the bathtub curve of Figure 9.1.

9.2.2.1 Exponential Distribution

The exponential distribution is a one parameter (λ) distribution that is often used in reliability analysis because is the distribution function in which the failure rate is constant (see Figure 9.2). If the time to failure of a product is exponentially distributed with the λ parameter, then the probability density function is:

$$f(t) = \lambda . e^{-\lambda t} \ t \geq 0 \qquad (9.6)$$

The *unreliability* function can be calculated from equation (9.6) as:

$$F(t) = \int_0^t \lambda . e^{-\lambda t} \mathrm{d}t = 1 - e^{-\lambda t} \qquad (9.7)$$

and the reliability function is:

$$R(t) = e^{-\lambda t}. \qquad (9.8)$$

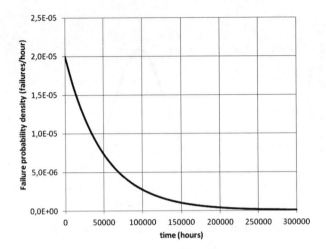

Figure 9.2 Exponential failure distribution function ($\lambda = 5000$ hours)

The failure rate function is calculated directly as:

$$\lambda(t) = \frac{f(t)}{R(t)} = \frac{\lambda.e^{-\lambda t}}{e^{-\lambda t}} = \lambda \qquad (9.9)$$

Therefore failure rate is constant and equal to the λ parameter of the exponential. From previous equations, MTTF can be calculated as:

$$MTTF = \int_{0}^{\infty} t.\lambda e^{-\lambda t} dt = \frac{1}{\lambda} \qquad (9.10)$$

and the reliability function at *MTTF* time is:

$$R(t = MTTF) = e^{-\lambda MTTF} = e^{-\lambda.\frac{1}{\lambda}} = e^{-1} = 0.368 \qquad (9.11)$$

The exponential distribution is widely used in electronic devices with low stress level because in many cases the wear out period starts much later than the end of the life of the product and therefore, the exponential distribution can be used.

9.2.2.2 Normal Distribution

The normal or Gaussian failure distribution is a symmetrical distribution meaning that failures are distributed in a symmetrical way around a key instant or mean time ($t = \mu$). The normal distribution is a two-parameter distribution (μ and σ) with the following failure probability density:

$$f(t) = \frac{1}{\sqrt{2\pi}\sigma} e^{\left[-\frac{(t-\mu)^2}{2\sigma^2}\right]}, \qquad -\infty < t < \infty \qquad (9.12)$$

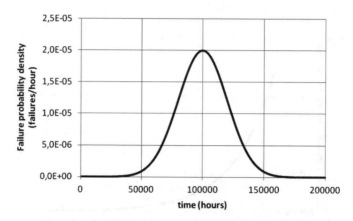

Figure 9.3   Normal failure distribution function for $\mu = 100\,000$ hours and $\sigma = 20\,000$ hours

μ being the mean time and σ the standard deviation. Figure 9.3 represents the failure probability density function as a function of time for $\mu = 100\,000$ hours and $\sigma = 20\,000$ hours. It can be seen how the failure probability is symmetrically distributed around μ.

The unreliability function can be calculated from equation (9.12) as:

$$F(t) = \int_{-\infty}^{t} \frac{1}{\sqrt{2\pi}\sigma} e^{\left[-\frac{(t-\mu)^2}{2\sigma^2}\right]} dt = \Phi\left[\frac{t-\mu}{\sigma}\right] \tag{9.13}$$

and the reliability function is:

$$R(t) = \int_{t}^{\infty} \frac{1}{\sqrt{2\pi}\sigma} e^{\left[-\frac{(t-\mu)^2}{2\sigma^2}\right]} dt = 1 - \Phi\left[\frac{t-\mu}{\sigma}\right] \tag{9.14}$$

while the failure rate function is expressed as:

$$\lambda(t) = \frac{f(t)}{R(t)} = \frac{\dfrac{1}{\sqrt{2\pi}\sigma} e^{\left[-\dfrac{(t-\mu)^2}{2\sigma^2}\right]}}{\displaystyle\int_{t}^{\infty} \dfrac{1}{\sqrt{2\pi}\sigma} e^{\left[-\dfrac{(t-\mu)^2}{2\sigma^2}\right]} dt} \tag{9.15}$$

Unreliability, reliability and failure rate for the normal distribution have no analytical solution.

Figure 9.4 presents the failure rate versus time. It can be seen that the failure rate increases with time. This means that if the product has a normal failure distribution it will be in the wear-out state of life. Normal failure distribution is often used to fit the failures of products that are in the wear-out period of life such as mechanical elements. The normal distribution is commonly used for reliability products in the wear out period of life due to its increasing failure rate.

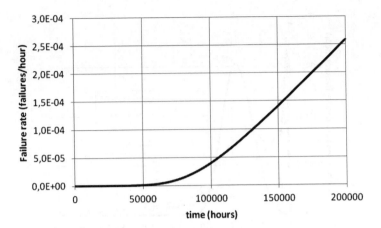

Figure 9.4 Failure rate function for the normal failure probability distribution function for $\mu = 100\,000$ hours and $\sigma = 20\,000$ hours

In normal distribution MTTF is the mean time of the distribution function, i.e. $MTTF = \mu$, and thus $R(MTTF) = 0.5$ as it was anticipated in section 9.2.1.

9.2.2.3 Weibull Distribution

The Weibull distribution is a two-parameter distribution (η and β) that is intensively used in reliability analysis, since it can fit the different failure distributions that are in the three periods of life of the *bathtub curve*. The failure probability density of the Weibull distribution is:

$$f(t) = \frac{\beta}{\eta^{\beta}} t^{\beta-1} \exp\left[-\left(\frac{t}{\eta}\right)^{\beta}\right] \qquad t \geq 0 \qquad (9.16)$$

β being the shape parameter, that takes values higher than 0, and η is the scale parameter. Figure 9.5 shows the Weibull failure probability density function for different shape parameter values.

Based on the probability density function, the Weibull unreliability function can be calculated as:

$$F(t) = 1 - \exp\left[-\left(\frac{t}{\eta}\right)^{\beta}\right] \qquad t \geq 0 \qquad (9.17)$$

and the failure rate function is:

$$\lambda(t) = \frac{\beta}{\eta}\left(\frac{t}{\eta}\right)^{\beta-1} \qquad t \geq 0 \qquad (9.18)$$

Figure 9.6 shows that the failure rate can show a very different trend depending on the shape parameter, β, fitting the different states of the life of the product:

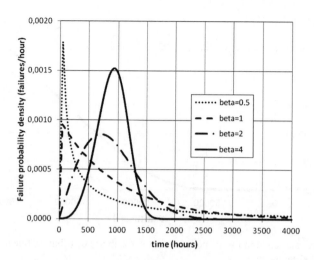

Figure 9.5 Weibull failure probability density function for $\eta = 1000$ hours and for different β values (0.5, 1, 2 and 4)

- If $\beta < 1$ the failure rate decreases over time fitting the life distribution of a product in the early period of life.
- If $\beta = 1$ the failure rate is constant, fitting the random failure period of life, because in this particular case the Weibull distribution coincides with exponential distribution.
- If $\beta > 1$ the failure rate increases with time and it fits the life distribution of a product in the wear-out period of life.

Therefore, the shape parameter, β, gives information about the state of the life of the product. On the other hand, the scale parameter, η, gives information about the time when failures occur. In particular, the unreliability at $t = \eta$ does not depend on the shape parameter and is:

$$F(t = \eta) = 1 - \exp\left[-\left(\frac{\eta}{\eta}\right)^{\beta}\right] = 1 - \exp(-1) = 0.632 \qquad t \geq 0 \qquad (9.19)$$

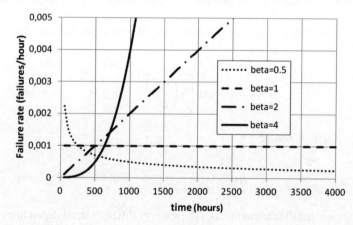

Figure 9.6 Weibull failure rate function for $\eta = 1000$ hours and for different β values (0.5, 1, 2 and 4)

Once the main reliability distribution functions have been explained, the next step is to describe the accelerated tests required to evaluate the failure distributions.

9.2.3 Accelerated Life Tests

The determination of the reliability of a product is an important target for any manufacturer. In order to do this, it is necessary to collect failure data over time, which subsequently allows the reliability analysis. The reliability of many electronic devices, and particularly that of solar cells, is very high so its evaluation in the field would need to collect data for decades. Therefore, getting failure data in a suitable timeframe requires the use of *accelerated life tests* (ALTs). In an accelerated life test devices are stressed to a level higher than the used in nominal operating conditions to shorten their failure times. After that, reliability at high stress level is extrapolated to operating conditions stress by means of life-stress models.

Accelerated life tests can be classified in two groups [4,5]:

- *Qualitatively accelerated life tests* are used primarily to reveal probable failure modes in a short period of time. A *highly accelerated life test* (HALT) [6] is a qualitatively accelerated life test designed to detect reliability weaknesses of a product in a short time with the objective of improving its reliability. In this type of tests a few samples are subjected to a gradually increasing stress until their failure occurs. These tests do not allow obtaining an assessment of the product reliability but give very important information about the stress limit and the reliability weaknesses of the products.

  Qualitatively accelerated life tests are also used as *qualification tests*. In this case the tests are used to predict in a short period of time whether the product will fail in working conditions. The tests are designed to cause the main known failures of the product. If the product passes the qualification tests, there will be a high confidence that during a long period of time, which is not specified and depends on working conditions, the product will not fail by means of the main known failure mechanisms.
- *Quantitative accelerated life tests* (QALT) are designed to evaluate the reliability of a product and therefore, the stress levels are not as high as in the HALT. Life-stress models are used to extrapolate the reliability to nominal conditions. To extrapolate working life from QALT requires the following steps:
  1. *To select the type of stress that will accelerate the product life.* In a QALT a stress will be used to accelerate the life of the product. In many cases temperature is used as the stressing magnitude to accelerate the life of the product but, depending on the product, other accelerated stresses must be used.
  2. *To define the stress values.* Stress values must be large enough to accelerate the product life and shorten the duration of the accelerated test but, at the same time, these values must not induce failures that do not appear at nominal operating conditions. At least three different stress values must be selected and they must be sufficiently spaced apart in order to estimate their influence on the product life.
  3. *To select the number of samples to be tested.* In order to estimate the product life of a product it is necessary to test at least 10 to 15 product units for each accelerated test.
  4. *To develop the accelerated tests.* Product units must be working during the accelerated test and failure times must be detected instantaneously. An instrumentation system that controls the working mode and detects the failures during the accelerated tests is required.

5. *To evaluate the product life in each accelerated test.* Once failures times have been precisely recorded, it is necessary to assess the life of the product using the distribution functions described in section 9.2.2. In most cases, a Weibull distribution function is used due its versatility but, depending on the product, other distribution functions could be more suitable.

6. *To apply life stress models to estimate the product life at nominal working conditions.* Once the product life for the different accelerated tests has been evaluated, it is necessary to extrapolate the product life at nominal working conditions. In order to do that, life stress models are required and they are described in the following section.

9.2.3.1 Life Stress Models

Product life depends on the electrical and environmental stresses. Product life can be accelerated by means of increasing one of the stresses. The main target of a QALT is to estimate the life distribution at nominal operation conditions. To do this it is necessary to use models that estimate the life at nominal conditions from the life at accelerated stress conditions.

The acceleration factor between two stress intensities is the ratio between the life, L, at the stress S_2 (which usually is the stress at the nominal conditions) and the life at the stress S_1 (usually being the stress intensity higher than that of nominal conditions). Therefore:

$$AF = \frac{L(S_2)}{L(S_1)} = \frac{t_{0.99}(S_2)}{t_{0.99}(S_1)} = \frac{t_{0.90}(S_2)}{t_{0.90}(S_1)} = \frac{t_{0.x}(S_2)}{t_{0.x}(S_1)} \tag{9.20}$$

Life is defined by a set of time parameters that describe the whole period of life of a population of devices, from the start of the test (when all the devices work properly) until the end of their lifetime when all the devices have failed. Usually, these time parameters are related to reliability, i.e. $t_{0.99}$, $t_{0.90}$, $t_{0.x}$ are the times at which the reliability values 0.99, 0,9 or 0.x are achieved. For instance, in a population of 100 samples $t_{0.90}$ is defined as the time at 10 samples have failed (that obviously depends on the applied stress, S). In life stress models, the acceleration factor is assumed to be independent on the selected reliability value and it is the same for the whole life, as equation (9.20) shows. Depending on the selected accelerated stress, there are different suitable life-stress models [5,7] whose brief description is stated below.

Temperature is widely used in ALTs because most of the failure mechanisms are accelerated with temperature. In many applications the dependence of life on temperature is well described by the Arrhenius model. The Arrhenius model assumes that the reaction rate of many failure modes is affected by temperature such as chemical processes are and thus it follows the equation:

$$r = B \exp\left[\frac{-E_A}{kT}\right] \tag{9.21}$$

B being a constant and E_A the activation energy, both of them depending on the failure mechanism; k is the Boltzmann constant; T is the temperature and r is the chemical reaction rate that causes a specific failure. The rate is assumed to be inversely proportional to the failure time. For example if an experiment is performed at two temperatures, T_1 and T_2, the acceleration factor (*AF*) ratio between the failure times at both temperatures can be described as:

$$AF = \frac{t_2}{t_1} = \frac{B\exp\left[\dfrac{E_A}{kT_2}\right]}{B\exp\left[\dfrac{E_A}{kT_1}\right]} = \exp\left[\frac{E_A}{k}\left(\frac{1}{T_2} - \frac{1}{T_1}\right)\right] \tag{9.22}$$

Therefore, once the activation energy of the main failure mechanism is known, it is possible to evaluate the acceleration factor between the accelerating test temperature and the working temperature. The main problem is that usually the activation energy is not known and several accelerated tests are required to determine it. Once E_A is determined, it is possible to evaluate the life of a product at working temperature.

Figure 9.7 shows the *AF* referred to 25 °C for different acceleration test temperatures and three different activation energies. It can be seen that *AF* is highly dependent on temperature. Accelerated test temperatures must be carefully selected and must not bring about failures different that those caused in nominal working conditions. The activation energy must be also be carefully determined because small deviations strongly affect *AF* and, consequently, the estimated life.

The inverse power law is also a frequently used to describe life as a function of different stresses. In the inverse power law, the product life (*L*) is calculated as follows:

$$L(S_1) = \frac{K}{S_1^{\ n}} \tag{9.23}$$

K being a constant; S_1 the applied stress and *n* a power factor (also constant). Therefore, the acceleration factor of a product working at two different stresses is:

$$AF = \frac{L(S_2)}{L(S_1)} = \left(\frac{S_1}{S_2}\right)^n \tag{9.24}$$

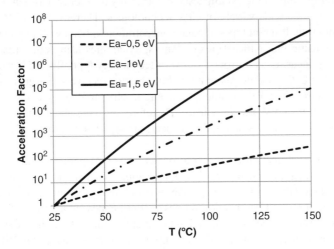

Figure 9.7 Acceleration factor referred to 25 °C as a function of accelerated test temperatures for three different activation energies

By measuring the life at two different stresses, the n power factor can be empirically calculated and from it, the *Life* at any stress.

The inverse power law is widely used with electrical stresses, either voltage or current, in electronic devices. As an example, the life of high power light emitting diodes (LEDs) follows an inverse power law, the stress being the injected current [8] with $n = 2$ (as we will see in section 9.3.5.2). Other life stress models that are also applied in QALT include Peck for humidity [9] and Coffin-Manson [10] for thermal cycling. These two can be considered as particular cases of inverse power law model where the stresses are relative humidity (RH%) in Peck's model and temperature increase (ΔT) in the Coffin-Manson model.

9.2.4 Reliability Versus Qualification

Since the late 1970s there have been a number of qualification test sequences in the PV industry. Nowadays, with huge growth and internationalization, the PV qualification standards from the International Electrotechnical Commission (IEC) are commonly accepted by manufacturers and buyers. For example, the IEC 61215 standard 'Crystalline silicon terrestrial photovoltaic modules – design qualification and type approval' [11] requires a sample of eight modules randomly taken from the production, which have to be subjected to a qualification sequence which contains different tests [12] such as UV exposure, thermal cycling, humidity-freeze cycling, damp-heat exposure, and outdoor exposure. If all the modules pass the tests, the module type gets the qualification. If two or more modules do not pass the tests, the module type does not get the qualification. If one module fails a given test, another two modules are taken from the production and both have to pass the test.

The qualification standard for CPV modules – namely IEC 62108 [13] – is a compendium of accelerated tests that identify long-term reliability problems observed in CPV modules and assemblies [14]. It is partially inspired by IEC61215 having a similar methodology: CPV modules follow a testing sequence and then are subjected to characterization tests. However, there are important differences since IEC62108 takes into account the special features of CPV receivers and modules so there are tests in the field and in the laboratory, high current density tests and high ramp temperature cycling tests. Another key difference between flat plate PV and CPV is that there are a wide variety of CPV technologies. Thereby, IEC 62108 groups CPV technologies into four different types of solar concentration: modules using point focus Fresnel lenses, modules using linear focus Fresnel lenses, assemblies using point focus mirrors and assemblies using linear focus mirrors. For every technology type, IEC 62108 defines a set of tests, not only for CPV modules/assemblies but also for receivers. Seven modules and two receivers are required for CPV module qualification; while nine receivers and seven primary optics are required for CPV assembly tests. Therefore, an issue about qualification tests that should be noticed is the very small and statistically insignificant number of products that are tested.

The purpose of qualification testing is to detect in a short period of time the presence of known failures or degradation modes that would appear in working conditions [15]. If the product passes the qualification test sequence, it is possible to assure with a high confidence that the known failures and degradation modes will not appear in a period of time that is not specified. However, the defects considered in the qualification tests are in some cases different than those appearing in real operation. For example, one controversial test in IEC 62108 is that of the thermal cycling that tries to evidence the failures that could appear when a cloud or another external agent blocks the direct solar radiation on a CPV receiver. The thermal cycling test is designed to determine the ability of receivers to withstand thermal mismatch, fatigue, and other stresses caused by repeated changes of temperature. Three different maximum temperatures (T_{max}) with different number of cycles can be used. The numbers of cycles are larger when T_{max} is lower. In the three cases the minimum temperature is $-40\,°C$ and the modules have an injected current of $1.25 \cdot I_{sc}$ when temperature is higher than $25\,°C$. However, there are studies [16] proving that thermal cycling at high temperature ramp rates may activate unrepresentative failure mechanisms and that the time to failure depends on the ramp rate.

The final goal of qualification tests is that almost all commercially available products pass the test sequence and, thereby, to pass the qualification tests means that the product meets a set of requirements. However, it is not possible to know the long term performance of different products succeeding the qualification tests. For example, to the question of how long will be the life time of a module passing the IEC61215 – the flat plate PV qualification standard –, some works suggest that the test time is equivalent to about 15–20 years in a soft climate [17,18]. These timeframes are lower than the warranty times offered by module suppliers so, many of them use their own tests in order to warrant 25 years [19,20]. Conversely, there are no estimates about an equivalent period of life for CPV modules when succeeding to pass the IEC 62108 standard because of the frequent changes in CPV technology and its still short history in the field (only a few years). Similarly to the IEC61215 case, to pass the IEC 62108 qualification tests does not assure that modules will achieve a certain reliability target. Therefore, the main limitation of qualification tests is their impossibility to give quantitative data about product lifetime in the field.

Accordingly, in order to quantify the life of a product it is necessary to make reliability tests. Reliability tests take much longer than qualification tests and require a significantly higher number of samples. A reliability test goes beyond a qualification test in order to cause failures in a substantial number of samples. The failure can be caused by increasing the test time or by increasing the stress level. Reliability tests analyze when, why and how a failure appears. The typical procedure of accelerated reliability testing needs to test three different lots of samples working at three different high stress levels. By obtaining the life distribution at three such stress levels, it is possible to evaluate the activation energy and from this, the life distribution at working conditions together with valuable information such as the reliability function, *MTTF*, failure rate function, power loss evolution, etc. Therefore, qualification tests are not reliability tests and the first should not substitute the latter.

9.3 Reliability of Solar Cells

9.3.1 Issues in Accelerated Aging Tests in CPV Solar Cells

As has been stated, an accelerated aging test stresses a given parameter (temperature, humidity, illumination, etc.), while the other stress parameters of the solar cells remain at nominal conditions (i.e. no stress is applied to the rest of the conditions). In particular, if the stressed solar cells do not operate at their nominal conditions, then the intended accelerated test would be a simple storage test that will not produce results about the solar cell reliability.

To apply the proper stress in temperature, humidity, cycling, etc. on concentrator solar cells operating under their nominal irradiance is a very difficult task. Assume a test following the procedure described in section 9.2.3 in which three temperatures have to be applied to three solar cell groups (each one with tens of cells) with a typical duration of several weeks or even months. How can nominal irradiance conditions be ensured in a constant mode? In other words, if temperature is stressed by using a climatic chamber, how can the solar cells be properly illuminated by the same irradiance (in terms of spectrum, intensity, uniformity etc.) that in addition has to be kept constant for the whole test duration? To the best of our knowledge, there is no answer by now. Therefore, the achievement of nominal irradiance conditions for concentrator solar cells during ALTs is a pending issue.

Perhaps the best way, by now, to emulate the electric performance of the solar cell under concentrated light is by forward bias. In other words, the solar cell voltage and current distribution that would be produced under concentrator operation are emulated by injecting a current in the absence of illumination. As can be anticipated by the discussions in chapter 2, both regimes (forward biasing and photogeneration) are not equivalent. In the following we discuss their similarities and differences.

Figure 9.8 shows a one-diode lumped model for a single junction solar cell operating under illumination and at forward bias. For the sake of simplicity, this model was to explain the different operation conditions (see section 9.3.5.2).

At normal operation, the solar cells of a CPV system will be paired with a maximum power point tracker (MPPT), so they will be operating around this bias point for most of their lifetime. In this situation, as Figure 9.8.a shows, the current delivered to the external load (I_{RS}) is slightly lower than the photogenerated current (I_L) so, the current flowing through the p-n junction which is forward biased, $I_D = I_L - I_{RS}$, is low. In a first approach, we may neglect the current flowing through the parallel resistance (R_p), since otherwise that would mean a faulty solar cell. In any case, the undesirable situation of a non-negligible R_p is also considered below. When the solar cell is forward biased (Figure 9.8.b) the diode (i.e. the p-n junction) is also forward biased, but most of the current injected into the solar cell flows through the junction diode ($I_{RP} \approx 0$, $I_D \approx I_{RS}$). Therefore, both at illumination and at forward bias the p-n junction is forward biased although it drains different current levels.

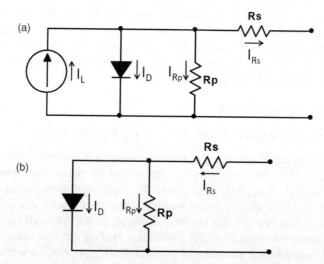

Figure 9.8 One diode model of a solar cell under illumination (a), and at forward bias (b). In this model, R_s and R_p are lumped parameters collecting all the resistive parts of the solar cell. For each circuit, the values of the currents are different. Source: Nuñez et al. 2015 [22]. Reproduced with permission of John Wiley and Sons, Ltd

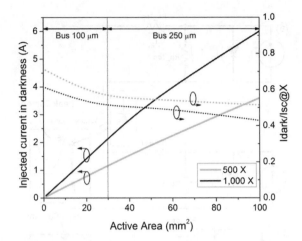

Figure 9.9 (Left axis) Calculated injected current in darkness, I_{dark}, in triple-junction solar cells of different sizes for emulating two nominal working concentrations: 500× (gray solid line) and 1000× (black solid line). (Right axis), Corresponding I_{dark} to $I_{sc@X}$ ratio for emulating two nominal working concentrations: 500× (gray dotted line) and 1000× (black dotted line). In the simulation, solar cells smaller than 30 mm² have bus bar widths of 100 μm while solar cells larger than 30 mm² have bus bar widths of 250 μm. Source: Nuñez *et al.* 2015 [22]. Reproduced with permission of John Wiley and Sons, Ltd

In the lumped diode model of Figure 9.8, when the solar cell is at open circuit conditions no current flows through the series resistance[1] ($I_{RS} = 0$) while almost all of the photocurrent flows through the diode and only a small fraction will flow through the parallel resistance. So, in a solar cell at open circuit the p-n junction drains the same current than at forward biasing, provided that the current injected under forward bias equals the photogenerated current under illumination. At short circuit, almost all photocurrent circulates through the series resistance and only a small fraction flows through the diode and parallel resistance. Therefore, in a solar cell under illumination, depending on its operating conditions, the stress will go either on the series resistance (i.e. electrical contacts, top cell emitter, etc.) or on the junction, but not on both at the same time. Conversely, when the solar cell is under forward biasing the stress is the same for the series resistance and for the pn junction [60]. Accordingly, the current injected to a solar cell when emulating illumination conditions has to be carefully calculated in order to avoid an overstress at any part of the cell [21]. This always results in an injected current lower than the photogenerated current [22] as can be seen in Figure 9.9 which guarantees that no part of the solar cell is overstressed, although someones are under-stressed. Therefore this is a conservative criterion.

An alternative criterion would consist of injecting a current which makes that the minimum current density is the photogenerated current density, but this will cause electrical overstress in some parts of the solar cell.

---

[1] In reality, as a result of distributed effects discussed in Chapter 2, a very small current flows through the horizontal series resistances of top cell emitter and front metal grid. For example, for a 3 × 3 mm² triple junction solar cell at 1000 suns the voltage loss in these resistances is of only 20 mV, i.e. 0.62% of the V_{oc} [22].

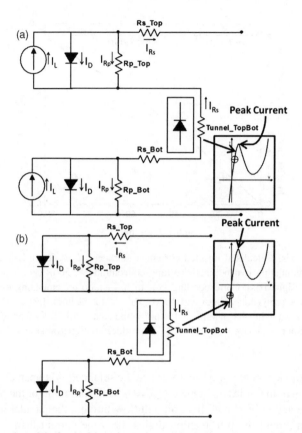

Figure 9.10 Lumped model of a dual-junction solar cell including, (a) tunnel diode under illumination and (b) at forward bias. Circles on the I–V curves of the tunnel junction indicate examples of operation points under illumination and at forward bias. Source: Nuñez *et al.* 2015 [22]. Reproduced with permission of John Wiley and Sons, Ltd

The simple model described above can be extended to a dual-junction solar cell by adding the tunnel diode (Figure 9.10). Under illumination the tunnel diode is forward biased and it is intended to work at the linear part of its I-V curve (Figure 9.10.a). On the other hand, if the solar cell is forward biased, the tunnel diode is reverse biased (Figure 9.10.b). The I-V curve of a tunnel diode at reverse bias is linear and has almost the same slope as in forward bias [22], as is shown in the insets of Figure 9.10. Because the maximum reverse current is much higher than the peak current (which limits the linear region in forward bias) no risk of damaging the tunnel diode is expected, when forward biasing the dual-junction solar cell [22]. The extension of the model from dual to triple junction solar cells does not add any additional qualitative difference between the illumination and biasing regimes.

The situation in which R_p is not high enough (so it can drain a non-negligible current) mainly arises from the appearance of defects in the semiconductor structure of the solar cell. The stress suffered by these defects will be similar both under illumination and at forward bias conditions, except in the case that defects are beneath the busbar. In this case, the defects would be overstressed at forward bias (because current flows mainly through them) while they would not have so high stress under illumination conditions (because almost no current would flow through them except at open circuit conditions). However, this difference between both regimes is not worrying because, in the end, these defects underneath the busbar will be the seed of failures excited by thermal-mechanical stress, etc. under operation conditions. Besides, the failures resulting from these defects are commonly regarded as infant failures when forward biasing and they appear in a very small proportion in commercial solar cells. This is one the reasons why ALTs must be carried out on a significant sample size.

Finally, there could be differences between the illumination and forward bias regimes in case of electromigration of the metal contacts due to the inverse current flow at both regimes. Previous works show that electromigration does not occur with the metals used in current commercial III-V concentrator multijunction solar cells and similar optoelectronic devices [23,24]. Even so, alternative tests would be desirable to dismiss failures due to electromigration using for example, forward/reverse bias on metallizations.

In summary, from the comparison between photogeneration and forward bias regimes experimented by a multijunction solar cell, the conclusions are [22]:

- The p-n junctions of a multijunction solar cell are forward biased both under illumination and under forward bias.
- The p-n junctions of a multijunction solar cell could be overstressed under forward bias, so a proper 3D simulation is required in order to avoid that the current flow at forward bias is higher than under illumination at any part of the cell.
- The tunnel junctions are reverse biased under forward bias while forward biased under illumination. However, in both regimes the tunnel junctions work at the linear part of their I-V curve. Because the maximum reverse current is much higher than the peak current, it seems there is no risk of damaging the tunnel diodes when the multijunction solar cell is forward biased.
- The inverse current flow under illumination and at forward bias could affect the evolution of electromigration in metal contacts. However, electromigration has not been observed on previous reliability tests using forward bias on commercial solar cells and other opto-electronic devices using similar metal stacks, so it is improbable for this to cause a different failure mechanism in commercial III-V cells under forward bias.
- Defects underneath the busbar (low shunt resistance) can cause failures with forward biasing that would not appear so early under illumination. However, this is uncommon and can be counterbalanced by using a significant sample size in ALTs, addressing those failures as infant mortality. Besides, these defects will be the seeds for failures under thermal-mechanical stress occurring under normal operation of the solar cells or under other qualification tests.

In consequence, the use of forward bias to emulate photogeneration does not result in additional overstresses if proper 3D modeling is carried out to control the current flowing at each part of the solar cell [22]. Other works have also shown that forward biasing itself (even at $4 \cdot I_{sc}$) does not degrade concentrator solar cells [25]. Accordingly, the use of forward bias seems to be a valid approach to emulate nominal conditions when developing ALTs for CPV solar cells.

9.3.2 Types of Failure

Failure is defined as the event or inoperable state in which any device does not perform as previously specified in its used environment [26]. The failure can be:

- catastrophic, when it causes an abrupt performance loss of the device;
- gradual, when due to degradation the device performance does not meet the specifications.

9.3.2.1 Catastrophic Failures

In the particular case of solar cells, a catastrophic failure results in an abrupt loss of power. This kind of failure is typically revealed within a few working hours in the field (infant mortality failure) and in the case of concentrator multijunction solar cells the infant mortality ranges from 0.5% [27] and up to 10% [28] in exceptional cases. These are non-negligible values which will have a high impact on the loss of energy production because of the series connection of the solar cells within a module [27]. Therefore, several studies have been carried out, as will be shown in section 9.3.3, with the purpose of determining the causes of these failures in order to try to prevent them.

9.3.2.2 Gradual Failures

Typically, it is considered that a flat plate module does not meet the specifications (gradual failure) when there is a decline in the output power rate higher than 1%/year (20% in 20 years) in its use environment [29]. However, in the case of CPV systems there is not a unified definition of failure as of today. According to several economic studies, for CPV to be cost-competitive, it has to fulfill the same reliability requirements as flat plate modules [30]. Then, all elements in the CPV system should have a very low yearly degradation rate; but what is the degradation which could be tolerated for each part of the system and in particular for concentrator solar cells? The response to this question is not straightforward. Núñez and co-workers proposed in [31] as a failure criterion a power loss in the solar cell of 2.5%. They opted for a conservative criterion in the degradation of the solar cell because they assumed that the degradation of other elements of the system, such as the optics, might be significant [32,33]. However, for practical reasons, if the degradation of the solar cells is monitored by their indoor characterization under a solar simulator, the failure criterion for the solar cell should be a loss of power higher than the uncertainty of such characterization technique. Typically, these concentration measurements are affected by a non-negligible uncertainty mainly resulting from the challenge of controlling the uniformity, the spectral match and the time stability of the irradiance. Such uncertainty could be in some systems higher than 2.5% absolute when measuring the solar cell efficiency [34]. This is why in subsequent works the failure criterion was increased within the range of 5–10% [21,35].

Finally, up to now there has not been any specific information about the gradual degradation of concentrator solar cells in the field. However, the few reports published about the evolution of the whole CPV systems during their first operating years [36–38] are very promising, showing a yearly degradation lower than 0.5% as will be shown in section 9.5.

9.3.3 Failures in Real Time Operation

As pointed out above, CPV companies and laboratories have reported minor degradation of the CPV systems in the field, at least for the first few years of operation. However, a non-negligible infant mortality of the solar cells has been observed. Several studies have been carried out in order to determine the origin of such infant mortality failures [27,28,33,39–41] with the aim of optimizing the design to avoid these failures and/or to define a proper screening characterization which allows the removal of weak solar cells from production. However, it is not always straightforward to discern if the infant mortality observed in the field is caused by failures inherent to the solar cell, its assembly, the packaging or even the optics.

The failure mode which has been observed more repeatedly in the field is a sudden drop of the open circuit voltage to 0 (the solar cell becomes a short-circuit [27,28,33]). Several studies have shown that the failure mechanism associated with this failure mode is thermal runaway. The evolution of thermal runaway consists in the concentration of the photo-generated current in a small area of the solar cell which increases the temperature of the solar cell locally causing thermal stress and finally leading to micro-cracks. The micro-crack causes a positive feedback process by making more current flow through it causing a further increase in temperature and finally the solar cell is shunted. That is, all the photo-generated current flows through a small region of the solar cells with a low shunt resistance. There are two different mechanisms which can cause thermal runaway:

- Voids in the epoxy die attach larger than 2.5% of the back area of the solar cell, as concluded by Bosco and co-workers after an exhaustive analysis of numerous receivers [27,39]. Then, in this case the origin is not related to the solar cell itself.
- Initial defects in the semiconductor structure which short-circuit the subcells [40,41]. In order to avoid thermal runaway in this case, Zimmerman proposes in [40] screening the solar cells by measuring their electroluminescence and dark I-V curve to remove the solar cells which reveal significant shunts. However, Zimmerman warns that these characterization techniques will reject more solar cells than will actually suffer infant mortality because these methods are not able to discern the dependence of the initial shunt with the temperature which is a key factor in its evolution towards an infant failure.

9.3.4 Accelerated Life Tests

Silicon flat modules have proven their reliability in the field for more than 20 years [42], III-V multijunction solar cells for space missions have also proven their stability and very similar devices such LEDs exhibit long life (see section 9.3.5) so, where does the skepticism regarding the reliability of concentrator solar cells come from? We could say that it has two primary origins; first the degradation of concentrator solar cells should be particularly low as noted in the previous section in order to achieve a whole CPV system degradation similar to flat-plate modules. Secondly, as we will show in section 9.3.5 the working conditions for CPV cells are very different from those of silicon solar cells, space cells and LEDs. In focusing on multijunction solar cells, the main issues when working under concentration are:

- The high photo-generated current densities which can promote defects.
- The high working temperatures which can also cause the evolution of defects.
- The non-uniform light profiles (in irradiance levels and in spectral content in refractive systems) caused by concentrator optics. These light profiles could potentially stress the solar cell in three

different ways: by causing non-uniform voltage profiles across the subcells which may force them to be draining (at least locally) a significant photo-generated current density, by making the tunnel junction to have some regions working in the first quadrant of the I-V curve and others in the third quadrant [43] and finally, by causing significant temperature gradients.

• The humidity and water condensation inside the CPV modules since typically they are not watertight [33].

In addition to these demanding working conditions, CPV is a technology with a low track record and in many cases, with open issues in the design of the different elements which constitute the CPV system that are in a testing phase. In particular, regarding the multijunction solar cells currently being used, there are several architectures with efficiencies above 40%. Therefore, different solar cell architectures and/or different CPV system designs (using different concentration levels, solar cell sizes, working temperatures, non-uniformity levels, preventing moisture mechanisms, etc.) could lead to very different reliabilities. Then, in order to have feedback on the reliability of the different solar cells and designs during their development, accelerated life tests are essential.

As it has been stated in section 9.2.3, there are two different accelerated tests that can be carried out: qualitative accelerated tests (mainly HALT) and quantitative accelerated life tests (QALT).

9.3.4.1 Qualitative Life Tests

Several studies have been addressed in order to determine failure modes in concentrator solar cells and to compare the performance of different semiconductor architectures. In the following tables (9.1–9.4) a review of the main studies reported in the literature is presented. None of them determine reliability functions/parameters. The tables group the studies according to the analysis of the main concerns regarding the degradation of concentrator solar cells: the window layers, since in the case of the top subcell it is the most exposed semiconductor layer (Table 9.1), the evolution of the threading dislocations in metamorphic architectures (Table 9.2), foreseeing the failure of standard lattice matched solar cells under the demanding concentrator working conditions (Table 9.3) and finally, to analyze the evolution and impact of initial defects on the semiconductor structure (Table 9.4). In order to exclude degradation analysis from semiconductor structures which could be outdated, Tables 9.1–9.4 only include the most recent studies (from the year 2000). Finally, it should be noted that some relevant studies carried out on solar cells used in space have been included in Tables 9.1–9.4 when the results obtained in the cited studies could be extrapolated to concentrator solar cells.

9.3.4.2 Quantitative Accelerated Life Tests

As pointed out above, the purpose of quantitative accelerated life testing (QALT) is to obtain the distribution of failures over time more quickly than under normal operating conditions. The philosophy of these tests is to stress at least one of the working parameters of the solar cell under test while leaving the rest of the parameters at nominal operating conditions. The high levels of stress force failures by accelerating the effects of natural aging. Then, once the solar cells under test have failed, a reasonable statistical model that relates the lifetime to the level of stress through an acceleration factor has to be determined and thus the life data from the

Table 9.1 Review of the main degradation works carried out in the study of the degradation mechanism of different window layers for concentrator solar cells. RH means relative humidity from now on.

Potential failure mode	Degradation of the window layer causing an increase of surface recombination
Solar cell structure	GaInP solar cell on GaAs substrate with n-AlInP window
Test conditions and duration	At open circuit condition, three cases:
	• 60 °C/60% RH/3×
	• 60 °C/60% RH/3× no UV
	• 60 °C/60% RH/0×
	2800 hours
Stress parameter	No stress
Evolution	No degradation in any case
Reference	[44]
Solar cell structure	Solar cells with different top cell windows: $(Al_{0.5}Ga_{0.5})_{0.51}In_{0.49}P$/AlInP
Test conditions and duration	95 °C and atmosphere saturated with moisture
	270 hours (for AlGaInP windows)/200 hours (for AlInP) windows
Stress parameter	Damp heat
Evolution	No significant evolution
Reference	[45]
Solar cell structure	p-on-n GaAs single junction solar cells with $GaInP_2$ or $Al_{0.85}Ga_{0.15}As$ windows
Test conditions and duration	Humid environment (RH 30%)
	8500 cycles of 40 minutes
Stress parameter	Thermal cycling (from −5 to 95 °C)
Evolution	Slight degradation of I_{sc} (<4%) and V_{oc} (<2%)
	Same behavior in both structures with different windows
Reference	[46]
Solar cell structure	p-on-n GaAs single junction solar cells with $GaInP_2$ or $Al_{0.85}Ga_{0.15}As$ windows
Test conditions and duration	Injection current of 1100 $I_{sc@1X}$ in darkness
	6000 hours
Stress parameter	Irradiance emulated
Evolution	Drop in Isc due to the appearance of circular defects on the surface
	More defects in the structures with $Al_{0.85}Ga_{0.15}As$ windows (I_{sc} drop ≈ 47%) than with $GaInP_2$ windows (I_{sc} drop ≈ 0%)
	Similar drop in V_{oc} for both structures ≈ 6% (hypothesis: increase of perimeter recombination)
Reference	[46]

accelerated life test can be used to extrapolate reliability information at nominal working conditions [29,55] as it has been explained in section 9.2.3.

The complexity of designing an ALT which leads to failure modes representative of those in the field together with the fact that they are usually time-consuming tests explains not only that the ALT reported on concentrator solar cells is scarce, but also that the level of sophistication in

Table 9.2   Review of the main degradation works focused on metamorphic concentrator solar cell structures

Potential failure mode	Penetration of misfit dislocations from the buffer layers into the active layers in lattice mismatched solar cells
Solar cell structure	Comparison between two solar cell structures: • 1 eV metamorphic GaInAs solar cells with $1-2 \cdot 10^6$ cm^{-2} threading dislocations • Lattice matched GaAs solar cells
Test conditions and duration	Open circuit/short circuit conditions 336 hours
Stress parameter	Temperature (125 °C) & irradiance (equivalent to 1300×, laser 808 nm)
Evolution	The same evolution of V_{oc} with time in both solar cell structures -0.53%/log(time) → projection: 2% V_{oc} loss after 30 years Perimeter degradation No extra failure mode due to threading dislocations
Reference	[47]
Solar cell structure	Lattice mismatched $Ga_{0.35}In_{0.65}P/Ga_{0.83}In_{0.17}As/Ge$ solar cells (0.0425 cm^2)
Test conditions and duration	3000 hours
Stress parameter	Damp heat (85 °C and 85% RH)
Evolution	No evolution of the electrical parameters Back side oxidation of the Ge subcell Change in the metallization color, gold wires and the adhesive (oxidation)
Duration	2500 hours
Stress parameter	Thermal Cycling (-30 °C to 130 °C), cycles of 2 hours
Evolution	No evolution
Duration	1600 hours
Stress parameter	Humidity freeze (-40 °C to 85 °C/85% RH), cycles of 24 hours
Evolution	Slight decrease in FF Back side oxidation of the Ge subcell Change in the metallization color, gold wires and adhesive (oxidation)
Duration	1500 hours
Stress parameter	High temperature (150 °C)
Evolution	Decrease of 4% in maximum power and I_{sc} after 500 hours No higher evolution at 1500 hours Change in the metallization color, gold wires and adhesive (oxidation)
Reference	[48]
Solar cell structure	Metamorphic $Ga_{0.35}In_{0.65}P/Ga_{0.83}In_{0.17}As/Ge$ concentrator solar cell
Test conditions and duration	500 hours
Stress parameter	Thermal cycles -40 °C $+ 90$ °C (2 hours cycle)
Evolution	Higher dark current density in the low voltage regime → decrease of the shunt resistance Strong reduction in FF and V_{oc} especially for low concentrations and more significant in small solar cells → Perimeter degradation
Reference	[49]
Solar cell structure	Lattice mismatched $Ga_{0.35}In_{0.65}P/Ga_{0.83}In_{0.17}As$ solar cells
Test conditions and duration	Injected current in darkness, ≈ 600 hours
Stress parameter	Irradiance (emulated 1000×)
Evolution	Dark spots in electroluminescence measurements → Sudden death Reduction V_{oc} → logarithmic increase with time of the saturation current density in the low voltage region → V_{oc} reduction of 0.5% after 20 years
Reference	[45]

Table 9.3 Review of the main degradation works carried out on lattice matched concentrator solar cells

Potential failure mode	To foresee failure modes in lattice matched concentrator solar cells
Solar cell structure	Lattice matched GaInP/GaInAs/Ge space solar cells
Test conditions and duration	11× of UV radiation
	≈ 2000 hours
Stress parameter	Temperature (197 °C, 227 °C)
Evolution	Bare cells themselves do not degrade
	When silicone adhesives are present in the packaging (top or underneath) degradation up to 20%, mainly because of subcell shunts reducing FF.
Reference	[50]
Solar cell structure	Lattice matched GaInP/Ga(In)As/Ge, solar cells 2.05×2.05 mm^2
Test conditions and duration	Injection current in darkness (equivalent to $500 \times I_{sc@1X}$),
	≈ 800 hours
Stress parameter	Temperature (140 °C)
Evolution	85% of the solar cells show a V_{oc} decrease (3–10%) and a I_{sc} decrease between (2–15%)
	15% of the solar cells present a higher degradation
Reference	[51]
Solar cell structure	n-on-p concentrator GaAs single junction solar cells (1 mm^2) with GaInP window
Test conditions and duration	Injection current ($1000 \times I_{sc@1X}$) in darkness
	≈315 hours
Stress parameter	Temperature step stress from 90 °C to 150 °C
Evolution	Gradual decrease in the perimeter parallel resistance
Reference	[52]
Solar cell structure	p-on-n concentrator GaAs single junction solar cells with Al$_{0.85}$Ga$_{0.15}$As window
Test working conditions	Irradiance equivalent to 1100× (laser 808 nm)
	Temperature 30 °C
	Solar cells at open circuit conditions
	57 hours
Stress parameter	None
Evolution	Increase of the saturation current in the low voltage region with the logarithm of time → Perimeter degradation
	Same evolution as tests with injection of current in darkness and humidity [46]
	Extrapolation to 20 years → 1% relative efficiency drop
Reference	[53]
Solar cell structure	p-on-n GaAs single junction solar cells with GaInP$_2$ or Al$_{0.85}$Ga$_{0.15}$As windows
Test conditions and duration	In a humid environment (RH 30%)
	≈5700 hours, cycles of 40 minutes
Stress parameter	Thermal cycling (from −5 to 95 °C)
Evolution	Slight degradation of I_{sc} (<4%) and V_{oc} (<2%) after 8500 cycles
	Same behavior in both structures with different windows
Reference	[46]

Table 9.4 Review of the main studies carried out to determine the origin and/or the impact of typical failures/defects detected on lattice-matched triple-junction solar cells

Potential failure mode	Evolution of initial shunts
Solar cell structure	Lattice matched GaInP/GaInAs/Ge
Test conditions and duration	6× and open circuit conditions
	5 minutes
Stress parameter	Temperature (250 °C)
Evolution	0.6% of the solar cells tested shunted after the test
Reference	[40]
Solar cell structure	GaInP/GaInAs/Ge space solar cells
Test conditions and duration	Inert atmosphere at ambient pressure
Stress parameter	Thermal cycling
	• From −175 °C to +86 °C
	• From −120 °C to +120 °C
Evolution	Low shunt resistance
	Grid finger interruption
Reference	[54]

ALT in the PV world has been relatively low [56]. Few QALTs have been carried out on concentrator lattice-matched [21,31] and lattice-mismatched [35] solar cells and they are usually controversial due mainly to the challenges of emulating nominal working conditions in terms of the spectrum impinging the solar cell, the concentration level and their stability throughout the time of the test (see section 9.3.1). Then, the approximations followed up to date to emulate working conditions lead to the question of whether they may be causing a failure mode which would not appear at nominal working conditions and/or they may not be accelerating a failure mode which would be revealed in field operation. Therefore, it is essential to corroborate the results and failure modes obtained in the accelerated life tests with the ones observed in field operation as we will see below.

Regarding the reliability data obtained up to date from this kind of tests, only two articles [21,31] report a fit of the concentrator solar cell failures subjected to different stress levels to a life model which takes into account their acceleration rate. Both studies [21,31] accelerate the solar cell's life by subjecting them to high temperatures. Therefore, the life-stress model used in both cases is the Arrhenius model (see section 9.2.3).

GaAs single-junction solar cells at three different temperatures were tested in [31]. The solar cells were divided into two groups and a different level of current was injected in darkness to each group in order to emulate the solar cells working at two different concentrations. The solar cells were covered with silicone in order to protect the perimeter and emulate the behavior of the solar cell with a secondary optic. A progressive degradation in the electrical behavior of the solar cells was reported. The activation energy obtained in that work was $E_A = 1.02$ eV and the injected current to emulate working conditions, that is the nominal concentration at which the solar cell is expected to be working, had a significant impact on the degradation of the solar cell after the working temperature. The reliability functions and parameters for the solar cells at different working temperatures and two different equivalent concentrations were extracted.

Considering that the solar cells work eight hours per day under concentration, they concluded that a 30-year warranty for these GaAs cells may be offered for both concentrations (700× and 1050×) if the solar cells were operated at a working temperature of 65 °C [31].

Commercial lattice-matched GaInP/GaInAs/Ge triple-junction solar cells were tested in [21]. The procedure followed was similar to [31]. Current was injected in darkness to emulate nominal working conditions and the solar cells were stressed at three different temperatures. All failures were catastrophic with $E_A = 1.59$ eV. By fitting the distribution of failures across the different test temperatures to an Arrhenius-Weibull model the reliability functions and parameters were extracted. For the nominal working conditions of the solar cells tested (820× and 80 °C), assuming that the concentration module works five hours per day, the warranty time for a failure population of 5% was 113 years [21]. Besides, the solar cell life was found to be very sensitive to the working temperature because of the large E_A value.

The failure analysis of the ALT [23] detected a severe deterioration of the silver grown electrolytically which made the busbar and fingers not equipotential. That would have led to the circulation of very high current densities into the semiconductor in very small areas of the cell structure during the emulation of nominal working conditions in the ALT. The confinement of these high current densities into small semiconductor regions would have favored the evolution of defects and cracks turning the solar cells into shunts (see Figure 9.11). This is in some extent a failure driven by thermal runaway. Thermal runaway failures caused by thermal fatigue were also found in the field although their roots are strongly dependent on the presence of moisture and the thermal cycling of the module [33], voids in the die attach [27] and perhaps in the semiconductor structure [28].

An additional similarity between the failures caused in an ALT [31,52] observed under temperature stress and the real ones occurring in the field by a combination of temperature, light bias and moisture [57] are the solar cell shunts at the perimeter. Therefore, it seems that the proposed ALTs are able to promote some of the failures appearing in the field. Even so, much more ALT proposals are needed to create a wide enough platform to recreate all real failures happening in the field.

Figure 9.11   SEM image of the evolution of the metallization of a faulty solar cell (left). SEM image of a crack propagation and the expected current flow (right). Source: Mercedes Gabás Pérez, Universidad de Málaga, 2015. Reproduced with permission of Mercedes Gabás Pérez

9.3.5 Reliability of Similar Devices

While the study of the reliability of CPV solar cells is going to be completed by means of both experience in the field and accelerated aging tests, it can be useful to look at other well proven and experienced PV cells and/or semiconductor devices, which share similar semiconductor materials, manufacturing techniques or operating conditions, namely, III-V space solar cells and light emitting diodes (LEDs).

9.3.5.1 III-V Space Solar Cells

The required high reliability of space systems forces them to have very demanding qualification standards such as the ECSS-E-20-OSA and AIAA S-111-2011 for III-V space solar cells [58,59]. The need of high efficiency solar cells showing a high radiation resistance spurred, in the end of 20th century, the deployment of GaInP/GaAs/Ge triple-junction space solar cells with efficiencies of 30% under AM0 spectrum as of 2015.

The materials and the structures of these space cells are almost the same than those of current commercial concentrator lattice-matched GaInP/GaAs/Ge triple-junction solar cells. However, the operating conditions are different, namely: a) in space there is an intense high energy particle radiation while in terrestrial applications there is not; b) in space the daily temperature variation is of some hundred degrees while in terrestrial operation is of only some tens of degrees; nevertheless, the thermal cycling on CPV cells is much more frequent because of the presence of clouds; c) the photogenerated current density in space solar cells is that produced at about 1.35 suns while that of CPV cells is that corresponding to some hundreds or even a thousand suns; and d) there is no humidity in space while in terrestrial applications there is.

Because of these meaningful differences in operating conditions, the high reliability experience of space cells cannot be directly applied to CPV cells, but their qualification standards can be used as a general framework for the development of qualification standards of CPV cells.

9.3.5.2 Light Emitting Diodes

LEDs are generally considered as very reliable devices. One of the promises of LED technology for illumination applications is a long service life higher than 100 000 h. The problem with this claim is that is not true under realistic conditions. A single, bare LED might last that long, but once integrated into a lighting system, the LED life could be far less. LED life estimation, especially lighting system life, is complicated by matters of integration with other components, including optics and housing (such as in the case of concentrator solar cells). Because of these constraints, the *MTTF* of LEDs spans from $2 \cdot 10^4$ to 10^5 h [60].

In order to analyze how to extrapolate the LED reliability experience to III-V concentrator solar cells, it is necessary to find out the role played by the semiconductor material and the working conditions on the device reliability. An analysis of these aspects can be found in [60]. The impact of the stress factors on LEDs which are not yet widely experienced on CPV cells can be summarized as follows:

- *Current density:* the need of a higher light output in LEDs forces a higher injection current density. However, the current density increase diminishes the LED life because of the activation of degradation mechanisms related to defects (see the inverse power law equation introduced in section 9.2.3.1 and now referred to current density in Table 9.5). As a consequence, most LED manufacturers have decided to limit the injection current densities from about 50 to 100 A/cm². On the other hand, concentrator solar cells manage photo-generated current densities around 15 A/cm² when operating at 1000 suns (7.5 A/cm² at 500 suns). Therefore, CPV solar cells operate at current densities 3–13 times lower than LEDs do. However, for a meaningful comparison not only the current density value has to be taken into consideration but also the current distribution in the device as it has been shown in section 9.3.1.

- *Humidity:* is a key degradation factor of LEDs in outdoor applications such as signage because of water penetration within the LED package. The influence of humidity in the device life can be described by the Peck's model (already introduced in section 9.2.3.1) with the equation shown in Table 9.5.
- *Light intensity:* LED light can degrade the surrounding encapsulant producing its yellowing. This degradation is accelerated by temperature and by short wavelength light (UV). Soft silicones are being used as encapsulants for LEDs because they are less sensitive to short wavelengths than conventional epoxies. Regarding solar cells, the degradation of silicones used as encapsulants is detailed in section 9.4.3. Concerning the photodegradation of concentrator solar cells themselves as a result of the high irradiance (hundred/thousand suns) they receive, there are few publications [53] and no one is devoted to concentrated UV.

A comparative summary of the abovementioned stress factors affecting the operational lifetime of LEDs and concentrator solar cells is presented in Table 9.5. By analyzing the similarities between concentrator solar cells and high-power LEDs it is possible to take advantage of the experience in LED reliability and apply it to CPV solar cells.

Table 9.5 Some stress factors affecting the reliability of LEDs that have not yet been extensively analyzed in CPV solar cells. Adapted from [60]

Stress factor	High power LEDs	III-V Concentrator Solar Cells	Operational device lifetime influence
Current Density	50–100 A/cm²	15 A/cm² (1000×)	$\frac{t_{J1}}{t_{J2}} = \left[\frac{J_2}{J_1}\right]^n$ $J \approx$ current density, $n \approx 2$ for LEDs
Humidity	Outdoor applications such as signage are exposed to humidity	Cells are exposed to water condensation inside the PV module.	$\frac{t_{RH_1}}{t_{RH_2}} = \left(\frac{RH_2}{RH_1}\right)^m$ • RH = relative humidity • m depends on failure type
Light Intensity	• Epoxy yellowing due to light emitted by LED. • Acceleration by temperature and UV light	• Solar cell light exposure is higher than LEDs. • Epoxy yellowing is higher under UV light.	• Radiation yellows encapsulation, reducing output light in LEDs and light impinging the cell. • No data about photodegradation of CPV cells themselves as a result of the high irradiance.

9.3.6 Links Among Degradation Studies, Reliability and Qualification Standards

It has been stated that there is not enough accumulated experience about the reliability of concentrator solar cells. The study of degradation is of great importance but it is important to keep in mind that reliability is a completely different issue. For instance, in terms of degradation, there are several studies describing tests in which a set of solar cells is introduced in a climatic chamber, the temperature is increased and the cells are biased to a specific current level, the number of failures is registered and the MTTF is calculated with the following expression (note the differences regarding Eq. (9.5))

$$MTTF = \sum_{i=1}^{N} \frac{t_{Fi}}{N_F} \qquad (9.25)$$

where t_{Fi} is the time of every failure and N_F is the number of failures. In such cases, the test is just a *degradation limited in time* study which only evaluates the solar cell power in a very aggressive and disperse way. The objective of these tests is to compare qualitatively in a short period of time the new generations of solar cells without any reliability analysis.

The problem is that this kind of tests does not establish a correlation between the test time and the lifetime of the device. With degradation limited in time, it is not possible to determine in which particular period of the classical bathtub curve the solar cell is working at. CPV has not the tradition of pure reliability including all its protocols, statistical methods, etc. Therefore, it is very common within the CPV field to find the use of the term 'reliability analysis' with the simple meaning, for example, of a set of degradation experiments in order to know if a given type of solar cell shows infant failures.

Therefore, reliability analysis of concentrator solar cells by means of QALTs must be carried out following the steps described in section 9.2.3. The accelerated tests allow shortening the time to evaluate the reliability of the solar cells, and have to be complemented with real time tests to determine if the stress factor introduces failure modes that would never happen in real operation.

Summarizing, the solar cell reliability testing should be directed not only to determine how long devices are going to live, but also the way in which these devices are going to live. In other words, reliability is also interested in knowing the probability distribution of failure in nominal conditions of operation [61]. This kind of reliability analysis should be carried out by solar cells manufacturers in order to know as much as possible of their own product and to supply the customers as much information as possible about the solar cell performance evolution.

Regarding the solar cells supplier-customer relationship, there is a need of a qualification standard in order to guarantee the customer that the cells have passed a given quality control. The IEC-62108 standard does not cover the qualification of solar cells so a new standard, namely IEC-62787 'Concentrator photovoltaic (CPV) solar cells and cell-on-carrier (CoC) assemblies – Qualification' is being developed. This requires the development of new procedures to determine the 'quality' of the different concentrator multijunction solar cells in a short period of time.

As of 2015, there are no reliability-based qualification tests on the IEC-62787 standard draft, but mainly storage based tests (i.e. the solar cells/CoCs are not under operating conditions during the tests), so, proper tests to predict the life of the devices are needed. Accordingly, it would be desirable for procedures to be simple and short in order to have an estimation of the reliability for any kind of concentrator solar cell architecture/CoC. That is, in addition to typical pass/fail tests being included in any qualification standard, tests that also determine if the solar cells/CoCs will pass their long-term nominal reliability target (in years) are required. Finally, not only tests on bare concentrator solar cells have to be developed but also tests considering the operation of cells in the module. As is shown in section 9.4, the way in which the die-attach and encapsulation is done plays a key role in the CPV solar cell life. So, reliability tests such as that described in [31] are compulsory.

9.4 Reliability of Modules

9.4.1 Introduction

Table 9.6 summarizes primary reliability concerns for CPV modules [32,62]. CPV modules are at risk to the same types of failures as flat-plate modules, but, in addition, will suffer catastrophic failure if thermal control is lost at the location of high solar flux and degradation in performance if the integrity of the optical path is compromised. The challenge with accelerated testing for CPV failure modes is that the modules often operate near or at their temperature or solar flux exposure limits. Below, each of the identified reliability issues are addressed and the current state of the art knowledge for their testing and evaluation is presented.

9.4.2 Die-attach

The die-attach has been identified as a key reliability concern, because failure of this thermal and electrical connection between the CPV cell and its assembly will result in a catastrophic failure of the cell. Therefore, reliability testing and modeling of the CPV die-attach has received the most attention in the current scientific work on CPV module reliability.

When in service under concentrated direct sunlight, the cell assembly in a CPV module will experience a complex temperature history. This history will contain many temperature reversals, or thermal cycles, that will impart stress to the die attach due to a mismatch in the coefficient of thermal expansion between the cell and its substrate. Thermal fatigue will lead to die-attach cracking which compromises the thermal conductivity of the bond and will result in thermal runaway and failure of the cell [39]. Thermal runaway is the feedback loop of Joule

Table 9.6 High-level summary of primary reliability concerns for CPV modules

Issue	Description
Die-attach	Die-attach and conductive epoxies, greases. Loss of thermal conductivity results in catastrophic cell failure.
Encapsulation	Chemical degradation or mechanical failure leading to optical absorption with catastrophic effect on cell operation.
Optics	Chemical or physical motivated reduction in transmittance leading to a gradual loss of module performance. Mechanical failure with catastrophic effect on module operation.

heating and current crowding that ultimately results in the formation of a current shunt, rendering the cell as an overall electrical short (see section 9.3.4.2).

The appropriate accelerated test to evaluate die-attach reliability is thermal cycling. Historically, empirical thermal fatigue equations have been used to calculate the fatigue life of the die-attach through accelerated thermal cycling tests. These equations include the Coffin-Manson and Engelmaier thermal fatigue models, among others. The Engelmaier equation calculates a fatigue life (N_f, number of cycles to failure) for a specific set of stress conditions [63]:

$$N_f = \frac{1}{2}\left[\frac{K_c}{\Delta T}\right]^{-1/c}$$

(9.26)

where the fitting constant K_c is a function of material properties and specimen geometry, ΔT the temperature range of the cycle and the exponent c depends on the cycle time and mean temperature. The Coffin-Manson model is also expressed in terms of fatigue life [64]:

$$N_f = K_{CM}\Delta T^{-2} f^{\frac{1}{3}} exp\left(\frac{E_A}{kT_{max}}\right)$$

(9.27)

This equation, demonstrated by experiment to be valid for ceramic and plastic packages attached with SnPb solders, computes N_f for a unique thermal cycle characterized by its ΔT, a new fitting constant K_{CM}, cycle frequency (f) and an Arrhenius dependence on the maximum cycle temperature (T_{max}) where an activation energy of $E_A = 0.12\,eV$ is typically used for SnPb solder, and k is Boltzmann's constant [65,66].

Numerical approaches using a 3D finite element model (FEM) of the CPV cell assembly have also been used to evaluate thermal fatigue in the CPV die-attach [67]. In this case, an accumulated energy approach is taken to evaluate the solder's lifetime. This approach considers the inelastic strain energy density as a damage indicator [68]. Accordingly, both crack initiation and growth are functions of the average inelastic strain energy density (plastic work) accumulated per thermal cycle. It is proposed that a typical cell assembly will fail after it has sustained a characteristic quantity of damage.

One study used the three options (TCA1, TCA2 and TCA3) of thermal cycle parameters, defined in the module qualification standard IEC 62108, as a basis to compare the empirical and numerical approaches [67]. The results presented were the simulated inelastic strain energy density (damage) accumulated during a single cycle of each type (right axis of Figure 9.12), normalized to unity for the most damaging cycle (left axis of Figure 9.12). To complete the comparison, the figure also includes the fits of the Engelmaier and Coffin-Manson (C-M) model for the same cycles. The Engelmaier fit had its constants adjusted to match the ratio of damage calculated from the FEM simulations of these TCA cycles ($K_c = 35$). The Coffin-Manson model fit the FEM well using the reported activation energy [67].

The same study also examined how the empirical approaches break down when trying to simulate the smaller, irregular cycles representative of a cell assembly's in-service temperature history. It was suggested that the numerical approach had the utility of simulating an actual temperature history to derive the amount of damage accumulated through service. The example presented was the measured cell temperature of one day in Golden, Colorado, which was used as input for the FEM (see Figure 9.13). The simulation suggested that, for this particular day,

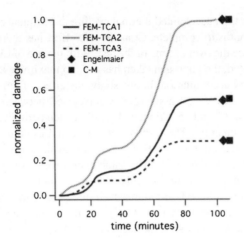

Figure 9.12 Comparison of empirical and numerical models to quantify rate of damage during thermal cycling described in IEC 62108

approximately 1 kPa of inelastic strain energy is accumulated in the die attach. Referring to Figure 9.12, the same amount of energy is accumulated during about one TCA3 cycle or about three of these particular days accumulate a similar amount of energy as one TCA2 cycle. This approach provides a critical step towards making the accelerated test a lifetime prediction tool, and thus is the subject of current research in this area [16,69].

Using this type of analysis, an acceleration factor between accelerated testing and service conditions may be determined. Considering this day in Golden accumulated an equivalent amount of inelastic strain energy as 1/3 of a TCA2 cycle, the acceleration factor is 3-days/ TCA2 cycle. Therefore, roughly 120 TCA2 cycles are equivalent to one year of exposure, if the weather on this particular day was repeated for an entire year.

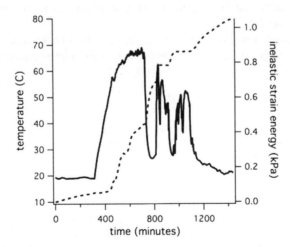

Figure 9.13 CPV cell temperature (solid) and simulated accumulation of inelastic strain energy within die-attach (dashed) over the period of one day in Golden, Colorado

While an effective test, an accelerated thermal cycling test sequence in IEC 62108 can still take on the order of months to complete. Considerable effort has therefore been put towards identifying how to reduce the overall time of the test. With the use of FEM simulations such as those previously presented, the approach taken has been to maximize the overall damage rate, or rate of inelastic energy accumulation. In one study, the effect of temperature ramp rate was examined [16,69]. Here, CPV cell assemblies were cycled between −40 and 110 °C with temperature ramp rates of 7.5 and 15 °C/min and the die-attach periodically evaluated for crack growth using ultrasonic imaging. The results were presented as cumulative crack growth on both a cycle and time basis and indicated that the crack growth rate was faster with the greater temperature ramp rate (see Figure 9.14). The same cycles were simulated with an FEM that indicated the faster cycle accumulated roughly 40% more inelastic strain energy. When the experimental results were evaluated for failure as a limit of crack area percent, the results were in good agreement with the model's ratio of damage accumulation rate. This study emphasized that while the faster cycle was only 40% more damaging on a per cycle basis, if the cycle times were also considered, it would accumulate crack growth 2.5× faster, producing a much more efficient accelerated test.

Accelerated thermal cycling is an effective test to evaluate thermal fatigue resistance of the CPV module's die-attach. Historical empirical equations for thermal fatigue work well for predicting the effect of various accelerated thermal cycles however are not useful in extrapolations that include the irregular cycles experienced while in service [67]. The employment of FEM simulations has made this extrapolation possible with an accumulated energy approach. One significant result from the FEM studies is the finding that increasing temperature ramp rate increases damage rate and provides for the most efficient accelerated thermal cycling test.

9.4.3 CPV Encapsulation

The encapsulation of CPV cells is a key reliability concern because failure of the encapsulant to protect the cell can lead to chemical or physical damage to the cell. Discoloration of the encapsulant will result in a catastrophic failure if the discoloured portion of the encapsulant is in the concentrated-light path.

Concentrating photovoltaic systems typically use a polymeric encapsulation to couple an optical component or coverglass to the photovoltaic cell. The encapsulation improves the

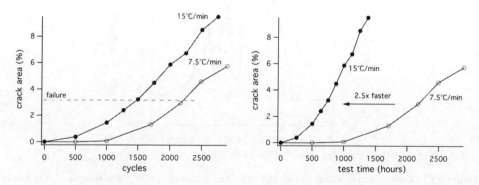

Figure 9.14 Crack growth vs. number of cycles (left). Crack growth vs. test time (right)

transmittance of concentrated optical flux by reducing reflection losses at interfaces, while protecting the cell from the field environment. The environment may corrode the gridlines (often composed of silver) or the window layer (such as AlGaAs [45] or AlInP [44]) or antireflective layers (such as Al_2O_3 or TiO_2) in multijunction cells. Poly(dimethylsiloxane) (PDMS) silicones find popular use in CPV, because of their known robustness in extreme environments [70]. While the use of hydrocarbon-based polymers may eventually prove riskier, these materials are sometimes used in CPV systems with a low (<10×) optical concentration. The durability of any CPV encapsulation, however, is not well established relative to the desired service life of 25 years.

Both hydrocarbon and silicone encapsulation are subject to degradation in response to stresses including UV, temperature and humidity. All of these stresses occur in a time-varied combination in the field environment, which may be further affected by the concentration of optical flux and the thermal management system (both unique to the module design). Degradation modes for CPV encapsulation examined in the literature include: discoloration, brittle fracture, fatigue crack growth, delamination at interfaces, void/bubble formation, optical haze formation, and densification. The modes and the corresponding methods of examination and aging will be examined in the following sections.

9.4.3.1 Discoloration

The topic of encapsulation discoloration was first examined for flat-panel PV [71], and has been observed for silicone encapsulation in CPV systems with a high (>1,000×) optical concentration [72,73]. Discoloration is much more detrimental in CPV than in flat-panel PV, because the formation of optically absorbing species within the encapsulation can facilitate thermal runaway and subsequent combustion of the encapsulation [70] subject to concentrated optical flux. The key parameters that may be used to predict encapsulation discoloration include the activation energy and action spectrum for the polymer. The activation energy, which depends on factors including temperature, UV, and humidity, can be applied in an Arrhenius representation. The activation spectrum is the convolution of the irradiation source (e.g., the direct solar spectrum) and the action spectrum [74,75] which characterizes the susceptibility for damage as a function of wavelength:

$$\Lambda(\lambda) = E(\lambda)s(\lambda) = E(\lambda)c_1 e^{-c_2\lambda} \tag{9.28}$$

In Eq. (9.28), given here for the international system of units (SI), Λ represents the activation spectrum ($W \cdot m^{-2} \cdot nm^{-1}$); E, the spectral irradiance ($W \cdot m^{-2} \cdot nm^{-1}$); s, the action spectrum (unitless); c_1 and c_2 represent empirical coefficients (unitless or nm^{-1}); and λ, the optical wavelength (nm). A power law representation is often used for the action spectrum.

The activation spectrum approach may be applied to distinguish representative CPV optical systems [76] (see Figure 9.15). For example, activation spectrum analysis readily identifies that 99% of the damaging UV radiation is attenuated using a poly(methyl methacrylate) (PMMA) Fresnel lens relative to a Silicone on Glass (SoG) lens or reflector-enabled (Ag or Al) optical system. Attenuation of the activation spectrum, even relative to a flat-panel module (FP-PV), qualitatively follows from the reduced area under the profiles for the PMMA lens containing systems in Figure 9.15.

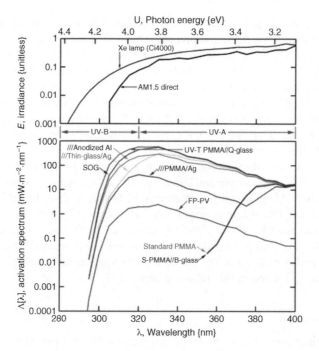

Figure 9.15 $\Lambda[\lambda]$ (at $C_g = 500\times$, except for FP-PV where $C_g = 1$) for key representative optical systems taken from [76]. The normalized spectral irradiance profiles for the direct solar resource (AM1.5 in ASTM G173) as well as a Xe lamp are provided for reference. The 'UV-A' and 'UV-B' bands are indicated between the two parts of the figure

The change in transmittance from discoloration, or any other mechanism facilitating optical absorption, may be examined using a spectrophotometer according to the method described in IEC 62788-1-4 [77]. Because a reduction in transmittance will likely correspond to catastrophic failure, more acute methods of diagnosis such as fluorescence [70,78], Fourier transform infrared (FTIR), or Raman [79,80] spectroscopy may be used to assess degradation. In the present IEC 62108 CPV module qualification standard [13], field deployment most rigorously queries the durability of the encapsulation material, particularly if the module is maintained at V_{oc} (where the cell temperature will be at its maximum). The CPV community may eventually prescribe an official standardized test condition (or sequence) that may be used to directly examine the UV durability of CPV encapsulation materials.

9.4.3.2 Brittle Fracture, Fatigue Crack Growth, and Delamination at Interfaces

Cohesive or interfacial fracture may result from the considerable difference in the thermal coefficient of expansion between silicone (often 200–300 ppm·°C^{-1}) and glass or thin film layers (on the order of 10 ppm·°C^{-1}). Cracking of rigid silicone encapsulation subject to a reduced temperature relative to the processing (cross-linking) temperature has been observed in a field deployed material specimen. Cracking has also been observed to result from the diurnal temperature variation through the process of fatigue. Cracking or delamination may not completely compromise the optical performance of a CPV module, but they do allow for rapid transport of water and the subsequent corrosion of the cell. Void formation within the encapsulation is undesirable because of the resulting optical loss, but also because such features may act as the weakest-link critical defects, facilitating mechanical fracture or delamination.

As in the die-attach, the issues of fracture, fatigue, and delamination may be understood through the methods of fracture mechanics. In the present IEC 62108 CPV module qualification standard [13], the thermal cycling and humidity freeze test sequences would both query the fracture or delamination of the CPV encapsulation. While these test sequences have not been analyzed specifically for encapsulation, they may stress the encapsulation in a manner consistent with the identification of infant mortality issues.

9.4.3.3 Void/Bubble Formation, Optical Haze Formation, Densification

Optical haze formation results from the scattering of light, where the corresponding optical scattering would reduce performance of the CPV module. Haze may follow from UV and/or thermal degradation of the encapsulation, including special additives [70]. Haze formation may also result from misuse (excessive application) of a primer, possibly making the encapsulation prone to delamination. Localized densification of field-deployed test specimens has been observed based on image distortion [70]. In addition to reduced optical performance believed to result from refractive index change, the localized mechanical strain present in densified regions could contribute to fracture, fatigue, and delamination.

To date, haze formation and densification have not proven problematic in fielded CPV modules and these issues may result only for experimental materials or the thick test specimens examined in [70]. Void formation, haze formation, and densification may all be examined using the same methods as those used for encapsulation discoloration, i.e., IEC 62788-1-4 [77]. For more rigorous field examination, a module may be deployed at V_{oc}, where cell temperature will be increased.

9.4.4 CPV Optics

CPV optical elements are a key reliability concern because loss of optical efficiency directly translates into reduced module efficiency.

Lens-based refractive designs (and not mirror-based reflective systems) currently dominate the CPV industry for modules using a substantial (>20×) geometric concentration of the sun's light. Monolithic Fresnel lenses composed of PMMA are a historically low-cost implementation. Such lenses may be manufactured by hot-embossing, casting, extruding, laminating, compression-molding, or injection-molding thermoplastic PMMA [81]. A laminated Fresnel lens consisting of discrete concentric prism elements composed of silicone, patterned on a glass superstrate is another lens implementation. Silicone on glass (SoG) lenses have gained popularity in CPV despite their greater weight and cost, based on their better optical performance and anticipated durability. SoG lens technology is presently more popular within the CPV industry for modules using a greater (≥500×) geometric concentration of the sun's light.

The relatively small deployment volume of CPV systems limits the available data to elucidate the most important degradation modes. A recent summary [82] of failures after for short-term deployment of CPV prototype modules reported breakage from hail, loss of mechanical integrity, and condensation on the inside of lenses. Full lifetime exposures of CPV systems are quite rare and may or may not reflect expected failures with today's newer materials.

9.4.4.1 PMMA Fresnel Lenses

If positioned well away from the receiver, a PMMA lens will experience field deployment close to the ambient conditions. Here, amorphous PMMA is subject to stresses including UV, temperature, humidity, and abrasion. The reliability of refractive CPV optics was recently reviewed in [83]. Degradation modes for PMMA Fresnel lenses include: surface crazing, brittle fracture, fatigue crack growth, delamination at interfaces, lens detachment, physical aging, solid erosion, wear, soiling, and discoloration. The modes and the corresponding methods of examination and aging will be examined in the following sections.

9.4.4.1.1 Surface Crazing, Brittle Fracture, Fatigue Crack Growth, Delamination at Interfaces, Lens Detachment

Crazing refers to the cracking of the base material, which may extend through the entire thickness [84], resulting in optical loss (haze) and contributing to fracture (as a weakest-link critical defect). Brittle fracture may result from the difference in the thermal coefficient of expansion between the PMMA lens ($70\,\text{ppm·}°\text{C}^{-1}$) and the module body (often aluminum, $\sim24\,\text{ppm·}°\text{C}^{-1}$). Brittle fracture may also result from impact, e.g., hail or wind-propelled-rocks. The toughness of PMMA allows crack growth from a micro-defect through fatigue to be analyzed using the log-linear Paris law [85]:

$$\frac{\partial a}{\partial N} = C \cdot \Delta K^m \tag{9.29}$$

where a represents the crack half-length (m); N, the number of cycles (cycles); ΔK, the range of stress intensity factor ($\text{MPa·m}^{0.5}$); C, a material-specific constant ($\text{m}^2\text{·Pa}^{-2}\text{·cycle}^{-1}$), and m, the Paris-law exponent (unitless, [86]). Delamination, which may be motivated by thermal misfit, may occur at the edges of the lens (where it is attached to the CPV module) or within the material (e.g., for a multilayer manufacturing process developed for PMMA [87]). The literature, however, only identifies loss of adhesion of the lens from the fielded module. Such delamination may occur from mechanical (wind) load, or the embrittlement of the adhesive material with age, combined with cold temperature, as in [82]. Of the issues motivating macroscopic fracture, the hail survivability of CPV modules may require the most design and verification effort. For example, in one hail storm at the NREL test site, many PMMA lenses and some SoG lenses were critically damaged on an assortment of prototype CPV modules, while no flat-panel PV modules were permanently damaged [82]. As in the die-attach, the issues of crazing, fracture, fatigue, and delamination may be understood through the methods of fracture mechanics.

The crazing of PMMA lenses might be examined using material specimens deployed in a hot-humid location, where both thermal- and hydro-expansion are occurring. The thermal cycling and humidity freeze test sequences of the IEC 62108 CPV module qualification standard [13] provide a very limited number of thermal or moisture cycles, not expected to invoke significant crazing. Regarding lens delamination, this issue is queried to some extent in the IEC 62108 thermal cycling and humidity freeze test sequences. There is yet no dynamical mechanical load test for CPV modules such that might simulate the rapid fluctuations occurring during transport to installation or from the wind at an installation site. Regarding lens fracture, this issue is queried in the thermal cycling and humidity freeze test sequences of IEC 62108.

Like lens delamination, fracture from mounting is queried in IEC 62108 based on CTE misfit, where the extent of examination would vary with the module design. Fracture from impact is presently examined in the hail impact test within IEC 62108.

9.4.4.1.2 Physical Aging

Physical aging refers to the densification of a polymer over prolonged time due to the consolidation of free volume within the material [88]. Sag (change in shape, e.g., in conjunction with gravity), embrittlement (contributing to fracture), and refractive index change (contributing to change in optical focus) may all result from physical aging. An applied stress ('creep test,' where creep compliance, J (Pa^{-1}), is proportional to $\log[t]$) or an applied strain ('relaxation test,' where creep modulus, E_c (Pa), is proportional to $\log[t]$) may be used to model the effects of physical aging. The double-logarithmic dependence for T (Equation (9.30)) –known as the Williams-Landel-Ferry (WLF) representation– may be used to obtain the time multiplier that may be applied to compare the rate of physical aging in different environments:

$$\log[a_t] = \log\left[\frac{\eta}{\eta_o}\right] = \frac{-C_1(T - T_r)}{C_2 + T - T_r} \tag{9.30}$$

Parameters in the equation include: a_t, the time multiplier (unitless); η, the dynamic viscosity ($Pa \cdot s$); η_o, a reference viscosity, typically the zero-shear viscosity ($Pa \cdot s$); C, the parameter-specific coefficients (unitless or K); T, the ambient temperature (K); and T_r, the reference temperature (K). Physical aging should not be confused with the outright melting of PMMA, which can occur near $\sim 105\,°C$. In comparison, field-deployed lenses typically rise $5–16\,°C$ above the ambient temperature [70], which can reach $\sim 55\,°C$ in hot desert locations like Riyadh, Saudi Arabia.

Temperature induced sag is one of the motivations for the options in the IEC 62108 protocol, including the thermal cycling test (maximum temperatures of 110, 85, or $65\,°C$) and damp heat test (conditions of $85\,°C/85\%$ relative humidity or $65\,°C/85\%$ RH). PMMA with a low molecular weight, including low quality or aged material, can have a melting temperature less than $80\,°C$. While there is no test standard for physical aging, it might be directly examined by subjecting lens or material specimens to elevated temperature or field deployment in a hot location. Because several of the effects of physical aging would only be manifest on a patterned lens, a module test (whether aged indoors or outdoors) or a lens test (which may require custom instrumentation) must be used.

9.4.4.1.3 Solid Erosion and Wear

Solid erosion and wear refer to the removal of material from soft polymer surfaces, e.g., sand-facilitated abrasion in the presence of wind. Erosion may also occur during the cleaning of the field facing surface of the lens, rendering an optical haze and corresponding reduction in optical transmittance. Surface damage may occur during maintenance when physical contact is used during cleaning, rather than a liquid spray. Erosion may facilitate mechanical fracture, if the material removal site was significant enough to act as the weakest-link for fracture. In one study [89], the cleaning of lens specimens using physical contact was found to damage the lens surface locally, and significantly deeper than wind-mediated solid erosion.

The standardized tests that most closely emulate the condition of field-induced solid erosion include the ASTM D 968 sand drop test [90] as well as the ASTM D658 and ASTM G76 air blast abrasion tests [91,92]. While the latter standard is presently withdrawn, the advantage of either test over methods using mechanical abrasion equipment or a stylus/indenter tip is the use of sand as the abrasive medium, presumably facilitating correlation to field results.

9.4.4.1.4 Soiling

The literature suggests that the accumulation of external contamination with time in the field, known as natural soiling, can reduce transmittance to the extent that might limit deployment of CPV technology [83,93]. Soiling may be reduced through the use of surface coatings [83,93], however, it remains to be established whether hydrophobic, hydrophilic, or anti-static surface functionalities perform adequately for CPV applications. The frequency and method(s) used for cleaning of CPV modules are also topics that have been studied to a limited extent [83,93]. The literature suggests, however, that no anti-soiling coatings and cleaning methods are perfect, i.e., a permanent loss of transmittance on the order of a few percent should be expected from soiling.

Regarding the deployment locations, soiling occurring in the south-western United States is considered more benign than in some Middle Eastern countries and India (in terms of the amount and tenacity of contamination) or China (where the desert environment can also be much more abrasive). Desert locations may result in the formation of cement-like surface layers; whereas more humid locations may favor the growth of organic material. Urban or industrial sites may have a very different species of contamination, e.g., soot from combustion or chemical vapor. Because the accumulation and consolidation of contamination requires precipitation and/or condensation cycles, soiling is difficult to simulate in a laboratory environment. Lacking a standard indoor test procedure that is well correlated with field data, deployment in challenging outdoor environments (or test sites) is recommended for the study of soiling. Test coupons may also be monitored in CPV installation sites prior to and after installation.

9.4.4.1.5 Discoloration

Discoloration, including a yellow or brown appearance, corresponds to optical absorptance and reduced module current generation. Examples of the effect of aging for the combined stresses of UV, temperature, and humidity are shown in Figure 9.16 for two different PMMA formulations [94]. The initial transmittance is shown to be reduced by aging for both materials,

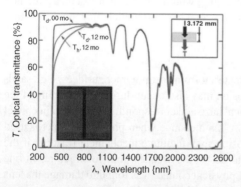 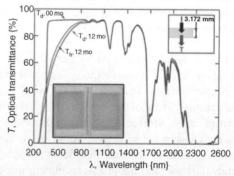

Figure 9.16 Measured optical transmittance (in air, compensated to the thickness of 3.175 mm) for the two different PMMA material specimens, taken from [94]. Both the hemispherical (T_h) and direct (T_d) measurements are shown for the specimens stressed for 12 months in addition to their visual appearance (photographed on black or white paper)

particularly in the visible spectrum. The material on the left in Figure 9.16 develops an optical haze, whereas the material on the right becomes discolored. Discoloration typically indicates the formation of a chromophore species and may correspond to molecular damage, i.e., chain scission. The mechanism of photolysis (including discoloration, crazing, and embrittlement enabled via UV-facilitated chain scission) is observed in the literature for PMMA subject to CPV field conditions, whereas the mechanism of thermal decomposition (indicated by mass loss via depolymerization) is not.

Regarding the indoor aging to examine discoloration, xenon lamps are recommended because they closely emulate the terrestrial UV solar spectrum [95]. A metal-halide lamp may also be used, however verification of the spectrum is advised (metal-halide bulbs and their companion filters are not nearly as standardized as for xenon). Test standards for Xe sources include ISO 4892-2 [95], ASTM G155 [96], which might also be applied for metal-halide equipment in addition to DIN 75220 [97]. Popular methods to achieve accelerated outdoor aging include ISO 877-3 [98] and ASTM G90 [99]. Because these procedures use a Fresnel apparatus to reflect up to 6× the intensity of the terrestrial sun, no spurious optical wavelengths can be introduced. The Fresnel reflector can be applied with temperature control and/or water spray at the specimen site.

Optical loss for PMMA lenses resulting from chemical mediated discoloration is often examined relatively early by those entering the CPV community. It remains to be established whether the optical loss associated with discoloration is more significant than the loss from surface crazing (occurring by optical scattering), solid erosion (also scattering), or soiling (both scattering and absorption). It also remains to be established if optical loss is a more prevalent problem, i.e., with a greater loss in energy production or need for maintenance, than mechanical damage, e.g., hail impact.

9.4.4.2 Silicone on Glass (SoG) Fresnel Lenses

Much of the discussion of the durability of SoG lenses is presently speculative, as unlike PMMA lenses, no 30-year-old SoG-enabled field installations exist. Degradation modes for SoG lenses include: physical aging, solid erosion and wear, discoloration, soiling, brittle fracture, fatigue crack growth, delamination at interfaces, lens detachment, focusing error, surface crazing, glass corrosion, and solarization. Some of these modes, including corrosion and solarization, uniquely apply to the glass superstrate. Several of the degradation modes for SoG lenses occur through the same underlying physical mechanisms or may be examined using the same methods as in PMMA lenses. For these modes, the literature describing PMMA lenses (discussed in the previous sections) also applies to SoG.

9.4.4.2.1 Physical Aging, Solid Erosion and Wear, Discoloration, and Soiling

The physical aging of SoG lenses is expected to be minimally relevant, because the glass transition temperature of the superstrate is well above ($>500\,°C$) that of the application environment and the silicone materials used for the lens facets are typically cross-linked. Because the glass transition temperature for silicone occurs well below ($<-120\,°C$) that of the application environment, physical aging or melting (typically occurring $\sim -40\,°C$) only becomes relevant in the case of inadequately cured material. The erosion of SoG lenses is expected to be greatly reduced for SoG because the glass superstrate is much harder than PMMA. Likewise, the discoloration of the lens facets is not anticipated for UV-robust silicone, aside from well-known exceptions like phenyl containing silicone [100,101]. Several studies suggest that an exposed glass surface will accumulate contamination at a lesser rate than PMMA [83,93]. Glass may, however, be more readily coated or functionalized to affect the rate of accumulation and cleaning of the SoG superstrate [83,93].

9.4.4.2.2 Brittle Fracture, Fatigue Crack Growth, Delamination at Interfaces, Lens Detachment, and Focusing Error

Considering that some of the silicone materials used for lens facets have self-healing characteristics, fracture and fatigue may only apply to the glass superstrate in SoG lenses. The superstrate may be tempered to improve its fracture, e.g., thermal misfit from mounting at the periphery, and impact, e.g., hail, resistance. Delamination between the glass superstrate and silicone facets is a potential weakness in SoG that is not possible in monolithic PMMA lenses. The significant difference in thermal coefficient of expansion between glass and silicone, in addition to consequences of failure, warrant study by those in the CPV community, including analysis as well as experimental verification of good adhesion for SoG lenses subject to CPV field conditions. Thermal misfit between the superstrate and lens facets can also result in a focusing error that varies with temperature [102,103]. This focal variation can be compensated with a secondary optic, and is therefore considered a performance issue rather than a reliability issue, as discussed in Chapter 5.

The fracture-related issues of lens cracking, delamination, and detachment for SoG are queried by CTE misfit in the thermal cycling and humidity freeze test sequences of IEC 62108 [13]. Like with fracture of PMMA lenses, the extent of examination facilitated by IEC 62108 varies with the module design. The addition of a dynamical mechanical load test to IEC 62108 could simulate the rapid fatigue facilitated by wind. Variation of the focal point with temperature may be understood through analysis and may be verified using a custom test fixture. Focusing error may also be examined using field deployment, provided the environment provides a temperature range that extends below the minimum temperature required for loss of optical flux.

9.4.4.2.3 Surface Crazing and Glass Corrosion

For SoG, the issue of traditional (polymer) surface crazing is limited to the silicone lens facets. Unlike PMMA, the silicone is a thin layer, limiting the depth of damage possible. Glass corrosion occurs through the leaching of alkali species present in glass (Na and Ca) in a basic environment, resulting in cracking, delamination, spalling, and pitting of the glass surface [104]. Like surface crazing, the roughened surface scatters light (reducing transmittance). The micrometer-scale defects created at the corroded surface can facilitate mechanical fracture under applied stress. Severely corroded glass may shatter upon handling.

The crazing of SoG lenses might be examined using component specimens deployed in a hot-humid location, which allows thermal- and hydro-expansion to contribute to crazing. As with any field-deployed SoG specimen, care must be taken to contain the faceted silicone surface, which will readily become covered with permanently adhered contamination in the field environment. Field studies may use a glass substrate, attached to the periphery of the SoG lens using UV durable polytetrafluoroethylene (PTFE) tape. The corrosion of some glass materials may be readily observed for hot and humid accelerated tests, including 1000 hours of 85 °C/85% relative humidity, 65 °C/85% RH, or 60 °C/60% RH. Corrosion, however, typically occurs at a much slower rate in the environment and there are no known examples of critically affected field-deployed CPV modules.

Rapid corrosion of glass can, however, occur readily in improper storage conditions – when wet glass sheets are stacked, a corrosive basic chemistry will develop between the glass pieces. The relevance for corrosion resulting from accelerated aging may be quite limited, i.e., a *substantial* site-specific acceleration factor or occurrence specific to manufacturing practices.

9.4.4.2.4 Solarization of the Superstrate

The redox balance between Fe^{2+} (ferrous iron) and Fe^{3+} (ferric iron) affects the transmittance of glass [83,105–107]. The transmittance spectra for the cerium-containing, low-iron-containing, soda-lime float glass (from the flat-panel PV industry) is shown as an example, before and after solarization (1000 hours of Xenon-arc lamp) in Figure 9.17. In the figure, the UV transmittance is reduced by solarization, which would reduce the power output for a CPV module, but may extend the life of the CPV encapsulation. Because solarization depends on the charge state of Fe, solarization may also occur with non-cerium-containing glass. The most overt result of solarization occurs at infrared wavelengths, however, the broad absorption band about 1,200 nm would minimally affect contemporary multijunction cells (the IR absorbing junction typically does not current-limit the cell in present triple-junction solar cells with germanium as the third junction, but could affect other architectures).

Like discoloration, haze-formation, and soiling, the change in transmittance from solarization may be examined using a spectrophotometer according to the method described in IEC 62788-1-4 [77]. Solarization will automatically be invoked during field deployment or through indoor aging with a Xe or metal-halide light source, e.g., ISO 4892-2 [95], ASTM G155 [96], or DIN 75220 [97]. The process of solarization will achieve completion in the environment, because it is enabled by a finite (ppm) concentration of chemical species. Rarely would a reducing atmosphere exist long enough in the field to reverse the reaction within the glass. In the case of nominal outdoor aging, the process of solarization may take as long as several years. For indoor aging, where the light intensity and temperature are often elevated, solarization may

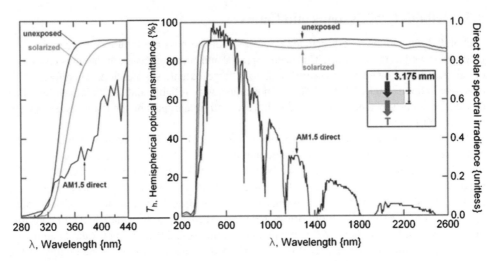

Figure 9.17 Measured optical transmittance (in air, compensated to the thickness of 3.175 mm) for the same glass specimen before and after solarization. The normalized direct solar spectral irradiance (AM1.5 in IEC 60904-3 [108]) is provided for reference

achieve completion within the period of an accelerated test, i.e., durations on the order of 1000 hours.

9.4.5 Other CPV Module Reliability Issues

Section 9.4 provides an abbreviated rather than a comprehensive treatment of all of the possible degradation and failure mechanisms for CPV modules. Issues related to the cells are treated in section 9.3 and issues that involve the tracker are covered in section 9.5. A few other CPV module issues are treated in this section.

A very common problem for CPV modules is moisture inside of the module. Most CPV module designs include a provision to equalize pressure inside and outside of the module so as to avoid deformation of the optical alignment as the atmospheric pressure changes. This transfer of air typically allows some transport of water vapor, resulting in the possibility of liquid water in the module. If the water touches the cells, the cells may be damaged, especially in the presence of light, eventually leading to complete exfoliation of the cells in some situations. In operation, the cell temperature increases, vaporizing the water and the water condenses on the coolest part of the module (usually the lenses). Fogging of the lenses decreases the output of the module. However, the performance may be completely recovered when the module dries if the cells have not been damaged. If the module seal is compromised, dirt may infiltrate the module. Any dirt in the light path will be problematic.

Additionally, CPV modules may experience ground faults. These faults are most likely to be found between the cell and the heat sink, since this connection is designed to have excellent thermal contact, motivating minimal thickness of the electrically insulating layers (see Chapter 7). Ground faults may be found at the time of manufacture or may appear in the field after months or years of use.

Other potential failure mechanisms include problems with the bypass diodes, junction box or connectors between modules.

9.5 Reliability of Systems and Plants

Section 9.4 constitutes a very detailed description of the possible failures occurring in CPV modules as well as their origins. This deep appraisal and comprehension of the physical causes of failures has allowed their control and, to a large extent, suppression. Therefore, well-developed commercial CPV modules are nowadays very reliable and we anticipate here that this section will show that module failures have a negligible impact on the unreliability of CPV systems and plants.

After more than eight years of experience (as of 2015) with CPV power plants in the world, several studies about power plant components performance and reliability have been published. Throughout this section, we will describe the degradation studies carried out in the different power plants at ISFOC (Puertollano, Spain) [109], a facility conceived as a test field for CPV technology where systems from the leading CPV manufactures have been operating –under exhaustive monitoring– for years (see Chapter 8, Annex 8-II). In addition, the most typical failures appeared in the first power plants and the very new qualification and aging tests which have been started to be performed in order to improve the quality of the CPV plant components will be also taken care of. As an example, we can say that Soitec's technology [110] has been installed in more than 20 locations in four continents and both the module and system technology have shown to be suitable for all conditions [111].

9.5.1 Performance Degradation in Power Plants

The experience of CPV regarding power plants is scarce. In fact, no many power plants are currently in operation for more than three years (as of 2015). In this section the experience of ISFOC, Soitec and Amonix power plants is presented as representative cases.

9.5.1.1 Experience at ISFOC

After almost three years of operation of the CPV plants at ISFOC, a study of their performance was carried out [112] looking for any sign of degradation. For this study only one CPV plant was selected, analyzing the evolution of efficiency as a function of time. To study the degradation three different methodologies were used:

9.5.1.1.1 Methodology 1: Comparison of the Efficiency of the Plants in Days with Similar Radiation

This methodology assesses losses by comparing efficiency of the plants during days with similar radiation in different seasons of different years. For instance, in Figure 9.18, we compare the efficiency of a CPV plant in summer 2009, 2010 and 2011, the results are very similar, with 21.4% in 2009, 21.5% in 2010 and 22.1% in 2011 as shown in Figure 9.18.

The comparison for winter, spring and autumn gives also similar results. At first sight, we can say that we cannot see a visible degradation of the performance of this plant during the first two years of operation using this methodology.

9.5.1.1.2 Methodology 2. Study the Performance of the Plant Versus DNI

The objective is to study the performance of the plant versus DNI. As illustrated by Figure 9.19, the energy produced by a CPV plant versus DNI yields a scatter plot that can be reasonably modeled by a linear fit, the slope of such line being proportional to plant efficiency. Therefore, if we represent these slopes for different years, the degradation factor could be analyzed.

Figure 9.19 shows the regression lines for the hourly energy production data versus DNI for different years of operation. The slopes of the regression lines are very similar and there is only a 0.84% difference between 2009 and 2011. Therefore, we can conclude that we cannot see any degradation with this methodology.

9.5.1.1.3 Methodology 3: Check the Power Generation of a Single System in the Plant

A final check to confirm the results obtained with Methodologies 1 and 2, is based on the analysis of the power generated by a single system (randomly chosen) of the plant, to verify if any degradation is observed. We have carried out different power measurements (I-V curve) of one concentrator system of the studied plant for several years. Table 9.7 shows the efficiency of the system in different dates (measurements carried out after cleaning the modules). The comparison should always be done in the same season, because the temperature conditions are key for the optical efficiency and therefore for the whole efficiency. We can observe that the efficiency is better in 2011 than in 2008. Of course we cannot say that the module has improved its efficiency, but the difference is probably due to the uncertainty in the measurement, different conditions and maybe some improvements in the tracking system. Anyhow, what can be said is that the degradation is not significant or, at least, it is not measurable.

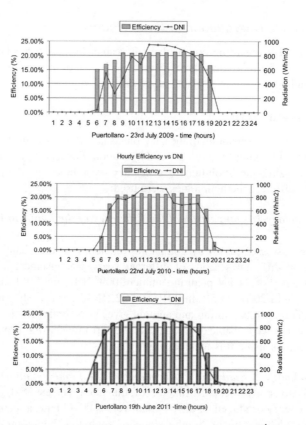

Figure 9.18 Hourly efficiency and DNI vs. time in Puertollano the 23rd July 2009 (top). Hourly efficiency and DNI vs. time in Puertollano the 22nd July 2010 (middle). Hourly efficiency and DNI vs. time in Puertollano the 19th June 2011 (bottom). Taken from [112]

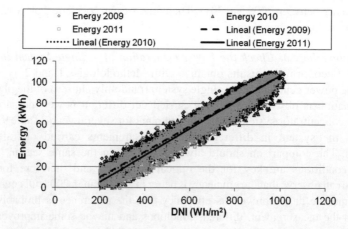

Figure 9.19 Hourly production vs. DNI during 2009, 2010 and 2011. Performance measurements of ISFOC for years 2009–2011.Taken from [112]

Table 9.7   Efficiency of one system of the CPV Puertollano plant in 2008 and 2011

Date	Efficiency
July 2008	23.08%
March 2011	22.72%
July 2011	23.70%

These negligible degradation rates are also indirectly confirmed with the information from Soitec, with zero module return from the field due to degradation after more than 5 years of experience and more than one million module operation months (number of modules times their time in operation, in months) [113].

In order to verify this degradation analysis, a module was removed from a CPV system operating in Puertollano (at ISFOC) for 3.5 years and then it was opened. It looked brand-new without neither signs of corrosion nor degradation. The optical transmission of the lens plate was measured and compared with a lens plate which was kept in the laboratory without using. Again no sign of degradation was found.

This study was done independently by ISFOC after the first 2.5 years of operation. Later on, Soitec has continued the study of degradation of this first power plant [114,115].

9.5.1.2   Experience at Soitec

The analysis carried out by Soitec about the performance and the degradation of its power plant at ISFOC in 2014 shows the results after 5 years of operation. The study is based in the AC performance ratio (*PR*) as defined in IEC 62670-2 (Draft) [116]:

$$PR = \frac{Y_f}{Y_r} = \frac{\dfrac{E_{net,AC}}{P_{CSTC}}}{\dfrac{E_{DNI}}{1\,kW/m^2}} \tag{9.31}$$

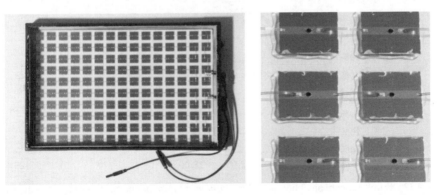

Figure 9.20   Module randomly chosen in the ISFOC power plant (left) that was open. Magnification of a part of the module containing six CoCs without any sign of degradation (right)

Figure 9.21 AC Performance Ratio for the first 5 years of operation of ISFOC's installation. Data filtered with DNI> 500 W/m². Source: Zech *et al.* 2014 [115]. Reproduced with permission of AIP Publishing LLC

Accordingly, Figure 9.21 shows the performance ratio of all the days during the first 5 years of operation.

For a more detailed comparison of the *PR*, similar days with clear sky and similar DNI and environmental conditions in the same season have been considered for the different years of operation and depicted in Figure 9.22. This figure shows very similar performance, independently of the year of operation.

Figure 9.23 shows the same days included in Figure 9.22, but only for the period between 12:00 and 13:00. As can be seen, *PR* values are again very similar for the 5 years.

All these results show no measurable difference in the performance over the 5 years of operation. This data confirm that no significant degradation could be measured at the power plant level. To confirm this result at the module level, Soitec has repeatedly measured a set of 7 modules installed at the ISFOC's power plant. Every 12–19 months the modules have been sent to Fraunhofer Institute in Freiburg to be flashed and compared with a reference module stored at controlled conditions. The comparison is shown in Figure 9.24 where there is no efficiency degradation detectable on any of the seven tested modules.

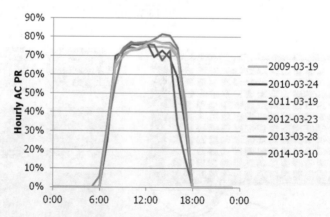

Figure 9.22 Daily comparison of hourly AC *Performance Ratio* for one clear-sky day of each year with comparable environmental conditions. Source: Zech *et al.* 2014 [115]. Reproduced with permission of AIP Publishing LLC

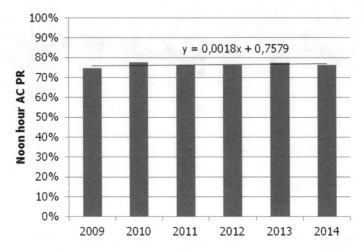

Figure 9.23 Noon hours (12:00–13:00 GMT) of AC *PR* for clear-sky days. The trend shows no change of the Performance Ratio. Source: Zech *et al.*, 2014 [115]. Reproduced with permission of AIP Publishing LLC

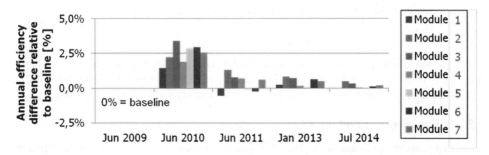

Figure 9.24 Annual efficiency differences of the seven Soitec CX-75-II CPV modules relative to the baseline efficiency. The relative difference converges to 0, because, every year the difference is divided by the number of years. Source: Zech *et al.*, 2014 [115]. Reproduced with permission of AIP Publishing LLC

In summary, following the different studies of this power plant there is no a noticeable degradation after the first 5 years of operation. This negligible degradation of CPV technology is not widely known and it should be one of its major advantages in comparison with other PV technologies.

9.5.1.3 Experience at Amonix

Amonix is one of the oldest CPV companies in the world. Founded in 1989, Amonix has installed systems in several climate conditions. They have analyzed the performance and reliability of their plants in several papers [117,118]. A work detailing the analysis of 5 years of field performance data gathered from five different sites totaling 570 kW of installed Amonix systems [119] shows the trend in performance, which can be obtained by dividing the monthly generated energy by the integrated direct incident sun irradiance for the month. Since it is desirable to be able to compare the performance trend of one system unit with other units of different size, this term is also divided by the rated power level. The resulting number is referred to as the monthly performance energy factor (MPEF). The MPEF for one unit is shown in Figure 9.25.

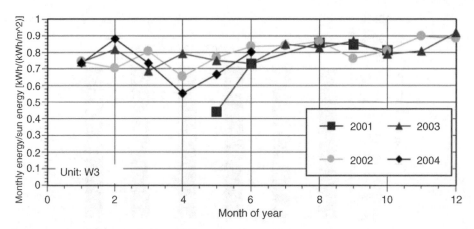

Figure 9.25   Monthly energy performance of one of the Amonix systems. Source: Stone *et al.*, 2006 [119]

Although the data presented in Figure 9.25 do not indicate any general degradation, there is significant scattering in the data points. This variation is the result of ambient temperature variation, wind speed variation, wind stow time, soiling rate, downtime, etc.

9.5.2   Failures of Components

In this section a description of the most typical failures of the components used in CPV plants is presented. These types of failures are not related with CPV technology itself but with bad designs of specific components of the plant or with quality failures of the components.

9.5.2.1   Experience at ISFOC

After two years of operation in the ISFOC's first installation, an analysis of the most frequent issues during the operation and maintenance (O&M) of the plant was reported [120]. In that study, the failure was identified as the intervention of the O&M team, without analyzing the impact in the availability of the plant (see Chapter 8, 8-II Annex). This study allowed to pinpoint reliability issues of some of the components and thus reduced the O&M works in the plant (Figure 9.26).

Figure 9.26 evidences that most of the O&M interventions were done in the trackers with around 72% of the tasks. All the other causes of failure have frequencies lower than 10%. The second in the rank is *protection and wirings*, the third is *inverters* and we need to go to the fourth position to find *modules*, followed by *software and communications* problems.

This remarkable accumulation of issues centered on trackers could be explained if we take into account that trackers are the only mechanical element of the installation; that ISFOC's plants were the first ones installed by most manufacturers and the tracker design was, maybe, not yet validated.

Figure 9.27 represents a breakdown of tracker issues occurred at ISFOC's demonstration plants between 2009 and 2010. The most frequent failures are related to the elevation mechanism (35%) and the control unit (30%). These failures were analyzed, together with system manufacturers, and it was concluded that some of them were design problems. That is the reason why the third most frequent failure is assigned as *design improvement*, namely, the periods of time in which CPV manufacturers were making the necessary replacements or improvements to solve the design failures.

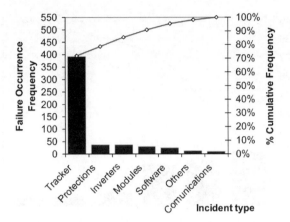

Figure 9.26 Pareto chart of the number of failures of each CPV component during 2009 and 2010 at ISFOC's demonstration plants (Puertollano, Spain). Source: Rubio *et al.*, 2011 [112]. Reproduced with permission of AIP

The second most frequent failure concerns the control unit. These types of failures are more difficult to detect at the design stage, but they can be improved during the commissioning period and, eventually, during the whole life of the power plant.

The rest of the issues in Figure 9.27 are not really failures, but really refer to internal maintenance, R&D activities, design improvements, calibration, cleaning, etc.

The design failures could have been solved with a good design validation. Therefore, the Working Group 7 of the IEC Committee has worked to develop characterization and qualification standards for trackers. As a consequence, the tracker standard IEC62817 was published in August 2014. A summary of IEC tests on trackers is described in Chapter 6.

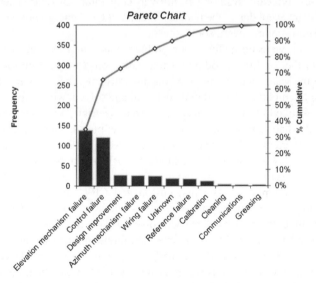

Figure 9.27 Pareto chart of the type of tracker failures during 2009 and 2010 at ISFOC's demonstration plants (Puertollano, Spain). Source: Rubio *et al.*, 2011 [112]. Reproduced with permission of AIP

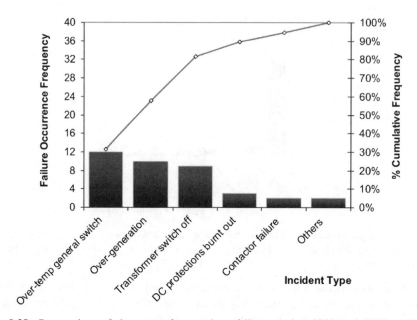

Figure 9.28 Pareto chart of the type of protections failures during 2009 and 2010 at ISFOC's demonstration plants (Puertollano, Spain). Source: Rubio *et al.*, 2011 [112]. Reproduced with permission of AIP

Back to Figure 9.26, the second general cause of failures in CPV systems is related to protections. Figure 9.28 represents a breakdown of these issues. Mostly, they are related to switches problems, typically resulting from over-temperature or over-generation situations. These early problems were due to the underestimation of CPV energy production as compared to that of PV. These failures were soon solved.

Module problems were normally minor problems such as breakage during assembly or maintenance with the forklift. In addition, some early non-watertight prototypes allowed the penetration of some water into the module housing. Anyhow, these few incidences (affecting less than 0.5% of the modules) were rapidly detected and solved, with defective modules being removed and substituted by the manufacturer so that electricity production was not affected.

The inverters also had some problems in the first power plants, mainly as a result of the use of off-the-shelf PV inverters with standard size, which were not adapted to the high electricity production of CPV. As is described in Chapter 8, CPV inverters have to be adapted to the situation of 0–100% (produced by no clouds–clouds cycles) and also need to recover very fast after a production plunge. The first customized CPV inverters also presented some manufacturing problems resulting in some water-tightness issues. Not being fundamental aspects, such manufacturing issues were soon solved. Again, qualification of new components stands out as a fundamental tool to overcome the aforementioned issues.

In summary, according to ISFOC's experience, the most important issue affecting the reliability of CPV systems and plants was related to the trackers and their influence could have been minimized with a good design validation. Therefore, qualification of power plant components emerges as a fundamental topic and, thereby, it will be discussed in section 9.5.3.

9.5.2.2 Experience at Amonix

After 5 years of operation in different locations [119], Amonix has analyzed the number of failures in the plant in order to identify the areas that need design attention. For the analysis, the plant was broken into an incident/failure tree composed of the subsystems and main components. The incident/failure tree is shown in Table 9.8. The table also shows the number of incidents, number of failures, the mean time between incidents (MTBI) and the mean time between failures (MTBF). The time base used in these calculations is on-sun generating time and not elapsed time or operating time.

As can be deduced from Table 9.8, the results of Amonix are quite similar to those of ISFOC. The first problem was the tracking system, which includes electronic control, hydraulics and drive with 55% of the issues when considering incidents and failures. The problems ranked second and third at ISFOC (see Figure 9.26), which were the protections and inverters, respectively, are jointly considered for Amonix within the category *AC/DC interface issues*.

Table 9.8 Unit incident and failure tree in Amonix plants (Taken from [119])

			Number of		Mean time between	
			Incident	Failure	Incident	Failure
	2.0 MegaModule	2.1 Misc	0	0	5857	5857
		2.2 Receiver plates	0	0	5857	5857
		2 3 Fresnel lens	0	0	5857	5857
		2.4 Connectors	0	0	5857	5857
		Total for subsystem 2.0 =	0	0	5857	5857
	3.0 AC/DC interface	3.1 Misc	0	0	5257	5257
		3.2 AC controls	3	1	1314	5257
		3.3 Inverter	19	4	263	1314
		3.4 DC controls	0	0	2176	5257
		Total for subsystem 3.0 =	22	5	236	1752
	4.0 Drive	4.1 Misc	0	0	5257	5257
		4.2 Azimuth ram	0	0	5257	5257
		4.3 Elevation ram	0	0	5257	5257
		Total for subsystem 4.0 =	0	0	5257	5257
1.0 System	5.0 Hydraulics	5.1 Misc	0	0	5257	5257
		5.2 Valves	0	2	5257	1752
		5.3 Sensors	0	0	5257	5257
		5.4 Pump/controls	0	0	5257	5257
		5.5 Hydraulic lines	0	0	5257	5257
		Total for subsystem 5.0 =	0	2	5257	1752
	6.0 Electronic controls	6.1 Misc	0	0	5257	5257
		6.2 Sun sensor	1	0	2629	5257
		6.3 Encoders	6	1	751	2629
		6.4 Wind sensors	0	0	5257	5257
		6.5 Control electronics/S	18	1	277	2629
		6.6 Tracking	4	0	1051	5257
		Total for subsystem 6.0 =	29	2	175	1752

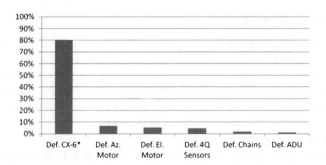

Figure 9.29 Most typical reliability problems of early Soitec technology. The codes used in the horizontal axis are: Def CX-6∗: Defects in CX-6 device: inverter + control unit; Def. Az. Motor: Defects in motor for azimuth movement; Def. El. Motor: Defects in motor for elevation movement; Def. 4Q sensors: Defect in sun sensor for tracking; Def. Chains: Defects in chain of the mechanism; Def. ADU: Defects in the Air Drying Unit used in Soitec technology

This type of issues accounted for 45% of the incidents and failures, the inverters being largely responsible for this amount. Finally, Amonix modules exhibited no problems, a situation very similar to what was described for ISFOC plants.

Therefore, the conclusions that can be drawn from Amonix experience are in full agreement with those of ISFOC plants. The specific elements of CPV (i.e. modules integrating highly efficient concentrating optics and multijunction solar cells) seem extremely reliable as a result of the efforts made in this field summarized in sections 9.3 and 9.4. It is the adaptation to CPV of existing technologies, namely, inverters and trackers, which is impacting the reliability of plants.

9.5.2.3 Experience at Soitec

After several years of operation, Soitec has identified the most typical reliability problems of their plants, as shown in Figure 9.29 and they have used this information as lesson learned to determine the solutions of Table 9.9.

Table 9.9 Identified on field reliability problems with identified solution by Soitec

Part	Part Description	# Failed Units	Identified Problem	Solution
CX-6 Inverter	The inverter converts the DC output of the modules into AC power	310	Integrated inverter and Tracker Control Unit (TCU) design. The CX-6 inverter design showed low reliability and led to limited flexibility regarding CPV system size.	Separate inverter and tracker control function and use of off-the-shelf inverters plus TCU
Azimuth (AZ)/ Elevation (EL) Motor	DC motor used to drive the tracker in AZ/EL direction	27/21	Relatively high number of mechanically brocken motors. The housing of the small DC motors used in the CX-P6 seem to be too weak under high (wind) loads.	AC motors with significantly more robust design used in actual CX-S530-II system

The statistics shows that the use of non standard key components, with low reliability, like the customized inverter of early Concentrix technology, could produce many problems on field. Therefore, the proposed solution was to use standard and well qualified inverters to be integrated in the field. To avoid this type of problems, Soitec developed a program to validate and qualify the inverters before validating them for a power plant installation. Again, qualification of power plant components emerges as a fundamental need to ensure the reliability of CPV installations and, thereby, it will be discussed in the next section.

9.5.3 Qualification Tests on Power Plants Components

9.5.3.1 Qualification Tests on Trackers and Drives

As we have already studied, many factors influence the reliability of a CPV plant, but the one which has the highest impact on O&M interventions is the tracker, including both the mechanical system and the control unit. Therefore, the CPV community started to work to improve the tracker quality. The tracker-related IEC standard is '*IEC 62817 Photovoltaic systems – design qualification of solar trackers*' [121]. One of the most important outcomes of this IEC standard is the description of the tests procedures for tracker qualification, which are described in the section 8 of the standard. The tests to be carried out are: a) visual inspection; b) functional validation tests; c) performance tests; d) mechanical testing; e) environmental testing; and f) accelerated mechanical cycling.

Some tracker manufacturers or CPV system developers have already started to use this standard (or equivalent internal standards) to qualify their trackers. For example, Sener [122] has used its own methodology to validate its tracker design [123]. First, Sener has carried out FEA and CFD computer simulations and then, they have carried out some wind tunnel tests with a scale model to verify the simulation results. Additionally, Sener has done some statistical and structural analysis and finally, it has done some measurements in the field, with a completely monitored tracker during one year, to characterize the behavior of the tracker. This type of characterization is very valuable to qualify the performance of the tracker system.

On the other hand, Soitec is developing an internal test procedure that details the methodology and equipment to be used in order to conduct a test that is compliant with the mechanical testing of the IEC 62817. Throughout the development of the standard, Soitec has followed the methods described in the IEC 62817 [121] to carry out the tests of pointing repeatability, backlash testing and moment testing under extreme wind loads, including a method involving the use of a hanging load with sand bags and metal weights which is shown in Figure 9.30.

Some other independent laboratories, like ISFOC or ITECAM, are preparing their facilities to carry out the different environmental tests detailed in IEC 62817. The overall objective of this testing is to induce failures or infant mortality associated with design that may occur as a result of accelerated environmental cycling of the drive system, control system, and associated wiring in a wide range of environmental conditions. ISFOC has already performed some of these tests [124] in their climatic chambers. The environmental tests are: temperature cycle with dust, temperature cycling, humidity freeze cycling and freeze/spray test.

In summary, the testing procedures described in IEC 62817 are very useful during the development of a solar tracking system as they provide the technical requirements that need to be fulfilled in order to ensure the reliability of such a key element in a CPV plant.

Figure 9.30 Setup for the moment testing under high wind loading

9.5.3.2 Qualification Tests on Other Plant Components

The IEC 62817 is not only a qualification standard for the tracker, but also for electronic equipment that can have separate failure mechanisms from that of the mechanical equipment associated with trackers. These electronic component system (ECS) consists of control electronics, power supplies, sensors, encoders and enclosures. These tests are covered in section 9 of IEC 62817, '*Design qualification testing specific to tracker electronic equipment*', where the following tests are described: visual inspection; functional test; protection against dust, water and mechanical impacts; robustness of terminals; surge immunity; shipping vibration shock; UV; thermal cycling; humidity-freeze and damp heat tests. Some of these tests have already been carried out by independent laboratories for different manufacturers.

Besides electronic components, other system components should be also qualified during the system validation. These elements are, among others, the limit switch assemblies, brackets, cable loom, cable clamps, tubing, diode connector, etc. These tests were mainly defined considering identified critical parameters, lessons learned and also from *failure mode and effects analysis* (FMEA). The validation which is carried out on this elements is mainly assembly testing (Figure 9.31 left), functional testing and UV tests (Figure 9.31 right).

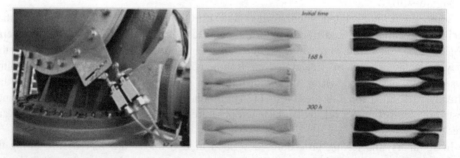

Figure 9.31 Elevation limit switch assembly validation in a prototype (left). UV tests in different cables of a CPV system (right)

Figure 9.32 Picture of dummy tracker test stand at the supplier factory. Source: Kinematics Manufacturing Inc. Reproduced with permission of Kinematics Manufacturing Inc

9.5.4 Aging Tests

Additionally, some CPV manufacturers have started to carry out special tests on the tracker with the double objective of checking the quality of the system and understanding the O&M needs of their CPV systems.

For example, Soitec has started aging tests in one of the most critical components of the tracker system, i.e. the drive. A test bench has been built to test the drive system under load (see Figure 9.32). An accelerated cycle has been defined to simulate one year of operation in only 15 days. After every simulated year the test stops to verify the play in the 3 axis x, y, z (see Figure 9.33) to study the degradation over the time. The test duration will emulate the complete lifetime of the tracker system, namely, 25–30 years. The backlash measurement results do not

Dialguage No.1   Dialguage No.2   Dialguage No.3

Figure 9.33 Torque in azimuth direction - Location of dial gauges. Source: Kinematics Manufacturing Inc. Reproduced with permission of Kinematics Manufacturing Inc

Table 9.10 Inspection results of the azimuth drive after aging test (simulating 25 years). Source: Kinematics Manufacturing Inc. Reproduced with permission of Kinematics Manufacturing Inc

	Item	New Part	After 25 Years
	Axial Run out (mm)	0.05	0.03-0.05
Slewing Ring	Radial Run out (mm)	0.07	0.03-0.07
	Teeth Thickness (mm)	7.50	7.18~7.32
Worm Shaft		OK	There is no obvious wear.
Bearing		OK	There is no obvious wear.

Figure 9.34 Inspection of the wear in an azimuth drive after a test simulating 25 years. Source: Kinematics Manufacturing Inc. Reproduced with permission of Kinematics Manufacturing Inc

show any significant increase of the backlash. After the disassembly of the drive, the inspection results (see Table 9.10 and Figure 9.34) do not show any significant wear or degradation which would affect the performance of the drive.

The objective of the tests is to understand the influence of the operation time on the performance of the drive as a function of time. In the analysis of the first aging tests of the drive, it was realized that the wind influence, which could be very high, was not included in the first set up of the test. Therefore, Soitec, together with some laboratories is trying to define an aging test which includes the effect of wind.

Other component manufacturers, like gears suppliers, have very efficient aging test methodologies to verify the wear of the gear in a very short time applying different load profiles. They can emulate 25 years lifetime in only six weeks.

All these aging tests will definitely help the CPV industry to understand better the reliability of the product and will improve the quality of the whole system.

9.6   Standards Development for CPV

The IEC 62108 CPV qualification standard has a provenance traceable back to early reliability studies by Jet Propulsion Laboratory (JPL) for non-concentrating PV modules in the 1970s, evaluation testing by Sandia National Laboratories in the 1980s and development of a US qualification standard by the IEEE in the 1990s before it was finally published in 2007. However, IEC 62108 is only the beginning. New test procedures need to be identified before a reliability standard can be published. Meanwhile, additional CPV standards and technical specifications are appearing to support the development of a mature CPV industry. In this section, all these aspects are described in detail.

9.6.1   Standards as the Mark of a Mature Industry

Consumers always look for evidence that the products they buy will work as expected. They may talk with friends who have bought the product or search the internet for product reviews such as test reports prepared by 'Consumer Reports'. Investors in large PV projects, however, need more convincing evidence of reliable product performance before funding projects requiring billions of dollars or euros. Investors will hire well-qualified engineering firms to conduct *due diligence* of the project in order to minimize risk. Investors may also look to government incentives to help recover original investment and reduce risk. However, incentives are being reduced and government incentives may still insist on meeting requirements for qualifying for the incentives. PV and CPV manufacturers are now facing requests for evidence that their products must last decades to be competitive with conventional electricity sources. Thus, it becomes essential that a CPV customer or investor can be given clear evidence that a product will be reliable.

Test standards are a primary vehicle for documenting reliability of a new product. If it has been shown that a standard test procedure is a strong indicator of field performance, then that test procedure provides a tool for helping the PV customer or investor decide whether a new product can be purchased with minimum risk. Other standards, e.g. performance standards, help customers understand what they are buying by documenting performance under a set of standard operating conditions. Both types of standards facilitate the launching of a new technology.

9.6.2   History of CPV Standards Development

9.6.2.1   The Jet Propulsion Laboratory *Block Buys*

The basis for all PV and CPV standards began within the Flat-Plate Solar Array (FSA) project initiated in 1975 at the Jet Propulsion Laboratory (JPL) and funded by the US Department of Energy (DOE) [125]. These activities built on JPL's earlier history for making precise measurements and conducting failure analyses within its space PV programs. The JPL project had three reliability activities that helped launch the early PV industry through 1) a series of module purchases (called *Block Buys*) supporting industry development of improved PV module designs; 2) reliability physics studies of module degradation and failure mechanisms; and 3) module performance and failure analyses directed toward developing standard qualification tests for PV modules.

Modules procured by JPL through five successive *Block Buys* were installed at 16 field sites with a variety of environments: extreme weather (e.g. Alaska), marine (e.g. Key West, Florida), high desert (Albuquerque, New Mexico), mountain (Colorado), urban coastal (New Orleans, Louisiana), Midwest (Indiana), upper great lakes (Michigan), northwest (Seattle, Washington) and urban southwest (Pasadena, California). When problems occurred during field tests, JPL performed an in-depth failure analysis to find the exact cause of each problem. A highly detailed failure report was prepared presenting the analysis results and recommendations for correcting module deficiencies. These reports were supplied to the module manufacturer and to JPL personnel responsible for modules development programs, including module qualification tests. During the various Block Buys (five in all), these analyses were instrumental in correcting design and processing problems to the extent that new modules incorporating the recom-mended changes were then successful in passing the qualification tests. The qualification tests evolved over a 10-year period resulting from information provided by the field test experience, module research and from the tests themselves. By 1985 JPL had developed a PV module qualification test that began with an initial evaluation for:

- electrical performance;
- visual and mechanical inspections;
- an electrical insulation test;
- nominal operating cell temperature (NOCT) determination;
- voltage and current temperature coefficient determinations.

Following the initial evaluation, a set of modules were subjected to the following test sequences:

- A thermal test cycling the module from −40 °C to 90 °C for 50 cycles (for Blocks I through Block IV) and an added extended test of 200 cycles for Block V. This test was followed by:
- A humidity-freeze test cycling initially from −23 °C to 40 °C (95% relative humidity) for 5 cycles (Block I through IV) and finalizing after Block V with cycling from 85 °C (85% relative humidity) to −40 °C for 10 cycles. This test was followed by:
- A mechanical loading cycle starting at 100 cycles of 2400 N/m^2 for Blocks I to III and 10000 cycles for Blocks IV and V. This test was followed by:
- A wind test (residential applications only) based on an Underwriters Laboratories standard. This test was followed by:
- A module twist test where a module corner was lifted. This test was followed by:
- A hail impact test introduced for Blocks IV and V using 20 mm diameter hail with a velocity of 23 m/s. This test was followed by:
- A final evaluation included electrical performance, visual inspection, insulation test and documentation of all changes.

In addition to this sequence of tests, the following tests were part of the qualification testing:

- An electrical isolation test specifying a maximum current of 50 µA at 1500 V for Blocks I to III; 2000 V for Block IV and 3000 V for Block V;
- A ground continuity test introduced for Blocks III to V of 50 mΩ minimum resistance to ground for exposed conductors; and

- A hot-spot endurance test introduced for the final Block V procurement specifying 100 hr endurance short circuited at normal operating cell temperature and 100 mW/cm^2 irradiance.

Over a 10-year period, JPL performed qualification tests on PV modules of about 150 designs. Studies of field performance showed reduced failure rates with progressive *Block Buys* culminated by the most dramatic improvement between Blocks IV and V [126]. Despite this impressive demonstration of the value of qualification testing, the final JPL report emphasized the following: '*Module and array field testing are required because module qualification testing has not been entirely successful in predicting and/or duplicating module field failures.*'

Because of these JPL results, PV and CPV module qualification testing have been considered to be strong qualifiers for successful field performance for a couple of years but not for 20 to 30 years of reliable field operation. As a result, PV and CPV qualification standards today have many tests resembling those in the JPL test sequence. However, later standards developers have made test modifications, added new tests and even eliminated some tests as more information became available or new technologies were developed. While evidence for PV and CPV reliability begins with qualification testing the strongest evidence for reliability still comes from long-term field deployments.

9.6.2.2 CPV Evaluation Testing at Sandia National Laboratories

In the late 1970s and 1980s, again with funding from DOE, Sandia National Laboratories conducted research and development of CPV technologies. In the late 1980s and as a prelude to a planned CPV initiative (later cancelled due to funding reductions) Sandia developed a test protocol for evaluating CPV technologies [127]. The Sandia evaluation tests had many similarities to the JPL Block 5 specification but recognized the fundamental geometrical differences between PV and CPV. Most flat-plate modules have a planar two-dimensional geometry, whereas concentrator modules have optics separated from the solar cells, forming a three-dimensional structure. The Sandia evaluation protocol had parallel test sequences for CPV receivers (two-dimensional assemblies of CPV cells including their electrical and thermal conduits) and CPV module (environmentally protected three-dimensional assemblies of receivers, optics and related components). Some CPV tests, such as the humidity-freeze test for receiver sections, were adapted directly from the JPL Block V specification. A few tests, such as an off-axis beam damage test, were developed specifically for CPV evaluation.

9.6.2.3 IEEE 1513–2001

In 1997, an IEEE (International Institute of Electrical and Electronics Engineers) standards working group began developing a qualification standard for CPV receiver sections and modules. The first draft was based on the evaluation tests developed at Sandia National Laboratories. It also followed the general outline of the IEEE standard 1262–1995 that, in turn, had been based on the earlier JPL Block V specification [128]. The work began with the support of the National Renewable Energy Laboratory as a direct result of a CPV company's concern that the lack of a CPV qualification standard was affecting their marketing, sales and ability to participate in state and national demonstration projects. IEEE 1513–2001, published in May 2001, was the world's first CPV qualification standard [129].

IEEE 1513–2001 began with a set of baseline tests for:

- electrical performance
- ground continuity
- electrical isolation
- wet insulation resistance
- visual inspections.

Following the baseline tests were several test sequences:

- *Sequence A* specified a test for bypass diodes and a thermal cycle test for two receivers.
- *Sequence B* specified different thermal cycle tests from that in *Sequence A*, humidity-freeze tests for two receivers and parallel tests for two modules involving thermal cycling, humidity-freeze, electrical isolation and terminations.
- *Sequence C* specified damp heat exposure for two receivers followed by a test for electrical isolation.
- *Sequence D* involves several module stress tests, including outdoor exposure, water spray, off-axis beam damage, hail impact and hot-spot endurance.
- *Sequence E* gave a specification for the special case of inaccessible by-pass diodes.

The final test and inspections included:

- visual inspections
- electrical performance
- electrical isolation
- wet-insulation resistance
- ground continuity.

A critical question after many stress tests is the performance of the CPV module or receiver section. In the case of flat-plate PV modules, a solar simulator and reference cell provide a means to verify performance. However, solar simulators couldn't meet the CPV requirement for collimated irradiance over a module area of about $1\,m^2$. Therefore IEEE 1513–2001 proposed three possible options to verify any degradation.

- The first option allowed the testing organization to conduct baseline outdoor performance testing under various temperature and solar irradiance conditions while measuring output currents and voltages to obtain an analytical expression for the module and receiver performance. This option required lots of time to determine the analytical equation for the module's performance and required outdoor testing after the stress tests to identify any degradation.
- The second option used a reference module, not subjected to any stress tests, whose performance was measured along with all other test modules. After a module is subjected to a test sequence, its performance is again compared with that of the reference module.
- The third option was the use of dark current-voltage (I-V) measurements before and after intermediate stress tests.

Almost simultaneously with the final publication of the IEEE 1513–2001 CPV standard, US CPV companies realized that international standards would soon be needed to participate in overseas projects. In the 1990s, US flat-panel PV companies had been participating in an IEC (International Electrotechnical Commission) technical committee to develop international PV standards. In 2000, NREL worked to create a new CPV working group within that technical committee. By 2001, Working Group 7 (WG7) had been approved within the PV Technical Committee 82 (TC82) to create international CPV standards. The first draft submitted to IEC for a new international CPV qualification standard was IEEE 1513–2001.

9.6.2.4 The First International CPV Standard: IEC 62108 Edition 1

One major problem in submitting IEEE 1513–2001 as a draft international standard was recognition that the US CPV experts were more knowledgeable about refractive optics than reflective optics. Soon after WG7 was formed, CPV experts with reflective optics experience from other countries, e.g. Australia and Spain, began participating in the development of a new international CPV qualification standard [13,130]. Further developments and availability of high efficiency multijunction solar cells provided additional impetus for a new IEC qualification standard. As a result there were differences between IEEE 1513–2001 and the later IEC drafts. The significant differences included the following:

* new definitions for receivers, modules and assemblies;
* test options for systems with active or passive cooling of receivers;
* new electrical isolation tests, since CPV receivers and modules have areas significantly different than those of flat-plate systems;
* eliminated the complex analytical procedure to verify degradation after a test sequence;
* incorporated the use of a control receiver or control module as the principal method for verifying degradation after IEC stress tests;
* added an additional current cycle during the high temperature dwell of the thermal cycle test.

9.6.2.5 Standards Beyond IEC 62108

The final IEC 62108 Edition 1, 'CPV modules and assemblies – design qualification and type approval', was published in December 2007 [13]. Since then WG7 membership has grown dramatically from approximately 20 members in 2007 to over 80 members in 2013. During this period of growth, WG7 initiated several projects for CPV standards and technical specifications. An IEC Technical Specification (IEC/TS) provides guidelines or recommendations without the pass/fail criteria of IEC standards. The following two specifications and two standards have been completed:

* IEC/TS 62727 Edition 1, Photovoltaic systems – Specifications for solar trackers.
* IEC/TS 62789 Edition 1, Specification of concentrator cell description.
* IEC 62670-1 Edition 1, Concentrator Photovoltaic (CPV) performance testing – standard conditions.
* IEC 62817 Edition 1, Solar trackers for photovoltaic systems—Design qualification.

IEC 62670-1 Edition 1 is very important to today's growing CPV market. It provides the following standard conditions for CPV performance testing:

a. CSOC – *Concentrator Standard Operating Conditions*
- Irradiance: 900 W/m^2 direct normal irradiance.
- Temperature: 20 °C ambient temperature.
- Wind speed: 2 m/s
- Spectrum: Direct normal AM1.5 spectral irradiance distribution consistent with conditions described in IEC 60904-3

b. CSTC – *Concentrator Standard Test Conditions*
- Irradiance: 1000 W/m^2 direct normal irradiance.
- Temperature: 25 °C cell temperature.
- Spectrum: Direct normal AM1.5 spectral irradiance distribution consistent with conditions described in IEC 60904-3.

Finally, IEC has approved as of April 2015 the development of the following:

- IEC/TS 62108-9 (62108 retest guidelines);
- IEC 62108 Edition 2 (CPV modules and assemblies—design qualification and type approval);
- IEC 62670-2 Edition 1 (CPV performance testing - energy measurement);
- IEC 62670-3 Edition 1 (CPV performance testing – performance measurements and power rating);
- IEC 62688 Edition 1 (module and assembly safety qualification);
- IEC/TS 62989 Edition 1 (concentrator optics technical specification);
- IEC 62787 Edition 1 (CPV cells and cell-on-carrier assemblies qualification);
- IEC 60904-3 Edition 3 (Edition 3 includes direct normal spectral irradiance distribution);
- IEC 62925 Edition 1 (thermal cycling test to differentiate increased thermal fatigue durability).

Acknowledgement

The development of IEC standards is primarily a group activity by individuals developing, testing or purchasing CPV systems. Hundreds of individuals have participated over the years and deserve recognition for their participation in dozens of IEEE and IEC meetings. Without their support, the final documents would not have been published.

References

1. O'Connor, Patrick, and Andre Kleyner. *Practical Reliability Engineering*. John Wiley and Sons. 2011.
2. Dimitri B. Kececioglu. *Reliability and Life Testing Handbook*, DEStech Publications.
3. Yang, Guangbin. *Life Cycle Reliability Engineering*. John Wiley and Sons. 2007.
4. Vassiliou, P., Mettas, A., and El-Azzouzi, T. (2008) Quantitative Accelerated Life-testing and Data Analysis. In *Handbook of Performability Engineering* (pp. 543–557). Springer London.
5. Escobar, L. A., and Meeker, W. Q. (2006). A review of accelerated test models. *Statistical Science*, **21**(4), 552–577.
6. Catelani, M., and Ciani, L. (2014, May). Highly Accelerated Life Testing for avionics devices. *First IEEE Metrology for Aerospace* (MetroAeroSpace), 29–30 May, Benevento, Italy, pp. 418–422. 2014.
7. Crowe, D., and Feinberg, A. (eds) *Design for Reliability*. CRC Press. 2010.
8. M. Vázquez, N. Nuñez and A. Borreguero Degradation of AlInGaP red LEDs under drive current and temperature accelerated life tests. *Microelectronics Reliability*, **50**, 1559–1562. 2010.
9. Peck, D. S. (1986, April). Comprehensive model for humidity testing correlation. In IEEE Reliability Physics Symposium, 1986. 24th Annual (pp. 44–50).

10. Blish, R. C. Temperature cycling and thermal shock failure rate modeling. Annual Proceedings, 35th IEEE International Reliability Physics Symposium, pp. 110–117. 1997.

11. International Electrochemical Commission. IEC 61215 Crystalline silicon terrestrial photovoltaic modules-design qualification and type approval. 1993.

12. Wohlgemuth, J., and Kurtz, S. Photovoltaic module qualification plus testing. 40th IEEE Photovoltaic Specialist Conference (PVSC), pp. 3589–3594. 2014.

13. International Electrochemical Commission. Working Group 7 within IEC Technical Committee 82 (Photovoltaics), IEC 62108 Edition 1, Concentrator photovoltaic (CPV) modules and assemblies- design qualification and type approval, 2007, https://webstore.iec.ch/publication/6469 (accessed November 2015).

14. Muñoz, E., Vidal, P. G., Nofuentes, G. CPV standardization: An overview. *Renewable and Sustainable Energy Reviews*, **14**(1), 518–523. 2010.

15. C. R. Osterwald and T. J. McMahon. History of accelerated and qualification testing of terrestrial photovoltaic modules: a literature review. *Progress in Photovoltaics Research and Applications*, **17**, 11–33. 2009.

16. Nick S. Bosco, Timothy J. Silverman, and Sarah R. Kurtz. On the effect of ramp rate in damage accumulation of the CPV die-attach. 2012 IEEE Photovoltaic Specialists Conference. Austin, Texas June 3–8. 2012.

17. H. Ossenbrink and T. Simple, Results of 12 years of module qualification of the IEC61215 standard and CEC specification 503. Proceedings of the 3rd World Conference on Photovoltaics Energy Conversion, p. 1882–1867. 2003.

18. J. H. Wohlgemuth, D. W. Cunningham, A. M. Nguyen and J. Miller, Long term reliability of PV modules. Proceedings of the 20th European Phtovotoltaic Solar Energy Conference, p. 1942–1946. 2005.

19. J. H. Wohlgemuth, Reliability testing of PV modules. Proceedings of the 23rd IEEE Photovoltaics Specialists Conference, p. 889–892. 1994.

20. Wohlgemuth JH. Long term photovoltaic module reliability, NCPV and Solar Program review meeting, Denver, CO, USA. National Renewable Energy Laboratory, NREL/CD-520-33586, pp. 179–183. www.nrel.gov/docs/fy03osti/33586.pdf (accessed November 2015). 2003.

21. P. Espinet-González, C. Algora, N. Núñez, *et al.* Temperature accelerated life test on commercial concentrator III-V triple-junction solar cells and reliability analysis as a function of the operating temperature. *Progress in Photovoltaics Research and Applications*, **23**(5), 559–569 DOI: 10.1002/pip.2461. 2015.

22. N. Nuñez, M. Vazquez, V. Orlando, P. Espinet-Gonzalez and C. Algora, Semi-quantitative temperature accelerated life test (ALT) for the reliability qualification of concentrator solar cells and cell on carriers, *Progress in Photovoltaics Research and Applications*, **23**, 1857–1866. DOI: 10.1002/pip.2631. 2015.

23. P. Espinet-Gonzalez, R. Romero, V. Orlando, M. Gabas, N. Nunez, M. Vazquez, *et al.,* "Case study in failure analysis of accelerated life tests (ALT) on III-V commercial triple-junction concentrator solar cells," in *Photovoltaic Specialists Conference (PVSC), 2013IEEE 39th*, Tampa, FL, 2013, pp. 1666–1671.

24. M. Fukuda, *Reliability and Degradation of Semiconductor Lasers and LEDs*. Artech House. 1991.

25. Kenji Araki, Hirokazu Nagai, and Kazuyuki Tamura, Fatigue failure of concentrator III-V solar cells - Does forward bias current injection really kill III-V CPV cells? *AIP Conference Proceedings* **1477**, 281 DOI: 10.1063/1.4753886. 2012.

26. Military Handbook Electronic Reliability Design Handbook, MIL-HDBK-338B. 1998.

27. N. Bosco, C. Sweet, M. Ludowise and S. Kurtz. An infant mortality study of III-V multijunction concentrator cells. *IEEE Journal of Photovoltaics*, **2**(4), 411–416. 2012.

28. K. Araki, O. A. Taher, H. Nagai, P. Hebert and J. Valles Are electro-luminescence defects in concentrator III-V cells responsible to thermal runaway and sudden death? 7th International Conference on Concentrating Photovoltaic Systems *AIP Conference Proceedings* 1407, 303–306. 2011.

29. T. J. McMahon. Accelerated testing and failure of thin-film PV modules. *Progress in Photovoltaics Research and Applications*, **12**, 235–248. DOI: 10.1002/pip.526. 2004.

30. B. Prior. Roadmap for CPV technology. GTM Research. October 31 2011.

31. N. Nuñez, J. R. González, M. Vázquez, C. Algora, P. Espinet., Evaluation of the reliability of high concentrator GaAs solar cells by means of temperature accelerated aging tests. *Progress in Photovoltaics Research and Applications*, **21**, 1104–1113. 2013.

32. S. Kurtz, J. Granata and M. Quintana. Photovoltaic-reliability R and D toward a solar-powered world. *Proceedings of the Society of Photographic Instrumentation Engineers* (SPIE) Solar Energy + Technology Conference San Diego, California August 2–6. www.nrel.gov/docs/fy09osti/44886.pdf. 2009.

33. M. Muller. Experience with CPV module failures at NREL. Reliability Workshop, Golden CO, USA. 2012. www.nrel.gov/docs/fy12osti/54838.pdf (accessed November 2015).

34. International Electrotechnical Commission (IEC) IEC 60904-9 Photovoltaic devices – Part 9: Solar simulator performance requirements. Geneva.

35. V. Orlando, P. Espinet, N. Nuñez, *et al.* Preliminary temperature accelerated life test (ALT) on lattice mismatched triple-junction concentrator solar cells-on-carriers. CPV-10 International Conference on Concentrating Photovoltaic Systems, 7–9 April, 2014 Albuquerque, NM (USA), *AIP Conference Proceedings*, 1616, 250–253. DOI: http://dx.doi.org/10.1063/1.4897072. 2014.

36. M. Martínez, F. Rubio, G. Sala, *et al.* CPV plants data analysis. ISFOC and NACIR projects results. AIP Conference Proceedings, 1477, 32316–18 April 2012, Toledo, Spain. 2012.

37. Gombert, A. Low cost reliable highly concentrating photovoltaics - a reality. 38th IEEE Photovoltaic Specialists Conference (PVSC), pp. 001651,001656, 3–8 June 2012 DOI: 10.1109/PVSC.2012.6317913. 2012.

38. K. Ghosal, D. Lilly, J. Gabriel, S. Seel, B. Fisher, S. Burroughs. Semprius module and system results. 40th IEEE Photovoltaic Specialist Conference, Denver, Colorado (USA), June 8–13, 2014.

39. N. Bosco, C. Sweet, T. Silverman and S. Kurtz. CPV cell infant mortality study. 7th International Conference on Concentrating Photovoltaic Systems. *AIP Conference Proceedings,* 1407. 2011.

40. C. G. Zimmermann. Thermal runaway in multijunction solar cells. *Applied Physics Letters*, **102**, 233506. 2013.

41. R. Campesato, G. Gori, G. Gabetta, M. Casale. CPV infant mortality and indoor characterization for high efficiency, reliable Solar Cells. 9th International conference on concentrator photovoltaic systems, April 15–17, Miyazaki, Japan. 2013.

42. D. C. Jordan and S. R. Kurtz. Photovoltaic degradation rates – an analytical review. *Progress in Photovoltaics: Research and Applications*, **21**, 12–29. 2013.

43. P. Espinet-González, I. Rey-Stolle, C. Algora and I. García. Analysis of the behavior of multijunction solar cells under high irradiance Gaussian light profiles showing chromatic aberration with emphasis on tunnel junction performance. *Progress in Photovoltaics: Research and Applications,* DOI: 10.1002/pip.2491. 2014.

44. R. France, M.A. Steiner, T. G. Deutsch, *et al.* Oxidation and characterization of AlInP under light-soaked, damp heat conditions. 35th IEEE Photovoltaic Specialists Conference (PVSC), pp. 002016–002020, 20–25 June 2010 DOI: 10.1109/PVSC.2010.5616977. 2010.

45. S. van Riesen and A. W. Bett. Degradation study of III–V solar cells for concentrator applications. *Progress in Photovoltaics: Research and Applications*, **13**(5), 369–380. 2005.

46. I. Rey-Stolle and C. Algora. Reliability and degradation of high concentrator GaAs solar cells. Proceedings of the 17th European Photovoltaic Solar Energy Conference and Exhibition, October 2001, Munich, Germany.

47. R. France and M.A. Steiner. High-irradiance degradation studies of metamorphic 1eV GaInAs solar cells. MRS Proceedings 1432. DOI: http://dx.doi.org/10.1557/opl.2012.1028. 2012.

48. F. Eltermann, M. Wiesenfarth, G. Siefer, J. Wilde, A. W. Bett. The effects of accelerated aging tests on metamorphic III-V concentrator solar cells mounted on substrates. 26th European Photovoltaic Solar Energy Conference and Exhibition, 5–9 September, pp. 163–168, Hamburg, Germany. 2011.

49. J. Schone, G. Peharz, R. Hoheisel, *et al.* Impact of thermal cycles on the material quality of metamorphic III-V solar cell structures. Proceedings of the 23rd European Photovoltaic Solar Energy Conference and Exhibition, 1–5 September, Valencia, Spain. 2008.

50. C. G. Zimmermann, C. Nomayr, M. Kolb and A. Rucki. A mechanism of solar cell degradation in high intensity, high temperature space missions. *Progress in Photovoltaics: Research and Applications*, **21**(4), 420–435. 2013.

51. S. Padovani, A. Del Negro, M. Antonipieri *et al.* Triple junction InGaP/InGaAs/Ge solar cells for high concentration photovoltaics application: Degradation tests of solar receivers. *Microelectronics Reliability*, **50** (9–11), 1894–1898. 2010. http://dx.doi.org/10.1016/j.microrel.2010.07.087

52. J. R. González, M. Vázquez, N. Núñez *et al.* Reliability analysis of temperature step-stress tests on III–V high concentrator solar cells. *Microelectronics Reliability*, **49**(7), 673–680. 2009. http://dx.doi.org/10.1016/j.microrel.2009.04.001.

53. I. Rey-Stolle and C. Algora. High-irradiance degradation tests on concentrator GaAs solar cells. *Progress in Photovoltaics: Research and Applications*, **11**(4), 249–254. 2003. http://dx.doi.org/10.1002/pip.487

54. C. G. Zimmermann. The impact of mechanical defects on the reliability of solar cells in aerospace applications. *IEEE Transactions on Device and Materials Reliability*, **6**(3), 2006.

55. W. Q. Meeker and L. A. Escobar. Review of recent research and current issues in accelerated testing. *International Statistical Review* **61**, 1, 147–168. 1993.

56. P. Hacke, R. Smith, K. Terwilliger, S. Glick, D. Jordan, S. Johnston, M. Kempe and S. Kurtz"Testing and analysis for lifetime prediction of crystalline silicon PV modules undergoing degradation by system voltage stress" Photovoltaic Specialists Conference (PVSC), Volume 2, 2012 IEEE 38th, pp.1,8, 3–8 June 2012, DOI: 10.1109/PVSC-Vol2.2012.6656746.

57. Bosco, N., and Kurtz, S. CPV cell characterization following one-year exposure in Golden, Colorado. 10th International Conference on Concentrator Photovoltaic Systems, CPV-10, 1616, 242–245. AIP Publishing.

58. ECSS European Cooperation for Space Standardization. Photovoltaic Assemblies and Components. 2004.

59. AIAA Standard — Qualification and Quality Requirements for Space Solar Cells (S-111-2005). 2005. https://www.aiaa.org/StandardsDetail.aspx?id=3921

60. Manuel Vázquez, Carlos Algora, Ignacio Rey-Stolle and Jose Ramón Gonzalez. III-V concentrator solar cell reliability prediction based on quantitative LED reliability data. *Progress in Photovoltaics: Research and Applications.* **15**, 477–491. 2007.

61. C. Algora. Reliability of III–V concentrator solar cells. *Microelectronics Reliability* **50**, 1193–1198. 2010.

62. S. Kurtz, J. Granata, and M. Quintana. Photovoltaic-reliability R and D toward a solar-powered world. pp. 74120Z-74120Z./ 2009.

63. W. Engelmaier. Generic reliability figures of merit design tools for surface mount solder attachments. *IEEE Transactions on Components, Hybrids, and Manufacturing Technology,* 16, 103–112. 1993.

64. K. C. Norris and A. H. Landzberg. Reliability of controlled collapse interconnections. *IBM Journal of Research and Development,* 13, 266–271. 1969.

65. A. E. Perkins and S. K. Sitaraman. *Solder Joint Reliability Prediction for Multiple Environments.* Springer, New York. 2009.

66. H. Cui. Accelerated temperature cycle test and Coffin-Manson model for electronic packaging. Annual Proceedings, Reliability and Maintainability Symposium, pp. 556–560. 2005.

67. N. Bosco, T. Silverman, and S. Kurtz. Modeling thermal fatigue in CPV cell assemblies. *IEEE Journal of Photovoltaics,* **1**, 242–247. 2011.

68. R. Darveaux. Effect of simulation methodology on solder joint crack growth correlation. Proceedings 50th Electronic Components and Technology Conference, pp. 1048–1058. 2000.

69. N. Bosco, T. Silverman, and S. Kurtz. Simulation and experiment of thermal fatigue in the CPV die attach. International Conference on Concentrating Photovoltaics (CPV-8), Toledo, Spain. 2012.

70. D.C. Miller, M.T. Muller, M.D. Kempe, *et al.* Durability of polymeric encapsulation materials for concentrating photovoltaic systems. *Progress in Photovoltaics: Research and* Applications, DOI: 10.1002/pip.1241.

71. A.W. Czanderna and F.J. Pern. Encapsulation of PV modules using ethylene vinyl acetate copolymer as a pottant: A critical review. *Solar Energy Materials and Solar Cells,* **43**(2), p. 101–181. 1996.

72. M.R. Winter, I. Aeby, J. Foresi. Performance and reliability of silicone polymers in 1000× concentration CPV applications. Proceedings PV Module Reliability Workshop. 2012.

73. M.R. Winter, I. Aeby, J. Foresi. Designing reliability into a 1000× concentration CPV system through targeted stress testing. Proceedings SPIE, 846805-1-12. 2012.

74. A.L. Andrady. Wavelength sensitivity in polymer degradation. *Advanced Polymer Science,* **128**, 47–94. 1996.

75. N.D. Searle. Activation spectra: Techniques and applications to stabilization and stability. ASTM STP 1385. 2000.

76. D.C. Miller, M.D. Kempe, C.E. Kennedy, S.R. Kurtz. Analysis of transmitted optical spectrum enabling accelerated testing of multi-junction CPV designs. *Optical Engineering,* **50**(1), 013003. 2011. DOI: 10.1117/1.3530092

77. International Electrotechnical Commission. IEC 62788-1-4: Measurement procedures for materials used in Photovoltaic Modules: Part 1 – Encapsulants. Measurement of optical transmittance and calculation of the solar-weighted photon transmittance, yellowness index, and UV cut-off wavelength. IEC, Geneva. In progress.

78. Silverstein RM, Bassler GC, Morrill TC. *Spectrometric Identification of Organic Compounds,* 5th edn. Chapter 7: ultraviolet spectrometry. John Wiley and Sons Inc., New York. 1991.

79. Röder B, Ermilov EA, Phillipp D, Köhl M. Observation of polymer degradation processes in photovoltaic modules via luminescence detection. Proceedings SPIE 7048, 70480F-1. 2008.

80. Pieke C, Kaltenbach T, Köhl M, Weiss KA. Lateral distribution of the degradation of encapsulants after different damp-heat exposure times investigated by Raman spectroscopy. Proceedings SPIE, 7730, 77730E-1. 2010.

81. R. Leutz, L. Fu, H. P. Annen., Stress in large-area optics for solar concentrators. Proceedings SPIE, 7412, 7412-06. 2009. DOI: 10.1117/12.827357

82. M.T. Muller, S.R. Kurtz, J. Rodriguez. Three years of observed failures and performance related issues associated with on-sun CPV module testing. *Proceedings International CPV Conference. American Institute of Physics Conference Proceedings* **1477**, 143–147. 2012. DOI: 10.1063/1.4753854

83. D.C. Miller, S.R. Kurtz. Durability of Fresnel lenses: A review specific to the concentrating photovoltaic application. *Solar Energy Materials and Solar Cells,* **95**(8), 2037–2068. 2011. DOI: 10.1016/j.solmat.2011.01.031

84. E.H. Andrews, L. Bevan. Mechanics and mechanism of environmental crazing in polymeric glass. *Polymer*, **13**, 337–346. 1972.

85. B. Mukherjee, D.J. Burns. Fatigue-crack growth in polymethylmathacrylate. *Experimental Mechanics*, **11**, 433–439. 1971.

86. N.H. Watts, D.J. Burns. Fatigue crack propagation in polymethylmethacrylate, *Polymer Engineering and Science*, **7**, 90–93. 1967.

87. M.J. O'Neill, A.J. McDanal. Photovoltaic manufacturing technology (PVMaT) improvements for ENTECH's concentrator module. National Renewable Energy Laboratory, NREL/TP-411-20277, pp. 1–74. 1995. www.nrel. gov/docs/legosti/fy96/20277.pdf (accessed November 2015).

88. R. Greiner, F.R. Schwarzl, Thermal contraction and volume relaxation of amorphous polymers, *Rheologica Acta*, **23**, 378–395. 1984.

89. D.C. Miller, L.M. Gedvilas, B. To, C.E. Kennedy, S.R. Kurtz. Durability of poly(methyl methacrylate) lenses used in concentrating photovoltaics. *Proceedings SPIE*, 7773-02. 2010.

90. American Society for Testing and Materials. ASTM D968: Standard Test Methods for Abrasion Resistance of Organic Coatings by Falling Abrasive. ASTM International, West Conshohocken, PA, USA, pp. 1–5. 2010.

91. American Society for Testing and Materials. ASTM D658: Standard Test Method for Abrasion Resistance of Organic Coatings by Air Blast Abrasive (Withdrawn 1996). ASTM International, West Conshohocken, PA, USA, pp. 1–3. 1991.

92. American Society for Testing and Materials. ASTM G76: Standard Test Method for Conducting Erosion Tests by Solid Particle Impingement Using Gas Jets. ASTM International, West Conshohocken, PA, USA, pp. 1–6. 2013.

93. T. Sarver, A. A. -Qaraghuli, L.L. Kazmerski. A Comprehensive review of the impact of dust on the use of solar energy: History, investigations, results, literature, and mitigation approaches. *Renewable and Sustainable Energy Reviews*, **22**, 698–733. 2013.

94. D.C. Miller, J.D. Carloni, D.K. Johnson, *et al.* An investigation of the changes in poly(methyl methacrylate) specimens after exposure to ultra-violet, heat, and humidity. *Solar Energy Materials and Solar Cells*, **111**, 165–180. 2013. DOI: dx.doi.org/10.1016/j.solmat.2012.05.043.

95. International Organization for Standardization. ISO 4892-2: Plastics - Methods of exposure to laboratory light sources - Part 2: Xenon-arc lamps. ISO, Geneva, pp. 1–13. 2013.

96. American Society for Testing and Materials. ASTM G155: Standard Practice for Operating Xenon Arc Light Apparatus for Exposure of Non-Metallic Materials. ASTM International, West Conshohocken, PA, USA, pp. 1–11. 2005.

97. Deutsches Institut Fur Normung E.V. DIN 75220: *Ageing of Automotive Components in Solar Simulation Units.* DIN, Berlin, pp. 1–6. 1992.

98. International Organization for Standardization. ISO 877-3: Plastics - Methods of exposure to solar radiation - Part 3: Intensified weathering using concentrated solar radiation. ISO, Geneva, pp. 1–11. 2009.

99. American Society for Testing and Materials. ASTM G90: Standard Practice for Performing Accelerated Outdoor Weathering of Nonmetallic Materials Using Concentrated Natural Sunlight. ASTM International, West Conshohocken, PA, USA, 1–11. 2010.

100. Department of Energy. Dow Corning Corporation Develop Silicone Encapsulation Systems for Terrestrial Silicon Solar Arrays. DOE/JPL-954995-80, pp. 1–57. 1979.

101. G. Deshpande, M.E. Rezac. The effect of phenyl content on the degradation of poly(dimethyl diphenyl) siloxane copolymers. *Polymer Degradation and Stability*, **74**, 363–370. 2001.

102. T. Schultz, M. Neubauer, Y. Bessler, P. Nitz, A. Gombert. Temperature dependence of Fresnel lenses for concentrating photovoltaics. Proceedings of the Concentrating Photovoltaic Workshop, Darmstat, Germany, 9–10 March 2009.

103. T. Hornung, A. Bachmaier, P. Nitz, A. Gombert. Temperature dependent measurement and simulation of Fresnel lenses for concentrating photovoltaics. Proceedings of the International Conference on CPV Systems, *AIP Conference Proceedings,* 1277, p. 85. 2010.

104. C.G. Pantano, D.E. Clark, L.L. Hench. *Corrosion of Glass, Books for Industry and the Glass Industry.* New York, 1979.

105. A.J. Faber, Optical properties and redox state of silicate glass melts, *Comptes Rendus Chimie*, **5**, 705–712. 2002.

106. S.B. Donald, A.M. Swink, H.D. Schreiber, High-iron ferric glass. *Journal of Non-Crystalline Solids*, **342** 539–543. 2006.

107. W. Thiemsorn, K. Keowkamnerd, S. Phanichphant, P. Suwannathada, H. Hessenkemper. Influence of glass basicity on redox interactions of iron-manganese-copper ion pairs in soda-lime-silica glass. *Glass Physics and Chemistry*, **34**, 19–29. 2008.

108. International Electrotechnical Commission. IEC 60904-3: Photovoltaic devices - Part 3: Measurement principles for terrestrial photovoltaic (PV) solar devices with reference spectral irradiance data. IEC, Geneva, pp. 1–62. 2008.

109. F. Rubio, P. Banda, J.L. Pachón and O. Hofmann. Establishment of the Institute of Concentration photovoltaics systems – ISFOC. Proceedings of the 22nd PVSEC, Milan. 2007.

110. S. van Riesen et al. Concentrix Solar's progress in developing highly efficient modules. Proceedings of the 7th International Conference on Concentrating Photovoltaic Systems CPV-7, Las Vegas. American Institute of Physics Conference Proceedings **1407**, pp. 235–238 (2011).

111. A. Gombert, T. Gerstmaier, M. Gomez, et al. Field experience of Concentrix Solar's CPV systems in different climatic conditions. AIP Conference Proceedings **1407**, p. 327. 2011.

112. F. Rubio, M. Martínez, D. Sánchez, R. Aranda, P. Banda., Two years operating CPV plants: Analysis and results at ISFOC. Proceedings of the 7th International Conference on Concentrating Photovoltaic Systems CPV-7, Las Vegas. American Institute of Physics Conference Proceedings **1407**, 323. 2011.

113. A. Gombert and F. Rubio. The importance of manufacturing processes and their control for the reliability of CPV systems. Proceedings of the 9th International Conference on Concentrating Photovoltaic Systems CPV-9, Miyazaki, Japan. American Institute of Physics Conference Proceedings **1556**, 279–283. 2013.

114. T. Gerstmaier, M. Röttger, T. Zech et al. Return of experience from 4+ years of field data: reliability and predictability of Soitec's CPV technology. 28th EU PVSEC, Paris, pp. 113–117. 2013.

115. T. Zech, T. Gerstmaier, M. Röttger, et al. Return of experience from 5 years of field data: long term performance reliability of Soitec's CPV technology. Proceedings of the 29th European PV Solar Energy Conference and Exhibition, Sep 22th –26, Amsterdam, Netherlands, pp. 2017–2019. 2014.

116. International Electrotechnical Commission. IEC 62670-2 (Draft): Concentrator Photovoltaic (CPV) Performance Testing - Part 2: Energy Measurement.

117. K.W. Stone, V. Garboushian, R. Hurt et al. Operation and performance of the Amonix high concentration photovoltaic system at the University of Nevada, Las Vegas during the second year of operation. ASME 2006 International Solar Energy Conference, pp. 131–138. American Society of Mechanical Engineers, 2006.

118. McConnell, R., et al. Assuring long-term reliability of concentrator PV systems. SPIE Solar Energy+ Technology. International Society for Optics and Photonics, 2009.

119. K.W. Stone, V. Garboushian, R. Boehme et al. Analysis of five years of field performance of the Amonix High Concentration PV system. Proceedings of the Power-Gen Renewable Conference. pp. 1–12. 2006.

120. Sánchez D., Martínez M., Gil E., et al. First experiences of ISFOC in the maintenance of CPV plants. 6th International Conference on Concentrating Photovoltaic Systems, Freiburg, April. American Institute of Physics Conference Proceedings, **1277**, p. 148. 2010.

121. International Electrotechnical Commission. IEC 62817, Photovoltaic Systems – design qualification of solar trackers (released at August 2014).

122. http://www.sener-power-process.com/EPORTAL_DOCS/GENERAL/SENERV2/DOC-cw499d8e0908599/ CPVCSPtwoaxessolartracker.pdf (accessed November 2015).

123. Rafael Rebolo, Jesús Lata, Javier Vázquez. Design of heliostats under extreme and fatigue wind loads. Proceedings of SolarPACES, Granada, Spain.

124. M. Martínez, G. Calvo-Parra, E. Gil, et al. Environmental testing results over a tracker drive. CPV10 Conference. American Institute of Physics Conference Proceedings, 1616, p. 228. April 2013.

125. E. Christensen. Flat plate solar array project: 10 years of progress. JPL 400-279, October 1985.

126. A.L. Rosenthal, M.G. Thomas, and S.J. Durand. A ten year review of performance of photovoltaic systems". Proceedings 23rd IEEE Photovoltaic Specialists Conference, pp. 1289–1291. May 1993.

127. J. Woodworth, M. Whipple. Evaluation test for photovoltaic concentrator receiver sections and modules. SAND09-0958, June 1992. Sandia NATIONAL Laboratories, Albuquerque, New Mexico, www.osti.gov/ scitech/servlets/purl/10161641 (accessed November 2015).

128. R. McConnell, S. Kurtz, W. Bottenberg, et al. Qualification standard for photovoltaic concentrator modules", Proceedings of the 16th European Photovoltaic Solar Energy Conference and Exhibition, (also available as NREL/ CP-590-28323, May 2000.

129. Institute of Electrical and Electronics Engineers CPV Working Group within IEEE Standards Coordinating Committee 21 (Fuel Cells, Photovolaics, Dispersed Generation and Energy Storage). Practice for Qualification of Concentrator Photovoltaic (PV) Receiver Sections and Modules. IEEE 1513–2001, May 2001.

130. R. McConnell, L. Ji, J. Lasich, R. Mansfield. Concentrator photovoltaic qualification standards for systems using refractive and reflective optics. Proceedings of the Nineteenth International European Photovoltaic Solar Energy Conference, pp. 2125–2128, June 2004.

10

CPV Multijunction Solar Cell Characterization

Carl R. Osterwald[1] and Gerald Siefer[2]
[1]*National Renewable Energy Laboratory, United States*
[2]*Fraunhofer-Institut für Solare Energiesysteme ISE, Germany*

10.1 Introduction

Characterization of solar cells can be divided into two types; the first is measurement of electro-optical semiconductor device parameters, and the second is determination of electrical conversion efficiency. Although there is of course overlap between the two, efficiency measurements are sufficiently important by themselves to justify this division. In this chapter, we will attempt to outline the best current practices available for characterizing high-efficiency semiconductor solar cells intended for CPV applications.

Efficiency measurements of conventional single-absorber cells, such as crystalline-Si, have been well established and standardized with respect to a reference spectral irradiance distribution (or reference spectrum). Using a reference spectrum is necessary because electrical performance is a strong function of the incident total and spectral irradiance, and because indoor solar simulators in general cannot reproduce solar spectra [1–3].

Standard reporting conditions (SRC) define the total and spectral irradiance (G and E), and the internal cell operating temperature, T, to which standard PV performance is measured and corrected. For conventional non-concentrating PV, these are $G_{ref} = 1000\,\text{Wm}^{-2}$, the hemispherical (global) reference spectrum (AM1.5g), and 25 °C [4–6]. A calibrated reference cell is used to determine the total irradiance, and measured cell currents are corrected to the reference spectrum by means of the spectral mismatch parameter, M [7–9].

The economics of CPV systems dictate that the solar converters employed have the highest efficiency possible; however, the inherent efficiency limits associated with single-absorber solar cells have led to the development of cells with multiple absorbers stacked on top of each other, i.e. subcells, where each subcell converts a different portion of the solar spectral

Handbook of Concentrator Photovoltaic Technology, First Edition. Edited by Carlos Algora and Ignacio Rey-Stolle.
© 2016 John Wiley & Sons, Ltd. Published 2016 by John Wiley & Sons, Ltd.

irradiance. Such multijunction devices have been realized in two different ways. The first is mechanical stacking, in which separate, complete cells are positioned on top of each other; the stacks generally have separate electrical connections to each constituent cell. The second uses monolithic crystal growth of multiple p-n junctions that are series-connected electrically with semiconductor tunneling diodes (or tunnel junctions). Collectively, these stacks are called multijunction devices. Further details on this topic can be found in Chapter 2 of the Handbook.

Mechanically stacked multijunction cells have been constructed only rarely, therefore we will not consider them further here and instead focus on monolithically stacked devices, other than to state that single-absorber characterization techniques can be used if all the component cells can be individually contacted. For the monolithically stacked case, the series connection topology greatly complicates characterization because the individual subcells cannot be measured separately. And for concentrator cells, testing at high irradiance adds additional challenges.

The chapter begins with a review of the multijunction concepts that are necessary for understanding CPV cell characterization techniques, and then proceeds to describe how CPV efficiency is defined and used. Next, several procedures for spectral irradiance adjustments of solar simulators, essential for multijunction measurements, are presented, followed by an overview of the light sources and optics commonly used in simulators for CPV cells under concentration. The final section gives details about the cell area, quantum efficiency (QE), and current-voltage (*I-V*) curve measurements that are needed to characterize cells as a function of irradiance (i.e. concentration).

10.2   Basic Concepts About Multijunction Solar Cells for Characterization Purposes

10.2.1   Review of Multijunction Solar Cell Theory

Multijunction solar cells were thoroughly discussed in Chapter 2 of this volume. As they are vital for understanding CPV multijunction characterization, a brief review of these concepts is also presented here. Figure 10.1 shows an equivalent circuit for a multijunction solar cell

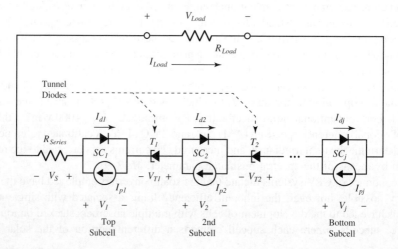

Figure 10.1   Equivalent circuit model of a *j*-subcell multijunction solar cell. Effects of parallel (shunt) resistance R_{sh} are not included because for well-behaved CPV cells, V/R_{sh} is small compared with the current density J_{sc}

composed of j subcells electrically connected in series. Light enters the top subcell and photons with energies higher than the bandgap energy of the subcell semiconductor material are absorbed; because it is transparent, photons with energies less than the bandgap are passed into the next subcell. In order to absorb as much light as possible, the subcell bandgaps are progressively smaller. Thus, the top subcell absorbs light of the shorter wavelengths while the bottom subcell absorbs the longest wavelengths.

Under illumination, each subcell generates an internal photocurrent I_{pj} equal to the convolution integral of its quantum efficiency (QE) with the incident spectral irradiance. The photocurrents are modeled as constant current sources in parallel with the subcell's diode. Because of the stacked structure, a subcell's QE usually spans the wavelength range between the bandgap of the subcell immediately above and its own bandgap. Each subcell therefore has a distinctive current-voltage (I-V) characteristic; the individual subcell I-V curves combine in a non-intuitive way to produce the external I-V characteristic of the multijunction cell.

If the load resistance is disconnected, i.e. $R_{Load} = \infty$, the load current $I_{Load} = 0$, all the photocurrents flow through the corresponding diodes, I_{p1} through I_{pj}, which puts the diodes into forward bias, and photovoltages ($V_1 - V_j$) are generated across each subcell. The photovoltages are closely related to the subcell bandgaps, thus the top subcell has the highest voltage while the bottom subcell the lowest. Because no current flows through the cell series resistance (R_{Series}) and the tunnel diodes, the voltages across these elements are zero, and the open-circuit voltage is:

$$V_{oc} = \sum_1^j V_i \tag{10.1}$$

when $R_{Series} = 0$, $V_{Load} = 0$ and I_{Load} is equal to the short-circuit current, I_{sc}. The photocurrents in the subcells cause internal voltages, which can be expressed as (see also Figure 10.1):

$$0 = -V_s + V_1 - V_{T1} + V_2 - V_{T2} + \ldots + V_j \tag{10.2}$$

The V_S and V_T terms are losses that cell designers try to minimize, and at high currents they are generally of the order of tens of millivolts. At I_{sc} these voltages are then distributed across the subcells, resulting in forward bias (dark) diode currents. The same external current must flow through each diode-current source pair, which means that the diode current will be the difference between the photocurrent and I_{sc}.

Consider now an ideal two-subcell device at I_{sc} with no series resistance or tunnel diode losses. If the two photocurrents are equal, i.e. $I_{p1} = I_{p2}$, no diode currents flow and I_{sc} is equal to the photocurrents. If one photocurrent is then increased slightly so that $I_{p1} = I_{p2} + \Delta I_p$, the extra photocurrent is forced to flow through its parallel diode because it is blocked by the reverse bias characteristic of the other cell, and the external current does not increase. Thus, the subcell with the smallest photocurrent limits the current through the external load. In this example, at I_{sc}, the subcell with the excess photocurrent is forced into forward bias, which then forces the other subcell into reverse bias.

For $0 \le R_{Load} \le V_{oc}$, a similar expression can be written:

$$V_{Load} = -V_s + V_1 - V_{T1} + V_2 - V_{T2} + \ldots + V_j, \tag{10.3}$$

Assuming that the photocurrents and diode I-V characteristics of each subcell are known, the load voltage and current can be calculated for a given value of R_{Load} using circuit theory and solving the simultaneous equations. By varying R_{Load}, the external I-V curve of the multi-junction cell can be calculated with numerical circuit simulations. The implications of the series-connected nature are such that special techniques are required for determination of both QE and I-V characteristics of multijunction cells (see Chapter 2).

10.2.2 Definition of CPV Cell Efficiency

As noted above, standardized efficiency measurements of non-concentrator (or flat-plate) solar cells are done with respect to the SRC, which are $1000 \, \mathrm{Wm}^{-2}$, the hemispherical (global) reference spectrum, and a cell temperature of $25 \, ^\circ\mathrm{C}$. These were established to enable comparisons of efficiency measurements between cells and performed by different laboratories with different equipment, especially solar simulators. Although it can be said that the SRC are arbitrary, especially considering that solar cells almost never operate at this relatively cold temperature, they provide a reliable basis for comparing one cell against another. For concentrator cells these issues remain, but the SRC have been adjusted for the ways CPV cells are used.

Concentrator cells operate under optics that focus sunlight, thereby intentionally raising the total irradiance to levels that are much greater than $1000 \, \mathrm{Wm}^{-2}$; the actual total irradiance is determined by the CPV system in which a cell is used. Therefore, it isn't possible to require efficiency measurements be done at a fixed total irradiance; instead, the efficiency is measured as a function of total irradiance, which is varied over a range appropriate for the particular cell under test. Because efficiency increases with total irradiance until series resistance losses cause it to decrease, this allows the maximum efficiency to be determined, along with the irradiance at which it occurs.

Optics for CPV systems increase irradiance by focusing sunlight onto a surface, and one of the design parameters is the geometrical concentration ratio (C_g), which can be specified as the entrance aperture area divided by the focused area. The concept is carried over to cell efficiency, but it is instead defined as an irradiance ratio rather than an area ratio (C). Because C is a ratio of homogenous quantities (two irradiances) it should be a dimensionless magnitude. However, the tradition in CPV is to use × or *suns* as the units for this ratio in order to highlight its physical meaning. A concentration ratio of one is called '1-sun', and by convention is equal to $1000 \, \mathrm{Wm}^{-2}$ [10,11].

Next, when sunlight is focused above about 50×, the optics produce an image of the sun on the cell front surface, and in so doing reduce the field-of-view from the hemisphere of flat-plate PV to a narrow cone of sky (concentrations less than 50× are not considered here because multijunction cells are typically not economically viable at low irradiances). When the system is pointed at the sun, the optics block collection of the diffuse sky irradiance from outside of the acceptance angle. This reduces the total irradiance available for conversion, and the spectral irradiance at the cell is effectively only that of the direct beam (or direct normal). Therefore, the hemispherical reference spectrum is not appropriate, and by convention CPV efficiency measurements are instead with respect to the direct reference spectrum (AM1.5d) [12], which is similar to the hemispherical spectrum but lacks the diffuse sky irradiance. The integrated total irradiance drops from $1000.1 \, \mathrm{Wm}^{-2}$ to $900.1 \, \mathrm{Wm}^{-2}$. Even so, 1-sun is still defined as $1000 \, \mathrm{Wm}^{-2}$, thus the direct reference spectrum is effectively scaled by a factor of 10:9.

When the *I-V* curve of a cell under test is traced from I_{sc} to V_{oc}, output power into a load resistance increases and then peaks at the maximum power point, P_{max}. Efficiency η is the power out divided by the power in:

$$\eta = \frac{P_{max}}{G_{in} \cdot A} = \frac{P_{max}}{G_{ref} \cdot C \cdot A} = \frac{V_{oc} \cdot I_{sc} \cdot FF}{G_{ref} \cdot C \cdot A} \qquad (10.4)$$

where G_{in} is the total irradiance, G_{ref} is the irradiance at one sun reference conditions (here 1000 Wm^{-2}), FF is the fill factor, and A is the cell area (see also section 10.5.1).

Eq. (10.4) points to a significant problem that must be solved for CPV efficiency measurements, which is determination of the concentration ratio, C. At 1-sun, for single absorber cells, irradiance is determined with a reference cell and spectral mismatch corrections to the reference spectrum. This technique is still available for testing above 1-sun, but reference cells usually are not designed to operate at high current levels. For those that are so designed, high currents can introduce errors if the cell's *I-V* curve has a non-negligible slope at I_{sc}. Another problem with reference cells is that the test plane of the solar simulator may not be large enough to illuminate a cell under test and a reference cell at the same time.

To get around such problems with reference cells, the following alternative has evolved. First, the cell under test is measured using a reference cell to obtain the I_{sc} at 1-sun, I_{1s}. Next, the cell is placed in a concentrator solar simulator and its *I-V* curve is measured with the total irradiance increased. Lastly, I_{sc} is assumed to vary linearly with irradiance, and that the cell's photocurrent is equal to its I_{sc}, which allows the concentration ratio to be calculated. Using I_C as the I_{sc} under concentration, Eq. (10.4) becomes:

$$\eta = \frac{P_{max}}{G_{ref} \cdot C \cdot A} = \frac{P_{max}}{G_{ref} \cdot \left(I_C / I_{1s} \right) \cdot A} \qquad (10.5)$$

In effect, with this procedure, the cell under test becomes its own reference cell after it is calibrated at 1-sun. The validity of the linearity assumption remains an active area of research that is beyond the scope of this chapter. However, it can be summarized in a general way by stating that Si-based cells are usually linear below about 100× and then gradually lose current, as much as 30–40% by 1000× (i.e. are sublinear). Other Si cells are superlinear [11,13–15]. III-V cells, especially GaAs, in contrast, have been described as linear up to 1000× [14,16]. Luminescent coupling can cause non-linear response in III-V multijunction cells, unless the cell is operating with the top subcell limiting the overall cell current (see section 10.3.4) [17]. It should be noted that the linearity assumption breaks down if the cell's *I-V* curve is poor, which can be caused by high series resistance or by subcells being forced into reverse bias in combination with a low shunt resistance or a reverse breakdown of the current limiting junction.

Lastly, the standard CPV cell temperature remains 25 °C.

10.2.3 Current-Voltage as a Function of Concentration

Using the definition of CPV efficiency, the task of characterization is then to determine how a solar cell under test behaves as the irradiance is varied over a range of interest. Figure 10.9 below shows a complete CPV characterization for a GaInP/GaAs/Ge cell, where the V_{oc}, FF, and η are plotted versus concentration; from these curves a number of important parameters are obtained, especially the concentrations at which the fill factor and efficiency reach their maximum values, and how rapidly the efficiency decreases past the maximum.

10.3 Spectral Matching and Adjustment

For any *I-V* measurement of a multijunction cell the sun simulator spectrum has to be adjusted in a way that all junctions generate the same photocurrent ratios with respect to each other as under reference conditions. Several methods for achieving these conditions are used at the moment. All of them are equivalent to each other in the sense of the intended result and they all rely on the usage of either spectrally adjustable sun simulators or multi-source simulators.

10.3.1 Isotype Method

This method is widely used in space applications and relies on the availability of so called isotype or component cells used as reference irradiance devices. Such devices correspond to structures similar to the actual multijunction solar cell to be measured however only one of the subcells is realized with an active p-n junction. In this way the absorption in the above subcells is replicated and consequently a single-junction cell with –in theory– identical EQE as the corresponding subcell in the multijunction stack is realized. The isotype method relies on this fact that the relative EQE of isotype cell and corresponding subcell are identical or at least as close to each other that any resulting spectral mismatch factor [7–9] is close to unity and thus negligible. Consequently this method neither requires the knowledge of the spectral distribution of the sun simulator nor the measurement of the EQE of the subcells of the multijunction cell to be tested. The sun simulator settings are varied empirically until all isotype cells simultaneously deliver their rated reference current under the simulator setting within a predefined tolerance range (typically ± 1 %). Then the multijunction cell is measured under these settings.

The big advantage of this method clearly is the fact that neither the EQE of the multijunction cell nor the sun simulator spectrum need to be known. However the method is only applicable in the case where matched isotype reference cells are available. Additionally, it has to be noted that the EQE of isotype or component cells is known to potentially be affected by luminescent coupling effects [18] thus the assumption above needs to be verified.

10.3.2 Reference Cell Method

The reference cell method can be considered as an enhancement of the isotype method. Here the differences in relative EQE between device under test (i.e. subcells of the multijunction cell) and reference cell(s) are considered [19,20]. Consequently this method requires both the knowledge of the EQE of the multijunction cell and of the simulator spectrum. For each of the subcells *i* in the multijunction cell the mismatch factor M_i with the corresponding reference cell (RC_i) is calculated (M_i is the product-ratio of four numeric convolution integrals of spectral irradiance and cell responsivity, see equation 10.11 below and [7–9]). The rated current at reference conditions of RC_i is referred to as $I_{RC_i}^{ref}$. Starting from the initial settings of the simulator the M_i values are calculated. For the final *I-V* curve measurement the currents $I_{RC_i}^{sim}$ of the reference cells RC_i under the simulator consequently need to be equal to:

$$I_{RC_i}^{sim} = I_{RC_i}^{ref}/M_i, \tag{10.6}$$

Under the present setting of the simulator the actual prevailing $I_{RC_i}^{sim}$ currents are measured and compared to their target values from Eq. (10.6). If the values coincide within the tolerated deviation (typically ± 1 %) the setting of the simulator is finished and the *I-V* curve of the

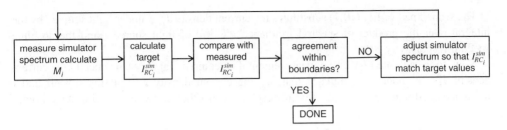

Figure 10.2 Flow chart of the reference cell method. The iterative process is repeated until measured and target $I_{RC_i}^{sim}$ match within the required boundaries (typically $\pm 1 \%$)

multijunction cell can be measured. However, typically this will not be the case and the simulator needs to be adjusted in a way that the values $I_{RC_i}^{sim}$ will match their target values. This however can only be achieved by changing the relative spectrum of the simulator, thus the spectral distribution needs to be re-measured and M_i values are re-calculated, starting again from the start. This iterative procedure is summarized in Figure 10.2

The advantage of this method compared to the isotype method is that differences between the EQE of device under test and reference cell(s) are considered. Thus this method is also practicable for multijunction cells where no corresponding isotype/component cells are available. The disadvantage however is related to the iterative nature of the method involving the re-measurement of the spectrum of the simulator which makes the method rather time consuming.

10.3.3 R_{ij} *Method and Linear Equation System Method*

The third type of method involves in fact two versions which are referred to as R_{ij} and linear equation system method. These kinds of methods rely on the knowledge of the simulator spectrum and the possibility to adjust different wavelength regions in the simulator spectrum independently of each other. Thus these methods are only suitable for multi-source sun simulators. Both methods require at least as many light sources as junctions in the device under test that can be adjusted in intensity independently of each other and show different spectral distributions. The spectral distributions of the light sources do not need to be completely disjoint. However the more they differ and the more the main irradiated power of one light source matches the response region of a subcell of the multijunction cell to be tested, the better applicable the procedures will be.

The linear equation system method requires the same number of independent light sources as subcells involved and is described in detail in [21]. The R_{ij} method corresponds to an enhancement of the linear equation system for the case where the number of independent light sources exceeds the number of subcells in the device under test [22].

Both methods are based on the superposition principle: The total current density $J_i^{sim,tot}$ of subcell i under illumination by a multi-source sun simulator with j light sources corresponds to the sum of the j currents densities $J_i^{sim,j}$ that are generated under illumination by the j single, individual light sources:

$$J_i^{sim,tot} = \sum_j J_i^{sim,j} = \sum_j \int E^{sim,j}(\lambda) \cdot SR_i(\lambda)d\lambda, \qquad (10.7)$$

The second part of Eq. (10.7) substitutes the current density $J_i^{sim,j}$ under light source j by the integral over the product of spectral irradiance $E^{sim,j}(\lambda)$ of light source j and the absolute spectral responsivity $SR_i(\lambda)$ of subcell i.

Now, for a calibrated I-V curve measurement it is required that for each subcell i the current density $J_i^{sim,tot}$ under the simulator matches the current density J_i^{ref} of this subcell under reference conditions. This means that for each subcell i the following equation must hold:

$$J_i^{sim,tot} = \sum_j J_i^{sim,j} = \sum_j \int E^{sim,j}(\lambda) \cdot SR_i(\lambda)d\lambda! J_i^{ref} = \int E^{ref}(\lambda) \cdot SR_i(\lambda)d\lambda, \quad (10.8)$$

Here the current density J_i^{ref} of subcell i under the reference spectrum has been substituted by the integral over the product of the spectral irradiance distribution E^{ref} of the reference spectrum and the absolute spectral responsivity $SR_i(\lambda)$ of subcell i. This leads to the following equation for each subcell i:

$$\sum_j A_j \int E^{sim,j}(\lambda) \cdot sr_i(\lambda)d\lambda = \int E^{ref}(\lambda) \cdot sr_i(\lambda)d\lambda, \quad (10.9)$$

Note that absolute spectral responsivity $SR_i(\lambda)$ of subcell i in capital letters has been replaced by the relative spectral responsivity $sr_i(\lambda)$ in small letters. This is due to the fact that the absolute spectral responsivity $SR_i(\lambda)$ occurs on both sides of the equation and thus any scaling factor will cancel out. Additionally, lamp scaling factors A_j have been introduced. Eq. (10.9) corresponds to a linear equation system with j equations and j unknowns A_j. Thus the lamp scaling factors Aj correspond to the solution of the linear equation system. Finally after determination of the A_j values calibrated reference cell(s) are used for setting the intensity of the j light sources in a way that:

$$J_{RC}^{sim,j} = A_j \int E^{sim,j}(\lambda) \cdot SR_{RC}(\lambda)d\lambda, \quad (10.10)$$

where $J_{RC}^{sim,j}$ is the target current density of the reference cell used for setting the intensity of light source j, A_j is the above mentioned lamp scaling factor that corresponds to the solution of the linear equation system and $SR_{RC}(\lambda)$ is the absolute spectral responsivity of the reference cell. Note that this procedure relies on the premise that the relative spectral distributions of the j light sources do not change when adjusting their intensity. If this assumption does not hold, e.g. when the current of a xenon or tungsten lamp is used for adjustment of intensity, the lamp adjustment factors A_j are required to be as close to unity as possible. In this way errors due to changes in relative spectral distribution of the light sources are minimized. In practice this can be realized by feeding several different spectral irradiance distributions $E^{sim,j}(\lambda)$ for each light source j determined at different intensity levels into the linear equation system represented by Eq. (10.9). From the solutions of Eq. (10.9) the set of A_j is chosen where all values of A_j only differ from unity by a few percent (typically $\ll 5\%$ is required).

The R_{ij} method [22] expands the above linear equation system concept to the case of multi-source sun simulators where the number of light sources is larger than the number of subcells involved. In principle the R_{ij} are defined as spectral mismatch factors following [7–9] for the

Table 10.1 Main advantages and disadvantages of the three introduced methods for spectral adjustment of a sun simulator for the measurement of multijunction cells

Method	Advantage	Disadvantage
Isotype	Potentially quick, no EQE and simulator spectrum required	Potential differences between EQE of isotype cell and corresponding subcell not taken into account
Reference cell	Takes into account differences between EQE of reference cell(s) and subcells	Iterative procedure, requires re-measurement of simulator spectrum
R_{ij}/linear equation system	As above No iterative procedure	Requires multi-source sun simulator, whereas others also work with spectrally variable simulators

case of subcell i being the reference cell and subcell j being the device under test:

$$R_{ij} = \frac{\int E^{ref}(\lambda) \cdot sr_i(\lambda)d\lambda}{\int E^{ref}(\lambda) \cdot sr_j(\lambda)d\lambda} \cdot \frac{\int E^{sim}(\lambda) \cdot sr_j(\lambda)d\lambda}{\int E^{sim}(\lambda) \cdot sr_i(\lambda)d\lambda}, \quad (10.11)$$

Again absolute responsivity has been replaced by the relative one in small letters, as any scaling factor will cancel out. The simulator spectrum $E^{sim}(\lambda)$ corresponds to the sum of the spectral distributions $E^{sim,k}$ of the k light sources. The requirement in respect to simulator spectrum for a calibrated I-V curve measurement now is that for all possible combinations of i and j R_{ij} equals unity. For the number of light sources equaling the number of subcells involved the outcome of the R_{ij} method will be the same as of the linear equation system method. However the R_{ij} method has been established especially for the case where the number of light sources is larger than the number of subcells involved. In that case the number of potential settings of the simulator fulfilling the $R_{ij} = 1$ criteria will be >1. This allows for e.g. choosing a solution for the simulator spectrum where besides the requirement for the currents of the subcells also additional constraints regarding e.g. similarity to the reference spectrum can be added.

The advantages and disadvantages of the three methods introduced above are summarized in the following Table 10.1.

10.3.4 Effects of Subcell Mismatching

Due to the internal series connection, the current of a multijunction cell will be limited by the subcell with the lowest current. As the subcells absorb light in different wavelength regions, this immediately leads to the conclusion that the spectral distribution of the light impinging the cell plays a more important role compared to the case of single-junction solar cells. Current limitation effect always will dominate the behavior of multijunction cell efficiency. However, there is a second effect of lower magnitude that goes in the opposite direction. Assuming a multijunction cell where subcell i is limiting the overall current of the device; between I_{sc} and I_{max} of the I-V curve of the multijunction cell all subcells except for subcell i will operate at voltages above the V_{max} on the individual subcell I-V curve. This in turn leads to the effect that the V_{max} of the multijunction I-V curve will be above the sum of the V_{max} from the subcell I-V curves. This leads to a more 'squared' I-V curve of the multijunction cell and consequently to an increase in fill factor FF. An elegant method for studying the effect of varying current ratios in a multijunction cell is the so called spectrometric characterization [21]. A spectrometric characterization corresponds to a systematic change of the simulator spectrum while recording I-V curves. The simulator spectrum is categorized using effective irradiances G_{eff}. In the case of multijunction solar cells the effective irradiance G_{eff}^i on subcell i under a given spectral distribution $E(\lambda)$ corresponds to the photocurrent density $J_{sc}^i(E(\lambda))$ of subcell i under illumination with $E(\lambda)$:

$$G_{eff}^i(E(\lambda)) = J_{sc}^i(E(\lambda)) = \int E(\lambda) \cdot SR_i(\lambda) \, d\lambda, \tag{10.12}$$

G_{eff}^i is defined for any given spectrum including the reference spectrum considered by replacing $E(\lambda)$ with $E_{ref}(\lambda)$ in Eq. (10.12).

Starting from reference conditions for two subcells i and j the effective irradiances are now varied in the following systematic way within a spectrometric characterization:

$$\frac{G_{eff}^i(E(\lambda))}{G_{eff}^i(E_{ref}(\lambda))} + \frac{G_{eff}^j(E(\lambda))}{G_{eff}^j(E_{ref}(\lambda))} = \frac{J_{sc}^i(E(\lambda))}{J_{sc}^i(E_{ref}(\lambda))} + \frac{J_{sc}^j(E(\lambda))}{J_{sc}^j(E_{ref}(\lambda))} = 2, \tag{10.13}$$

In other words, starting from reference conditions where:

$$\frac{G_{eff}^i(E(\lambda))}{G_{eff}^i(E_{ref}(\lambda))} = \frac{G_{eff}^j(E(\lambda))}{G_{eff}^j(E_{ref}(\lambda))} = 1, \tag{10.14}$$

the effective irradiances on subcells i and j are changed in a way that if effective irradiance (or current density) on subcell i is increased by x % the effective irradiance on subcell j is decreased by x %. The effective irradiances of additional subcells are kept constant equaling unity.

The following Figure 10.3 shows the example of a spectrometric characterization of a $Ga_{0.35}In_{0.65}P/Ga_{0.83}In_{0.17}As/Ge$ triple junction cell.

Spectral conditions on the left of Figure 10.3 correspond to 'red-rich' conditions. Here the effective irradiance on the $Ga_{0.35}In_{0.65}P$ top cell is decreased (values on bottom x-axis <1) whereas the effective irradiance on the $Ga_{0.83}In_{0.17}As$ middle cell is increased (values on top x-axis >1). Consequently spectral conditions on the right of Figure 10.3 correspond to 'blue-rich' conditions as here the effective irradiance on the $Ga_{0.35}In_{0.65}P$ top cell is increased (values on bottom x-axis >1) and at the same time the effective irradiance on the $Ga_{0.83}In_{0.17}As$ middle cell is decreased (values on top x-axis <1). The short circuit current of the triple-junction cell shows a triangular behavior with a maximum around $x = 0.98$ on the bottom x-axis. At these conditions both subcells (almost) generate the same current and the cell operates current matched. For x-values on the bottom x-axis >0.98 the middle cell limits the current (effective irradiance on top cell is increased and effective irradiance on middle cell decreased). For x-values on the bottom x-axis <0.98 the top cell is limiting the current (effective irradiance on top cell is decreased and effective irradiance on middle cell increased). The top and middle

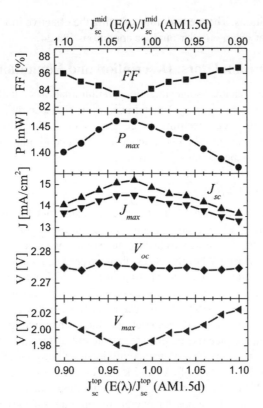

Figure 10.3 Spectrometric characterization of a $Ga_{0.35}In_{0.65}P/Ga_{0.83}In_{0.17}As/Ge$ triple junction cell using AM1.5d (ASTM G173-03, 1000 W/m²) as reference spectrum. *I-V* curve parameters on a vertical line correspond to a single *I-V* curve at spectral conditions categorized by the ratios of effective irradiances on the two *x*-axes

subcells of the triple-junction cell thus are current matched for spectral conditions slightly red rich compared to AM1.5d (*x* values 1.00 on both, bottom and top *x*-axis).

The current I_{max} at maximum power point follows the trend of the I_{sc}. Open circuit voltage V_{oc} is almost constant; however the voltage V_{max} at maximum power point shows an opposite behavior to I_{sc} due to the above mentioned effect under current mismatching between subcells. This in turn leads to an increase in fill factor with increasing current mismatch between the subcells – in other words: the minimum in fill factor can be used to identify spectral conditions where the investigated subcells operate current matched. This effect slightly weakens the decrease in P_{max} when increasing the current mismatch between top and middle subcells. Spectrometric characterizations can be a useful tool for the investigation of multijunction cells' response to changes in the spectrum. Moreover the spectrometric characterization nicely demonstrates the impact of spectra that generate a current balance between subcells that differs from the reference spectrum. The behavior found in Figure 10.3, particularly of short circuit current and fill factor, clearly highlights the importance of an accurate control and adjustment of the simulator spectrum in order to generate the same current balance between subcells as under reference conditions. The methodology of using the effective irradiances is also applied for CPV modules that operate under outdoor conditions [23,24].

10.4 Flash Solar Simulators: Description and Limitations

For CPV measurements, an ideal solar simulator can illuminate a cell under test with a variable total irradiance, over several orders of magnitude, while the spectral irradiance is fixed and equal to the direct reference spectrum. In general, such a simulator does not exist and performance measurements must be made under imperfect conditions, with additional procedures to overcome the imperfections [19,25].

10.4.1 Sources and Optics

At the moment there are only two viable light sources available for CPV cell testing: the sun itself and Xe arc lamps. Somewhat ironically, the sun isn't a good choice for a solar simulator because both total and spectral irradiance vary over the course of a day, and clouds will usually prevent testing. The total irradiance must be increased with optics, and to attain a concentration of 1000× over an area of 10 cm^2 requires an entrance aperture of 1 m^2, which must be pointed at the sun with an outdoor tracker. Keeping the cell temperature at 25 °C requires the illumination time be minimized with a fast shutter system. The most significant problem with using the sun as a solar simulator is the lack of spectral irradiance control, which is required for multijunction testing.

Xenon arc lamps can produce spectra that are reasonably close to the desired direct reference spectrum, with the exception of the prominent emission lines between 850 nm and 1000 nm. In the long-arc pulsed mode, a voltage of 1 kV to 3 kV is sourced across the lamp, then a very short trigger pulse of ~50 kV initiates the discharge. Flash durations can be anywhere from one to a hundred milliseconds. Short-arc lamps that operate in continuous mode can also be utilized, but require a fast shutter for momentary illumination.

Arc lamps are typically focused to high intensities with reflective optics, and most systems include secondary reflectors, baffles, or light pipes to improve spatial uniformity at the test plane.

10.4.2 Adjusting Total Intensity

Total intensity of an arc lamp is easily changed by changing the power into the lamp with the applied voltage, but doing so changes the temperature of the plasma during the flash, which causes the spectral irradiance to shift to higher or lower wavelengths. For CPV testing in which the spectrum needs to be constant as the total irradiance is varied, this spectral effect dictates that other means of varying total irradiance are necessary. One simple method is to change the distance from the lamp to the test plane, assuming the beam is divergent. Often, other factors make this unfeasible, especially the spatial uniformity at the test plane.

Another method that is commonly used is obscuration: by blocking part of the beam at a strategic point between the lamp and the test plane, total irradiance is reduced. Mesh screens with varying hole sizes are commonly used.

10.4.3 Irradiance Versus Time

Another important characteristic of arc lamps is the change of irradiance versus time during a flash. While simulator manufacturers strive to minimize such changes, they cannot be

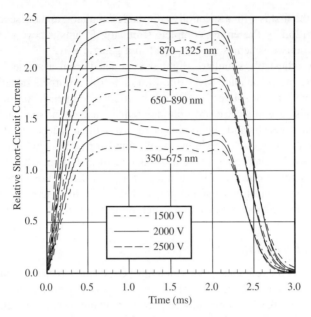

Figure 10.4 Relative short-circuit current versus time for three different isotype reference cells in a pulsed Xe solar simulator, for three lamp voltages (see text)

completely avoided. Figure 10.4 is a plot of the relative I_{sc} produced by three different isotype reference cells in a simulator with a 3 ms pulse duration, for three different values of lamp voltage. The wavelength ranges to which the cells respond are indicated; the I_{sc} data were scaled to group the three curves corresponding to each cell together. The absolute magnitudes are unimportant; instead consider the slopes of these curves between about 0.7 ms and 2.0 ms.

Looking at the 350–675 nm data, which are characteristic of a typical GaInP top subcell, it is evident that the lamp voltage increases the short-wavelength irradiance in the early portion of the flash when the arc plasma is hottest. In the GaAs-under-GaInP range (2nd subcell), the slope of the curves is less pronounced, and the irradiance in the 3rd subcell region (870–1325 nm), actually increases with time at a lamp voltage of 1500 V. Collectively, these curves show that the spectral irradiance during the flash changes as much as 10% in some cases.

These effects complicate multijunction testing because they limit the extent to which the spectral irradiance can be matched for a given multijunction cell – a 10% variation is enough to cause the limiting subcell to shift during a flash, which can cause artificial discontinuities in the I-V curve. The best mitigation is to restrict the portion of the flash during which I-V data is collected.

10.4.4 Spectral Irradiance Adjustment

Proper I-V measurements of multijunction cells requires the spectral irradiance to be adjusted so that for all subcells the same photocurrent ratios with respect to each other as under reference conditions are generated. In general, this requires changing the total irradiance in the spectral bands that correspond to the bands to which the subcells are sensitive. There are two ways to accomplish this; the first is with a spectrally adjustable simulator, and the second with a multi-source simulator [26,27]. Presenting complete simulator designs is beyond the scope of this chapter; instead the two schemes are described in a generic way.

In a spectrally adjustable simulator, light is subtracted from a single Xe flash lamp in different bands by means of multi-layer thin-film dielectric absorbers or reflectors equipped with adjustable apertures or shutters to control the irradiance in each spectral band. A multi-source simulator also uses dielectric absorption or reflection to define spectral bands, but with a separate flash lamp for each band; the irradiance of each lamp is independently controlled. The advantage of a multi-source over a single-source simulator is that light can be added to any spectral band by simply increasing the total irradiance of that lamp, whereas a single-source simulator can only remove light. This advantage comes at the cost of increased complexity, however, because each adjustment band requires what is essentially its own solar simulator, all of which must be optically combined, and synchronized in time.

10.4.5   Spectral Irradiance Measurement

Unless calibrated isotype reference cells which have QE characteristics that match those of the multijunction cell to be tested are available, so that the spectral mismatch factors (M_i) for all subcells are unity (see 10.3.1), correct subcell photocurrent tuning requires determination of the simulator's spectral irradiance. And once again, the nature of CPV cell I-V testing greatly complicates the requirements placed on any spectroradiometer used for flash spectral irradiance measurements.

Wavelength range – in order for the necessary convolution integrals to be valid, the instrument must capture the entire wavelength response range of a test cell. In terms of the CPV cells currently under development, this range is about 350–1800 nm. A single diffraction grating is unable to disperse light across this wide range of wavelengths, which forces the use of multiple monochromators.

Speed of measurement – Capturing the spectrum of a single flash in time periods of milliseconds or less places more constraints on the data acquisition system, which needs high-speed electronics and photodiode detector arrays, and synchronization of the multiple monochromators.

Calibration – Spectroradiometers are calibrated to convert photodiode currents to irradiance by comparison against specialized filament light sources that have known spectral irradiance versus wavelength curves at fixed distance from their exit apertures. The total irradiance of these standard lamps is low (less than 1-sun, about 300 Wm^{-2}) compared with the very high irradiances produced by CPV flash solar simulators. If the photodiode array detectors are not linear with irradiance, calibration errors can result that are difficult to quantify. Another issue with standard lamp calibration is the low irradiance available from filament lamps below 400 nm, which increases the measurement uncertainty in the UV. This problem however is also present in measurements of continuous light sources.

Another option for capturing the spectra of flash solar simulators does not attempt to measure an entire spectrum during a single flash. Using a specialized spectroradiometer, the spectrum is obtained at a single wavelength for each light flash, which means that it needs several hundred flashes to cover the wavelength range needed. This can be a time-consuming task that makes it unsuitable for iterative spectral adjustments, but it eliminates a number of error sources associated with diode array instruments [28].

10.5   Concentrator Solar Cell Characterization

10.5.1   Overview

Complete characterization of a single multijunction concentrator solar cell requires multiple measurements of a variety of parameters and apparatus. Figure 10.5 shows the sequence of these measurements; area is discussed in section 10.5.2, external QE in section 10.5.3, 1-sun calibration in section 10.5.4, and *I-V* under concentration in section 10.5.5.

10.5.2   Area Measurement

The measurement of the performance of photovoltaic devices is typically performed with sun simulators that have an illumination area larger than the area of the device under test. The power incident to the device which is needed for determination of device efficiency consequently is calculated from the power flux of the sun simulator and the area of the device under test.

The determination of the area of is commonly regarded as one of the most uncritical measurements to be performed on a photovoltaic device. However in the case of concentrator solar cells this is not necessarily true. In this context three different kinds of area specifications are typically used in PV [10]:

- total area (t.a.): full projected area of a device (identical to the area of its shadow);
- aperture area (a.a.): device under test is masked, but essential parts (like busbars, fingers, interconnects) are within masked area;
- designated (illumination) area (d.a.): portion of contacting parts of the device under test like busbars are excluded.

In the case of CPV cells designated area is typically referred to when stating solar cell efficiencies. This can be justified by the fact that in the application in a CPV module the busbar(s) of the cell will be outside of the illuminated area.

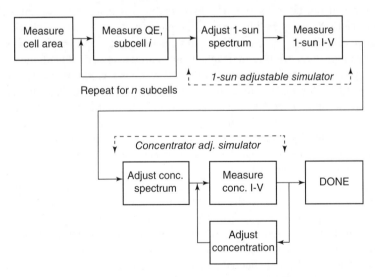

Figure 10.5   Flowchart for complete characterization of CPV cell efficiency versus concentration

An additional fact that is complicating the determination of the cell area for CPV cells is the fact that the chip size of the cell is typically larger than the active area of the solar cell. Mesa etching through the upper p-n junctions down into the substrate is usually used for definition of active cell area. This arises two challenges for the area measurement. Firstly these mesa edges hardly correspond to an easily identifiable vertical edge. Secondly even carriers that are generated outside the active cell area in the substrate where no active p-n junction is present might contribute to the current generation if diffusion length of carriers is long enough to reach the active p-n junction of the cell. This is particularly evident in the case of e.g. III-V solar cells realized on active silicon substrates and needs to be considered by e.g. additional masking [29].

10.5.3 External Quantum Efficiency

Main challenge in respect to the measurement of the external quantum efficiency (EQE) of multijunction devices is the fact that the individual subcells cannot be directly electrically contacted. As in the case of e.g. silicon single-junction cells the measurement of the EQE of multijunction cells is typically done using the differential spectral responsivity method [30]. In this method, quasi-monochromatic light that is modulated in intensity is used as test light and additional continuous bias light is used for putting the cell under test into the desired working conditions. Classically grating monochromator setups are used as they offer higher flexibility in respect to step width of measured wavelengths as well as bandwidth of the monochromatic light.

The principle procedures for the measurement of the EQE of multijunction cells are well-known [31]. This involves two points that are different from the case of single-junction cells:

- the continuous bias light has to be chosen in a way that the subcell to be measured will limit the overall current of the device;
- a bias voltage has to be applied in order to let the current limiting subcell operates at short circuit current conditions.

Consider subcell i to be measured: The first condition is mandatory for any subcell EQE measurement and is realized by illuminating the multijunction cell with high intensity in the response range excluding subcell i and with low intensity in the response region of subcell i. Several possibilities to realize this are available like either using broadband light sources (e.g. tungsten halogen lamps) and removing light in the response region of subcell i with optical filters or using LEDs for the bias light illumination. The necessity for applying an additional bias voltage arises from the fact that the subcells are connected in series and all subcells except subcell i to be measured operate with an excess current. The overall current of the multijunction cell however is limited by subcell i. Consequently subcells excluding subcell i will usually operate at a point on their individual I-V curve between open circuit voltage V_{oc} and voltage V_{max} at maximum power point. Considering Kirchhoff's law applying an external voltage of 0 V as in the case of single-junction cells will consequently lead to the limiting subcell i to operate at a negative voltage. In order to compensate for this fact and to allow the limiting subcell i to operate at its short circuit current a positive bias voltage is applied whose magnitude can be estimated from the expected V_{oc} of the other subcells. The measured EQE of typical GaInP and Ga(In)As subcells in present triple-junction devices hardly show any dependence on applied bias voltage. On the other hand the EQE of germanium subcells was found to react very sensitive on the magnitude of the applied bias voltage and artifacts in the EQE measurements

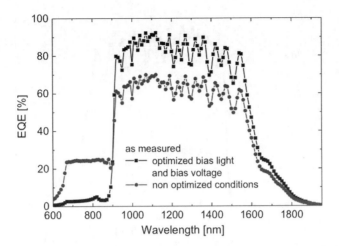

Figure 10.6 EQE measurements of the germanium bottom subcell of a lattice matched triple-junction solar cell. Besides the EQE measurement under optimized bias light and bias voltage conditions also a measurement under non-optimized conditions is shown

were reported [32]. Such an EQE measurement artifact can be identified by an unwanted measured EQE signal in the response region of another subcell and by a too lowly measured EQE of the subcell of interest, see Figure 10.6. There, two EQE measurements of the germanium subcell of a lattice matched $Ga_{0.50}In_{0.50}P/Ga_{0.99}In_{0.01}As/Ge$ triple cell are shown. In the optimized case –low EQE artifact– bias illumination and bias voltage have been adjusted in an iterative procedure maximizing the EQE signal in the response range of the Ge subcell and at the same time minimizing the EQE signal in the response range of the other subcells. The non-optimized case shows an EQE measurement using a different bias light spectrum and a non-optimized bias voltage. The occurrence of such an EQE measurement artifact could be explained through a low shunt resistance of the subcell under test [32].

The measurements in Figure 10.6 demonstrate the interdependence of height of unwanted EQE signal and lowering of EQE of subcell of interest: The EQE measurement under non optimized conditions in Figure 10.6 shows a significant EQE signal in the wavelength range below 900 nm. At the same time the absolute height of the measured EQE in the germanium subcell region above 900 nm is decreased compared to the measurement under optimized bias light and voltage conditions. Interestingly even the EQE measurement under optimized conditions in Figure 10.6 shows a small EQE artifact of 2–3 % measured EQE in the middle subcell region between app. 650 and 900 nm.

For the case of such an EQE measurement artifact being caused by a low shunt resistance in the subcell under test a correction procedure has been deduced in [33]: Consider the EQE measurement of subcell i to show an EQE measurement artifact in the response region of subcell j. The correction procedure is performed in two steps. Step one bases on scaling the actual measurement of subcell j to match the unwanted signal in measurement of subcell i and subtracting the so scaled EQE from the measured EQE of subcell i. The scaling factor sf_j to be applied to the EQE of subcell j is chosen in a way that the resulting EQE of subcell i after subtracting the scaled EQE of subcell j is zero at wavelengths where subcell i is not expected to

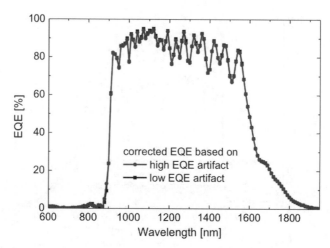

Figure 10.7 EQE measurements from Figure 10.6 after applying the above described EQE artifact correction

show any EQE signal. In the second step the resulting EQE of subcell i is divided by $(1-sf_j)$ in order to correct for the height of the EQE. Details of the correction procedure can be found in [33]. Figure 10.7 shows the EQE measurements from Figure 10.6 after applying the correction procedure described above.

Despite the fact that the EQE measurements in Figure 10.6 showed an EQE measurement artifact of different magnitude, after applying the EQE correction procedure both measurements are in perfect agreement, see Figure 10.7.

The above discussed EQE measurement artifacts are caused by a low parallel resistance of the measured subcell i. Similar effects can be observed for an early reverse breakdown of the measured subcell, often found in low bandgap subcells [34]. Recently it could be demonstrated that also strong radiative recombination in the above lying subcell j can cause a similar EQE artifact through luminescent coupling [35]. The correction procedure derived there is very similar to the one described above. The first step of removing the unwanted signal in the measured EQE of subcell i found in the response region of subcell j is identical. As above the EQE of subcell j scaled by a scaling factor sf_j is subtracted from the EQE of subcell i. However, the second step which is the correction of the height of the resulting EQE of subcell i is slightly different. Whereas in the above case of a low shunt resistance the EQE of subcell i is divided by $(1-sf_j)$ for luminescence coupling being the origin of the EQE artifact the EQE of subcell i is now multiplied with $(1+sf_j)$ [35]. An open issue is still how to identify the origin of the EQE artifact. A non-horizontal slope of the I-V characteristic of the multijunction cell under bias illumination for the EQE measurement might indicate a low shunt resistance to cause the EQE artifact. However, for most present multijunction cells probably more likely luminescence coupling will be the origin of the EQE artifact. For comparatively low EQE artifacts (i.e. small scaling factors sf_j) the difference between dividing by $(1-sf_j)$ and multiplying with $(1+sf_j)$ is low: for a scaling factor of 0.05 the difference between both methods is less than 0.3 %, whereas this difference increases to 1.1 % for a scaling factor of 0.1. In respect to the further calibration of the multijunction cell, this does not have any influence as relative EQEs only are used for all spectral correction procedures described in this chapter.

10.5.4 One-Sun Light I-V and One-Sun Short Circuit Current Calibration

The pure measurement of the light *I-V* curve of a multijunction cell is not critical at all. The complexity of this measurement is related to the correct setting of the simulator spectrum (see section 10.3).

One of the main outcomes of the one sun light *I-V* curve is the calibrated one sun short circuit current I_{1s} that is used for the determination of the concentration and thus calculation of cell efficiency for measurements under high light intensities (compare Eq. (10.5)).

In this context the shape of the one sun *I-V* curve is of particular interest. As mentioned above, at a terminal voltage of $V = 0$ V (i.e. short circuit current of the multijunction cell) the current limiting subcell operates at a negative voltage. In the case of a low shunt resistance present in the current limiting subcell this will lead to the fact that the measured short circuit current of the multijunction cell I_{1s} is significantly larger than the photocurrent of the limiting subcell. In such cases I_{1s} cannot be determined as measured I_{sc} from the *I-V* curve of the multijunction cell. This effect is illustrated in the following Figure 10.8.

The exemplary *I-V* curves Figure 10.8 demonstrate the effect of how the one sun short circuit current I_{1s} is overestimated in the case where the current limiting subcell suffers of a low shunt resistance that influences the slope of the *I-V* curve around the subcell's I_{sc}. In such cases I_{1S} from the one sun *I-V* curve cannot be used for further considerations which prevent a precise determination of the cell's efficiency under concentration. At the time being this prevents the accurate calibration of such devices under concentration. An alternative would be to determine the concentration level during measurements under high light intensities with appropriate isotype or component cells. However, isotype cells are not available in all cases and it can be expected that the uncertainty related to this approach will significantly increase (see also 10.2.2).

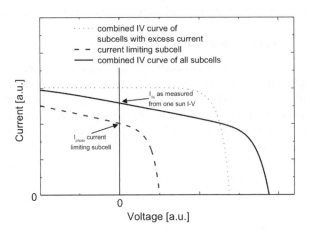

Figure 10.8 Illustration of the effect of overestimation of one sun current I_{1s} in the case of the presence of a low shunt resistance in the current limiting subcell. The dashed line corresponds to the *I-V* curve of the current limiting subcell with a low shunt resistance present. The dotted line represents the combined *I-V* curve of the subcells with excess current. At short circuit current of the multijunction cell's *I-V* curve ($V = 0$ V, straight line) the current limiting cell operates at a negative voltage and thus I_{1s} is significantly above the photocurrent of the limiting subcell at one sun ($V = 0$ V dashed line)

10.5.5 Concentration I-V

10.5.5.1 *I-V* Under Concentrated Light: Loads and Curve Sweeping

In order to avoid a potential heating of the cells, *I-V* curves under concentration are typically measured with flash simulators. This means that the *I-V* curve has to be measured within a time duration of less than one millisecond up to a few milliseconds. The *I-V* curve typically is switched using a quick enough bipolar power supply. For silicon solar cell based PV devices it is known since long that for *I-V* curve sweeping rates in the upper 100 Hz range transient effects are commonly observed [36,37]. The easiest way of verifying the presence or non-presence of transient hysteresis effects is to measure the *I-V* curve of the device under test twice; switching the *I-V* curve from short circuit current to open circuit voltage and vice versa from open circuit voltage to short circuit current. Transient hysteresis effects are observed around P_{max} and can be identified comparing the two measured *I-V* curves. If the difference found between the two *I-V* curves in P_{max} is below 1 % a simple averaging of the two *I-V* curves is recommended. However for larger deviations a comparison between averaged *I-V* curve and an additional measurement around P_{max} with increased effective measurement time is required. The effective measurement time can be increased either by measuring only sections of the *I-V* curve during one flash or by applying a point-by-point measurement. During the latter the voltage of the cell is kept constant during the whole flash (quasi static condition) and during each flash one $V - I$ pair is measured. If averaged *I-V* curve from above differs from the measurement with increased effective measurement time, the whole *I-V* curve is required to be re-measured. However, such transient hysteresis effects are rarely found in the case of III-V based multijunction cells.

In the case of *I-V* measurements under high intensities switching *I-V* curves in the above mentioned two directions has another benefit: The measured open circuit voltage values between the two curves can be compared in order to check for a potential heating of the cell. Although the measurement time is small, depending on the cell structure and mounting heating might occur even for measurement times of as low as 1 ms due the high light intensity.

10.5.5.2 V_{oc}, *FF*, η Versus Concentration

As the total irradiance on a solar cell is increased, the current scales linearly while the voltage increases logarithmically. Output power, the product of voltage and current, also increases approximately logarithmically until losses (especially series resistance) cause it to decrease at higher irradiance (see chapter 2). Thus, a plot of η versus G will show a peak value at a certain value of irradiance. An example measurement is shown in Figure 10.9 for a GaInP/GaAs/Ge multijunction cell in which the concentration was varied from 3× to >1000× under a Xe flash simulator; the critical parameters were extracted from the individual *I-V* curves and plotted versus the logarithm of concentration. Also shown are the 1-sun values measured separately in a 1-sun multi-source solar simulator (non-flash). The irradiance under concentration was determined from the I_{sc} ratio, as in Eq. (10.5).

10.5.6 Uncertainty Analysis

I-V measurements of CPV multijunction cells are based on techniques developed for single junction cells, for which example uncertainty analyses have been published and are applicable [39]; another example applied to thin-film tandem cells is in [40].

As discussed above, accurate measurements of multijunction solar cells require the incident spectral irradiance to be such that the ratios between the subcells' photocurrents are those that

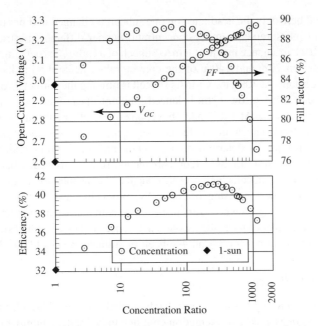

Figure 10.9 Open-circuit voltage, fill factor, and efficiency versus concentration for a GaInP/GaAs/Ge triple cell. The open circles are data points measured in a flash solar simulator, and the solid diamonds are the 1-sun calibration measured in a continuous multi-source simulator. Source: Data from [38]

are found under the reference spectral irradiance. Out-of-balance photocurrents affect the operating points of all subcells and cause errors in the measured *I-V* parameters (see section 10.3.4 and Figure 10.3); these effects are the largest error source for multijunction measurements when compared with single junctions.

An example of these effects is shown in Figure 10.10. In contrast to the spectrometric characterization from section 10.3.4 and Figure 10.3 where the effective irradiances on two

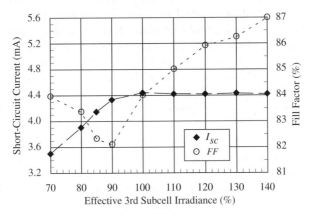

Figure 10.10 Short-circuit current and fill factor versus effective irradiance into the 3rd subcell of a GaInP/GaAs/GaInAs triple cell, measured in a continuous 1-sun multi-source simulator. Source: Data from [38]

subcells are varied here only the irradiance on one subcell is changed in order to correlate more to the situation of measurements under concentration under different spectra.

In Figure 10.10 the irradiance into the GaInAs subcell of a GaInP/GaAs/GaInAs multijunction cell is varied between 70% and 140% of the correct value, while the irradiance to the other subcells is held constant. At 70%, the GaInAs subcell is limiting the current through the external cell. As the irradiance increase toward the balanced condition, the I_{sc} increases in proportion, but a large drop in the FF is seen. At 100% and above, one of the other subcells is now the limiting subcell, so the I_{sc} is constant with irradiance. As at measurements under concentration the concentration ratio is typically determined based on the measured short circuit current I_C under concentration and the one sun current I_{1S} (compare section 10.2.2) this means that even though the irradiance on the GaInAs subcells increases the calculated concentration ratio will be constant. However, the FF is now increasing dramatically, producing a 4% error at 140%. In the 840–950 nm wavelength region where xenon spectra contain many emission lines, an unfiltered xenon simulator can easily produce 140% excess irradiance, and thus result in significant FF and efficiency measurement errors for 3- and 4-subcell multijunction cells [38,41,42].

10.5.7 Open Challenges

CPV multijunction metrology is not a completely settled science; although improvements in measurement techniques and instrumentation continue to be made, a number of challenges to obtaining accurate performance data have not yet been complete overcome.

Isotype/component reference cells – these remain difficult to obtain because they must be specially grown outside of cell manufacturers' normal production. This issue is compounded because CPV multijunction cells continue to evolve, rendering older isotype cells less useful.

Increasing numbers of subcells – the quest for the highest efficiencies has led to the current state-of-art cells composed of four subcells, and five- and six-subcell devices have been grown. Spectral irradiance adjustments for these cells can and will be difficult for measurement laboratories, specially if the existing wavelength adjustment bands in solar simulators do not correspond with those of the subcells. Changing or increasing the number of bands in simulators can be expensive as well as time-consuming.

Spectral irradiance measurements – as discussed above in section 10.4.5, accurate measurements of flash simulator spectral irradiance is complicated by the need for rapid data acquisition and by the very high total irradiance levels. Currently, spectroradiometers are calibrated against standard filament lamps that only provide irradiance levels of a few tenths of a sun, yet the instruments must measure levels several orders of magnitude greater. Spectroradiometer calibrations at high irradiance could be done at discrete wavelengths using high-power lasers, but would require an expensive investment in hardware and personnel.

Linearity – verification of the assumption that I_{sc} is linear with total irradiance (see section 10.2.2) is another difficult measurement. In particular, the effects related to luminescence coupling between the subcells have not yet been investigated sufficiently for high light intensities. It requires a detector that is itself linear over several orders of magnitude, but even if such a detector was available, verification is also complicated by the fact that I_{sc} is a function of spectral irradiance. Thus, linearity data is colored by any spectral response mismatch between the detector and the CPV cell being tested. Measuring EQE as a function of bias irradiance is a way to avoid the detector problem, but currently almost all EQE measurements of CPV multijunction cells are done with bias irradiances of 1-sun or less.

Acknowledgments

Carl Osterwald acknowledges the support by the U.S. Department of Energy under Contract No. DE-AC36-08-GO28308 with the National Renewable Energy Laboratory.

Gerald Siefer wishes to thank all members of the department III–V – Epitaxy and Solar Cells at Fraunhofer ISE for their contributions to the work presented here.

Glossary

List of Acronyms

Acronym	Description
AM1.5d	Reference spectrum for direct radiation
AM1.5d	Standardized reference spectral irradiance, direct normal
AM1.5g	Standardized reference spectral irradiance, hemispherical (global)
CPV	Concentrator photovoltaics
EQE	External quantum efficiency
I-V	Current-voltage curve
LED	Light emitting diode
PV	Photovoltaics
QE	Quantum efficiency
SR	Spectral response
SRC	Standard reporting conditions

List of Symbols

Typical units given in square brackets. If no units are given, variable is dimensionless.

Symbol	Description [Units]
A	Cell area [m^{-2} or cm^{-2}]
A_j	Light source j scaling factor [arbitrary]
C	Concentration ratio, irradiance
C_g	Concentration ratio, geometric
E or $E(\lambda)$	Spectral irradiance [$Wm^{-2}nm^{-1}$]
FF	Fill factor [dimensionless or %]
G	Total irradiance [Wm^{-2}]
G_{eff}	Total irradiance, effective [Wm^{-2}]
G_{ref}	Total irradiance, reference [$1000\,Wm^{-2}$]
I	Current [A]
i or j	Subcell number, beginning with topmost
I_{1S}	Short-circuit current at 1-sun irradiance [A]
I_C	Short-circuit current under concentration [A]
I_d	Subcell dark current [A]
I_p	Subcell photocurrent [A]
I_{max}	Current at maximum power point [A]
I_{sc}	Short-circuit current [A]
I_{RC}	Short-circuit current, reference cell [A]

(continued)

Symbol	Description [Units]
j or k	Light source number, multi-source simulator
J_{sc}	Short-circuit current density [Am^{-2} or Acm^{-2}]
$Load$	As a subscript, refers to load resistance
M	Spectral mismatch factor
P_{max}	Maximum power point, I-V curve [W]
T	Temperature, solar cell internal [°C]
R	Resistance [Ω]
R_{series}	Series resistance [Ω]
R_{sh}	Shunt resistance [Ω]
RC_i	Reference cell for subcell index i
R_{ij}	Spectral adjustment procedure, multi-source simulator
sf	Scaling factor, quantum efficiency
SR	Spectral responsivity, absolute [AW^{-1}]
sr	Spectral responsivity, relative [dimensionless]
V	Voltage [V]
V_{max}	Voltage at maximum power point [V]
V_{oc}	Open-circuit voltage [V]
V_S	Voltage across multijunction cell series resistance [V]
V_T	Voltage across multijunction cell tunnel diode [V]
η	Efficiency, solar cell [%]
λ	Wavelength [nm or μm]

References

1. C. R. Osterwald. Translation of device performance measurements to reference conditions. *Solar Cells*, **18**, 269–279, 1986.
2. K. A. Emery. Solar simulators and I-V measurement methods. *Solar Cells*, **18**, 251–260, 1986.
3. K. Emery. Measurement and characterization of solar cells and modules, in *Handbook of PV Sci. and Eng.*, 2nd edn, John Wiley & Sons, W. Sussex, UK, 2011, 797–840.
4. American Society for Testing and Materials. ASTM E948-15 - Standard test method for electrical performance of photovoltaic cells using reference cells under simulated sunlight. ASTM International, West Conshohocken, Pennsylvania, USA. 2015.
5. International Electrotechnical Commission. IEC 60891 ed2.0 - Photovoltaic devices–Procedures for temperature and irradiance corrections to measured I-V characteristics. IEC, Geneva. 2009.
6. International Electrotechnical Commission. IEC 60904-3 ed2.0 - Photovoltaic devices—Part 3: measurement principles for terrestrial photovoltaic (PV) solar devices with reference spectral irradiance data. Geneva: IEC, Geneva. 2008.
7. International Electrotechnical Commission, IEC 60904–7 ed3.0 - Photovoltaic devices – Part 7: Computation of the spectral mismatch correction for measurements of photovoltaic devices. IEC, Geneva. 2008.
8. American Society for Testing and Materials. ASTM E973-15 - Standard test method for determination of the spectral mismatch parameter between a photovoltaic device and a photovoltaic reference cell. ASTM International, West Conshohocken, Pennsylvania, USA. 2015.
9. C. H. Seaman. Calibration of solar cells by the reference cell method—the spectral mismatch problem. *Solar Energy*, **29**, 291–298. 1982.
10. M. A. Green, K. Emery, Y. Hishikawa, W. Warta, and E. D. Dunlop. Solar cell efficiency tables (version 39). *Progress in Photovoltaics: Research and Applications*, **20**, 12–20. 2012.
11. J. M. Gee and H. B.R. Photovoltaic concentrator cell measurement methods. *Solar Cells*, **18**, 281–288. 1986.
12. American Society for Testing and Materials. ASTM G173-03(2012) Standard tables for reference solar spectral irradiances: direct normal and hemispherical on 37° tilted surface. ASTM International, West Conshohocken, Pennsylvania, USA. 2012.

13. R. C. Dondero, T. E. Zirkle, C. E. Backus. Superlinear behavior of short circuit current in silicon n +/p/p+ solar cells. Proceedings of the 18th IEEE Photovoltaic Specialists Conference, Las Vegas NV, USA. 1985.

14. J. M. Gee. Characterization of the Isc versus irradiance relationship for silicon and III-V concentrator cells. Proceedings of the 19th IEEE Photovoltaic Specialists Conference, New Orleans, LA, USA. 1987.

15. G. Paternostera, M. Zanuccolib, P. Belluttia, *et al.* Fabrication, characterization and modeling of a silicon solar cell optimized for concentrated photovoltaic applications. *Solar Energy Materials and Solar Cells*, **134**, 407–416. 2015.

16. T. E. Zirkle, S. Peterson, K. Joardar, D. K. Schroder, C. E. Backus. Short-circuit current superlinearity phenomenon in GaAs solar cells. Proceedings of the 8th European PV Solar Energy Conference and Exhibition, Florence, Italy. 1993.

17. M. A. Steiner, J. F. Geisz. Non-linear luminescent coupling in series-connected multijunction solar cells. *Applied Physics Letters*, **100**, 2511–06. 2012.

18. C. Baur, M. Meusel, F. Dimroth, A. W. Bett. Investigation of Ge component cells. Proceedings of the 31st IEEE Photovoltaic Specialists Conference, Orlando, FL, USA. 2005.

19. K. Emery, M. Meusel, R. Beckert, *et al.* Procedures for evaluating multijunction concentrators. Proceedings of the 28th IEEE Photovoltaic Specialists Conference, Anchorage, AL, USA. 2000.

20. K. Heidler, A. Schoenecker, B. Mueller-Bierl, and K. Buecher. Progress in the measurement of multi-junction devices at ISE. Proceedings of the 22nd IEEE Photovoltaic Specialists Conference, Las Vegas, NV, USA. 1991.

21. M. Meusel, R. Adelhelm, F. Dimroth, A. W. Bett, and W. Warta. Spectral mismatch correction and spectrometric characterization of monolithic III-V multi-junction solar cells. *Progress in Photovoltaics: Research and Applications*, **10**, 243–255. 2002.

22. T. Moriarty, J. Jablonski, and K. A. Emery. Algorithm for building a spectrum for NREL's One-Sun Multi-Source Simulator. Proceedings of the 38th IEEE Photovoltaic Specialists Conference, Austin, TX, USA. 2012.

23. C. Domínguez, I. Antón, G. Sala, and S. Askins. Current-matching estimation for multijunction cells within a CPV module by means of component cells. *Progress in Photovoltaics: Research and Applications*, **21**, 1478–1488. 2013.

24. G. Peharz, G. Siefer, and A. W. Bett. A simple method for quantifying spectral impacts on multi–junction solar cells. *Solar Energy*, **83**, 1588–1598. 2009.

25. J. Kiehl, K. Emery, and A. Andreas. Testing concentrator cells: spectral considerations of a flash lamp system. Proceedings of the 19th European PV Solar Energy Conference and Exhibition, Paris, France. 2004.

26. G. F. Virshup. Measurement techniques for multijunction solar cells. Proceedings of the 21st IEEE PV Specialists Conference, Kissimimee, FL, USA. 1990.

27. F. Nagamine, R. Shimokawa, M. Suzuki, and T. Abe. New solar simulator for multi-junction solar cell measurements. Proceedings of the 23rd IEEE PV Specialists Conference, Louisville, KY, USA. 1993.

28. A. M. Andreas and D. R. Myers. Pulse analysis spectroradiometer system for measuring the spectral distribution of flash solar simulators. Proceedings of the SPIE Optics and Photonics 2008 Conference San Diego, CA, USA. 2008.

29. S. B. Essig, J., Schachtner, M., Wekkeli, A., Hermle, M., Dimroth, F. Wafer-bonded GaInP/GaAs//Si-solar cells with 30% efficiency under concentrated sunlight. *IEEE Journal of Photovoltaics*, **5**(3), 977–981. 2014. DOI: 10.1109/JPHOTOV.2015.2400212

30. J. Metzdorf. Calibration of solar cells. 1: The differential spectral responsivity method. *Applied Optics*, **26**, 1701–1708. 1987.

31. J. Burdick and T. Glatfelter. Spectral response and *I-V* measurements of tandem amorphous-silicon alloy solar cells. *Solar Cells*, **18**, 301–14. 1986.

32. M. Meusel, C. Baur, G. Létay *et al.* Spectral response measurements of monolithic GaInP/Ga(In)As/Ge triple-junction solar cells: Measurement artifacts and their explanation. *Progress in Photovoltaics: Research and Applications*, **11**, 499–514. 2003.

33. G. Siefer, C. Baur, and A. W. Bett. External quantum efficiency measurements of germanium bottom subcells: Measurement artifacts and correction procedures. Proceedings of the 35th IEEE Photovoltaic Specialists Confeernce, Honolulu, HI, USA. 2010.

34. E. Barrigón, P. Espinet-González, Y. Contreras, and I. Rey-Stolle. Implications of low breakdown voltage of component subcells on external quantum efficiency measurements of multijunction solar cells. *Progress in Photovoltaics: Research and Applications*, **23**(11), 1597–1607. 2015. DOI: 10.1002/pip.2597

35. M. A. Steiner, J. F. Geisz, T. E. Moriarty *et al.* Measuring IV curves and subcell photocurrents in the presence of luminescent coupling. *IEEE Journal of Photovoltaics*, **3**, 879–887. 2013.

36. J. Metzdorf, A. Meier, S. Winter, and T. Wittchen. Analysis and correction of errors in current-voltage characteristics of solar cells due to transient measurements. Proceedings of the 12th European PV Solar Energy Conference and Exhibition, Amsterdam, The Netherlands. 1994.

37. F. Lipps, A. Zastrow, and K. Bücher. I-V characteristics of PV-modules with a msec flash light generator and a 2 MHz data acquisition system. Proceedings of the 13th European PV Solar Energy Conference and Exhibition, Nice, France. 1995.

38. C. R. Osterwald, M. W. Wanlass, T. Moriarty, M. A. Steiner, and K. A. Emery. Effects of spectral error in efficiency measurements of GaInAs-based concentrator solar cells. National Renewable Energy Laboratory, NREL/TP-520-60748, Golden CO, USA. 2014.

39. K. Whitfield and C. R. Osterwald. Procedure for determining the uncertainty of photovoltaic module outdoor electrical performance. *Progress in Photovoltaics: Research and Applications*, **9**, 87–102. 2001.

40. Y. N. Qiu, T. R. Betts, G. R., and W. Herrmann. Uncertainties in the calibration of thin film based silicon based multi-junction devices using single source solar simulators. Proceedings of the 23rd European PV Solar Energy Confrence and Exhibition, Valencia, Spain. 2008.

41. C. R. Osterwald, M. W. Wanlass, T. Moriarty, M. A. Steiner, and K. A. Emery. Concentrator cell efficiency measurement errors caused by unfiltered xenon flash solar simulators. Proceedings of the 10th International Conference on Concentrator PV Systems, Albuquerque, NM, USA, 2014. http://scitation.aip.org/content/aip/proceeding/aipcp/1616

42. G. Siefer, C. Baur, M. Meusel, F. Dimroth, A. W. Bett, and W. Warta. Influence of the simulator spectrum on the calibration of multi-junction solar cells under concentration. Proceedings of the 29th IEEE PV Specialists Conference, New Orleans, LA, USA. 2002.

11

Characterization of Optics for Concentrator Photovoltaics

Maikel Hernández
LPI-Europe, S.L., Spain

11.1 Introduction

The objective of this chapter is to provide to non-optical specialists some guidelines about the measurements that can be done in a laboratory to perform a basic quality control on concentrator photovoltaic (CPV) optical elements. The optical concepts or architectures used in CPV are customized and optimized for each design, and therefore quality control of such optics has to be adapted to each system. This means that technical staff in charge of the evaluation and assembly of the optics should know as much as possible about the performance of the optical elements in order to adjust the measurement procedures as needed.

In general the optical architectures used in CPV are composed by more than one optical element. Actually, there are CPV systems that use more than two optical elements but using two is the most common situation because they keep the balance between performance and simplicity (see Chapter 4). These optical elements are usually named by its function into the optical system, for example: *primary optical element* (POE) and *secondary optical element* (SOE). The POEs are the optics that face the sun and bend the light the most. They could have continuous (smooth) or faceted geometries (i.e. parabolic mirrors or Fresnel lenses respectively). The SOEs are smaller than the POEs and they are placed near the solar cell. The combination of a POE and a SOE is typically referred to as the optical train of the CVP system. When we are going to perform quality control of the optical elements we should be aware of which are the possible loss sources in the optics: Fresnel losses, reflection losses, material absorption losses, surface and bulk scattering, vertex curvature radius, inactive faces and surface shape errors. The first four are called optical losses and the rest geometrical losses.

This chapter is organized in two main sections, namely, *geometrical characterization* and *optical characterization*. Section 11.2 describes how to measure the surface accuracy of optical

---

Handbook of Concentrator Photovoltaic Technology, First Edition. Edited by Carlos Algora and Ignacio Rey-Stolle.
© 2016 John Wiley & Sons, Ltd. Published 2016 by John Wiley & Sons, Ltd.

elements depending on the kind optical surface they have: faceted optics and non-faceted optics. Section 11.3 describes the most common procedures to carry out the characterization of optical performance of lenses and concentrators. In this section the reader will find the schematic set-ups and discussion on how to perform optical measures to evaluate the optical efficiency, acceptance angle and spectral irradiance distribution on the solar cell. Some examples are also shown in order to illustrate how to proceed depending of the kind of optics we are evaluating.

11.2 Geometrical Characterization

Geometrical characterization of optical parts consists of a set of measurements to evaluate how the manufactured optics geometrically deviates from theory. It includes the measurement of curvature radii and draft angles of faceted optics like Fresnel or *total internal reflection* (TIR) lenses.

The objective is to predict the impact of the manufacturing errors on the system performance with the help of simulation tools. The impact on the performance of the optical surface errors mainly depends on:

- Slope error: deviation of measured surface slope at the measurement point from the slope of the theoretical surface at the same point. It gives a measure of how much the light deviates from its theoretical path after bouncing in a surface. If we know the local acceptance angle of the surface at the point we are evaluating, we will be able to know if the ray will be collected or not.
- The type of optical surface, reflective or refractive. For refractive surfaces, the light incidence angle and the ratio of air and medium refractive indexes are also important.
- The local acceptance angle of the surfaces.

As a rule of thumb, first we should expect that reflective surfaces are in general more sensitive to slope errors than refractive surfaces because a slope error (δ_e) in a reflection deviates the ray from this original trajectory two times the error ($2\delta_e$). Second, POEs are more sensitive to slope errors than SOEs because SOE optics have higher local acceptance angle [1].

There are two groups of measurements to be carried out during the geometrical characterization: general dimension measurements and optical surface shape or form-factor measurements. The measurement of general dimensions – like diameter or thickness – is, in general, easy to perform and gives us a general overview of the quality of the optical parts. However, measuring accurately the surface shape is a difficult task that requires precise instruments as 3D contact or non-contact profilometers [2].

From acquired data on the optical elements we will be able to create optical models in order to simulate the system performance. This procedure will show the impact of imperfections in the optics on the performance. This is a reliable method that will reveal if the accuracy of the manufactured surfaces fulfills the system specifications.

Optics with faceted profiles [3] and non-faceted profiles (smooth profile) sometimes require different procedures for their geometrical characterization. The reason is because faceted optics (like Fresnel lenses) have small features with huge derivative variations that make it very difficult (sometimes impossible due to the small size of the facets) and time consuming to measure their profile accurately. Figure 11.1 shows the cross-section of a segment of a Fresnel

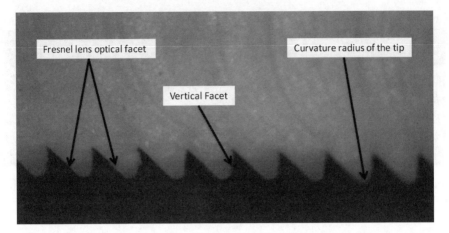

Figure 11.1 Detail of a Fresnel lens cross-section view under microscope

lens where the discontinuity of the surface is easy appreciated. Faceted optics also have two sources of geometrical losses that usually are measured at the same time: the one associated with the draft angles of the vertical walls and the one associated with the curvature radii of the facets tips. In Figure 11.1 we see the three types of surface that light has to cross in its path towards the solar cell: the optical facet, the vertical facet and the curved area resulting from the joint between the two aforementioned surfaces. However, for measuring a non-faceted or smooth surface – like the one of a parabolic mirror – a direct measurement of the shape or profile can be done by using traditional instruments as 3D contact or non-contact profilometers.

11.2.1 Faceted Optics

Measuring the section profile of faceted optics, like a Fresnel lens or a TIR lens, is a time and effort-consuming task. The reason is that it is not possible to use either a scanning system based on a laser nor a contact profile measurement system to acquire the lens profile. This is due to the large variations in the slope of the teeth forming the optical profile together with the depth of the facets (see Figure 11.1).

Since the source of the largest geometrical losses in a faceted optical element are the draft angle of the vertical facet and the curvature radii of the tips, we have to measure them if we detect any drop on the optical efficiency or if we are evaluating differences between the different optical manufacturing technologies of suppliers. The draft angles and tip radii are the ones that have the higher impact on the performance. Scattering is also an important source of losses; we will present later how to characterize it.

In order to measure the tip radius of the peak and valley, we have to cut the part into two halves or reproduce in silicone or in other material a *negative* of the optical part and cut it. In both situations we have to ensure that we are extracting the section that exactly coincides with a diameter in a rotationally symmetric optics or perpendicular to the extrusion direction in linear optics. Figure 11.2 shows the profile reproduction process for evaluating the accuracy of a Fresnel lens mold insert. The sequence is as follows: first we replicate the surface to be measured by filling it with a silicone rubber. After curing we extract the polymer part and finally we cut a diameter section of the surface.

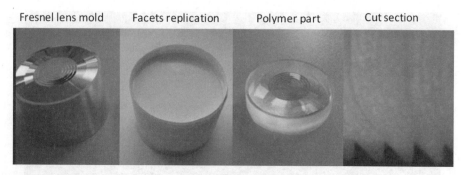

Figure 11.2 Figure showing the sequence to extract the section profile of a faceted optics in order to measure it

Once we have the cross-section of the optics, we will be able to scan its profile using, for instance, a microscope. From this data we can extract the draft angle and tips curvature radii information. Figure 11.3 shows an example of the calculation from the measured points of the curvature radii on the tips of the Fresnel facets.

In Figure 11.4 (left) we see an example of a TIR concentrator POE [4] being characterized using a microscope. Here the optical section was extracted by embedding the lens in a hard resin, cutting it in a half in a meridian plane and polishing. The microscope used to measure this profile was the *Unitron NMS Bi-9817* and the data acquisition was done with a *QuadraCheck QC-2210*, which allows to measure points along the lens profile. At the end, the outcome of the acquisition process is a set of points in text format that can be used for comparison with the theoretical profiles. In the right side of Figure 11.4 we show the detail of one of the TIR lens facet tips.

The expected theoretical efficiency losses due to the tip radii and draft angles depend on the design. Figure 11.5 shows an example of how the draft angle and tips curvature radii losses behave for a point focus Fresnel lens. For example, assuming that the lens was cut using a tool of 12 microns half radius and draft angles of 2°, the losses are around 1.7% and 3.4% respectively. This means that the maximum geometrical efficiency of the Fresnel lens will be around 94.6% when compared to a 'perfect' lens without geometrical losses.

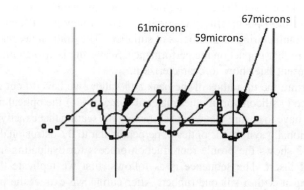

Figure 11.3 Curvature radii measurement of a Fresnel lens cross-section

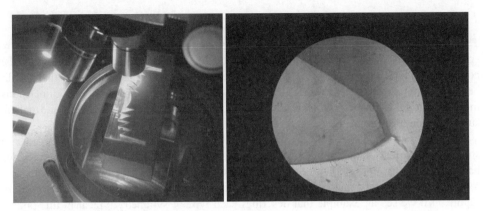

Figure 11.4 Right, detail of a facet tip of a TIR concentrator view under microscope. Left, cross-section of a TIR concentrator measured using a *Unitron NMS Bi-9817* microscope and a *QuadraCheck QC-2210* data acquisition system. Note that to obtain the cross-section, the concentrator has been embedded in a resin, cut in half and polished

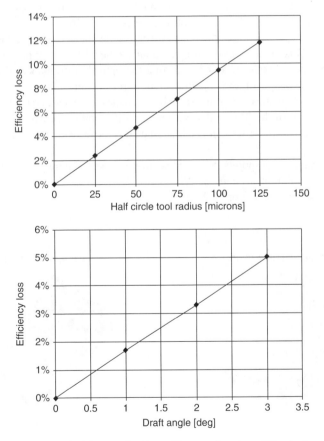

Figure 11.5 Relative geometrical losses of a Fresnel lens as function of the draft angle (top) and tips curvature radii (bottom)

11.2.2 Non-faceted Optics

The source geometrical losses in optics with non-faceted optical surfaces are mainly caused by errors in the shape (and assembly) of molds used for molding the optical parts or by the contractions of the optics after manufacturing. There are two sets of measurements that should be carried out to characterize the geometrical quality of the optics: general dimensions measurements (i.e. diameter, thickness, relative position between surfaces, etc.) and optical surface shape measurements.

In optics with rotational or linear symmetry a 2D profile could be measured in order to simplify the process. In this situation, as commented for the measurement of faceted optics, we have to ensure that the symmetry line being measured exactly coincides with a diameter in a rotationally symmetric optics or is perpendicular to the extrusion direction in a linear optics. In Figure 11.6 we see an example that describes each type of symmetry typical in this optical elements. On the left side of Figure 11.6 there is a rotational aspheric mirror created by revolving the 2D section around a corresponding axis. On the other hand, on the right side of Figure 11.6 we show an aspheric mirror with linear symmetry along one of the axis. In both cases the optics can be described or characterized by analyzing only a 2D section. Representative CPV Concentrators based on rotational and linear symmetric optics are the SMS XR Concentrator [5] and the DSMTS Concentrator, respectively [6].

After extracting the 2D section of the optics we can model the manufactured lens by using the measured section. This model could be used to simulate the CPV system performance based on the measured profiles. Another way of analyzing the optical performance of the optics is by calculating how the slope of the manufactured mirror deviates from the theoretical one. Figure 11.7 shows schematically a comparison between the theoretical and measured 2D sections of the aspheric mirror of the SMS XR Concentrator with rotational symmetry (Figure 11.8).

From this measured profile we can calculate the slope error using the following procedure:

1. Choose the number of points to be measured, N (see Figure 11.7). The slope error measurement resolution depends on N.
2. Loop from $k = 1$ to $k = N$. For each k:

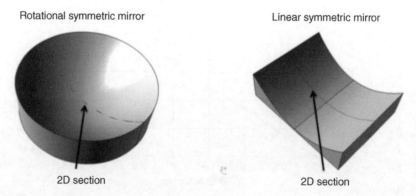

Rotational symmetric mirror   Linear symmetric mirror

2D section   2D section

Figure 11.6 Example of two symmetrical optics: on the left an aspheric mirror with rotational symmetry and on the right an aspheric mirror with lineal symmetry

Figure 11.7 Comparison between the theoretical and the measured 2D sections of an optical element with rotational symmetry

a. Calculate the unit vectors u_x and u_z:

$$u_x = \frac{(x_k - x_{k+1})}{\sqrt{(x_k - x_{k+1})^2 + (z_k - z_{k+1})^2}} \; ; u_z = \frac{(z_k - z_{k+1})}{\sqrt{(x_k - x_{k+1})^2 + (z_k - z_{k+1})^2}} , \quad (11.1)$$

b. Calculate the unit vectors v_x and v_z:

$$v_x = \frac{(x_k - x_{k+1})}{\sqrt{(x_k - x_{k+1})^2 + (f(x_k) - f(x_{k+1}))^2}} ; \quad (11.2)$$

$$v_z = \frac{(f(x_k) - f(x_{k+1}))}{\sqrt{(x_k - x_{k+1})^2 + (f(x_k) - f(x_{k+1}))^2}} , \quad (11.3)$$

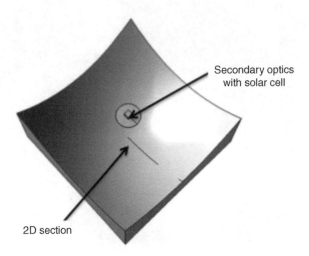

Figure 11.8 SMS XR Concentrator with rotational symmetry

c. Calculate the slope error in degrees (δ_e):

$$\delta_e = \arccos(u_x \cdot v_x + u_z \cdot v_z)\frac{180}{\pi}, \qquad (11.4)$$

The calculated slope error, δ_e, gives a measure of how much the light deviates from its theoretical path after bouncing off a surface. If we know the local acceptance angle of the surface at the point we are evaluating, we will be able to know if the ray will be collected or not, thus we will able to calculate the area of the optics which does not collect light. As we mentioned before, reflective surfaces are more sensitive to slope errors than refractive surfaces because a slope error (δ_e) in a reflection deviates the ray from this original trajectory two times the error ($2\delta_e$).

When the optical elements have free-form shape (i.e. with no prescribed symmetries as in the case of a Fresnel-Köhler POE or SOE) we have to measure the entire optical surface. From the acquired point cloud we can calculate the deviation from the theoretical surface and also we can create optical models to evaluate how the form-factor errors affect the system performance. Figure 11.9 shows an example of a free-form optical surface measurement and the calculation of the deviation from the theoretical shape. This measurement corresponds to the mirror of the SMS3D XR concentrator [7] shown in Figure 11.10. The procedure consists in measuring a set of points on the mirror surface and comparing them with the theoretical surface. The result is a deviation map along the optical surface. In the Figure 11.9 we show such deviation map where the darkest areas are the ones with larger deviations from the theoretical surface.

During the measurement of the shape of optical elements we have to guarantee that we do not introduce extra deformation in the optical part. This is visible when we measure large optical elements like, for instance, PMMA injected mirrors. If that happens, the results and the conclusions will be wrong. Effects like the gravity force can modify the shape of the optics

Figure 11.9   Measurement of a 3D point cloud of a free-form POE

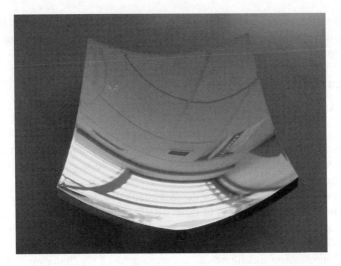

Figure 11.10   Image of the mirror PMMA injected of the SMS3D XR Concentrator

depending on its size and thickness, thus introducing measurement errors. Note that this effect will not appear when the optics are placed in the module because the housing is a structural element that helps to avoid extra deformations.

The measurement probe can be also another source of error when using measurement methods based on contact profile-meters in case the measuring tip touches the optics.

This means that we have to take these effects into account in order to design a good measurement procedure. Figure 11.11 shows the measurement of a thin optical element using a

Figure 11.11   POE measurement using a contact profilometer. Special positioners are used to minimize the effect of gravity force and tip contact

contact profile-meter. Errors introduced by the gravity force or probe touching effects are avoided by the use of specific designed gauges to fix the shape of the optical surface during the measurement to avoid extra deformation, as shown in the picture.

11.3   Optical Characterization

11.3.1   Measurement of the Optical Efficiency

In this section, some techniques to measure the optical losses of optical elements are shown. Here we will present how to perform efficiency measurements or transmission measurements. From the transmission measurements we can obtain the losses as the percentage of non-transmitted optical power.

Figure 11.12 shows the base set-up for the measurement of both optical efficiency of optical elements and the optical transmission of materials. This layout is general for monochromatic and spectral measurements. This means that we should be able to select a light source and detector pair matched to the spectral band under study.

The optical transmission measurements are more focused on characterizing the quality of bulk material and thin film coatings by selecting samples of the materials or sections of the optics where there are no geometrical losses. For example in a block with known thickness (d) light enters the part with intensity I_0 and leaves the part with a lower intensity I_1.

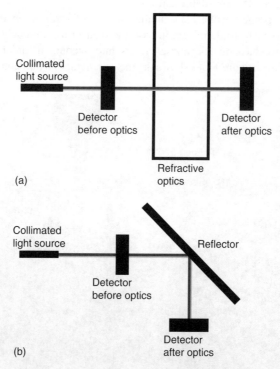

Figure 11.12   Schematic description of local optical efficiency measurements. The top set-up is the base layout for measuring the optical transmission for refractive optics; the bottom set-up is the base layout for reflective optics

The reduction in intensity is due to two factors:

- Fresnel reflections that decrease intensity by a transmission factor T_F.
- Absorption inside the material of thickness d given by $e^{-\alpha d}$, according to Beer–Lambert law.

The intensity at the other end of the part is then:

$$I_1 = T_F e^{-\alpha d}, \tag{11.5}$$

where α is the absorption coefficient of the material and T_F is the Fresnel losses transmission factor. In our case, there are two Fresnel reflections (when the light enters and leaves the transparent material). It should be also noted that T_F depends on the incidence angle of the light, the optical surface, the refractive index of the material and the wavelength. For the sake of simplicity, it is recommended to restrict the incidence angle of the light to normal incidence (i.e. use collimated light at normal incidence), in order to avoid the non-linear dependence of Fresnel losses with the incidence angle. The Fresnel losses transmission factor T_F can be calculated as:

$$T_F = \frac{16 n_1^2 n_2^2}{(n_1 + n_2)^4}, \tag{11.6}$$

Combining the two previous equations we can calculate the absorption coefficient as:

$$\alpha = -\frac{\ln(I_1/T_F)}{d} = -\frac{1}{d} \ln\left(\frac{I_1(n_1 + n_2)^4}{16 n_1^2 n_2^2}\right), \tag{11.7}$$

Making $n_1 = 1$ (if the transparent material is in air) and $n_2 = n$:

$$\alpha = -\frac{1}{d} \ln\left(\frac{I_1(1 + n)^4}{16 n^2}\right), \tag{11.8}$$

The optical transmission measurements along the solar spectrum are very important in CPV for selecting the materials and their thickness since the receiver (the solar cell) has a specific performance along the solar spectrum (see Chapter 2). For this measurement the use of a spectroradiometer is recommended as a sensor. Figure 11.13 shows the results of measuring the spectral transmission of a PMMA slab with a thickness of 3 mm.

The optical characterization at the laboratory is based on optical efficiency measurements performed locally or in an extended area of the optics. With the optical efficiency (η) we characterize the mirror reflectivity losses or the combination of Fresnel losses, absorption losses and geometrical losses. The optical efficiency simply is defined as the ratio of the power before the optics (i.e. impinging the optics) to the power after the optics (i.e. coming out of the optics):

$$\eta = \frac{P_{after\ Optics}}{P_{before\ Optics}}, \tag{11.9}$$

where $P_{after\ Optics}$ is the light power after the optics (POE or SOE) and $P_{before\ Optics}$ is the original light power before the optics. We have the choice of using a monochromatic source as laser or a solar simulator lamp as light source depending on whether we are performing monochromatic or broadband measurements, respectively.

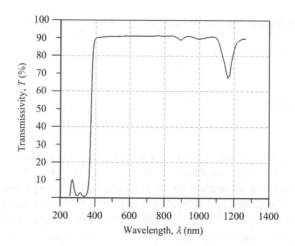

Figure 11.13 Spectral transmission measurement results of a 3 mm thick slab of PMMA

The detector selection depends on the spectral range we are planning to measure. Commonly, a silicon photodiode is used when there is no need to evaluate the optical efficiency as function of the wavelength. If an optical efficiency as function of the wavelength is required, the detector will be a spectral radiometer.

In addition, the optical efficiency measurements can be performed on the full area or, at least, in an extended area of the lens. This method consists in the use of a collimated beam to illuminate the entire optics or an extended part of it. In case we are measuring POE optics the angular extension of the light cone should be as wide as the acceptance angle of the concentrator. It is necessary that the light source will be stabilized in order to minimize measurement noise. This is an approximated method that is very useful for evaluating the quality of Fresnel lenses.

Figure 11.14 shows the equipment needed for the measurement. The collimated beam is created by the elements labeled in the picture as 'source' (stable LED or Lamp), d1 (a diaphragm), collimation lens and field stop. Using all these elements we are able to create the light beam with the required angular spread. The selection of the detector depends on many

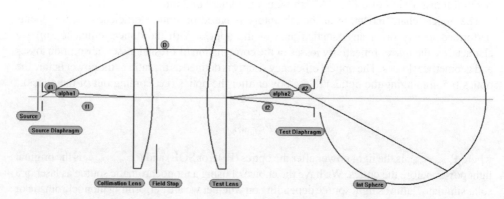

Figure 11.14 Schematic description of a set-up lens optical efficiency characterization

Figure 11.15 Set-up used for optical efficiency measurement of a Fresnel lens

factors as type of light source, spot size, signal level, etc. In this example, the selected detector is an integration sphere with a luxmeter.

This procedure allows measuring the relative efficiency of the optics under test (η_m), compared to a known optics. It contains all optical losses, transmission and geometrical losses. As defined in Eq. (11.10), η_m is the ratio of the sensor reading when the test lens is measured (I_f) to the reading when the reference lens is measured (I_c). It is necessary to guarantee that the solid angle collected by the sensor is the same when both lenses are measured, *alpha2* angle in Figure 11.14.

From this measurement and knowing the refractive index of lenses material it is possible to calculate the geometrical efficiency of the lens, $\eta_{g,f}$, to be tested:

$$\eta_m = \frac{I_f}{I_c} = \frac{\eta_{g,f} T_{F,f}}{\eta_{g,c} T_{F,c}} \rightarrow \eta_{g,f} = \eta_m \frac{T_{F,c}}{T_{F,f}}; \eta_{g,c} = 1, \tag{11.10}$$

where, $\eta_{g,c}$, is the geometrical efficiency of the reference lens; $T_{F,f}$ is the transmission at normal incidence of the tested lens; and $T_{F,c}$ is the transmission at normal incidence of the reference lens respectively.

As an example, we show in Figure 11.15 the measurement of direct cut Fresnel lens made of PMMA using as reference lens a plane-convex N-BK7 uncoated lens with 75 mm diameter. For the measurement we have used an integrating sphere and a luxmeter as sensor. A LED that illuminates an opal diffuser to make a Lambertian source composes the light source. In front of the opal filter there is a diaphragm to limit the emitting area to a small source in order to create the desired collimation angle.

In Figure 11.15 there are two plane-convex lenses, the first one (closer to the diffuser) is the collimation lens and the second one is the reference lens. At the end of the system there is a diaphragm with an aperture calculated before to maintain the same *alpha2* when interchanging the plane-convex lens with Fresnel lens. This diaphragm is used to reject the light that comes from facets draft angle, facets tips radii and scattering of the Fresnel lens.

11.3.2 POE Scattering Basic Measurements

Every manufactured optical element will inherently show some surface roughness. Characterizing the surface roughness [8,9] provides information on the amount of light that is scattered after hitting and optical surface. A laser-based measurement will help to find out if the optical system tolerates such high surface roughness. This is a local measurement using a laser to emulate a light ray. The laser parameters that we need to know are the divergence angle and the spot size, and they both should be as small as possible in order to avoid effects introduced by geometrical losses and surface curvature.

Figure 11.16 shows the schematic description of the measurement system. This procedure could be used for performing a local measurement of the scattering or along a section of an optical element. Essentially the set-up works as follows:

1. Initially, a detector is placed confronting the laser to calibrate its optical power.
2. The detector is removed, letting the laser beam impinge perpendicular to the POE (black thick line in Figure 11.16).
3. The laser light is refracted by a lens facet.

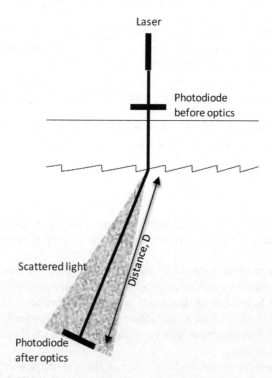

Figure 11.16   Schematic description of the used measured system

4. If the lens were perfect, the laser light would be refracted in a single ray at the designed angle (black thick line in Figure 11.16).
5. As a result of the roughness, the laser light is scattered producing a certain angular spread of the light (gray triangular area in Figure 11.16).
6. The photodiode acting as light detector after the optics, captures only a fraction of the laser power as a result of the angular dispersion of some scattered rays.

The laser used in the measurement must minimize the non-scattering losses, i.e.:

• Lens absorption must be negligible at the laser wavelength. It is recommended the use of laser beams in the visible range, of wavelength around 500 nm, where the typical materials used in CPV (glass, silicones, PMMA and mirror coatings) have a high optical transmission.
• In the situation of the POE (Fresnel or TIR lenses) the light must not hit on the vertices or on the quasi-vertical facets. For each lens facet the beam should be aimed perpendicularly to the aperture and pointing at the facet center.

In addition, there are some recommendations for the photodiode acting as the light sensor after the optics:

• As shown in Figure 11.16, the photodiode must be centered relative to the beam at the POE exit. This eliminates the contribution of the surface geometrical profile errors.
• In order to assess the impact of lens surface roughness of a real CPV set-up, the angular size of the photodiode must emulate the solar cell as seen through the SOE. This means that the photodiode must be placed precisely at the position at which it collects the light rays that would be collected by the set solar cell and SOE. Figure 11.17 illustrates such a position. The distance D at which the photodiode must be placed has to be calculated with a ray-tracing on theoretical profiles. Since the refractive index of the material depends on the radiation wavelength, the scattered light also depends on the wavelength. This means that the distance to place the photodiode has to be adjusted depending on the wavelength of the light source used.

11.3.3 Acceptance Angle Measurement

For the evaluation and quality control of the manufactured CPV optical elements the measurement of the acceptance angle is a good choice. This measurement will provide information on both optical efficiency and the acceptance angle, two of the merit functions that describe the optical system performance. This measurement can be used to obtain an absolute value of the acceptance angle of the system or to characterize acceptance angle variations compared to a reference value. The measurement is also suitable for doing the characterization of a single lens and receiver assembly as well as for characterizing a full module, of course depending on the dimensions of the testing facilities.

There are mainly two methods for measuring the acceptance angle: the direct method and the inverse method. In the direct method, either the optical train or the light source (solar simulator [10] or the sun) need to be mounted on a rotating platform in order to record the output signal of the detector as a function of the incident angle. In the case of CPV assemblies the natural choice for the detector is the solar cell itself, though other options could be also

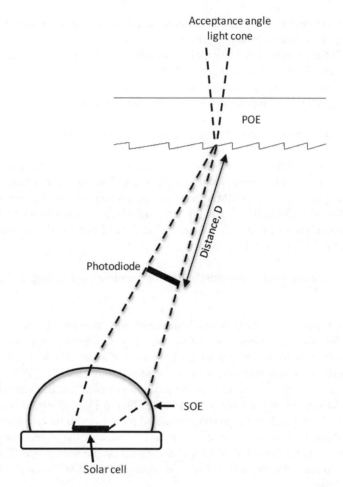

Figure 11.17 The photodiode will emulate the apparent size of the cell seen through the SOE

interesting in other applications (larger area, angular selectivity, spectral sensitivity, etc.). Figure 11.18 shows a schematic description of one of the possible set-ups for measuring directly the acceptance angle of a CPV receiver, which consist of a POE, a SOE and a solar cell. In Figure 11.18, it is assumed that the solar simulator position is fixed while the CPV receiver can turn on an axis located at the position labeled rotation center and perpendicular to the plane of the figure. In this sense, the CPV module can be deviated off the optical axis the desired angle. Figure 11.19 shows the measured photocurrent as function of the incidence angle for a CPV receiver formed by a point focus Fresnel lens (POE) and a dielectric truncated pyramid (SOE). The acceptance angle is calculated as the angle at which the photocurrent reaches 90% of the maximum photocurrent.

In the inverse method a Lambertian light source is placed at the solar cell position and the CPV module is used as collimator. As III-V semiconductors are efficient light emitters, it is possible to use the solar cell forward biased as the light source. This is a very interesting

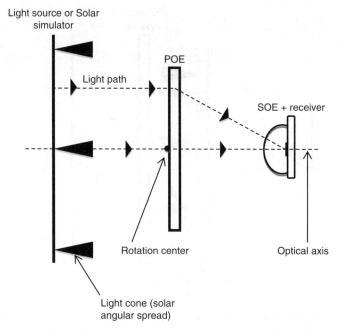

Figure 11.18 Schematic description of the measurement set-up used in the direct measurement of the concentrator acceptance angle

approach because it is highly suitable to be used in an in-line quality control process. In this measurement set-up the emitting diagram of the CPV module or assembly has to be projected on a screen by a projection system, as indicated in Figure 11.20. Finally, using a CCD camera the image on the screen is recorded and processed. This projection system has to be calibrated

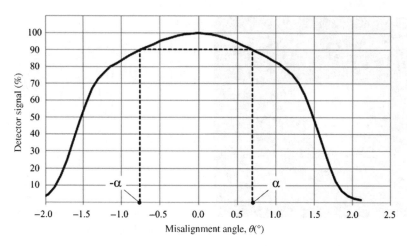

Figure 11.19 Photocurrent as function of an incidence angle measured using the direct method of a point focus Fresnel lens and truncated pyramid concentrator. The acceptance angle is labeled as α

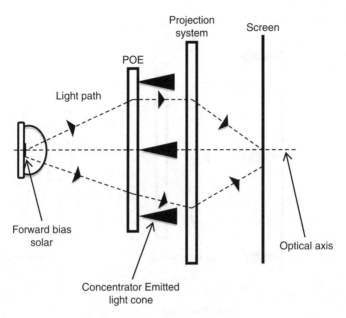

Figure 11.20 Schematic description of the measurement set-up used in the inverse measurement of the concentrator acceptance angle

and an image-processing step has to be done in order to calculate the real acceptance angle of the CPV system [11–14].

Figure 11.21 (left) shows the image captured by the CCD camera of the light emitted by a concentrator assembly in a set-up similar to Figure 11.20. On the other hand, Figure 11.21 (right) shows the iso-curves of the concentrator relative angular transmission obtained after

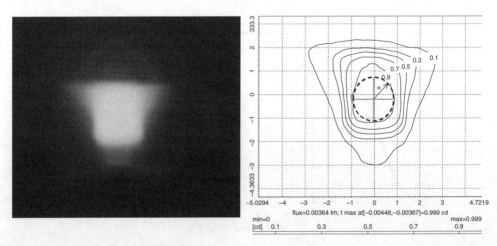

Figure 11.21 (Left) Emitting diagram of a concentrator assembly on a screen. (Right) Relative angular transmission obtained after data processing

processing. The acceptance angle of the concentrator will be the radius of the circle inscribed on the 0.9 iso-curve.

11.3.4 Spectral Irradiance Distribution Measurement at the Solar Cell Plane

The experimental set-up to measure the spatial and spectral irradiance distribution generated by the optics is based in the layout shown in Figure 11.22. For this measurement we change the solar cell for a Lambertian diffuser (transmitter in the picture) and an image proportional to the irradiance distribution on the cell is recorded using a CCD camera. In case of using an optical train without SOE, the measurement procedure is the same just placing the diffuser at the solar cell position.

In general, it is difficult to obtain an absolute measurement of the irradiance using a CCD camera because the sensibility of the camera depends on camera parameters like the exposition time and aperture. Also the bandpass filters adapted to the bandgaps of multijunction solar cell are not ideal (below we explain why filters are needed). However, the relative variations of the different areas of an irradiance pattern are valid. A more detailed description can be found in [14,15].

Optical filters will be used for the measurement of the spectral irradiance distribution. They should be placed between the Lambertian diffuser and the CCD camera. The filters will be monochromatic or adjusted to the spectral band of the subcells to be analyzed. It should be pointed out that virtually all CCD cameras use a silicon detector which means that wavelengths in the spectral band below the bandgap of silicon (e.g. the Ge bottom subcell of a GaInP/GaInAs/Ge triple-junction solar cell) cannot be detected using a conventional CCD camera.

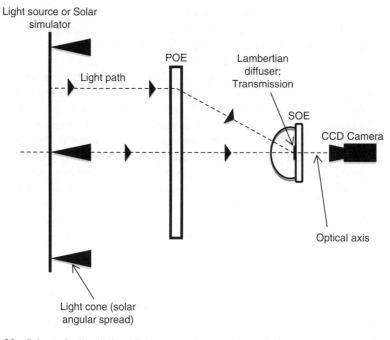

Figure 11.22 Schematic description of the measurement set-up used in the evaluation of the irradiance distribution at the solar cell plane

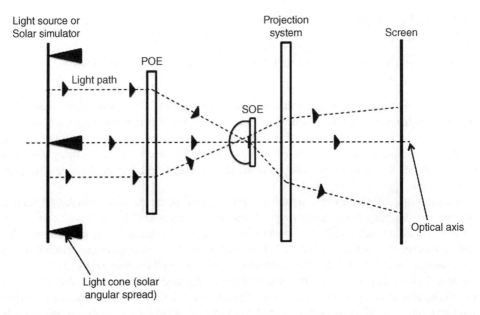

Figure 11.23 Schematic description of the measurement set-up used in the evaluation of the irradiance power distribution at the solar cell plane as function of the incidence angle

11.3.5 Angular Power Distribution at the Solar Cell Plane

The measurement of the angular power distribution on the solar cell plane could be based on projecting the light that passes through the optical system onto a screen and acquiring such image with a CCD camera. This is a very similar procedure to the one used in the acceptance angle inverse method characterization described in section 11.3.3.

Figure 11.23 shows the schematic description of one of the possible layouts to measure the power distribution as function of the incidence angle on the solar cells. For this measurement we just need the optical train (POE + SOE) without the solar cell, letting the light escape from the optics. The light is collected by the projection system generating an image on the screen. As indicated by Figure 11.23, the mission of the projection system is to deflect light impinging on it at a given angle to different positions of the screen. In other words, the projector turns angular spread into spatial spread and thus, the intensity of the CCD image will be mapping proportionally the angular power distribution on the solar cell. It should be noted that with this set-up we measure the angular power distribution without the solar cell in place. Therefore, to calculate the actual angular power distribution on the solar cell we need to transform the distribution measured in air to a distribution in the medium in contact with the solar cell.

11.3.6 In-line Characterization of Optics in Production

CPV manufacturers and integrators often face the challenge of having to characterize thousands of optical elements. Lenses are received from a variety of suppliers and routine measurements to ensure quality control are a must. For instance, POEs and SOEs may come from different vendors and use different materials, so tailored characterization procedures have to be implemented. In addition, this has to be done in a fast and effective way in order not to

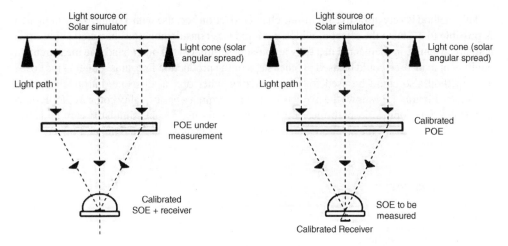

Figure 11.24 Schematic description of the measurement set-ups used in in-line characterization of CPV optical elements. In the left side is the layout proposed for POE characterization and in the right side the layout proposed of SOE characterization

incur in an undesired over cost. In this section we will review the optical characterization techniques most suited to perform such in-line quality control.

A complete quality control procedure for CPV should consider all possible sources of failures like optics deformation, optical surfaces quality (roughness), quality of materials (absorption, bulk scattering) and antireflective and mirror coatings. Performing all these measurements separately is not feasible in production due to the high cost and time consumption. In this scenario, the most practical approach is to perform relative functional measurements where the optical elements to be characterized are in a set-up that reproduces a concentrator receiver (see Figure 11.24). In the case of the POE the measurement procedure will consist of three steps:

1. Calibrate a reference system; in the case of the POE assembly of SOE and solar cell are calibrated under a POE which is known to be good (Figure 11.24 left).
2. Substitute the good POE for the POE to be checked, ensuring reproducibility and accuracy in the positioning.
3. Measure and compare the performance of the element under test with the reference one.

A similar process could be carried out to calibrate the SOE (Figure 11.24 right).

Finally, setting the pass/no-pass threshold (i.e. pass if short circuit current drop is lower than a certain percentage of the calibrated value) will allow the system to decide whether or not the optical element under inspection meets the quality requirements.

In the case of POE characterization we will need a calibrated SOE and receiver (solar cell) and a calibrated POE. As we mentioned before, after setting the acceptance threshold, we exchange the calibrated of reference POE and measure, typically, the short circuit current of the system. In case we are characterizing a SOE we will need a calibrated POE, SOE and receiver (notice that in this situation the SOE and receiver are not optically coupled). Here we swap the reference SOE for the one under measurement and compare the results.

This method is very attractive for in-line characterization because with a single measurement it is possible to evaluate the effect of all losses. It reduces considerably the characterization time; actually, it can be implemented in a fully automated set-up. On the other hand, the measurement equipment to be used could be used in other steps of the production line, thus reducing the costs. This method also could be used for characterizing other optical elements introducing small variations. Figure 11.24 shows the two set-ups for performing a basic quality control of POE (left) and SOE (right) optics. In both measurements we could use a solar simulator with emission spectrum AM1.5d and the solar angular spread (i.e. producing almost collimated light).

Glossary

List of Acronyms

Acronym	Description
2D	Two-dimensional
3D	Three-dimensional
CCD	Charge coupled device
CPV	Concentrator photovoltaics
DSMTS	Dielectric single-mirror two-stage concentrator
LED	Light emitting diode
PMMA	Polymethyl-methacrylate
POE	Primary optical element
SMS	Simultaneous multiple surfaces optical design method
SOE	Secondary optical element
TIR	Total internal reflection
XR	SMS-designed concentrator consisting of a mirror primary optics and a refractive secondary optics

List of Symbols

Typical units given in square brackets. If no units are given, variable is dimensionless.

Symbol	Description [Units]
d	Material thickness
N	Number of points in a lens profile
I_0, I_1	Incoming and outgoing light intensity (after traversing a lens)
I_f, I_g	Incoming and outgoing light intensity (after traversing a lens)
T_F	Fresnel losses transmission factor
u_x, u_z, v_x, v_z	Unit vectors
x, y, z	Cartesian coordinates
α	Absorption coefficient of a material or acceptance angle of optics
δ_e	Slope error
η	Optical efficiency
η_m	Relative optical efficiency (as compared to a reference lens)

References

1. Benítez, P., Mohedano, R., Miñano, J.C. (1997). Manufacturing tolerances for nonimaging concentrators. Proceedings SPIE Nonimaging Optics: Maximum Efficiency Light Transfer IV, 3139, p. 98–109.
2. Malacara D. (2007). *Optical Shop Testing*, Wiley, New Jersey.
3. Leutz, R. and Suzuki A. (2001). *Nonimaging Fresnel Lenses: Design and Performance of Solar Concentrators.* Springer, Berlin.
4. Hernandez, M., Alonso, J., *et al.* (2003). Sunlight spectrum on cell through very high concentration optics. 3rd World Conference on Photovoltaic Energy Conversion, Osaka, vol. 1, pp. 889–891.
5. Hernández, M., Benítez, P., Miñano, J.C., *et al.* (2007). Proceedings SPIE Nonimaging Optics and Efficient Illumination Systems II, (eds R. Winston and J. Koshel) vol. 6670 San Diego, USA.
6. Mohedano, R., Benítez, P., Pérez, F.J., and Miñano, J.C. (2000). Design of a simple structure for the D-SMTS concentrator. Proceedings 16th European Photovoltaic Solar Energy Conference, Glasgow, UK, 1–5 May, pp. 2563–2566.
7. Cvetkovic, A., Hernández, M., Benítez, P., *et al.* (2008). The SMS3D Photovoltaic Concentrator. Proceedings SPIE Optics and Photonics, vol. 7059-8, San Diego, USA.
8. Yoshizawa, T., (2009). *Handbook of Optical Metrology: Principles and Applications*, CRC Press.
9. Leach, R. (2011). *Optical Measurement of Surface Topography*, Springer, Berlin.
10. Domínguez, C., Antón, I., Sala, G. Solar Simulator for Concentrator Photovoltaic Systems. *Optics Express* **16**(19), **14** 894–901.
11. Herrero, R., Domínguez, C., Askins, S., Antón, I., Sala, G. (2010). Angular transmission characterization of CPV modules based on CCD Measurements. 6th International Conference on Concentrating Photovoltaic Systems. *AIP Conference Proceedings*, pp. 131–134, Freiburg, Germany.
12. Herrero, R., Victoria, M., Askins, S., Domínguez, C., Antón, I. (2010). Indoor characterization of Multijunction Solar Cells under non uniform light patterns. 6th International Conference on Concentrating Photovoltaic Systems. *AIP Conference Proceedings*, pp. 209–218, Freiburg, Germany.
13. Herrero, R., Askins, S., Sala, G., Antón, I. (2013). Evaluating Misalignments and Angular Transmission Function by the Luminescence Inverse Method. 9th International Conference on Concentrating Photovoltaic Systems, Miyazaki, Japan.
14. Zamora, P. (2014). Advanced Fresnel Köhler Concentrators for Photovoltaic Applications. PhD Thesis, Technical University of Madrid, Madrid.
15. Herrero, R., Victoria, M., Askins, S., *et al.* (2012). Concentration photovoltaic optical system irradiance distribution measurements and it effect on multijunction solar cells. *Progress in Photovoltaics: Research and Applications*, **20**(4), 423–430.

12

Characterization of CPV Modules and Receivers

César Domínguez, Rebeca Herrero, and Ignacio Antón
Instituto de Energía Solar, Universidad Politécnica de Madrid, Spain

12.1 Introduction

The purpose of this chapter is to present the key instruments and methods involved in the characterization of CPV modules, mono-modules and their sub-systems, *i.e.* concentrator optics and receivers.[1] We first present the main figures of merit that define their performance and then describe the state-of-the-art measurement equipment and test methods for measuring this performance both indoors and outdoors. A deep insight into practical instrumentation issues regarding the use of high-concentration optics and multijunction solar cells is provided. In most cases, characterization methods are described at the module level, but can be also applied to concentrator optics, receivers and solar cells. However, optics have to be characterized coupled to a suitable receiver. Special attention is given to relevant international standards throughout the chapter (most of them under development by the International Electrotechnical Commission (IEC) Technical Committee TC82, Working Group 7) in view of the need of the CPV industry for repeatable and reproducible tests. More details about CPV standards under development can be found at the end of Chapter 9.

---

[1] As a reminder, a receiver is the assembly of one or more solar cells and secondary optics (if present) that receives concentrated light from the primary optics and incorporates at least electrical terminals and the means for thermal transfer. A mono-module is a special construction of a particular module technology, where the minimum amount of optical and electrical elements are used to ensure similarity to the full size module in terms of linearity with irradiance and spectral and angular responses.

Handbook of Concentrator Photovoltaic Technology, First Edition. Edited by Carlos Algora and Ignacio Rey-Stolle.
© 2016 John Wiley & Sons, Ltd. Published 2016 by John Wiley & Sons, Ltd.

This chapter goes through the following topics:

- A brief description of the main specifications and figures of merit for reporting the performance of CPV modules and optics.
- A wide overview of the means and methods typically involved in the characterization of CPV performance indicators is presented through several sub-sections. Operating conditions relevant to CPV behavior are first introduced and their instrumentation described (irradiance, spectrum, circumsolar radiation, cell temperature and lens temperature). Tracker requirements and alignment procedures are also considered as they impact largely on CPV performance. Then, current methods being used for the power rating of CPV modules and the set of standard test conditions defined in the IEC 62670-1 are described. At the end of this section, a technique for the spectral characterization of CPV modules is presented, which can be also applied at the receiver and cell levels.
- Indoor characterization instruments and methods are considered in a separate section, with special consideration to solar simulators, the calibration and use of reference sensors and other precautions related to indoor measurements. A complete description of the physical meaning and characterization methods of the angular transmission curve is presented at the end of the section.

12.2 Figures of Merit of PV Concentrators

12.2.1 Reporting CPV Module Performance

The main figure of merit of a CPV module is its electrical efficiency, i.e. the magnitude of the maximum electrical power it provides divided by the light power incident on the area of aperture. This efficiency actually varies with the operating conditions, e.g. the characteristics of the sunlight, the module alignment or the operating temperature of its components, so it has to be given according to some agreed-upon standard measurement conditions. This is called a *power rating*. The maximum power P_{mpp} is extracted from the *I-V* characteristic, which provides the main description of the module electrical behavior.

Other module ratings are based on the energy generated over a particular period of time, which already accommodate the variability of the irradiance, spectral and ambient conditions with time, season or site. The system's *final yield* (Y_F), defined as the ratio of annual energy produced by the system (in kWh) to its nominal peak power (given in kW), is a rating indicator typically presented to compare the performance between different technologies, or between systems of the same technology installed in different sites. The units of Y_F are kWh per kW_p or simply hours. Therefore, Y_F can be seen as the number of hours per year that the system has been delivering its nominal power. In turn, the reference yield (Y_R) is defined as the ratio of integrated annual direct normal irradiance per unit area impinging on the system over a year (i.e. annual energy density received by the system in the form of DNI, in kWh/m^2) to the standard reference irradiance (kW/m^2) – more details on reference test condition in section 12.3.5.1. Finally, the *performance ratio* (*PR*) is obtained the ratio of Y_F to Y_R, so the parameter is more independent of the location, and is often used as a benchmark for performance.

As the ambient and operating conditions of a CPV module are constantly changing, the sensitivities of its *I-V* curve parameters to these variables are also important specifications. However, these ratings are more usually studied at array or system level, as they are strongly influenced by tracking errors, shading between arrays or inverter efficiency.

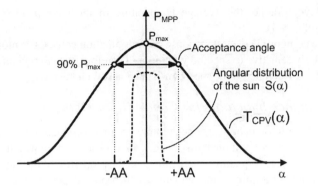

Figure 12.1 Example of an angular transmission curve. The acceptance angle of a concentrator *AA* is found where the angular transmission curve falls to 90% of its maximum

As discussed with solar cells (see Chapters 2 and 10), the spectral response (SR) – or its equivalent, the quantum efficiency – synthesizes the device response as a function of the wavelength of incident light. Similarly, this quantity could also be studied at the module level, for which we should calculate the ratio of the short-circuit current (I_{sc}) of the module to the spectral irradiance at a specific wavelength incident at the lens aperture (provided that the corresponding subcell is limiting the current). However, no instrumentation has been envisaged yet to measure SR of CPV modules or receivers. Therefore, the sensitivity of the module short circuit current to the incident spectrum is typically estimated through the product of the spectral transmittance of the optics and the cell spectral response [1], these two being measured or estimated separately.

Another major figure of merit of a CPV module is its *angular transmission curve* (also known as *angular transmittance* or *angular characteristic curve*). As shown by Figure 12.1, where an example of such angular transmission curve is included, this curve is given as the variation of the P_{mpp} (or sometimes the I_{sc}) as a function of the angle of misalignment between the incident light beam and the direction of perfect alignment. This ideal alignment is usually close to the normal to the module aperture but it is strictly defined as the centroid of the angular transmission curve (so it is only known *a posteriori*), which might not match the absolute maximum response (see Figure 12.4 in section 12.3.4). The angle for which the angular transmission curve falls to 90% of its maximum is called the acceptance angle of the concentrator ($\pm AA$)[2], although other percentages can be defined (and the symbol should be varied accordingly, e.g. $\pm AA97$, $\pm AA95$, etc.). Figure 12.1 also illustrates this calculation.

The temperature of a concentrator solar cell is primarily a function of the ambient temperature, the direct normal irradiance and the average wind speed. The *nominal operating cell temperature (NOCT)* is the mean solar cell junction temperature within a CPV module at which it operates when it is subject to some standard environmental conditions [2]. It is an indicator of the thermal performance of the module, as it gives info on the ability of the heat sink to dissipate the excess heat from the cell. This nominal temperature could also be defined

---

[2] In other texts $\pm \alpha$ is used as the symbol for acceptance angle (e.g. in Chapter 4 in this volume). We use $\pm AA$ instead here to avoid confusion with α being the variable describing angular displacement of the module optical axis.

at the module level [3], i.e. the average temperature of the back plate (at positions just underneath the solar cell).

Because of the composed spectral response of multijunction cells, CPV modules equipping them are particularly sensitive to spectral variations. Therefore, the *current mismatch* between the subcells under a reference light spectrum is also a key characteristic.

12.2.2 Performance Indicators for Concentrator Optics

The main figure of merit for an optical concentrator system is its *optical efficiency*, which might be simply stated as the fraction of the light power at the optical aperture which is available at its output (see Chapter 11). This quantity is dependent on the properties of the incident light, as the efficiency with which a light ray is manipulated by the optics depends on its wavelength and angle of incidence (whether through refraction, reflection, dispersion or absorption). However, no precise definition for this parameter has been yet standardized and many different ones can be found in the literature [4–6]. A norm on appropriate primary lens specifications is being prepared within the IEC TC82. Eq. (12.1) summarizes this:

$$\eta_{op}(E_{in}, S_{in}) = \frac{P_{out}}{P_{in}} = \frac{G_{out}A_{out}}{G_{in}A_{in}} = \frac{G_{out}}{G_{in}C_g}, \tag{12.1}$$

where P_{in} and P_{out} stand for the light power impinging on the entrance and output surfaces of the concentrator, respectively; G_{in} and G_{out} are the corresponding average irradiances, and A_{in} and A_{out} are their areas. The ratio of the aperture area to the output area is called *geometrical concentration*, C_g. As it has been said, this overall efficiency depends on the spectral and angular distributions of the incident light ($E_{in}(\lambda)$ and $S_{in}(\alpha)$, respectively). Figure 12.2 depicts the optical model of a concentrator.

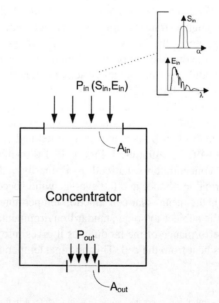

Figure 12.2   Diagram of a concentrator and fundamental quantities that determine the optical efficiency

The optical efficiency given as a function of the wavelength is called *spectral transmittance*, which still will depend on the angular distribution of the light source. The main interest of this quantity is to assess the impact of the optics on the effective spectral response of the solar cell used as receiver.

The *angular transmittance* described above – and depicted in Figure 12.1 – can be obtained also for the optical system alone. The *concentration-acceptance product* (CAP) of the optics (extensively discussed in Chapter 4) can be then derived as the product of the sine of the acceptance angle times the effective concentration ratio. The latter is given by the geometrical concentration ratio dimmed by the actual optical efficiency.

Another important feature of a concentrator is the homogeneity of the light profile that it produces on the cell, or irradiance map [7–9]. If the concentrator is not designed with specific uniformity prescriptions, at some points of the cell the local concentration level may exceed several times the average concentration, introducing a local overload of current density. This excess increases voltage losses due to series-resistance and might affect cell reliability due to thermal issues or trigger limitations of the tunnel junction. If the optics are refractive, some amount of chromatic aberration appears as well, which creates a different spectral distribution on every point of the cell surface and introduces distributed current-limitation issues [10,11].

All these figures of merit are in turn sensitive to operating temperature, especially in the case of refractive optics because refractive index varies with temperature. Besides, deformations induced by thermal dilatations and coefficient of thermal expansion (CTE) mismatch between construction materials might significantly alter optical performance. This is one of the main reasons of the large temperature sensitivity of silicone-on-glass lenses [11–15]. More details on temperature effects are given in Chapter 5.

12.3   Instruments and Methods for CPV Characterization

An accurate measurement of the figures of merit presented above requires appropriate instrumentation and reproducible procedures. Module *I-V* parameters have to be measured under well characterized operating conditions. This section gives an overview of the means and methods typically involved.

12.3.1 Indoors versus Outdoors

The characterization of a CPV module can be carried out indoors or under real sun. Both options present different advantages and disadvantages, which have to be taken into account in order to decide the appropriate characterization method in each case. In clear-sky days, outdoor measurements can provide very stable illumination within minutes (especially around noon), as well as truly spatially uniform irradiance. However, outdoor conditions are very dependent on the site, time and season due to weather variability, and they are affected by mounting and tracking errors, which hinder the repeatability and reproducibility of real sun measurements and increase the uncertainty. At the manufacturing line, this weather dependence leads to unpredictable time lags that are a bottleneck for module production. Moreover, despite that the best matching to the reference spectrum is achieved under the sun, it might be found only occasionally. On the other hand, outdoor measurements are the ultimate way to be sure about how the PV devices will perform under the sun. They are required for instance for the extraction of the module NOCT, which is the result of the actual heat dissipation capacity of the device under a particular set of realistic ambient conditions that can hardly be reproduced indoors.

Regarding indoor testing, the main drawbacks are the non-idealities of the light source with respect to the ideal solar illumination, which increase the errors and uncertainties of the measurement: spatial inhomogeneity of the irradiance, spectrum different from the reference one, inexact angular distribution of the light source etc. But then again, solar simulators facilitate readily reproducible procedures and environments that can be adjusted to the desired ranges, leading to more repeatable results.

12.3.2 Operating Conditions Relevant to CPV

The performance of multijunction solar cell based CPV systems is affected by a larger set of operating parameters than flat-plate PV [16,17]: intensity, spectral distribution and angular distribution of direct sun light, cell and lens temperature and module alignment are the most relevant factors. A good knowledge of these dependences is required for modeling CPV power output under any working condition [17], forecasting the energy yield for a particular period of time [18] and rating the systems with a low uncertainty [19].

12.3.2.1 Quantifying Spectral Effects

The measurement of the spectral irradiance of DNI is required for assessing its influence on concentrator performance or for assuring reference spectral conditions. A spectroradiometer is typically used for this purpose. The spectral range significant for conventional PV is 300–1100 nm, but it extends to 1800 nm when considering Ge-based triple-junction solar cells. The spectral response of the detector used defines the limits to the achievable range. Silicon detectors are used in conventional PV, but they have to be often improved through some boost of their UV response. If the spectral range to be achieved is too broad, two or more detectors with complementary spectral responses may be used. InGaAs or Ge detectors are often used to extend the sensitivity of the spectrometer to the near infrared range (mostly until 1700–2000 nm). If equipped with an irradiance probe beneath a collimator tube for limiting the field of view to 5°, the spectral irradiance of DNI can be measured. However, some practical limitations are typically found, namely outdoor reliability of this sophisticated instrument; the cumbersome amount of data to manage; the long acquisition times in the case of scanning monochromators, which will be affected by short-term variations of the atmospheric conditions; the low accuracy of the measurements in case of diffracting monochromators with diode arrays. Nevertheless, commercial equipment prepared for outdoor usage exist –like the pair EKO MS-710/MS-712–, which can be provided with collimator tubes for limiting their field of view to 5° (DNI). Together they cover the 350 nm to 1700 nm range and take measurements in 10 ms to 5 s.

The measured spectrum can then be compared to the some reference spectral distribution, e.g. AM1.5d defined in the ASTM G173-03 standard. This similarity can be quantified in different ways depending on the application and the PV technology. The simplest is a wavelength-by-wavelength comparison of the spectral irradiance distributions. For instance, the similarity according to IEC 60904-9 standard for the qualification a solar simulator is evaluated through the fraction of the total irradiance which is found at each spectral band. The minimum width of these bands is 100 nm [20]. Since the definition of the intervals is fixed (and implicitly given for the dominant crystalline Silicon technology), the resulting figure of similarity has a different weight on the measurement uncertainty depending on the spectral response of the test specimen.

When considering multijunction solar cells, the spectral similarity has to be evaluated first and foremost through the amount of light available for each subcell. The relationship between these values gives the spectral balance of the light. Any variations on this balance with respect to the reference spectrum vary the overall cell current and power even if the level of irradiance is kept constant. If a reference cell of a different technology is used for measuring it at the simulator, the error cannot be reduced simply by applying a correction through a spectral mismatch factor [21]. Distortions in the FF and the V_{oc} of secondary importance are also introduced. Thus, the measurement of this critical spectral balance can be carried out indirectly by measuring the spectrum with high resolution and estimating the photocurrents that may be obtained for each subcell, provided their spectral responses are known. Equivalently, a more straightforward measurement of this balance would be the use of component cells, i.e. single-junction cells with the same relative spectral response as the subcells in a multijunction solar cell [22–24]. A component cell is often grown using a similar stack of semiconductors as the multijunction cell, but where only one junction is a diode (the others are left electrically inactive by growing isotype heterojunctions, reason why these cells are often known as 'isotype' cells). The inactive layers work optically, nevertheless, leaving the active diode with the same spectral response as the subcell in the multijunction cell. Each component cell behaves as a sensor proportional to the irradiance within a particular spectral interval. The fact that the spectral response is not flat as in a thermal sensor implies a non-radiometric measurement, but this can also be seen as a weighted average with a significant weighting function (the one of the cells under study). The similarity between two spectra can now be evaluated through the ratios of the irradiance values obtained by these component cells. If the short-circuit current under reference spectrum (some standard test conditions[3]) $I_{sc,i-component}^{STC}$ is known for each component cell, the *effective irradiance* at the corresponding spectral band for each component cell i ($DNI_{i-component}$) is:

$$DNI_{i-component} = DNI^{STC} \frac{I_{sc,i-component}}{I_{sc,i-component}^{STC}}, \qquad (12.2)$$

where DNI^{STC} is the irradiance under standard test conditions that includes the reference spectral distribution. Now the spectral similarity of the incident spectrum $E_i(\lambda)$ to the reference AM1.5d can be evaluated through the ratio of every two values of component-cell DNI, known as the spectral matching ratio (SMR) [25]:

$$SMR(E_i(\lambda))_{j-subcell}^{i-subcell} = \frac{DNI_{i-subcell}}{DNI_{j-subcell}} = \frac{I_{sc,i-component}/I_{sc,i-component}^{STC}}{I_{sc,j-component}/I_{sc,j-component}^{STC}}, \qquad (12.3)$$

Note that SMR can also be seen as the current ratio between two component cells, divided by the same current ratio under reference conditions. When *SMR* is equal to *1*, the incident spectrum is *effectively* equivalent to the reference one in the spectral range of interest, i.e. the spectrum produces the same current ratio as AM1.5d. A value of *SMR*(top vs. middle) higher

---

[3] Typically a DNI of 1000 W/m², a cell temperature of 25 °C and a spectrum whose relative distribution is similar to the 'AM1.5d' spectrum defined in ASTM G173-03 standard. Only relative values are taken into account because the latter integrates to a value of 900.1 W/m².

than 1 implies that the photocurrent ratio between top and middle subcells is higher than that under the reference spectrum, and hence the spectrum is said to be 'blue rich.' On the contrary, a value of *SMR* lower than 1 indicates a 'red rich' spectrum. Less attention is paid to the bottom Ge subcell in conventional triple-junction GaInP/GaInAs/Ge solar cells, as this subcell has a typical excess of current of around 30% under AM1.5d. However, in more recent multijunction solar cell concepts (like those with a GaInAs bottom), an excess of illumination in the bottom cell can cause a large increase in the measured fill factor [26], so the *SMR*(top vs. bottom) or *SMR*(middle vs. bottom) need also to be considered. Moreover, even in the case of Ge-based triple-junction cells, variations of *SMR* in the bottom cell can affect measured performance: intensity variations in the bottom cell will vary the overall DNI while keeping cell *I-V* curve rather unaffected, thus disturbing electrical efficiency calculations.

The component cells have to be installed underneath collimator tubes with an opening angle around 5° (in order to match that of pyrheliometers) and aligned towards the sun in order to collect DNI. An advantage of the use of component cells is their fast light response, which allows a time-resolved measurement of the spectrum in pulsed solar simulators (limited by the sampling rate of the data acquisition system).

Commercial equipments are available for this purpose like the ICU-3J35 'spectral heliometer' from *Solar Added Value (SAV)* or the BPI-IT1 from *Black Photon Instruments*, both including calibrated lattice-matched triple-junction cells installed in weatherproof enclosures underneath collimator tubes for measuring direct normal irradiance (see Figure 12.3). A European spectral network for the continuous outdoor measurement of component-cells effective DNI has been setup recently within the SOPHIA project funded by the European Commission [27]. NREL has also recently set up this type of instrument for outdoor characterization purposes.

The main drawback of the use of component cells arises from the fact that there exist several multijunction solar cell architectures in the market and therefore their spectral responses vary significantly. This limitation in practice may be relaxed provided that a correlation can be found

Figure 12.3 Commercial spectral sensors based on component cells beneath collimator tubes. Solar Added Value *ICU-3J35* (left) and Black Photon Instruments *BPI-IT1* (right). Source: SAV and Black-Photon. Reproduced with permission of Black-Photon

between the spectral balance of the test specimen and the spectral balance defined by the component cells, which has been demonstrated to be usually the case [24].

12.3.2.2 Measurement of Irradiance

Thermal radiometric sensors are typically used to measure the irradiance outdoors. *Direct normal irradiance* (DNI) is measured by limiting the opening angle of the sensor to $5°$ and tracking the sun to an accuracy of at least $\pm 0.5°$. The sensitivity of a *normal incidence pyrheliometer* (NIP) is typically in the range of μV per $W \cdot m^{-2}$, so low-noise wiring is required. However, their slow dynamic response (a few seconds to reach the target value) makes them unpractical with pulsed solar simulators.

The measurement of a spectrally-corrected DNI based on component cells has also been proposed, as a means of accounting for the large spectral sensitivity of multijunction solar cell based CPV systems [17]. This approach tries to improve the linearity between DNI and module photocurrent by taking into account that different subcells will be limiting the overall current depending on the particular incident spectrum $E(\lambda)$. The spectrally corrected DNI for a particular incident spectrum $E(\lambda)$ can be defined as:

$$DNI_{3Jeq}(E(\lambda)) = DNI^{STC} \frac{\min(I^{E(\lambda)}_{sc,Top}, I^{E(\lambda)}_{sc,Middle}, I^{E(\lambda)}_{sc,Bottom})}{\min(I^{STC}_{sc,Top}, I^{STC}_{sc,Middle}, I^{STC}_{sc,Bottom})}, \qquad (12.4)$$

where $I^{E(\lambda)}_{sc,Top}$, $I^{E(\lambda)}_{sc,Middle}$ and $I^{E(\lambda)}_{sc,Bottom}$ are the short-circuit currents of the top, middle and bottom component cells under the incident spectrum, respectively, and $I^{STC}_{sc,Top}$, $I^{STC}_{sc,Middle}$ and $I^{STC}_{sc,Bottom}$ are their currents under the reference spectrum (standard test conditions).

12.3.2.3 Influence of Circumsolar Radiation

A CPV system has an acceptance angle significantly larger than the extent of the solar disc, so the light coming from the portion of sky within this field of view will be collected by the concentrator. This circumsolar radiation (see Chapter 1) is a function of the atmospheric conditions and the solar elevation, as it is created by the scattering of direct light into the region of the sky immediately surrounding the solar disk. For instance under hazy conditions the solar disc will appear diffuse and the fraction of circumsolar radiation will be enhanced. As conventional pyrheliometers have a field of view several times larger than typical acceptance angles of high-concentration PV systems, the light collected by the NIP will be higher. Therefore the DNI available for the concentrator will be overestimated and the calculated efficiency will be underestimated. This effect can be larger than a 10%, although the DNI will be very low in this case. The highest share of circumsolar radiation is found for large loads of aerosols or precipitable water vapor, or under thin cirrus clouds [28]. Other atmospheric and ambient factors influencing circumsolar radiation are ozone concentration, atmospheric pressure, ambient temperature or relative humidity.

The amount of circumsolar radiation can be measured directly using very sophisticated equipment (like the *Sun and Aureole Measurement System* from *Visidyne*) or be estimated upon the simultaneous measurements of two or more pyrheliometers with different penumbra functions [29]. Some experimental setups using charge coupled device (CCD) cameras have been proposed as well [30,31]. The *CSR 460* system from Black Photon Instruments consists of two different pyrheliometers with opening angles of $\pm 2.6°$ and $\pm 1.6°$ ($\pm 0.7°$ optionally). The *circumsolar ratio* (CSR) can be defined as the ratio of the circumsolar irradiance to sum of the disk and circumsolar irradiance (DNI). However, some further modeling effort is required to estimate the CSR or the circumsolar profile from the measurements taken by the *CSR 460* system.

Alternatively, we can take the ratio of direct normal irradiance to *global normal irradiance* (GNI) as a measure of the *clearness* of the sky, i.e. the share of the DNI in the total GNI. With no visible clouds, the higher this value, the less diffusing is the atmosphere. This is a quantity somewhat complementary to the CSR, although it only provides a qualitative estimation of this CSR. Nevertheless, these quantities can be readily measured or calculated by many meteorological stations worldwide, as only a NIP and a pyranometer are required. The requested GNI can be estimated from the global irradiance measured at a fixed plane.

In general, the effect of an increased circumsolar radiation will reduce the optical efficiency of the concentrator, which is reflected on a decrease of the I_{sc}/DNI ratio [32]. However, a very high value of DNI/GNI (low CSR) could affect the shape of the light spot on the cell negatively. For instance, high DNI/GNI conditions might produce a minimum-size spot on a point focus concentrator without any secondary optical element (SOE), thus increasing series resistance losses and shading factor at the metallization grid. The presence of a homogenizing SOE would minimize these effects.

12.3.2.4 Cell Temperature

It would be very useful to have a direct measurement of the cell temperature within the concentrator: it would allow the characterization of the thermal performance, the estimation of the module NOCT and the translation of the *I-V* curve into different temperatures. However, the solar cells within a CPV module cannot be directly contacted with temperature probes. A special assembly could be prepared to install a temperature transducer very close to the solar cell [33], but still a gradient to the real junction temperature might be found due to the very large power density cast on the cell. Also, an optical method for direct sensing of junction temperature based on the peak shift of the photoluminescence emission during normal operation has been recently proposed. However, its broad applicability for monitoring purposes is unlikely because of the need of a spectrometer and dedicated optical devices [34]. Owing to the aforementioned difficulties, several methods have been proposed to estimate cell temperature upon other variables easier to measure, namely module V_{oc}, backplane temperature, or ambient temperature and wind speed [35]. In either case, module current or DNI are used as a measure of the incident heat flux on the cell in order to improve the estimation accuracy.

Backplane temperature should be measured at the heat sink just underneath the solar cell, where the minimum gradient to the solar cell temperature is found. An average from several locations will increase the significance of the measurement, as the temperature is not uniform throughout a CPV module (differences larger than $10\,°C$ have been reported [33,36]). Small sensors with very low thermal inertia are preferable both having a faster response and not affecting local temperature. Resistance temperature detectors (RTD, e.g. Pt100) or thermocouples are typically used. For estimating cell temperature, a constant thermal resistance R_{th} [K/W] is assumed between the cell junction and the heat sink:

$$T_{cell} = T_{\text{heat sink}} + \dot{Q} \cdot R_{th}, \tag{12.5}$$

where $\dot{Q}$ [W] is the transfer rate of excess heat, which is given by the light power that is concentrated on the cell but is not converted into electrical power to the load. This heat is thus a function of the cell electrical efficiency ($\eta_{electrical}$) and the optical efficiency of the optical system ($\eta_{optical}$):

$$\dot{Q} = DNI \cdot A_{lens} \cdot \eta_{optical} \cdot (1 - \eta_{electrical}), \tag{12.6}$$

where A_{lens} is the area of the optical aperture. This was the approach envisaged by UPM for the specifications of the ISFOC[4] call for tenders [37]. The main drawback of this method is that it assumes that the thermal resistance between the cell and the heat sink is known. However, this information is neither usually specified by CPV module manufacturers nor easy to characterize. Instead, it has been pointed out that for measurements under small variations of the DNI, the thermal gradient between the cell and the heat sink may be assumed to be constant. In this case, module temperature can be directly used for those power rating procedures where stringent data filtering is carried out *a priori* in order to select very repeatable illumination conditions. This is the case for the Concentrator Standard Operating Conditions (CSOC) rating procedure in the current draft of the upcoming IEC 62670-3 norm, to be commented in detail in section 12.3.5.

Another group of methods takes advantage of the quasi-linear relationship between the V_{oc} and cell junction temperature (see Chapter 5). Thus, the V_{oc} thermal coefficient β of the cell has to be known for at least one concentration level, as it is actually a function of concentration (it approximately decreases with the logarithm of the concentration level [38], as discussed in Chapter 5). The popular 'V_{oc}-I_{sc}' method [39, 40] normalizes every V_{oc} measured to a known reference concentration level by means of the I_{sc} ratio between measurement and reference conditions. If the cell temperature is known at these reference conditions, β can be used to translate the voltage difference into the variations of temperature:

$$T_{cell} = \frac{V_{oc} - V_{oc,STC} + N_s \beta\, T_{cell,STC}}{N_s \dfrac{nk}{q} \ln\left(\dfrac{I_{sc}}{I_{sc,STC}}\right) + N_s \beta}, \tag{12.7}$$

where I_{sc} *and* V_{oc} are the measured short-circuit current and open-circuit voltage; $I_{sc,STC}$ and $V_{oc,STC}$ are said current and voltage under some reference conditions; T_{cell} is the cell temperature to estimate; $T_{cell,STC}$ is the temperature under the reference conditions; β is the V_{oc} thermal coefficient for a single cell (in the range 4–6 mV/K for conventional III-V triple-junction cells), n is the ideality factor (between 3 and 4 typ.), k is the Boltzmann constant, q is the charge of the electro and N_s is the number of solar cells in series in the module.

Here the assumption usually made is that the V_{oc} thermal coefficient and ideality factor of the module are those given by the solar cell specifications, which might not be necessarily true. Values either higher [41] or lower [33] have been reported, although the latter is more frequently found in the authors' experience. Thus, these parameters may be characterized *a priori* for the module itself, for which a solar simulator would be in principle required [41, 42]. Alternatively, the so-called shutter method could be used to extract module β outdoors, but only if a very fast unmasking of the module (20–30 ms) can be assured [43].

---

[4] Instituto de Sistemas Fotovoltaicos de Concentración

IEC 60904-5 standard [44] proposes the calculation of the *Equivalent Cell Temperature* through an analogous method, but where the ratio of $I_{sc}/I_{sc,STC}$ is substituted by the ratio of the incident irradiance to the reference irradiance (DNI/DNI^{STC}). The main advantages of the methods based on V_{oc} are that the whole module is described by a single temperature (rather than the average for several locations) and that this value varies instantaneously, i.e. V_{oc} does not suffer from the thermal inertia of temperature sensors. The main drawback is that results are very dependent on the set of parameters chosen for β and n. Muller and co-workers found differences up to 10 °C between assuming β to be constant or variable with irradiance [43]. An improved version of the IEC 60904-5 procedure has been proposed to account also for the variations of the ideality factor with irradiance level and temperature [45].

12.3.2.5 Lens Temperature

The optical properties of a concentrator depend also on lens temperature, mainly due to its effects on the overall shape or the refractive index (see Chapter 5). This sensitivity is especially relevant in silicone-on-glass (SoG) primary Fresnel lenses, as the silicone experiments a strong variation of its refractive index with temperature and a large CTE mismatch with its glass superstrate [12–15] which produces changes in the intensity and size of the light spot on the receiver [42]. This seems to be the reason for the large I_{sc} thermal coefficients found outdoors for modules with SoG primary lenses. Apparent coefficients of 0.8%/K versus ambient temperature and of 0.15%/K versus backplane temperature have been reported in the literature for a module with SoG Fresnel primary lens and no SOE [46,47]. These coefficients are up to one order of magnitude higher than those reported for the bare cell alone (see Chapter 5). The losses on the annual energy yield due to this effect can be very significant and much larger for SoG than for PMMA lenses [48, p. 2] so it is essential to assess this influence for every CPV technology. This large sensitivity of SoG Fresnel lenses can be counterbalanced by smarter designs of the Fresnel facets [49] or as a wider approach, by the inclusion of a secondary optic that accounts for the variations in size of the concentrated spot [50]. A detailed study on the sensitivity of module current to lens temperature for different CPV architectures has been recently presented [32].

12.3.3 Tracker Requirements

Outdoor characterization of CPV modules requires their installation on a two-axis tracker. High tracking accuracy is required in order to remove any interference of pointing errors on measured module performance. Real state-of-the-art high-concentration PV modules feature P_{mpp} acceptance angles as narrow as 0.4°, so tracking accuracy should be well below this figure (±0.1° accuracy is commonly specified by CPV tracker manufacturers, yet typically optimistic). However, the required accuracy should be rather given by the acceptable error in the *I-V* parameter of interest, which is found through the angular transmission curve of the particular module to be tested. Moreover, the mounting structure or tracker table should offer a large flexural rigidity in order to avoid differential tracking errors throughout the test bed or variable bending when a module is installed or removed.

Nevertheless, the uncertainty introduced by tracking errors can be limited if a pointing sensor is used to monitor sun-tracking accuracy simultaneously with module measurements. *I-V* data taken under a misalignment larger than a given threshold are discarded, thus reducing the overall uncertainty. High-resolution sun tracking sensors are available based either on

Position Sensitive Detectors (PSD) like the *BPI-TA1* [51] or on CCD detectors like the *Trac-Stat SL1* [52]. In principle PSD sensors enable faster readings (real-time scope output), while CCD allows for increased reading robustness through integrated image processing algorithms. The accuracy of the *TA1* and *SL1* is claimed to be in the $\pm 0.02°$–$\pm 0.03°$ range.

12.3.4 Alignment Procedures

One of the largest contributions to measurement uncertainty in CPV modules is related to the initial alignment of the module to the light beam. Thus, a reproducible alignment process that leads to a repeatable position (high precision) is desired. This is also central to inter-lab measurement comparisons. There are three main alternative procedures for aligning a module to the solar beam:

- visual alignment
- I_{sc} optimization
- P_{mpp} optimization or I_{mpp} optimization.

If the light spot cast by the primary optics can be seen on any of the subsequent optical stages (e.g. SOE) or the cell, then the alignment can be carried out using visual acuity: the operator has to ensure that the light spots are projected at the center of the SOE or the cell entrance (welding goggles or shading clothes are required to protect operators' eyes). The alignment precision using this visual method can be better than $0.1°$ for high-concentration optics, although it has some important drawbacks. First, not every module technology features this spot visibility. Second, it depends on the expertise of the operator. And third, the center position of the spot might vary between the different lenses of the module. In this case, the operator has to find a global (subjective) compromise: e.g. the position where most of the spots are well centered, or the position where none of the spots are outside the collecting surface. Nevertheless, this method provides quick and simple module installations when experienced operators are involved.

Alignment using *I-V* parameters allows for more reproducible procedures. The objective is to find the centroid of the 2D angular transmission curve of the CPV module (recall Figure 12.1) in terms of either I_{sc} or P_{mpp}. This does not necessarily mean *maximizing* the parameter. The centroid of the angular transmission might differ from the maximum response orientation either due to manufacturing non-idealities or to the particular optical architecture. Thus, the maximum response might be found in a position where the angular tolerance is narrower and hence is more sensitive to energy losses due to tracking errors. The main drawbacks of *I-V* based alignments are that they require stable operating conditions during the whole adjustment process and portable instrumentation.

In order to find the centroid, the *I-V* parameter of choice is monitored while performing angular misalignment sweeps across at least two orthogonal rotation axes.[5] The alignment is then defined by the mean of the edge angles for which the power falls to some percentage of the maximum (97% or 95% for example – see Figure 12.4). Practical alignments require in practice portable or handheld instrumentation close to the tracker on which the module is installed (and hence probably different from the *I-V* bench that will be used for module *I-V* curve monitoring) so that the operator can readily have a feedback of the misalignment steps. While I_{sc} optimization requires a simple digital multimeter, P_{mpp} optimization would require a fast

---

[5] everal points below the acceptance angle claimed by the manufacturer should be swept.

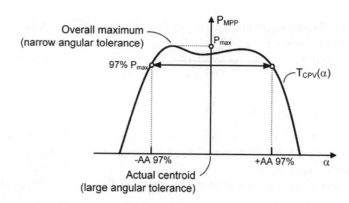

Figure 12.4   Power angular transmission curve where the position for maximum electrical efficiency does not match the centroid of the curve (optimal alignment for field operation)

handheld power multimeter[6] (e.g. *Metra HIT Energy* from *Gossen Metrawatt*[7]) together with a rheostat for biasing the module close to V_{mpp}. Alternatively, the expensive power multimeter can be substituted for two conventional multimeters: in the *pseudo-I_{mpp}* optimization process, the module voltage is fixed to a value close to 0.8 times the V_{oc} using the rheostat, and then the variations of current (close to I_{mpp}) are monitored with a separate multimeter (ideally, using the voltage of a shunt resistor adapted to the module under test). Angular misalignments can be carried out using either an offset-adjustable tracker (a single module can be tested at a time, though) or, more generally, an adjustable mounting structure installed on the tracker frame. Precise angular steps can be simply carried out using fixation elements like the number of turnings on screws or nuts (spring loaded). Sun-tracking sensors attached to the mounting frame or the module chassis would also allow precise quantification of misalignments.

Anyhow, this process can also be carried out in a more time-consuming way by taking the complete *I-V* curve – with an appropriate load – and DNI at every misalignment step, provided irradiance, ambient temperature and wind speed are reasonably stable. This is usually ensured by using the shortest time possible around noon. Then the centroid is found *a posteriori* by processing P_{mpp} divided by DNI (and the module is oriented accordingly).

12.3.5 Rating CPV Module Performance

Determining the power rating of a CPV module can support many different purposes: providing a nameplate value (datasheet), performing a comparison between technologies (benchmarking), acceptance of a device, sizing a power plant and balance-of-system (BOS) components, providing the basis for energy rating parameters or yield modeling etc. Therefore a standardized rating process is a priority for the CPV industry and the successful deployment of this technology [53]. However, due to the higher complexity, rapid evolution and wider variety of designs being pursued by CPV manufacturers compared to flat-plate PV, standardization efforts have not yet issued an IEC norm on CPV module rating methods. Working

---

[6] Portable *I-V* curve meters might be hardly suitable because their precision is typically in the range of the optimization being carried out and they provide slow feedback to the operator when varying module alignment.
[7] Metra HIT Energy: https://www.gossenmetrawatt.com/english/produkte/metrahitenergy.htm

Table 12.1 Concentrator Standard Reference Conditions as defined in IEC 62670-1:2013 norm

Parameter	CSOC	CSTC
Irradiance	$900\,\text{W}\cdot\text{m}^{-2}$	$1000\,\text{W}\cdot\text{m}^{-2}$
Temperature	20 °C (ambient)	25 °C (cell)
Wind speed	$2\,\text{m}\cdot\text{s}^{-1}$	*N/A*
Spectrum	AM1.5d	AM1.5d

Group 7 of the IEC Technical Committee 82 is currently developing a number of different CPV standards including power and energy rating methods, both of which are in a very advanced state. The first part of the IEC 62670 series 'Photovoltaic concentrators (CPV) – Performance testing' has already been approved and published, which defines the set of *standard conditions for assessing the power produced by CPV systems and their photovoltaic subcomponents* [54].

12.3.5.1 Standard Measurement Conditions: CSTC and CSOC

Power rating tries to offer a representative figure of the module power under some agreed-upon operating conditions. There are two sets of standard conditions currently defined in the IEC 62670-1 norm: on the one hand, some concentrator standard *operating* conditions (CSOC) representing module performance outdoors under representative operating conditions; on the other hand, some concentrator standard *test* conditions (CSTC) to be assured in a readily reproducible environment. They both are summarized in Table 12.1:

The 'AM1.5d' standard spectral irradiance distribution for direct beam light will be defined within the IEC 60904-3 in its upcoming third edition, although a similar version is already defined in the ASTM G173-03(2012) standard [55]. The distribution is generated using version 2.9.2 of the SMARTS model[8] fed with representative atmospheric and environmental data and assuming a field of view of 5.8° [56]:

- air mass (AM): 1.5
- U.S. Standard Atmosphere with CO_2 concentration of 370 ppm, a rural aerosol model, and no pollution
- surface pressure: 1013.25 mbar
- aerosol optical depth at 500 nm: 0.084
- precipitable water: 1.4164 cm
- ozone content: 0.3438 atm-cm.

The resulting direct spectral irradiance integrates to $900.14\,\text{W/m}^2$, so it has to be scaled to $1000\,\text{W/m}^2$ when considering CSTC (in this case the distribution is labeled as AM1.5d to point out that only the relative distribution is kept). The irradiance level of $1000\,\text{W/m}^2$ is often referred to as '1 sun'.

As an important note, especially when considering low-concentration PV systems, the norm specifies that the DNI has to be corrected for solar elevation over the plane of the array for

---

[8] Available at http://www.nrel.gov/rredc/smarts/

devices using single-axis tracking. This quantity is sometimes referred to as *direct plane-of-array irradiance* (DPOA).

12.3.5.2 Power Rating Procedures

The power rating procedures for the future norm IEC 62670-3 are still under discussion. Research labs and manufacturers are evaluating different methods publicly available through their application on different module technologies [37,20,18,56,57]. The objective is to come up with the procedures producing the highest reproducibility among labs and different times of the year, while keeping time and cost of completion and complexity to a minimum. Due to the wide variety of CPV designs available, the applicability of the ratings to most technologies is also considered. The SOPHIA project funded by the European Commission has initiated the widest round robin to date on CPV module ratings (both indoors and outdoors), with non-European laboratories like NREL getting also involved.

There are two landmark rating procedures on which most experience is based: the ASTM E2527-09 norm and the ISFOC call for tenders' procedure. The first one is based on a simple multi-linear regression of the maximum power point with respect to DNI, ambient temperature (T_{amb}) and wind speed (w_s), as developed in the PVUSA rating procedure in the early 90 s [59]:

$$P_{mpp} = DNI \cdot (a_1 + a_2 \cdot DNI + a_3 \cdot T_{amb} + a_4 \cdot w_s), \qquad (12.8)$$

where a_1 to a_4 are fitting coefficients extracted from long-term outdoor measurements of the CPV module. The independent variables are then fed with the desired standard reporting conditions to find the rated power P_{ASTM} ($DNI = 850$ W·m^{-2}, $T_{amb} = 20$ °C and $w_s = 4$ ms^{-1} for this ASTM norm). Spectrum influence is neglected, which is a major weakness of this method – rating variability in the range of 10% has been reported because of seasonal effects [19].

On the other hand, the *ISFOC procedure* was prepared as an acceptance test for the installed power of CPV plants. It is based on the translation of the measured *I-V* curve in terms of DNI and cell temperature using a lumped diode model [37]. The main difficulties linked to this method are the need of estimating cell temperature (see section 12.3.2.4) and the fact that model parameters might be unknown. Its main flaw is again the lack of an appropriate consideration of some operating parameters relevant to CPV like spectrum or lens temperature. In order to appropriately take them into account, two approaches appear: either the dataset is filtered for narrow ranges of the most relevant quantities (i.e. measurement conditions as close as possible to CSOC), or the dependence of *I-V* parameters on them is explicitly added to the translation and regression expressions involved in the process. Recent studies on rating methods try to find the best tradeoff between the two [58,60].

Both translation and regression methods are being considered within the draft of the IEC 62670-3 norm. It is generally accepted that regression methods perform better when long periods of valid data are available [53]. Translation methods on the contrary might arrive to repeatable ratings using data from few days [37]. Anyhow, the accuracy of both methods is affected by the similarity between the actual outdoor conditions during the measurement and CSOC.

I-V translations are considered in current 62670-3 draft (*Approved New Work* with reference 82/919/NP in April 2015) through simplified translations of the maximum power point, instead

Table 12.2 Data filtering requirements in 62670-3 draft

Parameter	Valid range
DNI	700–1100 W/m^2
Air temperature	0 °C–40 °C
DNI/GNI	> 0.8
Tracking error	< half DUT acceptance angle
SMR (top/middle)	0.975–1.025
SMR (top/bottom)	0.975–1.025
SMR (middle/bottom)	0.975–1.025
5-min average wind speed	0.5–3.5 m/s
DNI fluctuation during previous 10 min	< 10%
Ambient temperature fluctuation during previous 30 min	< 5 °C

of separate current and voltage corrections – Eq. (12.9):

$$\eta_{CSOC} = f_{V,CSOC}\left\{\eta_{I-V} - \delta\left[(T_{amb} - 20) + f_{DNI}(DNI - 900)\right]\right\}, \tag{12.9}$$

where η_{CSOC} is the rated electrical efficiency at CSOC (from which the rated power is directly obtained), $f_{V,CSOC}$ is a correction factor for the voltage as a function of irradiance – Eq. (12.11), η_{I-V} is the measured electrical efficiency – Eq. (12.10), δ is the thermal coefficient of efficiency with ambient temperature (T_{amb}) and f_{DNI} is a temperature correction factor based on DNI:

$$\eta_{I-V} = \frac{P_{mpp}}{DNI \cdot A_{in}}, \tag{12.10}$$

$$f_{V,CSOC} = 1 + N_S \frac{nkT_{cell}}{qV_{oc}} \ln\left(\frac{900}{DNI}\right), \tag{12.11}$$

Although spectral variability is not corrected for in these expressions, it is taken into account within the norm draft (as well the dependence on other relevant quantities) through filtering measurement data to a narrow range of SMR values (2.5% around $SMR = 1$) before performing the rating. Main requirements currently defined are summarized in Table 12.2.

Recently, some alternative expressions and sets of filters considering the influences of spectrum, lens temperature or sky clearness have been evaluated [58]. The main conclusion is that power conversion methods show best reproducibility when the data is filtered for a narrow range of spectrum and lens temperature. If lens temperature is not available, ambient temperature filtering might be used instead. Another method based on the lumped diode model of a multijunction solar cell [61] and the use of the spectrally-corrected DNI defined in Eq. (12.4) has shown to model accurately the seasonal (spectral) variability of CPV array performance after appropriate restriction of lens temperature [17].

Regarding energy ratings, although the IEC 61724 norm already offers general guidelines for PV systems in general, the IEC TC82 WG7 is preparing a specific norm (IEC 62670-2) on the energy rating of CPV modules and assemblies, based on extended monitoring.

12.3.6 Spectral Characterization of CPV Modules and Receivers

The current response of multijunction cell-based CPV modules cannot be linearly correlated to DNI variations due to their composed spectral response. However, this relationship is typically used in power rating procedures, performance modeling or energy yield estimations. Thus, the characterization of module sensitivity to spectral variations is frequently desired. This is typically studied through long-term outdoor monitoring of the module I-V curve together with spectral conditions [61,62]. Spectrum can be either directly measured or estimated as described earlier in this chapter (see section 12.3.2.1), and its impact in the main I-V parameters is analyzed, especially on I_{sc} and P_{mpp}. Since solar irradiance and spectral balance are highly correlated, I_{sc} is usually divided by DNI to better analyze current mismatch losses. Peak I_{sc}/DNI values and FF minima signal spectral conditions for which the cells are current matched [63,64] (see Chapter 10 for a discussion on the impact of current matching on FF). However, whether this optimum current matching between subcells correspond to the maximum electrical efficiency of the module depends on the particular CPV technology considered, as other factors like series resistance losses might play a counterbalancing role.

Methods for establishing direct relationships between module current and spectral conditions have been proposed using component cells DNI [65,66] or AM [15], as the latter is the main driver of daily spectral variations. Component cells can also be used to estimate the module current mismatch (more precisely, the current mismatch between the subcells likely to limit module current) under any spectral conditions [67]. This enables a deeper understanding of module performance variability and allows more accurate modeling of the module power output over time.

A very elegant way to assess how spectral variations affect CPV module performance can be developed using component cells. This method is based on the so called *subcell-limitation graphs*, which constitute a visual summary of subcell normalized current as a function of the spectral content of the incident spectrum. Figure 12.5 shows an example of such subcell-

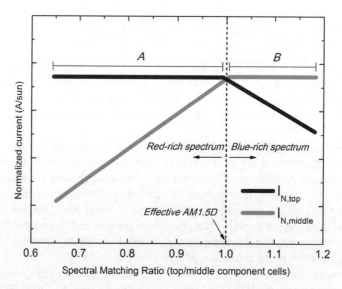

Figure 12.5 Sample subcell-limitation graph for the top and middle subcells of a Ge-based triple-junction solar cell optimized to be current-matched under reference spectrum

limitation graphs for a Ge-based triple-junction solar cell. This method is particularly simple for modules using this type of multijunction solar cells, where the Ge junction is overexcited and only the spectral mismatch between GaInP and GaInAs subcells plays a role. Therefore, in Figure 12.5, the short circuit current produced by the device under test (DUT) – which can be either a CPV receiver, a module or an individual multijunction solar cell – normalized to the DNI 'seen' by GaInP top cell – Eq. (12.2) and Eq. (12.13) – is plotted as a function of the SMR (thick black line in Figure 12.5). The same is done normalizing to the DNI 'seen' by GaInAs middle cell (thick gray line in Figure 12.5). The aforementioned normalization is calculated as follows:

$$I_{N,i-component} = \frac{I_{sc,DUT}}{DNI_{i-component}},$$ (12.12)

where $I_{N,i-component}$ [A/suns] is the multijunction solar cell current normalized to the DNI as measured by the i-component cell from Eq. (12.2), and $I_{SC,DUT}$ is the short circuit current produced by the specimen under test. When the DUT is limited by a particular subcell, its short-circuit current will be linear with that of the corresponding component cell. Therefore, $I_{N,i-component}$ will appear constant throughout the range of SMR in which the subcell limits. Note that the quantity is given in amperes per *sun* –where *1 sun* refers to 1000 W/m^2 – or, alternatively, in A/(W/m^2).

In the spectral zone labeled *A*, the DUT short-circuit current normalized to the top-cell DNI remains constant (the DUT is said to *follow* the top component cell), so in this zone $I_{sc,DUT} \approx I_{L,top-subcell}$. This knowledge is used to 'calibrate' the top-subcell current sensitivity $I_{N,top}*$ as a function of the top-component DNI, as in Eq. (12.13). Analogously, in zone *B*, the middle subcell is limiting the multijunction solar cell current, and its calibration constant $I_{N,middle}*$ can be determined, Eq. (12.14):

$$\{Zone\ A\} : I^*_{N,top} = \frac{I_{sc,DUT}}{DNI_{top-component}}\bigg|_{I_{sc,DUT} \approx I_{L,top}},$$ (12.13)

$$\{Zone\ B\} : I^*_{N,middle} = \frac{I_{sc,DUT}}{DNI_{middle-component}}\bigg|_{I_{sc,DUT} \approx I_{L,middle}},$$ (12.14)

These two calibration values can be used to determine the overall current-mismatch *CM* under the reference spectrum AM1.5d, as in Eq. (12.15):

$$CM^{top}_{middle}(E_{AM1.5D}(\lambda)) = \frac{I^*_{N,top}(E_{AM1.5D}(\lambda))}{I^*_{N,middle}(E_{AM1.5D}(\lambda))},$$ (12.15)

Moreover, the current photogenerated at each subcell can be estimated for any arbitrary irradiance-spectrum condition by means of the component-cells DNI:

$$I_{L,i-subcell} = DNI_{i-component} \cdot I^*_{N,i-subcell},$$ (12.16)

The current-matching ratio under any particular illumination spectrum $E_i(\lambda)$ can be directly obtained through the value of SMR, as shown in Eq. (12.17):

$$CM^{top}_{middle}(E_i(\lambda)) = \frac{I_{L,top}(E_i(\lambda))}{I_{L,middle}(E_i(\lambda))} = \frac{DNI_{top}(E_i(\lambda))}{DNI_{middle}(E_i(\lambda))} \cdot \frac{I^*_{N,top}(E_{AM1.5D}(\lambda))}{I^*_{N,middle}(E_{AM1.5D}(\lambda))}$$
$$= SMR^{top}_{middle}(E_i(\lambda)) \cdot CM^{top}_{middle}(E_{AM1.5D}(\lambda)) \qquad (12.17)$$

12.3.6.1 Non-Idealities in Subcell-Limitation Diagrams

The preceding analysis allows the prediction of the multijunction solar cell current under varying working conditions, e.g., throughout a day, as long as the DNI of the component cells is available. However, the subcell-limitation graph may be obscured and deviate from this ideal analysis for two primary reasons. First, even in the case in which perfect component cells are available, concentrator optical systems alter the spectrum according to their spectral transmittance $T_{CPV}(\lambda)$. Thus, the external spectral responses of the subcells under the concentrator become somewhat mismatched to those of the component cells, and they may not continue to have the same proportionality with the illumination spectrum. However, because the spectral response bounds do not change, it can be expected that the proportionality would be maintained for conventional concentrators, although with some amount of mismatch. This *spectral mismatch factor, M,* in the ratio of photocurrents can be estimated for two different spectra $E_1(\lambda)$ and $E_2(\lambda)$ as shown in Eq. (12.18). This is analogous to the standard formula for the error in the measured I_{sc} of a test specimen under a solar simulator when the reference cell used has a spectral response that is mismatched to the device under test (see Chapter 10). This concern is also applicable in the case where the CPV module is equipped with multijunction solar cells from a different technology than the component cells:

$$M(E_1(\lambda), E_2(\lambda)) = \frac{\int SR(\lambda)\, T_{CPV}(\lambda)\, E_2(\lambda)\, d\lambda}{\int SR(\lambda)\, E_2(\lambda)\, d\lambda} \cdot \frac{\int SR(\lambda)\, E_1(\lambda)\, d\lambda}{\int SR(\lambda)\, T_{CPV}(\lambda)\, E_1(\lambda)\, d\lambda}, \qquad (12.18)$$

A second possible reason for not finding a subcell-limitation zone that is flat arises from spectral non-homogeneities over the cell surface. Chromatic aberration of refractive concentrators typically produces an inhomogeneous spatial distribution of the current mismatch over the cell. This implies that both the magnitude of the current mismatch and the identity of the limiting subcell vary throughout the cell surface, introducing larger current losses defined by the overall current mismatch. This situation is signaled in the subcell-limitation graph by a non-flat zone in the normalized current, typically found as a transition between the zones of perfect subcell limitation.

Figure 12.6 illustrates a general subcell-limitation graph that includes the situations described above –again for CPV module using Ge-based triple-junction solar cells–, which may be found in a real refractive CPV system. In zone A, the module short circuit current perfectly follows the top-component DNI, indicating that the top subcell is limiting the module current for most of the cell area – i.e. $CM < 1$ according to the definition of Eq. (12.17). In zone B, the SMR and the overall CM increase, but the spatial non-uniformity of the spectrum over the solar cells causes the top and

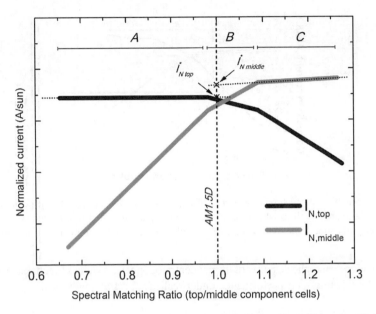

Figure 12.6 Sample subcell limitation graph for a refractive concentrator module using Ge-based triple-junction solar cells, where the current-matching is not uniform throughout the cell surface and there is some spectral mismatch between the DUT middle subcell and the middle component cell

middle subcells to limit the current simultaneously in different parts of the cell. Finally, in zone C, the middle subcell is limiting the module current for most of the cell area. In this region, a quasi-horizontal monotonic trend was found for $I_{N,middle}$, meaning that, in this case, some mismatch exists between the spectral response of DUT middle subcells and the middle component cell caused by the non-ideal transmittance of concentrator optics $T_{CPV}(\lambda)$. The calculation of the current mismatch under this situation requires a linear extrapolation of this monotonic trend to the point of $SMR = 1$ in order to subtract the mismatch error bias and obtain $I_{N,middle}*$.

12.3.6.2 Analysis Beyond a Saturated Bottom Cell

Assuming that the bottom subcell does not limit the overall current may not be valid for several reasons. On the one hand, an increasing number of high efficiency triple-junction cells with alternative substrates that produce a lower excess current in the bottom cell are being proposed. This tighter match is also found in next generation concepts like the recent four-junction cell. On the other hand, a particular concentrator optical design may produce a large unbalanced absorption in the near-infrared region such that the bottom subcells does not saturate anymore. In either case, the analysis of the subcell limitation diagrams should be extended beyond top and middle subcells ($I_{N,bottom}$). In this case, plotting the subcell limitation diagram versus other SMR (middle to bottom, for instance) might be useful. Figure 12.7 shows a sample subcell limitation diagram for a record-efficiency triple-junction solar cell with a very slight excess current in the bottom cell (not based on a Germanium susbtrate). Note that for very blue-rich spectra (*SMR(top/middle) > 1.1*) the bottom cell limits DUT current. Because *SMR(middle/bottom)* is highly correlated to *SMR(top/middle)*, the bounds to each spectral range can be clearly identified. This

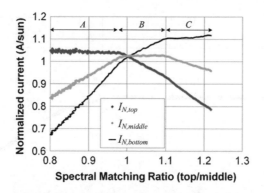

Figure 12.7 Sample subcell limitation graph of $I_{N,top}$, $I_{N,middle}$ and $I_{N,bottom}$ for a metamorphic high-efficiency triple-junction cell in which three spectral ranges are revealed where each subcell limits the overall current (top subcell in zone *A*, middle subcell in zone *B* and bottom subcell in zone *C*)

information is critical to the designer of the concentrator optics, as well as for characterizing these devices and predicting their performance under realistic spectral conditions.

12.3.7 Angular Transmission Curve

The angular transmission curve (discussed in section 12.2.1) is obtained through the measurement of the module *I-V* curve under varying misalignment angles. The whole *I-V* curve needs to be measured since the angular transmission curve will be calculated plotting the P_{mpp} (or sometimes the I_{sc}) of each curve vs. misalignment angle. The misalignment angle can be defined for any rotation direction from the center, although it is typically given only along one or two axes for simplicity. If a tracker with programmable offset angles is available, the whole 2D map of angle deviations (as a function of misalignment angle α and its azimuth direction ϕ as in Figure 12.8) can be covered, although the measurement might take too long to keep stable

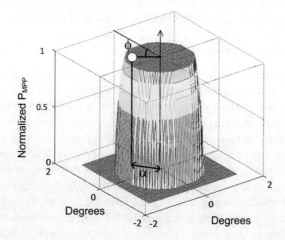

Figure 12.8 Normalized 2D map of P_{mpp} as a function of the misalignment angle α and the azimuth direction ϕ of said misalignment

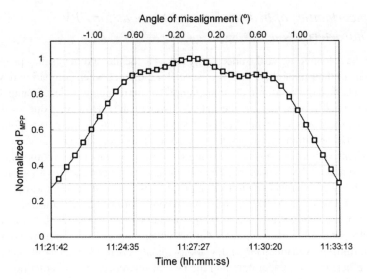

Figure 12.9 Outdoor measurement of the P_{mpp} angular transmittance curve of a concentrator: the tracker is aligned a few minutes ahead from the actual position of the sun and then stopped. The time at each sample (bottom abscissa) and the solar ephemeris are used to calculate the misalignment angle (top abscissa)

operation conditions throughout the whole measurement process (and hence introducing larger uncertainties).

The simplest method for the outdoor measurement of the angular transmission curve simply requires the tracker to be stopped at any moment close to noon (where the slowest change in solar elevation is found). The module I-V curve is measured at short intervals as the sun deviates from the vector of optimal alignment. The time under which every measurement has been carried out is then used to calculate the angular deviation from the initial position (assumed to be the center of the curve), using solar ephemeris.[9] To be sure that the center of the curve is found, the tracker can be manually aligned a few minutes ahead from the actual position of the sun and then stopped (see Figure 12.9).

Since irradiance and spectrum are not constant with time, the registered I-V data has to be corrected for DNI or spectrally-corrected DNI (e.g. use only middle-component effective DNI if the middle sub-cell is known to be limiting module current during the whole measurement) following some translation models like those defined in IEC 60891. Temperature variations can be neglected provided irradiance, wind speed and ambient temperature were reasonably stable throughout the whole measurement. Then the corrected P_{mpp} or I_{sc} are plotted versus the calculated angles to obtain the angular transmission curve.

Indoor characterization of this angular characteristic requires a more profound description of the quantities involved (see section 12.4.4).

---

[9] Free software is available for this purpose like the SOLPOS code developed by NREL. Available at: http://www.nrel.gov/midc/solpos/solpos.html

12.3.8   Uncertainties of Instruments and Methods for CPV Characterization

Laboratories have to investigate the uncertainty linked to the measurements they provide in order to give an idea of their reliability. This is of special importance when comparing devices (is the difference lower than the measurement uncertainty?) or for reducing project financial risks (how much can the rated power vary in the actual power plant?). However, due to the fact that measurement procedures have not yet been standardized and the large variability of CPV architectures, the measurement uncertainty linked to the power rating of CPV devices has not been comprehensively assessed yet. Nevertheless, a few international intercomparison exercises involving most major CPV characterization labs have already been carried out in order to assess the reproducibility of current characterization procedures. The SOPHIA round-robin on outdoor power rating of Soitec CX-M400 modules has recently showed an agreement in rated power between labs better than 4% for labs fulfilling the requirements on prevailing conditions [68]. However, labs performing measurements during winter months (cold ambient temperature and red-rich spectrum) showed much larger deviations, mainly as a result of the large sensitivity of the lens optical efficiency on temperature and the lack of reference spectral conditions,

Figure 12.10 shows the main sources of uncertainty affecting the characterization of CPV devices outdoors. Some of them are specifically linked to module power rating, like the

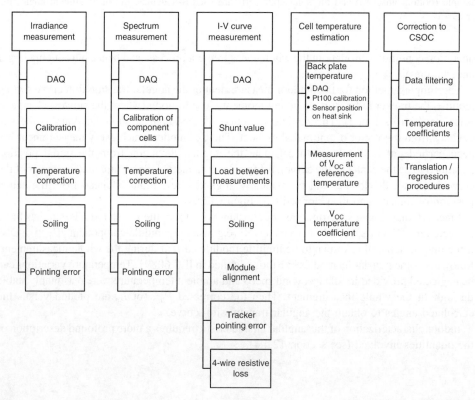

Figure 12.10   Main sources of uncertainty for the outdoor power rating and characterization of CPV modules

contributions derived from correcting to CSOC. Some of them are common to PV characterization – pyrheliometer calibration, data acquisition (DAQ), resistive losses or Pt100 calibration – so the calculation of their typical uncertainty is known [69]. However, most of them have neither been carefully evaluated yet nor quantified specifically for CPV devices and their particular measurement procedures (which in turn are not yet standardized). The main sources of uncertainty linked to outdoor measurement of CPV modules are related to the specificities of multijunction cells and concentrator optics described above, i.e. spectral sensitivity, narrow angular acceptance, large sensitivity to lens temperature and large cell operating temperature (large and variable gradient between heat sink and cell, which complicates its estimation).

Measurement procedures under discussion try to reduce the largest contributions to uncertainty by introducing an appropriate filtering of the working conditions. Soiling can be simply avoided by cleaning the primary lenses before acquiring data. Pointing errors can be avoided by filtering out data outside certain bounds (linked to the acceptance angle of the device). Translation to CSOC requires an estimate of the cell temperature, which is highly sensitive to uncertainty in the measurement of the V_{oc} under reference temperature. Two methods are currently provided in the IEC 62670-3 norm draft: dark *I-V* and simulator based measurements. Another large contribution to uncertainty is linked to a good knowledge of the temperature coefficients for efficiency and short circuit current for the translation to CSOC.

Once the standard uncertainty of every source is quantified, their combined uncertainty in the measurement is linked to the sensitivity of the measurand to each component. The sensitivity coefficients are typically found empirically: how a change in a particular quantity propagates to the measurand. In many cases, this sensitivity is unity, when the error in the input component fully propagates to the measurand. For instance, short circuit current measurements are directly affected by errors in the measurement of irradiance. The correlation between the different uncertainty components has to be known as well, although they are typically assumed to be uncorrelated [70]. The combined uncertainty u_c is then calculated as root mean square of the uncertainties of every component u_{Xi} with the appropriate sensitivity coefficient c_i as in Eq. (12.19):

$$u_c^2(y) = \sum_i^n (c_i u_{Xi})^2 \tag{12.19}$$

In the calculation of the combined uncertainty, the standard uncertainty of a component with a rectangular probability distribution is equal to the distribution half width divided by $\sqrt{3}$. This is assumed to be the case for instance for shunt or Pt100 calibration or module misalignment. A coverage factor of two is used in the final expanded combined uncertainty in order to obtain a 95% coverage interval [69].

12.4   Indoor Measurements of CPV Modules

12.4.1   Solar Simulators for CPV Modules

12.4.1.1   Introduction

The purpose of a solar simulator is to force the device under test to produce the same performance as under some predefined reference conditions, and then allow for its measurement. In an idealistic approach, one would construct a light source that reproduces accurately the characteristics of the sun under those reference conditions: standard irradiance level with negligible spatial and temporal variation over the whole prescribed area, a standard spectrum with negligible spectral mismatch between reference unit and device under test, and the angular distribution of the average solar disk. However, the fabrication of such a system might be either not feasible or prohibitively expensive. In a practical approach we could try to build a light source *relatively* similar to the sun such that makes the PV device generate the same current as under the reference conditions. Thus, reference conditions would be verified by the effect they produce on a light sensor with the same sensitivities as the device under test to irradiance, spectrum, temperature and angular distribution of the radiation. Such a reference sensor accommodates the non-idealities of the light source. Any mismatches between the responses of the reference sensor and the test specimen might require the application of additional correction factors (such as the four-integral spectral mismatch correction factor M, Eq. (12.18)). For instance, the use of an amorphous silicon solar cell as a reference sensor for testing crystalline silicon cells can lead to errors around 10% even for top-class simulators [71].

The main applications of a solar simulator are:

- general *characterization* of a device for the prediction of in-sun performance;
- the *rating* of a device for comparison with other technologies;
- the comparison between a series of devices of the same technology for *sorting* purposes, e.g. classification of modules out of a production line;
- *qualification*, i.e. to compare the performance of a device before and after some ageing or reliability tests.

The quality standards required for the solar simulator depend on the target of the measurements. Sorting or qualifying might only require the measurement conditions to be very repeatable, but not to resemble too accurately the real sun, while module rating for intercomparison of technologies requires an accurate measurement with respect to some well-defined reference conditions.

12.4.1.2 Specification of Requirements for a CPV Solar Simulator

International standards exists for classifying the quality of a solar simulator according to three characteristics: the spatial uniformity of the irradiance, the match to a reference spectral irradiance distribution and the temporal stability of the irradiance during a measurement [21,68,69]. Nevertheless, no classification parameters are set for the angular distribution of the light beam over the receiver. However, the limited angular acceptance of concentrators implies a high sensitivity to the angular distribution of the light source. Therefore a solar simulator for CPV modules should fulfill the fundamental requirement of reproducing the angular distribution of direct light, i.e. the solar disc of $\pm 0.28°$ plus some amount of circumsolar radiation within a field of view of at least $\pm 2.5°$, if one wants to reproduce in-sun behavior. However, conventional flat-plate solar simulators do not produce a beam of parallel rays. The only requirement for solar simulators in current specification standards regarding the angular size of the light is to provide a rough description of it, to be included in the simulator report from the manufacturer. In ASTM E927 standard specification for solar simulators, the manufacturer has to indicate the fraction of total light power that falls within a 30° field of view at any point of the test area [72,73]. We may define a *collimation angle* as the maximum angle in which 90% of the irradiance is contained. What is an acceptable collimation angle depends on the angular transmittance of the device under test (the wider the acceptance angle, the more tolerant is to poor collimation angles), the type of testing to be carried out and the uncertainty limits required. Nevertheless, recommendations or requirements should be given for the indoor testing of CPV modules in relation to the collimation angle. These recommendations could be either absolute (specific bounds for the valid collimation angle) or technology-dependent: either by requiring that the collimation angle is at least half of the acceptance angle, or by requiring that the angular transmission curves obtained indoors and outdoors be close enough (and defining metrics for this similarity). Nevertheless, in our experience, a collimation angle in the range of $\pm 0.4°$ is suitable for most state-of-the-art high-concentration systems.

Since present-day high-concentration PV modules employ monolithic series-connected multijunction cells, their characterization has to take into account the balance of photocurrents between subcells. The spectral match to a reference spectrum in current standards is defined through the similarity in spectral irradiance throughout several fixed wavelength intervals of 100 and 200 nm, which is not sufficient for this purpose. Instead, we could use the aforementioned SMR – calculated using Eq. (12.3) – to compare the expected current matching under the simulator with that under a reference spectrum.

In concentrator systems and terrestrial multijunction cells, the AM1.5d spectral distribution defined in the ASTM G173-03 standard is used as reference spectrum [74]. As already commented in this chapter, for common triple-junction cells on Ge substrate, where the bottom subcell produces an excess of current under reference spectrum, the relevant spectral balance is the *top-to-middle* SMR to AM1.5d. An SMR of 1 indicates that the simulator spectrum is equivalent to AM1.5d in relative shape, throughout the spectral range in which those subcells have significant sensitivity. The ability to adjust this SMR to 1 is therefore required for CPV solar simulators. Depending on the way that the simulator spectrum can be adjusted (filtering, mixture of multi-sources, flash lamp triggering voltage etc.), there may be only some types of the available multijunction cell technologies for which the SMR can be set to 1. Therefore, the information about which multijunction cell technologies the simulator is adjustable for should be a basic specification to appear in the simulator labeling.

The spatial non-uniformity of the irradiance is calculated through the maximum and minimum light intensities found over the defined test area. To do this, the irradiance is measured throughout a matrix of equally spaced test positions over the test plane. ASTM E927 gives detailed recommendations about the minimum proportion of the area that should be checked and imposes the use of a light sensor with 'appropriate' linearity and time response for the characteristics of the solar simulator. This appropriateness actually depends on the type of device to be tested (linearity with irradiance, spectral and angular responses etc.), so this requirement is vague. In the end, the non-uniformity is calculated as:

$$\text{Irradiance non-uniformity}(\%) = \pm \frac{\max\{DNI(x, y)\} - \min\{DNI(x, y)\}}{\max\{DNI(x, y)\} + \min\{DNI(x, y)\}} \cdot 100 \qquad (12.20)$$

where $\max\{DNI(x,y)\}$ and $\min\{DNI(x,y)\}$ are the maximum and minimum irradiance found at any location (x,y) on the test plane area.

The temporal instability of the irradiance during a single measurement and between measurements has to be evaluated as well. The characterization of this property depends on the type of simulator used (continuous, single-pulse or multi-flash). An example is given, for

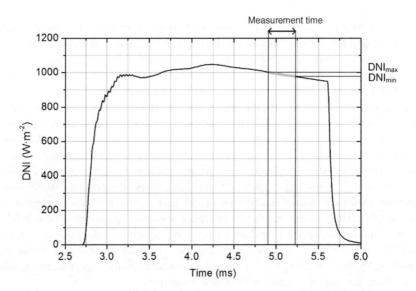

Figure 12.11 The temporal instability of the irradiance is evaluated throughout the *I-V* curve measurement time

a flash based simulator, in Figure 12.11. In the case of continuous and single-pulse simulators, the irradiance stability has to be checked without interruption throughout the whole measurement time frame. It is calculated as follows:

$$Temporal\ instability\ (\%) = \pm \frac{\max\{DNI(t)\} - \min\{DNI(t)\}}{\max\{DNI(t)\} + \min\{DNI(t)\}} \cdot 100 \qquad (12.21)$$

On the contrary, in multi-flash measurements triggered by an irradiance sensor, the temporal instability is checked only throughout the acquisition of the *I-V* pair and the corresponding irradiance level. Again, the better the linearity and the spectral and angular responses of the irradiance sensor are matched to the device under test, the better the results.

The characteristics of a solar simulator shall be checked periodically or whenever there is a change in the solar simulator (including aging) that could affect these characteristics beyond acceptable limits. Aging is a big concern particularly in continuous solar simulators, whose lamp and optical components may degrade much faster than pulsed ones due to the higher energies involved. A frequent checking of the spectral matching characteristics is encouraged especially when the test specimen uses multijunction solar cells.

12.4.1.3 Types of Illumination Systems

The illumination system of a solar simulator typically comprises an optical assembly of one or more light bulbs and other optical elements that guide, collimate, mix and/or homogenize the light sources to create the prescribed light beam over a particular plane and area. Gas discharge lamps achieve the highest radiances, especially if they are pulsed. Flash lamps filled with noble gases can be triggered to conduct very large current densities for a short time, producing radiances that resemble blackbody radiation with an equivalent temperature even larger than that of the sun (i.e. higher than 5800 K). Xenon-filled flash lamps produce the best similarity to solar white light. The maximum intensity of flash lamps can be controlled varying the discharge voltage, which determines the maximum current density. Continuous Xe–arc lamps are also used, but produce very large spectral emission lines that have to be carefully filtered. Moreover, their spectrum shows a red-shift with ageing, i.e. the proportion of power in the longer wavelengths increases with bulb use [75]. Their intensity can be controlled through their load current. As a rule of thumb, for the same irradiance level and illuminated area, a continuous illumination system requires much more complexity and provides poorer spatial uniformity of the light. Alternatively, tungsten halogen lamp arrays are also used, but they cannot provide the solar spectrum alone as they feature a low color temperature (around 3000 K). Infrared light-rejecting filters (dichroic filters, or 'hot mirrors') are used to improve the similarity to the solar spectrum. Thus, they are frequently used as complementary lights for multi-source solar simulators, in order to precisely adjust the combined spectrum or even perform spectral sweeps useful for the characterization of multijunction devices [76]. However, the spectrum of tungsten lamps alone may be good enough for rough PV testing with low-grade simulators. Their intensity, as well as their spectral distribution, depends on the load current. Whether a simulator light is acceptable or not depends on the level of accuracy required for the measurement or on the goal of the testing. Main uncertainties in indoor measurement of PV devices arise from the differences between artificial light sources and the 'reference sun'. A calibration laboratory requires the highest possible quality, while poor simulators may be acceptable for sorting tests at the production line. An arrangement of several light sources, spectral filters and homogenizing lenses is usually required to create a continuous light beam with the necessary quality for a calibration laboratory [77].

12.4.1.4 Collimated Light Beams

The optical design of a solar simulator for CPV modules is fundamentally different from flat-plate PV simulators because of the requirement of providing light with a very small angular size. Such a collimated light beam can be created by very distant sources. Allowing a great distance between a point source and the receiver can create a small angular size, but at the cost of a very low irradiance. If a module has an aperture of 1 m in side, a small source would have to be located almost 100 m far from it to be seen within a field of view lower than ±0.3° at every point of the module. The alternative is the use of a collimator element, i.e. a device that transforms a divergent light beam into a collimated beam or plane wave front. A point light source at the focus of an optical collimator is imaged to infinity. This can be done either in reflection with mirrors or by refraction with lenses. Refractive collimators suffer from chromatic aberration due to the wavelength dependence of refractive index, and their transmittance varies with the angle of incidence due to the angular dependence of Fresnel reflection losses. The result is a considerable non-uniformity of irradiance, spectrum and collimation angle. A mirror, on the contrary, lacks these aberrations.

In a parabolic mirror, the rays coming from the focus of the parabola are reflected to infinity. Thus, if a reflective paraboloid of revolution is illuminated by small lamp located at its focus, it will produce in reflection a quasi-collimated light beam of the same size as the mirror cross-section. The angular size of this light will be that of the lamp seen by the mirror, as depicted in Figure 12.12, so a small lamp and a long focal distance are needed to produce a good

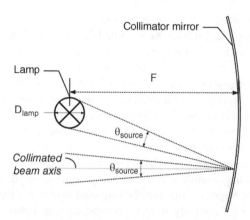

Figure 12.12 The angular size of the light beam is that of the lamp seen by the collimator element

collimation. For a light bulb diameter D_{lamp} of 6 cm (typical of a high intensity commercial Xenon flash bulb), a focal distance F longer than 6 m is required to produce the angular size of the sun, i.e. $\theta_{source} \approx \pm 0.28°$– see Eq. (12.22):

$$\theta_{source} = \tan^{-1}(D_{lamp}/F) \tag{12.22}$$

The aperture area of most current CPV modules presents some linear dimension in the range of 1 meter. The cross-section of the collimator has to be large enough to accommodate such extensive modules, but the manufacturability and accuracy of mirrors is more difficult to attain as the size of the mirror raises.

The CPV solar simulators available in the market use either approach. The *Helios 3198* from *Solar Added Value*[10] features a large 2 m diameter collimator mirror and flash light, while the *Neonsee* simulator uses continuous light and transmission optics to create a light beam up to 50 cm wide.[11]

12.4.2 Reference Sensor

Light sensors are used at the simulator to identify the desired illumination conditions. In pulsed simulators, thermal sensors cannot be used so radiometric measurements are difficult. Moreover, as the angular size and spectrum of the light differ from that of the sun, it may be convenient to somehow outweigh the part of the light intensity coming from wavelengths or angles irrelevant to CPV. This is to say, we are interested in identifying the realistic illumination conditions that make the PV device perform the same as if it were illuminated by those ideal reference conditions. For instance, an artificial light with a radiometric irradiance of 1100 W/m^2 and a non-ideal spectrum might produce exactly the same photocurrent and current matching on a test specimen as if it were illuminated under 1000 W/m^2 and the AM1.5d spectrum. For this reason, reference sensors matched to the test specimen in angular and spectral responses should be used. This is the idea behind the classical usage of reference cells

---

[10] Helios 3198 solar simulator: http://solaraddedvalue.com/en/category/productos/helios-3198/
[11] Neonsee Solar Simulator: www.neonsee.com/en/cpv/-iv-measurement/

in the indoor measurement of flat PV modules, for which it is demanded that the monitor cell has a similar spectral response and linearity with irradiance as the test specimen. When the light characteristics change, a sensor that is not matched to the device under test varies its response incoherently with it and provides biased estimations of the irradiance level. This might be corrected, to some extent, through the calculation of mismatch factors, but nevertheless the uncertainty of the measurement is increased [78].

International standards do not specify that the test specimen and the reference cell should have a similar sensitivity to the angular distribution of the light source. This is obviated because flat PV devices typically feature very wide angular responses. At most, the norms demand to specify the fraction of the light that is falling within a reasonable field of view (30° in ASTM E927). However, the narrow angular response of concentrators implies a high sensitivity to the angular distribution of the light source, and therefore the reference sensor has also to accommodate the non-idealities of the collimation. A temptation here is the use of monitor cells in collimator tubes to reduce their field of view; however, this does not make a matched sensor for two reasons: first, the angular transmission curve of a concentrator has not a flat-top shape as that of the collimator, and second, the concentrator optics do change the effective spectral response of the cell.

Thus, the natural choice for a CPV reference unit is a module of the same technology as the test specimen. However, in order to reduce the area that needs to be illuminated by the simulator and for the sake of simplicity, the reference module may be constructed including only the minimum optical-electrical unit within the module that produces the same spectral and angular sensitivities. This minimum unit of a particular technology is referred to as 'mono-module' (see Figure 12.13). Such sensors can be fabricated from existing components available to all CPV manufacturers. When a mono-module is not available, another module of the same technology can be used as CPV reference unit, after proper outdoor calibration [25].

To prepare a mono-module for use as a light sensor an outdoor calibration is performed, whereby its short-circuit current under some reference conditions is obtained (I_{sc}^{STC}). These conditions should include irradiance, spectrum and lens temperature because of their large influence on module current. These conditions can be ensured either by interpolation, translation or filtering of the measured values to the desired levels (analogous to power rating

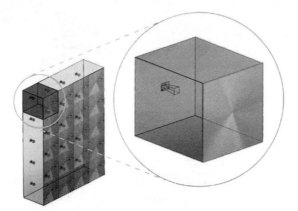

Figure 12.13 A mono-module is the minimum optical-electrical unit of a CPV module that features the same spectral and angular responses as the whole module. Source: © Black Photon Instruments GmbH, 2015. Reproduced with permission of Black Photon Instruments GmbH

procedures). SMR given by component cells can be used to identify reference spectrum conditions.

An important note is that the same alignment used on the tracker should be reproduced at the simulator with high precision in order to keep the validity of the calibration.

12.4.2.1 Procedure for Outdoor Calibration of Mono-Module

The objective of the calibration is to obtain the short-circuit current of the mono-module under some standard conditions, which may be chosen to be those of CSOC: an irradiance of 900 W/m^2, AM1.5d spectrum and 20 °C ambient temperature. As in CSOC power rating, several clear-sky days would be required in which AM values close to 1.5 are found. The occurrence of the AM1.5d spectrum is detected through the SMR for the relevant subcells of the solar cell used as receiver in the mono-module (typically lattice-matched Ge-based triple-junction cells). An indication on how to measure the SMR is given in the next section. The mono-module has to be placed on a solar tracker, and properly aligned to the sun (this alignment has to be repeated with great precision for its usage in the simulator). If the angular transmission curve is known to have an absolute maximum at the center of the transmission curve, the unit is simply aligned by maximizing the short-circuit current.

Mono-module I_{sc} is recorded simultaneously with the DNI (through a pyrheliometer) and the SMR to AM1.5d (calculated for the least generating subcells of the receiver cell) for the whole calibration period. Then the current normalized by the DNI is calculated for every specimen under test as:

$$I_N(t) = \frac{I_{sc}(t)}{DNI(t)} \tag{12.29}$$

which varies as a function of the SMR, as shown in Figure 12.14. A linear regression can be used to interpolate the I_N for the sensor at $SMR = 1$ (effective AM1.5d), which will be the calibrated

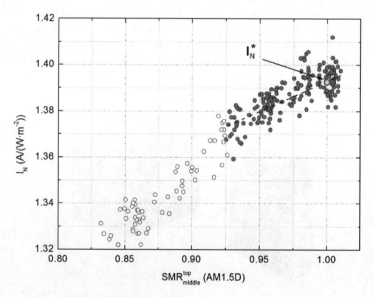

Figure 12.14 Sample calibration plot for a CPV reference mono-module. Note that since two different I_N trends vs. SMR were found (the subcell limiting the overall current changes), only filled dots should be used for interpolating the I_N^* calibration value

constant I_N^* [A/(W·m^{-2})] used for the sensor when it is placed in the solar simulator. The effective standard irradiance of 1000 W/m^2 for CSTC measurements at the simulator is achieved when the mono-module is generating a current equal to 1000 W/m$^2 \times I_N^*$. This procedure is still under development by the IEC TC82 WG7 for the IEC 62670-3 norm.

12.4.2.2 Monitoring Spectral Conditions

The most relevant measurement of the spectrum for CPV modules is the balance between the spectral bands corresponding to each subcell. A deviation in this balance translates into a change in the current mismatch between subcells with respect to reference conditions. Thus, module subcells will work under different intensities of effective irradiance. If the photocurrent of non-limiting subcells is higher than under reference conditions, the excess current will contribute to a higher V_{mpp} and FF, thus overestimating P_{mpp}. Fortunately, voltage is proportional to the logarithm of the concentration, so this effect may be negligible for moderate spectral errors.

When pulsed solar simulators are considered, the time resolved measurement of the spectrum might not be possible using spectroradiometers. Instead, calibrated component cells can be used to monitor the effective irradiance at each spectral band (average weighted by the SR of the cells) as described in section 12.3.2.1. For their use in the simulator, the component cells have to be installed in collimator tubes with an opening angle of about ±2.5° and aligned towards the light beam in order to collect only direct normal irradiance. The current of the mono-module and the component-cells short circuit current – and therefore DNI through Eq. (12.2) – are measured simultaneously throughout the flash pulse. Then the SMR can be calculated and plotted as a function of the mono-module DNI. With these data we can identify the irradiance ranges for which the simulator produces a spectrum sufficiently matched to reference spectrum (see Figure 12.15). This valid range will vary with each technology considered.

This large spectral sweep provided by pulsed solar simulators can be used to characterize the sensitivity of CPV module performance to spectrum, similarly as done in outdoor operation.

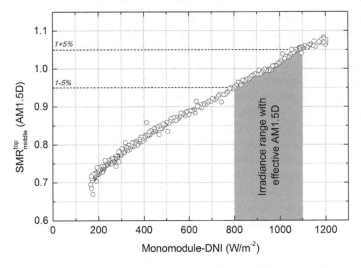

Figure 12.15 Spectral matching ratio as a function of the effective DNI measured by the mono-module of a particular CPV technology. The range of irradiance featuring the reference spectrum is indicated

However, indoor spectral characterization allows for much more stable and repeatable conditions regarding module alignment or lens and cell temperatures.

12.4.3 Caveats on Indoor Measurements

In order to measure the I-V characteristic of a PV device, solar simulators employ a variable load to bias the device at different working points across the I-V curve. Then the voltage and current across the device are measured for every working condition, simultaneously with the incident irradiance. If the linearity of the current with irradiance is known for the particular device under test and reference sensor, corrections can be made to account for the variability of the irradiance between the measured points, translating all the current measurements to a single reference irradiance level. Voltage can also be corrected for light fluctuations if the *effective* ideality factor of the PV device is known, as the open-circuit voltage increases with the logarithm of the photocurrent, but weighted by the ideality factor. The changes in the voltage losses due to the series resistance depend also on the photocurrent.

The aforementioned dependence of PV devices on temperature is also taken into account. Algebraic methods like the one in the IEC 60891 standard use linear temperature coefficients to correct both current and voltage [79]. Analytical modeling of the device (like in the lumped diode model) can also be used to extract translation equations for the voltage [80]. In order to carry out these corrections, the temperature of the device has to be measured. In pulsed solar simulators with reasonably short light pulse lengths, cell temperature can be considered to be equal to room temperature. As a rule-of-thumb, a pulse can be considered to be short if its duration is lower than 10 ms, although the precise temperature rise depends on the actual energy delivered by the pulse and the heat capacity of the particular cell receiver configuration.

There are many different measurement schemes that can be used to produce the required sweep of voltage and current across the device. The particular strategy used may be determined by the type of illumination used (continuous or pulsed) or by the electric characteristics of the specimen under test. Resistive and capacitive loads are the simplest passive loads and are typically chosen for very high power systems or when custom measurement systems are required. They cannot operate beyond the power quadrant of the I-V curve, though. Automatic 'source-meters' can actively bias the PV device to reverse or forward voltages, i.e. beyond the I_{sc} and V_{oc}, but usually their limitations in maximum current or power do not allow the measurement of large modules. Programmable bipolar power supplies allow the same measurement scheme, but pieces of equipment for much higher powers are commercially available. Capacitive loads and bipolar power supplies have also the advantage of providing very fast voltage sweeps, so they may be the only option for solar simulators with pulsed illumination.

Solar simulators can be classified into three types according to the combination of illumination system and measurement scheme: *continuous* or steady-state simulators, *single–pulse* simulators and *multi–flash* simulators. Continuous and single-pulse simulators measure the whole I-V curve at one stroke, while multi-flash simulators measure only one I-V point per flash pulse. Single-pulse simulators require a light plateau while the voltage sweep is carried out in order to assure stable irradiance, so the energy of the pulse is much higher than in multi-flash testing. Moreover single-pulse measurements can be affected by transient-related artifacts due to the fast voltage sweep. These issues depend on the particular CPV technology being measured, but they are especially important in high-efficiency silicon devices [81]. Multi-flash and continuous simulators on the contrary can usually assume that the device is

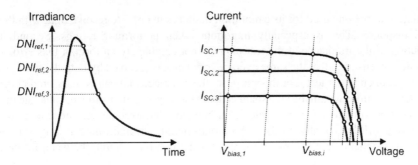

Figure 12.16 Scheme of multi-flash *I-V* measurement. A single voltage bias $V_{bias,I}$ is applied at each pulse to obtain the *I-V* points for several irradiance levels $DNI_{ref,I}$ (*left*). A family of *I-V* curves at different irradiance levels is obtained at the end of the measurement (*right*)

under electrical equilibrium. Continuous simulators are affected by thermal issues because they need accurate temperature sensing and active cooling of the test bed. This is especially a problem in concentrated light simulators due to the enormous heat flux that has to be removed from the receiver plate to maintain a particular temperature. Both continuous and single-pulse simulators require checking the temporal instability of the irradiance, or else correct the *I-V* curve proportionally. On the contrary, multi-flash measurements that are triggered by the DNI given by a reference sensor (see three of these threshold levels marked as $DNI_{ref,I}$ in Figure 12.16) may not need any irradiance correction.

12.4.3.1 Alignment of DUT and Reference Sensor

A final key issue in the indoor measurement of CPV modules is the procedure used to align the device under test to the light beam. All comments made to this respect in section 12.3.4 'Alignment procedures' are again applicable here. For a proper alignment at the simulator, a holding structure at the receiver plane should be prepared for the module allowing small angular rotations across two orthogonal axes (e.g. azimuth and elevation) in steps shorter than a tenth of a degree typically. This structure shall be rotated mechanically with respect to the light path in order to align the module. Simple visual alignment can be used in pulsed solar simulators if a steady lamp is located at the same position as the flash lamp, as in the *Helios 3198* solar simulator from *Solar Added Value*. However, the method based on P_{mpp} maximization leads to lower uncertainty and depends less on user experience. Its application is much simpler than outdoors due the stable operation conditions indoors. Estimated alignment accuracy can be translated into measurement uncertainty through the angular transmission curve of the module under test.

If a series of identical modules have to be measured, two possibilities arise: either a single module is aligned following any of the above procedures and then the alignment is preserved for the rest of them through the fixation of the holding structure; or the alignment process is repeated for every module. The choice depends on the objective of the measurement, e.g. if the modules out of a production line will be ultimately mounted on the same array without individual alignment in the field, then taking all measurements against a fixed datum plane gives more relevant info.

The reference devices like the mono-module have to be fixed and aligned at a separate structure. There are two main precautions. First, the mono-module has to be installed on the same fixture as the one used for its calibration outdoors in order to receive light with the same overall response. This is especially important when performing ratings. Second, as the distribution of the irradiance at the test plane is not completely uniform, a bias error should be avoided by ensuring that the same average irradiance is received by both the mono-module and the specimen under test. For this purpose, the reference unit can be measured at different equidistant locations within the area spanned by the test specimen in order to calculate the mean value. Then a correction factor is calculated between that average and the irradiance at the position that will be actually used for the mono-module. In multi-flash testing, this process requires an auxiliary light sensor to account for the non-repeatability between flashes.

12.4.4 Angular Transmission Curve: Direct and Inverse Methods

The angular transmission curve was introduced in section 12.2.2. In a more rigorous way, the angular transmission curve of a concentrator describes the fraction of light rays (of wavelength λ and incidence direction angles (α, ϕ) with respect to the concentrator optical axis) that are focused at its output while illuminating uniformly its aperture area. This function usually is normalized to range from 0 to 1 which means that absolute optical losses are not included. This function depends on the angular and spectral characteristics of the light distribution $S(\alpha,\phi,\lambda)$. In an analogy with linear systems in signal processing, an impulse-response angular transmission function $H(\alpha,\phi,\lambda)$ can be defined as the angular transmittance of the module while it is illuminated by a perfectly parallel ray beam, i.e. $S(\alpha,\phi)=\delta(\alpha,\phi)$. For a given light source, the angular transmittance $T_S(\alpha,\phi,\lambda)$ can be obtained through the convolution of the two functions:

$$T_S(\alpha, \varphi, \lambda) = \int\limits_0^{2\pi} \int\limits_0^{\pi/2} H(\rho, \omega, \lambda) \cdot S(\alpha - \rho, \varphi - \omega, \lambda) \cdot \mathrm{d}\rho \cdot \mathrm{d}\omega \qquad (12.24)$$

In the case of a non-monochromatic light source, the angular transmittance $T_S(\alpha,\phi)$ is obtained after integrating over the whole spectral range. In practice, the angular transmittance function $T_S(\alpha,\phi)$ related to a light source not perfectly collimated $S(\alpha,\phi)$ is the one normally measured both outdoor (the sun angular size is close to 0.27 degrees) and indoor (perfect collimated light can be prohibitively complicated and expensive to be reproduced at large aperture areas).

Two different kinds of illumination methods can be distinguished for the indoor characterization of the angular transmission curve of concentrator optical systems and concentrator photovoltaic modules, referred to as *direct* or *inverse*, depending on whether the concentrator is irradiated from the input or output aperture, respectively [82–85]. If all possible optical paths from the input to the output aperture of the concentrator and *vice versa* are investigated, some of them are totally reflected backwards or absorbed inside the concentrator optical system while others connect the two apertures. In the hypothesis that these light beams go through only reversible processes, the illumination can be said to comprise only non-polarized light without experimenting diffusion or diffraction phenomena. As established by the reversibility principle [86], the transmission factors in direct and inverse illumination match because of the invariance attenuation suffered by light while reversing the optical path. This implies that the normalized angular distribution of the inverse radiance, $L_{inv,out}(\alpha,\phi)$ (related to the light distribution at the aperture of the concentrator when irradiated from its output) is equal to the normalized impulse-response angular transmittance H (α,ϕ) function (obtained when the concentrator is irradiated from its input).

12.4.4.1 Direct Methods

For direct illumination (i.e., direct methods), the basic procedure of measuring the angular transmittance function of a concentrator consists of recording the transmitted flux trough the input to the output aperture area of the concentrator while varying the direction of the light source $S(\alpha,\phi)$ with respect to the normal of the concentrator. This is similar to what was described for outdoor measurements of the angular transmission curve in section 12.3.7. If the concentrator does not have rotational symmetry, an exhaustive characterization of the angular properties of the concentrator requires the measurement of the 2-dimensional (2D) angular transmittance function $T_S(\alpha,\phi)$. In practice, the angular transmission is typically described by 1-dimensional (1D) angular transmittance function $T_S(\alpha)$ related to a fixed azimuth angle ϕ (that defines the meridian plane in which the incident angles α are contained).

For a CPV module, the angular transmittance function $T_S(\alpha,\phi)$ related to a given light source $S(\alpha,\phi)$ is calculated as the ratio of the maximum output power P_{mpp} for a given incident light direction relative to the maximum output power P_{max} at the best module alignment (which might not match the normal to the module):

$$T_S(\alpha, \phi) = \frac{P_{mpp}(\alpha, \phi)}{P_{max}} \qquad (12.25)$$

The measured angular transmission curve actually depends on the particular angular distribution of the incident light $S(\alpha,\phi)$, so the comparison between the angular transmittance curves taken both indoors and outdoors provides an assessment of the resemblance of the simulator light's collimation to that of the sun. Such a comparison is nevertheless a little challenging because it may be difficult to pass through the same meridian of the 2D misalignment-angle map in both setups and, most specially, because the measurement conditions can hardly be kept throughout the whole outdoors measurement process. Figure 12.17 shows the power angular

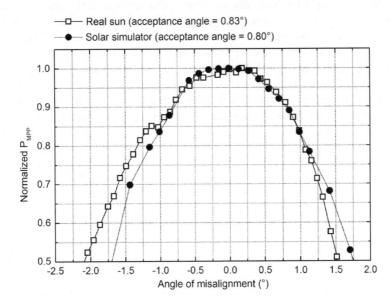

Figure 12.17 Indoor and outdoor measurements of the angular transmittance of a concentrator

transmission curves measured for a 500× concentration module both under the sun and at a *SAV Helios 3198 CPV* solar simulator. The lower acceptance measured in the simulator may be explained by the wider angular size of the source.

12.4.4.2 Inverse Methods

For inverse illumination (i.e., inverse methods), several measurement procedures have been proposed to evaluate the angular transmittance curve of a concentrator module based on the reverse optical path. In these setups, the exit of the concentrator (i.e. the place at which the solar cell is located) is illuminated with a Lambertian light and the illumination transmitted through the concentrator aperture is measured. A Lambertian distribution obeys the so-called Lambert's cosine law, which states that the radiance L (W·m^{-2} sr^{-1}) scattered from ideal Lambertian surface follows a cosine relationship with the polar angle of the scattered light with respect to the normal of the surface.

For concentrators of large field of view (i.e., medium-low geometrical concentration), the evaluation of the transmitted flux can be performed by recording several photographs of the transmission pattern of radiation emanating through the entrance aperture while a Lambertian source is placed at the exit of the concentrator, and a CCD camera is displaced along a single axis [87]. To inspect the transmitted flux at different directions (e.g., α_1 or α_2), one image has to be taken at different positions of the CCD camera with respect to the optical axis of the concentrator. Instead of recording the emitted pattern at the exit of the concentrator with a CCD camera, a Lambertian target can be included far away and aligned with the aperture of the concentrator. An image can be taken with a CCD camera to measure the light emitted by the concentrator and diffused by the Lambertian target [6]. With this measurement scheme, the angular dependence of concentrator with small aperture areas can be resolved by taking only one image with a CCD camera [88].

When characterizing CPV modules in inverse illumination, they cannot be studied as isolated optics (in which a Lambertian source is included at its output aperture) but including all of the defects resulting from coupling to the cell and assembly of the module. In this regard, the solar cell can be forward biased to reproduce the Lambertian light beams (at 680 nm and 890 nm for triple-junction GaInP/GaAs/Ge solar cells) by electro-luminescence [89–92]. Therefore, the performance of the actual manufactured concentrator can be studied based on the principle of reversibility in optics. If the whole cell area emits in all possible ray directions with the same intensity, then the light rays exiting the module in a given angular direction (α, ϕ) will have an intensity that is proportional to the direct angular transmittance for that direction when the concentrator is illuminated by uniform, collimated radiation (i.e., the impulse response angular transmission function $H(\alpha, \phi, \lambda)$). The registration of near field luminescent patterns was first proposed in the 90's to identify defective solar cells in the module or misalignments between optical system-cell units with respect to the module optical axis. To perform this last measurement, a CCD camera must be placed at a large distance from the concentrator to have a field of view several times smaller than the angular size of the sun or of the acceptance angle of the module. For modules with large aperture area, it can highly reduce the spatial resolution of the measurement, and still the angular inspection can be not as accurate as required. By rotating the module at different angular directions with respect to the camera, a visual inspection can be performed to observe the angular transmittance [93].

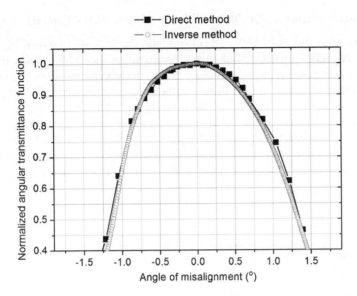

Figure 12.18 Direct and indirect measurements of the angular transmittance curve of a sample CPV module

Figure 12.18 presents the comparison of two angular transmittance curves for a sample CPV module obtained through both direct and indirect methods.

Likewise, instead of rotating the module at different positions with respect to a CCD camera as presented above, a new method (so-called luminescence-inverse (LI) method [94]) is proposed for capturing the light of large CPV aperture areas with high spatial resolution. A large collimator mirror or any other concentrating element is used to focus the backward light with a given angular direction (α,ϕ) to the same point in the focal plane of the mirror. Therefore, module rotation is unnecessary nor placing the camera far away from the module to have good accuracy in the angular inspection. In the classical configuration of the LI method, a Lambertian screen is placed at the focal plane of a collimator mirror, and the light emitted from the module, reflected by the mirror, and scattered by the Lambertian surface is captured by a CCD camera. The image created on the Lambertian surface at the focal plane of the collimator mirror is a representation of the inverse radiance $L_{inv,out}(\alpha,\phi,\lambda)$ of the module, and consequently of the impulse-response angular transmittance function $H(\alpha,\phi,\lambda)$.

Not only the angular transmittance function but also the evaluation of misalignments within a module can be a valuable tool for identifying manufacturing or assembly errors in CPV modules [95]. The misalignments between the optical system-cell units in a CPV module (understood as the differences in the optimum alignment between units) can be obtained from the set of angular transmittance functions of the units if given for the same datum angle. This measurement, based on direct illumination, requires long time for acquisition of output power values at different angular positions of the module, and also a high consume of light resource if measured indoors. By indirect methods, the so-called *Module Optical Analyzer* (MOA) from *Solar Added Value* address this issue evaluating the angular transmittance function and misalignments of a module within seconds, and without the need of any light source.

12.4.5 Uncertainties in the Indoor Measurement of I-V Curves

Owing to the wide variety of CPV technologies and the limited amount of laboratories and companies that count on indoor characterization capabilities for CPV modules, the cumulated knowledge on best practices regarding solar simulator measurements are even scarcer than for outdoor procedures (section 12.3.8). Therefore, the uncertainty in indoor CPV measurement is higher and, in any case, it has been less studied in the literature. Even though part of the procedures are similar, the use of an artificial non-ideal pulsed light source, essentially modifies the traceability chain and the magnitude and components of the uncertainty. Namely, the use of mono-modules and component cells as reference devices, the spatial non-uniformity of the irradiance and the non-ideal collimation and spectrum of simulator light are the main sources of uncertainty.

Glossary

List of Acronyms

Acronym	Description
AM	Air mass
AM1.5d	Reference spectrum for direct radiation
CAP	Concentration-acceptance product
CSOC	Concentrator solar cell operating conditions
CSR	Circumsolar ratio
CSTC	Concentrator standard test conditions
CTE	Coefficient of thermal expansion
DAQ	Data acquisition
DNI	Direct normal irradiance
DPOA	Direct plane-of-array irradiance
GNI	Global normal irradiance
ISFOC	Instituto de Sistemas Fotovoltaicos de Concentración
NIP	Normal incidence pyrheliometer
NOCT	Nominal operating cell temperature
PMMA	Poly methyl methacrylate
PR	Performance ratio
PSD	Position sensitive detectors
RTD	Resistance temperature detector
SMR	Spectral matching ratio
SOE	Secondary optical element
SoG	Silicone-on-glass (primary lens)
SR	Spectral response

List of Symbols

Typical units given in square brackets. If no units are given, variable is dimensionless.

Symbol	Description [Units]
AA	Acceptance angle [°]
A_{in}	Aperture area of primary lens or module [m^2]

A_{out}	Output area of concentrator optics [m^2]
C_g	Geometrical concentration [suns or ×]
CM	Current mismatch
$DNI_{i\text{-}component}$	Effective DNI measured by i component cell [W/m^2]
$E(\lambda)$	Spectral irradiance [W·m^{-2} nm^{-1}]
F	Focal distance [cm]
G	Irradiance [W/m^2]
$H(\alpha,\phi)$	Impulse-response angular transmittance
$I_{sc,STC}$	Short circuit current at standard test conditions [A]
$I_{L,i\text{-}subcell}$	Photogenerated current at subcell i [A]
$I_{L,i}*$	Current sensitivity of subcell I [A/(Wm^{-2})]
I_{mpp}	Current at maximum power point [A]
I_N	Current normalized to DNI [A/(Wm^{-2})]
L	Radiance [W·m^{-2}·sr^{-1}]
M	Spectral mismatch factor
n	Ideality factor
$\dot{Q}$	Transfer rate of excess heat [W]
R_{th}	Thermal resistance [K/W]
$S(\alpha)$	Light angular distribution
$T(\alpha,\phi)$	Angular transmittance
$T(\lambda)$	Spectral transmittance
T_{amb}	Ambient temperature [°C or K]
T_{cell}	Equivalent cell temperature [°C or K]
V_{mpp}	Voltage at maximum power point [V]
$V_{oc,STC}$	Open circuit voltage at standard test conditions [V]
u_i	Uncertainty of source i
w_s	Wind speed [m/s]
Y_F	Final yield
Y_R	Reference yield
α	Misalignment angle [°]
β	Open-circuit voltage temperature coefficient [%/°]
δ	Efficiency temperature coefficient [%/°]
η	Conversion efficiency [%]
θ_{source}	Angular size of solar simulator light beam [°]
ϕ	Azimuth direction of misalignment angle [°]

References

1. G. S. Kinsey, K. M. Edmondson. Spectral response and energy output of concentrator multijunction solar cells. *Progress in Photovoltaics: Research and Applications*, **17**(5), 279–288. 2009.
2. International Electrotechnical Commission. IEC, 61215 Crystalline Silicon Terrestrial Photovoltaic (PV) Modules-Design Qualification and Type Approval. IEC, Geneva. 2005.
3. M. Muller, B. Marion, J. Rodriguez. Evaluating the IEC 61215 Ed.3 NMOT procedure against the existing NOCT procedure with PV modules in a side-by-side configuration. 38th IEEE Photovoltaic Specialists Conference (PVSC), pp. 697–702. 2012.
4. I. Anton, D. Pachon, G. Sala. Characterization of optical collectors for concentration photovoltaic applications. *Progress in Photovoltaics: Research and Applications*, **11**(6), 387–405. 2003.
5. M. Brogren. Optical efficiency of low-concentrating solar energy systems with parabolic reflectors. Doctoral thesis. Uppsala, Sweden. 2004.

6. A. Parretta. Optical efficiency of solar concentrators by a reverse optical path method. *Optics Letters*, **33**(18), 2044. 2008.

7. M. Victoria, R. Herrero, C. Dominguez. *et al.* Indoor characterization of non uniform light distribution due to concentration optics and its effects on solar cell performance. 25th European Photovoltaic Solar Energy Conference. WIP Renewable Energies, 143–146. 2010.

8. M. Victoria, R. Herrero, C. Domínguez, *et al.* Characterization of the spatial distribution of irradiance and spectrum in concentrating photovoltaic systems and their effect on multi-junction solar cells. *Progress in Photovoltaics: Research and Applications*, **21**(3), 308–318. 2011.

9. I. García, C. Algora, I. Rey-Stolle, and B. Galiana. Study of non-uniform light profiles on high concentration III–V solar cells using quasi-3D distributed models. 33rd IEEE Photovoltaic Specialists Conference, PVSC '08. pp. 1–6. 2008.

10. S. R. Kurtz and M. J. O'Neill. Estimating and controlling chromatic aberration losses for two-junction, two-terminal devices in refractive concentrator systems. Twenty Fifth IEEE Photovoltaic Specialists Conference, pp. 361–364. 1996.

11. V. D. Rumyantsev, N. Y. Davidyuk, E. A. Ionova, et al. Thermal regimes of Fresnel Lenses and cells in 'all-glass' HCPV modules. *AIP Conference Proceedings*, **1277**(1), 89–92. 2010.

12. T. Schult, M. Neubauer, Y. Bessler, P. Nitz, and A. Gombert. Temperature dependence of Fresnel lens for concentrating photovoltaics. Proceedings 2nd International Workshop on Concentrating Photovoltaic Optics and Power.

13. T. Hornung, A. Bachmaier, P. Nitz, and A. Gombert. Temperature dependent measurement and simulation of Fresnel lenses for concentrating photovoltaics. Proceedings 6th International Conference on Concentrating Photovoltaic Systems. *AIP Conference Proceedings*, **1277**(1), 85–88. 2010.

14. S. Askins, M. Victoria, R. Herrero, *et al.* Effects of temperature on hybrid lens performance. *AIP Conference Proceedings*, **1407**(1), pp. 57–60. 2011.

15. K. Araki, Y. Kemmoku, and M. Yamaguchi. A simple rating method for CPV modules and systems. 33rd IEEE Photovoltaic Specialists Conference, PVSC '08, pp. 1–6. 2008.

16. C. A. Gueymard. Spectral circumsolar radiation contribution to CPV. *AIP Conference Proceedings*. **1277**, 316–319. 2010.

17. I. Antón, M. Martínez, F. Rubio, *et al.* Power rating of CPV systems based on spectrally corrected DNI. *AIP Conference Proceedings*, **1477**(1), 331–335. 2012.

18. T. B. Young, A. Chan, N. J. Ekins-Daukes, *et al.* Atmospheric considerations when estimating the energy yield from III-V photovoltaic solar concentrator systems. 25th European Photovoltaic Solar Energy Conference and Exhibition/5th World Conference on Photovoltaic Energy Conversion, Valencia, Spain, pp. 106–109. 2010.

19. M. Muller, B. Marion, J. Rodriguez, and S. Kurtz. Minimizing variation in outdoor CPV power ratings. *AIP Conference Proceedings*, **1407**(1), 336–340. 2011.

20. International Electrotechnical Commission. IEC, 60904-9 Photovoltaic devices - Part 9: Solar simulator performance requirements. IEC, Geneva, Switzerland. 2007.

21. K. Emery. Measurement and characterization of solar cells and modules. *Handbook of Photovoltaic Science and Engineering*, (eds A. Luque and S. Hegedus), John Wiley & Sons Ltd, Chichester, England, pp. 797–840. 2011.

22. C. Domínguez, I. Antón, G. Sala. Solar simulator for concentrator photovoltaic systems. *Optics Express*, **16**(19), 14894–14901. 2008.

23. T. Gerstmaier, S. van Riesen, J. Schulz-Gericke, *et al.* Spectrally resolved DNI measurements: Results of a field comparison of spectroradiometers, component cells and the SOLIS satellite model. *AIP Conference Proceedings*, **1477**, 131–134. 2012.

24. J. Jaus, T. Missbach, S. P. Philipps, G. Siefer, A. W. Bett. Spectral measurements using component cells: examinations on measurement precision. 26th European Photovoltaic Solar Energy Conference and Exhibition, Hamburg, Germany, pp. 176–181. 2011.

25. C. Domínguez, S. Askins, I. Anton, and G. Sala. Characterization of five CPV module technologies with the Helios 3198 solar simulator. 34th IEEE Photovoltaic Specialists Conference (PVSC), **2009**, pp. 001004–001008. 2009.

26. C. Osterwald, M. Wanlass, T. Moriarty, M. Steiner, K. Emery. Effects of spectral error in efficiency measurements of GaInAs-based concentrator solar cells. National Renerwable Energy Laboratory, Technical Report TP-5200-60748. 2014.

27. P. Malbranche. The new European initiative on photovoltaic research infrastructures: an improved support to the scientific community, industry and markets. 26th European Photovoltaic Solar Energy Conference and Exhibition, Hamburg, Germany, pp. 4497–4499. 2011.

28. G. Peharz, L. Bugliaro, G. Siefer, and A. W. Bett. Evaluation of satellite cirrus data for performance models of CPV Modules. *AIP Conference Proceedings*, **1277**, 312–315. 2010.
29. S. Wilbert, R. Pitz-Paal, and J. Jaus. Circumsolar radiation and beam irradiance measurements for focusing collectors. ES1002, Cost Wire Workshop, Risö, Denmark. 2012.
30. A. Neumann, A. Witzke, S. A. Jones, and G. Schmitt. Representative terrestrial solar brightness profiles. *Journal of Solar Energy Engineering*, vol. **124**, no. 2, pp. 198–203. 2002.
31. M. Schubnell. Sunshape and its influence on the flux distribution in imaging solar concentrators. *Journal of Solar Energy Engineeering*, **114**(4), 260–266. 1992.
32. C. Domínguez and P. Besson. On the sensitivity of 4 different CPV module technologies to relevant ambient and operation conditions. *AIP Conference Proceedings*, **1616**, 308–312. 2014.
33. Y. Ota, H. Nagai, K. Araki, and K. Nishioka. Temperature distribution in 820X CPV module during outdoor operation. *AIP Conference Proceedings*, **1477**, 364–367. 2012.
34. E. Menard, M. Meitl, and S. Burroughs. Indirect temperature measurement of CPV solar cells using wavelength shift of the sub-cells luminescence emission peaks. Proceedings 27th Eur. Photovoltaic Solar Energy Conerence, pp. 189–193. 2012.
35. D. L. King, W. E. Boyson, and J. A. Kratochvil. Photovoltaic array performance model. Sandia National Laboratories, SAND2004-3535, Aug. 2004.
36. J. Jaus, R. Hue, M. Wiesenfahrt, G. Peharz, and A. W. Bett. Thermal management in a passively cooled concentrator photovoltaic module. Fraunhofer Institut für System-und Innovationsforschung, Karlsruhe, 2008.
37. F. Rubio, M. Martínez, J. Perea, D. Sanchez, and P. Banda. Comparison of the different CPV rating procedures: Real measurements in ISFOC, 34th IEEE Photovoltaic Specialists Conference (PVSC), 2009, pp. 000800–000805.
38. H. Helmers, M. Schachtner, and A. W. Bett. Influence of temperature and irradiance on triple-junction solar subcells. *Solar Energy Materials and Solar Cells*, **116**, 144–152. 2013.
39. E. Sánchez and G. L. Araujo. Correction of the temperature effect on the solar cell Isc–Voc characteristic. *Solid-State Electronics*, **25**(8), 817–819. 1982.
40. D. L. King. Photovoltaic module and array performance characterization methods for all system operating conditions. *AIP Conference Proceedings*, **394**, 347–368. 1997.
41. G. Peharz, J. P. Ferrer Rodríguez, G. Siefer, and A. W. Bett. Investigations on the temperature dependence of CPV modules equipped with triple junction solar cells. *Progress in Photovoltaics: Research and Applications*, **19**(1), 54–60. 2011.
42. S. Askins, C. Domínguez, I. Antón, and G. Sala. Indoor performance rating of CPV modules at multiple temperatures and irradiance levels. *AIP Conference Proceedings*, **1277**, 209–212. 2010.
43. M. Muller, C. Deline, B. Marion, S. Kurtz, and N. Bosco. Determining outdoor CPV cell temperature. *AIP Conference Proceedings*, **1407**, 331–335. 2011.
44. International Electrotechnical Commission. IEC, 60904-5 Photovoltaic devices - Part 5: Determination of the equivalent cell temperature (ECT) of photovoltaic (PV) devices by the open-circuit voltage method. IEC, Geneva, Switzerland. 2011.
45. G. H. Yordanov, O.-M. Midtgard, and T. O. Saetre. Equivalent cell temperature calculation for PV modules with variable ideality factors. 38th IEEE Photovoltaic Specialists Conference (PVSC) pp. 000505–000508. 2012.
46. I. Anton, C. Dominguez, M. Victoria, *et al*. Characterization Capabilities of solar simulators for concentrator photovoltaic modules. *Japanese Journal of Applied Physics*, **51**(10), p. 4.
47. G. Peharz, J. P. Ferrer Rodríguez, G. Siefer, and A. W. Bett. Investigations on the temperature dependence of CPV modules equipped with triple-junction solar cells. *Progress in Photovoltaics: Research and Applications*, **19**(1), 54–60. 2011.
48. T. Hornung, M. Steiner, and P. Nitz., Estimation of the influence of Fresnel lens temperature on energy generation of a concentrator photovoltaic system. *Solar Energy Materials and Solar Cells*, **99**, 333–338. 2012.
49. S. van Riesen, A. Gombert, E. Gerster, *et al*. Concentrix Solar's progress in developing highly efficient modules. *AIP Conference Proceedings*, **1407**(1), 235–238. 2011.
50. M. Victoria, S. Askins, R. Nuñez, *et al*. Tuning the current ratio of a CPV system to maximize the energy harvesting in a particular location. *American Institute of Physics Conference Proceedings*, **1556**, 156–161. 2013.
51. T. Missbach and J. Jaus. New sensor for measuring tracking accuracy, tracker vibration, and structural deflection, *American Institute of Physics Conference Proceedings*, **1477**, 262–266. 2012.
52. M. Davis, J. Lawler, J. Coyle, A. Reich, and T. Williams. Machine vision as a method for characterizing solar tracker performance. 33rd IEEE Photovoltaic Specialists Conference, PVSC '08, pp. 1–6. 2008.

53. S. Kurtz, M. Muller, B. Marion, *et al.* Considerations for how to rate CPV. *AIP Conference Proceedings*, **1407**, 25–29. 2011.

54. International Electrotechnical Commission. IEC, 62670-1 Photovoltaic concentrators (CPV) - Performance testing - Part 1: Standard conditions, 1st edn. IEC, Geneva, Switzerland. 2013.

55. American Society for Testing and Materials International. Tables for Reference Solar Spectral Irradiances: Direct Normal and Hemispherical on 37 Tilted Surface. ASTM International, 2012.

56. C. A. Gueymard, D. Myers, and K. Emery. Proposed reference irradiance spectra for solar energy systems testing. *Solar Energy*, **73**, 443–467. 2002.

57. P. Trujillo, J. P. Ferrer, M. Martínez, and F. Rubio. Influence of ambient conditions in the power rating of CPV modules. *AIP Conference Proceedings*, **1477**, 135–138. 2012.

58. P. Besson, C. Domínguez, and M. Baudrit. Contributions to reproducible CPV outdoor power ratings. AIP Conference Proceedings, Albuquerque, New Mexico. 2014.

59. R. N. Dows and E. J. Gough. PVUSA procurement, acceptance, and rating practices for photovoltaic power plants. Pacific Gas and Electric Co., San Ramon, Department of Research and Development; Bechtel Corp., San Francisco, CA, USA, DOE/AL/82993-21, Sep. 1995.

60. M. Muller, S. Kurtz, and J. Rodriguez. Procedural considerations for CPV outdoor power ratings per IEC 62670 *AIP Conference Proceedings*, 2013, vol. **1556**, pp. 125–128.

61. C. Domínguez, I. Antón, and G. Sala. Multijunction solar cell model for translating I–V characteristics as a function of irradiance, spectrum, and cell temperature. *Progress in Photovoltaics: Research and Applications*, **18**(4), 272–284. 2010.

62. K. Araki and M. Yamaguchi. Influences of spectrum change to 3-junction concentrator cells. *Solar Energy Materials and Solar Cells*, **75** (3–4), 707–714. 2003.

63. M. Muller, B. Marion, S. Kurtz, and J. Rodriguez. An Investigation into spectral parameters as they impact cpv module performance. *AIP Conference Proceedings*, **1277**(1), 307–311. 2010.

64. W. E. McMahon, K. E. Emery, D. J. Friedman, *et al.* Fill factor as a probe of current-matching for GaInP2/GaAs tandem cells in a concentrator system during outdoor operation. *Progress in Photovoltaics: Research and Applications*, **16**(3), 213–224. 2008.

65. G. Peharz, G. Siefer, and A. W. Bett. A simple method for quantifying spectral impacts on multi-junction solar cells. *Solar Energy*, **83**(9), 1588–1598. 2009.

66. G. Peharz, G. Siefer, K. Araki, and A. Bett. Spectrometric outdoor characterization of CPV modules using isotype monitor cells. 33rd IEEE Photovoltaic Specialists Conference, PVSC '08 pp. 1–5. 2008.

67. C. Domínguez, I. Antón, G. Sala, and S. Askins. Current-matching estimation for multijunction cells within a CPV module by means of component cells. *Progress in Photovoltaics: Research and Applications*, **21**(7), 1478–1488. 2012.

68. G. Siefer, M. Steiner, M. Baudrit, C. Dominguez, I. Antón, and F. Roca. Evaluation of the SOPHIA CPV Module Round Robin on Power Rating at CSOC. 11th International Conference on Concentrator Photovoltaics (CPV-11), Aix-les-Bains, France. 2015.

69. H. Müllejans, W. Zaaiman, and R. Galleano. Analysis and mitigation of measurement uncertainties in the traceability chain for the calibration of photovoltaic devices. *Measurement Science and Technology*, **20**(7), pp. 075101. 2009.

70. D. Dirnberger and U. Kraling. Uncertainty in PV module measurement—Part I: calibration of crystalline and thin-film modules. *IEEE Journal of Photovoltaics*, **6**(3), 1016–1026. 2013.

71. K. Emery, D. Myers, and S. Rummel. Solar simulation-problems and solutions. Conference Record of the Twentieth IEEE Photovoltaic Specialists Conference, vol. **2**, pp. 1087–1091. 1988.

72. American Society for Testing and Materials International. E927-10 Standard Specification for Solar Simulation for Terrestrial Photovoltaic Testing. ASTM International, West Conshohocken, PA, United States. 2010.

73. Japanese Standards Association. Japanese Industrial Standard JIS, C 8912:1998/Amendment 1:2005 Solar simulators for crystalline solar cells and modules (Amendment 1). 2005.

74. American Society for Testing and Materials International. G173-03 Standard Tables for Reference Solar Spectral Irradiances: Direct Normal and Hemispherical on 37° Tilted Surface. ASTM International, West Conshohocken, PA, United States. 2008.

75. K. A. Emery. Solar simulators and I-V measurement methods. *Solar Cells*, **18**, 251–260. 1986.

76. R. Adelhelm and K. Bücher. Performance and parameter analysis of tandem solar cells using measurements at multiple spectral conditions. *Solar Energy Materials and Solar Cells*, **50** (1–4), 185–195. 1998.

77. K. Emery. Measurement and characterization of solar cells and modules, in *Handbook of Photovoltaic Science and Engineering*, (eds A. Luque and S. Hegedus), John Wiley & Sons Ltd, Chichester, England, pp. 701–752. 2004.

78. J. Hohl-Ebinger and W. Warta. Uncertainty of the spectral mismatch correction factor in STC measurements on photovoltaic devices. *Progress in Photovoltaics: Research and Applications*, 19. 2011.

79. International Electrotechnical Commission, IEC 60891 Photovoltaic devices - Procedures for temperature and irradiance corrections to measured I-V characteristics. IEC, Geneva, Switzerland. 2009.

80. G. Sala. Cooling of solar cells. *Solar Cells and Optics for Photovoltaic Concentration*, (ed. A. Luque), Adam Hilger, Bristol, pp. 239–245. 1989.

81. H. A. Ossenbrink, W. Zaaiman, and J. Bishop. Do multi-flash solar simulators measure the wrong fill factor? Conference Record of the Twenty Third IEEE Photovoltaic Specialists Conference, pp. 1194–1196. 1993.

82. A. Parretta, L. Zampierolo, and D. Roncati. Theoretical aspects of light collection in solar concentrators. Imaging and applied Optics Congress, Optical Society of America Technical Digest (CD) paper STuE1. 2010.

83. A. Parretta, G. Martinelli, A. Antonini, D. Vincenzi, and C. Privato. Direct and inverse methods of characterization of solar concentrators. *Optics for Solar Energy*, p. STuA1. 2010.

84. E. Yablonovitch, O. D. Miller, and S. R. Kurtz. The opto-electronic physics that broke the efficiency limit in solar cells. 38th IEEE Photovoltaic Specialists Conference (PVSC) pp. 001556–001559. 2012.

85. R. Winston. Thermodynamically efficient solar concentrators. *Journal of Photonic Energy*, 2(1), 025501–1. 2012.

86. R. C. Jones. On reversibility and irreversibility in optics. *Journal of the Optical Society of America*, **43**(2), 138–143. 1953.

87. A. Timinger, A. Kribus, H. Ries, T. Smith, and M. Walther. Optical assessment of nonimaging concentrators. *Applied Optics*, **39**(31), 5679–5684. 2000.

88. A. Parretta, A. Antonini, G. Martinelli, and M. Armani. Optical characterization of CPC concentrator by an inverse illumination method. 22nd EPSEC, Milan, Italy. 2007.

89. V. D. Rumyantsev and M. Z. Shvarts. A luminescence method for testing normal operation of solar modules and batteries based on AlGaAs solar cells with radiation concentrators. *Geliotekhnika*, **28** p. 5. 1992.

90. V. M. Andreev, V. A. Grilikhes, and V. Rumiantsev. *Photovoltaic Conversion of Concentrated Sunlight*. John Wiley & Sons. 1997.

91. J. L. Álvarez, M. Hernández, P. Benítez, and J. C. Miñano. RXI concentrator for 1000 X photovoltaic energy conversion. SPIE'99 Conference Nonimaging Optics: Maximum Efficiency Light Transfer V, vol. **3781**, pp. 30–37. 1999.

92. R. Herrero, C. Domínguez, S. Askins, I. Antón, and G. Sala. Two-dimensional angular transmission characterization of CPV modules. *Optics Express*, **18**(S4), A499–A505. 2010.

93. V. A. Grilikhes, V. D. Rumyantsev, and M. Z. Shvarts. Indoor and outdoor testing of space concentrator AlGaAs/GaAs photovoltaic modules with Fresnel lenses. Twenty Fifth IEEE Photovoltaic Specialists Conference, pp. 345–348. 1996.

94. R. Herrero, C. Domínguez, S. Askins, I. Antón, and G. Sala. Luminescence inverse method For CPV optical characterization. *Optics Express*, **21**(S6), A1028–A1034. 2013.

95. I. Antón and G. Sala. Losses caused by dispersion of optical parameters and misalignments in PV concentrators. *Progress in Photovoltaics: Research and Applications*, **13**(4), 341–352. 2005.

13

Life Cycle Analysis of CPV Systems

Vasilis Fthenakis

Center for Life Cycle Analysis, Columbia University, United States; and Photovoltaics Environmental Research Center Brookhaven National Lab., United States

13.1 Introduction

The market for photovoltaics is growing fueled by drastic price reductions and concerns about climate change. Although most of the drastic price reductions have happened in c-Si and CdTe photovoltaic (PV) technologies, there is considerable interest in high efficiency solar cells to overcome area use constraints in specific applications. Particularly attractive are solar cells based on III-V semiconductors such as GaAs and GaInP/GaAs, which are capable of utilizing a wider spectrum of sunlight than other solar cells. Due to their high cost, use of III-V solar cells has been initially limited to space applications. Terrestrial applications emerged as a viable option with advances in sun-tracking concentrating photovoltaic (CPV) technologies that focus sunlight onto these cells, reducing the required cell area by two to three orders of magnitude. Choices are available on concentration ratio and design, optics, cell types, cooling, and tracking, and these are described in depth in other chapters of this volume. The current chapter focuses on the life cycle assessment of the most common design of CPV that is *point focus high concentration* (HCPV) systems, employing Fresnel lenses and large modules on two-axis trackers.

Life cycle assessment (LCA) is an analytical methodology used to measure material- and energy-inputs, transformations, and outputs throughout the entire life-cycle of a product or process, thus from cradle to grave, or in the case that materials are recycled from cradle to cradle. The energy and material inputs used in LCA are most commonly derived from process data from each of the processes in the various life-stages. These include extraction and processing of materials, manufacturing of solar cells and the other components of a

Handbook of Concentrator Photovoltaic Technology, First Edition. Edited by Carlos Algora and Ignacio Rey-Stolle.
© 2016 John Wiley & Sons, Ltd. Published 2016 by John Wiley & Sons, Ltd.

photovoltaics system, construction, operation, maintenance and eventual decommissioning and disposal and/or recycling. These data are typically grouped in *life cycle inventory* (LCI) databases. The most common LCA metrics of the environmental footprint of a product or process are *energy payback times* (EPBT), *energy return on investment* (EROI), and greenhouse and toxic gas emissions. Precise definitions of these concepts, which are almost self-evident, will be given in section 13.6 and 13.7. Also of interest to energy technologies are land use and water use per unit of energy output. These are described in the Methodology section below. The chapters that follow present as a case study a detailed LCA of an Amonix CPV system installed in the southwest of the United States, with comparisons of the estimated environmental footprint of this system with other reported in the literature.

13.2 Case Study Description

In the following we describe the methodology and results for applying LCA to the most common design of CPV that is point focus high concentrator employing III-V cells, Fresnel lenses, assembled into modules, and mounted on two-axis trackers (see Figure 13.1). The methodology is applied to the Amonix 7700 CPV system for which a complete LCI is available.

The Amonix 7700 CPV system consists of seven concentrating module units called *MegaModules* mounted on a two-axis tracker (see Figure 13.1). Sunlight is concentrated on 7560 solar cells at a concentration ratio of 500 suns. This system uses multijunction GaInP/GaInAs/Ge cells grown on germanium substrates, rated at 37% under the test condition of $50.0 \, W/cm^2$ ($500 \, kW/m^2$), 25°C, and AM 1.5d spectrum. The total area of the system is $328 \, m^2$; its aperture area is $267 \, m^2$. The rated power of the Amonix 7700 CPV system when it was

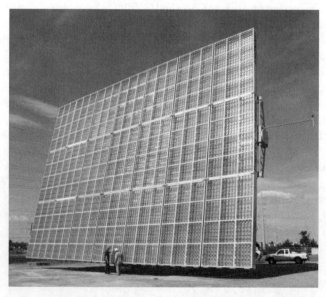

Figure 13.1 Amonix 7700 System. Source: Amonix Inc, 2012. Reproduced with permission of Amonix Inc

introduced was $53\,kW_p$ at standard test conditions. Amonix reported that with efficiency increases in the cells, lenses, heat sink, interconnect and inverter, the more advanced version of the same unit has produced up to 63 kW under the same test conditions. It is now assumed that the system has an average rated capacity of $60\,kW_p$.

13.3 Methodology

The most common and widely tested LCA approach uses process-based data, giving detailed energy and materials inputs in each of the many process in the life-cycle of a photovoltaic. For example, the manufacturing of solar cells entails multiple material deposition processes alternated with scribing, etching, and cleaning processes This is delineated by ISO 14040 and International Energy Agency PVPS Guidelines [1,2]. Sometimes cost data are used as indicators of material and energy intensities but this approach, called economic input/output (EI/O) is not well developed for use in PV life cycles. Sets of process-based LCI data for commercial photovoltaics can be found elsewhere [3]. Constructing a LCI starts by determining the input- and output- physical flows for each stage from the inventory of materials, fuel- and electricity-usage, and operational data from Amonix Corporation. The spatial system boundary for this LCA study is the United States for which the major energy- and emissions-flows are evaluated. The results were extrapolated to the south European condition in section 13.9.1 Comparison with other CPV Systems.

The life cycle of the CPV system starts by acquiring the materials, and encompasses the production of materials, the manufacturing of components, their assembly/installation, operation/maintenance, decommissioning at the end of its useful life, and disposal or recycling (Figure 13.2). The upstream inventory data, (i.e., energy and materials inputs and outputs during the materials acquisition and production stages), were derived from commercial databases including those of Franklin [4] and Ecoinvent [5], along with open-data sources from the National Renewable Energy Laboratory (NREL, the Aluminum Industry Association, and Brookhaven National Laboratory [6]). The mass inventory data were provided by Amonix and Spectrolab, the producer of the III-V multijunction solar cells used in the system; these are representative data for point focus high concentration systems, employing Fresnel lenses and large modules on two-axis trackers.

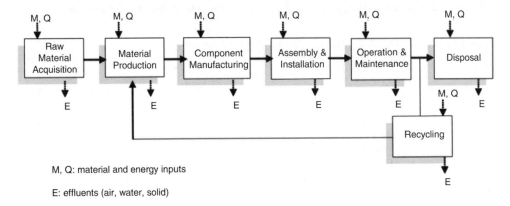

M, Q: material and energy inputs

E: effluents (air, water, solid)

Figure 13.2 Life-cycle stages of CPV

The recycling and disposal stages were modeled by scenario analyses on the basis of industrial practices (e.g., recycling automobiles). Energy- and materials-inputs for transportation, (for example, delivering products and removing solid waste between life-cycle stages), were estimated based on the distance between Amonix Inc. manufacturing site in Torrance CA and the installation sites in Phoenix AZ, assuming equal volume transportation by trucks and by track-trailers, as reported by Amonix.

13.4 Life-Cycle Inventory Analysis

13.4.1 Production of Materials and Associated Emissions

Table 13.1 shows down the materials and components that comprise the Amonix 7700 system. The *MegaModules* (35%) and tracker (63%) account for the majority of the mass of the components; they comprise mainly steel (72%) and aluminum (10%), with concrete (12%) used in their foundation, and a minor share of other materials (6%). Fuel- and electricity-usages for installing the unit and operating the tracker were measured directly in the field. The energy usages for building and operating a trailer office for monitoring and maintaining the system are estimated from data associated with the Tucson Electric Power PV plant in Springerville AZ, for which we have detailed data [7]. Additional materials (not used in Springerville), include water used for cleaning of the CPV system, assuming a frequency of cleaning four times per year with 2000 liters per wash, resulting in 8000 l/year. The flow of materials and emissions are shown in Figure 13.3.

The end-of-life stage of the Amonix 7700 CPV components consists of dismantling, transporting, and shredding. Dismantling operations are documented in section 13.4.4. As decommissioning has not yet occurred, the impacts were conservatively determined based on dismantling the associated components with the same amount of energy use as the installation stage, carrying those components a distance of 100 km by heavy truck, shredding and separating them.

The recycling and disposal stages were modeled by scenario analyses on the basis of industry practice, e.g. automobile recycling. The metals are recycled at the end-of-life stage, but we do not consider here the processes that occur after shredding; we assumed an open-loop recycling scheme for this analysis.

We used the LCA software 'Simapro' [8] to compile materials and fuel inputs, and to calculate the associated energy and emissions from the whole Amonix 7700 system.

Table 13.1 Material breakdown of the Amonix 7700 CPV system

Element	Components	Materials	Mass (kg)	Fraction (%)
MegaModules	Frame	Steel	1780	6.6
	Fresnel lenses	Acrylic	1261	4.6
	Heat sink	Aluminum	2557	9.4
	Inside structure	Steel	3866	14.3
Tracker	Foundation	Concrete	3126	11.5
	Hydraulic drive	Steel	2724	10.0
	Pedestal and torque tube	Steel	11 260	41.6
Power components	Cells/cables/inverter/Transformer/ controller	Several	555	2.0

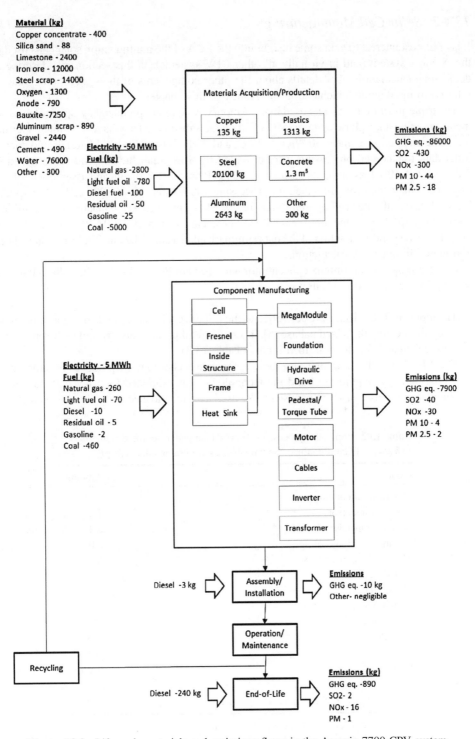

Figure 13.3 Life-cycle materials and emissions flows in the Amonix 7700 CPV system

13.4.2 Solar Cell Manufacturing

It is of utmost interest to gain some insight into the LCA of the multijunction solar cells used in the Amonix system (and in virtually all other CPV systems). In this section, we describe in detail such calculations. For details about the other components of the system, the reader is referred to the diagrams included in the Annex of this chapter.

The triple-junction III-V cells used in the Amonix systems are produced by Spectrolab using metal-organic vapor-phase epitaxy (MOVPE) on Ge substrates. They have a nominal aperture area efficiency of 37% under 50 W/cm^2, 25°C, and AM1.5D conditions. Materials and energy input data for the cell processing were provided by Spectrolab Inc [9]. and summarized in Table 13.2. A 10% loss of inputs during the solar cell production/dicing and assembly has been assumed. The inventories are scaled for processing 1000 wafers into 50 000 cells, corresponding to 4.95 m^2 of solar cell aperture area and 915 kW$_p$ of rated DC capacity. As mentioned above, only 7560 multijunction solar cells are needed in an Amonix 7700 CPV system and thus only the corresponding fraction of the data summarized in Table 13.2 and Table 13.3 have been considered in our LCA calculations.

In order to model the Amonix concentrator with the GaInP/GaInAs/Ge solar cells, additional assumptions were made as follows.

- The inputs in Table 13.2 were converted to the full life cycle energy and emissions inventory using the Ecoinvent LCA database [5]. Then, the total primary energy and GHG emissions were calculated for the described inputs (Table 13.3).
- The EPBT, i.e. the time in years for a renewable technology to produce the amount of primary energy required to build the technology, was estimated for III-V cells and for a complete Amonix concentrator system. We used a primary to electricity conversion factor of

Table 13.2 Inputs for processing 1000 4-inch wafers, corresponding to 915 kW$_p$[a] of GaInP/GaInAs/Ge terrestrial concentrator solar cells [9]

Inputs	Amount
Materials for components	(kg)
Wafer/precursors	21.7
Contact metals	3.4
Anti-reflection coating	0.02
Materials use for process	(kg)
Hydrogen	57.2
Nitrogen	0.2
Photoresist	2.3
Solvents	1073.4
Acids	255.8
Bases	161.3
Electricity	(kWh)
MOVPE	2365
Other[b] (gas scrubbing and cell processing)	470

[a] 37% rated solar cell efficiency at 500 suns.
[b] estimated from Mohr *et al.*, (2007) [10].

Table 13.3 The environmental impacts from 1000 4-inch wafers,, corresponding to 915 kW$_p$[a] of GaInP/GaInAs/Ge terrestrial concentrator solar cells

Inputs	Primary Energy (MJ)	%	GHG (kg)	%
Materials for components				
Wafer/precursors	2787	13.3	130.4	14.6
Contact metals	323	1.6	26	2.9
Anti-reflection coating	17	0.1	1	0.1
Materials used for process				
Hydrogen	556	2.7	13	1.5
Nitrogen	0.3	0.0	0.01	0.0
Photoresist	21	0.1	1	0.1
Solvents	10 035	49.3	331	37.4
Acids	648	3.2	32	3.6
Bases	949	4.7	47	5.4
Electricity				
MOVPE	4209	20.7	253	28.6
Other (gas scrubbing and cell processing)	842	4.1	51	5.7
Total	*20 368*	*100*	*885*	*100*

[a] 37% rated solar cell efficiency at 500 suns.

0.29 corresponding to the average US grid mix. The GHG emissions were calculated with a 100-year timeframe, taking into account CO_2, CH_4, N_2O, and chlorofluorocarbons employing the global warming factors defined by the IPCC [11]. The results are presented in Table 13.3.

As shown in Table 13.2, in the MOVPE-based cell production at Spectrolab, the use of solvents accounts for the greatest impacts followed by the electricity consumptions. The total impacts were seven and nine times higher than the mono-crystalline Si cell, in terms of primary energy and GHG emissions, respectively [5,6]. However, compared with the total impacts from the Amonix CPV system, they are relatively small, i.e. 2.4% and 1.6% of the total impacts correspondingly (Table 13.4). This reduced impact is the result of the high concentration ratio (500×) and thus the small area of the solar cells used in the system.

Table 13.4 EPBT and GHG emissions for the GaInP/GaInAs/Ge solar cell and the total Amonix CPV under the DNI of 2500 kWh/m^2/year

Metrics	Multijunction solar cells	Total Amonix system life-cycle	Fraction of cells (%)
Primary energy (GJ)	20.4	835	2.4
EPBT[a] (years)	0.02	0.9	2.5
GHG[a] (g CO$_2$-eq/ kWh)	0.4	25.6	1.6

[a] These are hypothetical values assuming that the cells are part of a system that generates AC power.

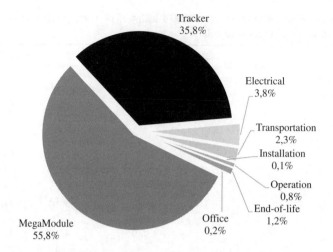

Figure 13.4 Breakdown of the life cycle primary-energy demand

13.4.3 Primary Energy Demand

Figure 13.4 and Table 13.5 present the breakdown of the primary-energy demand during the life cycle of the 7700 Amonix CPV system. The component-manufacturing stage dominates this demand, accounting for 95.3% of the primary energy. Energy flows are shown in Figure 13.5 and additional details are shown in the diagrams included in the Annex of this chapter.

Table 13.5 Breakdown of primary energy from cradle-to-grave of an Amonix 7700 system

Stage	Energy (MJ)	%	GHG (kg CO$_2$-eq.)	%
Parts	1 470 633	88.3	102 108	92.4
Cells	14 562	0.9	615	0.6
Foundation	2341	0.1	430	0.4
Frame	234 218	14.1	16 836	15.2
Fresnel lenses	171 974	10.3	9086	8.2
Heat sink	440,089	26.4	31 194	28.2
Tracker (pedestal and tube)	427 106	25.7	31 420	28.4
Inverter	33 395	2.0	2130	1.9
Transformer	11 973	0.7	566	0.5
Hydraulic drive	117 972	7.1	8912	8.1
Motor	2 056	0.1	113	0.1
Cables	5 278	0.3	265	0.2
Controller	8 907	0.5	498	0.5
Anemometer and sensor	762	0.05	43	0.04
Assembly/Installation	162	0.01	12	0.01
Operation/Maintenance	111 830	6.7	2463	2.2
Transportation	61 364	3.7	4480	4.1
End-of-Life	20 745	1.2	1512	1.4
Total	*1 664 733*	*100*	*110 575*	*100*

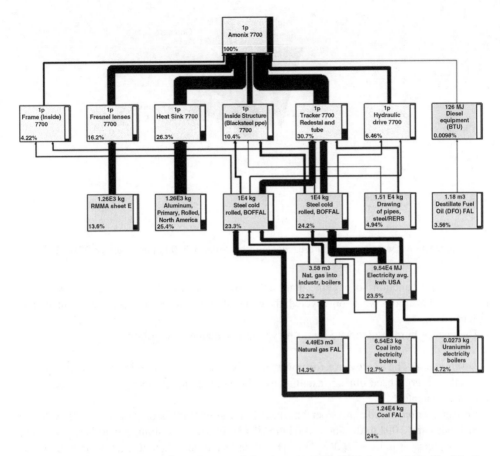

Figure 13.5 Life-cycle energy flow diagram of the Amonix 7700 CPV System (cut-off = 5%; thus energy contributions smaller than 5% of the total are not shown in this figure). Detailed diagrams for each system component are included in Annex 13-I

13.4.4 End-of-Life Processing

It is assumed that the metal parts of the Amonix 7700 will be recycled. Steel and aluminum parts which constitute 81% of the weight of an Amonix 7700 system (see breakdown in Figure 13.6), are easily recyclable. The balance of the materials, i.e. concrete, acrylic, and other substances, needs additional processes like sorting and separation before entering a recycling circuit. We describe the prospective recycling stage of an Amonix 7700 unit based on the automobile recycling practices. After decommissioning, some parts will be separated for reuse (e.g., sensors) and others will be transported and shredded. If the foundation is removed, then would undertake crushing and screening to be used as aggregates for new concrete production. After shredding, ferrous materials will be sent to steel smelters while the rest will be further divided into non-ferrous metals and non-metallic materials. The non-ferrous metals will be individually extracted depending on the market value. The final residue, including plastics, glass, and so on is considered non-recoverable waste material and is sent to landfills for

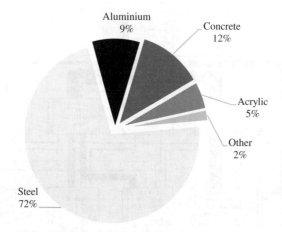

Figure 13.6 Materials weight breakdown of an Amonix 7700 system (total = 27100 kg)

disposal. Multijunction solar cells, made from III-V compound semiconductors, would be disposed as hazardous waste or recycled; the end-of-life fate of the cells is not accounted in the current LCA.

The energy burden associated with recycling is calculated as follows:

- *Dismantling*: The energy use for power tools is estimated to be ~35 kWh (assumed to be equal to energy used during installation). This corresponds to using 2.7 kg of diesel in a diesel generator.
- *Transportation*: 3000 tonnes per km, based on the total mass of Amonix 7700 system of 30 tonnes and 100 km distance. 100 km × 30 t = 3000 tkm. Assuming that the use of diesel use in heavy trucks is 0.068 l/tkm (Franklin database), then the transportation of the components of a decommissioned CPV system will require 204 l (174 kg) of diesel.
- *Shredding*: 0.34 MJ/kg × 30 tonnes = 10 200 MJ of diesel = 222 kg of diesel.

Thus, a total of 396 kg (464 l) of diesel fuel would be required to the power tools and the trucks used in decommissioning and transporting the CPV metal parts to recycling warehouses 100 km away.

13.5 System Performance Data and Estimates

The electricity generated by an Amonix 7700 CPV system is approximated from measurements of the performance of a smaller system, the Amonix 7500, operated in Las Vegas over a period of six months (January 31 to July 31, 2009) as data for the 7700 were not available. Figure 13.7, provided by Amonix, displays the electricity generated per day as a function of the direct normal solar radiation for the Amonix 7500 CPV. From this data the slope is determined using a least-squares-fit and has units of kWh/(kWh/m^2) that can be used for estimating energy production for periods longer than six months.

This slope serves as an energy performance rating for the Amonix 7500. The calculation for annual energy generation of the 7500 is the product of the annual average direct normal radiation for a site (7.14 kWh/m^2 per day for Las Vegas). The 7700 has seven *MegaModules*

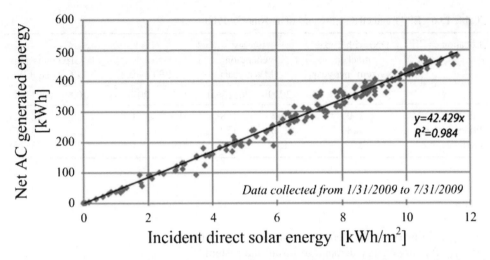

Figure 13.7 AC electricity Production of an Amonix 7500 CPV system in Las Vegas as a function of direct normal irradiation on the system, during six months (January–July 2009). Source: Amonix Inc, 2012. Reproduced with permission of Amonix Inc

while the 7500 has five, since the Amonix 7700 would produce 7/5 times more electricity than the 7500. Using the annual average direct solar radiation in Las Vegas with the analysis above, leads to the ideal annual energy generation by Amonix 7700 in this location:

$$Y = 42.429[\text{kWh}/(\text{kWh}/\text{m}^2)] \times 365[\text{days}/\text{year}]$$
$$\times 7.14[(\text{kWh}/\text{m}^2)/\text{day}] \times 7/5 = 154,803.9[\text{kWh}/\text{year}]$$
(13.1)

Amonix hired a third-party engineering firm, Black and Veatch, to conduct an analysis of field performance losses of 7700 units operating in three locations. This analysis identified the following field performance losses: Additional soiling (2%), AC wiring and transformer losses (2%), availability for maintenance (1%), wind stow (0.5%), shading (0.5%) and elevation angle limit losses (0.8%). Therefore the ideal energy generation of a single unit is reduced by 6.8% for fields of 7700 Amonix CPV systems, and thus the expected energy generation in the Las Vegas area would be:

$$Y = 154,803.9 \times 0.932 = 144,277[\text{kWh}/\text{year}]$$
(13.2)

This includes 5 kWh/day of parasitic energy consumed in operating the system (mainly associated to the tracker motor).

13.6 Energy Payback Time

The energy payback time (EPBT) is defined as the period required for a PV system to generate the same amount of energy as that used in the PV system from cradle to grave. For a PV system it is quantified as follows:

$$EPBT = (E_{comp} + E_{inst} + E_{EOL})/(E_{a.gen} - E_{a.oper})$$
(13.3)

Table 13.6 EPBT and GHG emissions of an Amonix 7700

Location	DNI with 2-axis tracker (kWh/m^2 per year)	Energy generation (MWh/year)		EPBT (years)		GHG emissions (g CO_2-eq./kWh)	
		2009	2011(est)	2009	2011(est)	2009	2011(est)
Las Vegas, NV	2600	144	168	0.9	0.8	26	22 (16)
Phoenix, AZ	2480	136	159	0.9	0.8	27	23 (16)
Glendale, AZ	2570	139	163	0.9	0.8	27	23 (16)

where

- E_{comp}: Primary-energy demand to produce the components
- E_{inst}: Primary-energy demand to install the system
- E_{EOL}: Primary-energy demand for end-of-life management
- $E_{a.gen}$: Annual electricity-generation (primary-energy equivalent)
- $E_{a.oper.}$: Annual energy-demand for operation and maintenance (primary-energy equivalent).

Calculating the primary-energy equivalent requires country-specific, energy conversion parameters for the fuels and technologies used to produce energy and feedstock. The annual electricity generation ($E_{a.gen}$) is represented as primary energy, based on the efficiency of electricity conversion at the demand side. The United States average energy-mixture efficiency of 0.29 was adopted in converting the electricity generated into primary energy.

We estimated the electricity generation of the Amonix 7700 CPV from Eq. (13.2) at the locations where the units have been operating; the energy payback times for those sites are listed in Table 13.6.

13.7 Greenhouse and Toxic Gas Emissions

13.7.1 Emissions in the Life-Cycle of Amonix 7700

The greenhouse gas (GHG) emissions during the life cycle stages of the 7700 Amonix CPV unit are estimated as an equivalent of CO_2 using an integrated time-horizon of 100 years (GWP_{100}). Emissions considered for calculating GHG emissions include CO_2, CH_4, N_2O, and chlorofluorocarbons. Unlike fixed, standard PV configurations in which the emissions mostly are linked to producing solar cells, the tracking and concentrating equipment contributes the majority of the GHG emissions from this system.

After normalizing for the electricity generated, this system generates 26–27 g CO_2-eq./kWh during its 30-year life-cycle (Table 13.6) in its current operating locations. Extending the system's life to 50 years would reduce life-cycle GHG emissions to approximately 16 g CO_2-eq./kWh (Table 13.6, numbers in parenthesis).

Fifty years might look as a surprising duration but the fact of the matter is that the life expectancy of the CPV system could be extended to 50 years by replacing every 25 years the

Table 13.7   GHG, particle and toxic gas emissions of an Amonix 7700 system

Elements	CO_2 eq (kg)	PM_{10}^a (kg)	$PM_{2.5}^a$ (kg)	SO_2 (kg)	NO_x (kg)	Hg (kg)
III-V Multijunction Cells	659	0.1	0.3	2.5	2.3	1.2E-04
Foundation	801	8.1	0.9	0.5	1.8	3.2E-05
Frame (Inside)	4303	0.8	N/A	25.4	14.0	4.2E-05
Fresnel Lenses	10 869	0.7	0.4	44.5	38.5	2.1E-04
Heat Sink	25 883	26.4	9.5	67.8	97.2	9.8E-04
Inside Structure (Black steel pipe)	10 581	1.7	0.4	58.1	32.5	1.5E-04
Tracker (2 axis)	31 057	5.1	1.8	170.0	95.4	4.9E-04
Inverter (30 kW)	2064	0.2	3.1	29.0	11.5	7.3E-05
Transformer	517	0.1	1.4	12.0	4.4	1.9E-05
Hydraulic actuator	6585	1.2	N/A	38.9	21.4	6.4E-05
Motor	101	0.0	0.5	4.1	1.1	5.4E-06
Cables	227	3.1	1.6	12.3	3.2	1.6E-05
Controller	170	0.4	0.1	1.0	0.6	2.3E-05
Anemometer and Sensor	15	0.0	0.0	0.1	0.0	2.0E-06
Welding, arc, steel	3.4E-01	1.7E-03	8.6E-04	1.1E-03	1.2E-03	1.7E-07
Diesel equipment (kcal)	2.5	0.002	N/A	0.001	0.05	2.0E-09
Total	*93 842*	*47.8*	*20.1*	*466.2*	*324.3*	*2.2E-03*

^{a} PM_{10} and PM_{10} are two indicators of particle pollution. PM_{10} corresponds to particle sizes of up to 10 microns, whilst $PM_{2.5}$ corresponds to particle sizes of up to 2.5 microns.

III-V solar cells and the Fresnel lenses in the field, assuming that the metal structure would last those 50 years. Replacing the cells would not add significantly to the EPBT and GHG burden because the cells account for only about 1% of the total energy burden. However, for a detailed analysis of this scenario, replacing the Fresnel lenses, potential losses from breakage of cells in the field and increased maintenance and labor requirements associated with a longer than 30 years useful life, would have to be taken into account.

The emissions of GHG, NO_x, SO_x and Hg are mainly associated with burning fossil fuel; these are shown in Table 13.7.

13.7.2   Reduction of Emissions from PV Replacing Electricity from the Grid

Photovoltaic electricity generation technologies are the most sustainable and efficient pathways for reducing emissions of CO_2. The whole 'cradle-to-grave' life-cycle of the Amonix 7700 CPV system, that includes the extraction of raw materials from the ground, their processing and purification, production of solar cells, modules and balance-of-system (BOS) components, installation, operation and end-of-life processing, result in a mere 16 g CO_2-eq. emissions per kWh of electricity produced over an assumed life-expectancy of 50 years. These emissions are higher than those from CdTe PV but lower than those from 2010 production crystalline-Si PV life cycles. In any case, all PV technologies have GHG emissions that are 30 to 50 times lower than those of fossil-fuel based generation and ~50% lower than those from the nuclear life

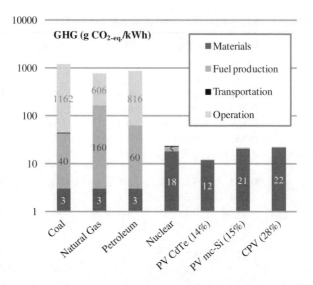

Figure 13.8 Greenhouse emissions in the life-cycle of various power plants operating in the U.S. Southwest. Sources: coal, natural gas and petroleum (Kim and Dale, 2005 [15]); nuclear (Fthenakis and Kim, 2007 [16]); PV and CPV (Fthenakis, adjusted for 2014 efficiencies and SW US irradiation). Adapted with permission

cycle in the United States (Figure 13.8). The Amonix 7700 system can reduce emissions from the average US grid by $(754-16) = 738$ g/kWh, accounting for approximately a 97.8% emissions reduction. As shown in Table 13.8, substituting conventional grid electricity with solar electricity will also result in drastic reductions in the emissions of criteria pollutants and mercury. It is noted that these estimates do not account for price-induced fuel substitutions and for the impact of a large penetration of renewable energy into the life-cycle emissions of PV. Detailed analysis of the GHG and criteria emissions from a 'reformed' grid requires energy-environmental-economic models like MARKAL and NEMS [12] and full cost accounting should considerate the external costs of electricity [13]. An analysis of mercury emissions and associated emissions is given elsewhere [14].

Table 13.8 Avoided emissions from wide-scale deployment of the Amonix CPV Systems

	Generation[a] (MWh/year)	Emissions (g/kWh)					
		PM_{10}	$PM_{2.5}$	SO_2	NO_x	Hg	GHG[b]
US average grid mix		0.078	0.053	3.94	1.43	1.6E-05	753.6
Amonix – Phoenix, AZ	159	0.010	0.0043	0.10	0.069	4.7E-07	20.9
Amonix - Las Vegas, NV	168	0.010	0.0040	0.09	0.065	4.5E-07	19.8
Amonix – Glendale, AZ	163	0.010	0.0042	0.10	0.067	4.6E-07	20.4
Emissions reduction (%)		87–88%	~92%	97–98%	~95%	~97%	~97%

[a] Expected –near-term
[b] CO_2 eq

13.8 Land and Water Use in CPV Systems

We assessed the life cycle usages of land and water for an Amonix 7700 unit and other power systems, considering the entire life cycle stages, i.e., extracting resources, generating the electricity, disposing of the waste, and both direct- and indirect- transformations. Detailed descriptions of resource-use indicators associated with electricity generation technologies are given in previous studies [17,18]. The land requirements are diverse within a technology, depending on regional conditions and the technologies adopted. The photovoltaic-fuel cycle transforms the least amount of land per GWh of electricity generated among the renewable technologies we assessed. Notably, over a 30-year timeframe, the land areas transformed by PV life cycles in US-SW operations, are comparable to those of the natural-gas fuel cycle and less than most coal-fuel cycles coupled with surface mining.

For our study, the land transformation (m^2) and water withdrawal (m^3) indicators were estimated for both upstream and on-site usages. For upstream usages, we took information from the Ecoinvent database [5]; for on-site data, we used actual field data. The currently operating Amonix 7700 units occupy $16\,000–24\,000\,m^2$ per MW of AC power, which translates into $266\,m^2$ of direct land-transformation per GWh of electricity in Phoenix. The indirect land transformation that is linked to the energy- and materials-inputs to build an Amonix 7700 unit corresponds to only $32\,m^2/GWh$.

In contrast, for water usages the indirect component dominates. Direct water withdrawal for cleaning an Amonix 7700 unit was reported to be only 26 l/MWh under the solar radiation of Phoenix, while, correspondingly, indirect water withdrawal for materials and energy inputs was estimated to be 682 l/MWh. We compare in Figure 13.9 and Figure 13.10 these usages of resource across PV technologies under the solar radiation of Phoenix, i.e., $2370\,kWh/m^2/year$ for optimal tilt flat PV and $2480\,kWh/m^2/year$ of DNI for collectors with 2-axis tracker. We assumed that flat PV is installed with the BOS of Tucson Electric Power Springerville power plant [7]. Accordingly, the land use for an Amonix 7700 system is about the same as that of multi-Si PV and CdTe PV (Figure 13.9).

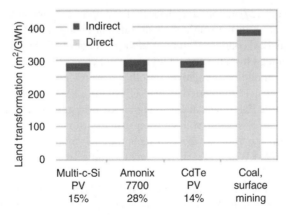

Figure 13.9 Comparison of land transformation. PV technologies are assessed under the environmental conditions of Phoenix, AZ, while the coal case refers to surface mining in the Eastern US (case 1, Table 1 in Ref. [17])

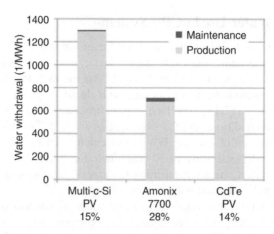

Figure 13.10 Comparison of water withdrawal across PV technologies under the environmental conditions of Phoenix, AZ

As shown in Figure 13.10, the water use in the life-cycle of the Amonix 7700 system is lower than that of multi- and mono-PVs, and comparable to that of CdTe PV. This is the result of relatively high water use in Si wafer manufacturing. We point out that estimates of land and water usage carry a larger uncertainty than those for emissions and energy payback since the availability and quality of data is poorer for the former.

13.9   Discussion and Comparison with Other CPV and PV Systems

13.9.1   Comparison with Other CPV Systems

There are many LCA studies on flat panel PV systems in the literature, but most relate to c-Si PV and CdTe PV technologies [19–26]. Our literature review shows only a few CPV LCA studies and these are summarized below

- Peharz and Dimroth [27] reported an EPBT of 0.7–0.8 years for a prototype FLATCON CPV system operating in the south of Spain (annual energy in DNI = 2093 kWh/m^2).
- Reich-Weiser et al. [28] conducted an LCA of the SolFocus CPV based mainly on EI/O LCI data; this was excluded from the current review as not compatible with the standard process-based LCI data methodology employed in other peer-reviewed LCA studies.
- De Wild-Schotten et al. [29] reported on the LCA of the CPV technologies developed by the Apollon collaborative EU project, namely a point focus system from SolarTec using Fresnel lenses and a dense array system from CPower using mirrors. They estimated EPBTs of 1.5 and 1.9 years corresponding to the two technologies under Sicilian irradiation (annual energy in DNI = 1794 kWh/m^2-year). The corresponding green-house gas (GHG) emissions were 37 g CO$_2$-eq./kWh.
- Fthenakis and Kim [30–32] estimated the EPBT and GHG emissions of an Amonix 7700 system operating in Phoenix, AZ (annual energy in DNI = 2480 kWh/m^2-year) to be 0.9 years and 27 g CO$_2$-eq//kWh correspondingly.
- Nishimura et al. [33] estimated the EPBT of a Sharp CPV assuming operation in the Gobi desert (annual energy in DNI = 1513 kWh/m^2-year) to be 2.0 years.

The performance of the four systems was normalized for the solar irradiation of Catania, Sicily and Las Vegas, Nevada and comparisons of their EPBT and carbon footprints are shown in Table 13.9 below.

The systems for which LCA was performed had DC efficiencies in the 22–24% range. More recent results on CPV efficiency from Semprius 3.5 kW being tested at Sandia (Albuquerque, New Mexico) show efficiencies of 32.6% [34]. These modules employ III-V cells and an optical concentration ratio of 1111 suns. Higher efficiencies are presumably the result of reducing optical and thermal losses and to the degree that the LCI do not change, they would result to lower EPBT and GHG emissions than the ones reported herein.

13.9.2 Comparison with Other PV Systems

Fthenakis and Alsema (2006) [35] and Fthenakis *et al.* (2008) [22] reported the EPBT and GHG emissions of crystalline Si and CdTe PV technologies. Fthenakis *et al.* (2009) [24] reported the environmental footprint metrics of thin-film CdTe PV. These values were updated for average 2014 efficiencies, assuming that the LCI have not changed, and listed in Table 13.10. EPBT and GHG emissions of 14% to 15% efficient multi-crystalline silicon, ground-mount PV modules produced in the United States, would correspond to 1.1 years and 19 g CO_2-eq./kWh, respectively, when operating in south-facing latitude tilt under the optimal angle insolation of Phoenix (2370 kWh/m^2 per year). In the same location, the EPBT and GHG emissions of a CdTe PV system would be 0.5 years and 10 g CO_2-eq./kWh respectively. The Amonix 7700 CPV system has an advantage over the considered multi-crystalline silicon solar cell in terms of EPBT. Also, if the deployment life of the 7700 CPV is extended to 50 years, then its life-cycle GHG emissions could be reduced to 16 g CO_2-eq./kWh (Table 13.6), lower than those of the considered multi-crystalline silicon life cycles.

Table 13.9 Harmonized life cycle assessment results. Assumed Operating Locations: Catania, Sicily, Italy (annual energy in DNI = 1794 kWh/m^2); Las Vegas, Nevada, Unite States (annual energy in DNI = 2600 kWh/m^2)

Technology	Cell type	Module actual efficiency (%)	Module Area (m^2)	EPBT (years)		Carbon footprint (g CO_2-eq./kWh)	
				Catania	Las Vegas	Catania	Las Vegas
Amonix 7700	III-V	28.3[a]	328	1.5	1.0	32	22
FLATCON	III-V	24.1	34	0.8	0.6	18	12
CPower	Mono-c-Si	21.8	9	1.5	1.0	37	25
SolarTec	Mono-c-Si & III-V	25.3	9	1.9	1.3	45	31
Sharp	III-V	18.8	11	1.7	1.2		

[a] Amonix reported a rated power of 60 kW AC under DNI of 850 W/m^2. This corresponds to an efficiency of 21.5% over the total tracker area of 328 m^2. Accounting for performance losses of 6.8% (described in 13.5) brings the efficiency of the MegaModule to 28.3%. It is noted that the efficiency of the FLATCON system assumes a DNI of 1000 W/m^2; this information is not available for the other systems listed in the table.

Table 13.10   Comparison of life cycle parameters and performances across PV Technologies (normalized for Phoenix insolation)

PV System	Amonix CPV, 60 kW$_p$	Multi-c-Si[a] ground-mount	CdTe[b] ground-mount
Module DC Efficiency	28%	14–15%	14%
Total System Loss	18%	20%	20%
Insolation (kWh/m^2 per year)	2480[c]	2370[d]	2370[d]
EPBT (years)	0.9	1.1	0.5
GHG (g CO$_2$-eq./kWh)-30 years life[e]	24	19	10
GHG (g CO$_2$-eq./kWh)-50 years life[e]	16		

[a] Adapted from Fthenakis and Alsema (2006) and Fthenakis *et al.* (2008) and adjusted for Phoenix insolation and 2014 average efficiencies.
[b] Adapted from Fthenakis *et al.* (2008) and Fthenakis *et al.* (2009) and adjusted for Phoenix insolation.
[c] Direct normal insolation with 2-axis tracker.
[d] South facing, longitude optimal.
[e] 100 years of integrated time horizon for the Global Greenhouse Potential (GWP).

Glossary

List of Acronyms

Acronym	Description
AM1.5d	Reference spectrum for direct radiation
CPV	Concentrator photovoltaics
DNI	Direct normal irradiance
EPBT	Energy payback time
EROI	Energy return on investment
GHG	Greenhouse gas
HCPV	High concentrator photovoltaics
IPCC	Intergovernmental panel on climate change
LCA	Life cycle analysis
LCI	Life cycle inventory
MOVPE	Metalorganic vapor phase epitaxy
NREL	National Renewable Energy Laboratory

List of Symbols

Typical units given in square brackets. If no units are given, variable is dimensionless.

Symbol	Description [Units]
DNI	Direct normal irradiance [W/m^2]
EPBT	Energy payback time [years]
PM$_{10}$	Mass of emitted particles of 2.5 to 10 microns in diameter [kg]
PM$_{2.5}$	Mass of emitted particles below 2.5 microns in diameter [kg]
Y	Annual energy generation [kWh]

References

1. E. Alsema, D. Fraile, R. Frischknecht, V. Fthenakis. Methodology guidelines on life cycle assessment of photovoltaic electricity. International Energy Agency, 2009.
2. V. Fthenakis, R. Frischknecht, M. Raugei *et al.* Methodology guidelines on life cycle assessment of photovoltaic electricity. International Energy Agency, 2011.
3. V. Fthenakis, H. C. Kim, R. Frischknecht *et al.* Life Cycle inventories and life cycle assessments of photovoltaic systems. International Energy Agency, 2011.
4. LCI Database Documentation. Franklin Associates USA, 1998.
5. Ecoinvent data v1.1. Final reports ecoinvent 2000 No. 1–15. Ecoinvent Centre, Swiss Centre for Life Cycle Inventories, 2004.
6. V. Fthenakis, W. Wang, H. C. Kim. Life cycle inventory analysis of the production of metals used in photovoltaics. *Renewable and Sustainable Energy Reviews*, **13**, 493–517. 2009.
7. J. E. Mason, V. M. Fthenakis, T. Hansen, H. C. Kim. Energy payback and life-cycle CO_2 emissions of the BOS in an optimized 3.5 MW PV installation. *Progress in Photovoltaics: Research and Applications*, **14**, 179–190. 2006.
8. Simapro (PRé Consultants). Available: http://www.pre-sustainability.com/simapro
9. R. King (Spectrolab)., Personal communication. 2008.
10. N. J. Mohr, J. J. Schermer, M. A. J. Huijbregts, A. Meijer, L. Reijnders. Life cycle assessment of thin-film GaAs and GaInP/GaAs solar modules. *Progress in Photovoltaics: Research and Applications*, **15**, 163–179. 2007.
11. Intergovernmental Panel on Climate Change. IPCC Fourth Assessment Report (AR4), IPCC Working Group I. 2007.
12. S. C. Morris, G. A. Goldstein, V. M. Fthenakis. NEMS and MARKAL-MACRO models for energy-environmental-economic analysis: a comparison of the electricity and carbon reduction projections. *Environmental Modeling and Assessment*, **7**, 207–216. 2002.
13. C. Sener and V. Fthenakis. Energy policy and financing options to achieve solar energy grid penetration targets: Accounting for external costs. *Renewable and Sustainable Energy Reviews*, **32**, 854–868. 2014.
14. V. M. Fthenakis, F. W. Lipfert, P. D. Moskowitz, L. Saroff. An assessment of mercury emissions and health risks from a coal-fired power plant. *Journal of Hazardous Materials*, **44**, 267–283. 1995.
15. S. Kim, B. Dale. Life cycle inventory information of the United States electricity system. *The International Journal of Life Cycle Assessment*, **10**, 294–304. 2005.
16. V. M. Fthenakis, H. C. Kim. Greenhouse-gas emissions from solar electric- and nuclear power: A life-cycle study. *Energy Policy*, **35**, 2549–2557. 2007.
17. V. Fthenakis, H. C. Kim. Land use and electricity generation: A life-cycle analysis. *Renewable and Sustainable Energy Reviews*, **13**, 1465–1474. 2009.
18. V. Fthenakis, H. C. Kim. Life-cycle uses of water in U.S. electricity generation. *Renewable and Sustainable Energy Reviews*, **14**, 2039–2048. 2010.
19. V. Fthenakis. Overview of potential hazards, in *Practical Handbook of Photovoltaics: Fundamentals and Applications*, (eds T. Markvart and L. Castañer), Elsevier, 2003, pp. 857–868.
20. V. M. Fthenakis, E. A. Alsema, M. J. de Wild-Scholten. Life cycle assessment of photovoltaics: perceptions, needs, and challenges. *Conference Record of the 31st IEEE Photovoltaic Specialists Conference*, pp. 1655–1658. 2005.
21. V. M. Fthenakis, H. C. Kim. CdTe photovoltaics: Life cycle environmental profile and comparisons. *Thin Solid Films*, **515**, 5961–5963. 2007.
22. V. M. Fthenakis, H. C. Kim, E. Alsema. Emissions from photovoltaic life cycles. *Environmental Science & Technology*, **42**, 2168–2174. 2008.
23. V. M. Fthenakis, H. C. Kim. Photovoltaics: Life-cycle analyses. *Solar Energy*, **85**, 1609–1628. 2011.
24. V. Fthenakis, H. C. Kim, M. Held, M. Raugei, J. Krones. Update of PV energy payback times and life-cycle greenhouse gas emissions. Proceedings of 24th European Photovoltaic Solar Energy Conference, Hamburg, Germany, 21–15 September, 2009.
25. V. Fthenakis, D. O. Clark, M. Moalem *et al.* Life-cycle nitrogen trifluoride emissions from photovoltaics. *Environmental Science and Technology*, **44**, 8750–8757. 2010.
26. V. Fthenakis. Solar cells: energy payback times, photovoltaic (PV) energy payback time (EPBT) and environmental issues, in *Encyclopedia of Sustainability Science and Technology*, (ed. R. Meyers), Springer New York, 2012, pp. 9432–9448.
27. G. Peharz, F. Dimroth. Energy payback time of the high-concentration PV system FLATCON®. *Progress in Photovoltaics: Research and Applications*, **13**, 627–634. 2005.

28. C. Reich-Weiser, S. Horne, D. A. Dornfeld. Environmental metrics for solar energy. Proceedings of 23rd European Photovoltaic Solar Energy Conference and Exhibition, Valencia, Spain, 2008.

29. M. de Wild-Scholten, M. Sturm, M. A. Butturi, *et al.* Environmental sustainability of concentrator PV systems: preliminary LCA results of the Apollon Project. Proceedings of the 25th European Photovoltaic Solar Energy Conference and Exhibition/5th World Conference on Photovoltaic Energy Conversion, Valencia, Spain, 2010.

30. H. C. Kim, V. M. Fthenakis. Life cycle energy demand and greenhouse gas emissions from an Amonix high concentrator photovoltaic system. *Conference Record of the 2006 IEEE 4th World Conference on Photovoltaic Energy Conversion*, pp. 628–631. 2006.

31. R. McConnell, V. Fthenakis. Concentrated photovoltaics, in *Third Generation Photovoltaics*, (ed. V. Fthenakis), InTech, 2012.

32. V. M. Fthenakis, H. C. Kim. Life cycle assessment of high-concentration photovoltaic systems. *Progress in Photovoltaics: Research and Applications*, **21**, 379–388. 2013.

33. A. Nishimura, Y. Hayashi, K. Tanaka *et al.* Life cycle assessment and evaluation of energy payback time on high-concentration photovoltaic power generation system. *Applied Energy*, **87**, 2797–2807. 2010.

34. B. King, D. Riley, C. Hansen, M. Erdman, J. Gabriel, K. Ghosal. HCPV characterization: Analysis of fielded system data. *AIP Conference Proceedings*, **1616**, 276–279. 2014.

35. V. Fthenakis, E. Alsema. Photovoltaics energy payback times, greenhouse gas emissions and external costs: 2004–early 2005 status. *Progress in Photovoltaics: Research and Applications*, **14**, 275–280. 2006.

13-I Annex

Energy Flow Diagrams for Amonix 7700 System Components

In section 13.4.3, Figure 13.5, the total energy flow diagram for a single Amonix 7700 was depicted. In this annex, we include the detailed energy flow diagrams for all the elements that constitute an Amonix 7700, namely, 1) frame (Figure 13-I.1); 2) Fresnel lenses (Figure 13-I.2); 3) heat sink (Figure 13-I.3); 4) tracker (Figure 13-I.4); 5) hydraulic drive (Figure 13-I.5). All these flow diagrams were produced using the Simapro software [8].

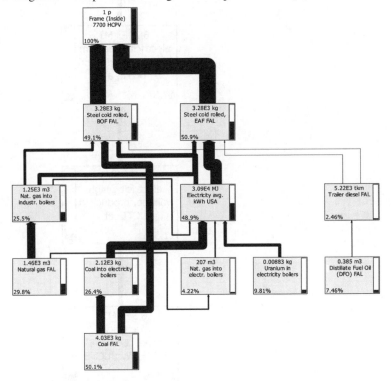

Figure 13-I.1 Frame: Life-cycle primary energy (diagram cut-off = 3%; thus energy contributions smaller than 3% of the total are not shown in this figure)

Handbook of Concentrator Photovoltaic Technology, First Edition. Edited by Carlos Algora and Ignacio Rey-Stolle. © 2016 John Wiley & Sons, Ltd. Published 2016 by John Wiley & Sons, Ltd.

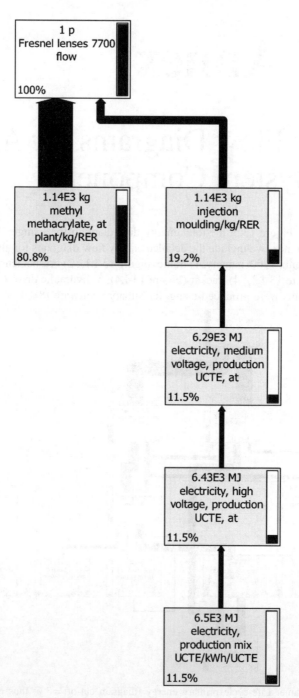

Figure 13-I.2   Fresnel lenses: Life-cycle primary energy distribution (diagram cut-off = 5%; thus energy contributions smaller than 5% of the total are not shown in this figure)

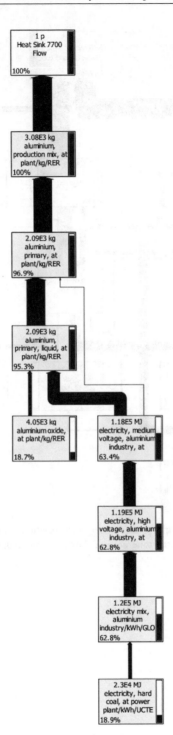

Figure 13-I.3 Heat sink: Life-cycle primary energy distribution (diagram cut-off = 17%; thus energy contributions smaller than 17% of the total are not shown in this figure)

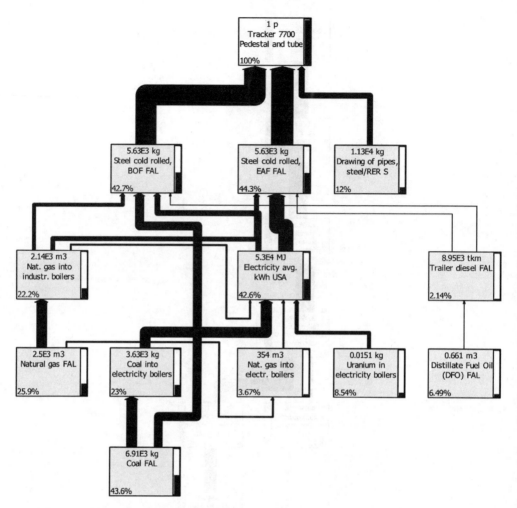

Figure 13-I.4 Trackers: Life-cycle primary energy distribution (diagram cut-off = 3%; thus energy contributions smaller than 3% of the total are not shown in this figure)

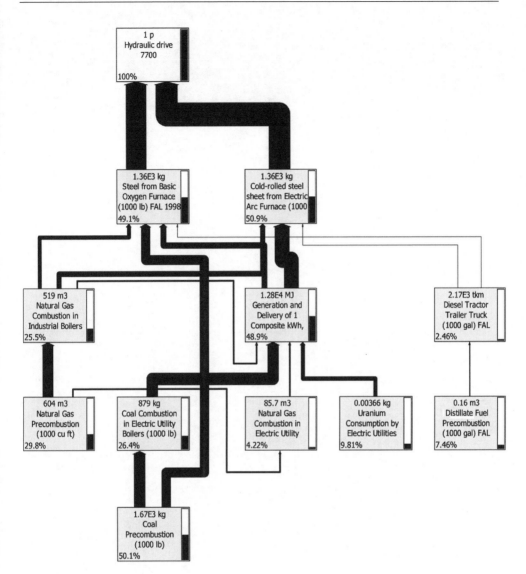

Figure 13-I.5 Hydraulic drive: Life-cycle primary energy distribution (diagram cut-off = 3%; thus energy contributions smaller than 3% of the total are not shown in this figure)

14

Cost Analysis

Carlos Algora,[1] Diego L. Talavera,[2] and Gustavo Nofuentes[2]

[1]*Instituto de Energía Solar, Universidad Politécnica de Madrid, Spain*
[2]*Grupo de Investigación y Desarrollo de Energía Solar, Universidad de Jaén, Spain*

14.1 Introduction

Economics is not a pure science because although it yields theories and explains the past, frequently is unable to make correct predictions. Therefore, although in this chapter we will make forecasts about the cost and price of CPV, the main purpose is to describe the concepts, to introduce the mathematic tools used and to show the complete process to carry out a cost calculation. In this way, if the assumptions considered through this chapter turn out to be inaccurate, the chapter will still be valuable in the sense of showing the main tendencies and in providing the reader the tools for doing his own calculations based on new guesswork. Even so, in order to mitigate the possibility of inaccurate predictions, sensitivity analyses of several key parameters and functions are carried out throughout the chapter. We have adopted this approach, rather than presenting rigid predictions based on fixed parameters with only marginal chances of providing the right guess for coming years. Accordingly, this chapter provides the required economic concepts (section 14.2) to carry out the profitability analysis of a project for a CPV plant (section 14.3). After that, the cost of CPV technology (section 14.4) is analyzed including both the cost of the CPV system (€/W_p, $/$W_p$, etc.) and the *levelized cost of electricity* (€/kWh, $/kWh, etc.) resulting in a forecast about CPV grid parity.

14.2 Basic Concepts of Cost and Profitability Analysis

In this section, some concepts and elements of cost and economic analysis are reviewed, namely, elements of the investment, present and future worth (i.e. value) of sums, the discount rate, effect of the inflation, impact of taxation and financing.

Handbook of Concentrator Photovoltaic Technology, First Edition. Edited by Carlos Algora and Ignacio Rey-Stolle.
© 2016 John Wiley & Sons, Ltd. Published 2016 by John Wiley & Sons, Ltd.

14.2.1 Elements of the Investment

From a business point of view, an investment may be stated as the acquisition of assets in order to obtain benefits which are expected to be generated over the coming years. From a financial standpoint, an investment may be stated as the flows of revenue and costs generated since the beginning of this investment until its liquidation. Some of the factors to be taken into account in the analysis of the investment, from a financial point of view, are defined below:

- Initial investment cost (CPV_{IN}) is the initial payment of the total amount of funds which are set aside for the acquisition of assets of the investment project. In other words, CPV_{IN} is the total turnkey installed cost of the project.
- Revenue and costs that are originated along the life cycle of the investment and generated by the exploitation of the assets object of the investment.
- *Net cash flow* (NCF) is the difference between revenue and costs generated along the life cycle of the investment project. NCF is usually calculated on an annual basis.
- *Salvage value* (S_V), is a salvage value received by the sale (settlement) of the assets of the investment. This inflow will be obtained at the end of the life cycle of the investment.
- Analysis period (N_{ap}) is the period of time for which the evaluation is carried out. N_{ap} is usually equal to the life cycle of the investment project. The estimation of N_{ap} is often difficult owing to the number of variables influencing its quantification.

Therefore, an investment project may be depicted over time (on the *x*-axis) by means of its cash flows as depicted in Figure 14.1. In year zero, cash outflows (such as the initial investment cost) are shown as negative in this axis, while net cash flows are shown as positive in this axis over the remaining years.

14.2.2 Present and Future Worth of Sums. The Impact of Inflation

If a sum is received from a debtor at the present time, it is said to have a present worth (i.e. present value), Q. If the reception of this sum is postponed n years, the debtor should pay more than Q for

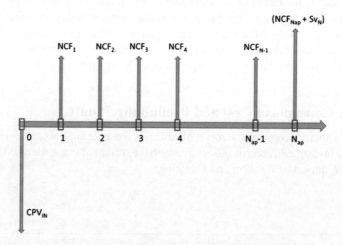

Figure 14.1 Financial dimension of the investment

two reasons. First, this sum could have been safely invested at a given interest rate – by buying government bonds, for example – so that the investor would have obtained Q plus the corresponding interests after this period. Second, the purchase power of Q is bigger at the present time than that after n years, due to the impact of inflation. Bearing this in mind, the debtor should pay more in order to preserve this purchase power. Hence, if i and g are the annual interest and inflation rates, respectively, a sum Q received at the present time will have a future worth (i.e. future worth) of $Q \cdot [(1+i) \cdot (1+g)]^n$ after n years. Consequently, a sum Q to be received after n years has a present worth of $Q/[(1+i) \cdot (1+g)]^n$. In other words, receiving $Q \cdot [(1+i) \cdot (1+g)]^n$ after n years equals receiving Q at the present time. Parameter i is usually termed real discount rate, so that the nominal discount rate – or simply discount rate – is stated as d, by means of $1+d=(1+i)(1+g)$. Last, parameter i is also noted as d_r.

The amount Q may be expressed either in current currency or constant currency. Current or nominal currency takes into account the effect of inflation. Constant or real currency does not consider this effect.

14.2.3 The Discount Rate

Choosing an appropriate figure for d is a controversial issue. The appropriate selection of the value of d for the analysis of a given investment project should be the rate of return for an investment of comparable risk that would be made instead. Commonly, government bonds are considered as a suitable alternative investment.

As previously explained, estimating the discount rate proves difficult. Usually, in conditions of certainty the figure of the discount rates must be the largest number between the following: a) opportunity cost of capital, understood as the profitability foregone on the next most attractive investment and b) explicit cost of capital, or cost incurred by the investor when using capital drawn from the financial market.

Opportunity cost of capital is based on opportunities arising from the financial market. Therefore, the value of the discount rate would be equivalent to the price of money determined by the free interplay of supply and demand. It may also represent the rate of return on the best alternative investment available (i.e.: government bonds).

Explicit cost of capital derives from the capital sources or funds borrowed to finance the investment. Consequently, the investment should provide a return that at least covers the cost of capital of the investment. In this case, the value of the discount rate should be equivalent to the *weighted average cost of capital* (WACC). WACC is the cost that the owner or investor of the project must pay to use capital sources in order to finance the investment. A widespread practice is to use a discount rate equal to the organization's weighted average cost of capital [1].

14.2.4 Effect of Inflation

Inflation influences all the factors that determine an investment project, such as the current revenues and cost, cash flows and discount rate, among others. The impact of inflation on the discount rate has already been presented in section 14.2.2. The effect of inflation on net cash flows of the investment is going to be dealt with below.

An annual escalation rate (m) – i.e. the percentage at which the yearly change in the price of the goods or services produced by the investment is expected to occur – of the net cash flows may be assumed as a consequence of the effect of the annual inflation rate (g). Then, if g equals m, inflation does not influence the profitability of the investment. If $g > m$, inflation affects this profitability in a negative manner. Conversely, $g < m$ leads to a positive impact on profitability.

14.2.5 Impact of Taxation

The impact of taxation may cause a noticeable variation of the profitability of an investment project. In general, most existing tax laws consider that every investment owner must pay a yearly amount, mostly attributable to the earnings of the previous year. This amount depends on the law defined tax rate (T), investment revenue, the annual operation and maintenance cost, debt interest paid on funds borrowed to finance the investment, tax depreciation, etc. Tax depreciation is a means of recovering the cost of property, by means of income tax deduction. The method used in the calculation of the depreciation may differ from country to country. The amplitude and variability of tax regulations makes it complex to encompass taxation in our study. Therefore, only some general indications will be provided, aimed to include the effect of taxation on factors that influence the profitability of an investment such as NCF, SV and the cost of capital.

The effect of taxation on NCF – net cash flow before tax ($NCF_{(before-tax)}$) – is determined by applying the tax rate on some of the components that make up the net cash flow, namely: revenue, operation and maintenance cost (O&M cost) and tax depreciation (DEP). Then, the net cash flows after-tax ($NCF_{(after-tax)}$) may be written as:

$$NCF_{(before-tax)} = \text{Revenue-Costs}, \tag{14.1}$$

$$NCF_{(after-tax)} = NCF_{(before-tax)} - (NCF_{(before-tax)} - DEP) \cdot T, \tag{14.2}$$

The taxation impact on the salvage value – salvage value before-tax ($SV_{(before-tax)}$) – is derived from the fact that the Administration registers any increase of net worth derived from the sale of assets. On the other hand, if a decrement of net worth takes place, it implies a tax deduction. In this chapter, the salvage value is assumed equal to zero, so these considerations do not apply.

The influence of taxation on funds borrowed to finance an investment is exerted on the paid debt interest – $Interest_{(before-tax)}$ – since it is tax deductible. Accordingly, taxation leads to a decrease in the cost of capital. Hence, interests after paying taxes ($Interest_{(after-tax)}$) are given by:

$$Interest_{(after-tax)} = Interest_{(before-tax)}(1 - T), \tag{14.3}$$

14.2.6 Financing

The project investment is usually financed by means of debt or equity capital. Some forms of debt include long-term loans, bonds and debentures. Other forms of financing include common stock and preferred stock (equity capital). Paid dividends (return on equity) from both common and preferred stock are not tax deductible, while paid debt interest is tax deductible, as commented above. Therefore, the cost of debt is usually lower than that of equity capital [1]. Each one of these financial mechanisms poses some advantages and drawbacks. Making a decision on the most suitable financing option of the investment project implies choosing a specific one or, more commonly, a combination of them.

14.3 Review of Profitability Analysis

Prior to introducing some classical criteria intended to carry out the profitability analysis of a project investment, some helpful concepts are going to be reviewed, namely, the life-cycle cost of the system together with the present worth of cash inflows related to a CPV plant. These concepts are essential to apply the profitability analysis to these systems with their own peculiarities. The most common criteria aimed when measuring the profitability of the project, such as the *net present value* (NPV), *benefit-to-cost* (B-C), *profitability index* (PI), the *discounted payback time* (DPB) and the internal rate of return (IRR) are presented hereafter. Criteria based on NPV, B-C, PI and IRR, are addressed when measuring profitability, while DPB is based on the measurement of the liquidity of the investment.

The core idea that lies in the profitability assessment of investment projects may be broadly summarized as follows: finding a suitable parameter, or a set of them, in which all considered financial factors are taken into account so that a profitability estimate of the investment is provided.

14.3.1 The Life Cycle Cost of a CPV System

The life cycle cost of the CPV system (LCC) may be calculated by adding the initial investment cost of the CPV system (CPV_{IN}) plus the present worth of its operation and maintenance cost ($PW[CPV_{OM}(N)]$) over the system life cycle (N):

$$LCC = CPV_{IN} + PW[CPV_{OM}(N)], \tag{14.4}$$

Life cycle operation and maintenance cost, $PW[CPV_{OM}(N)]$ may be written as:

$$PW[CPV_{OM}(N)] = \left(CPV_{AOM} \cdot \frac{q(1-q^N)}{1-q}\right), \tag{14.5}$$

where CPV_{AOM} is the annual operation and maintenance cost, assumed constant over the system life cycle, while $q(1-q^N)/(1-q)$ is the present value interest factor. Factor q is equal to $1/(1+d)$. Assuming an annual escalation rate of the operation and maintenance cost of the CPV system (ε_{CPVOM}), implies writing $PW[CPV_{OM}(N)]$ as follows:

$$PW[CPV_{OM}(N)] = \left(CPV_{AOM} \cdot \frac{K_{PV} \cdot (1-K_{PV}^N)}{1-K_{PV}}\right), \tag{14.6}$$

where $K_{PV} = (1+\varepsilon_{CPVOM})/(1+d)$ and ε_{CPVOM} is considered since the first year.

If taxation is taken into account, Eq. (14.6) might be rewritten as follows:

$$PW[CPV_{OM}(N)] = \left(CPV_{AOM}(1-T) \cdot \frac{K_{PV} \cdot (1-K_{PV}^N)}{1-K_{PV}}\right), \tag{14.7}$$

The share of external and equity financing may be included in the analysis explicitly through the weighted average cost of capital over the nominal discount rate. Therefore, CPV_{IN} may be financed through long-term debt or/and by means of equity capital. Usually, the amount financed through debt is greater than that financed by equity capital in renewable energy projects, as the cost of debt is usually lower than that of equity capital, as detailed in section 14.2.6. If CPV_{IN} is financed partially through debt (CPV_l,) and the remainder by means of equity capital (CPV_s), then $CPV_{IN} = CPV_l + CPV_s$. Therefore, CPV_l is borrowed at an annual loan interest (i_l) to be paid back in N_l years, which correspond to the loan duration. CPV_s corresponds to the share of equity (common stock and preferred stock), with an annual retribution in form of dividends (d_s) (return on equity) and it must be paid in full at the end of the life cycle of the project (N years).

Concerning CPV_l, the payment of each year (PY_{CPVl}) can be set equal so that:

$$PY_{CPVl} = CPV_l \cdot \frac{i_l}{1 - (1 + i_l)^{-N_l}}, \tag{14.8}$$

Regarding CPV_s, the payment of each year (PY_{CPVs}) over the life cycle of the system is shown in Eq. (14.9). It must be kept in mind that CPV_s is payable in full after N years:

$$PY_{CPVs} = d_s \cdot CPV_s, \tag{14.9}$$

Considering the previous financing scheme, the initial investment cost of the CPV system can be expressed as:

$$CPV_{IN} = CPV_l \cdot \frac{i_l}{1 - (1 + i_l)^{-N_l}} \cdot \frac{q(1 - q^{N_l})}{1 - q} + d_s \cdot CPV_s \cdot \frac{q(1 - q^N)}{1 - q} + CPV_s \cdot q^N, \tag{14.10}$$

If the present value interest factor of year k, i.e. $q(1 - q^k)/(1 - q)$, is noted as $PVIF(k)$, and taking into account the impact of taxation on financing – tax deduction applies to interest payments of loan – the initial investment cost of the CPV system may be rewritten as:

$$CPV_{IN} = CPV_l \cdot \frac{i_l(1 - T)}{1 - (1 + i_l(1 - T))^{-N_l}} \cdot PVIF(N_l) + d_s \cdot CPV_s \cdot PVIF(N) + CPV_s \cdot q^N,$$

$$\tag{14.11}$$

It is worth mentioning that the left-hand side of Eq. (14.11) only equals its right-hand side if the selected value for d is set equal to $WACC$. As shown in the paragraph above, d is involved in the calculation of $PVIF(k)$ given that $q = 1/(1 + d)$. Consequently, CPV_{IN} may be expressed by the right-hand side of Eq. (14.11) provided that $d = WACC$ in the latter equation.

If an initial investment subsidy (CPV_{IS}) – in the form of grants, rebates, buy down subsidies, among others – is available, the amount $CPV_{IN} - CPV_{IS}$ is to be paid by the owner of the system. This amount may be financed through debt and equity, so that $CPV_{IN} - CPV_{IS} = CPV_s + CPV_l$, in this case. It should be noted that CPV_{IS} is taxable in a given period of time over which it is amortized. Thus, the initial investment cost of the CPV system may be expressed as:

$$CPV_{IN} = CPV_l \cdot \frac{i_l(1 - T)}{1 - (1 + i_l(1 - T))^{-N_l}} \cdot PVIF(N_l) + d_s \cdot CPV_s \cdot PVIF(N) + CPV_s \cdot q^N$$

$$+ \frac{CPV_{IS}}{N_{IS}} \cdot T \cdot PVIF(N_{IS}) \tag{14.12}$$

where N_{IS} is the period of time over which an initial investment subsidy is amortized. Accordingly, taxation of CPV_{IS} takes place over N_{IS}.

As remarked above, the left-hand side of Eq. (14.12) only equals its right-hand side if the selected value of d is set equal to $WACC$. However, it should be noted that $WACC$ in Eq. (14.12) is lower than that obtained by means of Eq. (14.11). Obviously, less funding is needed to be borrowed to finance the investment in Eq. (14.12) than that needed in Eq. (14.11). As mentioned in section 14.2.3, a widespread practice consists of setting the discount rate equal to the organization's $WACC$ [1,2].

Finally, if the impact of taxation is taken into account by assuming that tax depreciation is deductible [1]:

$$LCC = CPV_{IN} + PW[CPV_{OM}(N)] - PW[DEP(N_d)] \cdot T, \qquad (14.13)$$

where PW[$DEP (N_d)$] is the present worth of tax depreciation and N_d is the period of time over which an investment is depreciated for tax purposes. The method used in the tax depreciation may differ from one country to another. Usually, the *modified accelerated cost recovery system* (MACRS) is used as the tax depreciation system in the United States [3]. Anyway, readers must refer to national taxation laws. For example, if tax depreciation is assumed linear over a given period of time, the present worth of the tax depreciation may be estimated by:

$$PW[DEP(N_d)] = DEP_y \cdot PVIF(N_d), \qquad (14.14)$$

where DEP_y is the annual tax depreciation for the CPV system. The consideration of a linear tax depreciation implies that $DEP_y = CPV_{IN}/N_d$.

14.3.2 The Present Worth of the Cash Inflows Generated by a CPV System

The present worth of the cash inflows generated by a CPV system is linked to government regulations concerning economic relations between the utility and the user. In those countries where net or bill metering exists, a fraction of the annual CPV electricity generated is used for self-consumption (E_{CPVs}) and consequently saved, instead of buying it from the grid at a given price (p_s), which is set at the retail price of the market. The remaining annual electricity generation (E_{CPVg}) is fed into the grid, which may be compensated at a different price (p_g). The most general case would take into account different prices applied for both self-consumption – or saved – and for feeding electricity production into the utility. Bearing in mind all these considerations, the present worth of the cash inflows from a CPV system (PW[CIF(N)]) may be written as:

$$PW[CIF(N)] = p_s E_{CPVs} \frac{K_{ps}\left(1 - K_{ps}^N\right)}{1 - K_{ps}} + p_g E_{CPVg} \frac{K_{pg}\left(1 - K_{pg}^N\right)}{1 - K_{pg}}, \qquad (14.15)$$

where K_{ps} and K_{pg} stand for two factors related to the annual increase rate of the electricity price and the annual decrease rate of the CPV system peak power. These factors are expressed by:

$$K_{ps} = (1 + \varepsilon_{ps}) \cdot (1 - \varepsilon_{pl})/(1 + d), \qquad (14.16)$$

$$K_{pg} = (1 + \varepsilon_{pg}) \cdot (1 - \varepsilon_{pl})/(1 + d), \qquad (14.17)$$

where ε_{ps} and ε_{pg} stand for the annual increase rate of the electricity price that is consumed and fed to the grid, respectively. Factor ε_{pl} is the annual degradation rate of the efficiency of the CPV system. Both factors are considered since the first year.

As stated above, this applies to the most general case. However, if the annual CPV generated electricity is totally fed into the grid, Eq. (14.15) is simplified:

$$PW[CIF(N)] = p_g E_{CPV} \frac{K_{pg}\left(1 - K_{pg}^{\ N}\right)}{1 - K_{pg}}, \tag{14.18}$$

where $E_{CPV} = E_{CPVs} + E_{CPVg}$.

If taxation is taken into account, Eq. (14.18) might be rewritten as follows:

$$PW[CIF(N)] = p_g E_{CPV}(1 - T) \frac{K_{pg}\left(1 - K_{pg}^{\ N}\right)}{1 - K_{pg}}, \tag{14.19}$$

A later sale or any further use of the parts of the CPV system has been considered improbable in our analysis, so a zero salvage value of the system is assumed.

14.3.3 Assessment of the Profitability of a CPV System

14.3.3.1 Net Present Value

The net present value of an investment project is the sum of present values of all cash inflows and outflows related to the investment [4], which is an estimate of the wealth increase that the project will produce. Therefore, NPV equals the present worth of the cash inflows from the system minus the life cycle cost, so it may be written, in general, as follows:

$$NPV = PW[CIF(N)] - LCC, \tag{14.20}$$

The economic meaning of NPV equals the total net profit of the investment, once the cash inflows have paid the life cycle cost of the CPV system. An investment project is assumed profitable when $NPV > 0$, it is inacceptable when $NPV < 0$ and indifferent or neutral when $NPV = 0$. When some alternative investments exist, these have to be ranked so that the project with the highest NPV is to be selected. However, this method is unsuitable in order to choose between two projects with the same NPV but different initial investment cost requirements and life cycles.

14.3.3.2 Profitability Index and Benefit-to-Cost Ratio

The profitability index (PI) of an investment project is defined as the ratio between the net present value of the system and the initial investment cost. If this criterion is used, the project is profitable when PI is positive, while negative values of this parameter represent unprofitable investments:

$$PI = \frac{NPV}{CPV_{IN}}, \tag{14.21}$$

The benefit-to-cost (B-C) ratio of an investment project is defined as the ratio between the present worth of its cash inflows and the project life cycle cost:

$$B - C = \frac{PW[CIF(N)]}{LCC},$$

(14.22)

The use of this criterion for assessing the profitability implies assuming that the project is feasible when B–C is greater than one, so that if some projects are to be assessed, the one with the highest B–C should be preferable.

Both PI and B–C solve the problem of how to prioritize investment projects with the same NPV but different initial cost requirements. However, PI and B–C fail when used to prioritize independent investments and mutually exclusive projects with different life cycles. Both IP and B–C are closely related and provide similar information to the investor/user.

14.3.3.3 Discounted Payback Time

The discounted payback time (DPB) is the period – typically expressed in years – it takes to recover the project cost of an investment. Therefore, to determine DPB, the present worth of the cash inflows minus the present worth cash outflows, are added year by year until the total sum equals zero:

$$0 = PW[CIF(DPB)] - (CPV_{IN} + PW[CPV_{OM}(DPB)] - PW[DEP(N_d)] \cdot T)$$

(14.23)

where PW[CIF (DPB)] is the present worth of the cash inflows over DPB, PW[$CPV_{OM}(DPB)$] is the present worth of the operation and maintenance cost over DPB and PW[$DEP(N_d)$] is obtained by means of Eq. (14.14) when the time over which the investment is depreciated, for tax purposes, is shorter than DPB. On the other hand, if the period over which the investment is depreciated is longer than DPB, the discounted payback time is calculated by:

$$PW[CIF(DPB)] = CPV_{IN} + PW[CPV_{OM}(DPB)] - PW[DEP(DPB)] \cdot T,$$

(14.24)

where PW[$DEP(DPB)$] $= DEP_y \cdot PVIF(DPB)$. An investment project is viewed favorably if the DPB is shorter than the system life cycle, given the considered discount rate. Shorter figures of DPB are obviously preferable. Although easily understandable and straightforward, this method does not consider the cash flows that are produced after the DPB. No conclusions concerning the economic profitability can be drawn from this criterion, so it must be used combined with others.

14.3.3.4 Internal Rate of Return

The internal rate of return (IRR) of an investment project is the value of the discount rate d that leads to $NPV = 0$. Also, the IRR for a given project can be defined, as the actual interest rate at which the project initial investment should be lent during its useful life cycle to achieve the same profitability [5]. Therefore, IRR is the value of d that satisfies:

$$0 = PW[CIF(N)] - LCC,$$

(14.25)

It should be noted that Eq. (14.25) leads to the calculation of a 'gross' internal rate of return. However, since most projects use financial mechanisms, the net internal rate of return (IRR_n) provides a more realistic assessment. Thus, IRR_n is obtained by subtracting $WACC$ from IRR as calculated by means of Eq. (14.25), this is: $IRR_n = IRR - WACC$. It is worth noting that, if the parameters involved in the calculation of IRR in Eq. (14.25) are considered in current currency, nominal IRR is obtained. On the other hand, if these parameters are considered in constant currency real IRR is obtained.

From an economic point of view, an investment on a CPV system is acceptable if a) IRR exceeds a profitability threshold fixed by the future investor/owner or b) IRR is above $WACC$. In this sense, this method is very important for investors since it provides a meaningful estimate of the rate of return of their investment.

14.3.4 Sensitivity Analysis on the Profitability of CPV Systems

In this section, we want to show the usefulness of the concepts defined above. As detailed in previous sections, many factors are involved in the calculation of the different criteria aimed at assessing the profitability of CPV systems when considered as investment projects. This section introduces some other factors that influence many of those criteria.

Defining a base case is a good starting point to look into the deviations caused on its related profitability criterion by the variations in the values of the factors that define this base case. Of course, such base case should be representative of state-of-the-art CPV systems, and some realistic assumptions must be taken into account when defined. The impact of the variation of a given factor on the values of a specific profitability criterion obviously depends on the interval within which this factor is assumed to range. Thus, most of these intervals considered hereafter vary between 50% and 150% of the base case value. This allows us to analyze the dependence of a specific criterion on the factor under study. Consequently, sound conclusions may be drawn regarding the sensitivity of the former to variations of the latter. Anyway, it must be kept in mind that when studying the influence of the variations of one of these factors, the values of the remaining factors will be assumed equal to those that configure the base case.

Bearing in mind all the above comments, the base case is defined below. It should be noted that the figures corresponding to this section and referring to costs, incentives and operation and maintenance costs are all them normalized-per-W_p. The symbols used for these factors are the same for those not normalized, except that they are shown in brackets and with the subscript 'W_p'.

Due to the present small size of the CPV market – when compared to that of flat-plate modules [6] – values for the normalized-per-W_p initial investment cost ($[CPV_{IN}]_{W_p}$) are quite scattered. Nowadays, this initial investment cost per W_p in CPV systems of around one megawatt might vary from €1.4/W_p to €2.2/W_p [7–9]. A value of $[CPV_{IN}]_{W_p} = €2.05/W_p$ may be assumed for the base case, as it will be calculated in section 14.4.1.2.

The available annual direct normal irradiation highly depends on the location of the site. This is a crucial factor since the AC electricity annual yield (E_{CPV}) of a CPV system is given by:

$$E_{CPV} = DNI_a \cdot P^* \cdot PR / DNI_{CSOC} \qquad (14.26)$$

where DNI_a is the annual direct normal irradiation; DNI_{CSOC} is the direct normal irradiance at CSOC (0.9 kW/m^2); P^* is the power of the CPV system rated at CSOC (Concentrator Standard

Operating Conditions, see Chapter 8). Setting $P^* = 1\,kW_p$ in Eq. (14.26) implies that the normalized-per-kW_p annual CPV electricity yield ($[E_{CPV}]_{kWp}$, in kWh/(kW_p·year)) may be written as:

$$[E_{CPV}]_{kWp} = DNI_a \cdot PR/DNI_{CSOC} \qquad (14.27)$$

A value for DNI_a equal to 2135 kWh/(m^2·year) has been assumed for the base case. This specific figure coincides with the value of DNI_a for Almeria, Spain (Longitude: 2.464 W, Latitude: 36.834 N) [10]. As discussed in Chapter 8, the performance ratio (PR) accounts for losses derived from operating module temperature, spectral effects, power conditioning, wiring losses, impact of dirt and soil, etc. Values of PR, ranging from 0.8 to 0.9 may be considered for the base case, assuming it is a well designed state-of the art CPV system [11,12].

The electricity unitary price, p_u, of the CPV generated electricity paid to the owner – fed to the grid, or saved by the owner, *in situ* self consumption – can be fixed at wholesale or retail price of the market. However, this price can be set by law in some countries with incentives for CPV generation. Values of p_u varying between €0.07/kWh and €0.30/kWh comprise the range of most market prices and present generation-based incentives for PV in different countries –Germany, Italy, France, USA, Greece, UK, among others [13–17]–. Some examples of incentives for flat-plate PV installations include: Germany offers from €0.1102/kWh for free standing facilities; Italy from €0.106/kWh to €0.176/kWh as a function of rated power plus premium for personal consumption; France from €0.0818/kWh to €0.3159/kWh as a function of rated power; and in the United States net metering is regulated by law in most states, but state policies vary widely [18]. The feed-in tariff values for HCPV in Italy, according to the Ministerial Decree 05/07/2012, are as follows: from €0.215/kWh for installations with rated power from 1–200 kW_p; €0.201/kWh for installations with rated power from 200.01–1,000 kW_p; and €0.174/kWh for plants larger than 1000 kW_p [17]. For a HCPV system with a rated power greater than 1 MW_p, a reasonable value for the assumed base case is given by p_u = €0.115/kWh, what means that all the electricity is assumed to be fed into the grid.

The annual increase rate of the CPV electricity unitary price, ε_{pu}, linked to the evolution of electricity markets, is always difficult to forecast. In the coming years, electricity prices could experience an annual increase that may range from 2% to 6.7% yearly, depending on the country [19]. In our analysis, ε_{pu} is set to 2.5%. Regarding inflation, values of $g = 2.12\%$ for the Euro area [20,21] and $g = 2.41\%$ for the United States [20] result from obtaining averages of historical data related to annual inflation rates (period 2004–2013) for some countries and states of these two top PV areas. Thus, g is assumed equal to 2.2%, in the base case.

Regarding CPV_{IS}, in the United States, some states, utilities and a few local governments offer direct cash incentives to promote the deployment of PV systems [22]. These upfront incentives encourage PV installations by reducing the initial investment costs. In some other countries – China, Italy, Germany, Japan, among others – an initial investment subsidy is used as a supporting measure of PV systems [13,14,16]. However, this measure is being progressively phased out, so $[CPV_{IS}]_{Wp}$ is assumed to be zero in this study.

In our base case, the initial investment cost of the CPV system is financed through long-term debt and equity capital. Thus, 80% of this amount (share of debt) is borrowed at an annual loan interest equal to 6% and a 20-year loan term [23,24]. The remaining 20% of the amount is financed by means of equity capital, with an annual retribution in the form of dividends (d_s, return on equity) equal to 13% [1,25] and are payable in full at the end of the life cycle of the project. This life cycle is set equal to 25 years. Investors tend to feel that CPV projects pose a higher risk than that of projects related to some other renewable technologies, so cost of debt and equity capital – loan interest rate (i_l) and dividends (d_s) – are higher.

Nominal discount rate (d) is assumed equal to *WACC* in order to calculate the different profitability criteria and the subsequent analysis [1]. Of course, the value of *WACC* varies depending on how the capital resources are chosen to finance the initial investment. In the base case, after-tax *WACC* is equal to 6.7%, given the assumptions stated in the preceding paragraph.

Values of the income tax rate for the owner of the CPV system may vary according to each country regulations. In some studies, *T* is assumed equal to 34% [2], 18% [26] and 30% [27]. Therefore, *T* is assumed equal to 30% for the base case. The method used in the tax depreciation, is a general method, which considers a maximum linear coefficient of 5%, with a tax life for depreciation of 20 years [28–30].

The annual PV electricity yield generated by the system is assumed to decrease every year. A typical annual degradation rate in the efficiency of flat-plate PV modules equals 0.5% [31,32]. This value of ε_{pl} leads to 18.3%-decrease of the PV system efficiency after 25 years of operation. Nowadays, flat-plate PV systems have a life cycle of around 25 years and more. In our analysis, CPV systems are assumed to have the same life cycle and the same value of ε_{pl}. This is in fact a quite conservative assumption since much lower yearly decreases in efficiency have been measured so far in CPV plants although for operating periods lower than 10 years (as discussed in Chapter 9).

The normalized-per-W_p annual operation and maintenance costs $[CPV_{AOM}]w_p$ equals 2% of $[CPV_{IN}]w_p$ for the CPV system [11]. Further, the annual escalation rate of the operation and maintenance costs is set equal to the value of g, so that $\varepsilon_{CPV_{AOM}} = 2.2\%$.

In summary, the values assumed for every factor that defines base case are gathered in Table 14.1.

Table 14.1   Values of factors assumed for the base case of section 14.3.4

Factors	Base case values	Units
$[E_{CPV}]kW_p$	2032	kWh/(kW$_p$·year)
DNI_{COSC}	0.9	kW/m^2
DNI_a	2135	kWh/(m^2·year)
PR	0.86	n.a.
$[CPV_{IN}]w_p$	2.05	€/W$_p$
ε_{pl}	0.5	%
p_u	0.115	€/kWh
ε_{pu}	2.5	%
$[CPV_{AOM}]w_p$	2.0[a]	%
$\varepsilon_{CPV_{AOM}}$	2.2	%
T	30	%
d	6.7	%
g	2.2	%
$[CPV_{IS}]w_p$	0	€/W$_p$
i_l	6.0	%
N_l	20	years
d_s	13	%
N	25	years

[a] This value should be considered as the percentage of $[CPV_{IN}]w_p$ spent on operation and maintenance tasks on an annual basis.

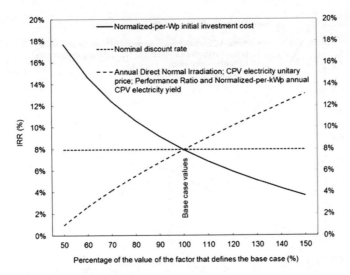

Figure 14.2 *IRR* as a function of the percentage of the value of some factors that define the base case. As regards to performance ratio, only percentage values below 120% of the base case (0.86) should be considered, given that annual *PR* cannot exceed the ideal value of 1

Solving the equations introduced in section 14.3.3 by using the values shown in Table 14.1 paves the way to the estimation of the profitability criteria of the base case. Accordingly, the influence that deviations from the values listed in Table 14.1 have on *IRR*, *DPBT* and *PI* analyzed below. Figures 14.2–14.6 show the impact of deviations from the values of these factors as provided to define the base case. Most variations range from 50% to 150% of the value of the factor assumed for the base case.

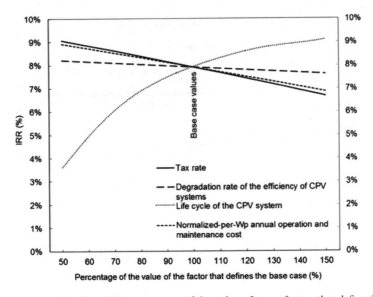

Figure 14.3 IRR as a function of the percentage of the value of some factors that define the base case

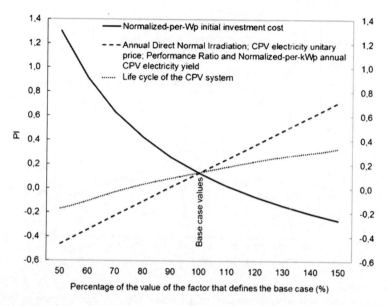

Figure 14.4 Profitability index as a function of the value of some factors that define the base case. As regards to performance ratio, only percentage values below 120% of the base case (0.86) should be considered, given that annual *PR* cannot exceed the ideal value of 1

Figure 14.2 and Figure 14.3 show the influence of the variation of different factors considered on *IRR*, while Figure 14.4 and Figure 14.5 show the impact of such variations on *PI*. Last, Figure 14.6 show both *DPB* and *PI* as a function of the variation of the values of the initial investment subsidy.

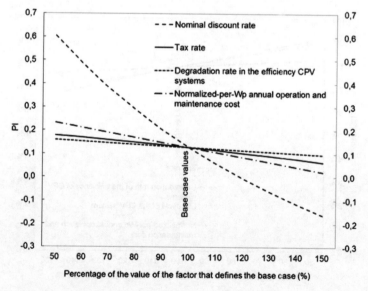

Figure 14.5 Profitability index as a function of the value of some factors that define the base case

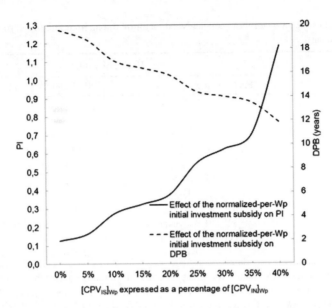

Figure 14.6 Profitability index and discount pay-back time as a function of percentage changes of the normalized-per-W_p initial investment subsidy (base case). Altering the initial investment leads to an implicit variation of both equity and debt which are involved in funding the project. This causes the sinuous shape of both two curves

As shown in Figure 14.2 and Figure 14.4, percentage changes in annual direct normal irradiation, CPV electricity unitary price, performance ratio, and normalized-per-kW_p annual PV electricity yield exert a similar impact on *IRR* and *PI*. These factors, together with the normalized-per-W_p initial investment cost, influence *IRR* and *PI* in a very noticeable way. They cause a bigger impact on these two profitability criteria than those due to changes in the remaining factors that define the base case; i.e. useful life of the CPV system, normalized-per-W_p annual operation and maintenance cost, tax rate, degradation rate of the CPV systems efficiency and weighted average cost of capital – which is set equal to nominal discount rate, as previously commented. It seems from Figure 14.3 that variations in the useful life cycle of the CPV system exert a moderate effect on *IRR*, while that normalized-per-W_p annual operation and maintenance cost, tax rate and degradation rates exert a relatively small influence on *IRR*.

Figure 14.4 and Figure 14.5 clearly show how variations in the weighted average cost of capital – which is set equal to nominal discount rate – exert a noticeable influence on *PI*, while *PI* shows a smaller sensitivity to the life cycle of the system. Further, normalized-per-W_p annual operation and maintenance cost, tax rate and degradation rate in the CPV system efficiency exert less influence on this profitability index. Figure 14.6 shows values of discount pay-back time and profitability index, as a function of percentage changes of the normalized-per-W_p initial investment subsidy ($[CPV_{IS}]w_p$). In this figure, variations in the factor $[CPV_{IS}]w_p$ exert a moderate effect on discount pay-back time and profitability index. This impact is similar to those exerted by other factors such as the useful life of the system and the weighted average cost of capital.

Table 14.2 summarizes the impact of the analyzed factors on the values of internal rate of return and profitability index. The values shown in this Table have been obtained from

Table 14.2 Effect of the variation of each analyzed factor on the internal rate of return and profitability index of the base case

Factor	Range of value		
	Factor value	IRR (%)	PI
d	$3.4 \div 10\%$	No variation[a]	$0.60 \div -0.16$
N	$12.5 \div 37.5$ years	$3.6 \div 9.7$	$-0.17 \div 0.33$
T	$15 \div 45\%$	$9.1 \div 6.7$	$0.18 \div 0.06$
$[CPV_{IN}]_{Wp}$	$1.025 \div 3.075$ €/W$_p$	$17.7 \div 3.7$	$1.30 \div -0.26$
$[CPV_{IS}]_{Wp}$	$0 \div 40\%$	Not studied	$0.13 \div 1.19$
ε_{pl}	$0.25 \div 0.75\%.$	$8.2 \div 7.6$	$0.16 \div 0.10$
$[CPV_{AOM}]_{Wp}$	$1 \div 3\%$[b]	$8.9 \div 6.9$	$0.23 \div 0.02$
DNI_a	$1067 \div 3202$ kWh/m^2	$0.97 \div 13.9$	$-0.46 \div 0.71$
$[E_{CPV}]_{kWp}$	$1016 \div 3048$ kWh/(kW$_p$·year)	Same as above	
p_u	$0.06 \div 0.17$ €/kWh		
PR	$0.43 \div 1.0$	$0.97 \div 9.76$	$-0.46 \div 0.32$

[a] It should be understood that IRR stands for 'gross' internal rate of return, so that this profitability criterion remains constant irrespective of the value of $WACC$. However, as commented in section 14.3.3.4, the net internal rate of return (IRR_n) is directly influenced by $WACC$, since $IRR_n = IRR - WACC$.
[b] These values should be interpreted as the percentage of $[CPV_{IN}]_{Wp}$ that is spent on operation and maintenance tasks on an annual basis.

Figures 14.2–14.6. In Table 14.2, column 2 depicts the variation range of the factors configuring the base case, while columns 3–4 show the variation range of the values of each considered profitability criterion.

From this sensitivity analysis of both the internal rate of return and the profitability index of a CPV system, some interesting conclusions can be derived. First, tax rate, annual degradation rate of the CPV system efficiency and normalized-per-W$_p$ annual operation and maintenance cost, ordered from the lowest to the highest impact, exert a moderate influence on the two profitability criteria studied. The useful life of the CPV system and weighted average cost of capital also exert a moderate, but more noticeable, influence on PI, while WACC has no effect on IRR –'gross' IRR– due to the reasons detailed above. On the other hand, the normalized-per-kW$_p$ initial investment subsidy, performance ratio, normalized-per-kW$_p$ annual CPV electricity yield, annual direct normal irradiation, CPV electricity unitary price and normalized-per-W$_p$ initial investment cost, ordered from smallest to biggest impact, are highly influential on PI and IRR. It is worth noting that $[E_{CPV}]_{kWp}$, DNI_a, PR and p_u produce the same influence on both profitability criteria.

Let us now consider a future scenario, namely, that CPV technology reaches a cumulative installed CPV power of 1 GW$_p$, as described hereafter in section 14.4.2.1. It is expected that in this scenario, $[CPV_{IN}]_{Wp}$ would be around €0.95/W$_p$, while the perceived risk of CPV projects would be similar to that of some other renewable technologies. Under these circumstances, cost of equity capital and debt should be lower – by means of dividends and loan interest rate, respectively – than the values set for the base case analyzed according to Table 14.1, so that $d_s = 5.5\%$ and $i_l = 3.5\%$. Accordingly, CPV electricity unitary price would be around €0.05/kWh – pool price [33,34] – as no generation-based incentives for CPV would be available. The values of all other factors should remain the same to those of the base case of Table 14.1.

Table 14.3 Values for the base case corresponding to a future scenario, where cumulative installed CPV power reaches 1 GW$_p$, that differ from those shown in Table 14.1

Factors	Base case values	Units
$[CPV_{IN}]$w$_p$	0.95	€/W$_p$
p_u	0.05	€/kWh
d	3.4	%
i_l	3.5	%
N_l	20	years
d_s	5.5	%

The new values assumed for the factors that define this future scenario (and consequently, differ from those gathered in Table 14.1) are shown in Table 14.3.

With these assumptions, a sensitivity analysis for *IRR* is carried out whose results are shown in Figure 14.7. The variations of the intervals considered in this analysis range from 80% to 120% of nominal value of the factor set for the future scenario. It should be noted that annual direct normal irradiation, performance ratio, CPV electricity unitary price and normalized-per-kW$_p$ annual PV electricity generation, exert the same effect on *IRR*. Therefore, Figure 14.7 shows *IRR* as a function of normalized-per-W$_p$ initial investment cost for variations of such factors.

The high sensitivity of *IRR* to annual direct normal irradiation, performance ratio, CPV electricity unitary price and normalized-per-kW$_p$ annual PV electricity generation can be easily noticed in Figure 14.7. It is worth mentioning that the span between the maximum and minimum

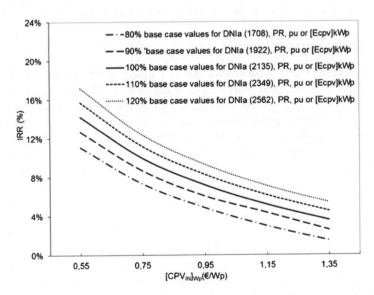

Figure 14.7 *IRR* values as a function of normalized-per-W$_p$ initial investment cost for variations of annual direct normal irradiation, performance ratio, CPV electricity unitary price and normalized-per-kW$_p$ annual CPV electricity yield. For each depicted line, all other factors are set to their default values in Table 14.1 and Table 14.3

value of *IRR* for each case obtained by means of varying DNI_a, p_u, *PR* or $[E_{CPV}]_{kWp}$ (from 80 to 120%) is around 10% (absolute).

14.4 The Cost of CPV

In section 14.3, we have analyzed several economic criteria to assess the economic viability, profitability, etc. of a project for a CPV plant. These criteria need as input data the CPV costs. Accordingly, this section discusses the installed cost and levelized cost of electricity (*LCOE*) for CPV. Firstly, we present the results of the installed cost of a system and its breakdown into component costs. After that, we analyze the *LCOE* as a function of several technical and financial aspects and finally, we determine the conditions for CPV grid parity *LCOE*.

14.4.1 The Cost of CPV Systems

The cost of the CPV system coincides with the initial investment cost (CPV_{IN}) which has shown its impact in several equations and figures of section 14.3. However, it is more useful to use the normalized-per-W_p cost of the CPV system $[CPV_{IN}]_{Wp}$, expressed in €/W_p, \$/W_p, etc. which is widely used to compare competing technologies because it is an indicator for investors of the initial outlay. When supplied by system manufacturers, the cost of the CPV system is a simple metric which only considers the cost of a system divided by its peak output rating. However, this simplicity could hide some unquoted figures. The real $[CPV_{IN}]_{Wp}$ should include not only the cost of the system itself but also the costs of transport, land preparation and installation. Therefore, all these aspects will be considered in our analysis. Since $[CPV_{IN}]_{Wp}$ is highly impacted by the cost of the components of the CPV system, we individually analyze their influence below. Through this chapter we will assume an exchange rate of 1 € = 1.25 \$.

14.4.1.1 Installed CPV System Cost

For this work, several industry players and EPCCs (Engineering, Procurement and Construction Contractors) have been surveyed in order to know the current CPV system costs, including the costs of the balance-of-system components needed for the power plant. To date, as a result of its low volume, CPV does not constitute a real market. Therefore, prices are very changing and depend on the type of offer, request for bid, etc. At this still early commercial stage, there is a big distance between the lowest and highest prices (from around 1 to €3/W_p). On one hand, the reasons for offering the lowest prices are to sell systems –even at a loss– to attain significant production volumes to subsequently bring prices down and/or for wining a certain bid. On the other hand, the highest prices are mainly related to new market entrants with very low volume productions and lack of commercial experience. Conversely to prices, costs are much more constant. Therefore, we have preferred to calculate the system cost from their components and to add a commercial margin of 5% for the EPCC.

The calculations have been carried out following the procedure described in [35]. The result for 1 MW_p plant is shown in Figure 14.8 which presents the impact on $[CPV_{IN}]_{Wp}$ of both solar cell efficiency and concentration. These values can be considered valid also for installations up to a maximum of a few megawatts. As compared to the values published for installations in the range of tens of MW_p, the $[CPV_{IN}]_{Wp}$ values of Figure 14.8 seem higher but to an extent which is difficult to quantify because of the opacity of the largest CPV companies.

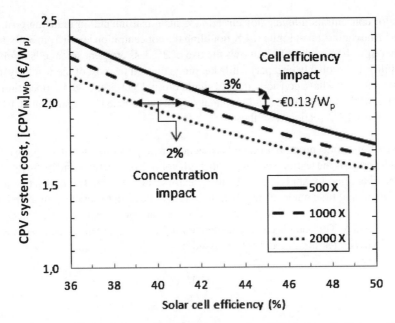

Figure 14.8 $[CPV_{IN}]_{wp}$ of a 1 MW$_p$ complete CPV plant as a function of solar cell efficiency for different concentration levels. Both the impact of solar cell efficiency (top right) and concentration level (center left) are explained in the text

As it will be shown in section 14.4.1.2, the cost of the module is responsible of about 58% of the whole system cost. Although module efficiency is key, we have preferred to show the solar cell efficiency in Figure 14.8 because of its widest span regarding the rest of the efficiencies in the module. The module efficiency is calculated as the product of solar cell efficiency, optics efficiency and module manufacturing efficiency. We have assumed for installations deployed in 2014 an optics efficiency of 0.85 and a module manufacturing efficiency of 0.95. These values are kept constant for Figure 14.8 because small improvements are envisaged in the short term. So, the assumption of a solar cell efficiency of 40% results in a module efficiency of 32.3% (ASTM G173-03, CSTC: 1000 W/m^2, 20°C) in agreement with that of state-of-the-art commercial modules [36,37], although some module prototypes have already surpassed 35%.

Figure 14.8 shows two main impacts:

- Impact of solar cell efficiency: an efficiency increase of 3% produces an average price reduction of €0.13/W$_p$ for the three concentrations considered as it is shown on top right. This means a reduction of €0.043/W$_p$ when solar the cell efficiency increases 1%. This value aligns fairly well with previously published ones ($0.047/W$_p$ [38] and $0.045/W$_p$ [39] reduction for 1% efficiency gain). From a comparative point of view, this means that if top-class efficiency solar cells are purchased by a module manufacturer at a given price, 1% less efficient solar cells would need to be purchased at 25% cheaper price for its impact on the CPV system to be cost neutral when operating at 1000×. Conversely, solar cells with 1% lower efficiency should be a 13% cheaper at 500× to be cost neutral (assuming a price for 40% solar cells of $0.188/W$_p$ and $0.355/W$_p$ for 1000× and 500×, respectively calculated from [40]).

- Impact of concentration level: concentration is also instrumental in reducing cost. As it is shown in the center-left of Figure 14.8, doubling the concentration level (from 500× to 1000× and also from 1000× to 2000×) allows the use of 2% less efficient solar cells (with highest availability and probably cheaper) while keeping constant the cost of the whole system. For this calculation, we have considered that a single Fresnel lens is enough for operation at 500× (€48/m^2), while a Fresnel lens and secondary optics are needed for operation at 1000× and 2000×, as 14.4.1.2 section describes.

However, if the concentration level and efficiency are key factors for reducing cost, the most important leverage is the learning coming from high volume productions and maturity [41]. This fact can be quantified by applying the historic learning rate model. The cost reduction is not simply a function of time but is related to cumulative production experience impacted by R&D advances, the amount of technology transfer and the rate of investment in advanced manufacturing processes. This later is closely related to market development, which is influenced by cost reduction. The learning model is expressed by:

$$\frac{C}{C_0} = \left(\frac{M}{M_0}\right)^{-L} \tag{14.28}$$

where C_0 is the cost of a given cumulative production, M_0. So, if the learning elasticity parameter, L, is known, the new cost, C, of another cumulative production, M, can be calculated.

PV learning curves are often presented at the module level only, thereby neglecting balance of system (BOS) cost, although the reduction of PV system cost associated to BOS components is in turn of crucial importance. Therefore, we will take into consideration the learning rates of both the module and the system. The cost reduction rate (LR) of c-Si modules each time the production is doubled ($LR = 1 - 2^{-L}$) is 16.9% [42] which results in a learning elasticity parameter of $L_{c\text{-}Si} = 0.267$. These values were extracted from the period 1988 to 2010. Nevertheless, from 2006 an acceleration in the learning rate is occurring for all PV technologies as it is shown in Figure 14.9 with an average LR about 27% which means $L = 0.454$. The history of CPV is more recent and an overall LR around 20% ($L = 0.322$) has been determined [7] without paying attention to the learning rates of the different components of the CPV system. Therefore, based on the learning curve approach, we calculate the CPV system cost taking into account the following assumptions:

- The world cumulative installed CPV power, as of the end of 2013, was 160 MW$_p$ [43]. These data will be used as starting value for learning curve calculations. In addition, estimates (which include data collected from public presentations, press releases, or website announcements) for a cumulative installed power of about 330 MW$_p$ by the end of 2014 have been recently published [44].
- Two important milestones for cumulative installed CPV capacity have been considered in our calculations, namely, 500 MW$_p$ and 1 GW$_p$. The first value is taken as a demonstration of CPV maturity being able to survive the market pressure of Si-PV. The second value could be achieved around 2020 [45] if CPV would eventually overcome current barriers such as the lack of track record, lack of bankability, lack of awareness, lack of project finance for a heightened perceived risk, etc. so CPV would become in a real competing technology.

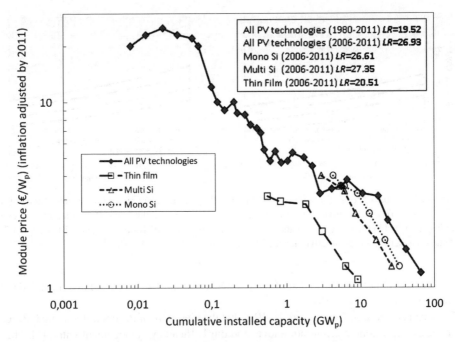

Figure 14.9 Learning curve of different PV technologies (adapted from [46]). *LR* is the cost reduction (%) when production (or cumulative installed capacity) is doubled

- Module price evolution considers, on the one hand, the trend in current prices of solar cells depending on the customer requested volume (see details in section 14.4.1.2) and, on the other hand, the price evolution of the rest of the components assuming two scenarios: $L = 0.267$ (equal to that of c-Si modules in the period 1988–2010) and $L = 0.454$ (equal to that of c-Si modules from 2006).
- The rest of the system excluding module (tracking, inverter, land and preparation, etc) follows a cost evolution ruled by $L = 0.322$ according to [7].
- Analyzed concentration levels: 500× and 1000×.
- Considering the improvements expected in technology, optics efficiency increases from its current value of 0.85 to 0.86 for 500 MW$_p$ and to 0.87 for 1 GW$_p$. Similarly, module manufacturing efficiency does not change from its current value of 0.95 for 500 MW$_p$ while it increases to 0.96 for 1 GW$_p$.

The results of the learning impact can be seen in Figure 14.10. This figure clearly evidences that concentrations of 1000× result in lower costs than 500×. In the case of 1000×, the learning influence allows going down the €1/W$_p$ barrier for solar cell efficiencies higher than 42% and once the 1-GW$_p$ cumulative installed CPV power is reached; while for 500×, solar cell efficiencies higher than 46% are required. Besides, for CPV plants of some tens of MW$_p$ that cost would be even lower than the €0.85/W$_p$ shown in Figure 14.10 left. On the other hand, PV could reach a cumulative installed power of about 500 GW$_p$ by 2020 [47] resulting in a flat plate Si-PV cost of about €0.7/W$_p$ assuming an average learning rate $LR = 22\%$. Therefore, CPV

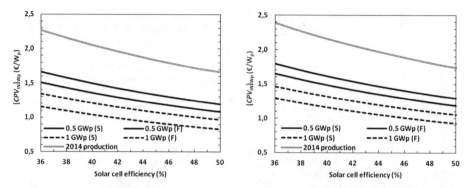

Figure 14.10 Cost of a multi-megawatt complete CPV plant as a function of solar cell efficiency for $0.5\,GW_p$ and $1\,GW_p$ cumulative installed CPV power, each one with the two different learning rates for modules described in the text (S, slow and F, fast). Concentration is of 1000× (left) and 500× (right). The baseline considering the cumulative installed CPV power in 2014 is also shown

system cost would be similar to that of flat plate Si-PV if the 1-GW_p cumulative installed CPV power is reached around 2020.

What can be deduced from Figure 14.10 is that the hierarchy of factors influencing CPV cost is first learning, second concentration level and third efficiency in agreement with [41]. This is the classic chicken-and-egg situation, where higher efficiency and concentration are needed to configure an attractive product with greater demand that can deliver the cost reduction needed to experience the learning for increasing efficiency and concentration. Figure 14.10 shows that after the initial period in which high efficiency and high concentration are the main levers allowing some cost reduction, learning takes the lead when 0.5 and $1\,GW_p$ cumulative installed CPV power are reached. In other words, equivalent increases in solar cell efficiency will produce lower price reductions as cumulative installed CPV capacity increases (in Figure 14.10 the slope of the different lines becomes less steep for increasing cumulative installed power). The reason for the aforementioned slowing down is the reduced impact that the solar cell cost is reaching in the price breakdown of a CPV system, as we show below.

14.4.1.2   Component Cost Breakdown of CPV Systems

The cost breakdown of a turnkey CPV system in 2014 is represented in Figure 14.11. In Figure 14.12, the same is shown after the level of $1\,GW_p$ cumulative installed CPV power is achieved. Figure 14.11 shows that the module takes a 57.8% of the whole CPV system cost whose breakdown is 9.6% for the solar cells, 12% for the optics and 36.1% for the rest of the module (frame, front glass, heat sinks, receivers, module assembly and calibration). The low impact of the solar cells (9.6%, i.e. €0.188/W_p) is because of the operation at 1000× since at 500× their impact increases up to 19.4%, i.e. €0.355/W_p (see section 14.4.1.1). We are considering a solar cell size of 5.5 × 5.5 mm with an average efficiency of 40% and including the expenses associated with dicing and testing [40]. It should also be noted that the solar cell cost (€0.188/W_p) reflects order volumes for $1\,MW_p$ while for a $10\,MW_p$ order, the price goes down to €0.176/W_p (which would be equivalent to $L = 0.029$) and for a $25\,MW_p$ order, the price would decrease to about €0.158/W_p (which would be equivalent to $L = 0.054$). The link between

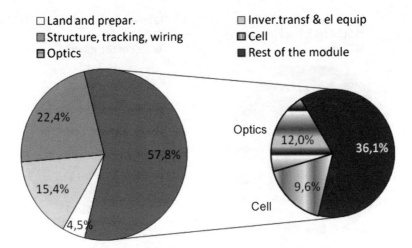

Figure 14.11 Component cost breakdown of a turnkey system of one megawatt in 2014. Cost of transport, installation, etc. are included in each category (left). Module cost breakdown where numbers indicate percentage of the whole CPV system cost which is €2.05/W_p (right). Concentration level is 1000× and module efficiency is 32.3% based on 40% efficient solar cells. A benefit of 5% for the EPCC is included

L and order volumes is because learning encompasses a variety cost-reducing mechanisms. They include the fall in costs that occurs as manufacturing ramps up, and engineers involved with this introduce new approaches for cutting costs, the trimming of costs that result from the use of mass production plants and the negotiation of attractive discounts from suppliers (such as high order volumes), among others. Nowadays, the production capacity of the main solar cell suppliers is over 100 MW_p/year each and considering all of them it is estimated to be about 500 MW_p/year. Of course, this capacity was untapped in 2014.

On one hand, the efficiency of commercial solar cells is assumed to be 40% until the 0.5 GW_p cumulative installed CPV power is reached, since most probably triple-junction (lattice-matched and lattice mismatched) solar cells will be the dominant architecture used. The corresponding solar cell price based on learning for this situation is assumed to be the one for typical orders of 10 MW_p. On the other hand, the efficiency of commercial solar cells is assumed to reach 44% for the 0.5–1.0 GW_p range thanks to the availability of upcoming solar cell architectures (inverted metamorphic, wafer bonded, dilute nitrides, etc., see Chapter 2). Because there is no reliable data about the production cost of these cells as of 2015, we have assumed the lowest price of current triple-junction solar cells, i.e. that for order volumes of 25 MW_p and higher. The reason for that is because the expected solar cell price reduction by learning would be counterbalanced by the higher cost associated to the higher manufacturing complexity of these upcoming solar cell architectures.

The cost calculation based on the abovementioned assumptions, shows that it is likely that the ratio of system costs attributed to solar cells will increase from about 10% (as of 2014, Figure 14.11) to 15% (until 1 GW_p is reached, Figure 14.12). The reason for that is that the solar cell cost decreases 26%, namely from €0.188/W_p (in 2014) to €0.139/W_p (once 1 GW_p is reached) while the overall system cost would decrease 54%, namely from €2.05/W_p (in 2014) to €0.95/W_p (once 1 GW_p is reached).

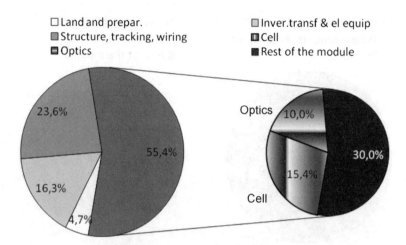

Figure 14.12 Component cost breakdown of a turnkey system of a few megawatts after a cumulative installed CPV power of $1 GW_p$ assuming the learning conditions of section 14.4.1.1 and $L = 0.454$ for module. Cost of transport, installation, etc. are included in each category. The whole CPV system cost is $€0.95/W_p$ (left). Module cost breakdown with numbers indicating percentage of the whole CPV system (right). Concentration level is 1000× and module efficiency is 36.7% based on 44% efficient solar cells. A benefit of 5% for the EPCC is included

With regard to optics and according to [48], we assume that the cost of flat Fresnel lenses depends on the technology (hot-embossed PMMA films or plates, SoG, i.e. silicone on glass). In volume (>20 000 m^2/year) their cost is expected to reach the $€40–56/m^2$ range and it does not depend on concentration level (for its use in 500× or 1000× systems).

For concentrations above 500×, a secondary optic element (SOE) is highly recommended in order to avoid tight tolerances. The SOE cost contribution to the whole system cost depends on the concentration level (number of SOEs/m^2) and also on the solar cell size (volume of optics material). For instance, given a solar cell size the SOE cost contribution reduces when concentration increases. In our case, at 1000× and for solar cells of 5.5 × 5.5 mm, glass SOEs in volume are expected to be in the range of $€24–32/m^2$ (of Fresnel lens aperture), so the primary plus secondary optics cost would add up to $€64–88/m^2$ [48].

The structure, tracking, assembling and DC wiring bring the second highest contribution to the system cost as of 2014 (22%, i.e about $€0.44/W_p$) as Figure 14.11 shows. However, as trackers are also being used in flat-plate and solar-thermal applications there is market to reduce cost since they share some similar components with their higher accuracy counterparts used in CPV. In addition, significant innovative tracker designs are now under development. Therefore, the cost of the structure, tracking, assembling and DC wiring is likely to go down to $€0.22/W_p$ although its contribution when $1 GW_p$ is reached will be also significant as Figure 14.12 shows (23%).

The third component to cost contribution is that of inverter, transformer and electric equipment which brings 15% as of 2014 and 16% when $1 GW_p$ is reached although with costs of $€0.31/W_p$ and $€0.15/W_p$, respectively. Finally, the land and preparation takes 4–5% as of 2014 and when $1 GW_p$ is reached with costs of $€0.09/W_p$ and $€0.05/W_p$, respectively.

The summary of the CPV system cost improvements expressed in $€/W_p$ is presented in Figure 14.13. Starting from a current cost of $€2.05/W_p$ a reduction of $€1.11/W_p$ will be achieved

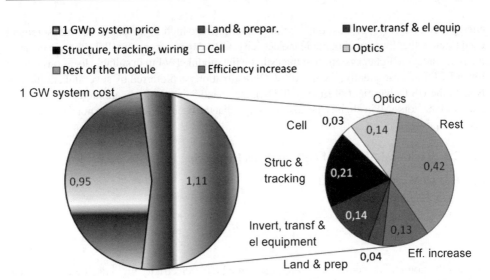

Figure 14.13 CPV installed system cost and cost reductions at 1000× expressed in €/W$_p$. Starting from a cost of €2.05/W$_p$ (the left pie chart), a final cost of €0.95/W$_p$ (left sector of the left pie chart) would be achieved once 1 GW$_p$ is installed thanks to a cost reduction of €1.11/W$_p$ (right part of the left pie chart). The breakdown of this cost reduction is shown in the pie chart on the right

for a multi-MW$_p$ installation once 1 GW$_p$ is installed, down to a cost of €0.95/W$_p$. In order of importance, the cost reductions will be 1) module €0.42/W$_p$; 2) structure, tracking and DC wiring €0.21/W$_p$; 3) inverter, transformer and electric equipment €0.14/W$_p$; 4) optics €0.14/W$_p$; 5) efficiency increase €0.13/W$_p$; 6) land and preparation €0.04/W$_p$; and 7) solar cells €0.03/W$_p$.

It is worth mentioning the fact that efficiency increase ranks fifth. Of course, making the systems more efficient increases the power output per square meter, in turn reducing the number of modules, trackers, etc. needed for an installation. However, our expectations within the period required to reach 1 GW$_p$ installed capacity, consider an efficiency increase for modules from about 32% to 37% as an average value in the industry. Accordingly, and in agreement with Figure 14.10, the key element to reduce cost in that period and considering a given concentration will be the learning well over the efficiency increase.

14.4.2 Levelized Cost of Electricity (LCOE) of CPV

The *levelized electricity cost* (LEC) or *levelized cost of electricity* (LCOE) is defined as the cost of a unit of electricity in current monetary units. As a tool for system analysis, LCOE is particularly suitable thanks to its projection of costs over the useful life of the system. It incorporates the normalized-per-W$_p$ initial investment or cost of the system [CPV_{IN}]$_{Wp}$ described in section 14.4.1 and adds operation and maintenance costs. For CPV, LCOE incorporates the available annual DNI and system lifetime to obtain a total energy output. It then discounts the future revenue from the system to present values, reflecting inflation and costs of financing. The cost is therefore calculated as a constant over the whole life of the system, expressed in €/kWh, $/kWh, etc.

Therefore, LCOE allows comparisons between costs of the electricity generated by different technologies It ontains more information that the system cost, CPV$_{IN}$. For example, nuclear energy has a huge system cost but a competitive LCOE which in the end is the measurement used to show the 'low price' of nuclear electricity. CPV experiences a parallel situation because it has nowadays a higher system cost than silicon flat module PV but in suitable locations it has a lower LCOE. Consequently, the use of LCOE estimations is preferred by several institutions – NREL, the US Department of Energy (DOE), etc. – and is increasingly adopted as a metric by many solar actors. All in all, it is the cost of energy that the user is interested in, not the cost of the installed power.

14.4.2.1   Technical Aspects of CPV Systems Influencing LCOE

$LCOE$ of CPV can be expressed as:

$$\sum_{n=1}^{N} \frac{E_{CPVn}}{(1+d)^n} \cdot LCOE = [LCC]_{Wp}, \tag{14.29}$$

being d the nominal discount rate, E_{CPVn} the annual CPV electricity yield during the n-th year and $[LCC]_{Wp}$ the normalized-per-W_p life cycle cost of the CPV system whose calculation (see Eq. (14.4)) involves $[CPV_{IN}]_{Wp}$ and $[CPV_{OM}]_{Wp}$. The summation calculation does not start from $n=0$ but from $n=1$ since: a) LCC includes the initial investment cost of the project, b) electricity generation, degradation, etc. are considered at the end of the first year.

Since LCOE is a constant value per year, Eq. (14.29) can be written as follows:

$$LCOE = \frac{[LCC]_{Wp}}{\sum_{n=1}^{N} \dfrac{E_{CPVn}}{(1+d)^n}}, \tag{14.30}$$

The LCOE value derived from Eq. (14.30) can be expressed in either current or constant currency. This expression of the LCOE is determined by the type of the discount rate used in the denominator of Eq. (14.30). If the nominal discount rate is used in the denominator, the resulting LCOE is expressed in current (or nominal) values (see section 14.2.2). Current units are often preferred when expressing LCOE for making predictions concerning grid parity and they will be the used throughout this section.

If we assume that E_{CPVn} remains constant over life cycle and taking into account that ε_{pl} is the annual decrease rate of the power of the CPV system because of degradation, Eq. (14.30) can be written as follows:

$$LCOE = \frac{[LCC]_{Wp}}{E_{CPV} \sum_{n=1}^{N} \dfrac{(1-\varepsilon_{pl})^n}{(1+d)^n}}, \tag{14.31}$$

In order to do a proper forecast of LCOE, the values of the main parameters involved have to be determined together with a LCOE sensitivity analysis on those parameters. When Eq. (14.31) is analyzed thoroughly, it is seen that: 1) its numerator depends on initial system cost, O&M, concentration level and discount rate; and 2) its denominator depends on efficiency (AC), irradiation, degradation rate, system life and discount rate.

It must be noted that initial system cost, concentration level and efficiency are not independent but they are linked among them as we described in section 14.4.1. Our choice for the ranges of values of the parameters influencing Eq. (14.31), which constitute the base case for LCOE calculations, is described in the following paragraphs.

The financeable life for a solar PV system is usually coincident with the manufacturer's guarantee period which is often 20–25 years. Accordingly, we assume as base case $N = 25$ years. However, field experience has shown that the life of solar PV systems spans well beyond 25 years even for the older Si-flat plate technologies, and current ones are likely to improve lifetime further. Therefore, a 30–40 year lifetime is becoming expected. This lifetime improvement obviously would increase the energy yield delivered over this enlarged useful life of the system. Additionally, longer system lifetimes would allow loan terms greater than 25 years paving the way to shrink WACC. Therefore, increased energy generation together with lower financial costs would yield lower values of LCOE than those calculated hereafter. However, a conservative criterion has been adopted, by assuming a system lifetime of 25 years and loan terms shorter than this period (20 years).

For the initial system cost, we assume as base case the value of 2.05 €/W_p calculated in section 14.4.1.2 for a one MW_p installation in the conditions of 2014 (1000×; module efficiency of 32.3% based on 40% efficient solar cells together with a benefit of 5% for the EPCC, see Figure 14.11). In LCOE calculations, further reductions in system cost down to 0.8 €/W_p – achievable after 1 GW_p cumulative installed CPV capacity – will be also taken into account. Regarding O&M cost, for flat-plate PV this only comprises the expense of cleaning panels and replacing the inverter, in most cases. A typical assumption might be to schedule the replacement of the inverter once during the system's lifetime, with this still representing the majority of the O&M cost. Anyhow, for CPV the tracker maintenance is an additional issue (see Chapter 9 on Reliability). The accuracy of O&M cost data is not very high because of the relatively young companies producing CPV systems and it will increase as experience grows. An assumption that is frequently made in the absence of meaningful operating data is that O&M cost adds up to 2% [11] of the normalized-per-W_p initial investment cost, giving a figure in €/W_p, $/$W_p$ and are assumed to remain constant over time. Accordingly, we adopt an annual 2% for O&M costs with a variation ranging from 0.5% (which considers an improved and mature technology) to 3.0% (resulting from unsolved problems with the tracker performance together with a more frequent cleaning), for the sensitivity analysis.

The choice of a discount rate value comes with ample uncertainty and is dealt with the sensitivity analysis carried out in section 14.4.2.2. The concept of discount rate puts a value on time preference on money, which varies by circumstance, location and the period considered. Besides, some investors vary their discount rate between PV technologies to reflect their perception of their financial risks. In this sense, CPV is discriminated against. Therefore, we have chosen $d = 8\%$ in comparison with a typical 4% for flat plate PV [25]. As stated in section 14.3.1, d is assumed equal to $WACC$.

PV systems are often financed based on an assumed 0.5–1.0% per year degradation rate although 1% per year is used based on warranties. This rate is faster than many historical data given for silicon PV exhibiting 0.2–0.5% per year degradation rate [32] with a median value of 0.5% [31]. Therefore, and in agreement with the very low CPV plant power degradation reported in Chapter 9, we consider 0.5% for the base case with a variation ranging from 0.2 to 1.0%, for the sensitivity analysis.

Table 14.4 Values for the base case and for the sensitivity analysis in the *LCOE* calculation

Parameter	Symbol	Units	Base case value	Variation range of the base case value
Useful life of the CPV system	N	years	25	25–40
Nominal discount rate	d	%	8	3–10
Annual degradation rate of the CPV system efficiency	ε_{pl}	%	0.5	0.2–1.0
Normalized-per-W$_p$ initial investment system cost	$[CPV_{IN}]_{Wp}$	€/W$_p$	2.05	0.80–2.50
Normalized-per-W$_p$ annual operation & Maintenance cost[a]	$[CPV_{AOM}]_{Wp}$	%	2.0	0.5–4.0
Concentration level	C	×	1000	500
CPV system efficiency (AC)	$\eta_{CPV,\,AC}$	%	26.8	24–35
Annual direct normal irradiation	DNI_a	kWh/m^2	2135 (Almería, Spain)	1800–2600

[a]This value should be interpreted as the percentage of $[CPV_{IN}]_{Wp}$ that is spent on operation and maintenance tasks on an annual basis.

Regarding commercial module efficiency we have considered as the base case the current typical value of 32% as it was described in section 14.4.1.1 with variations from 29% (current lower efficiency modules) to 40% (expected to be reached well after 1 GW$_p$ cumulative installed power). The whole CPV system efficiency (AC) results of applying a correcting factor of around 0.83–0.85 to module efficiency (resulting from interconnection losses; module mismatch losses; tracking & wiring losses; high operating temperature, spectrum variations; soiling, and, finally, DC-AC conversion efficiency). Therefore, the 32% module efficiency results in 27% system efficiency (AC) while the subsequent variations for the sensitivity analysis range from 24 to 35%.

Finally, the DNI base case value is associated to Almería (Spain) where a pioneer concentrated solar power (CSP) plant of 50 MW$_p$ was installed (Plataforma Solar de Almería) in the early 1980s, thanks to an annual DNI of 2135 kWh/m^2 [10]. Variations from 1800 (in order to consider less sunny places such as Madrid, Austin, Nice, etc.) to 2600 (in order to take into account the highest insolation sunbelt places) will be analyzed in the sensitivity analysis.

In summary, the values assumed for the base case and their variation range for the sensitivity analysis are included in Table 14.4.

Figure 14.14 shows the LCOE forecast for a one-MW$_p$ system as a function of the system efficiency (AC) for concentration levels of 500× and 1000×. The system cost is 2.05 €/W$_p$ for a system efficiency (AC) of 26.8% (corresponding to a module efficiency of 32.3%) and 1000× in agreement with the cost breakdown described in Figure 14.11. The rest of the points of the lines in Figure 14.14 have a different system cost because of the links among efficiency, concentration and system cost described in section 14.4.1.2. The other parameters for LCOE calculation are those of the base case of Table 14.4. CPV systems operating at 1000× are around €c0.5/kWh cheaper than those operating at 500×. A decrease of about €c4/kWh results from an increase of system efficiency from 24% to 35%.

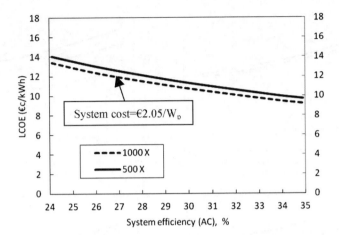

Figure 14.14 LCOE calculation for a one-MW$_p$ system as a function of the system efficiency (AC) for concentrations levels of 500× and 1000×. The system cost of €2.05/W$_p$ only applies for 1000× and 26.8% system efficiency point as indicated by the arrow (see details in the text)

In order to know the influence of several parameters on the *LCOE*, it is interesting to consider the case in which an EPPC would deploy multi-MW$_p$ CPV systems at a cost of €2.05/W$_p$ operating at 1000× and he would look for the influence of other parameters such as DNI$_a$, O&M, d and N. The calculations are presented in Figure 14.15 which shows that the most influencing levers are the discount rate and O&M cost.

The low influence of the useful life of the CPV system on LCOE seems surprising. The reason is because the discount rate of the base case, $d = 8\%$, is very high so it attenuates any parameter influence when time increases. This is shown in Figure 14.16 which shows the LCOE as a function of DNI_a for different system lifetimes in the case of $d = 8\%$ (previously shown in Figure 14.15 bottom right) and $d = 4\%$. As can be seen in this figure, if CPV would have a discount rate similar to that of silicon flat plate PV (4%), the influence of lifetime would increase (i.e. the three curves in the case of $d = 8\%$ are quite close whilst for $d = 4\%$ are further apart). However, what is more important, is that a LCOE ranging from 6 to €c11/kWh (depending on the location) would be achievable for current CPV systems. These LCOE values would be cost competitive, as of 2015, with conventional electricity sources in several countries.

Therefore, our LCOE calculation shows that the main efforts in current CPV should be oriented towards facing the lack of bankability because of a limited awareness about CPV. With CPV there is a feeling of risk that the amount of electricity that the financier is paying for will not be delivered. Operating data that will stand up to intense scrutiny and due diligence is needed to reassure financiers that this electricity will be delivered. While some of the CPV firms claim to have adequate data to have passed the 'bankability milestone', their technology must still withstand close analysis from each financier dealt with [11]. In addition, some high-profile and potentially lucrative 'agreements' involving CPV companies announced over the last years have been canceled or have not yielded the expected results so far. Such announcements are likely to make financiers more skeptical, potentially raising the risk premium they assign to CPV power plants. Therefore, while CPV is perceived as riskier than conventional PV by potential investors and customers, the lever of technical improvements will not be strong enough neither to substantially decrease LCOE nor to improve its deployment.

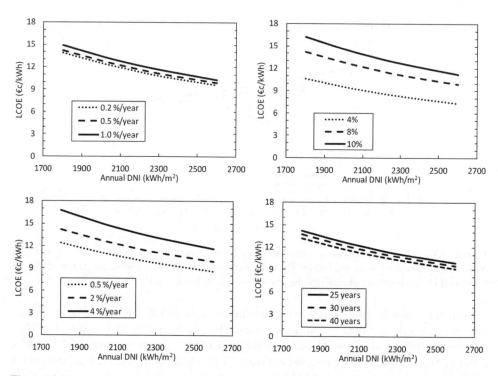

Figure 14.15   *LCOE* calculation for a one-MW$_p$ plant with a system cost of €2.05/W$_p$ operating at 1000×
with a system efficiency (AC) of 26.8% as a function of annual DNI (DNI$_a$) for variations of degradation
rate (top left), discount rate (top right), O&M cost (bottom left) and useful life of the CPV system (bottom
right). For each figure the rest of parameters are those of Table 14.4

If these barriers were overcome, then technological improvements and cost reduction by
learning would play a key role in CPV *LCOE* decrease. Accordingly, we have calculated the
LCOE for a 1 GW$_p$ cumulative installed power. In this scenario, the base case will be the same of
Table 14.4 but with the changes shown in Table 14.5. The resulting changes arise from assuming
a reduced initial system cost (€0.95/W$_p$) and a higher module efficiency of 36.7% (resulting in an
AC system efficiency of 30.5%) as calculated in section 14.4.1.2 and Figure 14.12. Besides,
once the 1 GW$_p$ milestone is reached, bankability will not be then a problem and discount rates

Table 14.5   Values for the base case parameters and for the sensitivity analysis in the LCOE calculation
used for the 1 GW$_p$ scenario that change with regard to Table 14.4

Parameter	Symbol	Units	Base case value	Variation range of the base case value
Nominal discount rate	d	%	4	2–6
Normalized-per-W$_p$ initial investment system cost	$[CPV_{IN}]_{Wp}$	€/W$_p$	0.95	0.70–1.50
CPV system efficiency (AC)	$\eta_{CPV,\,AC}$	%	30.5	24–35

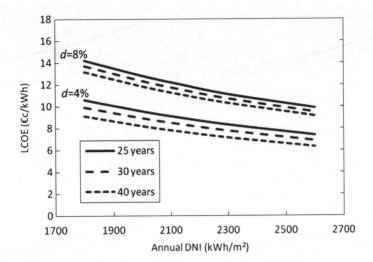

Figure 14.16 *LCOE* calculation for a one-MW$_p$ plant as a function of annual DNI for system lifetimes of 25, 30 and 40 years and considering discount rates of 4 and 8%. For each line the rest of parameters are those in Table 14.4

similar than PV (4%) will become widespread. The results of these calculations are shown in Figure 14.17 and Figure 14.18.

As can be seen, LCOE in the range of €c3–6/kWh will be achievable for many situations. These values are cost competitive with other electricity sources such as nuclear, wind, coal, etc. in many countries and they are below those of silicon flat plate PV (as of 2015). Besides, the

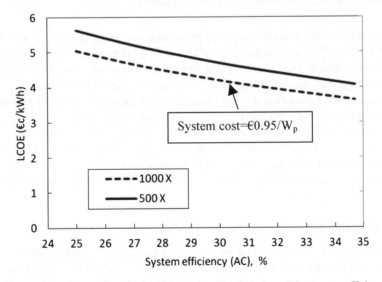

Figure 14.17 *LCOE* calculation for a multi-MW$_p$ plant as a function of the system efficiency (AC) for concentrations levels of 500× and 1000×. The system cost of €0.95/W$_p$ only applies for 1000× and 30.5% system efficiency. That point is indicated by the arrow (see details in the text)

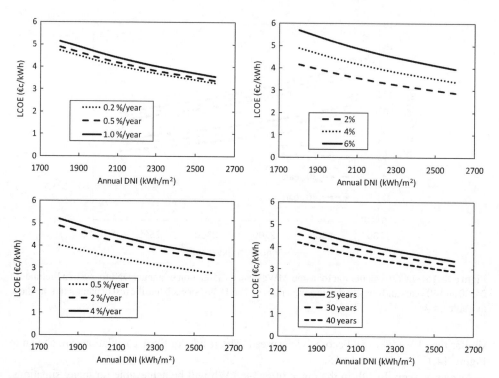

Figure 14.18 *LCOE* calculations for a few megawatts system with a system cost of €0.95/W$_p$ operating at 1000× with a system efficiency (AC) of 30.5% as a function of annual DNI for variations of degradation rate (top left), discount rate (top right), O&M costs (bottom left) and system life (bottom right). For each figure the rest of parameters are those of base case in Table 14.5

system cost of €0.95/W$_p$ was calculated for plants of a few megawatts so, as it was stated in section 14.4.1, for CPV plants in the range of tens of MW$_p$ that cost would be lower. LCOE of €c2-3/kWh were already predicted [35] and there is even more room for LCOE reductions if optimum values for all the parameters involved were simultaneously achieved.

The influence of system efficiency and concentration is shown in Figure 14.17. Operation at 1000 suns results in a reduction of about €c0.5/kWh with regard to 500 suns, while an efficiency increase from 25 to 35% produces an LCOE reduction of about €c1.5/kWh.

In addition to the concentration and efficiency impacts, it is interesting to see in Figure 14.18 that the hierarchy of factors governing LCOE is: 1) discount rate; 2) O&M cost; 3) system lifetime; and 4) degradation rate, assuming that the rest of parameters of the base case remain unaltered.

14.4.2.2 Including Financial Aspects in the Calculation of LCOE

In addition to the technical aspects dealt with in section 14.4.2.1, it is worth trying to ascertain the influence on LCOE of the following financial aspects: tax rate, inflation rate, external and equity financing, together with the escalation rate of the O&M cost. The latter factor was stated in section 14.3.1, (see definition of K_{PV} in equation (14.6)) being the percentage ratio by which this cost annually increases.

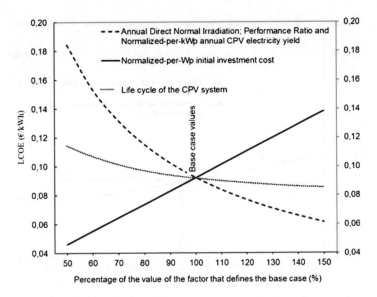

Figure 14.19 LCOE as a function of the percentage of the value of some factors that define the base case. As regards to performance ratio, only percentage values below 120% of the base case (0.86) should be considered, given that annual PR cannot exceed the ideal value of 1

The abovementioned factors were previously described in section 14.3 and their values were summarized in Table 14.1. The influence of some technical aspects analyzed in section 14.4.2.1 is revisited here in order to have an overall description when technical and financial aspects are jointly considered. Besides, the technical factors span used in this section is larger than in section 14.4.2.1 thus contributing to the consideration of more possible LCOE scenarios.

The base case stated in Table 14.1 is a good starting point to look into how LCOE varies when the value of a specific factor is changed while the values of the rest of the factors remain unaltered. Figure 14.19, Figure 14.20 and Figure 14.21 show the impact of such deviations from the base case. Most deviations range from 50% to 150% of the value of the base case.

Table 14.6 summarizes the impact of the analyzed factors on the values of LCOE. The values shown in this Table have been drawn from Figure 14.19, Figure 14.20 and Figure 14.21. In this Table, column 2 depicts the variation range of the factors studied the base case, while column 3 depicts the variation range of the value of LCOE.

It stems from Table 14.6 that for the base case values, annual normal irradiation, performance ratio and normalized-per-kW$_p$ annual PV electricity yield are at the top of the hierarchy of factors that rule LCOE. These three factors exert a similar impact on LCOE. Normalized-per-W$_p$ initial investment cost is also a crucial parameter for LCOE, to which it is also very sensitive but to a lesser extent. It must be pointed out that these conclusions consider $d = 6.7\%$ for the base case. However, if d would reach higher values, its influence would be key and would mask those of the aforementioned parameters, as has been stated in section 14.4.2.1.

If a future scenario of 1 GW$_p$ cumulative installed CPV power – described previously in section 14.4.2.1 – is considered, $[CPV_{IN}]$w$_p$ may be set equal to 0.95 €/W$_p$. Further, in this scenario the risk of CPV projects is perceived as similar as that of some other renewable technologies.

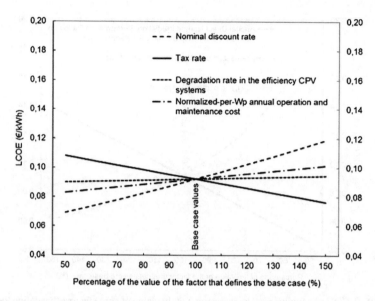

Figure 14.20   LCOE as a function of the percentage of the value of some factors that define the base case

Consequently, costs of both equity capital and debt would shrink, so that dividends and loan interest rates might lower down to $d_s = 5.5\%$ and $i_l = 3.5\%$, respectively. Let us assume that the values of all other factors remain as those in Table 14.1. To summarize, the figures now assumed for factors that differ from those of Table 14.1 are gathered in Table 14.7.

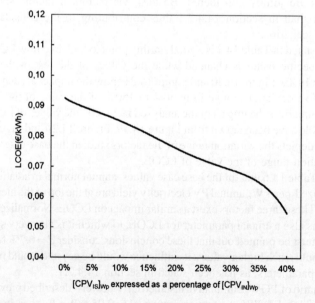

Figure 14.21   LCOE as a function of percentage changes of the initial normalized-per-W_p investment subsidy (base case)

Table 14.6 Effect of the variation of each analyzed factor on LCOE of the base case

Factor	Range of values	
	Factor value	LCOE (€/kWh)
$d = WACC$	3.4÷10%	0.069÷0.119
N	12.5÷37.5 years	0.114÷0.086
T	15÷45%	0.108÷0.077
$[CPV_{IN}]_{Wp}$	1.035÷3.075 €/W$_p$	0.046÷0.138
$[CPV_{IS}]_{Wp}$	0÷40%	0.092÷0.055
ε_{pl}	0.25÷0.75%	0.072÷0.076
$[CPV_{AOM}]_{kWp}$	1÷3%[a]	0.083÷0.102
DNI_a	1067÷3202 kWh/(m^2·year)	0.184÷0.062
$[E_{CPV}]_{kWp}$	1016÷3048 kWh/(kW$_p$·year)	0.184÷0.062
PR	0.43÷1.0	0.184÷0.079

[a] This value should be interpreted as the percentage of $[CPV_{IN}]_{kWp}$ that is spent on operation and maintenance tasks on an annual basis.

A sensitivity analysis regarding LCOE in this future scenario is carried out hereafter. The impact of variations of the factors that influence most LCOE from those values that define the future scenario are shown in Figure 14.22. These variations range from 80% to 120% of the value of the factor stated for this new scenario. It should be borne in mind that annual direct normal irradiation, performance ratio and normalized-per-kW$_p$ annual PV electricity yield exert a similar effect on LCOE.

Figure 14.22 shows values of LCOE as a function of normalized-per-kW$_p$ initial investment cost for variations of such factors. It is clear that a future scenario – optimistic but also realistic – brought by the 1 GW$_p$ of cumulative installed power would turn this technology into a cost-competitive electricity generation technique, when compared to conventional alternatives. This would be true for a wide variety of sites, regardless of their solar resource up to a large extent. In this sense, it should be noted that Figure 14.22 deals with values of annual direct normal irradiation ranging from 1,708 to 2,563 kWh/m^2. This Figure also considers noticeable positive deviations from the normalized-per-W$_p$ initial investment cost assumed in this scenario. Thus, even if the worst case is assumed i.e. $[CPV_{IN}]_{Wp} = 1.35$ €/W$_p$ and $DNI_a = 1708$ kWh/m^2, then values for LCOE would stay below €c6/kWh. Indeed, it should be kept in mind that taxation decreases the cost of financing by means of lowering the WACC –assumed equal

Table 14.7 Values for factors of the base case that differ from those shown in Table 14.1. These different values correspond to a scenario of 1 GW$_p$ cumulative installed CPV power

Factors	Base case values	Units
$[CPV_{IN}]_{Wp}$	0.95	€/W$_p$
d	3.4	%
i_l	3.5	%
N_l	20	years
d_s	5.5	%

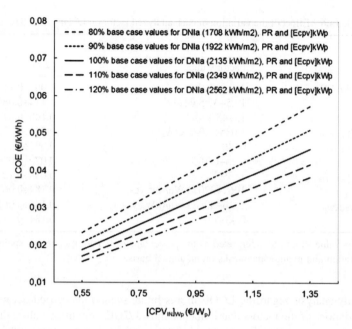

Figure 14.22 LCOE values for the future 1-GW$_p$ CPV installed power scenario as a function of initial investment cost for variations of annual direct normal irradiation, performance ratio and normalized-per-W$_p$ annual CPV electricity yield. For each line, the values of all other parameters are drawn from the base case as stated in Table 14.1 and in Table 14.7

d–, which in turn leads to smaller values of LCOE. The LCOE-competitiveness of CPV is analyzed in detail in the next section.

14.4.3 Towards the CPV Grid Parity

In order to compare the LCOE of CPV, the prices of electricity resulting from the energy mix in several countries have to be considered. The LCOE evolution of CPV is based on the analysis of section 14.4.2. For the evolution of electricity prices, we have assumed that its growth trend bases on historical growth rates and is therefore rather conservative. There are two main markets determining the electricity prices which separate the potentially competitive functions of generation and retail from the natural monopoly functions of transmission and distribution; and establish a wholesale electricity market and a retail electricity market. The role of the wholesale market is to allow trading between generators, retailers and other financial inter-mediaries both for short-term delivery of electricity (spot price) and for future delivery periods (forward price). The retail electricity market exists when end-use customers can choose their supplier from competing electricity retailers. Therefore, wholesale electricity prices are closer to the generation prices than retail ones while retail prices happen to be better metrics to compare with self-consumption electricity prices. Accordingly, wholesale and retail electricity prices are analyzed in this section.

Figure 14.23 shows a forecast of the normalized-per-W$_p$ initial investment cost of CPV systems expressed in current monetary units as a function of the cumulative installed CPV power. Figure 14.23 has been produced by taking into account many of the assumptions made in

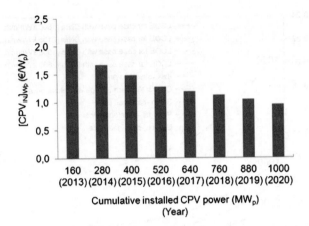

Figure 14.23 Normalized-per-W_p initial investment cost of CPV forecast as a function of the cumulative installed power. A rough estimate of the time evolution (years) in the horizontal axis has been done by assuming that a cumulative installed power of $1\,GW_p$ could be reached by 2020 [45]

section 14.4.1.1, starting from a world total installed capacity of CPV power of $160\,MW_p$ by the end of 2013 [43] up to a capacity target of $1\,GW_p$ (which can be accomplished around 2020 [45]). A value of $[CPV_{IN}]w_p = €2.05/W_p$ is assumed for the end of 2013, as detailed in section 14.4.1.1. Values of $[CPV_{IN}]w_p$ for future cumulative installed CPV power are derived from Eq. (14.28) by means of applying the learning ratios discussed in section 14.4.1.

The calculation of LCOE for CPV systems as a function of the cumulative installed CPV power to be presented hereafter results from assuming the base case scenario as configured by Table 14.1, with the exception of $[CPV_{IN}]w_p$ which is assumed to vary according to Figure 14.23.

14.4.3.1 Comparison Between Wholesale Electricity Prices and LCOE of CPV

Average prices of wholesale electricity for some countries in the European Union between 2010 and 2013 are shown in Table 14.8. These prices could increase at an annual escalation rate ranging from 3.8% to 6.7%, depending on the country during the period 2014–2020 [19]. Consequently, in this study an annual average escalation rate of 4% is assumed.

Figure 14.24 depicts the forecast of wholesale electricity average prices in the selected EU countries over the period 2013–2020 together with the LCOE of CPV systems. The wholesale

Table 14.8 Average wholesale electricity prices (€/kWh) for the selected countries of the European Union between 2010 and 2013. Taxation not included [33]

Country	2010	2011	2012	2013 (first semester)
Central Western Europe: Austria, Belgium, Germany, France, Netherlands, Switzerland (baseload price)	0.046	0.053	0.044	0.041
Italy (baseload price)	0.057	0.067	0.069	0.061
Spain (baseload price)	0.042	0.051	0.049	0.039
Portugal (baseload price)	0.041	0.050	0.048	0.036
Greece (baseload price)	0.053	0.066	0.047	0.032

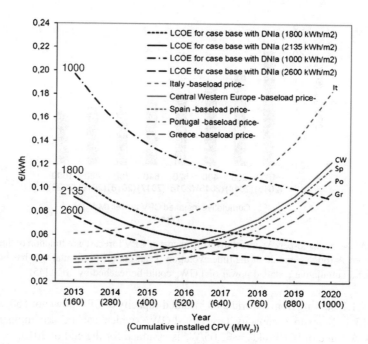

Figure 14.24 Forecast of wholesale electricity average prices in some selected EU countries and predicted development of LCOE of CPV. The relation between the time evolution (years) and the cumulative installed CPV power evolution, MW_p, (shown in brackets) follows the equivalence shown in Figure 14.23

electricity price evolution is calculated from 2013 prices (see Table 14.8) and by considering an annual average escalation rate of 4%. Regarding the LCOE, our evolution forecast is based on the cumulative installed power and not in years. However, in order to build Figure 14.24 (and the following), we have related the time evolution (years) with the cumulative installed power evolution (GW_p) by using the equivalence shown in Figure 14.23. For this reason, in the following figures the horizontal axis includes both the cumulative installed CPV power and the year.

It would not be earlier than 2017-2018 (or for a cumulative installed power around 700 MWp) when the cost of CPV-generated electricity would be lower than the price of wholesale electricity by means of conventional methods for sunny sites in this geographical area. Sites with a lower solar resource, namely, those with $1000\,kWh/m^2 < DNI_a < 1800\,kWh/m^2$, would not produce electricity at a lower cost than that paid by grid companies until the end of the decade or when a cumulative installed power around 1 GWp is achieved.

The average wholesale prices of electricity in the United States according to the specific company are listed on Table 14.9 during the period 2010–2013. Wholesale electricity prices are expected to grow in the remaining years of the decade (2014–2020). Using a conservative criterion, an annual escalation rate of 2.4% might be assumed for these prices, given that this rate is set equal to predicted annual inflation rate. This predicted value has been obtained by means of averaging the historical data related to annual inflation rates during the period 2004–2013 [20].

Figure 14.25 shows the forecast of wholesale electricity average prices in the specific electricity hubs and *LCOE* of CPV systems over the period 2013–2020. The same assumptions

Table 14.9 Selected hubs for the wholesale electricity average price ($/kWh) in the United States for the period 2010–2013 [34]

Region	Electricity Hub	2010	2011	2012	2013
New England	Mass Hub	0.056	0.053	0.042	0.066
PJM	PJM West	0.054	0.052	0.041	0.046
Midwest	Indiana Hub	0.041	0.041	0.034	0.038
Northwest	Mid-C	0.036	0.029	0.023	0.037
Northern California	NP-15	0.040	0.036	0.033	0.044
Southwest	Palo Verde	0.039	0.036	0.030	0.037
Southern California	SP-15	0.040	0.037	0.036	0.048

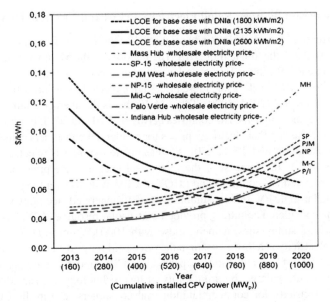

Figure 14.25 Forecast of average wholesale electricity prices in some specific Hubs in the United States and predicted development of LCOE of CPV. The relation between the time evolution (years) and the cumulative installed CPV power evolution, MW$_p$, (shown in brackets) follows the equivalence shown in Figure 14.23

for relating time evolution with cumulative installed power than for Figure 14.24 have been considered. Undoubtedly, achieving grid parity for CPV in the United States within the coming years would mean a tough competition between this technology and conventional electricity generation techniques.[1] According to this forecast, the cost of CPV-generated electricity in sunny sites – $DNI_a > 1800\,kWh/m^2$ – would at last be equal or lower than the price paid by grid companies by the end of the decade.

---

[1] It has to be taken into account that an exchange rate of 1 € = 1.25 $ has been assumed for the cost of CPV.

Table 14.10 Average retail electricity prices (€/kWh) in the residential segment for some selected EU countries between 2010 and 2013. Taxation not included [50,51]

Country	Year			
	2010	2011	2012	2013
Germany	0.138	0.141	0.151	0.149
Italy	0.125	0.142	0.148	0.150
Spain	0.145	0.160	0.178	0.163
France	0.097	0.099	0.100	0.110
United Kingdom	0.135	0.137	0.170	0.171
Portugal	0.108	0.102	0.120	0.124
Greece	0.097	0.100	0.105	0.119

14.4.3.2 Comparison Between Retail Electricity Prices and LCOE of CPV

The average increase of the retail electricity price for the whole of the EU-28 was 6.0 % between the first half of 2012 and the first half of 2013. The majority of the Member States followed this upward development [49]. In the European Union, retail electricity prices for consumers have steadily grown from 2011 to 2013 and these prices are expected to maintain this growing trend until the end of the decade (2014–2020). Table 14.10 shows the average retail electricity prices in the residential segment in some EU countries between 2010 and 2013. If these historical data are considered, forecasting that electricity prices for residential consumers could increase at an annual average rate of around 2% [19,50] is a reasonable hypothesis which is assumed here.

Figure 14.26 depicts the forecast of the retail electricity average prices in the residential segment for the selected EU countries and LCOE of CPV systems over the period 2013–2020. As happens with flat-plate PV, self consumption of CPV-generated electricity would already be profitable for the considered countries, provided that a solar resource of $DNI_a > 1800\,kWh/m^2$ were available. Less sunny sites, namely, those with $1000\,kWh/m^2 < DNI_a < 1800\,kWh/m^2$, would produce electricity at a lower cost than the price paid for grid electricity in the residential segment before 2017. Cost reductions expected for CPV and the growing trend of retail electricity prices would enlarge the profitability of this technology as year 2020 approaches.

The price of electricity for commercial/industrial consumers rose in the EU-28 by 4.1% between the first half of 2012 and the first half of 2013 [49]. Table 14.11 shows the average retail

Table 14.11 Average retail electricity prices (€/kWh) in the commercial/industrial segment for some selected EU countries between 2010 and 2013. Taxation not included [50,51]

Country	Year			
	2010	2011	2012	2013
Germany	0.092	0.09	0.082	0.091
Italy	0.092	0.117	0.112	0.112
Spain	0.108	0.108	0.114	0.114
France	0.066	0.072	0.068	0.066
United Kingdom	0.095	0.094	0.091	0.116
Portugal	0.088	0.09	0.101	0.101
Greece	0.087	0.094	0.098	0.102

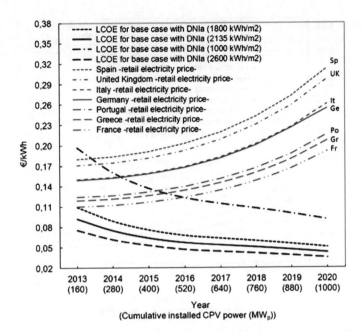

Figure 14.26 Forecast of retail electricity average prices in the residential segment in the selected EU countries and predicted development of LCOE of CPV. The relation between the time evolution (years) and the cumulative installed CPV power evolution, MW$_p$, (shown in brackets) follows the equivalence shown in Figure 14.23

electricity prices in the commercial/industrial sector in some EU countries in the period from 2010 to 2013. These prices are expected to keep growing until the end of the decade (2014–2020). By means of exploring historical data, electricity prices for the commercial/industrial segment could be reasonably hypothesized to increase at an annual average rate of around 2.2% [50].

Figure 14.27 depicts the forecast over the period 2013–2020 of the retail electricity average prices in the commercial/industrial segment for the selected EU countries and LCOE of CPV systems. It would not be earlier than 2016 (or when a cumulative installed power higher than around 500 MWp is achieved) when the cost of CPV-generated electricity drops to become competitive in this segment for sunny sites in this geographical area with low retail prices in this segment (i.e. France). For other countries such as Spain, Italy or Portugal, grid parity in this segment could have been already reached. Sites with a lower solar resource, namely, those with $1000 \, \text{kWh/m}^2 < DNI_a < 1800 \, \text{kWh/m}^2$, would not produce electricity at a lower cost than the retail price paid for grid electricity in this segment in France before the end of the decade.

As regards to another top PV area, the retail prices of electricity for ultimate customers in United States are detailed below. Residential consumers had to pay \$0.121/kWh in average during 2013 [52]. Besides, electricity prices for residential consumers are expected to grow in the upcoming years (2014–2040) at an annual average rate of around 2.2% [53]. Commercial consumers had to pay \$0.103/kWh in average in 2013 [52]. An annual increase rate of around 2.2% is expected within the coming years (2014–2040) [53]. Last, electricity prices for industrial consumers during 2013 averaged \$0.068/kWh [52]. These prices for industrial

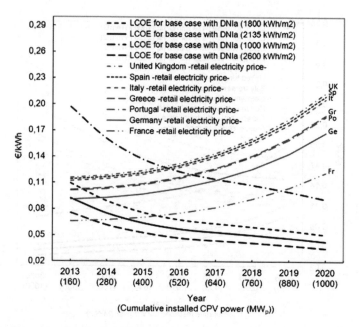

Figure 14.27 Forecast of retail electricity average prices in the commercial/industrial segment in the selected EU countries and predicted development of LCOE of CPV. The relation between the time evolution (years) and the cumulative installed CPV power evolution, MW_p, (shown in brackets) follows the equivalence shown in Figure 14.23

consumers could increase at an annual average rate of around 2.6% [53]. Table 14.12 gathers the average retail electricity prices in the residential, commercial and industrial sectors in the United States from 2010 to 2013.

Figure 14.28 shows the forecast over the period 2013–2020 of the retail electricity average prices in the residential, commercial and industrial segments in the United States and LCOE of CPV systems. It would not be until 2018 when CPV-generated electricity became competitive in the three segments in US sunny sites ($DNI_a > 1800\ kWh/m^2$). North American locations with a low irradiation profile, namely, those with DNI_a around $1000\ kWh/m^2$, would not produce electricity at a lower cost than the retail price paid for grid electricity until the end of the decade.

Table 14.12 Average retail electricity prices ($/kWh) in the residential, commercial and industrial segments in the United States between 2010 and 2013. Taxation not included [52]

Segment	Year			
	2010	2011	2012	2013
Residential	0.115	0.117	0.119	0.121
Commercial	0.102	0.102	0.101	0.103
Industrial	0.068	0.067	0.068	0.068

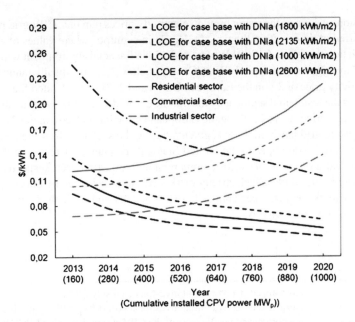

Figure 14.28 Forecast of retail electricity average prices in the residential, commercial and industrial segment in the United States and predicted development of LCOE of CPV[2]. The relation between the time evolution (years) and the cumulative installed CPV power evolution, MW_p, (shown in brackets) follows the equivalence shown in Figure 14.23

14.4.3.3 Grid Parity of LCOE of CPV in a Nutshell

Figures shown throughout section 14.4.3 turn out to be valuable tools to estimate when parity in both wholesale and retail markets is expected to be achieved by CPV power plants. It should be understood that the estimates of LCOE refer to the base case. Thus, factors involved in the calculations of LCOE may vary from one country to another. Anyway, the following conclusions may be drawn from the analysis carried out in this section:

- Everything suggests that the rising trend in electricity prices will persist. Economies of scale and learning curves will lead to a steady decrease of CPV system costs. Therefore, parity in residential segment is likely to be achieved by the middle of the present decade, whereas parity in commercial and industrial segments would not be a fact until the second half of this decade. Grid parity in the wholesale electricity market may be reached by the end of the decade.
- Parity achieved in the residential segment under a net metering scheme may boost the CPV market, thus helping to achieve the parity in the wholesale electricity market before the end of the present decade.
- Countries with high electricity prices and levels of annual direct normal irradiation above $1800\,kWh/m^2$ may prove themselves as suitable areas to turn CPV into a competitive alternative to conventionally-generated electricity. Besides, such levels of solar resource ensure competitiveness of CPV in case of no so expected electricity prices increases.

---

[2] It has to be taken into account that an exchange rate of 1 € = 1.25 $ has been assumed for the cost of CPV.

In conclusion, as long as the rising trend in electricity prices remains, CPV grid parity will be reached at different times within the present decade in countries where values of annual direct normal irradiation that exceed $1800\,kWh/m^2$ and with a financial environment within the after-tax weighted average cost of capital (WACC) stays below 7%. Specifically, annual escalation rates of electricity price of 2% in the European Union and 2.2% in the United States would lead to grid parity in the residential segment by the middle of this decade. Likewise, annual escalation rates of electricity price such as 2.2% in the European Union for both commercial and industrial uses would ensure grid parity in these segments during the second half of the present decade. The same forecast would apply to the United States if the annual increases of the electricity price of 2.2% and 2.6% took place for commercial and industrial uses, respectively. Finally, CPV grid parity in the wholesale electricity market would be achieved by the end of the 2010s if prices in such market experienced annual increases of 2.4% and 4% in the United States and in the European Union, respectively.

Glossary

$[CPV_{IS}]_{Wp}$	Normalized-per-W_p initial investment subsidy ($€/W_p$)
$[CPV_{IN}]_{Wp}$	Normalized-per-W_p initial investment cost ($€/W_p$, $\$/W_p$, etc.). It coincides with the normalized per-W_p cost of the CPV system including EPCC benefit
$[CPV_{AOM}]_{Wp}$	Normalized-per-W_p annual operation and maintenance cost of the CPV system ($€/W_p$; $\$/W_p$)
$[E_{CPV}]_{kWp}$	Normalized-per-kW_p annual CPV electricity yield (kWh/(kW_p·year)).
$B\text{-}C$	benefit-to-cost
C	Cost of a given cumulative production M ($€/W_p$; $\$/W_p$)
C_0	Cost of a given cumulative production M_0 ($€/W_p$; $\$/W_p$)
CPV_{AOM}	Annual operation and maintenance cost of the CPV system ($€$, $\$$, etc.)
CPV_{IN}	Initial investment cost of the CPV system ($€$, $\$$, etc.). It coincides with the cost of the CPV system including EPCC benefit
CPV_{IS}	Initial investment subsidy ($€$, $\$$, etc.)
CPV_l	Amount equal to the portion of the initial investment financed with loan ($€$)
CPV_s	Amount equal to the portion of the initial investment financed with equity capital ($€$)
DEP	Tax depreciation for the CPV system ($€$, $\$$, etc.)
DEP_y	Annual tax depreciaton for the CPV system ($€$, $\$$, etc.)
DNI_a	Annual direct normal irradiation (kWh/m^2 year)
DNI_{CSOC}	Direct normal irradiance at CSOC (concentrator standard operating conditions) (kW/m^2)
DPB	Discounted payback time (years)
d	discount rate, nominal discount rate (%)
d_r	Real discount rate (%)
d_s	Annual dividends (%)
E_{CPV}	Annual CPV electricity yield (kWh)
E_{CPVg}	Annual CPV electricity generated injected into to the grid (kWh)
ε_{CPVAOM}	Annual escalation rate of the operation and maintenance cost of the CPV system (%)
E_{CPVn}	Annual CPV electricity yield during the n-th year
E_{CPVs}	Annual CPV electricity generated is used for self-consumption (kWh)

ε_{pg}	Annual increase rate of the electricity price into to the grid by user (%)
ε_{pl}	Annual degradation rate in the efficiency of the CPV system (%)
ε_{ps}	Annual increase rate of the electricity price self-consumption or saved by user (%)
ε_{pu}	Annual increase rate of the CPV electricity unitary price (%)
$\eta_{CPV,\,AC}$	CPV system efficiency (AC)
g	Annual inflation rate (%)
i	Annual interest rate of money (%)
i_l	Annual loan interest (%)
IRR	Internal rate of return
IRR_n	Net internal rate of return
K_{pg}	Factor equal to $((1+\varepsilon_{pg})(1-\varepsilon_{pl}))/(1+d)$
K_{ps}	Factor equal to $((1+\varepsilon_{ps})(1-\varepsilon_{pl}))/(1+d)$
K_{PV}	Factor equal to $(1+\varepsilon_{CPVOM})/(1+d)$.
L	Learning elasticity parameter
LCC	Life - cycle cost of the CPV system (€, $, etc.)
$[LCC]_{Wp}$	Normalized-per-W_p LCC (€/W_p; $/$W_p$)
$LCOE$	Levelised cost of electricity (€/kWh, $/kWh, etc.)
LR	Cost reduction rate (%)
m	Annual escalation rate of NCF (%)
M	Cumulative production assigned to a cost C (kW$_p$, MW$_p$, etc.)
M_0	Cumulative production assigned to a cost C_0 (kW$_p$, MW$_p$, etc.)
N	Useful life of the CPV system, equal analysis period (years)
NCF	Net cash flow (€, $, etc.)
N_d	Period of time over which an investment is depreciated for tax purposes (years)
N_{IS}	Period of time over which an initial investment subsidy is amortized (years)
N_l	Time duration of loan (years)
NPV	Net present value (€, $, etc.)
p_g	CPV electricity unitary price into to the grid (€/kWh, $/kWh, etc.).
PI	Profitability index
PR	Performance ratio of a CPV system
p_s	CPV electricity unitary price self-consumption (€/kWh, $/kWh, etc.)
p_u	CPV electricity unitary price (€/kWh, $/kWh, etc.)
$PVIF(k)$	Present value interest factor of year k equal to $q(1-q^K)/(1-q)$
$PW[CIF(DPB)]$	Present worth of the cash inflows from a CPV system over DPB (€, $, etc.)
$PW[CIF(N)]$	Present worth of the cash inflows from a CPV system over N (€, $, etc.)
$PW[CPV_{OM}(DPB)]$	Present worth of the CPV system operation and maintenance cost over DPB (€, $, etc.)
$PW[CPV_{OM}(N)]$	Present worth of the CPV system operation and maintenance cost over N (€, $, etc.)
$PW[DEP(DPB)]$	Present worth of tax depreciation over DPB (€, $, etc.)
$PW[DEP(N_d)]$	Present worth of tax depreciation over N_d (€, $, etc.)
PY_{CPVl}	Payments of each year of debt (€, $, etc.)
PY_{CPVs}	Payments of each year of equity capital (€, $, etc.)
q	Factor equal to $1/(1+d)$
S_V	Salvage value of the system at the end of their life cycle
T	Income tax rate (%)
$WACC$	Weighted average cost of capital (%)

Through this chapter we assume an exchange rate of 1 € = 1.25 $

References

1. Short W, Packey DJ, Holt T. A manual for the economic evaluation of energy efficiency and renewable energy technologies 1995; NREL/TP-462-5173, National Renewable Energy Laboratory: 1–120.
2. Swift KD. A comparison of the cost and financial returns for solar photovoltaic systems installed by businesses in different locations across the United States. *Renewable Energy* 2013; **57**: 137–143.
3. Internal Revenues Service United States, Department of the Treasury. Figuring Depreciation Under MACRS. 2013. http://www.irs gov/publications/p946/ch04 html#d0e4318 (accessed November 2015).
4. Lasnier F, Ang T. *Photovoltaic Engineering Handbook*, Adam Hilger, Bristol, England and New York, 1990, pp 371–399.
5. Chabot B. From costs to prices: Economic analysis of photovoltaic energy and services. *Progress in Photovoltaics Research and Applications* 6: 55–68. 1998.
6. Fraunhofer Institute for Solar Energy Systems Ise. Photovoltaics Report November 7, 2013; www.ansfans.org/documenti/corsiEM/RN_1_scenari?ISE_Photovoltaics-2013-11-07.pdf (accessed November 2015).
7. Haysom J, Jafarieh O, Anis H, Hinzer K. Concentrated photovoltaics system costs and learning curve analysis. *AIP Conference Proceedings*, **1556**: 239–243, 2013; DOI: 10.1063/1.4822240.
8. Fraisopi F. *The CPV Market: An Industry Perspective*. GTM Research, Intersolar München. June 2013;
9. NREL. Energy Technology Cost and Performance Data for Distributed Generation 2013 (August); 2014. http://www.nrel.gov/analysis/tech_lcoe_re_cost_est.html (accessed November 2015).
10. Meyer R, Torres Butron J, Marquardt G, Schwandt M, *et al*. Combining solar irradiance measurements and various satellite-derived products to a site-specific best estimate. Proceedings of the 14th SolarPACES Symposium, Las Vegas March 2008: 1–8.
11. Extance A, Márquez C. The Concentrated Photovoltaics Industry Report. CPV Today, 2010. www.dsireusa.org (accessed November 2015).
12. King C. Site data analysis of CPV plants. IEEE Photovoltaic Specialists Conference, 2010, pp. 3043–3047.
13. International Energy Agency (IEA). Trends 2013 in Photovoltaic Application: Survey Report of Selected IEA Countries Between 1992 and 2012. 2013. Report IEA-PVPS T1-23: 2013.
14. Yamada H, Ikki O. National Survey Report of PV Power Applications in Japan 2012. International Energy Agency (IEA), Task 1 Exchange and dissemination of information on PV power systems. 2013.
15. Fang L, Honghua X, Sicheng W. National Survey Report of PV Power Applications in China 2012. International Energy Agency (IEA), Task 1 Exchange and dissemination of information on PV power systems. 2013.
16. Castello S, De Lillo A, Guastella S, Paletta F. National Survey Report of PV Power Applications in Italy 2012. International Energy Agency (IEA), Task 1 Exchange and dissemination of information on PV power systems. 2013.
17. Campoccia A, Dusonchet L, Telaretti E, Zizzo G. An analysis of feed-in tariffs for solar PV in six representative countries of the European Union. *Solar Energy*, **107**: 530–542. 2014.
18. DSIRE. Solar policy and information. Database of states of incentives for renewables. 2014. www.dsireusa.org (accessed November 2015).
19. European Photovoltaic Industry Association. Solar Photovoltaics Competing in the Energy Sector: On the road to competitiveness 2011. Available at: http://www.epia.org/news/publications/ (accessed November 2015).
20. Global rates.com. Inflation - summary of current international inflation figures 2013. http://www.global-rates.com/economic-indicators/inflation/inflation.aspx (accessed November 2015).
21. European Central Bank. Inflation in the Euro area 2014. Available at: http://www.ecb.europa.eu/stats/prices/hicp/html/inflation.en.html (accessed November 2015).
22. US Department of Energy. Database of states incentives for renewables and efficiency (DSIRE); 2013. www.dsireusa.org (accessed November 2015).
23. European Central Bank. MFI interest rates on euro-denominated deposits from and loans to euro area residents 2013; http://sdw.ecb.europa.eu/reports.do?node=100000173 (accessed November 2015).
24. Global rates.com. Central banks - summary of current interest rates 2013; http://www.global-rates.com/interest-rates/central-banks/central-banks.aspx (accessed November 2015).
25. Fraunhofer Institute for Solar Energy Systems ISE. Levelized cost of electricity renewable energy technologies. November 2013.
26. Colmenar-Santos A, Campíñez-Romero S, Pérez-Molina C, Castro-Gil M. Profitability analysis of grid-connected photovoltaic facilities for household electricity self-sufficiency. *Energy Policy*, **51**: 749–764. 2012.

27. Talavera DL, Nofuentes G, De La Casa J, Aguilera J. Sensitivity analysis on some profitability indices for photovoltaic grid-connected systems on buildings: The case of two top photovoltaic European Areas. *Journal of Solar Energy Engineering, Transactions of the ASME* 2013, pp. 135.

28. Ministry of Economics, Spain. Royal Decree 1777/2004, Ministry Economic. RD 1777/2004 2004; BOE number 189:28377–28429.

29. Ministry of Economics, Spain. Royal Decree 1793/2008, Ministry Economic. RD 1793/2008 2008; BOE number 278:45770–45786.

30. Thonson Reuters. Consulta A.E.A.T. 128308. IS. Central fotovoltaica. Amortización. 2014. http://portaljuridico. lexnova.es/doctrinaadministrativa/JURIDICO/77405/consulta-aeat-128308-is-central-fotovoltaica-amortizacion

31. Jordan DC, Kurtz SR. Photovoltaic degradation rates - An analytical review. *Progress in Photovoltaics Research and Applications*, **21**: 12–29. 2013.

32. Branker K, Pathak MJM, Pearce JM. A review of solar photovoltaic levelized cost of electricity. *Renewable and Sustainable Energy Reviews*, **15**: 4470–4482. 2011;

33. European Commission DE. Quarterly Report on European Electricity Markets 2013;Volume 6, issue 2: Second quarter 2013.

34. U.S. Energy Information Administration. Wholesale Electricity and Natural Gas Market Data 2014; http://www.eia. gov/electricity/wholesale/index.cfm (accessed November 2015).

35. Algora C. The Importance of the very High Concentration in Third-Generation Solar Cells. In: *Next Generation Photovoltaics*, (eds Martí, A. and Luque, A), pp 108–139 IoP Publishing. 2004.

36. Soitec. Solar Energy 2015. http://www.soitec.com/en/technologies/concentrix/ (accessed November 2015).

37. Suncore Photovoltaic Technology Company Limited. 2015. http://www.suncorepv.com/index.php? m=content&c=index&a=lists&catid=40 (accessed November 2015).

38. GTM Research. Concentrating Photovoltaics 2011: Technology, Costs and Market 2014; https://www .greentechmedia.com/research/report/concentrating-photovoltaics-2011 (accessed November 2015).

39. Jeff Allen. Pushing the Boundaries of CPV Efficiency, CPV USA, 23–24 September, 2013, San Jose, California. 2013.

40. Prívate communications from a solar cell supplier 2014.

41. Algora C. Very high concentration challenges of III-V multijunction solar cells. In: *Concentrator Photovoltaics* (eds A. Luque & V. Andreev), Springer, Germany, pp. 89–111. 2007.

42. Friederike Kersten, Roland Doll, Andreas Kux, *et al*. PV learning curves: past and future drivers of cost reduction. 26th European Photovoltaic Solar Energy Conference, September 2011, WIP, Hamburg, Germany 2011.

43. Kinsey G. Welcome to CPV-10. Verbal presentation at CPV10 Conference, Albuquerque, USA. 2014.

44. Philipps SP, Bett AW, Horowitz K, Kurtz S. Current status of concentrator photovoltaic (CPV) technology January 2015; CPV Report. TP-6A20-63916. www.nrel.gov/docs/fy15osti/63196.pdf (accessed January 2015).

45. IHS. Concentrated Photovoltaic Solar Installations Set to Boom in the Coming Years 2013;Report. Available at: www.ihs.com (accessed January 2015).

46. Weber ER, Schleicher-Tappeser R. Financing Future Innovations in Photovoltaics. 28th EU-PVSEC Paris October 1, 2013. www.ise.fraunhofer.de/de/veroeffntlichungen/vortaege-prof-weber/20131001.pdf (accessed November 2015).

47. Solar Power Europe (formerly known as EPIA). Global Market Outlook for Solar Power 2015–2019. 2015, p. 15.

48. Mohedano R. Personal communication. LPI, www lpi-llc com (2014).

49. European Commission e. Energy price statistics 2014; http://epp.eurostat.ec.europa.eu/statistics_explained/index. php/Energy_price_statistics#Electricity_prices_for_household_consumers (accessed November 2015).

50. European Commission e. Electricity and natural gas price statistics 2014; http://epp.eurostat.ec.europa.eu/ statistics_explained/index.php/Electricity_and_natural_gas_price_statistics (accessed November 2015).

51. WIP – Renewable Energies (WIP). PV parity project. Electricity prices scenarios until at least the year 2020 in selected EU countries. Deliverable 22 January 2012; (IEE/10/307/SI2.592205).

52. U.S. Energy Information Administration. Short-Term Energy Outlook. Table 2. U.S. Energy Price 2014; http:// www.eia.gov/forecasts/steo/tables/?tableNumber=8#startcode=2005 (accessed November 2015).

53. U.S. Energy Information Administration. Annual Energy Outlook 2014.Table A3, Energy prices by sector and source 2014; http://www.eia.gov/forecasts/aeo/pdf/tbla3.pdf (accessed November 2015).

Index

3D model, 537
3TIER, 496

A

AA, *see* acceptance angle
Abengoa, 406–412, 504
absolute cavity radiometers, 3, 16
absorption and emission
   of phonons, 141
   of photons, 141, 144, 160, 164
absorption coefficient, 157, 160, 173, 175, 176,
   549, 554, 559, 561, 625, 636
absorption of light, 138, 158, 163, 166–169
   by nanostructures, 173
accelerated life tests (ALT), 529, 539, 540, 544
   highly accelerated life test (HALT), 529, 540
   humidity accelerated life test, *see* life stress
      models
   qualification tests, 529, 532–533, 537, 548,
      573, 574, 577, 578
   quantitative accelerated life tests (QALT), 529,
      540
   temperature accelerated life test, *see* life stress
      models
acceleration factor, AF 530–531, 540, 551, 561
acceptance angle, 349, 351, 352, 359–364, 367,
   398, 402, 616, 622, 626, 629, 630–634,
   636, 641, 643, 647, 650, 663–665, 676,
   678
acceptance half-angle, 190, 208, 226, 229
accuracy, 2–4, 11–25, 29, 33, 38
activation energy, 530–531, 533, 544, 550, 553
active cooling, *see* active dissipation

active dissipation, 275, 285
adhesive, 278
ADU: Air Drying Unit, 421, 424
aerosol, 8, 22–27, 30, 32, 33, 43, 47, 50, 52
aerosol optical depth, 653
aging, 533, 540, 546, 553, 556–562,
   575, 576
air mass, 2, 7, 8, 22–24, 26, 27, 33, 49, 50, 52
albedo, 7, 28
alignment, 11, 20, 640, 641, 644, 651, 661, 670,
   672, 673, 675, 677
alumina, 278–282
aluminum, 347, 348, 356, 380, 386–388, 390,
   391, 393–395
AM1.5d, direct reference spectrum, 592, 599,
   600, 611, 644, 645, 646, 653, 657, 665,
   668, 670, 678
AM1.5g global hemispherical reference
   spectrum, 589, 611
ambient, 340, 341, 343, 346, 356, 375, 377,
   379–382, 386–388, 390, 402, 403
ambient conditions, *see* standard test conditions
American Society for Testing and Materials, 4
   ASTM E2527–09, 654
   ASTM E927, 664, 665
   ASTM G173–03, 644, 645, 653, 665
Amonix, 353, 355, 358, 563, 567, 571, 572,
   686–702
analysis, 495, 496, 497, 498
angle of incidence, 642, 667
Angstrom exponent, 8
Angstrom's equation, 8
angular acceptance, *see* acceptance angle

Handbook of Concentrator Photovoltaic Technology, First Edition. Edited by Carlos Algora and Ignacio Rey-Stolle.
© 2016 John Wiley & Sons, Ltd. Published 2016 by John Wiley & Sons, Ltd.